KU-281-446

1997–2001 Recomm[...] [...]ary Allowances (RDA) and Adequate Intakes (AI)

Vitamins		Minerals											
Vitamin E RDA (mg/day)[e]	Vitamin K AI (µg/day)	Calcium AI (mg/day)	Phosphorus RDA (mg/day)	Magnesium RDA (mg/day)	Iron RDA (mg/day)	Zinc RDA (mg/day)	Iodine RDA (µg/day)	Selenium RDA (µg/day)	Copper RDA (µg/day)	Manganese AI (mg/day)	Fluoride AI (mg/day)	Chromium AI (µg/day)	Molybdenum RDA (µg/day)
4	2.0	210	100	30	0.27	2	110	15	200	0.003	0.01	0.2	2
5	2.5	270	275	75	11	3	130	20	220	0.6	0.5	5.5	3
6	30	500	460	80	7	3	90	20	340	1.2	0.7	11	17
7	55	800	500	130	10	5	90	30	440	1.5	1.0	15	22
								40	700	1.9	2	25	34
								55	890	2.2	3	35	43
								55	900	2.3	4	35	45
								55	900	2.3	4	35	45
								55	900	2.3	4	30	45
								55	900	2.3	4	30	45
								40	700	1.6	2	21	34
								55	890	1.6	3	24	43
								55	900	1.8	3	25	45
								55	900	1.8	3	25	45
								55	900	1.8	3	20	45
								55	900	1.8	3	20	45
								60	1000	2.0	3	29	50
								60	1000	2.0	3	30	50
								60	1000	2.0	3	30	50
								70	1300	2.6	3	44	50
								70	1300	2.6	3	45	50
								70	1300	2.6	3	45	50

[...]Academy Press. Copyright 1997, [...]ademy Press, Washington, D.C.

[...]take Levels (UL)

							[...]oron (mg/day)	Nickel (mg/day)	Vanadium (mg/day)
							—	—	—
							—	—	—
							3	0.2	—
							6	0.3	—
23	600	280	5000	6	10	1100	11	0.6	—
34	900	400	8000	9	10	1700	17	1.0	—
40	1100	400	10,000	11	10	2000	20	1.0	1.8
40	1100	400	10,000	11	10	2000	20	1.0	1.8
34	900	400	8000	9	10	1700	17	1.0	—
40	1100	400	10,000	11	10	2000	20	1.0	—
34	900	400	8000	9	10	1700	17	1.0	—
40	1100	400	10,000	11	10	2000	20	1.0	—

NOTE: An Upper Limit was not established for vitamins and minerals not listed and for those age groups listed with a dash (—) because of a lack of data, not because these nutrients are safe to consume at any level of intake. All nutrients can have adverse effects when intakes are excessive.

SOURCE: Adapted with permission from the *Dietary Reference Intakes* series, National Academy Press. Copyright 1997, 1998, 2000, 2001, by the National Academy of Sciences. Courtesy of the National Academy Press, Washington, D.C.

Nutrition Now

Third Edition

Judith E. Brown
University of Minnesota

WADSWORTH
THOMSON LEARNING™

Australia · Canada · Mexico · Singapore · Spain ·
United Kingdom · United States

WADSWORTH
THOMSON LEARNING™

Dedication

To instructors everywhere who love to teach nutrition,
and to students who find learning about nutrition to be both enriching and fun.

PUBLISHER: Peter Marshall
DEVELOPMENT EDITOR: Elizabeth Howe
ASSISTANT EDITOR: John Boyd
EDITORIAL ASSISTANT: Andrea Kesterke
MARKETING MANAGER: Becky Tollerson
MARKETING ASSISTANT: Adam Hofmann
ADVERTISING PROJECT MANAGER: Brian Chaffee
PROJECT MANAGER: Sandra Craig
PRINT/MEDIA BUYER: Barbara Britton
PERMISSIONS EDITOR: Joohee Lee
PRODUCTION, ART, AND TEXT DESIGN: Ann Borman
COVER DESIGN: Bill Stryker
PHOTOGRAPHIC ILLUSTRATION: Stan Maddock
COPY EDITOR: Pat Lewis
PROOFREADING: Heather Jones
INDEXER: Terry Casey
PHOTO RESEARCH: Ann V. Swift
COVER PHOTOGRAPH: Photo Disc
COMPOSITOR: Parkwood Composition Service
PRINTER: Transcontinental Interglobe

Printed in Canada
3 4 5 6 7 05 04 03 02

Library of Congress Cataloging-in-Publication Data
Brown, Judith E.
Nutrition now / Judith E. Brown. —3rd ed.
p. cm.
Includes Bibliographical references and index.
ISBN 0-534-58005-X
1. Nutrition. I. Title.

QP141 .B873 2002
613.2—dc21 2001026541

Wadsworth/Thomson Learning
10 Davis Drive
Belmont, CA 94002-3098
USA

For information about our products, contact us:
Thomson Learning Academic Resource Center
1-800-423-0563
http://www.wadsworth.com

International Headquarters
Thomson Learning
International Division
290 Harbor Drive, 2nd Floor
Stamford, CT 06902-7477
USA

UK/Europe/Middle East/South Africa
Thomson Learning
Berkshire House
168-173 High Holborn
London WC1V 7AA
United Kingdom

Asia
Thomson Learning
60 Albert Street, #15-01
Albert Complex
Singapore 189969

Canada
Nelson Thomson Learning
1120 Birchmount Road
Toronto, Ontario M1K 5G4
Canada

About the Author

Judith E. Brown is a professor of nutrition and public health at the University of Minnesota. She received her Ph.D. in human nutrition from Florida State University and her M.P.H. in public health nutrition from the University of Michigan. She has received competitively funded research grants from the National Institutes of Health, the Centers for Disease Control and Prevention, and the Maternal and Child Health Bureau and has over 80 publications in the scientific literature including the *New England Journal of Medicine*, the *Journal of the American Medical Association*, and the *Journal of the American Dietetic Association*. A recipient of the Agnes Higgins Award in Maternal Nutrition from the March of Dimes, Dr. Brown teaches Human Nutrition and Health and Maternal and Infant Nutrition at the University of Minnesota. A registered dietitian, Brown is the successful author of *Everywoman's Guide to Nutrition* and *Nutrition for Pregnancy*.

Contents in Brief

Contents

Unit 5
Nutrition, Attitudes, and Behavior 5-1

Unit 6
What's a Healthful Diet? 6-1

Unit 7
How the Body Uses Food 7-1

Unit 8
Calories! Food, Energy, and Energy Balance 8-1

Unit 9
The Highs and Lows of Body Weight 9-1

Unit 10
Weight Control: The Myths and Realities 10-1

Unit 11
Disordered Eating: Anorexia Nervosa, Bulimia, and Pica 11-1

Unit 12
Useful Facts about Sugars, Starches, and Fiber 12-1

Unit 13
Alcohol: The Positives and Negatives 13-1

Unit 14
Diet and Dental Health 14-1

Unit 15
Proteins and Amino Acids 15-1

Unit 16
Vegetarian Diets 16-1

Unit 17
Food Allergies and Intolerances 17-1

Unit 18
Fats and Cholesterol in Health 18-1

Unit 19
Nutrition and Heart Disease 19-1

Unit 20
Vitamins and Your Health 20-1

Unit 21
Phytochemicals and Genetically Modified Food 21-1

Unit 22
Diet and Cancer 22-1

Unit 23
Good Things to Know about Minerals 23-1

Unit 24
Herbal and Other Dietary Supplements 24-1

Unit 25
Water Is an Essential Nutrient 25-1

Unit 26
Nutrition and Physical Fitness for Everyone 26-1

Unit 27
Nutrition and Physical Performance 27-1

Unit 28
Good Nutrition for Life: Pregnancy, Breastfeeding, and Infancy 28-1

Unit 29
Nutrition for the Growing Years: Childhood through Adolescence 29-1

Unit 30
Nutrition and Health Maintanence for Adults of All Ages 30-1

Unit 31
Nutrient-Gene Interactions in Health and Disease 31-1

Unit 32
The Multiple Dimensions of Food Safety 32-1

Unit 33
Aspects of Global Nutrition 33-1

Preface

We are pleased to present to instructors and students the third edition of *Nutrition Now.* Thoroughly updated, this text remains dedicated to complementing the instructional efforts of faculty who teach a one- to four-credit course on introductory nutrition to nonmajors. In line with the previous editions, *Nutrition Now* retains its scientific, student-focused, and practical approach to nutrition. It continues to be directed at the goals of increasing student understanding of basic nutrition concepts and developing skills for making healthful decisions about personal nutrition now and in the future. For many students, the course supported by *Nutrition Now* represents a happy balance between meeting an educational requirement and learning useful information about a subject of particular interest.

Readers of this preface will discover that the third edition covers new topics of growing importance to nutrition, and that the text's unique didactic features have been strengthened. There are now 33 "stand-alone" units in *Nutrition Now.* As before, subsets of units that match instructors' and students' interests and needs can be ordered and presented in class in any desired arrangement. Eleven of the units cover core nutrition topics, including nutrition concepts, nutrition and health relationships, the nutrients, healthy diets, and digestion. Other units address topics that touch students' lives regularly and that relate to students' professional interests. Topics covered in these 22 units include body weight, weight control, eating disorders, heart disease, cancer, herbal and other dietary supplements, sports nutrition, life cycle nutrition, food safety, and global nutrition. Exciting discoveries related to genes, genetic variation, and nutrition and health relationships receive first-time coverage in the nutrient-gene interactions unit.

In the third edition of *Nutrition Now,* you will find:

- **Expanded coverage** of herbal supplements (Unit 24), diabetes (Unit 12), genetically modified foods (Unit 21), food safety (Unit 32), Canadian dietary guides (Unit 6), the fetal origins hypothesis (Units 19 and 28), and lactose maldigestion (Units 12 and 17), as well as an expanded glossary.
- **Dietary Reference Intake values** released through 2001.
- **2000 Dietary Guidelines for Americans.**
- **New Nutrition Up Close** activities and other student activities.
- **A table of measurement equivalents.**
- **Additional Health Action features.**

Pedagogical Features

Nutrition Now continues to incorporate a number of unique approaches to student learning:

- Each unit begins with a feature called ***Nutrition Scoreboard*** that presents a set of true-false questions designed to test students' knowledge of various topics addressed in the unit. Answers to the questions appear on the following page.
- A **summary of the key concepts** covered is presented on the second page of each unit.
- **Every major point in each unit is accompanied by an illustration** that helps to drive the point home.

- ***Health Action* boxes** provide examples and guidelines for making healthful decisions about nutrition, food handling, and more.
- ***Nutrition Up Close* boxes** provide opportunities for students to apply nutrition concepts in their daily lives.
- ***WWW Links*** provide URLS and descriptions of high-quality nutrition resources.
- **Margin definitions** and an **end-of-text glossary** help students master the material presented.
- **Every version of the text includes a resource-rich appendix.** Materials include a table of food composition, the food exchange systems for the United States and for Canada, a table of intentional food additives, supplemental information related to the basic structure and function of cells, and a glossary.

A number of high-quality, useful aids to instruction are available to faculty adopting *Nutrition Now:*

- An ***Instructor's Manual*** is replete with class outlines, student activities, test questions and answers, and transparency masters of the Nutrition Scoreboard for each unit.
- A set of 80 **full-color transparencies** of key illustrations in the text.
- **CNN Videos** available to qualified adopters. These news clips are a good way to introduce nutrition topics and ignite class discussion.
- **Nutrilink CD-ROM** presentation software supplement. This software contains illustrations from this and other introductory nutrition textbooks. Instructors can download images, photographs, and animations from Nutrilink into Power Point or other electronic presentation programs.
- **Diet Analysis** software and a **Nutrition Interactive CD-ROM** are available for students to supplement their learning. These are sold individually or may be bundled with the text for a reduced price.
- ***Nutrition Now* on CD-ROM** provides the full text with student quizzes, animations, and video clips to enhance learning.

Acknowledgments

Nutrition Now has been and continues to be the product of a group of creative, kind, and highly skilled people. From rough drafts to published copy, the atmosphere of shared enthusiasm made my part of the effort entirely fun. For making that possible, and for his inside-out knowledge of the ingredients of high-quality textbooks, I thank Pete Marshall, Publisher, Wadsworth Publishing. Special thanks go to the rest of the editorial team for their work on all facets of production. A pat on the back goes to Becky Tollerson, Marketing Manager, for her effort in marketing this book. The creative and talented mind behind the art work in *Nutrition Now* belongs to Ann Borman. (She is the reason I traveled to a little UPS box in a dark alley late at night to drop off chunks of the manuscript.) The text has been thoughtfully reviewed for author's mistakes by Patricia Lewis. Her terrific copy-editing job is much appreciated. Once again a miracle has happened. Batches of typed pages, taped-on notes, and ideas have been converted into a text to be proud of. My thanks to you all.

Textbooks don't get to see the light of day unless instructors know they are available and have a chance to hear about them and review them. I am fortunate to have representation of *Nutrition Now* placed in the highly capable hands of Wadsworth sales representatives and managers. I hope you enjoy reading this new edition.

The Instructor's Manual for *Nutrition Now* was developed by Judy Kaufman. Thank you for expanding the usefulness of the text.

The second edition of *Nutrition Now* received a thorough going-over by faculty who teach introductory nutrition. Their comments always make me sit up and pay attention. And you will note in the third edition that I did pay attention. To the reviewers, I send my thanks for your time and helpful comments.

Sara Long Anderson
Southern Illinois University, Carbondale

Raga Bakhit
Virginia Polytechnic Institute

Becky Black
Long Beach Community College

Elsa Ramirez Brisson
County of Monterey Department of Nutrition Services

Sandra Cobb
North Georgia College

Paula Cochrane
Albuquerque TVI Community College

Joyce Davis
Eastern New Mexico University

Bruce Evans
Huntington College

Gary Fosmire
The Pennsylvania State University

Catherine Jen
Wayne State University

Noelle Kehrberg
Western Carolina University

Andrea Koepke
Anderson University

John Leary
State University of New York, Cortland

Karen Lowry-Gordon
Kent State University

Shortie McKinney
Drexel University

Glenn McNeil
Fort Hays State University

Marcia Magnus
Florida International University

Suzanne Martin
College of the Ozarks

Dennis Miller
Cornell University

Stella Miller
Mount San Antonio College

Connie Mueller
Illinois State University

Lisa Sasson
New York University

Joanne Spaide
University of Northern Iowa

Janet Walhberg-Rankin
Virginia Polytechnic Institute

Marsha Read
University of Nevada, Reno

Roger Sargent
University of South Carolina

Joanne Steiner
Indiana University of Pennsylvania

Anne VanBeber
Texas Christian University

Joy Ward
Mount San Antonio College

Unit 1

Key Nutrition Concepts and Terms

Photo Disc

Nutrition Scoreboard

	TRUE	FALSE
1. Calories are a component of food.		
2. Nutrients are substances in food that are used by the body for growth and health.		
3. The Dietary Reference Intakes (DRIs) specify minimal levels of nutrients people should consume in their diet each day.		
4. Both high and low intakes of nutrients threaten health.		
5. Foods can basically be divided into two groups: those that are "good" for you and those that are "bad" for you.		

(Answers on next page)

Key Concepts and Facts

- → At the core of the science of nutrition are concepts that represent basic "truths" and serve as the foundation of our understanding about normal nutrition. (They are listed in Table 1.4 on p. 1-19.)
- → Most nutrition concepts relate to nutrients.

Answers to

Nutrition Scoreboard

	TRUE	FALSE
1. Calories are a measure of the amount of energy supplied by food. They're a property of food, not a component of food.		✓
2. That's the definition of nutrients.	✓	
3. There are no "minimum dietary intake" standards. The DRIs represent nutrient intake levels appropriate for nearly all healthy people.		✓
4. Excessive as well as inadequate intake levels of vitamins, minerals, and other nutrients can be harmful to health.	✓	
5. There are *no* good or bad foods, but there are healthy and unhealthy diets.		✓

The Meaning of Nutrition

To be surprised, to wonder, is to begin to understand.
—José Ortega y Gasset

What is nutrition? It can be explained by situations captured in photographs as well as by words. This introduction presents a photographic tour of real-life situations that depict aspects of the study of nutrition.

Before the tour begins, take a moment to make yourself comfortable and clear your mind of clutter. Take a careful look at the photographs, pausing to mentally describe in two or three sentences what each shows.

PhotoEdit, David Young-Wolff

Photo Researchers, Lawrence Midgale

S.I.I.D.E. Phototake

Sygma, D. Aubert

Photo Researchers, Ulrike Welsch

Photo Disc

PhotoEdit, Michael Newman

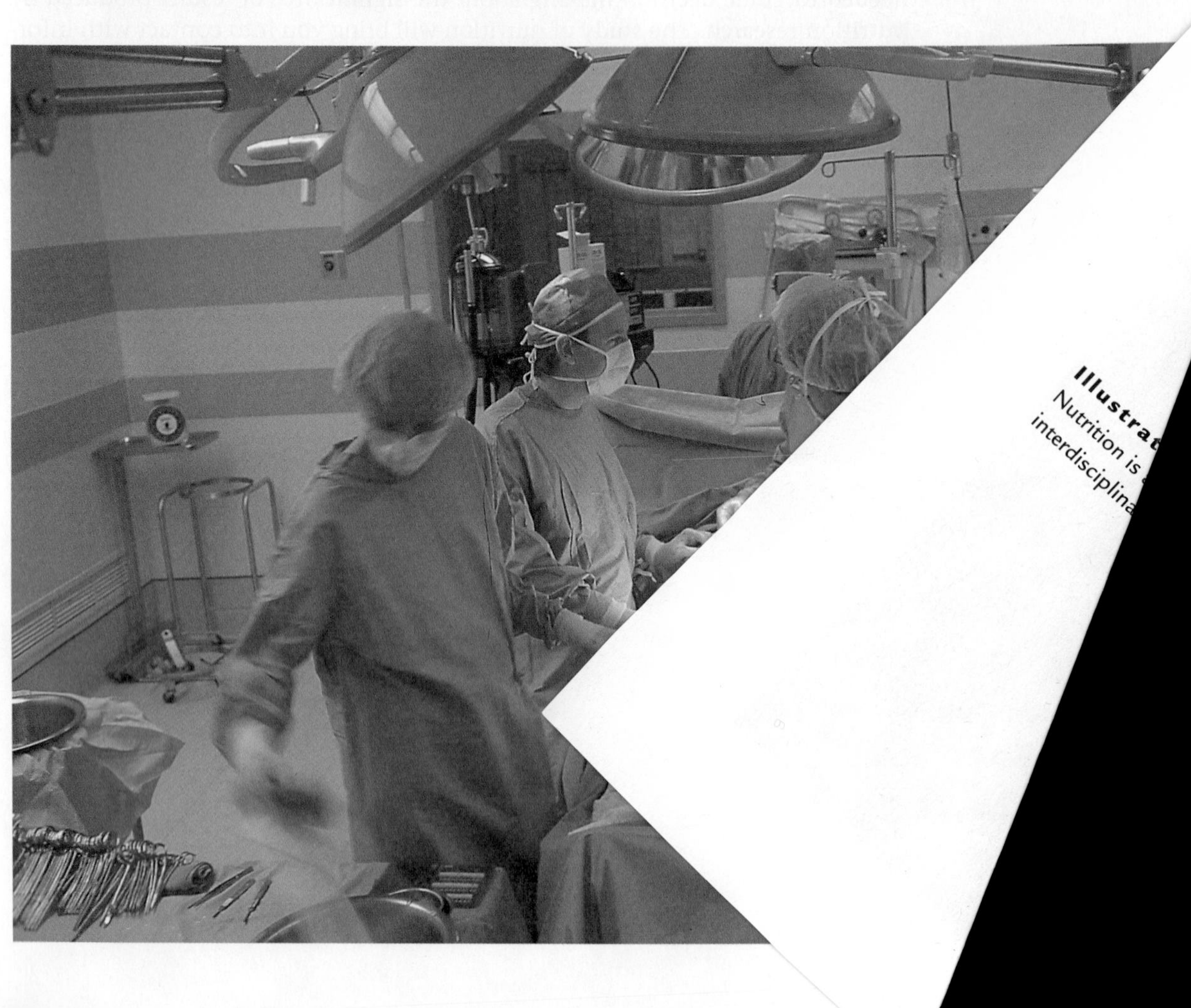

Illustrat

Nutrition is

interdisciplina

Not everyone who looks at the photographs will describe them in the same way. Reactions will vary somewhat due to personal experiences, interests, attitudes, and beliefs. An individual trying to gain weight will probably react differently to the photograph of the person on the scale than someone who is trying to lose weight. If you grew up in a family that farmed for a living, the picture of pesticides being sprayed on a crop may mean increased food production to you. But another person's reaction may be that pesticide residues on foods are harmful to health. Although knowledge about nutrition is generated by impersonal and objective methods, it can be a very personal subject.

Nutrition Defined

nutrition The study of foods, their nutrients and other chemical constituents, and the effects of food constituents on health.

In a nutshell, ***nutrition*** is the study of foods and health. It is a science that centers on foods, their nutrient and other chemical constituents, and the effects of food constituents on body processes and health. The scope of nutrition extends from food choices to the effects of specific components of foods on health.

Nutrition Is a "Melting Pot" Science

The broad scope of nutrition makes it an interdisciplinary science. Knowledge provided by the behavioral and social sciences, for example, is needed in studies that examine how food preferences develop and how they may be changed. Information generated by the biological, chemical, physical, and food sciences is required to propose and explain diet and disease relationships. Methods developed by quantitative scientists such as mathematicians and statisticians are needed to guide decision making about the significance of results produced by nutrition research. The study of nutrition will bring you into contact with information from a variety of disciplines (Illustration 1.1).

As you study this science, you will discover answers to a number of questions about your own diet, health, and eating behavior. Is obesity primarily due to eating behaviors, physical inactivity, or your genes? How do you know whether new information you hear about nutrition is true? Can sugar harm more than your teeth? Can the right diet or supplement give you a competitive edge? What is a healthful diet, how do you know if you have one, and, if improvements seem warranted, what's the best way to go about changing your diet for the better? These are just a few of the questions that nutrition and your course of study into it will address. If all goes well, you will take from this learn-

ion 1.1
n
ry science.

PhotoEdit, David Young-Wolff

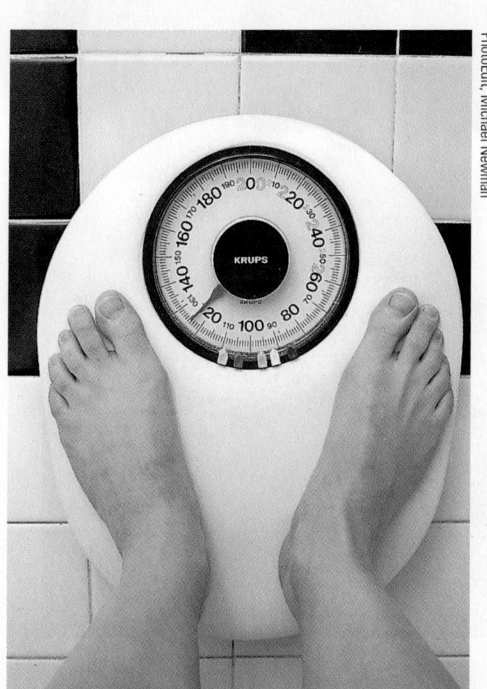

PhotoEdit, Michael Newman

Photo Disc

ing experience not only knowledge about nutrition and health, but skills that will keep the information and insights working to your advantage for a long time to come.

Foundation Knowledge for Thinking about Nutrition

Common sense requires a common knowledge.

You don't have to be a *bona fide* nutritionist to think like one. What you need is a grasp of the language and the basic concepts of the science. It's the purpose of this unit to give you this background. The essential topics covered here are explored in greater depth in units to come, and they build upon this foundation knowledge. With a working knowledge of nutrition terms and the concepts that provide the foundation of our understanding about nutrition, you will have an uncommonly good sense of nutrition.

NUTRITION CONCEPT **#1** **Food is a basic need of humans.**

Humans need enough food to live and the right assortment of foods for optimal health. In the best of all worlds, the need for food is combined with the condition of ***food security.*** People who experience food security have access at all times to a sufficient supply of safe, nutritious foods. They are able to acquire acceptable foods in socially acceptable ways—without having to scavenge or steal food, for example, in order to eat or to feed their families. ***Food insecurity*** exists whenever the availability of safe, nutritious foods, or the ability to acquire them in socially acceptable ways, is limited or uncertain (Illustration 1.2).[1]

Food insecurity exists in about 12% of U.S. households. It is most likely to occur in female-headed households with young children living in inner city areas. Compared to whites, African Americans are one and a half times more likely to experience food insecurity, and Hispanics are twice as likely.[2]

➡ **food security**
Access at all times to a sufficient supply of safe, nutritious foods.

➡ **food insecurity**
Limited or uncertain availability of safe, nutritious foods.

Photo Disc

Photo Disc

Illustration 1.2
"It is possible to go an entire lifetime without knowing about people's experiences with hunger."
—Meghan LeCates, Capitol Area Food Bank

NUTRITION CONCEPT #2 **Foods provide energy (calories), nutrients, and other substances needed for growth and health.**

People eat foods for many different reasons. The most compelling reason is that we need the ***calories, nutrients,*** and other substances supplied by foods for growth and health.

➡ **calorie**
A unit of measure of the amount of energy supplied by food. (Also known as a kilocalorie, or the "large Calorie" with a capital "C.")

➡ **nutrients**
Chemical substances found in food that are used by the body for growth and health. The six categories of nutrients are carbohydrates, proteins, fats, vitamins, minerals, and water.

➡ **phytochemicals (*phyto* = plants; pronounced fight-toe-chemicals)**
Chemical substances in plants, some of which perform important functions in the human body. Phytochemicals give plants color and flavor, participate in processes that enable plants to grow, and provide plants with protection against insects and diseases.

➡ **antioxidants**
Chemical substances that prevent or repair damage to cells caused by exposure to oxidizing agents such as environmental pollutants, smoke, ozone, and oxygen. Oxidation reactions are also a normal part of cellular processes.

Calories A calorie is a unit of measure of the amount of energy in a food—and of how much energy will be transferred to the person who eats it. Although we often refer to the number of calories in this food or that one, calories are not a substance present in food. And, because calories are a unit of measure, they do not qualify as a nutrient.

Nutrients Nutrients are chemical substances present in food that are used by the body (Illustration 1.3). Essentially everything that's in our body was once a nutrient in food we consumed.

There are six categories of nutrients (Illustration 1.4), and each category (except water) consists of a number of different substances used by the body for growth and health. The carbohydrate category includes simple sugars and complex carbohydrates (starches and dietary fiber). The protein category includes 20 amino acids, the chemical units that serve as the "building blocks" for protein. A number of different types of fat are included in the fat category. Of primary concern are the saturated fats, unsaturated fats, essential fatty acids, and cholesterol (read more about them in Illustration 1.4). The vitamin category consists of 13 vitamins, and the mineral category includes 15 minerals. Water makes up a nutrient category by itself.

Carbohydrates, proteins, and fats supply calories and are called the "energy nutrients." Although each of these three types of nutrients performs a variety of functions, they share the property of being the body's only sources of fuel. Vitamins, minerals, and water are chemicals needed for the conversion of carbohydrates, proteins, and fats into energy and for the building and maintenance of muscles, blood components, bones, and other parts of the body.

Other Substances in Food Food also contains many other substances, some of which are biologically active in the body. The ***phytochemicals*** are a major type of these substances, and there are over 2000 of them in plants. Examples of plant foods that are particularly rich sources of phytochemicals are shown in Illustration 1.5.

Phytochemicals provide plants with color, give them flavor, and foster their growth and health. A specific example of how one phytochemical works is described in the On the Side box (p. 1–10). In humans, consumption of phytochemicals in diets high in vegetables and fruits is strongly related to a reduced risk of developing certain types of cancer, heart disease, infections, and other disorders.[4]

Specific phytochemicals have names that are often hard to pronounce and difficult to remember. Nevertheless, here are a few examples. Plant pigments, such as lycopene (like-o-peen), which helps make tomatoes red, anthocyanins (an-tho-sigh-an-ins), which give blueberries their characteristic blue color, and beta-carotene (bay-tah-karo-teen), which imparts a dark yellow color to carrots, are phytochemicals that act as ***antioxidants.*** They protect plant cells and, in some cases, human cells, too, from damage that can make them susceptible to disease. Various types of sulfur-containing phytochemicals are present in cabbage, broccoli, cauliflower, brussels sprouts, and other vegetables of the same family. These substances may help prevent a number of different types of cancer.

Photo Disc

Illustration 1.3
Foods provide nutrients. "Please pass the complex carbohydrates, thiamin, and niacin . . . I mean, the bread!"

Illustration 1.4
Nutrients are grouped into six categories. Here the major types of nutrients in each category are listed, along with a description of each.

CARBOHYDRATES

Carbohydrates are substances in food that consist of a single sugar molecule or of multiples of them in various forms.

Simple sugars are the most basic types of carbohydrates such as glucose (blood sugar), sucrose (table sugar), and lactose (milk sugar).

Starches are complex carbohydrates consisting primarily of long, interlocking chains of glucose units.

Dietary fiber consists of complex carbohydrates found principally in plant cell walls. Because dietary fiber cannot be broken down by human digestive enzymes, it cannot be used for energy.

PROTEINS

Proteins are substances in food made of amino acids.

Amino acids are specific chemical substances from which proteins are made. Of the 20 amino acids, 9 are "essential," or a required part of our diet.

FATS

Fats are substances in food that are soluble in fat and not water.

Saturated fats are found primarily in animal products, such as meat, butter, and cheese, and in palm and coconut oil. Diets high in saturated fat may elevate blood cholesterol.

Unsaturated fats are found primarily in plant products, such as vegetable oil, nuts, and seeds. Unsaturated fats tend to lower blood cholesterol levels.

Essential fatty acids are two specific types of unsaturated fats that are required in the diet.

Cholesterol is a fat-soluble, colorless liquid found in animals but not plants. It can be manufactured by the liver.

VITAMINS

Vitamins are chemical substances found in food that perform specific functions in the body. Humans require 13 different vitamins in their diet.

MINERALS

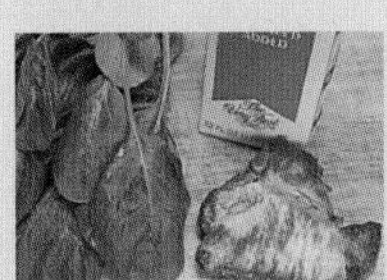

Minerals are chemical substances that make up the "ash" that remains when food is completely burned. Humans require 15 different minerals in their diet.

WATER

Water Most individuals need about 10 cups of water each day from food and fluids.

Illustration 1.5
Examples of good food sources of phytochemicals.

Photo Disc

➡ **essential nutrients** Substances required for normal growth and health that *cannot* be produced, or produced in sufficient amounts, by the body; they must be obtained in the diet.

➡ **nonessential nutrients** Nutrients required for normal growth and health that *can* be manufactured in sufficient quantities by the body from other components of the diet. We do not need a dietary source of nonessential nutrients.

Some Nutrients Must Be Provided by the Diet

Many nutrients are required for growth and health. The body can manufacture some of these from raw materials supplied by food, but others must come assembled. Nutrients that the body cannot make, or make in sufficient quantity, are referred to as ***essential nutrients.*** Here "essential" means "required in the diet." Vitamin A, iron, and calcium are examples of essential nutrients. Table 1.1 lists all the known essential nutrients.

Nutrients used for growth and health that can be manufactured by the body from components of food in our diet are considered ***nonessential.*** Cholesterol, creatine, and glucose are examples of nonessential nutrients. Nonessential nutrients are present in food and used by the body, but they are not required parts of our diet because we can produce them.

Both essential and nonessential nutrients are required for growth and health. The difference between them is whether we need a dietary source. A dietary deficiency of an essential nutrient will cause a specific deficiency disease, but a dietary lack of a nonessential nutrient will not. People develop scurvy (the vitamin C–deficiency disease), for example, if they don't consume vitamin C. But you could have zero cholesterol in your diet and not become "cholesterol deficient" because your liver produces cholesterol.

On the Side

The distinctive odor garlic produces when chopped or crushed is actually this plant's defense against insect predators. Garlic cloves contain an odorless, sulfur-containing compound called "alliin." When the clove is disrupted, alliin is released and reacts with an enzyme located in neighboring cells that converts the alliin to the odoriferous "allicin." Allicin is garlic's bug repellant—and the "people repellant" that makes many individuals shy about eating it.[3]

Photo Disc

Our Requirements for Essential Nutrients

The amount of essential nutrients humans need each day varies a great deal, from amounts measured in cups to micrograms. (See Table 1.2 to get a notion of the amount represented by a gram, milligram, and other measures. Illustration 1.6 helps give a sense of the scale.) Generally speaking, humans need 8 to 10 cups of water from fluids and foods, nine tablespoons of protein, one-fourth teaspoon of calcium, and only one-thousandth teaspoon (a 30-microgram speck) of vitamin B_{12} each day.

Table 1.1
Essential nutrients for humans: A reference table.

Energy Nutrients	Vitamins	Minerals	Water
Carbohydrates	Biotin	Calcium	Water
Fats[a]	Folate	Chloride	
Proteins[b]	Niacin (B_3)	Chromium	
	Pantothenic acid	Copper	
	Riboflavin (B_2)	Fluoride	
	Thiamin (B_1)	Iodine	
	Vitamin A	Iron	
	Vitamin B_6 (pyroxidine)	Magnesium	
		Manganese	
	Vitamin B_{12}	Molybdenum	
	Vitamin C (ascorbic acid)	Phosphorus	
		Potassium	
	Vitamin D	Selenium	
	Vitamin E	Sodium	
	Vitamin K	Zinc	

[a]Fats supply the essential nutrients linoleic and alpha-linolenic acid.
[b]Proteins are the source of 9 "essential amino acids": histidine, isoleucine, leucine, lysine, methionine, phenylalanine, threonine, trptophan, and valine. The other 11 amino acids are not a required part of our diet; they are considered "nonessential."

We all need the same nutrients, but not always in the same amounts. The amounts needed vary among people based on:

- Age
- Sex
- Growth status
- Body size
- Genetic traits

and the presence of conditions such as:

- Pregnancy
- Breastfeeding
- Illnesses
- Drug use
- Exposure to environmental contaminants

Table 1.2
Units of measure commonly employed in nutrition.

Measure	Abbreviation	Equivalents
Kilogram	kg	1 kg = 2.2 lb = 1000 grams
Pound	lb	1 lb = 16 oz = 454 grams = 2 cups (liquid)
Ounce	oz	1 oz = 28 grams = 2 tablespoons (liquid)
Gram	g	1 g = 1/28 oz = 1000 milligrams
Milligram	mg	1 mg = 1/28,000 oz = 1000 micrograms
Microgram	mcg, µg	1 mcg = 1/28,000,000 oz

1 egg = 50 grams or 1¾ oz; 212 milligrams (0.2 grams) of cholesterol in yolk

1 slice of bread = 1 oz = 28 grams

1 nickel = 5 grams

1 teaspoon of sugar = 4 grams, 1 grain of sugar = 200 micrograms

Photo Disc

Photo Disc

Illustration 1.6
The magnitude of a million. An olympic-sized swimming pool is said to contain about one million drops of water. One gram (approximately ¼ teaspoon of sugar) equals 1 million micrograms. Most nutrients are measured in grams, milligrams, or micrograms.

COMMENTARY

Dietary Reference Intakes:
Vitamin A, Vitamin K, Arsenic, Boron, Chromium, Copper, Iodine, Iron, Manganese, Molybdenum, Nickel, Silicon, Vanadium, and Zinc

Food & Nutrition Board

Illustration 1.7
The final set of Dietary Reference Intake levels on calories, protein, fat, and carbohydrate is scheduled for release in 2002.

Each of these factors, and others, can influence nutrient requirements. General recommendations for diets that provide all the essential nutrients usually make allowances for the major factors that influence the level of nutrient need, but they cannot allow for all of the factors.

The Recommended Dietary Allowances (RDAs) are the most widely used standard for identifying desired levels of essential nutrient intake in healthy people. Different RDAs are provided based on age and gender and according to whether a woman is pregnant or breastfeeding.

The first edition of the RDAs was published in 1943. It was prompted by the high rejection rate of World War II recruits due to underweight and nutrient deficiencies. The most recent update, which is taking place between 1997 and 2002, is giving the RDAs a whole new look—and a new name: Dietary Reference Intakes (DRIs) (Illustration 1.7).

The Revolutionary New DRIs Previous editions of the RDAs were based on levels of essential nutrients that protected people from deficiency diseases such as scurvy and rickets. As the science of nutrition advances, it is becoming abundantly clear that other components of food besides essential nutrients affect health. It is also becoming apparent that levels of nutrient intake associated with deficiency disease prevention may be too low to help prevent cancer, heart disease, and osteoporosis. Excessively high intakes of nutrients from fortified foods and supplements are another problem that was not considered when previous editions of the RDAs were prepared.

RDAs for the essential nutrients and other biologically active components of food are now being updated, and safe, upper levels of intake are being developed. The recommended daily levels of intake will not only meet the nutrient needs of almost all healthy people, but will also help reduce the risk of chronic disease.[5] Recommended intakes of some nutrients, such as calcium, vitamin D, and fluoride, for which too little conclusive evidence on disease prevention exists, are being represented by an "Adequate Intake," or AI, level. (Table 1.3 identifies the new terms and their abbreviations and includes a graphic representation of their meanings.) Regardless of which label is attached, the RDAs and AIs represent the best estimates of nutrient intake levels that promote optimal health. These values are represented in the main DRI tables presented on the inside front cover of this book. The term "Estimated Average Requirement" (EAR) has been introduced in the new dietary intake standards. It represents the intake level of a nutrient that is estimated to meet the requirement of 50% of the individuals within a group.

Estimates of safe, upper limits of nutrient intake are called "Tolerable Upper Intake Levels," abbreviated ULs. The ULs do not represent recommended or desired levels of intake. Instead, they represent total, daily levels of nutrient intake from food, fortified food products, and supplements that should not be exceeded. To reflect the fact that recommended nutrient intakes now include the RDAs, AIs, EARs, and ULs, the name of the recommendations has been broadened to Dietary Reference Intakes.

The new DRIs are also revolutionary in that they are being developed by experts from the United States and Canada, and will apply to the people in both countries. Updates on the latest available DRI reports can be obtained by visiting the web site listed at the end of the unit.

NUTRITION CONCEPT #3

Health problems related to nutrition originate within cells.

The main employers of nutrients are cells (Illustration 1.8). This is because all body processes required for growth and health take place within cells and the fluid that surrounds them. The human body contains more than one hundred trillion (100,000,000,000,000) cells. (Which type is the most

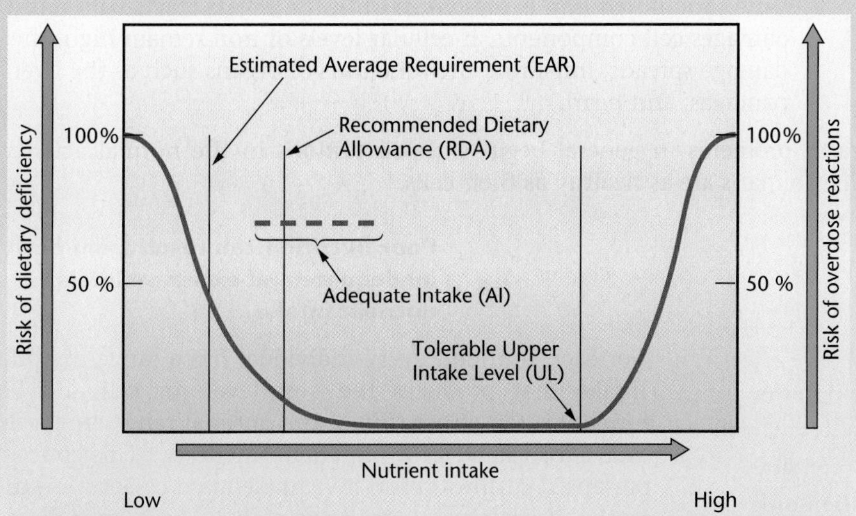

Table 1.3
Terms and abbreviations used in the 1997–2002 DRIs and a graphic representation of their meaning. (Source: Adapted from Yates et al. J Am Diet Assoc 1998; 98:702)

- **Dietary Reference Intakes (DRIs).** This is the general term used for the new nutrient intake standards for healthy people.
- **Recommended Dietary Allowances (RDAs).** These are levels of essential nutrient intake judged to be adequate to meet the known nutrient needs of practically all healthy persons while decreasing the risk of certain chronic diseases.
- **Adequate Intakes (AIs).** These are "tentative" RDAs. AIs are based on less conclusive scientific information than are the RDAs.
- **Estimated Average Requirements (EARs).** These are nutrient intake values that are estimated to meet the requirements of half the healthy individuals in a group. The EARs are used to assess adequacy of intakes of population groups.
- **Tolerable Upper Levels of Intake (ULs).** These are upper limits of nutrient intake compatible with health. The ULs do not reflect desired levels of intake. Rather, they represent total, daily levels of nutrient intake from food, fortified foods, and supplements that should not be exceeded.

common? See the On the Side insert on p. 1–14.) The functions of each cell are maintained by the nutrients it receives. Problems arise when a cell's need for nutrients differs from the amount of nutrients supplied.

Cells are the building blocks of tissues (such as muscles and bones), organs (such as the kidneys, heart, and liver), and systems (such as the respiratory, reproductive, circulatory, and nervous systems). Normal cell health and functioning are maintained when a nutritional and environmental utopia exists within and around the cells. Disruptions in the availability of nutrients, or the presence of harmful substances in the cell's environment, initiate diseases and disorders that eventually affect tissues, organs, and systems. Here are two examples of how cell functions can be disrupted by the presence of low or high levels of nutrients:

1. Folate, a B vitamin, participates in cellular reactions that keep the level of homocysteine (an animo acid–like substance) normal. When too little folate is available, homocysteine accumulates in cells and spills into the blood. Once in the blood, high levels of homocysteine enhance plaque formation in arteries and increase the risk of heart disease.

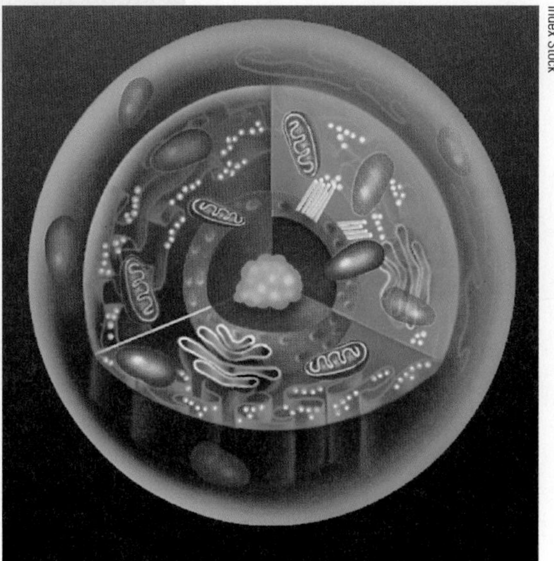

Index Stock

Illustration 1.8
Schematic representation of the structure of a human cell.

2. When too much iron is present in cells, the excess reacts with and damages cell components. If cellular levels of iron remain high, the damage spreads, impairing the functions of organs such as the liver, pancreas, and heart.

Health problems in general begin with disruptions in the normal activity of cells. Humans are as healthy as their cells.

On the Side

What's the most common cell in the body? It's the red blood cell. The human body contains more than 30 billion of them.

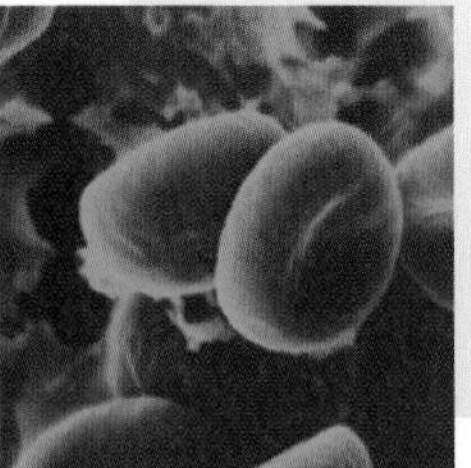

Phototake

NUTRITION CONCEPT #4 **Poor nutrition can result from both inadequate and excessive levels of nutrient intake.**

For each nutrient, every individual has a range of optimal intake that produces the best level for cell and body functions. On either side of the optimal range are levels of intake associated with impaired functions.[6] This concept is presented in Illustration 1.9. Inadequate essential nutrient intake, if prolonged, results in obvious deficiency diseases. Marginally deficient diets produce subtle changes in behavior or physical condition. If the optimal intake range is exceeded, mild to severe changes in mental and physical functions occur, depending on the amount of the excess and the nutrient. Overt vitamin C deficiency, for example, produces bleeding gums, pain upon being touched, and a failure of bone to grow. A marginal deficiency may cause delayed wound healing. On the excessive side, high intakes of vitamin C may contribute to the development of diarrhea and kidney stones.[7] Nearly all cases of vitamin and mineral overdose result from the excessive use of supplements—they are almost never caused by foods. For nutrients, "enough is as good as a feast."

Steps in the Development of Nutrient Deficiencies and Toxicities

Poor nutrition due to inadequate diets generally develops in the stages outlined in Illustration 1.10. To help explain the stages, this illustration includes an example of how vitamin A deficiency develops. After a period of deficient intake of an essential nutrient, tissue reserves of the nutrient become depleted. Blood levels of the nutrient then decrease because there are no reserves left to replenish the blood supply. Without an adequate supply of the nutrient in the blood, cells get shortchanged. They no longer have the supply of nutrients needed to

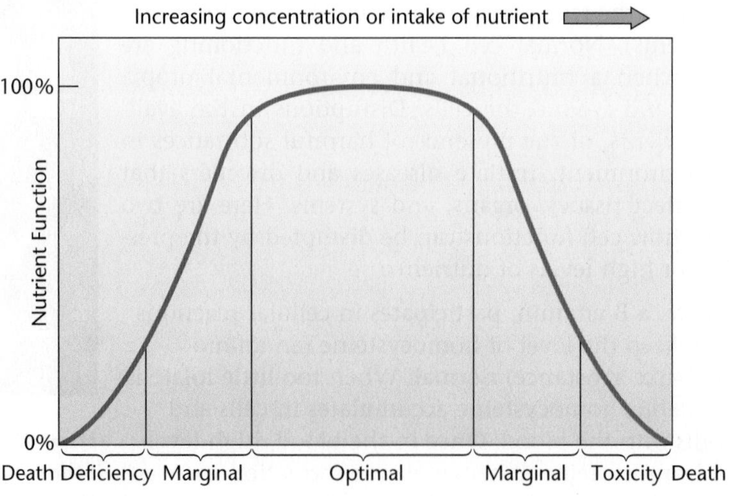

Illustration 1.9
For every nutrient, there is a range of optimal intake that corresponds to the optimal functioning of that nutrient in the body.

maintain normal function. If the dietary deficiency is prolonged, the malfunctioning cells cause sufficient impairment to produce physically obvious signs of a deficiency disease. Eventually, some of the problems produced by the deficiency may no longer be repairable, and permanent changes in health and function may occur. In most cases, the problems resulting from the deficiency can be reversed if the nutrient is supplied before this final stage occurs.

Excessively high intakes of many nutrients such as vitamin A and selenium produce toxicity diseases. The vitamin A toxicity disease is called "hypervitaminosis A," and the disease for selenium toxicity is called "selenosis." Signs of the toxicity disease stem from increased levels of the nutrient in the blood and the subsequent oversupply of the nutrient to the cells. The high nutrient load upsets the balance needed for normal cell function. The changes in cell functions lead to the signs and symptoms of the toxicity disease.

For both deficiency and toxicity diseases, the best time to correct the problem is usually at the level of dietary intake, before tissue stores are adversely affected. In that case, no harmful effects on health and cell function occur—they are prevented.

Nutrient Deficiencies Are Usually Multiple Most foods contain many nutrients, so poor diets will affect the intake level of more than one nutrient. Inadequate diets generally produce a spectrum of signs and symptoms related to multiple nutrient deficiencies. For example, protein, vitamin B_{12}, iron, and zinc are packaged together in many high-protein foods. The protein-deficient children you may see in news reports on television are rarely deficient just in protein. They likely have iron, zinc, and other deficiencies in addition to protein.

The "Ripple Effect" Dietary changes affect the level of intake of many nutrients. Switching from a high-fat to a low-fat diet, for instance, generally results in a lower intake of calories, cholesterol, and vitamin E as well. So, dietary changes introduced for the purpose of improving the intake level of a particular nutrient produce a "ripple effect" on the intake of other nutrients.

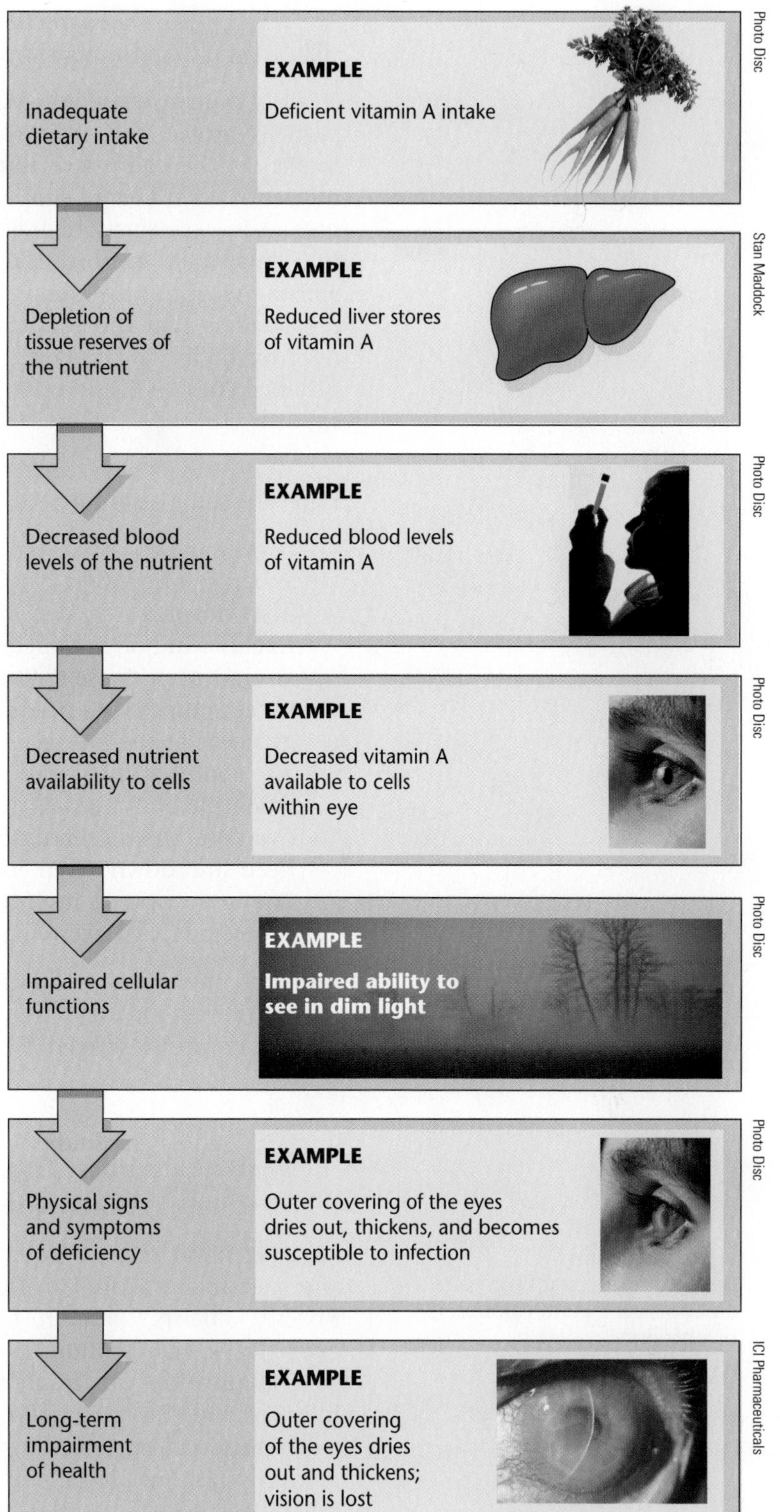

Illustration 1.10
The usual sequence of events in the development of a nutrient deficiency and an example of how vitamin A deficiency develops.

Source: Adapted with permission: Goodman AS. Vitamin A and retinoids in health and disease. N Engl J Med 1984;310:1023–31. Copyright 1984. Massachusetts Medical Society. All rights reserved.

NUTRITION CONCEPT #5

Humans have adaptive mechanisms for managing fluctuations in nutrient intake.

Healthy humans are equipped with a number of adaptive mechanisms that partially protect the body from poor health due to fluctuations in dietary intake. In the context of nutrition, adaptive mechanisms act to conserve nutrients when dietary supply is low, and to eliminate them when they are present in excessively high amounts. Dietary surpluses of energy and some nutrients, such as iron, calcium, vitamin A, and vitamin B_{12}, are stored within tissues for later use. In the case of iron and calcium, absorption is also regulated so that the amount absorbed changes in response to the body's need for them. The body has a low storage capacity for other nutrients, for instance vitamin C and water, and excesses are eliminated through urine or stools.

Here are some examples of how the body adapts to changes in dietary intake:

- When caloric intake is reduced by fasting, starvation, or dieting, the body adapts to the decreased supply by lowering energy expenditure. Declines in body temperature and the capacity to do physical work also act to decrease the body's need for calories. When caloric intake exceeds the body's need for energy, the excess is stored as fat for energy needs in the future.
- The ability of the gastrointestinal tract to absorb dietary iron increases when the body's stores of iron are low. To protect the body from iron overdose, the mechanisms that facilitate iron absorption in times of need shut down when enough iron has been stored.
- The body protects itself from excessively high levels of vitamin C from supplements by excreting the excess in the urine.

Although these built-in mechanisms do not protect humans from all the consequences of poor diets, they do provide an important buffer against the development of nutrient-related health problems.

Malnutrition can result from poor diets and from disease states, genetic factors, or combinations of these causes.

➡ **malnutrition**
Poor nutrition resulting from an excess or lack of calories or nutrients.

Malnutrition means "poor" nutrition and results from both inadequate and excessive availability of calories and nutrients in the body. Vitamin A toxicity, obesity, vitamin C deficiency (scurvy), and underweight are examples of malnutrition.

Malnutrition can result from poor diets and also from diseases that interfere with the body's ability to use the nutrients consumed. Diarrhea, alcoholism, cancer, bleeding ulcers, and HIV/AIDS, for example, may be primarily responsible for the development of malnutrition in people with these disorders.

In addition, a percentage of the population is susceptible to malnutrition due to genetic factors. For example, people may be born with a tendency to produce excessive amounts of cholesterol, absorb high levels of iron, or utilize folate poorly. Some cases of obesity and underweight appear to be related to a combination of a genetic predisposition and dietary factors.

NUTRITION CONCEPT #7 **Some groups of people are at higher risk of becoming inadequately nourished than others.**

Women who are pregnant or breastfeeding, infants, growing children, the frail elderly, the ill, and those recovering from illness have a greater need for nutrients than people who fit into "none of the above." As a result, they are at higher risk of becoming inadequately nourished than other people. In cases of widespread food shortages, such as those induced by natural disasters or war, the health of these nutritionally vulnerable groups is compromised the soonest and the most.

Within the nutritionally vulnerable groups of pregnant and breastfeeding women, children, the frail elderly, and the ill, certain people and families are at particularly high risk of malnutrition. These are people and families who are poor and least able to secure food, shelter, and high-quality medical services. The risk of malnutrition is not shared equally among all persons within a population.

NUTRITION CONCEPT #8 **Poor nutrition can influence the development of certain chronic diseases.**

Poor nutrition does not result only in nutrient deficiency or toxicity diseases. Faulty diets play important roles in the development of heart disease, hypertension, cancer, osteoporosis, and other ***chronic diseases.*** Diets high in animal fat, for example, are related to the development of heart disease, low vegetable and fruit diets to cancer, low calcium diets to osteoporosis, and high sugar diets to tooth decay. The harmful effects of negative dietary practices on the development of certain diseases and disorders may take years to become apparent.

➡ **chronic diseases** Slow-developing, long-lasting diseases that are not contagious (e.g., heart disease, diabetes, and cancer). They can be treated but not always cured.

NUTRITION CONCEPT #9 **Adequacy, variety, and balance are key characteristics of a healthy diet.**

Healthy diets contain many different foods that together provide calories and nutrients in amounts that promote the optimal functioning of the body. A variety of food is required to obtain all of the different essential nutrients we need in our diet. No one food (except breast milk for young infants) contains all of the nutrients we need, and most single foods don't come close.

Many different combinations of foods can make up a healthy diet. It could include milk, broccoli, chicken, and apples, or it could include ants, snails, sea cucumbers, and burdock root. What's important is that the contributions of a variety of individual foods add up to an adequate and balanced diet.

Nutrient Density

Adequate diets are most easily achieved by including foods that are good sources of a number of nutrients but not packed with calories. Foods that provide multiple nutrients in appreciable amounts relative to calories are considered ***nutrient-dense.*** Those that provide calories and low amounts of nutrients are called ***empty-calorie*** foods. Illustration 1.11 shows an example of a nutrient-dense food and an empty-calorie food. It is obviously easier to build an adequate diet around nutrient-dense foods such as fruits, vegetables, breads, cereal, lean meats,

➡ **nutrient-dense foods** Foods that contain relatively high amounts of nutrients compared to their calorie value. Broccoli, collards, bread, cantaloupe, and lean meats are examples of nutrient-dense foods.

➡ **empty-calorie foods** Foods that provide an excess of calories in relation to nutrients. Soft drinks, candy, sugar, alcohol, and fats are considered empty-calorie foods.

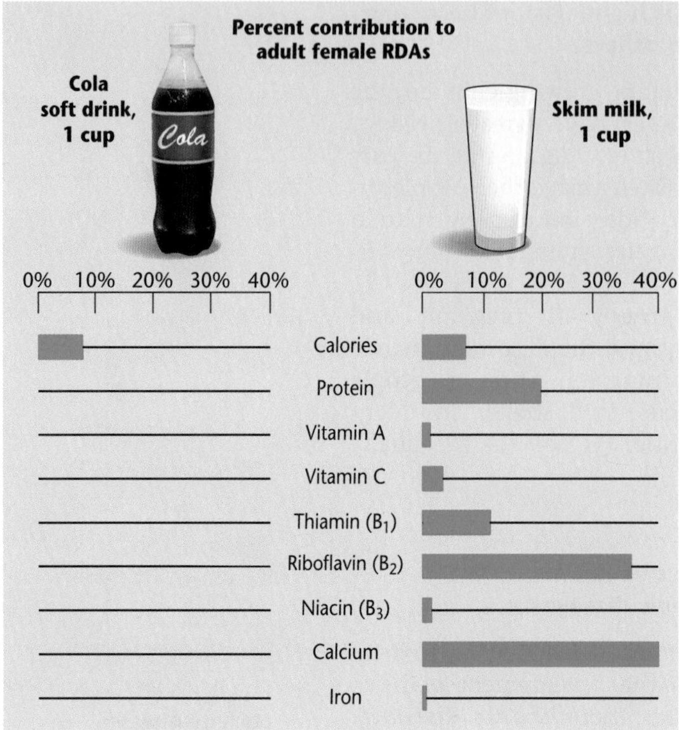

Illustration 1.11
An example of a food with a high nutrient density and a food with a low nutrient density. Percentages given represent percent contributions to adult female RDAs.

Photo Disc

Illustration 1.12
"Basic foods" are nutrient-dense.

and milk (Illustration 1.12) than around foods such as soft drinks, candy, pastries, chips, and alcoholic beverages.

NUTRITION CONCEPT #10 There are no "good" or "bad" foods.

All things in nutriment are good or bad relatively. —Hippocrates

Nearly 8 of 10 U.S. adults believe there are "good foods" and "bad foods."[8] Unless we're talking about spoiled stew, poison mushrooms, or something similar, however, there are no foods that are either good or bad. There are, however, combinations of foods that add up to a healthy or unhealthy diet. Consider the case of an adult who eats only foods thought of as "good": raw broccoli, apples, orange juice, boiled tofu, and carrots. Although all the foods consumed are nutrient-dense, they do not add up to a healthy diet because these foods don't supply a wide enough variety of the nutrients we need. Or, take the case of the teenager who occasionally eats fried chicken, but otherwise stays away from fried foods. The occasional fried chicken isn't going to knock his or her diet off the track. But the person who eats fried foods every day, with few vegetables or fruits, and loads up on soft drinks, candy, and chips for snacks has a bad diet. It's not any single food that defines a bad diet; it's the combination of foods generally consumed.

The basic nutrition concepts presented here are listed in Table 1.4. It may help you to remember the concepts and to start "thinking like a *bona fide* nutritionist," if you go back over each concept and give several examples related to it. If you understand these concepts, you will have gained a good deal of insight into nutrition.

1. Food is a basic need of humans.
2. Foods provide energy (calories), nutrients, and other substances needed for growth and health.
3. Health problems related to nutrition originate within cells.
4. Poor nutrition can result from both inadequate and excessive levels of nutrient intake.
5. Humans have adaptive mechanisms for managing fluctuations in nutrient intake.
6. Malnutrition can result from poor diets and from disease states, genetic factors, or combinations of these causes.
7. Some groups of people are at higher risk of becoming inadequately nourished than others.
8. Poor nutrition can influence the development of certain chronic diseases.
9. Adequacy, variety, and balance are key characteristics of a healthy diet.
10. There are no "good" or "bad" foods.

Table 1.4
Nutrition concepts.

Nutrition UP CLOSE

Nutrition Concepts Review

FOCAL POINT:
Nutrition concepts apply to diet and health relationships.

Write the number of the nutrition concept from Table 1.4 that applies to the situation described. Use each concept and do not repeat concept numbers in your responses.

Nutrition Concept Number	Situation
1. _____	The Irish potato famine caused thousands of deaths.
2. _____ and _____	Otis mistakenly thought that as long as he consumed enough calories, protein, vitamins, and minerals, he would stay healthy no matter what he ate.
3. _____	I feel guilty every time I eat potato chips. I wish they weren't bad for me.
4. _____	Phyllis was relieved to learn that her chronic diarrhea was due to the high level of vitamin C supplements she had been taking.
5. _____	A low amount of iron in Tawana's red blood cells was the reason for her loss of appetite and low energy level.
6. _____	Far more young children than soldiers died as a result of the 10-year civil war in the Sudan.
7. _____	For the past 20 years, Don's idea of dinner was a big steak and potatoes. His recent heart attack changed his view of what's for dinner.
8. _____	During the two weeks they were backpacking in the Netherlands, Tomás and Ozzie ate very few vegetables and fruits. Their health remained robust, however.
9. _____	Zhang wasn't aware that he had the inherited condition hemochromatosis until he began taking iron supplements and developed iron overload symptoms.

FEEDBACK (answers to these questions) can be found at the end of Unit 1.

WWW links

www.eatright.org
The American Dietetic Association's web site lets you know how to find a dietitian and provides information on hundreds of nutrition resources.

www.nal.usda.gov/fnic/foodcomp
Visit this USDA site if you need food composition data.

www.usda.gov.cnpp
Source of the 2000 Dietary Guidelines for Americans.

www.iom.edu/fnb
Get updates and reports on the DRIs from this web site.

www.navigator.tufts.edu
A reliable, comprehensive search engine for scientifically solid sites.

www.healthfinder.gov
A gateway to reliable nutrition and health information.

www.pueblo.gsa.gov
Tons of consumer nutrition information.

REFERENCES

1. Anderson SA. Core indicators of nutritional state for difficult-to-sample populations. J Nutr 1990;120:1598.
2. Kendall A, et al. Position of the American Dietetic Association: domestic food and nutrition security. J Am Diet Assoc 1998;98:337–42.
3. Herber D. The stinking rose: oregano-sulfur compounds and cancer. Am J Clin Nutr 1997;66:425–6.
4. Craig WJ. Phytochemicals: guardians of our health. J Am Diet Assoc 1997;97(suppl 2):S199–S204.
5. Standing Committee on the Scientific Evaluation of Dietary Reference Intakes (Food and Nutrition Board, National Academy of Sciences). Washington, DC: National Academy Press; 1997.
6. Based on Mertz W. The essential trace elements. Science 1981;213:1332–8.
7. Hodges, RE. Vitamin C. In: Human nutrition, a comprehensive treatise. vol. 3B Nutrition and the adult. New York: Plenum; 1980: pp 73–96.
8. Nutrition and you: trends 2000 survey results released. ADA Courier 2000; 39:1,3.

Nutrition

UP CLOSE

Nutrition Concepts Review

Feedback for Unit 1

The nutrition concepts apply to the situations as follows:

1. 1
2. 2, 9
3. 10
4. 4
5. 3
6. 7
7. 8
8. 5
9. 6

Unit 2

The Inside Story about Nutrition and Health

Stock Photo America, Eising

Nutrition Scoreboard

	TRUE	FALSE
1. How long people live and how healthy they are depend on four factors: lifestyle behaviors, the environment to which people are exposed, genetic makeup, and access to quality health care.		
2. Diet is related to the top two causes of death in the United States.		
3. The body of modern humans was designed over 40,000 years ago.		

(Answers on next page)

Answers to

Nutrition Scoreboard

	TRUE	FALSE
1. There *are* no secrets to a long, healthy life.	✓	
2. Diet is associated with the development of heart disease and cancer (which, by the way, cause over half of all deaths in the United States).	✓	
3. Hairstyles may be different, but our bodies are the same as they were 40,000 years ago.	✓	

Key Concepts and Facts

- Health and longevity are affected by diet. Other lifestyle behaviors, genetic makeup, the environment to which we are exposed, and access to quality health care also affect health and longevity.
- Dramatic changes in the types of foods consumed by modern humans compared with early humans are related in some ways to the development of today's leading health problems.
- The health status of a population changes for the better or worse as diets change for the better or worse.
- The diets and health of Americans are periodically evaluated by national studies.

Nutrition in the Context of Overall Health

Think of your body as a machine. How well this machine performs depends on a number of related factors: the quality of its design and construction, the appropriateness of the materials used to produce it, and how well it is maintained.

A photocopying machine designed to produce 10,000 copies a day will break down sooner if it is used to make 20,000 copies a day. The repair call will, in all probability, come earlier if the machine is overused *and* poorly maintained, or if it has a part that doesn't work well. On the other hand, chances are good the copy machine will function at full capacity if it is free from design flaws, skillfully constructed from appropriate materials, properly used, and kept in good shape through regular maintenance.

Although much more complex and sophisticated, the human body is like a machine in some important ways. How well the body works and how long it lasts depend on a variety of interrelated factors. The health and fitness of the human machine depend on genetic traits (the design part of the machine), the quality of the materials used in its construction (your diet), and regular maintenance (your diet, other lifestyle factors, and health care).

The Neighborhood

Half way through his "Hearty Man" breakfast, Dwayne thought he heard some of his smaller arteries slamming shut.

Lifestyles exert the strongest overall influence on health and longevity (Illustration 2.1).[1] Behaviors that constitute our lifestyle, such as diet, smoking habits, illicit drug use or excessive drinking, level of physical activity or psychological stress, and the amount of sleep we get, largely determine whether we are promoting health or disease.[2] Of the lifestyle factors that affect health, our diet is one of the most important.[3] In a sense, it is fortunate that diet is related to disease development and prevention. Unlike age, gender, and genetic makeup, our diets are within our control.

People have an intimate relationship with food—each year we put over 1000 pounds of it into our bodies! (See the types and amounts of food eaten in a year by a family of four in Illustration 2.2.) Food supplies the raw materials the body needs for growth and health, and these, in turn, are affected by the types of food we usually eat. The diet we feed the human machine can hasten, delay, or prevent the onset of an impressive group of today's most common health problems.

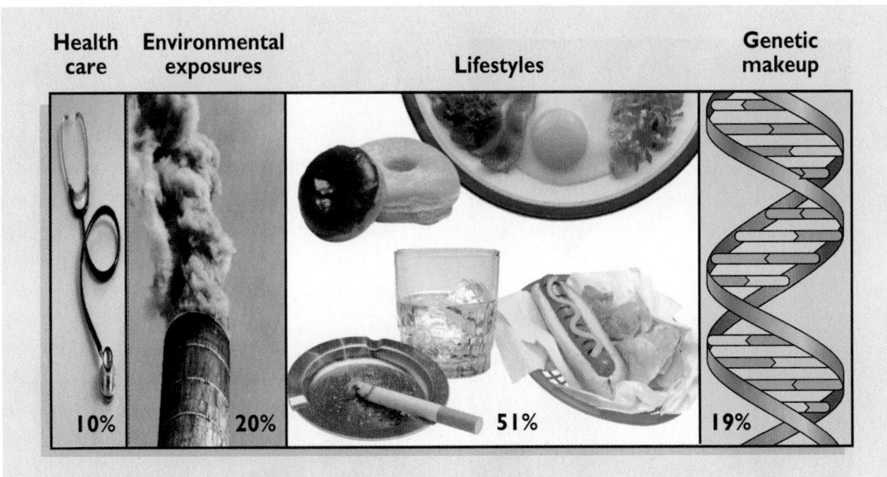

Photo Disc (1-3)

Illustration 2.1
Conditions that contribute to death among adults under the age of 75 years in the United States. Health care refers to access to quality care; environmental exposures include the safety of one's surroundings and the presence of toxins and disease-causing organisms in the environment; lifestyle factors include diet, exercise, smoking, and alcohol and drug use; and genetic makeup consists of inherited health traits.[1]

The Nutritional State of the Nation

Whosoever was the father of a disease, an ill diet was the mother.
—George Herbert, 1660

Since early in the twentieth century, researchers have known that what we eat is related to the development of vitamin and mineral deficiency diseases, to compromised growth and impaired mental development in children, and to the body's ability to fight off infectious diseases. Seventy years ago in the United States, widespread vitamin deficiency diseases filled children's hospital wards and contributed to serious illness and death in adults (Illustration 2.3). Now, however, dietary excesses are filling hospital beds and reducing the quality of life for millions of Americans.

➡ **chronic diseases**
Slow-developing, long-lasting diseases that are not contagious (e.g., heart disease, cancer, diabetes). They can be treated but not always cured.

Today, the major causes of death in Americans are slow-developing, lifestyle-related ***chronic diseases.*** As Illustration 2.4 indicates, the majority of Americans die from heart disease or cancer. Together these diseases account for 54% of all deaths. Diets high in saturated fat (the type found mainly in animal fat) and

Illustration 2.2
An American family of four surrounded by 5000 pounds of food representing what they eat in a year.

Source unknown

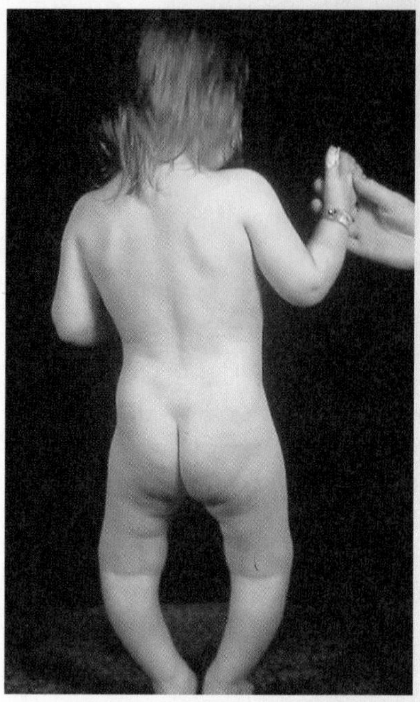

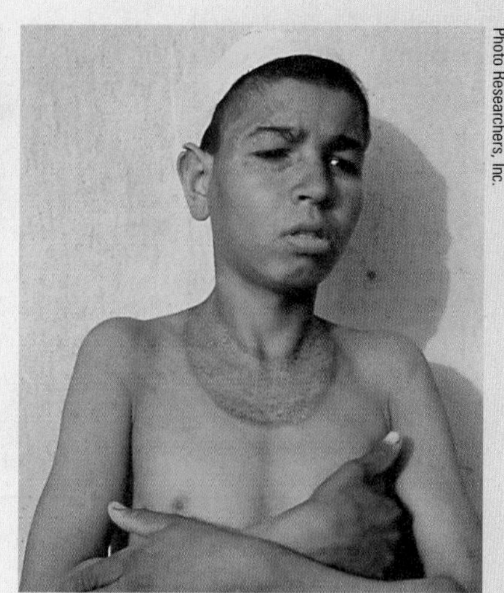

Illustration 2.3
Vitamin D deficiency (rickets) and niacin deficiency (pellagra) were leading causes of hospitalization of children in the United States in the 1930s.

➡ **diabetes**
A disease characterized by abnormally high levels of blood glucose.

➡ **stroke**
The event that occurs when a blood vessel in the brain suddenly ruptures or becomes blocked, cutting off blood supply to a portion of the brain. Stroke is usually associated with "hardening of the arteries" in the brain. (Also called a *cerebral vascular accident.*)

➡ **cirrhosis of the liver**
Degeneration of the liver, usually caused by excessive alcohol intake over a number of years.

cholesterol are linked to the development of heart disease. Certain types of cancer, such as breast and colon cancer, are related to low fruit and vegetable consumption and excessive levels of body fat.[4,5]

Diet is also related to three other leading causes of death: ***diabetes, stroke,*** and ***cirrhosis of the liver.*** Excessive body fat is closely associated with the development of diabetes in middle-aged adults, and excessive levels of alcohol intake contribute to stroke, accidents, and cirrhosis of the liver.[4]

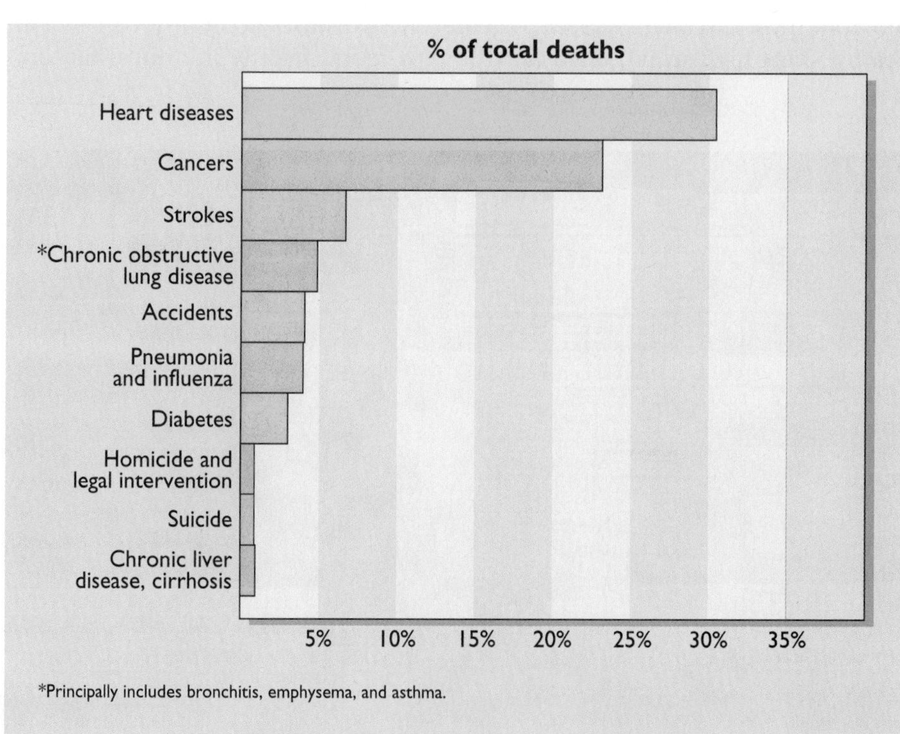

Illustration 2.4
Percentage of total deaths for the top 10 leading causes of death in the United States, 1998

Source: Health, United States, 2000, www.cdc.gov/nchs

That's not all. The types of food people eat are also associated with the risk of developing hypertension (high blood pressure), iron-deficiency anemia, tooth decay and gum disease, ***osteoporosis,*** and excessive levels of body fat:[5]

- High sodium intakes are associated with hypertension in some adults. Hypertension is one of the most common disorders in adults in the United States, and it is closely linked to heart disease and stroke.
- Low iron intake is a common problem among women and young children in the United States. Consequently, iron deficiency remains a public health problem in the United States today.
- The frequent consumption of sugars, especially in the form of "sticky sweets," is related to the development of tooth decay and gum disease. Adequate fluoride intakes help prevent tooth decay.
- Low intake of calcium, especially during the teen and young adult years, is associated with the development of osteoporosis and bone fractures. Nearly half of all women in the United States fail to get enough calcium in their diets.
- Approximately one-half of all adults in the United States are overweight or obese, and one in four children is overweight, too. Being overweight, or, more correctly, being overly fat, shortens life expectancy by increasing the likelihood that diabetes, heart disease, hypertension, and certain types of cancer will occur.[6]

➡ **osteoporosis**
A condition in which bones become fragile and susceptible to fracture due to a loss of calcium and other minerals.

The diet and health relationships overviewed here are summarized in Table 2.1. This discussion, however, doesn't cover all of the ways diet affects health. These and other ways diet affects health are examined in more detail in other units in this book.

The Importance of Food Choices People are not born with a compass that directs them to select a healthy diet—and it shows.[7] If given access to a food

Table 2.1
Examples of diseases and disorders linked to diet.

Disease or Disorder	Dietary Connections
Heart disease	High animal (saturated) fat and cholesterol intakes; low intakes of certain vitamins and minerals and vegetables and fruits; excessive body fat
Cancer	Low vegetable, fruit, and fiber intakes; excessive body fat and alcohol intake
Diabetes (primarily in adults)	Excessive body fat; low vegetable and fruit intake; high saturated fat intake
Cirrhosis of the liver	Excessive alcohol consumption; poor overall diets
Hypertension (high blood pressure)	High sodium (salt) and alcohol intake; excessive levels of body fat
Iron-deficiency anemia	Low iron intake
Tooth decay and gum disease	Excessive and frequent sugar consumption; inadequate fluoride intake
Osteoporosis	Inadequate intakes of calcium and vitamin D
Obesity	Excessive calorie intake and high-fat diets

supply like that available in the United States, people show a marked tendency to choose a diet that is high in meat and fat, and low in complex carbohydrates (Illustration 2.5). Such a diet also tends to include processed foods that are often high in salt or sugar and low in dietary fiber, vegetables, and fruits. It is this type of diet that poses the greatest risks to the health of Americans.

Diet and Diseases of Western Civilization

Why is the U.S. diet—a "Western" style of eating—hazardous to our health in so many ways? What is it about a diet that is high in animal fat, salt, and sugar and low in vegetables, fruits, and fiber that promotes certain chronic diseases? A good deal of evidence indicates that the chronic diseases now prevalent in the United States and other Westernized countries have their roots in dietary and other changes that have taken place over centuries.

Our Bodies Haven't Changed

Biological processes that control what the body does with food today were developed more than 40,000 years ago. These hardwired, evolution-driven processes exist today because they are firmly linked to the genetic makeup of humans and genes change very little over great spans of time.

For the first 300 centuries of existence, humans survived by hunting and gathering (Illustration 2.6). They were constantly on the move pursuing wild game or following the seasonal maturation of fruits and vegetables. Meat, berries, and many other plant products obtained from successful hunting and gathering journeys spoiled quickly, so they had to be consumed in a short time. Feasts would be followed by famines that lasted until the next successful hunt or harvest.

The bodies of modern humans, adapted to exist on a diet of wild game, fish, fruits, roots, vegetables, and grubs, to survive periods of famine, and to sustain a physically demanding lifestyle, are now exposed to a different set of circumstances. The foods we eat bear little resemblance to the foods available to our early ancestors. Sugar, salt, alcohol, food additives, and fats, such as oils, mar-

Photo Disc

Illustration 2.5
Lopsided, all-American food choices.

Photofile-Anthro

Photofile-Anthro

Illustration 2.6
These African tribespeople live very much as the primitive hunter-gatherers did, expending at least 3000 calories a day in search of wild game and plants.

garine, and butter, were not a part of their diets. These ingredients came with Western civilization. Furthermore, we do not have to engage in strenuous physical activity to obtain food, and our feasts are no longer followed by famines.

Then and Now The human body developed other survival mechanisms that are not the assets they used to be. Mechanisms that stimulate hunger in the presence of excess body fat stores, conserve the body's supply of sodium, and confer an innate preference for sweet-tasting foods, as well as a digestive system that functions best on a high-fiber diet, were advantages for early humans. They are not advantageous for modern humans, however, because diets and lifestyles are now vastly different.

Although the human body has a remarkable ability to adapt to changes in diet, health problems of modern civilization such as heart disease, cancer, hypertension, and diabetes are thought to result partly from diets that are greatly different from those of our early ancestors.[8] It appears that the human body was built to function best on a diet that is low in fat, sugar, and sodium, contains lean sources of protein, and is high in fiber, complex carbohydrates, and vegetables and fruits (Illustration 2.7). Strong evidence for this conclusion is provided by studies that track how disease rates change as people adopt a Western style of eating.

Illustration 2.7
The Western diet includes many processed foods that are high in fat, sugar, or sodium (left). The human body was designed to work best on a diet that includes whole grains, vegetables and fruits, and other basic foods (right).

Table 2.2
Life expectancy at birth for countries with high life expectancies.

Source: Health, United States, 2000. www.cdc.gov/nchs

Country	Life Expectancy (Years)
Japan	79.6
Sweden	79.0
Switzerland	78.5
Canada	78.3
Australia	78.0
France	77.9
Greece	77.9
Norway	77.8
Italy	77.7
Netherlands	77.5
Spain	77.0
United Kingdom	77.0
Austria	76.8
Finland	76.5
New Zealand	76.5
Belgium	76.4
Germany	76.3
United States	75.9
Luxembourg	75.8
Iceland	75.6
Denmark	75.2
Portugal	75.1

Illustration 2.8
A typical Japanese meal. Compare this to the "all-American" food choices shown in Illustration 2.5.

Changing Diets, Changing Disease Rates

Many countries are adopting the Western diet and the pattern of disease that accompanies it. People in Japan, for example, live longer than anyone else in the world—until they move to the United States (Table 2.2). In Japan, diets consist mainly of rice, vegetables, fish, shellfish, and meat (Illustration 2.8).[9,10] When Japanese people move to the United States, their diets change to include, on average, more meat and fat and fewer complex carbohydrate. Japanese living in the United States are much more likely to develop diabetes (Illustration 2.9), heart disease, breast cancer, and colon cancer than people who remain in Japan.[11]

Dietary habits in Japan are rapidly becoming similar to those in the United States. Hamburgers, fries, steak, ice cream, and other high-fat foods are gaining in popularity. Rates of diabetes, heart disease, and cancer of the breast and colon are on the rise in Japan.[9,12] Similarly, the "diseases of Western civilization" are occurring at increasing rates in Russia, Greece, Israel, and other countries adopting the Western diet.[8,12,13]

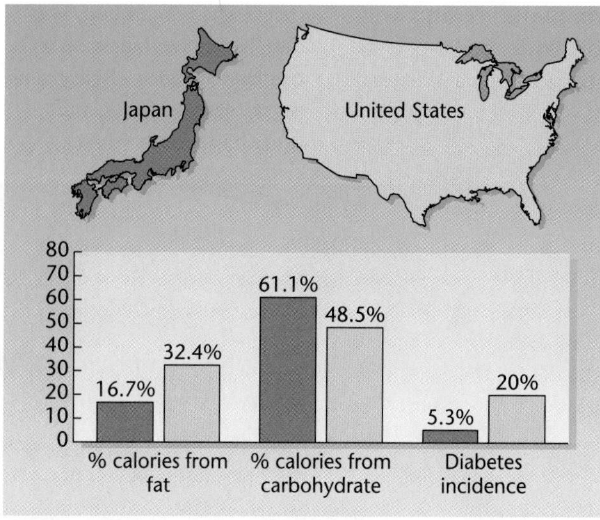

Illustration 2.9
An increased rate of diabetes in Japanese men immigrating to Seattle corresponds to dietary changes.[14]

The Power of Prevention

Heart disease, cancer, and other chronic diseases are not the inevitable consequence of Westernization. High-fat diets and lifestyle behaviors (such as smoking) that promote chronic disease can be avoided or changed. Although heart disease is still the leading cause of death in the United States, its rate has declined by 45% in the last 35 years. Reductions in animal fat and cholesterol intake, along with reductions in the proportion of people who smoke, are largely credited with this striking improvement in deaths from heart disease.[15]

Improvements in Health Status	Improvements in Health Practices	Improvements in Food Services, and Programs
Reduce rates of: · Overweight · Heart disease and stroke · Hypertension · Cancer · Osteoporosis · Diabetes · Congenital (in-born) abnormalities · Growth retardation · Iron deficiency · Baby bottle tooth decay · Food allergy deaths	*Increase rates of:* · Breastfeeding · Safe and effective weight-loss practices (diet and exercise) · Healthy weight gain in pregnancy *Reduce dietary intake of:* · Fat · Saturated fat · Sodium *Increase dietary intake of:* · Vegetables · Fruits · Grain and grain products (especially whole grains) · Calcium	· Improve food safety practices · Increase food security · Increase nutrition education in schools · Improve nutritional quality of school meals and snacks · Increase work site programs in nutrition and weight management · Increase nutrition counseling for patients with heart disease, diabetes, and other nutrition- and diet-related conditions

Table 2.3
A summary of nutrition objectives for the nation from *Healthy People 2010.*

Improving the American Diet

Many efforts are underway to improve the diet and health status of Americans. Like some other countries with high rates of "Western" diseases, the United States has set national health goals and implemented programs aimed at improving health. Goals for changes in health status in the United States are presented in the report, *Healthy People 2010: Objectives for the Nation.* Its objectives for improvements in nutrition are outlined in Table 2.3.

Attempts to improve the U.S. diet have led to the development of a set of guidelines for healthy eating. Known as the "Dietary Guidelines for Americans" (Illustration 2.10), this document identifies 10 key areas in need of improvement and provides practical suggestions for implementing dietary changes. The Dietary Guidelines, which are updated every five years, are a very useful tool for planning and evaluating dietary intake.

In addition to the Dietary Guidelines, recommendations for the selection of foods that contribute to a healthy diet are available. These recommendations are based on the food groups presented in the "Food Guide Pyramid." The Food Guide Pyramid is an updated version of the Basic Four Food Groups. It is shown in Illustration 2.11 (p. 2–10).

Illustration 2.10
Dietary Guidelines for Americans.

Nutrition Surveys: Tracking the American Diet

The food choices people make and the quality of the American diet are regularly evaluated by national surveys. Table 2.4 on p. 2–10 summarizes five major surveys that are conducted periodically by the federal government. The first of these surveys, the National Household Food Consumption Survey, was started in 1936. It was conducted in conjunction with the first national program aimed at reducing hunger, poor growth in children, and vitamin and mineral deficiency diseases in the United States.

Results of nutrition surveys are used to identify problem areas within the food supply and characteristics of diets

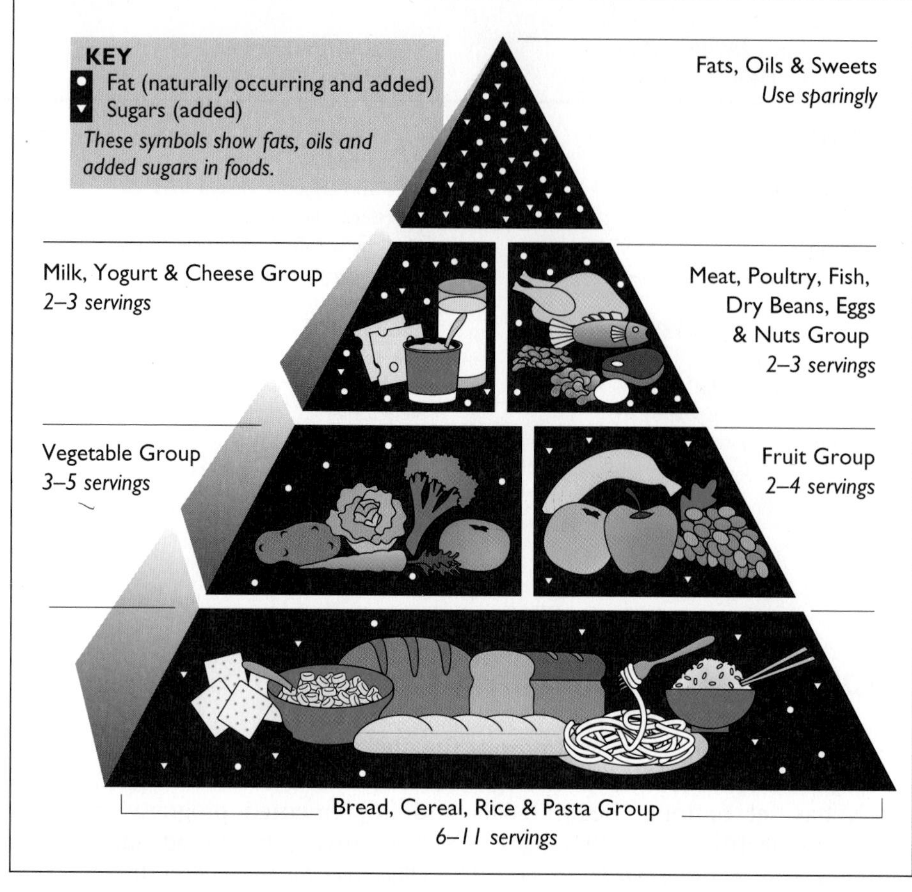

Illustration 2.1
Making daily f
the USD
Pyramid.

consumed by people in the United States. They provide information ranging from the amount of lead in certain foods to the adequacy of diets of low-income families. Together with the results of studies conducted by university researchers and others, they provide the information needed to give direction to food and nutrition programs and efforts aimed at improving the availability and quality of the food supply.

Table 2.4
Periodic surveys of diet and health in the United States conducted by federal health agencies and the Department of Agriculture.

1. **National Household Food Consumption Survey (NFCS)**
   - Conducted about every 10 years
   - Assesses the availability of foods and the level of nutrients in families' food supplies
2. **Continuing Survey of Food Intake by Individuals**
   - Conducted nearly every year
   - Assesses food intake of individuals
3. **Total Diet Study**
   - Conducted each year
   - Monitors the level of contamination of foods with certain industrial chemicals and pesticides in addition to nutrient content
4. **National Health and Nutrition Examinations Survey (NHANES)**
   - Conducted about every 10 years
   - Comprehensively assesses dietary intake and health of a sample of the U.S. population
5. **Nutrition monitoring**
   - Ongoing collection of data on the food supply and the diet and health of the U.S. population
   - Assesses certain characteristics of the nutritional health of low-income women and children

Nutrition UP CLOSE

Rate Your Nutrition Knowledge

FOCAL POINT: How well-informed are you? Compare your knowledge of nutrition against that of the typical American consumer.

Researchers at the Centers for Disease Control contacted a sample of U.S. adults and asked them a set of questions about diet and health relationships. If you answer all of these questions correctly, you know more than most.

1. In your opinion, which of these are the two best ways to lose weight?
   a. Do not eat at bedtime.
   b. Eat fewer calories.
   c. Take diet pills.
   d. Increase physical activity.
   e. Do not eat any fat.
   f. Eat a grapefruit at each meal.
2. Eating a diet high in animal fat ______ the risk of heart disease.
   a. increases
   b. decreases
   c. does not affect
3. A high blood cholesterol level ______ the risk of heart disease.
   a. increases
   b. decreases
   c. does not affect
4. Which of the following substances in food is most often associated with high blood pressure?
   a. cholesterol
   b. sugar
   c. sodium (or salt)
5. Which of the following aspects of diet is related to the development of osteoporosis?
   a. not enough dietary fiber
   b. too much fat
   c. not enough calcium
6. Which of the following aspects of diet is related to the development of certain types of cancer, such as colon cancer?
   a. too much sodium (or salt)
   b. not enough dietary fiber
   c. too much sugar

FEEDBACK (answers to these questions) can be found at the end of Unit 2.

Source: Adapted from material from the U.S. Government—Centers for Disease Control and Prevention.

WWW links

www.healthfinder.gov
Gateway for nutrition and health information.

web.health.gov/healthypeople
Provides access to *Healthy People 2010* objectives.

www.healthcanada.gov
Health status and goals for Canada.

www.nutrition.gov
One-stop shopping for information on government-sponsored nutrition programs, health statistics, and more.

www.iom.edu/fnb
Directs you to updates and reports on the Dietary Reference Intakes (DRIs).

www.mayohealth.org
Provides reliable health and medical information, an audio pronunciation guide of medical terms, and qualified staff to answer questions.

www.ncbi.nlm.nih.gov/PubMed
Access to scientific journal articles on nutrition topics via Medline.

REFERENCES

1. Centers for Disease Control, Disease Prevention and Health Promotion Unit. Atlanta; 1989.
2. Promoting Health: Intervention Strategies for Social and Behavioral Research, Institute of Medicine, National Academy of Sciences, July 2000.
3. Sauberlich HE. Implications of nutritional status on human biochemistry, physiology, and health. Clin Biochem 1984;17:132–42.
4. Weisburger JH. Approaches for chronic disease prevention based on current understanding of underlying mechanisms. Am J Clin Nutr 2000;71:1710S–4S; discussion 1715S–9S.
5. Krauss ML, et al. AHA Dietary Guidelines, Revision 2000. Circulation 2000;102:2296–311.
6. Deckelbaum RJ, Fisher EA, Winston M, et al. Summary of a scientific conference on preventive nutrition: pediatrics to geriatrics. Circulation 1999;100:450–6.
7. Story M, Brown JE. Do young children instinctively know what to eat? N Engl J Med 1987;316:103–6.
8. Milton K. Hunter-gatherer diets—a different perspective [editorial]. Am J Clin Nutr 2000;71:665–7.
9. Sugano M, Hirahara F. Polyunsaturated fatty acids in the food chain in Japan. Am J Clin Nutr 2000;71:189S–96S.
10. Matsuzaki T. Longevity, diet, and nutrition in Japan: epidemiological studies. Nutr Rev 1992;50:355–9.
11. Kato H, Haybuchi H. Study of the epidemiology of health and dietary habits of Japanese and Japanese Americans. Japan Journal of Nutrition 1989;47:121–30.
12. Cooper R, Schatzkin A. Recent trends in coronary risk factors in the U.S.S.R. Am J Public Health 1982;72:431–40.
13. Trichopoulou AD, Efstathiadis PP. Changes of nutrition patterns and health indicators at the population level in Greece. Am J Clin Nutr 1989;49(suppl):1042–7.
14. Tsunehara CH, Leonetti DL, Fujimoto WY. Diet of second-generation Japanese-American men with and without non-insulin-dependent diabetes. Am J Clin Nutr 1990;52:731–8.
15. Ernst ND, Sempos CT, et al. Consistency between US dietary fat intakes and fetal serum cholesterol concentrations. Am J Clin Nutr 1997;66(suppl):965S–72S.

Nutrition **UP CLOSE**

Rate Your Nutrition Knowledge

Feedback for Unit 2

The number in parentheses indicates the percentage of U.S. adults in the sample who responded correctly.

1. b and d (73%)
2. a (80%)
3. a (86%)
4. c (58%)
5. c (43%)
6. b (25%)

Unit 3

Ways of Knowing about Nutrition

Stock Food America, Eising

Nutrition Scoreboard

	TRUE	FALSE
1. It is illegal to convey false or misleading information about nutrition in magazine and newspaper articles and on television.		
2. Knowledge about nutrition is gained by scientific studies.		
3. "Double-blind" studies are used to diminish the "placebo effect."		

(Answers on next page)

Key Concepts and Facts

- For the most part, nutrition information offered to the public does *not* have to be true or even likely true.
- Nutrition information offered to the public ranges in quality from sound and beneficial to outrageous and harmful.
- Science is knowledge gained by systematic study. Reliable information about nutrition and health is generated by scientific studies.
- Misleading and fraudulent nutrition information exists primarily because of financial interests and personal beliefs and convictions.

Answers to

Nutrition Scoreboard

	TRUE	FALSE
1. Freedom of speech applies to information about nutrition in articles, speeches, pamphlets, and broadcasts. However, it is illegal to make false or misleading claims about nutrition in advertisements or on product labels and packaging.		✔
2. The "way of knowing" for the field of nutrition is science.	✔	
3. In a double-blind study, neither the research staff nor the research subjects know which subjects are getting the real treatment and which are receiving the fake treatment. This reduces the placebo ("sugar pill") effect, or changes in health that are due to the expectation that a specific treatment will have a particular impact on health.	✔	

How Do I Know if What I Read or Hear about Nutrition Is True?

Photo Disc

Maria, a college student: "You really ought to try this new Herbal Melt Down Diet I found on the Internet. I used the herbs for a week and lost five pounds!"

Max, a 50-year-old engineer: "If I hadn't been following a low-fat diet for the last 10 years, I would have had a heart attack by now—just as my dad did when he was 48."

Jessie, a premed student: "If I were you, I'd treat that urinary tract infection with cranberry juice."

Newspaper headline: "Eating Cauliflower Daily Prevents Breast Cancer!"

TV advertisement: "Lose fat, not muscle! Eat Complete Cereal every morning!"

How do you know if what you read or hear about nutrition is true? With only the information given in these examples, you couldn't know. In reality, however, this is how much of the information we receive about nutrition comes to us—in bits and pieces. The nutrition information offered to the public is a mix of truths, half-truths, and gossip. The information does not have to meet any standard of truth before it can be represented as true in books, magazines, newspapers, TV and radio reports and interviews, pamphlets, the Internet, and speeches.

Opinions expressed about nutrition are protected by the freedom of speech provisions of the U.S. Constitution. Although a tabloid article announcing that 19 foods have negative calories is misleading and fraudulent, it is not illegal. The promotion of nutritional remedies that are not known to work, such as amino acid supplements for hair growth and beef extract for the treatment of cancer, is likewise protected by freedom of speech. It is illegal, however, to put false or misleading information about a product or service on a product label, in a product insert, or in an advertisement. In addition, the U.S. and Canadian mail systems cannot be used to send or to receive payments for products that are fraudulent.

With so much misinformation available, it can be difficult to know what to do when we hear or read something about nutrition that may benefit us personally or perhaps help a friend or relative. Why does such a mix of nutrition sense and nonsense exist? How can you separate sound information from the highly questionable? Where does nutrition information you can trust come from? These questions are addressed in this unit.

Photo Researchers, David Fasier

Why Is There So Much Nutrition Misinformation?

In addition to protections afforded by freedom of speech, there are other reasons why consumers are bombarded with nutrition misinformation and ineffective or untested products and services. These reasons can be grouped into two basic categories: profit, and personal beliefs and convictions. The first and the more important reason, not surprisingly, is the profit motive.

Motivation for Nutrition Misinformation #1: Profit As long as consumers seek quick and easy ways to lose weight, build muscle, slow aging, and reduce stress, goals that cannot be achieved quickly or easily, there will continue to be a huge market for nutritional products and services that offer assistance. The financial incentives are in place:

> . . . [join] a special company with 22nd century breakthrough nutritional products and unequaled compensation plan!
>
> —*USA Today* business advertisement

> Teaming up nutritional science with network marketing makes an awesome twosome!
>
> —Promotional article for "Nutritional System" in an advertising magazine

Not everyone who is in the business of nutrition has the goal of maintaining or improving people's health. Many are in the business to make money from people who are willing to believe their advertisements and buy products or services just in case they work.

People believe nutrition nonsense and buy fraudulent nutrition products for a number of reasons:

- Many people believe what they want to hear.
- People tend to believe what they see in print.
- Promotional materials are made to sound scientific and true.
- The products offer solutions to important problems that have few or no solutions in orthodox health care.
- Promotional materials often appeal to people who are disenchanted with traditional medical care and the side effects of many medications, or who want a "natural" remedy.
- Laws that govern truth in advertising are often not enforced.
- New bogus products and services find a ready market because existing bogus products and services don't work.

Instead of evidence and facts, profit-oriented companies use paid testimonials ("It worked for me, it will work for you!"), "medical experts," and Hollywood stars and sports heroes to promote their products. Their advertisements use terms such as "a wonder to science" or "miraculous" that appeal to some people's inclinations toward the mysterious or the divine. Real remedies aren't advertised with such terms. Nor are they portrayed as being effective because they come from Europe, the Ecuadoran highlands, the ancient Orient, or organic algae ponds.

Illustration 3.1 offers a formula for developing and marketing a fraudulent nutrition product. Try following the steps and make up your own miraculous nutritional cure. Once you've devised your own product, you'll find it easy to detect when other people are doing it for real.

Illustration 3.1

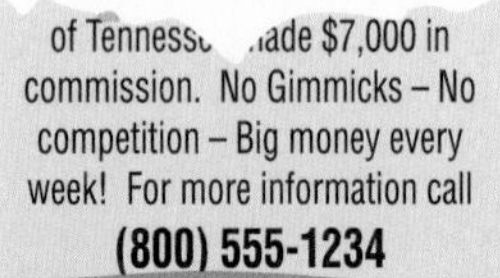

Create Your Own Fraudulent Nutrition Product

A Step-by-Step Guide

1. Identify a common problem people really want fixed that cannot easily or quickly be fixed another way.

   Examples: Obesity, low energy, weak muscles.

2. Make up a nutritional remedy and connect it to a biological process in the body. Try to think of a remedy that probably won't harm anyone. While you are at it, create a catchy name for your product.

   Examples: An herb that burns fat by speeding up metabolism, vitamins that boost the body's supply of energy, protein supplements that go directly to your muscles.

3. Develop a scientific-sounding explanation for the effect the product has on the body. Refer to the results of scientific studies.

   Examples: "Research has shown that the combination of herbs in this product stimulates the production of chemicals that trigger the energy-production cycle in the body." "As nutritional scientists have known for a century, the body requires B vitamins to form energy. The more you consume of this special formulation of B vitamins, the more energy you can produce and the more energy you will have!" "The unique combination of amino acids in this product is the same as that found in the jaw muscles of African lions. It is well known that an African lion can lift a 600-pound animal in its teeth."

4. Dream up testimonials from bogus previous users of the product, before-and-after photos, or "expert" opinions to quote. Use terms like "magical," "miraculous," "suppressed by traditional medicine," "secret," or "natural" as much as possible.

   Examples: Jane Fondu, Indianapolis: "I didn't think this product was going to work. The other products I tried didn't. My doctor couldn't help me lose weight. Thank God for [insert name of product]! I miraculously lost 20 pounds a week while eating everything I wanted." Dr. J. R. Whatsit, DM, NtD, Director of Nutritional Research: "We discovered this secret herb in our laboratories after years of looking for the substance in plants that keeps them from getting fat. Voilà! We found it. This discovery may win us a Nobel Prize." Or, show photos of a skinny person and a beefed-up person. (The photos don't have to be of the same person; the people only need to look similar.)

5. Offer customers a money-back guarantee.

Controlling Profit-Motivated Nutrition Frauds The Federal Trade Commission (FTC) has authority to remove from the airwaves and Internet advertisements that make false claims. It has exerted its authority in the past by removing blatantly false and misleading advertisements including those for the "European Weight Loss Patch," shark cartilage, and bee pollen. Nevertheless, misleading and inaccurate advertisements still appear because enforcement efforts are weak and are concentrated on very dangerous products. The FTC and other federal agencies do respond when a number of consumers register complaints about a nutritional product or advertisement.

Science for Sale Here's a true story:

> *Friday, 2:00* P.M. *A call came into the Nutrition Department from a man who wanted to talk to a nutrition expert. He had heard on last night's news that zinc lozenges were good for treating cold symptoms. Since he had a cold and didn't think the zinc would hurt, he purchased and consumed a whole roll of lozenges. Now he had a horrible, metallic taste in his mouth that he couldn't get rid of, no matter how often he brushed his teeth or gargled with mouthwash. He was worried that the taste was going to stay in his mouth forever. The nutrition expert assured him that it wouldn't. He had overdosed on zinc and the taste would go away slowly.*

This man was not the only one who heard the newscast about zinc and consumed too many lozenges the next day. Sales of zinc lozenges skyrocketed after the newscast, and as it happened, the author of the research study profited handsomely from the increased sales. After completing the study, the author bought shares of stock in the company that made the lozenges.[1] Such a financial tie should have been reported in the article. It is possible that a financial incentive could have influenced how he discussed the study's results.

Many researchers have a vested interest in their research results (Illustration 3.2). A study of 1000 Massachusetts scientists who had published research

On the Side

We should react continually to misinformation to combat quackery.
—Sara Short, Syracuse University

Have you seen an advertisement for a nutrition product that you suspect made false claims, or have you tried a product that didn't work or made you ill? Register your concerns by writing or e-mailing the following agency:

Health Care Fraud
Bureau of Consumer Protection, Federal Trade Commission
Washington, DC 20580;
or www.ftc.gov

"Dietol-7 Users Have Lost Over 5 Million Pounds!"
Available at most nutrition centers and other fine stores

HEALTH

Study links researchers to money

By Tim Friend
USA TODAY

SEATTLE — A study of 1,000 scientists who published their findings in major scientific journals shows at least a third had a financial stake in their research.

The study reported ... American As- ...

... ing affiliated with a biotechnology company involved in commercializing it. Although the study looked only at Massachusetts researchers, Krimsky says the situation is comparable to other stat... biomedical research and ...

"No one ...

Researcher Made Profit After Study

Invested in Maker of Cold Treatment

By PHILIP J. HILTS

A Cleveland researcher whose study showed that zinc lozenges might be an effective cold treatme... said yesterday that he had invest... in the company that makes the ... enges before the paper announci... the results was published. Once ... paper appeared, the company's stock soared, and he sold

Illustration 3.2
Researchers may have a financial interest in their research.

articles found that one-third held a patent for the product tested, were paid industry consultants, or had another form of financial stake in the research.[1]

Nutrition research is often conducted by people or companies that have a stake in the results. Although this doesn't mean the research results are invalid, it does mean that the study design should be carefully scrutinized before it is published in a scientific journal and broadcast on the news. People and companies with vested interests in the results of studies tend not to report findings that reflect negatively on a product, or their promotional materials may neglect to mention studies that produced different results. For these reasons, it is important to consider whether the study results or other presentations of nutrition facts may be tainted by a financial motive.

A Checklist for Identifying Nutrition Misinformation Illustration 3.3 lists some of the common features of fraudulent information for nutrition products and services. If you find any of these characteristics in articles, advertisements, Web sites, or pamphlets or hear them on infomercials or TV and radio interviews, beware of the product or service. People offering legitimate information are cautious and don't exaggerate nutritional benefits to health. They usually aren't selling anything but rather are trying to inform people about new findings so they can make better decisions about optimal nutrition.

A checklist for identifying nutrition misinformation

1. Is something being sold? ____ Yes ____ No
2. Does the product or service offer a new remedy for problems that are not easily or simply solved (e.g., obesity, cellulite, arthritis, poor immunity, weak muscles, low energy, hair loss, wrinkles, aging, stress)? ____ Yes ____ No
3. Are such terms as "miraculous," "magical," "secret," "detoxify," "energy restoring," "suppressed by organized medicine," "immune boosting," or "studies prove" used? ____ Yes ____ No
4. Are testimonials, before-and-after photos, or expert endorsements used? ____ Yes ____ No
5. Does the information sound too good to be true? ____ Yes ____ No
6. Is a money-back guarantee offered? ____ Yes ____ No

Illustration 3.3

The Business of Nutrition News Newspapers, TV stations, magazines, books, and other sources of information make money when the number of viewers or readers is high. To increase viewers or readers, the media attempt to present information that will pique people's interest. Nutrition breakthroughs tend to do that. The media have access to a large number of nutrition studies: more than 8000 nutrition-related research articles are published each year (Illustration 3.4).

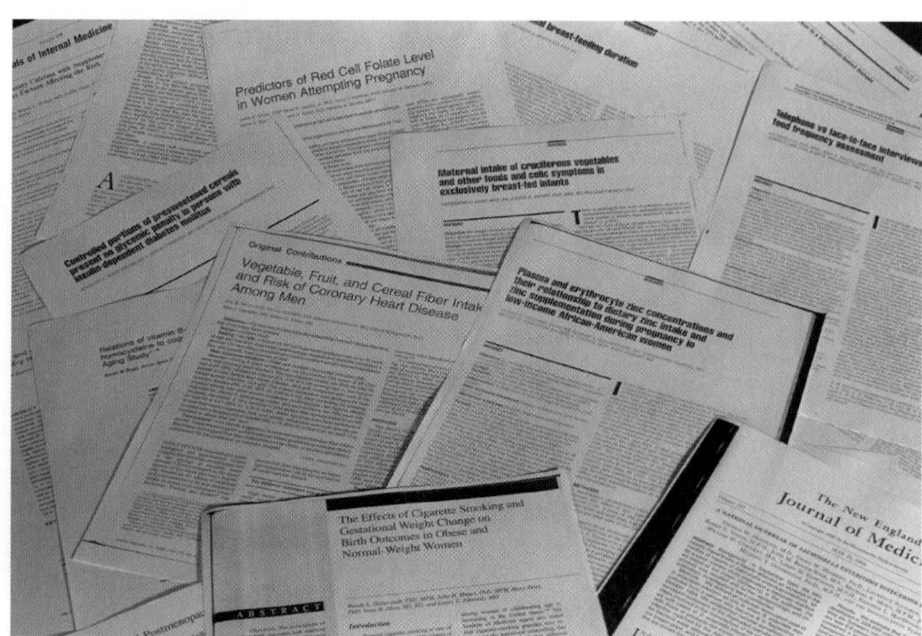

Illustration 3.4
More than 8000 research reports about nutrition appear in the scientific literature each year, or approximately 22 reports each day on average.

Studies reporting new results related to obesity, cancer, cholesterol, heart disease, vitamins, and food safety are particularly hot topics. To keep people interested, the media may sensationalize and oversimplify nutrition-related stories. They may report on one study one day and describe another study with opposite results the next. A classic example of headlines about nutrition that totally confused the public is shown in Illustration 3.5. Such study-by-study coverage of nutrition news leaves consumers not knowing what to believe or what to do. A Nutrition Trends Survey by the American Dietetic Association uncovered a strong consumer preference (81% of those sampled) for learning about the latest nutrition breakthroughs only *after* they had been generally accepted by nutrition and health professionals.

Nutrition studies are complex, and the results of one study are almost never enough to prove a point. Decisions about personal nutrition should be based on accumulated evidence that is broadly supported by nutritionists and other scientists.

Motivation for Nutrition Misinformation #2: Personal Beliefs and Convictions Numerous alternative health practitioners (such as nutripaths, irridologists, electrotherapists, scientologists, and faith healers) use nutrition remedies. These practitioners' approaches to and philosophies about health vary, but their belief in the benefits of what they do is often very strong. These firm beliefs may be enough to "talk" some people out of their problems—problems that might have gone away anyway.

There's no way to say whether the majority of remedies used by alternative nutrition practitioners work. (Some of the unproven remedies employed are shown in Illustration 3.6.) Strong belief in these remedies, rather than proof that they work or a desire for profits, may rule these practitioners' distribution of nutrition advice and products. Because the nutritional cures offered are not based on science and have not been scientifically evaluated, it is not possible to

Illustration 3.5
How to confuse the public: A lesson delivered by the headlines about oats and blood cholesterol. A consensus of scientific opinion about the effects of oat bran on blood cholesterol levels has now been reached: regular consumption of oat bran or oatmeal lowers LDL cholesterol (the so-called "bad" cholesterol) 5% or more.[2]

Illustration 3.6
Supplements that may be used to treat a variety of health problems. The safety and effectiveness of many of these products are yet to be proved.

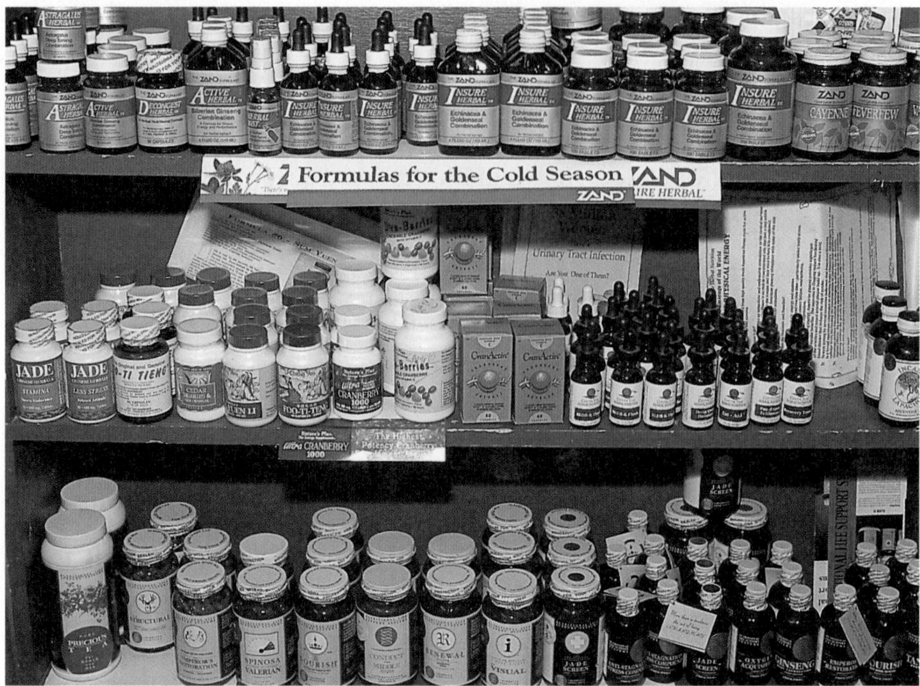

know whether they are safe or effective. Until they are evaluated, the logical conclusion is that they are neither.

Professionals with Embedded Beliefs Professionals who work in health care and research are not immune to the pull of deeply rooted convictions about diet and health relationships. Although scientists, they sometimes remain wedded to theories about diet and health relationships after they have been disproved. The nurse or doctor who holds onto the belief that salt intake should be restricted in pregnancy and the university professor who is convinced that pesticides on foods pose no risk to health (Illustration 3.7) are two examples of professional sources of misinformation. The failure of such professionals to give up erroneous convictions about diet and health adds to the flow of nutrition misinformation, and to consumer confusion and increased health risk.

Illustration 3.7

Source: © Sidney Harris

How Nutrition Truths Are Identified

The true method of knowledge is experiment.
—William Blake, 1788

Where does nutrition information you can count on come from? There is only one way to identify sound nutrition information—you have to put it to the test and see if it survives the dispassionate, systematic examination dictated by science.

Science is a unique and powerful explanatory system. It produces information based on facts and evidence. Our understanding of nutrition is based on scientifically determined facts and evidence obtained from laboratory, animal, and human studies. These types of studies provide information that qualifies for use when developing public policies about nutrition and health. (Illustration 3.8). Science delivers information eligible for inclusion in text-

The American Heart Association can help you choose heart-healthy foods when grocery shopping. Just look for products with our special, red heart-check mark. You'll find hundreds of them throughout your grocery store. Visit www.americanheart.org/FoodCertification/ for a list of certified foods. For more AHA Web sites, cookbooks and other helpful resources, see the list on pages 24 and 25.

American Heart Association
Meets American Heart Association food criteria for saturated fat and cholesterol for healthy people over age 2.

How Can I Use This Plan?

This booklet lists the basic food groups. It gives you the number of servings per day from each food group, serving sizes and suggested food choices. It's important to select a wide variety of foods low in saturated fat and cholesterol within each food group. Desserts, snacks and beverages are also listed.

Eating the lower number of servings from each food group gives you enough protein, vitamins and minerals — nutrients that your body needs each day. Eat moderate amounts of foods from the meat, fish, poultry, egg and fat groups. You may choose more servings of foods from the other groups if you don't need to lose weight or you want to gain weight.

The secret to success is balance. If you sometimes have a high-fat dish for dinner, balance it with lower-fat foods such as steamed vegetables or a fruit for dessert. The same idea applies to high-fat or salty snacks. Eat small amounts, and balance them with foods low in fat, cholesterol and sodium, such as fruits, vegetables and fat-free milk products.

The American Heart Association suggests this plan for all healthy Americans and children more than 2 years old. Growing children and teenagers have special needs. They must get enough energy (calories) and nutrients each day. Women who are pregnant or breast-feeding — or people who have a medical disorder such as diabetes, insulin resistance or obesity — should talk to their healthcare provider, a Registered Dietitian, or a licensed dietitian or nutritionist about their special dietary needs.

Eating Plan Tips

To control the amount and kind of saturated fat and dietary cholesterol you eat...

- Eat up to 6 ounces (cooked) per day of lean meat, fish and skinless poultry.
- Try main dishes featuring pasta, rice, beans and/or vegetables. Or create "low-meat" dishes by mixing these foods with small amounts of lean meat, skinless poultry or fish.
- Use cooking methods that require little or no fat — boil, broil, bake, roast, poach, steam, sauté, stir-fry or microwave.
- Use about 5 to 8 teaspoon servings of fats and oils per day for cooking and baking, and in salad dressings and spreads.
- Trim off the fat you can see before cooking meat and poultry. Drain off all fat after browning. Chill soups and stews after cooking so you can remove the hardened fat from the top.

4 5

American Heart Assn.

Illustration 3.8
Facts and evidence provided by scientific studies are used to develop nutrition information and guidelines that affect the public's health.

books about nutrition! (The information in nutrition textbooks is reviewed for scientific accuracy by many faculty members before it appears in print.)

Sources of Reliable Nutrition Information

Reliable sources of information about nutrition use the standards of proof required by science. They report decisions about nutrition and health relationships that are based on multiple studies and arrived at by "scientific consensus." These decisions represent the majority opinion of scientists knowledgeable about the particular nutrition topic.

Nutrition recommendations made to the public, such as the Dietary Guidelines for Americans and the Dietary Reference Intakes, are based on the consensus of scientific opinion. The information is made available to the public not to sell a product or to pass on an ideology, but to inform consumers honestly about nutrition and to help people use the information to maintain or improve their health. Organizations and individuals who offer reliable nutrition information on the Internet and in publications are listed in Table 3.1 (p. 3–10).

Who Are Qualified Nutrition Professionals? Many people refer to themselves as "nutritionists," but only a portion are qualified based on education and experience. These individuals are registered, licensed, or certified dietitians or nutritionists who meet qualifications established by national and state regulations. They have demonstrated a mastery of knowledge about the science of nutrition and appropriate clinical practices. Table 3.2 (p. 3–10) describes the qualifications of those who legitimately use the title dietitian or nutritionist. The specific titles vary somewhat depending on state regulations and laws governing the practice of nutrition and dietetics.

It is difficult for anyone to have all the information needed to make sound decisions about every nutrition question or issue that arises. As you try to make sense out of the information you read or hear, don't hesitate to get help. Visit the Web sites at the end of this unit, check out the index of this book for the relevant topic, or call your local health department or area Food and Drug

Table 3.1
Reliable sources of nutrition information.*

1. **Voluntary health organizations** such as the American Heart Association, the American Cancer Society, the National Kidney Foundation, and the American Red Cross.
2. **Scientific organizations** such as the National Academy of Sciences, the American Society for Clinical Nutrition, and the American Society of Nutrition Scientists.
3. **Professional organizations** such as the American Dietetic Association, the Society for Nutrition Education, and the American Heart Association.
4. **Government publications** such as nutrition, diet, and health reports from the National Institutes of Health, the Surgeon General, the Food and Drug Administration, the Centers for Disease Control, and the U.S. Department of Agriculture.
5. **Registered dietitians** in hospitals, public health departments, the Extension Service, and universities.
6. **Nutrition textbooks** used in college and university nutrition courses and the nutrition faculty of accredited universities.

*See Appendix B for additional information on reliable sources of nutrition information.

Administration office. The more solid the information you have, the better the decision you make will be.

The Methods of Science

Does the term "scientific method" conjure up an image of beady-eyed, white-coated scientists working diligently in windowless basement laboratories (Illustration 3.9)? Although we tend to assume that scientists are busy advancing our knowledge of nutrition and many other fields, the methods of science are a mystery to most people. They should not be. Consumers are big users of nutrition information. To make sound judgments about nutrition and health, people need to know how to distinguish results produced by scientific studies from those generated by personal opinion, product promotions, and bogus studies.

Table 3.2
Who's who in nutrition and dietetics: laws and regulations governing practice.[3]

Title	Qualifications
Registered dietitian	Individual who has acquired knowledge and skills necessary to pass a national registration examination and participates in continuing professional education. Qualification is conferred by the Commission of Dietetic Registration.
Licensed dietitian/nutritionist	Individual (usually a registered dietitian) who is qualified to practice nutrition counseling based on education and experience. Nonlicensed individuals who practice nutrition counseling can be prosecuted for practicing without a license. Licenses are conferred by state law; 40 states have such laws.
Certified dietitian/nutritionist	Individual who meets certain educational and experience qualifications. Unqualified individuals may practice nutrition counseling but have to call themselves something else (e.g., "nutritional counselor"). Certification is established by state regulation.
Dietitian/nutritionist	Individuals with any type of education or experience may register with the state of California as a dietitian or nutritionist.

There are many established methods of science. The specific methods employed vary from study to study depending on the type of research conducted. Nevertheless, all types of scientific studies have one feature in common: they are painstakingly planned. Planning is the *most* important, and often the most time-consuming, part of the entire research process.

Photo Disc

Illustration 3.9
The scientific method is not mysterious at all, but a carefully planned process for answering a specific question.

Developing the Plan

The first part of the planning process entails clearly stating the question that will be addressed—and, one hopes, answered—by the research. For the purposes of illustration, let's say our research will address the effects of vitamin X supplements on hair loss (Table 3.3 and Illustration 3.10). The idea for the research came from a study that hinted that users of vitamin X supplements had increased hair loss. In that study, adults were given 25,000 international units (IU) of vitamin X for three months to test its safety. Although the study found that vitamin X had no adverse effects on health and offered some benefits, a number of subjects who had received the supplements complained of hair loss. Accordingly, our study poses the question, "Do vitamin X supplements increase hair loss?"

Table 3.3
Overview of a human nutrition research study: vitamin X supplements and hair loss.*

A. **Pose a clear question:** "Do vitamin X supplements increase hair loss?"
B. **State the hypothesis to be tested:** "Vitamin X supplements of 25,000 IU per day taken for three months increase hair loss in healthy adults."
C. **Design the research:**
   1. *What type of research design should be used?* "In this study, a clinical trial will be used. The supplement and placebo will be allocated by a double-blind procedure."
   2. *Who should the research subjects be?* "The study will exclude subjects who may be losing or gaining hair due to balding, hair treatments, medications, or the current use of vitamin X supplements."
   3. *How many subjects are needed in the study?* "The required sample size is calculated to be 20 experimental and 20 control subjects."
   4. *What information needs to be collected?* "Information on hair loss, conditions occurring that might affect hair loss, and the use of supplements and placebos will be collected."
   5. *What are accurate ways to collect the needed information?* "The study will use the 'measure the hairs in a square inch of scalp' technique."
   6. *What statistical tests should be used to analyze the results?* "Appropriate tests identified."
D. **Obtain approval for the study from the committee on the use of humans in research:** "Approval obtained."
E. **Implement the study design:** "Implemented."
F. **Evaluate the findings:** "Subjects receiving the 25,000 IU of vitamin X for three months lost significantly more hair than subjects receiving the placebo. Hypothesized relationship found to be true."
G. **Submit paper on the research for publication in a scientific journal or other document.**

*This is a fictitious study used for illustrative purposes only.

Illustration 3.10
Do high doses of vitamin X increase hair loss?

Table 3.4
A short glossary of research terms.

Association	The finding that one condition is related to the occurrence of a particular event, such as a disease or disorder. For example, diets low in vegetables are associated with breast cancer. Associations do *not* prove that one condition (such as a diet low in vegetables) *causes* an event (such as breast cancer). They indicate that a statistically significant relationship between a condition and an event exists.
Cause and effect	A finding that demonstrates that a condition causes a particular event. For example, vitamin C deficiency causes the deficiency disease scurvy.
Clinical trial	A study design in which one group of randomly assigned subjects (or subjects selected by the "luck of the draw") receives an active treatment and another group receives an inactive treatment, or "sugar pill," called the placebo.
Control group	Subjects in a study who do not receive the active treatment or who do *not* have the condition under investigation. Control periods, or times when subjects are not receiving the treatment, are sometimes used instead of a control group.
Double blind	A study in which neither the subjects participating in the research nor the scientists performing the research know which subjects are receiving the treatment and which are getting the placebo. Both subjects and investigators are "blind" to the treatment administered.
Epidemiological studies	Research that seeks to identify conditions related to particular events within a population. This type of research does *not* identify cause-and-effect relationships. For example, much of the information known about diet and cancer is based on epidemiological studies that have found that high-fat diets are associated with the development of certain types of cancer.
Experimental group	Subjects in a study who receive the treatment being tested or have the condition that is being investigated.
Hypothesis	A statement made prior to initiating a study of the relationship sought to be proved (found to be true) by the research.
Meta-analysis	An analysis of data from multiple studies. Results are based on larger samples than the individual studies and are therefore more reliable. Differences in methods and subjects among the studies may bias the results of meta-analyses.
Peer review	Evaluation of the scientific merit of research or scientific reports by experts in the area under review. Studies published in scientific journals have gone through peer review prior to being accepted for publication.
Placebo	A "sugar pill," an imitation treatment given to subjects in research.
Placebo effect	Changes in health or perceived health that result from expectations that a "treatment" will produce an effect on health.
Statistically significant	Research findings that likely represent a true or actual result and not one due to chance.

The Hypothesis: Making the Question Testable

The question is then transformed into an explicit *hypothesis* that can be proved or disproved by the research. ("Hypothesis" and other research terms used in this unit are defined in Table 3.4.) The results of the research must provide a true or false response to the hypothesis, and not an explanatory sentence or paragraph. So, in this example, "vitamin X increases hair loss" wouldn't do as a hypothesis because it leaves too many questions about the relationship unanswered. The effect of vitamin X on hair loss may depend on the amount of vitamin X given, how long it is taken, or health, age, and other characteristics of the research subjects. The hypothesis "vitamin X supplements of 25,000 IU per day taken for three months by healthy adults increase hair loss" is concrete enough to be addressed by research.

The Research Design: Gathering the Right Information

Poor research design, or the lack of a solid plan on how the research will be conducted, is often a weak link that renders studies useless (Illustration 3.11). It's the "oops, we forgot to get this critical piece of information!" or the "how did all the measurements come out wrong?" at the end of a study that can ruin months or even years of work. Each step in the research process must be thoroughly planned. In research, there are no miracles (Illustration 3.12). If something can go wrong because of incomplete planning, it probably will.

Research designs are often based on the answers to the following questions:

- What type of research design should be used?
- Who should the research subjects be?
- How many subjects are needed in the study?
- What information needs to be collected?
- What are accurate ways to collect the needed information?
- What statistical tests should be used to analyze the findings?

What Type of Research Design Should Be Used? A number of different research designs can be used to test hypotheses. An *epidemiological* study design could be used to determine if hair loss is more common among people who take vitamin X supplements than among people who do not. This type of

Illustration 3.11

Source: © Sidney Harris.

"We don't get involved with things like double-blind tests and peer review. We're just a little Mom-and-Pop laboratory."

Illustration 3.12

Source: © Sidney Harris.

"I think you should be more explicit here in step two."

study is commonly employed to identify conditions that are related to specific health events in humans. Animal studies that determine hair loss in supplemented and unsupplemented animals could be used to provide preliminary evidence. Or, another design, such as a *clinical trial,* could be used. Since this design would work well for the proposed hypothesis about vitamin X supplements, the rules for conducting a clinical trial will be followed in the example. (You can see a map of the research design used in this hypothetical study in Illustration 3.13.)

The purpose of clinical trials is to test the effects of a treatment or intervention on a specific biological event. In our example, the effect of vitamin X supplements on hair loss would be tested.

The Experimental and Control Groups Clinical trials, as well as other research studies that address questions about nutrition, require an *experimental* group (the group of subjects who receive vitamin X in this example) and a *control* group (the comparison group that receives a *placebo* and not vitamin X). Never trust the results of a study that didn't employ both! Here's why. How do you know the effect of a certain treatment isn't due to something *other* than the treatment? What if you gave a group of adults vitamin X supplements and you found they lost hair? Does this mean the vitamin X caused the hair loss? Could it be that the subjects lost no more hair during the treatment period than they would have lost without the vitamin X? You can't know whether vitamin X sup-

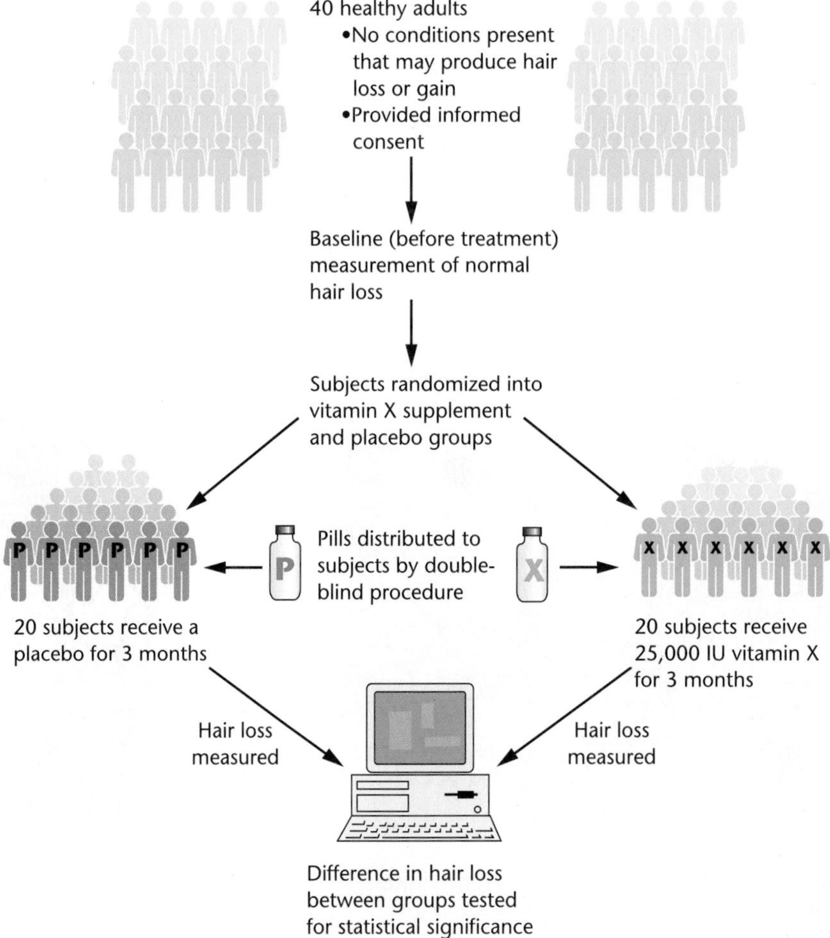

Illustration 3.13
Map of the design of the vitamin X supplement study.

plements produce hair loss if you don't know how much hair is lost without vitamin X.

After usual hair loss is measured, people in the study are randomly assigned (or chosen by the "luck of the draw") to serve in either the experimental or the control group. Individuals assigned to the control group will be given pills that look, taste, and feel like the vitamin X supplements, but have no effect on hair loss or gain. Neither the research staff who will be assessing the hair loss nor the people in the study will know who is getting vitamin X and who is getting the placebo (Illustration 3.14). This *double-blind* procedure is used because knowing which is which may affect people's expectations and change the results. This important topic deserves additional attention.

Illustration 3.14
Which is the vitamin X supplement? Neither the subject nor the investigator should know which is vitamin X and which is the placebo.

The "Placebo Effect" The *placebo effect* can cause a good deal of confusion in research. That's because people tend to have expectations about what a treatment will do, and those expectations can influence what happens.

A good example of the placebo effect occurred in a study that tested the effectiveness of a medication in reducing binge eating among people with bulimia.[4] After the usual number of binge-eating episodes was determined, 22 women with bulimia were given either the medication or a placebo in a double-blind fashion. The number of binge-eating episodes was then reassessed. At the end of the study period, a whopping 78% reduction in binge-eating episodes was found among women taking the medication. But binge-eating episodes dropped by *70%* in the control group. Was the medication effective? No. Was the expectation that a medication would help effective? Yes, at least for the time covered by the study. Due to the placebo effect, a greater than 90% reduction in episodes of binge eating would have been needed to conclude that the medication had a real effect on binge eating.

Who Should the Research Subjects Be? Although expedient, it is not enough to say, "Well, let's use 10 faculty members in the study." The type and number of subjects employed by the research are important considerations.

The type of subjects involved in research is important because you need to exclude people who have conditions that might produce the problem the research is examining. For example, hair loss may result from periodic balding, a bad perm, the use of mousse or hair spray, or illnesses or medications. If people with these conditions and circumstances were included in the study, it would be difficult to determine whether hair loss was due to vitamin X or something else. In addition, it is important to exclude from the study people who already take vitamin X supplements or are bald. An inappropriate or biased selection of subjects could make the results useless or, in the worst case, make the study's results come out in a predetermined way.

How Many Subjects Are Needed in the Study? The number of subjects needed for a study is a mathematical question, and we won't go into the formulas here. The number of subjects is based upon the number needed to separate true differences between the experimental and control groups from differences that are due to chance (Illustration 3.15). Suppose that hair loss normally varies from 40 to 80 strands per week. You would need to include enough subjects in the study to make sure differences in hair loss between the groups were not due to normal fluctuations in hair loss. This point is important because studies employing too few subjects lead to inconclusive results and a great deal of controversy about nutrition. One or a few subjects are never enough to prove a point about nutrition.

Let's assume that the mathematical formulas applied to the vitamin X study showed that 20 people are needed in both the experimental group and the control group.

Photo Disc

PhotoEdit, Amy C. Etra

Illustration 3.15
How many subjects are needed in the study? It's important to get the number right.

What Information Needs to Be Collected? Information recorded by researchers must represent an evenhanded approach to discovering the facts. Both findings that may support the hypothesis *and those that may refute it* must be identified. In order to know if vitamin X supplements increase hair loss, you would have to know how much hair is lost, whether the subjects faithfully took the supplement, and whether something came up during the study (such as a bad perm) that might alter hair loss. The presence of such conditions should be determined by methods known to provide accurate results.

What Are Accurate Ways to Collect the Needed Information? "Garbage in—garbage out." If the information obtained on and from subjects is inaccurate, then the study is worthless. To avoid this problem, good research employs methods of collecting information known to produce accurate results.

Let's consider the problem of measuring hair loss. How could you do that accurately? First, you look for a method that has already been demonstrated by research to be accurate. Otherwise, you first have to prove that the method you use to measure hair loss is accurate. Assume an accurate method is used.

What Statistical Tests Should Be Used to Analyze the Findings? Before the first piece of information is collected, appropriate tests for identifying *statistically significant* results of the research are selected. Many different statistical tests can be used to identify significant differences between the findings from the experimental and control groups. The trick here is to make sure the appropriate tests to prove or disprove the hypothesis are selected. Without such tests, it is often very hard to decide what the findings mean. What if the study shows that the vitamin X–supplemented group lost an average of 5% more hair than the control group? Is that 5% difference due to the vitamin X or to something else, such as chance fluctuations in normal hair loss? Well-chosen statistical tests will tell you whether the differences between groups are in all probability real or due to chance occurrences or coincidence.

Human Subjects' Approval

An important step in the planning process is applying for approval to conduct the proposed research on human subjects. Universities and other institutions that conduct research have formal committees that review plans to make sure they are consistent with the rules that govern research on human subjects. For this study, you would have to show to the committee's satisfaction that the level of vitamin X employed is safe. As a part of this process, the committee usually requires that subjects consent, in writing, to participate in the study.

Implementing the Study

With the design in place and the appropriate approvals obtained, it is time to implement the study. Assume that, due to the study's solid design and importance, a grant needed to fund the research has been awarded. Subjects are recruited for the study and enrolled if they are eligible and consent to participate. Measurements of hair loss are performed according to schedule, use of the pills by subjects is monitored, results are checked for errors and entered into a computerized data file, and statistical tests are applied. This process usually ends with computer printouts that exhibit the findings (Illustration 3.16).

Making Sense of the Results

Let's say subjects who received the supplement lost 30% more hair than the control subjects did. Having applied the appropriate statistical test, you find that 30% is a highly significant difference. Does that mean that vitamin X supplements *cause* hair loss? Cause is a strong word in research terms. It implies that a *cause-and-effect* relationship (that the vitamin X supplements caused the hair loss) exists. But many factors can contribute to hair loss, and since many causes may be unknown or not measured by the study, it is difficult to conclude with absolute certainty that vitamin X by itself caused the hair loss. Assume 4 of the 20 subjects in the experimental group who were known to have taken the vitamin X supplement as indicated lost less hair than usual. The vitamin X supplements *didn't* cause them to lose hair.

Illustration 3.16
Computer printout of statistical tests.

One could conclude from this research (if, as a reminder, it had actually been done) that vitamin X supplements, given to healthy adults at a dose of 25,000 IU per day for three months, are strongly *associated* with hair loss. The term "associated" as used here means that the vitamin X supplements were strongly *related* to hair loss but may not have *caused* the hair loss. The hypothesized relationship between supplemental vitamin X and hair loss would be found to be true. Although the research strongly indicates a cause-and-effect relationship, additional studies would be needed to prove the relationship exists in other groups of people and at different doses of vitamin X.

After the results are known, a paper describing the study and its results is written and submitted for publication in a scientific journal. If judged to be acceptable after review by other scientists, the paper is published.

Complete coverage of how nutrition questions are addressed through research would take a three-course sequence. Nevertheless, it is hoped that the information presented here makes it easier to judge the likely accuracy of reports about nutrition studies in the popular media. There is a lot more to a "Vitamin X Supplements Linked to Hair Loss" story than meets the eye.

Science and Personal Decisions about Nutrition

Science is based on facts and evidence. The grounding ethic of scientists is that facts and evidence are more sacred than any other consideration. These characteristics of science and scientists are strong assets for the job of identifying truths.

Although imperfect because they are undertaken by humans in an environment of multiple constraints, only scientific studies produce information about nutrition you can count on. Evidence is the single best ingredient for decision making about nutrition and your health.

Nutrition UP CLOSE

Checking Out a Fat-Loss Product

FOCAL POINT: How to make an informed decision about the truthfulness of an advertisement for a fat-loss product.

Read the following advertisement for a fat-loss product. Then, check out the information using the checklist for identifying nutrition misinformation.

1. Is something being sold?
   ____ Yes ____ No
2. Is a new remedy for problems that are not easily or simply solved being offered?
   ____ Yes ____ No
3. Are terms such as "miraculous," "magical," "secret," "detoxify," "energy restoring," "suppressed by organized medicine," "immune boosting," or "studies prove" used?
   ____ Yes ____ No
4. Are testimonials, before-and-after photos, or expert endorsements used?
   ____ Yes ____ No
5. Does the information sound too good to be true?
   ____ Yes ____ No
6. Is a money-back guarantee offered?
   ____ Yes ____ No

Optional: Repeat the activity using a nutrition-related advertisement from the Internet.

FEEDBACK (answers to these questions) can be found at the end of Unit 3.

P.O. Box 281752,
Miami, FL

Lose Over 2 Inches In 3 Weeks

A Scientifically Proven Fat Reduction Cream That Actually Works!

A scientific, double-blind, placebo-controlled research study of both sexes demonstrated that LIPID MELT fat reduction cream reduces inches from the thigh area. Subjects lost an average of 2 inches from each thigh in only 3 weeks!

Subjects from a preliminary pilot study reported that LIPID MELT cream reduced inches from their abdomen. Both studies reported no skin rashes, discomfort or sensitivity. Formulated with patented liposome technology, LIPID MELT contains no drugs, only natural active ingredients. Initially sold only in salons and spas.

Six-week supply: $25 + $4.95 s/h.
We guarantee that you will lose inches or your money will be refunded.

www.altmedicine.com
Directs readers to reliable information on alternative medicine and diets, herbs, supplements, and other nutrition topics.

www.eatright.org
The American Dietetic Association's Web site is an excellent source of information about diet, health, and qualified dietetics personnel in your area.

www.nlm.nih.gov/medlineplus
Maintained by the National Library of Medicine, this site is a sure bet for authoritative information on food and nutrition, ongoing clinical trials, and other information. Also offers a medical dictionary and glossary.

www.altmed.od.nih.gov/
Connects to the alternative medicine site of the National Institutes of Health.

www.navigator.tufts.edu
Scientifically sound nutrition information, search engine.

www.vm.cfsan.fda.gov/list.html
How to contact the FDA about nutrition misinformation and quackery; FDA regulations.

www.quackwatch.com
Connects to hyperlinks regarding quackery.

REFERENCES

1. Who's minding the lab? Tufts University Health and Nutrition Letter 1997, May: 4–5.
2. Schneeman BO. Building scientific consensus: the importance of dietary fiber. Am J Clin Nutr 1999;69:1.
3. Pathways to excellence. Chicago: The American Dietetic Association; 1997.
4. Alger SA, Schwalberg MD, Bigaouette JM, Michalek AV, Howard LJ. Effect of a tricyclic antidepressant and opiate antagonist on binge-eating behavior in normal weight bulimic and obese, binge-eating subjects. Am J Clin Nutr 1991;53:865–71.

Nutrition

UP CLOSE

Checking Out a Fat-Loss Product

Feedback for Unit 3

The product advertised earns five "yes" responses, making it highly unlikely that the product works. One or two "yes" responses provide a strong clue that advertised products may not work.

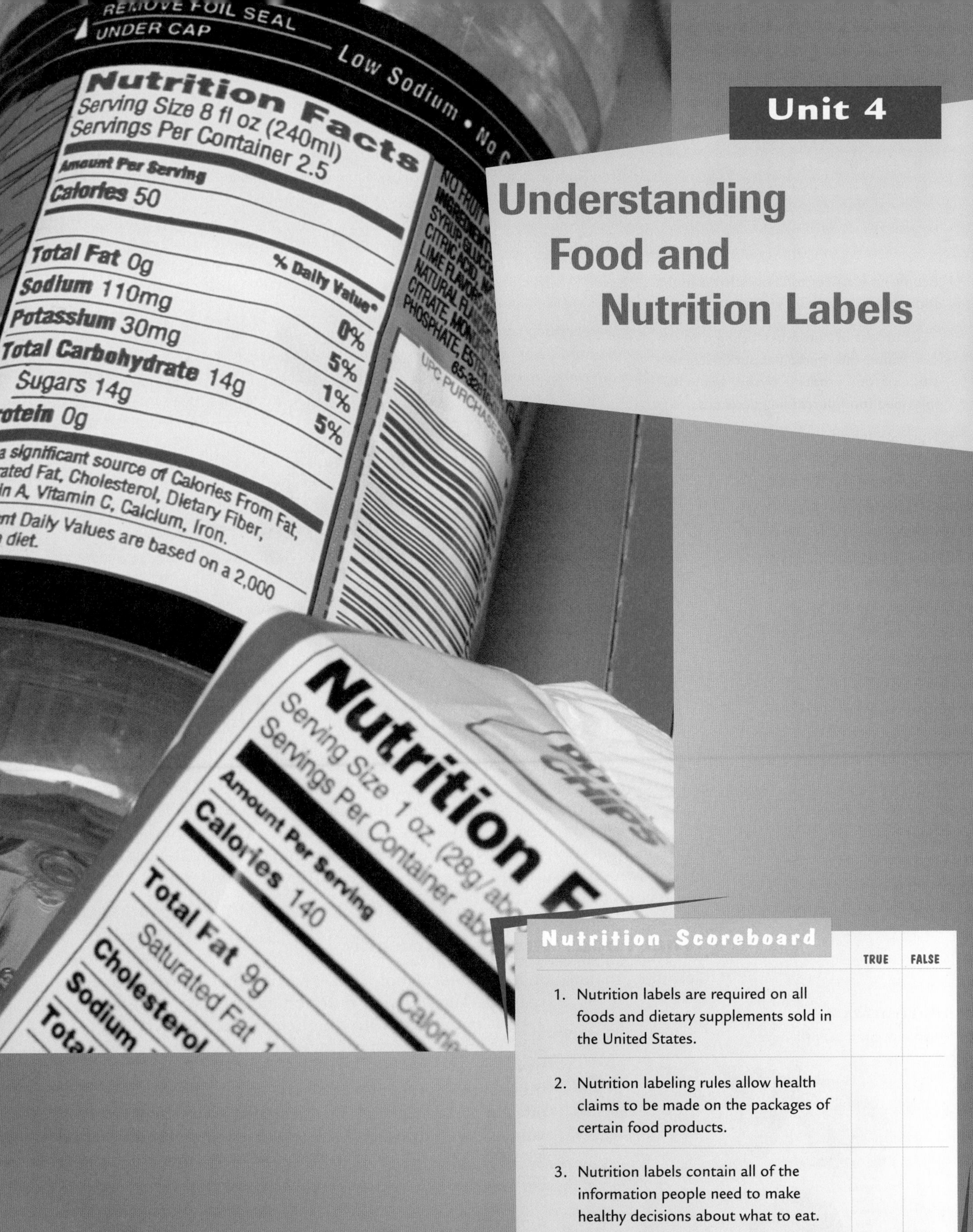

Unit 4

Understanding Food and Nutrition Labels

Nutrition Scoreboard

	TRUE	FALSE
1. Nutrition labels are required on all foods and dietary supplements sold in the United States.		
2. Nutrition labeling rules allow health claims to be made on the packages of certain food products.		
3. Nutrition labels contain all of the information people need to make healthy decisions about what to eat.		

(Answers on next page)

Key Concepts and Facts

- People have a right to know what is in the food they buy.
- The purpose of nutrition labeling is to give people information about the composition of food products so they can make informed food-purchasing decisions.
- Nutrition labeling regulations cover the type of foods that must be labeled and set the standards for the content and format of labels.

Answers to

Nutrition Scoreboard

	TRUE	FALSE
1. Labeling is now required for almost all processed foods and for dietary supplements, but remains largely voluntary for fresh fruits and vegetables and raw meats.[1]		✓
2. Health claims for food products are allowed on many food packages. The claims must be truthful and adhere to FDA standards.	✓	
3. Nutrition labels are necessarily short and can't tell the whole story about food and health. They help people make several key decisions about a food's composition.		✓

Nutrition Labeling

Misleading messages, hazy health claims, and the slippery serving sizes that characterized food labels in the past led to the recent revolution in nutrition labeling. Consumers, especially those who needed to watch their diets for health reasons, made it clear they wanted the mystery about what's in many foods to end. Passage of the 1990 Nutrition Labeling and Education Act by Congress indicated that their concerns had been heard, but another three years of lobbying, debate, and delays passed while the specific rules implementing the legislation were drawn up. Finally, in 1993, new rules for nutrition labeling were published, and revisions and implementation are ongoing.

What Foods Must Be Labeled?

According to nutrition labeling regulations, almost all multiple-ingredient foods must be labeled with nutrition information (Illustration 4.1). (Compliance with this requirement is very high—approximately 97% of packaged or processed food products have nutrition labels.)[1] Labeling of fresh vegetables and fruits and raw meats is voluntary because these foods are often sold individually and without packaging. It is recommended, however, that supermarkets make nutrition labeling information available to consumers through appropriately placed charts or pamphlets (Illustration 4.2).[2] Nutrition labeling is not required on restaurant menus, foods sold by local bakeries, and foods with packaging that is too small to fit a label (bite-sized candy bars, for example).[1]

Illustration 4.1
Read any good labels lately?

Photo Researchers, Jeff Greenberg

What's on the Nutrition Label?

The nutrition information label is known as the "Nutrition Facts" panel. The panel, shown in Illustration 4.3 (p. 4-4),

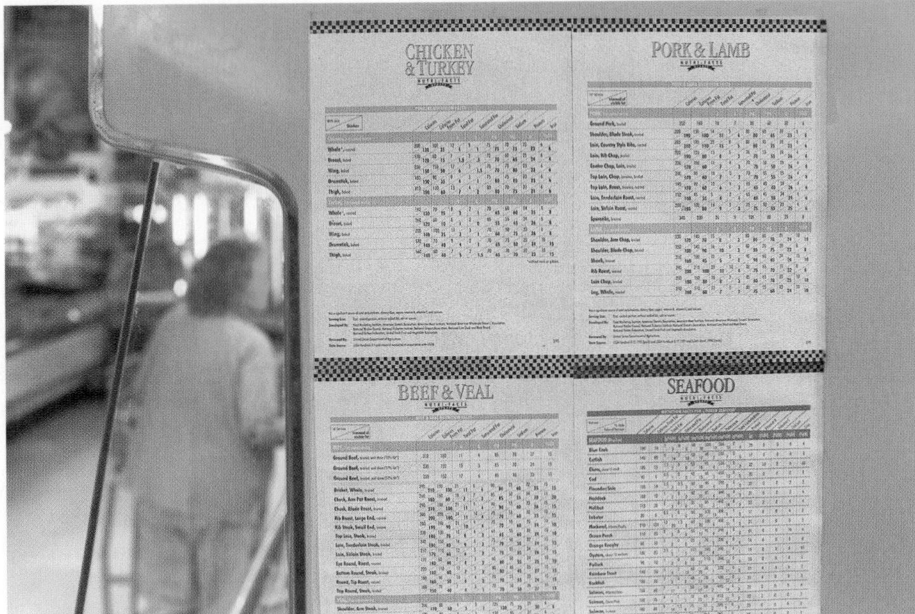

Illustration 4.2
Many supermarkets make nutrition information available in posters, cards, or pamphlets in areas where produce and meats are sold.

highlights a product's content of fat, saturated fat, cholesterol, sodium, dietary fiber, two vitamins (A and C), and two minerals (calcium and iron). The food's content of these nutrients must be based on a standard serving size as defined by the Food and Drug Administration (FDA). Standard serving sizes, which represent portion sizes generally consumed, are established for 131 types of food. For example, a serving of fruit juice is 6 ounces (¾ cup), and a serving of cooked vegetables is ½ cup.

All labeled foods must provide the information shown on the Nutrition Facts panel in Illustration 4.3 (p. 4-4). Additional information on specific nutrients can be added to the panel on a voluntary basis (Table 4.1 on p. 4-5). However, if the package makes a claim about the food's content of a particular nutrient not on the "mandatory" list, then information about that nutrient *must* be added to the Nutrition Facts panel.

Nutrition labels contain a column headed ***% Daily Value (%DV).*** Figures given in this column are intended to help consumers answer such questions as "Does a serving of this macaroni and cheese contain more fat than the other brand?" and "How much fiber is in this cereal?"

Daily Values are standard levels of dietary intake of nutrients developed specifically for use on nutrition labels. They are based on earlier editions of the Recommended Dietary Allowances and scientific consensus recommendations.[2] The %DV figures listed on labels represent the percentages of the standard nutrient amounts obtained from one serving of the food product. Standard values for total fat, saturated fat, and carbohydrate are based on a daily intake of 2000 calories. The %DV for total fat intake is based on 30% of total calories from fat (65 grams), the saturated fat standard on 10% of total calories (20 grams), and the standard for carbohydrate intake on 60% of total calories (300 grams). If, for example, a serving of a food product contained 10 grams of saturated fat, the %DV listed on the nutrition label would be 50% because the standard for saturated fat intake is 20 grams.

Food labels that make a claim about a particular nutrient or ingredient naturally contained in the food or added to it must include that nutrient or ingredient on the label. Other rules for including nutrition and health claims on food product labels exist.

On the Side

A 1997 survey of 300 foods by the FDA found that nutrient amounts listed on labels are correct 90% of the time.

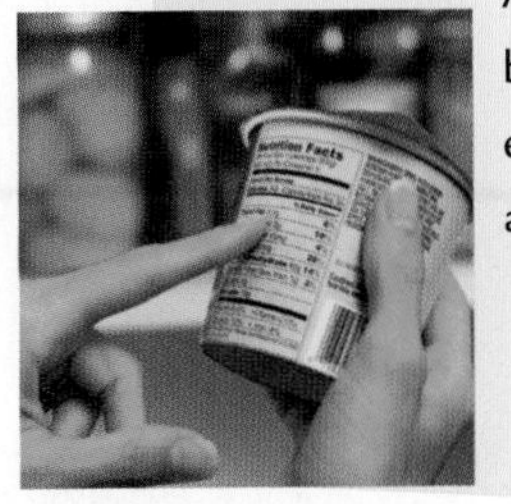
Photo Disc

% Daily Value (%DV) Scientifically agreed-upon standards of daily intake of nutrients from the diet developed for use on nutrition labels. The "% Daily Values" listed in nutrition labels represent the percentages of the standards obtained from one serving of the food product.

Daily Values (DVs) Scientifically agreed-upon daily dietary intake standards for fat, saturated fat, cholesterol, carbohydrate, dietary fiber, and protein intake compatible with health. DVs are intended for use on nutrition labels only and are listed in the Nutrition Facts panel. "% Daily Value" on Nutrition Facts panels is calculated as the percentage of each DV supplied by a serving of the labeled food.

Illustration 4.3
What are you having for dinner? The Nutrition Facts panel lets you know.

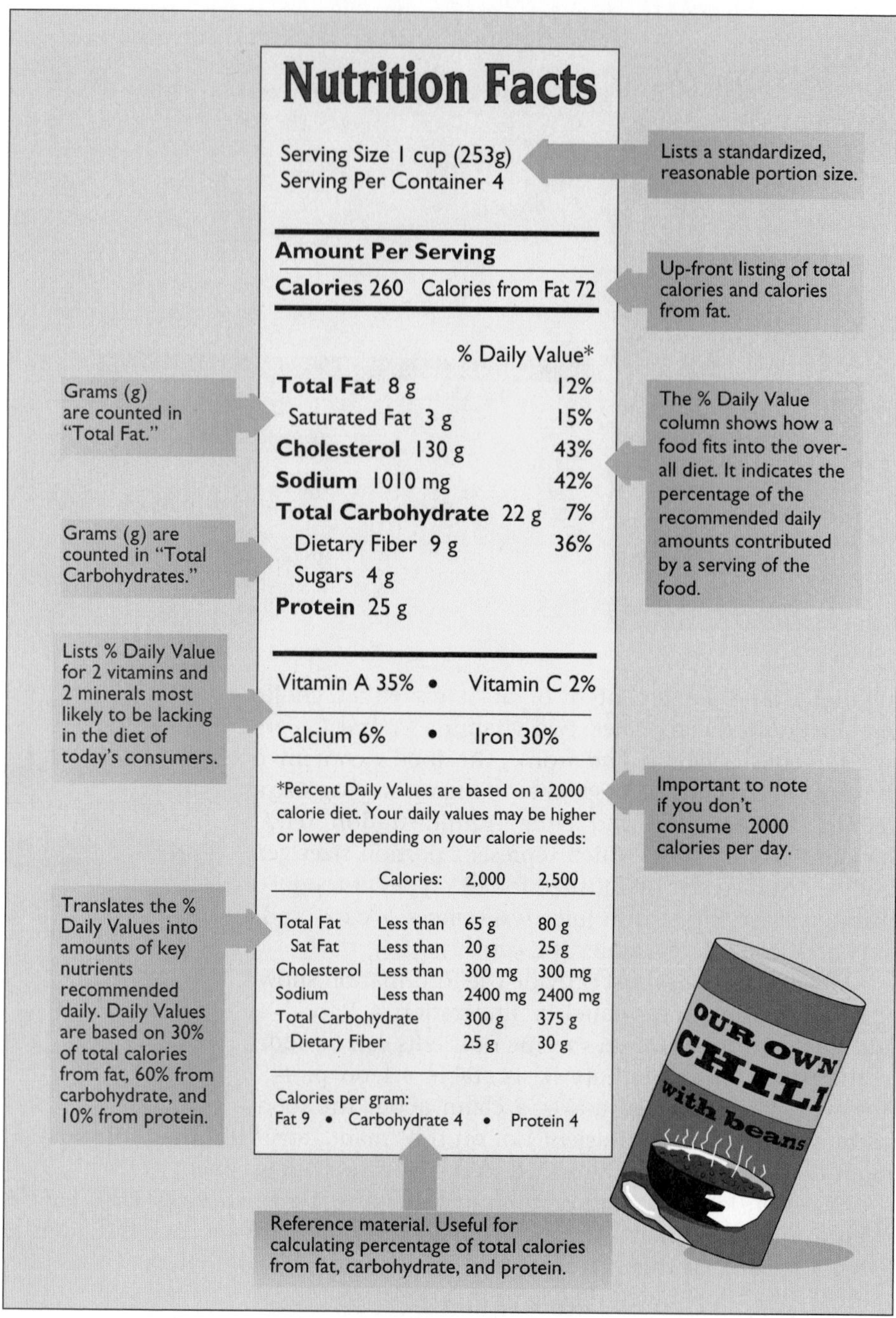

Nutrition and Health Claims Must Be Truthful Approximately 40% of food products sold in grocery stores make one or more nutrition claims on the packaging (Illustration 4.4 on p. 4–7).[1] Can these claims be trusted? If they are among the nutrition claims approved by the FDA, the answer to this question is almost always "yes."

If foods are to be labeled with terms such as "High fiber" and "Healthy," they must conform to standard definitions (Illustration 4.5 on p. 4–7). For example, foods labeled "low fat"—the most popular nutrition claim made on food packages[1]—must contain 3 grams of fat or less per serving. Low-fat foods can be labeled with a "percent fat free" label, such as "98% fat free" turkey. This label means that the product contains approximately 2% fat on a weight basis. Meat products labeled "lean" must contain less than 10 grams of fat, 4.5 grams of sat-

Table 4.1
Mandatory and voluntary components of the Nutrition Facts panel and assigned Daily Values (DVs). Components are listed in the order in which they must appear on the nutrition panel; unapproved components may not be listed.

Mandatory	DV	Voluntary	DV
Total calories	—	Calories from saturated fat	
Calories from fat	30%	Polyunsaturated fat	
Total fat	65 g	Monounsaturated fat	
Saturated fat	20 g	Stearic acid	
Cholesterol	300 mg	Potassium	
Sodium	2400 mg	Soluble fiber	
Total carbohydrate	300 g	Insoluble fiber	
Dietary fiber	25 g	Sugar alcohols (xylitol, mannitol, sorbitol)	
Sugars	—		
Protein	—	Other carbohydrates	
Vitamin A	5000 IU	Vitamin D	400 IU
Vitamin C	60 mg	Vitamin E	30 IU
Calcium	1000 mg	Vitamin K	80 mcg
Iron	18 mg	Thiamin	1.5 mg
		Riboflavin	1.7 mg
		Niacin	20 mg
		Vitamin B_6	2.0 mg
		Folate	400 mcg
		Vitamin B_{12}	6 mcg
		Biotin	300 mcg
		Pantothenic acid	10 mg
		Phosphorus	1000 mg
		Iodine	150 mcg
		Magnesium	400 mg
		Zinc	15 mg
		Selenium	70 mcg
		Copper	2 mg
		Manganese	2 mg
		Chromium	120 mcg
		Molybdenum	75 mcg
		Chloride	3400 mg
		Potassium	3500 mg

*If the food package makes a claim about any of the voluntary components, or if the food is enriched or fortified with any of them, they become a mandatory part of the nutrition panel.

urated fat, and 95 milligrams of cholesterol per serving. Consequently, many meats and meat products do not qualify for the "percent fat free" label.

A variety of approved nutrition claims are in use (see the Health Action on p. 4–6), and this list of claims will grow longer. A key feature of many of the approved claims is that they cannot be used to make certain foods appear to be healthier for you than they really are. Foods often have to meet several standards for nutritional value before they qualify for a claim.

You may see some food products labeled with claims like "pure," "10% fat," or "all natural" that are not listed in the Health Action section (Illustration 4.4). Because these claims have not been approved, they may not represent benefits and have no standard definition. These are the types of claims that are hard to trust.

Health Claims Upon approval by the FDA, foods or food components with scientifically agreed-upon benefits to disease prevention can be labeled with a health claim (Table 4.2 on p. 4–9). The health claim used on qualifying product labels, however, must be based on the FDA's "model claim" statements. For instance, scientific consensus holds that diets high in fruits and vegetables may lower

On the Side

The "Jelly Bean Rule" Foods high in fat, saturated fat, cholesterol, or sodium and low in certain vitamins, minerals, or fiber cannot be labeled with a nutrition claim. That means food manufacturers can't add, say, calcium to jelly beans, soft drinks, or similar foods to earn a claim that implies the product is "healthy."[3]

Photo Disc

HEALTH ACTION

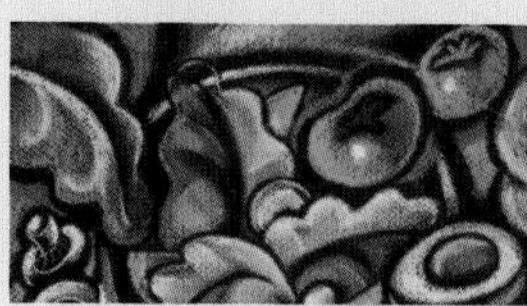

What "Front of the Package" Nutrition Terms Must Mean

Term	Examples	Means That a Serving of the Product Contains:
Extra lean	Extra-lean pork, extra-lean hamburger	Fewer than 5 grams of fat, fewer than 2 grams of saturated fat and trans fat combined, *and* fewer than 95 milligrams of cholesterol (applies to meats only).
Extra, More	Bread with added fiber, fortified foods	At least 10% more of the Daily Value of a nutrient per serving than in a similar food.
Fat free	Skim milk, no-fat salad dressing	Less than 0.5 grams of fat.
Free	Sugar-free, sodium-free	No or negligible amounts of sugars, sodium, or fat.
Good source	Good source of fiber, good source of calcium	10 to 19% of the Daily Value for a particular nutrient.
Healthy	Healthy burritos, canned vegetables	No more than 60 milligrams of cholesterol, 3 grams of fat, and 1 gram of saturated fat; and more than 10% of the Daily Value of vitamin A, vitamin C, iron, calcium, protein, or fiber. "Healthy" foods must also contain 360 milligrams or less sodium.
High	High in iron, high in vitamin C	20% or more of the Daily Value for a particular nutrient.
Lean	Lean beef, lean turkey	Fewer than 10 grams of fat, fewer than 4.5 grams of saturated fat and trans fat combined, *and* fewer than 95 milligrams of cholesterol (applies to meats only).
Less	Less saturated fat, less cholesterol	25% less of a nutrient than a comparable food.
Light or lite	Light in sodium, lite in fat, light brown sugar, light and fluffy	33% fewer calories or half the fat as the regular product, or 50% or less sodium than usual in a low-calorie, low-fat food. "Light" can also be used on labels to describe the texture or color of a food.
Low calorie	Low-calorie cookies, low-calorie fruit drink	40 calories or fewer.
Low cholesterol	Low-cholesterol egg product	20 milligrams or less cholesterol (applies to animal products only)

continued on next page

HEALTH ACTION (continued)

Term	Examples	Means That a Serving of the Product Contains:
Low fat	Low-fat cheese, low-fat ice cream	3 grams or less fat.
Low saturated fat	Low–saturated fat pancake mix, low–saturated fat eggnog	1 gram or less saturated fat and 0.5 grams or less trans fat.
Low sodium	Low-sodium soup, low-sodium hot dogs	140 milligrams or less sodium.
Percent fat free	95% fat free, 98% fat free	The specified percentage of fat on a weight basis (only low-fat foods can use this label).
Reduced	Reduced calorie/cholesterol/saturated fat bacon	25% less calories, cholesterol, or saturated fat and trans fat than the regular product.
Trans fat free	Trans fat–free margarine	Less than 0.5 grams trans fat and less than 0.5 grams saturated fat.

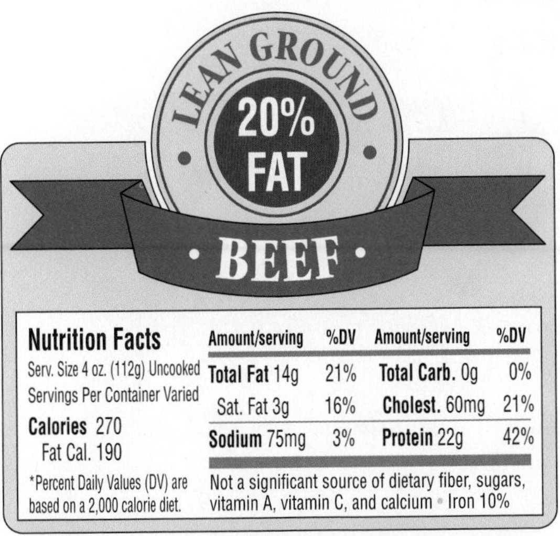

Illustration 4.4
Two examples of labels claiming "lean" ground beef. Only the 7% fat beef is actually lean. "20% fat" ground beef is far from lean, actually providing 22 grams and 75% of total calories from fat per serving. It is labeled incorrectly.

Illustration 4.5
Eye-catching nutrition claims on food labels must conform to standard definitions.

Topics Covered by Approved Claims	Topics of Claims Not Approved/Pending
1. Calcium and osteoporosis	1. Zinc and infection
2. Fats and cancer	2. Dietary fiber and colorectal cancer
3. Saturated fat and cholesterol and heart disease	3. 800 mcg folic acid more effective for reducing risk of neural tube defects than lower amounts
4. Fruits and vegetables and heart disease, cancer	4. Calcium and hypertension
5. Sodium and hypertension	
6. Folate or folic acid and neural tube defects (newborn malformations such as spina bifida)	
7. Oat bran and oatmeal and heart disease	
8. Soluble fiber in oats and psyllium seed husks and heart disease	
9. Sugar alcohols (e.g., xylitol, sorbitol) and tooth decay	
10. Soy protein and heart disease	
11. Whole grain foods and heart disease and certain cancers	
12. Plant stanols and sterols (used in spreads such as Take Control and Benecol) and heart disease	
13. Potassium and high blood pressure and stroke	
14. Antioxidant vitamins and cancer	

Table 4.2
Approved nutrition claims for disease prevention, and claims not approved.

the risk of cancer, and a health claim to this effect is allowed. The FDA's model claim for labeling fruits and vegetables is "Low fat diets rich in fruits and vegetables may reduce the risk of some types of cancer, a disease associated with many factors." Health claims are only approved by the FDA for food products that are not high in fat, saturated fat, cholesterol, or sodium. Examples of model health claims approved by the FDA are given in Table 4.3.

➡ **enrichment**
The replacement of thiamin, riboflavin, niacin, and iron lost when grains are refined.

➡ **fortification**
The addition of one or more vitamins and/or minerals to a food product.

Labeling Foods as Enriched or Fortified The vitamin and mineral content of foods can be increased by ***enrichment*** and ***fortification.*** Definitions for these terms were established more than 50 years ago. Enrichment pertains only to refined grain products, which lose vitamins and minerals when the germ and bran are removed during processing. Enrichment replaces the thiamin, riboflavin, niacin, and iron lost in the germ and bran. By law, producers of bread, cornmeal, pasta, crackers, white rice, and other products made from refined flour must enrich these foods. Beginning in 1998, federal regulations mandated that folate (a B vitamin) in the form of folic acid be added to refined

Table 4.3
Examples of model health claims approved by the FDA for labels of foods that qualify based on nutrient content. Model claims are often abbreviated on food labels.

- ***Whole grain foods and heart disease, certain cancers.*** Diets rich in whole grain foods and other plant foods and low in fat, saturated fat, and cholesterol may reduce the risk of heart disease and certain cancers.
- ***Soy protein and heart disease.*** Diets low in saturated fat and cholesterol that include 25 grams of soy protein a day may reduce the risk of heart disease.
- ***Sugar alcohols and tooth decay.*** Frequent between-meal consumption of food high in sugars and starches promotes tooth decay. The sugar alcohols in this food do not promote tooth decay.
- ***Saturated fat, cholesterol, and heart disease.*** Development of heart disease depends on many factors. Eating a diet low in saturated fat and cholesterol and high in fruits, vegetables, and grain products that contain fiber may lower blood cholesterol levels and reduce your risk of heart disease.
- ***Calcium and osteoporosis.*** Regular exercise and a healthy diet with enough calcium help maintain good bone health and may reduce the risk of osteoporosis later in life.
- ***Fruits and vegetables and cancer.*** Low-fat diets rich in fruits and vegetables may reduce the risk of some types of cancer, a disease associated with many factors.
- ***Folate and neural tube defects.*** Women who consume adequate amounts of folate daily throughout their childbearing years may reduce their risk of having a child with a neural tube birth defect.

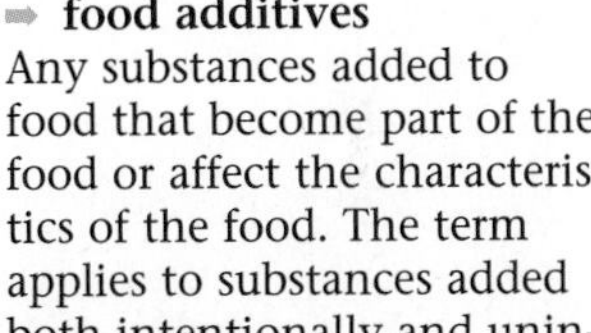

food additives
Any substances added to food that become part of the food or affect the characteristics of the food. The term applies to substances added both intentionally and unintentionally to food.

grain products. This new regulation was put into effect to help reduce the risk of a particular type of inborn, structural problem in children (neural tube defects) related to low blood levels of folate early in pregnancy.

Any food product can be fortified with vitamins and minerals—and many are. One of the few regulations governing the fortification of foods is that the amount of vitamins and minerals added to the food must be listed in the Nutrition Facts panel. Illustration 4.6 shows some examples of enriched and fortified foods.

Food enrichment and fortification began in the 1930s to help prevent deficiency diseases such as rickets (vitamin D deficiency), goiter (from iodine deficiency), pellagra (niacin deficiency), and iron-deficiency anemia. Today, foods are increasingly being fortified for the purpose of reducing the risk of chronic diseases such as osteoporosis, cancer, and heart disease.

Illustration 4.6
Examples of enriched and fortified foods.

The Ingredient Label

More useful information about the composition of food products is listed in ingredient labels. The label of any food that contains more than one ingredient must list the ingredients in order of their contribution to the weight of the food (Illustration 4.7 on p. 4–10). The ingredient that makes up the greatest portion of the product's weight must be listed first. Ingredients such as milk solids, peanuts, sulfites, or egg whites that may cause allergic reactions in some people should be listed on the ingredient label, too.[1]

Food Additives on the Label Specific information about ***food additives*** must be listed on the ingredient label. Nearly 3000 chemical substances are added to food to enhance its flavor, color, texture,

Illustration 4.7
All foods with more than one ingredient must have an ingredient label. Here are the rules.

All foods with more than one ingredient must have an ***ingredient label.*** Here are the rules.

1 Ingredients must be listed in order of their contribution to the weight of the food, from highest to lowest.

2 Beverages that contain juice must list the percentage of juice on the ingredient label.

3 The terms "colors" and "color added" cannot be used. The name of the specific coloringr edients must be given (for example, caramel color, yellow dye #5).

4 Cow's milk protein-based food additives or other ingredients to which some people are allergic must be listed on the label. (For example, if the protein casein is used in coffee lightener, the ingredient label must state that the casein was derived from cow's milk.)

15% Juice
Apple Juice Drink
Ingredients sugar, corn syrup, fruit juice (apple, orange, and pineapple) from concentrate, red dye #6, benzoates (used as a preservative).

Our Own Nutri•Bar
INGREDIENTS: ENRICHED WHEAT FLOUR, CORN SYRUP, WHOLE OATS, PARTIALLY HYDROGENATED SOYBEAN OIL, BLUEBERRY PRESERVES (BLUEBERRIES, CORN SYRUP), SUGAR, GLYCERIN, MALTODEXTRIN, HONEY, NATURAL AND ARTIFICIAL FLAVORINGS, NONFAT DRY MILK, SODIUM ALGINATE, WHEAT BRAN, MODIFIED CORN STARCH, SALT, POTASSIUM BICARBONATE, CITRIC ACID, CALCIUM PHOSPHATE, XANTHAN GUM, MALIC ACID, CASEIN CONCENTRATE (DERIVED FROM COW S MILK), BLUE #1, RED #40.

Illustration 4.8
Solving the mystery of food label terms. (Ingredients are listed in order by weight.)

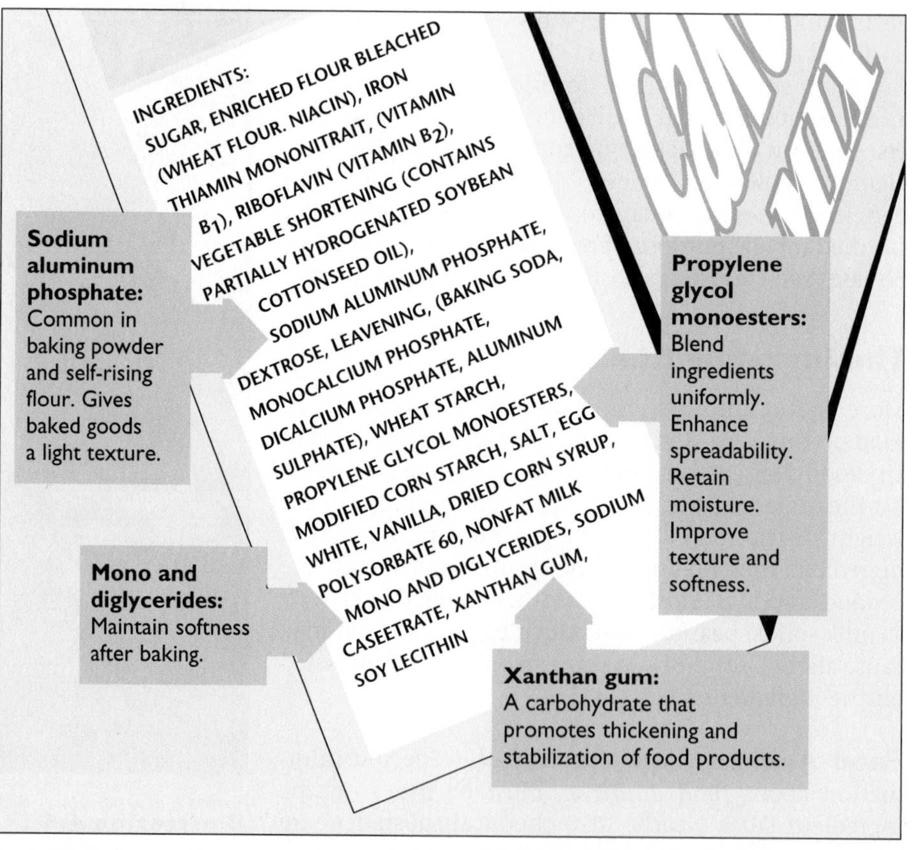

cooking properties, shelf life, or nutrient content. Food additives must be on the list of those considered safe by the FDA, however, before they can be added to food. An example of an ingredient label that shows the product's content of food additives is provided in Illustration 4.8.

The most common food additives are sugar and salt, but trace amounts of polysorbate, potassium benzoate, and many other additives that are not so familiar are also included in foods (Illustration 4.9). Although the new labeling regulations are more comprehensive than the old ones, they don't help consumers understand what some additives do. Appendix D lists many of the most common additives and indicates their function in foods.

Trace amounts of substances such as pesticides, hormones and antibiotics given to livestock, fragments of packaging materials such as plastic, wax, aluminum, or tin, very small fragments of bone, and insects may end up in foods. These are considered "unintentional additives" and do not have to be included on the label.

the NOOZ by Peter Kohlsaat

Meet Phil, the supermarket's in-house chemist.

Illustration 4.9 *Source: Tribune Media Services, Inc. All Rights Reserved. Reprinted with permission.*

Nutrition Labeling in Canada Food labeling standards apply in Canada. Appendix F shows an example food label and points out required components on Canadian food labels.

Dietary Supplement Labeling

The wide assortment of ***dietary supplements*** available on the market must be labeled and are regulated by a common set of rules. Dietary supplements differ from drugs in that they do not have to undergo vigorous testing and obtain FDA approval before they are sold. In return, dietary supplement labels cannot claim the products treat, cure, or prevent disease.[3]

According to FDA regulations, dietary supplements must be labeled as such and include a "Supplement Facts" panel that lists serving size, ingredients, and percent Daily Value (%DV) of essential nutrient ingredients (Illustration 4.10). Products can be labeled with a health claim, such as "high in calcium" or "low fat," if the product qualifies according to nutrition labeling regulations.

dietary supplements Any products intended to supplement the diet, including vitamin and mineral supplements; proteins, enzymes, and amino acids; fish oils and fatty acids, hormones and hormone precursors, herbs and other plant extracts. Such products must be labeled "Dietary Supplement."

Structure or Function Claims Supplements can be labeled with "structure or function" health claims such as "improves circulation," "supports the immune system," and "helps maintain mental health." If a function claim is made on the label or package inserts, the label or insert must acknowledge that the FDA does not support the claim. That's why you can see *"This statement has not been evaluated by the FDA. This product is not intended to diagnose, treat, cure, or prevent any disease."* printed on the labels of dietary supplements that make such claims. Structure and function claims don't have to be approved before they can appear on product labels. Dietary supplements labeled with misleading or untruthful information, and those that are not safe, can be taken off the market and the manufacturers fined.

Illustration 4.10

Dietary supplements labeled.

number of servings of antioxidant rich fruits and vegetables from your diet, yo
your diet with Nature Made Antioxidant Formula.
†This statement has not been evaluated by the Food and Drug Administr
intended to diagnose, treat, cure

Nutrition Facts
Serving Size 1 softgel

Amount Per Softgel	% Daily Value
Vitamin A 10,000 I.U.	200%
100% as Beta Carotene	
Vitamin C 250 mg	417%
Vitamin E 200 I.U.	667%
Zinc 7.5 mg	50%
Selenium 15 mcg	21%
Copper 1 mg	50%
Manganese 1.5 mg	75%

SUGGESTED USE: As a dieta
softgel daily, with a meal. Keep
in a cool, dry place, out of reach
Do not use if imprinted seal under
Our pure gelatin shell has no arti
you may notice a natural color va
INGREDIENTS: Ascorbic Aci
Acetate, Gelatin, Cottonseed Oil
Calcium Phosphate, Partially H
and Soybean Oils, Lecithin Oil, Z
Beeswax, Manganese Sulfate,
Selenate.
1-800-276-2878 M-F 7am-4pm P.
Oil- and Water-soluble Vitamins with M
Mfd. By Nature Made Nutritional F
Mission Hills, CA 91346-9606 Mad

Recently, manufacturers of dietary supplements are being allowed to label products with health claims supported to some extent by scientific studies. For example, EPA, DHA, and omega-3-fatty acid supplements (types of fat found primarily in fish and fish oils) can be labeled with a health claim that suggests these substances can reduce the risk of heart disease.[4]

Irradiated Foods

Food irradiation is an odd example of a food additive. It is actually a process that doesn't add anything to foods. Irradiation exposes foods to gamma rays that fragment DNA, thus destroying insects, bacteria, molds, and other microorganisms. Food irradiation enhances the shelf life of food products and decreases the risk of food-borne illness.[5] It must be performed according to specific federal rules.

On the Side

Irradiation is sometimes referred to as "cool pasteurization" because it destroys microorganisms in food without heat. Pasteurization was considered highly suspect by many consumers in the late 19th century, just as irradiation is now.[5]

Photo Disc

There is little evidence that irradiation adversely affects food quality, flavor, or safety.[6] Nevertheless, the notion of beaming gamma rays through foods may conjure up all sorts of concerns, in particular, the fear that radioactive particles stay in irradiated foods and end up in your body if you eat them. Unlike foods exposed to radioactive fallout (such as that released after the nuclear power plant accident at Chernobyl in 1987), irradiated foods retain no radioactive particles. The process is like having your luggage X-rayed at the airport. Your luggage doesn't become radioactive nor has anything in it changed. The process leaves no evidence of having occurred. Actually, this lack of change creates a challenge because inspection agencies can't determine whether a food has been irradiated or not. All irradiated foods except spices added to processed foods are required to display the international "radura" symbol and to indicate in writing that the food has been irradiated (Illustration 4.11).

Irradiated foods have a long history but have not yet won the trust of many American consumers. Today, irradiated products face many of the same hurdles microwaved foods confronted in the 1960s. The desire to limit radioactive processes of all sorts as well as radioactive waste products is widespread, and many people fear anything that involves radioactivity, especially when it comes to food.[5] Acceptance of irradiated foods and even a preference for them tend to develop as consumers learn more about the safety and benefits of the process.[6] Irradiation is approved for use on chicken, turkey, pork, beef, lamb, grains, fresh fruits and vegetables, and spices in the United States,[6] but currently the process is used mainly to sterilize food for astronauts and for hospital patients who easily get infections, and also to sterilize paper and plastic medical supplies, tampons, and germ-free surgical products. Food irradiation, along with processes that kill microorganisms with electricity or pressure, is increasingly viewed as an important part of efforts to control outbreaks of *E. coli* bacteria and other foodborne illnesses.[5]

Illustration 4.11
The "radura," as the symbol is called, must be displayed on irradiated foods. The words "treated by irradiation, do not irradiate again" or "treated with radiation, do not irradiate again" must accompany the symbol.

Labeling Organic Foods

For many years, consumers and producers of organic foods urged Congress to set criteria for the use of the term "organic" on food labels (Illustration 4.12). Consumers wanted to be assured that foods for which they might be paying premium prices were really organic, and producers wanted to keep the business honest. Standards were needed because you can't distinguish organically produced foods from others by looking at them or tasting them. The U.S.

Illustration 4.12
Organic foods are growing in popularity.

Department of Agriculture (USDA) has established standards for organic foods that make it illegal to sell or label a product as organic unless it meets specific criteria.[7]

Rules that qualify foods as organic are highlighted in Table 4.4. Consumers will be able to identify foods that meet these standards by a USDA organic food logo. Many other "certified organic" seals are placed on food, but they may not reflect the same standards for the food as will the USDA seal. Consumers are protected from "cheating" by producers, wholesalers, or retailers because violations of the organic standards carry penalties.

It should be pointed out that the term "organic" does not indicate that foods are healthier, more nutritious, or superior to conventionally produced foods. There is no scientific evidence that organic foods are more nutritious or better for health than the foods you buy in conventional supermarkets.[7]

Beyond Nutrition Labels: We Still Need to Think

> Even with the new food labels, consumers can't be stupid.
>
> —Max Brown

Understanding and applying the information relayed in nutrition labels calls for more, not less, nutrition knowledge on the public's part. People need to know a

Table 4.4
Rules that qualify foods as "organic."[7]

1. Plants
   - Must be grown in soils not treated with synthetic fertilizers, pesticides, and herbicides for at least three years
   - Cannot be fertilized with sewer sludge
   - Cannot be treated by irradiation
   - Cannot be grown from genetically modified seeds, or contain genetically modified ingredients
2. Animals
   - Cannot be raised in "factory-like" confinement conditions
   - Cannot be given antibiotics or hormones to prevent disease or promote growth

good bit about nutrition before they can understand nutrition labels and incorporate labeled foods appropriately into an overall diet.

Good diets include more than just foods with nutrition labels. The ice cream cone from the stand in the mall, the orange, potato, or fish we buy in the store, the pizza delivered to the dorm, and the salt and margarine or butter we add to foods at home are unlabeled parts of many diets. We need to know enough about the composition of unlabeled foods to fit them into a healthy diet. Nutrition labels often list only two vitamins and two minerals, but many more are required for health, so it's particularly important for people to know if their diet is varied enough to supply needed vitamins and minerals. Use of the Food Guide Pyramid should go hand-in-hand with label reading and food choices.

Healthy diets consist of a wide variety of foods that may include high-fat or high-sodium foods on occasion. Not every food we eat has to have the "right" label profile. Serving mostly low-fat or low-calorie foods to children, for example, might have unintended, unhealthy effects. Young children need calories and fat for growth and development. If diets are severely restricted, growth and development will be impaired. Nutrition labels are an important tool for helping people make informed food-purchasing decisions. However, labels do not now, nor will they ever, provide all the information needed to make wise decisions about food. Only people who are well-informed about nutrition can do that.

www.cyberdiet.com
Includes an interactive site for learning about nutrition labels.

www.navigator.tufts.edu
Search terms such as nutrition labeling, food additives, organic foods, food irradiation.

vm.csfan.fda.gov
Provides nutrition and dietary supplement labeling information and updates.

www.nal.usda.gov/fnic
Search "food irradiation" and "food labeling."

www.exnet.iastate.edu
Find food irradiation information plus much more.

REFERENCES

1. Brecher SJ et al. Status of nutrition labeling, health claims, and nutrient content claims for processed foods. J Am Diet Assoc 2000;100:1057–62.
2. Pennington JA, Hubbard VS. Derivation of daily values used for nutrition labeling. J Am Diet Assoc 1999;97:1407–12.
3. Ross S. Functional foods: the Food and Drug Administration perspective. Am J Clin Nutr 2000;71(suppl):1735S–8S.
4. FDA announces decision on another health claim for dietary supplements. Press release, 2000 Nov 1.
5. Shea KM. Technical report: irradiation of food. Pediatrics 2000;106:1505–9.
6. Position of the American Dietetic Association: food irradiation. J Am Diet Assoc 2000;100:246–52.
7. New organic foods standards announced by USDA. USDA press release, 2000 Dec 21.

Nutrition UP CLOSE

Comparison Shopping

FOCAL POINT: Comparing label information can help you choose the most nutritious product for your money.

Rate the nutrition value of these two cereals by completing the second paragraph below using information from the Nutrition Facts panels. The first paragraph has been done for you.

Whole-Grain Cereal

Nutrition Facts
Serving Size 1 cup (50g)
Serving Per Container 10

Amount Per Serving	
Calories 170 Calories from Fat 5	
	% Daily Value*
Total Fat 0.5 g	1%
Saturated Fat 0 g	0%
Cholesterol 0 g	0%
Sodium 0 mg	0%
Total Carbohydrate 41 g	14%
Dietary Fiber 5 g	21%
Sugars 0 g	
Protein 5 g	
Vitamin A 0%	Vitamin C 0%
Calcium 2%	Iron 8%

*Percent Daily Values are based on a 2000 calorie diet. Your daily values may be higher or lower depending on your calorie needs:

		Calories: 2,000	2,500
Total Fat	Less than	65 g	80 g
Sat Fat	Less than	20 g	25 g
Cholesterol	Less than	300 mg	300 mg
Sodium	Less than	2400 mg	2400 mg
Total Carbohydrate		300 g	375 g
Dietary Fiber		25 g	30 g

Health Granola Cereal

Nutrition Facts
Serving Size 1 cup (50g)
Serving Per Container 10

Amount Per Serving	
Calories 210 Calories from Fat 30	
	% Daily Value*
Total Fat 3 g	5%
Saturated Fat 0 g	0%
Cholesterol 0 g	0%
Sodium 120 mg	5%
Total Carbohydrate 43 g	14%
Dietary Fiber 3 g	12%
Sugars 16 g	
Protein 5 g	
Vitamin A 2%	Vitamin C 0%
Calcium 2%	Iron 10%

*Percent Daily Values are based on a 2000 calorie diet. Your daily values may be higher or lower depending on your calorie needs:

		Calories: 2,000	2,500
Total Fat	Less than	65 g	80 g
Sat Fat	Less than	20 g	25 g
Cholesterol	Less than	300 mg	300 mg
Sodium	Less than	2400 mg	2400 mg
Total Carbohydrate		300 g	375 g
Dietary Fiber		25 g	30 g

The **Whole-Grain Cereal**, providing 170 **calories** per serving, contributes 5 **calories from fat** (or 1% of the % Daily Value). Each serving also provides 0 mg of **sodium** (or 0 % of the % Daily Value), 5 g of **dietary fiber** (or 21% of the % Daily Value), and 0 g of **sugars**.

The **Health Granola Cereal**, providing _____ **calories** per serving, contributes _____ **calories from fat** (or _____ % of the % Daily Value). Each serving also provides _____ mg of **sodium** (or 5% of the % Daily Value), _____ g of **dietary fiber** (or _____ % of the % Daily Value), and _____ g of **sugars**.

Which cereal provides the better nutritional value?

FEEDBACK (answers to these questions) can be found at the end of Unit 4.

Nutrition

UP CLOSE

Comparison Shopping

Feedback for Unit 4

The **Health Granola Cereal,** providing 210 **calories** per serving, contributes 30 **calories from fat** (or 5% of the % **Daily Value**). Each serving also provides 120 mg of **sodium** (or 5% of the **% Daily Value)**, 3 g of **dietary fiber** (or 12% of the **% Daily Value**), and 16 g of **sugars.** The **Whole-Grain Cereal** provides the better nutritional value because it is lower in calories, total fat, sodium, and sugar and higher in fiber than the **Health Granola Cereal.** Don't be fooled by an attractive-sounding product. The proof of nutritional value is in the label, not the name.

Unit 5

Nutrition, Attitudes, and Behavior

Nutrition Scoreboard

	TRUE	FALSE
1. Food preferences are genetically determined.		
2. There are three things in life that never change: death, taxes, and food habits.		
3. Skipping breakfast does not affect school performance in well-nourished children.		

(Answers on next page)

Answers to

Nutrition Scoreboard

	TRUE	FALSE
1. Although genetics plays a role in food preferences, the predominant influences are environmentally determined.[1]		✓
2. True, true, false. Death and taxes are for certain, but the idea that food habits don't change is a myth.		✓
3. Well-nourished children who don't eat breakfast tend to score lower on problem-solving tests than children who eat breakfast.[2] (If you got this one wrong, answer one more question: Did you skip breakfast this morning?)		✓

Key Concepts and Facts

- Most food preferences are learned.
- The value a person assigns to eating right has more effect on dietary behaviors than does knowledge about how to eat right.
- Food habits can and do change.
- The smaller and more acceptable the dietary change, the longer it lasts.
- Behavior and mental performance can be affected by diet.

Why Do People Eat the Foods They Do?

Horse meat is a favorite food in a large area of north-central Asia. Pork, which is widely consumed in North and South America, Europe, and other areas, is rigidly avoided by many people in Islamic countries. Bone-marrow soup and sautéed snails are delicacies in France, while kidney pie is traditional in England. Dog is a popular food in Borneo, New Guinea, eastern Asia, west-central Africa, and the Philippines. In a number of countries, people enjoy insects but consider corn and soybeans fit only for animal feed. And then there are steamed clams and raw oysters—food passions for some but absolutely disgusting to others.[3]

When did you first think "yecck!"? The food choices just described would elicit that response among people from a variety of cultures. But they would not be responding to the same foods.

Illustration 5.1
This woman from the Libinza tribe of Zaire is preparing grubs, which will be cooked and eaten for dinner.

Photo Researchers, Jacques Jangoux

Why do people eat what they do? People learn from the society they live in which animals and plants are considered food and which are not (Illustration 5.1). Once items are identified as food, they develop a legacy of strong symbolic, emotional, and cultural meanings. Comfort foods, health foods, junk foods, fun foods, organic foods, soul foods, fattening foods, mood foods, and pig-out foods, for example, have been identified in the United States. And all cultures have their "super food." In Russia and Ireland, it's potatoes. In Central America, corn and yucca (a starchy root, also called manioc) are super foods, and in Somalia it's rice. The designation refers to the cultural significance of the food and doesn't imply "superior nutritional value."[3]

In countries where a wide variety of foods are available and people have the luxury of selecting which foods they will eat, food choices are influenced by a wide range of factors (Illustration 5.2). Note that heredity is but one of many factors that influence food choices. Food choices, and therefore diet quality, are primarily a consequence of learning.[1]

People Are Not Born Knowing What to Eat

The food choices people make are not driven by a need for nutrients or guided by food selection genes. People deficient in iron, for example, do not seek out iron-rich foods. If we're overweight, no inner voice tells us to reject high-calorie

Photo Disc

Illustration 5.2
Factors influencing food selection and dietary quality. Each of these sets of factors interacts with the others.[7,10]

foods. Women who are pregnant don't instinctively know what to eat to nourish their growing fetuses. No evidence indicates that young children, if offered a wide variety of foods, would select and ingest a well-balanced diet.[4]

Humans are born with mechanisms that help them decide when and how much to eat, however.[5] An inborn attraction to sweet-tasting foods, a dislike for bitter foods, and the response of thirst when water is needed influence food and fluid intake to an extent (Illustration 5.3).[6] There is evidence to suggest that people deficient in sodium experience an increased preference for salty foods.[7] Most people have a strong aversion to foods they think "smell bad," which are likely spoiled. Whether this response is inborn or learned is not clear.

Inborn traits that affect food choices are not very fine-tuned. People may overdo (consume excessive sugar and salt, for example) and underdo (for instance, drink too little water in response to thirst) their response to internal cues. The strongest influence on choices is food preference.[7]

Illustration 5.3
Newborn infants respond to different tastes: (a) Baby at rest, (b) tasting distilled water, (c) tasting sugar, (d) tasting something sour, and (e) tasting something bitter.[6]

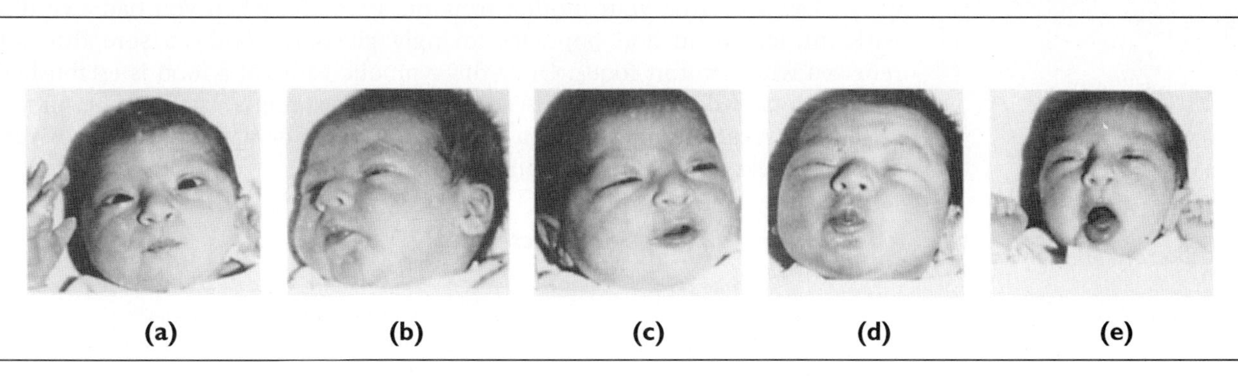

Source unknown

Why Do We Prefer the Foods We Do?

"You're going to eat that! Do you have any idea what's in that hot dog?"

"Yeah, I do. There's barbecue in the backyard, ball games with my mom and dad, and parties at my friend's house. The memories taste great!"

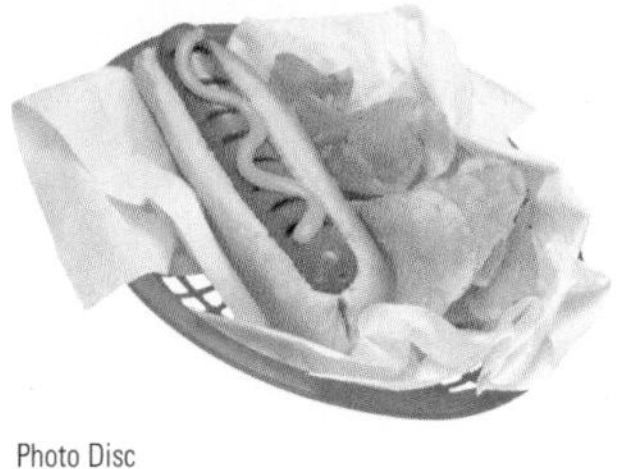

Photo Disc

The strong symbolic, emotional, and cultural meanings of food come to life in the form of food preferences. We choose foods that, based on our cultural background and other learning experiences, give us pleasure. Foods give us pleasure when they relieve our hunger pains, delight our taste buds, or provide comfort and a sense of security. We find foods pleasurable when they outwardly demonstrate our superior intelligence, commitment to total fitness, or our pride in our ethnic heritage. We reject foods that bring us discomfort, guilt, and unpleasant memories and those that run contrary to our values and beliefs.

The Symbolic Meaning of Food Food symbolism, cultural influences, and emotional reasons for food choices are broad concepts that may become clearer with concrete examples. Here are a few examples to consider.

Status Foods Vance Packard in *The Status Seekers* provided a memorable example of the symbolic value of food:

> *As a lad, this man had grown up in a poor family of Italian origin. He was raised on blood sausages, pizza, spaghetti, and red wine. After completing high school, he went to Minnesota and began working in logging camps, where—anxious to be accepted—he soon learned to prefer beef, beer, and beans, and he shunned "Italian" food. Later, he went to a Detroit industrial plant, and eventually became a promising young executive. . . . In his executive role he found himself cultivating the favorite foods and beverages of other executives: steak, whiskey, and seafood. Ultimately, he gained acceptance in the city's upper class. Now he began winning admiration from people in his elite social set by going back to his knowledge of Italian cooking, and serving them, with the aid of his manservant, authentic Italian treats such as blood sausage, spaghetti, and red wine!*[8]

Comfort Foods

The first thing I remember liking that liked me back was food.
—Rita Rudner, comedian

Ice cream, apple pie, chicken noodle soup, boxed chocolates, meatloaf and mashed potatoes: these are the most popular comfort foods in the United States.[9] The feeling of security and love that came along with the tea and honey or chicken soup that your mother or father gave you when you had a cold or with the ice cream and popsicles lovingly given to soothe a sore throat is renewed with comfort foods. Once the symbolic value of a food is established, its nutritional value will remain secondary.[10] Food status is a strong determinant of food choices, and after all, as a noted nutritionist once said, "life needs a little bit of cheese cake" (Illustration 5.4).[11]

"Discomfort Foods" Are Rejected

"When I was a kid, my parents made me eat the brussels sprouts they put on my plate. I finally ate them, but then I threw up. Nobody has ever made me eat brussels sprouts again."

Illustration 5.4

Source: Reprinted by permission: Tribune Media Services.

Memories of bad experiences with food and expectations that certain foods will harm us in some way contribute to our learning about food and affect our food preferences. Eating a piece of blueberry pie right before an attack of the flu hits or overdosing on sweet pickles or olives, for example, may take these foods off your preferred list for a long time.

Cultural Values Surrounding Food

> *A team of scientists observed that the diet of certain groups in the Chin States of Upper Burma was seriously deficient in animal protein. After considerable study a way was found to improve the situation by cross-breeding the small, local black pigs raised by the farmers with an improved strain to obtain progeny giving a greater yield of meat. The entire operation, however, completely failed to benefit the nutrition of the population because of one fact which had been viewed as irrelevant. The cross-bred pigs were spotted. And it was firmly believed—as firmly as we believe that to eat, say, mice, would be disgusting—that spotted pigs were unfit to eat.*[3]

Dietary change introduced into a culture for the purpose of improving health can only be successful if it is accepted by the culture. Cultural norms are not easily modified.

Other Factors Influencing Food Choices Food preferences and selections are also affected by the desire to consume foods considered healthy. The rewards and pleasures of reducing fat intake, eating more fruits and vegetables, or cutting down on sweets take such forms as weight loss and maintenance of the lost weight, an end to constipation, lower blood cholesterol level, and a newly discovered preference for basic foods.

Food Cost and Availability Food choices are also affected by the cost and availability of food. Researchers found that college students eating in dining halls have better diets when they prepay for their meals for the entire term rather than paying at each meal.[12] Grocery shoppers tend to select more low-fat and high-fiber foods when presented with a wide selection of those foods rather than just a few.[13]

Genetic Influences The primary role genetics plays in food preference development relates to taste. Some people are born with a strong sensitivity to bitterness and may reject foods such as brussels sprouts and broccoli (especially

On the Side

Photo Disc

Men tend to like olives and anchovies more than women do.[14]

if they are overcooked!), bitter-tasting teas and wines, and tonic water. Some people with this genetic trait will continue to eat strong-flavored vegetables—particularly after they cover up the bitterness with seasonings or sauces.

Food Choices Do Change

Who says old dogs can't learn new tricks? Most Americans aren't eating the way they used to. Per person consumption of pasta has more than doubled since 1970, while fresh egg intake has fallen from an average of 5⅓ eggs per week to 3½. Over the same period, low-fat milk sales have risen 300%, and there has been a wholesale switch to diet sodas and light beer. Beef consumption dropped 15% between 1970 and 1998.[15] Broccoli consumption is skyrocketing. Per person consumption of broccoli has risen 940% since 1970—a bigger gain than any other food. Americans are also eating more cheese and premium ice cream.[16] For more examples of changes in food choices, see Illustration 5.5.

Food choices are largely learned and do change as we learn more about foods. Perhaps your food choices have changed over time. How do the food choices you make now compare with the choices you made five years ago?

Illustration 5.5
Some changes in Americans' food choices in the last decade. Consumption of these foods is increasing: mixed fruit juice, low-fat and skim milk, frozen dinners, lasagna noodles, hamburger, pizza, roasted chicken, chili, pretzels (left). Consumption of these foods is decreasing: fried fish, chicken, and steak, shrimp and shellfish, eggs, whole milk, alcohol (right).

How Do Food Choices Change?

What are the ingredients for change in food choices? Why do some people succeed in improving their food choices while other people find that very hard to do? Nutrition knowledge, attitudes, and values have a lot to do with changing food choices for the better.

Nutrition Knowledge and Food Choices

Sound knowledge about good nutrition necessarily precedes the selection of a healthful diet. But is knowledge enough to ensure that healthy changes in diet will be made? The answer is "yes" for some people and "no" for others. Students at James Madison University in Virginia improved their diets as a result of a course in nutrition! At the end of the course, students reported eating, on average, less fat and sodium than at the beginning of the course (Illustration 5.6).[17] For some Americans, news that animal fats raise blood cholesterol levels was enough to convince them to cut down on beef and whole milk. Knowledge that diets containing an adequate amount of fiber reduce the risk of colon cancer led many Americans to switch to high-fiber breakfast cereals.

DAILY TRIBUNE

Wednesday, September 2, 2001 Daily 35 Cents

Nutrition Course Improves Diets of College Students

The diets of college students improved after taking a basic nutrition course

Results indicated that students decreased the amount of total fat from 82 to 68 grams and made other changes putting their diets more in line with the Dietary Guidelines.

Illustration 5.6
A true story.[17]

Information about good nutrition leads some people to modify their eating behavior some of the time and is more likely to change the food choices of women than men.[18] Knowledge alone is often not enough, however.

When Knowledge Isn't Enough Many people know far more about the components of a good diet than they put into practice. The problem is that between knowledge and practice lie multiple beliefs and experiences that act as barriers to change (Illustration 5.7).

Change of any type is most likely to succeed when the benefits of making the change outweigh the disadvantages. This makes changes in food choices a very individual decision. Individuals decide whether a change is in their best interests. But what sorts of circumstances, in addition to increased knowledge, make changes in food choices worthwhile for individuals—and even highly desired?

Illustration 5.7
Why knowledge about a good diet may not be enough to improve food choices.

"I feel guilty about eating the foods I like."

"Eating right is too expensive."

"I tried eating better, but I didn't stick with it."

"I don't have the time to eat right."

"The vegetables I like aren't available."

"I'm healthy now. Why should I worry about my diet?"

Photo Disc

Table 5.1
Factors that enhance food changes.

- Attitude that nutrition is important
- Belief that diet affects health
- Perceived susceptibility to diet-related health problems
- Perception that benefits of change outweigh barriers to change

Nutrition Attitudes, Beliefs, and Values The value individuals place on diet and health is reflected in the food choices they make. A survey of restaurant patrons found that food choices varied according to the consumer's perceptions of the importance of diet to health:[19]

- "Unconcerned" consumers—people who are unconcerned about the connection between diet and health and who tend to describe themselves as "meat and potato eaters"—select foods for reasons other than health.
- "Committed" consumers believe that a good diet plays a role in the prevention of illness. They tend to consume a diet consistent with their commitment to good nutrition.
- "Vacillating" consumers—people who describe themselves as concerned about diet and health but who do not consistently base food choices on this concern—tend to vary their food choices depending on the occasion. These consumers are likely to abandon diet and health concerns when eating out or on special occasions, but generally adhere to a healthy diet.

Avoiding illness and curing or diminishing current health problems are likewise strong incentives for changing food choices (Table 5.1).[20] One study found that nurses and dietitians based dietary changes primarily on the benefits of good diets to future health.[21] In almost all instances, the key to lasting improvements in diet is to make changes you can live with—changes that bring more pleasure than inconvenience or discomfort. Painful changes, such as those that require willpower to maintain, rarely last.

Successful Changes in Food Choices

> Perhaps the key in making dietary changes is to determine which changes are the easiest.[22]

The primary reason most efforts to improve food choices fail is that they are too drastic. Improvements that last tend to be the smallest acceptable changes needed to do the job.[23]

The Process of Changing Food Choices Assume you need to lose weight and want to keep it off by modifying your food choices. A promising plan to accomplish this would begin by identifying food choices you would like to change and lower-calorie food options you would be willing and able to eat (Table 5.2). Then you could make the plan more specific by identifying the changes that would be easiest to implement.

For example, assume the low-calorie foods you like and would eat include frozen nonfat yogurt and oranges. You might decide to eat yogurt or an orange in place of your usual bedtime snack of ice cream. A specific dietary change such as this is much easier to implement than a broad notion, such as "eat less." Although weight loss will take a while, such a small acceptable change has a much better chance of working than a drastic change in diet.

On the Side

Photo Disc

According to a population-based survey in 1997, 91% of U.S. consumers reported that they had changed their food-buying habits in the last two years.

Planning for Relapses When making a change in your diet, be prepared for relapses. Relapses happen for a number of reasons, and they don't mean the attempt has failed. People often return to old habits because the change they attempted was too drastic or because they tried to make too many changes at once. If the change undertaken doesn't work out, rethink your options and make a midcourse correction.

Table 5.2
Changing food choices for the better.[24]

The Process	An Example
1. Identify a healthful change in your diet you'd like to make.	1. I'd like to lower my fat intake.
2. Identify food choices you'd like to change.	2. I eat at fast-food restaurants five times a week. I always have a large order of fries, and several times a week I eat fried chicken.
3. Identify specific, acceptable options for more healthful food choices.	3. Options identified: · Order tossed salad instead of fries. · Eat grilled chicken sandwiches once a week instead of fried chicken. · Eat at a Mexican fast-food restaurant more often.
4. Decide which option is easiest to accomplish and requires the smallest change to get the job done.	4. I love Mexican food. It would be easy to eat it more often.
5. Plan how to incorporate the change into your diet.	5. Mondays and Fridays, I'll eat Mexican.
6. Implement the change. Be prepared for midcourse corrections.	6. Midcourse correction: On Fridays, when I'm with my friends, it's easier to eat at the restaurant they like. I'll order the grilled chicken sandwich and a salad.

Does Diet Affect Behavior?

Our behaviors most certainly influence our diet, but does it work the other way? Can what we eat affect our mental performance, mood, or activity level? The answer is "yes" for certain types of diets and "no" for others. Behaviors shown to be affected by diet, as well as behaviors where the effects of diet are unsubstantiated but nevertheless widely believed to exist, are listed in Table 5.3.

Table 5.3
Relationships between diet and behavior.[25–28]

***Established* relationships between diet and behavior:**

- Severe protein-calorie malnutrition early in life and impaired mental development
- Breakfast skipping and reduced problem-solving skills in children
- Fetal exposure to high levels of alcohol and mental retardation and behavioral problems
- Iron-deficiency anemia and reduced motor skills, attention span, and problem-solving skills
- Overexposure to lead and poor academic performance
- Folate deficiency and depression

***Unsubstantiated* claims about diet and behavior relationships:**

- Sugar intake, hyperactivity, and criminal behavior
- Vitamin supplements and increased intelligence
- Meat and aggressive behavior
- Oysters and sexual prowess (there are *no* food aphrodisiacs)

Malnutrition and Mental Performance

The World Health Organization estimates that 40 to 60% of the world's children suffer mild to moderate malnutrition.[29]

Like growth and health, mental development and intellectual capacity can be affected by diet. The effects range from mild and short term to serious and lasting, depending on when the malnutrition occurs, how long it lasts, and how severe it is. The effects are most severe when malnutrition occurs while the brain is growing and developing.

Brain growth in humans is concentrated during fetal growth and early childhood. Consequently, malnutrition that occurs for months during pregnancy or infancy produces the most serious and lasting effects on mental development.

Severe deficiency of protein, calories, or both early in life leads to growth retardation, low intelligence, poor memory, short attention span, and social passivity.[30] When the nutritional insult is early and severe, some or all of these effects may be lasting (Illustration 5.8). Children in Barbados, for example, who experienced protein-calorie malnutrition in the first year of life did not fully recover even with nutritional rehabilitation. Growth improved but academic performance did not. Compared to well-nourished children, those experiencing protein and calorie deficits during infancy were more likely to drop out of school and were four times more likely to have symptoms of hyperactivity (attention deficit disorder).[30]

Protein-calorie malnutrition that occurs later in childhood, after the brain has developed, produces behavioral effects that can be corrected with nutritional rehabilitation. Correction of other deficits that often accompany malnutrition, such as the lack of educational and emotional stimulation and harsh living conditions, hastens and enhances recovery.[30]

Protein-calorie malnutrition severe enough to cause permanent delays in mental development rarely occurs in the United States. When it does, malnutrition is usually due to neglect or inadequate caregiving. More common dietary events that impair learning in U.S. children are breakfast skipping, fetal exposure to alcohol, iron deficiency, and lead toxicity.

The Picture Cube, Michael Simpson

Illustration 5.8
This child in a Rwandan refugee camp may never fully recover from the effects of early malnutrition.

Does Breakfast Help You Think Better? Short-term fasting, such as skipping breakfast, reduces the late-morning problem-solving performance of otherwise well-nourished children.[30] About one in seven young children in the United States doesn't eat breakfast regularly.[31] No one has yet studied the effects of breakfast skipping on college students' midterm exam scores. . . .

Early Exposure to Alcohol Affects Mental Performance Mental development can be permanently delayed by exposure to alcohol during fetal growth. Although growth is also retarded, the most serious effects of fetal exposure to alcohol are permanent delays in mental development and behavioral problems associated with them.[32] Women are advised not to drink if they are pregnant or may become pregnant.[33]

Iron Deficiency Impairs Learning

> On a global basis, an estimated 20% of men, 35% of women, and 40% of all children are anemic, primarily due to iron deficiency.[34]

Most cases of iron-deficiency anemia in children result from inadequate intake of dietary iron. Iron-deficiency anemia in children is a widespread problem in developed and developing countries and likely is the most common single nutrient deficiency.[34] The potential impact of iron-deficiency anemia on the functional capacity of humans represents staggering possibilities.

Until recently, it was thought that the effects of iron-deficiency anemia on intellectual performance were short term and could be corrected by treating the anemia. It now appears that some of the effects may be lasting. Five-year-old children in Costa Rica treated for iron-deficiency anemia during infancy scored lower on hand-eye coordination and other motor skill tests than similar children without a history of anemia.[35] Studies in the United States have detected shortened attention span and reduced problem-solving ability in iron-deficient children.[36]

Overexposure to Lead Our concern about exposure to lead has recently increased, due in large measure to new information indicating that lead exposure has long-term behavioral effects. There are many opportunities for overexposure to lead. Approximately 84% of U.S. houses built before 1980 contain some lead-based paint.[27] Children living in or near these houses may eat the paint flakes (they taste sweet to children), or the old paint may contaminate the soil near the houses (Illustration 5.9). Lead also ends up in soil from industrial and agricultural chemicals, in water from lead-based pipes and solder, and in the air from the days when leaded gas was used.

Although the use of lead in cans, pipes, and gasoline has decreased dramatically (Illustration 5.10), lead remains in the environment for long periods. It also stays in the body, stored principally in the bones, for a long time—20 years or more. The effects of excessive exposure to lead include increased absenteeism from school, longer reaction times, impaired reading skills, and higher dropout rates. Blood lead levels in children have dropped substantially in recent years. Despite this drop, however, one million young children in the United States have elevated blood lead levels.[27]

Illustration 5.9
There are many opportunities for overexposure to lead. Young children are especially vulnerable.

PhotoEdit, Felicia Martinez

Food, Serotonin, and Behavior

Serotonin is a chemical messenger produced in the brain and other parts of the body from the amino acid tryptophan. It plays a role in

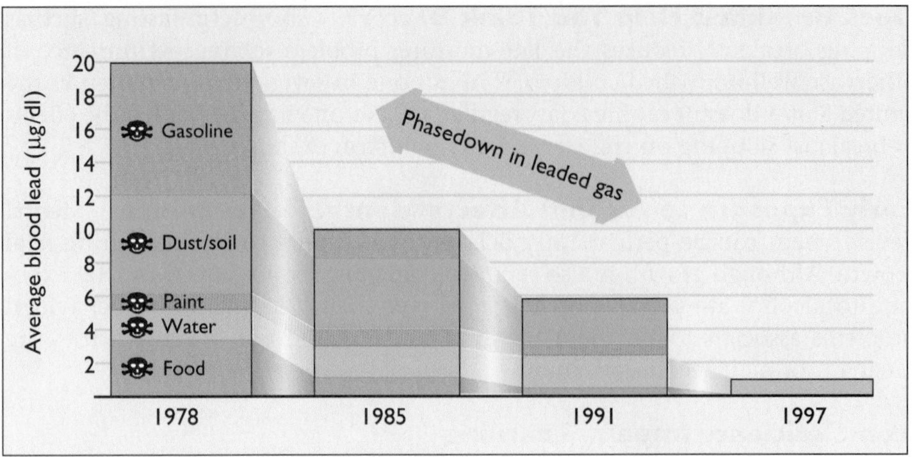

Illustration 5.10
Trends in children's lead exposure. A phasedown of leaded gasoline from 1976 to 1985 and near-elimination of leaded food cans have contributed to a reduction in children's lead levels.

Source: Environmental Protection Agency, 1991; FDA 1997.

appetite regulation (among many other functions) and consequently is a popular target for the action of some weight-loss drugs. Foods high in tryptophan, such as most meats and milk, increase brain levels of serotonin, especially if eaten with carbohydrates. High levels of serotonin tend to decrease calorie and fat intake and induce sleepiness. Brain serotonin levels may also modify a preference for foods high in protein, but this conclusion is controversial.[28]

Consuming foods that increase brain levels of serotonin may have short-term effects on appetite and food intake.[28] It may also be the reason people tend to get sleepy after the Thanksgiving turkey dinner, or why warm milk helps some people sleep.

Sugar, Hyperactivity, and Criminal Behavior

Sugar has been wrongly blamed for all sorts of behavioral problems. A number of studies have investigated the relationship of sugar intake to hyperactivity[25] and criminal behavior,[37] and none has shown an effect. An intriguing possibility was uncovered by one study that indicated sugars and other carbohydrates may *decrease* activity level in children.[38]

The myth that sugar causes hyperactivity in children lingers on in the minds of many parents, however. The excitement that often accompanies high-sugar eating occasions, such as Halloween and birthday parties, or the expectation that children will be more active after they eat sugary foods, may sustain the myth.[39]

It has also been claimed that food additives, especially artificial food colors, cause hyperactivity in children. Well-controlled studies do not support this claim, either.[40]

The Future of Diet and Behavior Research

Identifying the effects of nutrition on behavior is a tricky business. Many factors in addition to diet influence behavior, making it difficult to separate dietary from social, economic, educational, and genetic influences. We still have much to learn, and many assumptions about diet and behavior must await confirmation through research.

Nutrition
UP CLOSE

Improving Food Choices

FOCAL POINT:
Developing a plan for healthier eating.

Identify a change in your diet that you would like to make. Then develop a plan for making the change by thinking through and responding to each element of the dietary change process listed. (Refer to Table 5.2 for examples of responses.)

Dietary Change Process	Your Response
1. Identify a healthful change in your diet you'd like to make.	**1.** ______________
2. Identify food choices you'd like to change.	**2.** ______________
3. Identify specific, acceptable options for more healthful food choices.	**3.** ______________
4. Decide which option is easiest to accomplish and requires the smallest change to get the job done.	**4.** ______________
5. Plan how to incorporate the change into your diet.	**5.** ______________

FEEDBACK (answers to these questions) can be found at the end of Unit 5.

WWW links

www.nal.usda.gov/fnic
Provides ethnic and cultural resources related to food and nutrition and to nutrition and learning/behavior information.

www.nih.nimh.gov
The National Institute of Mental Health provides information on attention deficit hyperactivity disorder.

www.healthfinder.gov
Use this site to find reliable information about iron deficiency, fetal alcohol syndrome, and other disorders presented in this unit.

www.ncemch.org
Connects to the National Center for Education in Maternal and Child Health; provides information on lead poisoning and searchable databases.

www.ent.iastate.edu.edu/Misc/InsectsAsFood.html
Insect recipes from Iowa State University.

Nutrition **UP CLOSE**

Improving Food Choices

Feedback for Unit 5

There are no right or wrong answers to this Nutrition Up Close. Plans for dietary change that are specific, easy to accomplish, and highly acceptable are most likely to work in the long run.

REFERENCES

1. van den Bree MBM et al. Genetic and environmental influences on eating patterns of twins aged ⩾50 y. Am J Clin Nutr 1999;70:456–65.
2. Murphy JM et al. The relationship of school breakfast to psychosocial and academic functioning. Arch Pediatr Adol Med 1998;152:899–907.
3. Pyke M. Man and food. New York: McGraw-Hill; 1972.
4. Story M, Brown JE. Do young children instinctively know what to eat? The studies of Clara Davis revisited. N Engl J Med 1987;316:103–6.
5. Birch LL, Johnson SI, Jones MB, Peters JC. Effects of a nonenergy fat substitute on children's energy and macronutrient intake. Am J Clin Nutr 1993;58:326–33.
6. Steiner JE. The gustofacial response: observation on normal and anencephalic newborn infants. In: Bosma JF, ed. Fourth symposium on oral sensation and perception: development in the fetus and infant. Bethesda (MD): National Institute of Dental Research, DHEW Publication No. 73-546, U.S. Dept. HEW, NIH; 1973:254.
7. Factors influencing food choices in humans. Nutr Rev 1990;48:442.
8. Packard V. The status seekers. New York: Random House, Inc.; 1959; p 146.
9. Yankelovich Survey results reported in USA Today, 2000 Feb 7:D1.
10. Parraga IM. Determinants of food consumption. J Am Diet Assoc 1990;90:661–3.
11. Zallen DT, Brown ML. Food fights: deciding about diet and disease. Am J Clin Nutr 1990;52:944–5.
12. Beerman KA. Variation in nutrient intake of college students: a comparison by students' residence. J Am Diet Assoc 1991;91:343–4.
13. Cheadle A, Psaty BM, Curry S, et al. Community-level comparisons between the grocery store environment and individual dietary practices. Prev Med 1991;20:250–6.
14. Ullrich N, Tepper BJ. Food preferences are influenced by gender and genetic taste sensitivity to 6-n-propylthiouracil (PROP). Institute of Food Technology Annual Meeting, 2000, www.ift.confex.com/ift/2000/techprogram/paper_3781.htm, 1/01.
15. www.usda.gov/news/pubs/book98/chla.htm, 1999; accessed Dec 2000.
16. What we eat in America. Agricultural Research Service survey, Human Nutrition Center, USDA, 1997.
17. Brevard PB. Nutrition education improves the diet of college students (abs.). J Am Diet Assoc 1991, abs. A-47.
18. Ankeny K, Oakland MJ, Terry RD. Dietary fat: sources of information used by men. J Am Diet Assoc 1991;91:1116–7.
19. Shepherd R, Stockley L. Nutrition knowledge, attitudes, and fat consumption. J Am Diet Assoc 1987;87:615–9.
20. Becker MH, Maiman LA, Kirscht JP, Haefner DP, Drachman RH. The health belief model and prediction of dietary compliance: a field experiment. J Health Soc Behav 1977;18:348–66.
21. Holdt CS, Gates GE, Lassa S. Knowledge and behaviors of allied health professionals regarding meat (abs.). J Am Diet Assoc 1991;91S, abs. A-13.
22. Mitchell SJ. Changes after taking a college basic nutrition course. J Am Diet Assoc 1990;90:955–61.
23. van Beurden E, James R, Christian J, Church D. Dietary self-efficacy in a community-based intervention: implications for effective dietary counselling. Austr J Nutr Diet 1991;48:64–67.
24. Green GW, Rossi SR. Stages of change for reducing fat intake over 18 months. J Am Diet Assoc 1998;98:529–34.
25. Wolraich ML, et al. The effect of sugar on behavior or cognition in children: a meta-analysis. JAMA 1995;274:1617.
26. deAndraca I, et al. Psychomotor development and behavior in iron-deficient anemic infants. Nutr Rev 1997;55:125.
27. Child Health USA, 1997. US Maternal and Child Health Bureau, PHS, Washington, DC, p 29.
28. Fernstrom JD. Can nutrient supplements modify brain function? Am J Clin Nutr 2000;71(suppl):1669S–73S.
29. Bellamy C. The state of the world's children. Oxford (UK): Oxford University Press; 1996.
30. Allen LH. Functional indicators of nutritional status of the whole individual or the community. Clin Nutr 1984;3:169–75.
31. Nicklas TA, Boa W, Webber LS, Berenson GS. Breakfast consumption affects adequacy of total daily intake in children. J Am Diet Assoc 1993;93:886–91.
32. Tittman HG. What's the harm in just a drink? Alcohol Alcoholism 1990;25:287–91.
33. Dietary Guidelines for Americans, U.S. Dept. of HHS and USDA, 2000.
34. Lazoff B. Nutrition and behavior. Am Psychologist 1989;44:231–6.
35. Lazoff B, Jimenez E, Wolf AW. Long-term developmental outcome of infants with iron deficiency. N Engl J Med 1991;325:687–94.
36. Hurade EK et al. Early childhood anemia and mild or moderate mental retardation, Am J Clin Nutr 1999;69:115–9.
37. Gans DA, Harper AE, Bachorowski J-A, Newman JP, Shrago ES, Taylor SL. Sucrose and delinquency: oral sucrose tolerance test and nutritional assessment. Pediatrics 1990;86:254–62.
38. Ferguson HB, Stoddart C, Simeon JG. Double-blind challenge studies of behavioral and cognitive effects of sucrose-aspartame ingestion in normal children. Nutr Rev 1986;44:144–50.
39. Regalado M. Busting the sugar-hyperactivity myth. WebMD, www.webMD.com, 1999, accessed 1/01.
40. Background on food additives. www.ificinfo.health.org/backgrnd/bkgr9.htm. Accessed 1/22/01.

California Fruit Tree Agreement

Unit 6

What's a Healthful Diet?

This Is Just to Say

I have eaten
the plums
that were in
the icebox

and which
you were probably
saving
for breakfast

Forgive me
they were delicious
so sweet
and so cold

—William Carlos Williams

Nutrition Scoreboard

	TRUE	FALSE
1. Three out of four U.S. adults consume the recommended five or more servings of vegetables and fruits a day.		
2. The basic food groups include a "healthy snack" food group.		
3. Healthy diets include foods that together provide less than 30% of total calories from fat.		

(Answers on next page)

Answers to

Nutrition Scoreboard

	TRUE	FALSE
1. Only one in four U.S. adults consumes "five a day."[1]		✔
2. Does not.		✔
3. Diets should provide less than 30% of total calories from fat. High-fat foods can be part of a healthy diet if they are balanced with low-fat foods.	✔	

Key Concepts and Facts

- → Healthy diets are characterized by adequacy and balance.
- → There are many types of healthy diets.
- → There are no good or bad foods, only diets that are healthy or not healthy.
- → The "Dietary Guidelines for Americans" and the "Food Guide Pyramid" provide foundation information for healthy diets.

California Tree Fruit Agreement

Stock Food America, Eising

PhotoEdit, David Young-Wolff

Photo Disc

Photo Researchers, Lionela F. Stevenson

Illustration 6.1

Can you smell it? Can you taste it?

A plump, golden peach. It's so ripe that juice spurts from it and drips down your chin when you take a bite.

A golden brown turkey just taken out of the oven. The wonderful smell fills the kitchen.

A steaming loaf of homemade bread just set out to cool.

A perfect ripe tomato just picked from the garden. It melts in your mouth.

This Is Just to Say

I have eaten the plums that were in the icebox
and which you were probably saving for breakfast
Forgive me they were delicious so sweet and so cold

—William Carlos Williams

Healthy Eating: Achieving the Balance between Good Taste and Good for You

May I have your attention please? For a moment, think about the foods in Illustration 6.1. If your mouth is watering and you're ready to go out and buy some ripe peaches, you have found the balance between good taste and good for you. Who said foods that taste good aren't good for you?

Characteristics of Healthy Diets

Healthy diets come in a variety of forms. They may be based on bread, olives, nuts, fruits, beans, vegetables, lamb, and chicken (as in Greece); rice, vegetables, and small amounts of fish and other meats (as in China); or black beans, rice, meat, and tropical fruits (as in Cuba and Costa Rica). Examples of the diverse foods that can be part of a healthy diet are presented in Illustration 6.2.

Although the types of foods that go into them can vary substantially, healthy diets all share two characteristics: they are adequate and balanced. ***Adequate diets*** include a wide variety of foods that together provide sufficient levels of calories and ***essential nutrients.*** What's sufficient? For calories, it's the number that maintains a healthy body weight. For essential nutrients, sufficiency corresponds to intakes that are in line with recommended intake levels represented by the ***Recommended Dietary Allowances (RDAs)*** and ***Adequate Intakes (AIs).*** (These are shown in the Dietary Reference Intake tables on the inside, front cover of this book.) Recommended amounts of essential nutrients should be obtained from foods to reap the benefits offered by the variety of naturally occurring substances in foods that promote health. Following the pattern of food intake recommended in the Food Guide Pyramid helps ensure that both the need for essential nutrients and the need for other beneficial components of food will be met.

Balanced diets provide calories, nutrients, and other components of food in the right proportion—neither too much nor too little. Diets that contain too much sodium or too little fiber, for example, are out of balance. Americans are much more likely to consume an adequate diet than a balanced diet.

Why Is the American Diet Out of Balance?

Illustration 6.3 (p. 6–4) shows why the American diet is out of balance. For starters, it contains too much fat! The proportion of total calories supplied by total fat and ***saturated fat*** is too high. Protein and ***simple sugars*** are overrepresented in the American diet, but ***complex carbohydrates*** (mainly supplied by grain products and vegetables) don't appear often enough (Illustration 6.4 on p. 6–5).

Imbalanced Selection of Basic Foods Unfortunately, Americans on the whole make an imbalanced selection of foods from the Food Guide Pyramid. Average intakes of foods from four of the groups in the pyramid are below recommendations (Table 6.1 on p. 6–5). Only 33% of adults consume foods from all food groups, and only 1% consume the desired number of servings from each group.[2] Vegetables, fruits, and dairy products are most likely to be missing from the American diet, whereas meats, breads, and cereals are least likely to be

adequate diet
A diet consisting of foods that together supply sufficient protein, vitamins, and minerals and enough calories to meet a person's need for energy.

essential nutrients
Substances the body requires for normal growth and health but cannot manufacture in sufficient amounts; they must be obtained in the diet.

Recommended Dietary Allowances (RDAs)
Intake levels of essential nutrients that meet the nutritional needs of practically all healthy people while decreasing the risk of certain chronic diseases.

Adequate Intakes (AIs)
Provisional RDAs developed when there is insufficient evidence to support a specific level of intake.

balanced diet
A diet that provides neither too much nor too little of nutrients and other components of food such as fat and fiber.

saturated fats
The type of fat that tends to raise blood cholesterol levels and the risk for heart disease. They are solid at room temperature and are found primarily in animal products such as meat, butter, and cheese.

simple sugars
Basic types of carbohydrates such as sucrose (table sugar), fructose (fruit sugar), and glucose (blood sugar).

complex carbohydrates
The form of carbohydrate found in starchy vegetables, grains, dried beans, and many types of dietary fiber.

Illustration 6.2
Foods that contribute to healthy diets in different countries.

The main meal of the day ("la comida") in Mexico might include mole (chicken with chili sauce), rice, beans, tortillas, a side dish of chiles, and lemonade.

Dinner in Italy might be linguine primavera (pasta with vegetables).

A typical Japanese summer light meal includes soumen (very thin noodles) and tempura (Japanese fritters of seafood or vegetables).

PhotoEdit, Felicia Martinez

Photo Researchers, Inc.

Photo Researchers, Hiroshi Harada

Illustration 6.3
Why the U.S. diet is out of balance.

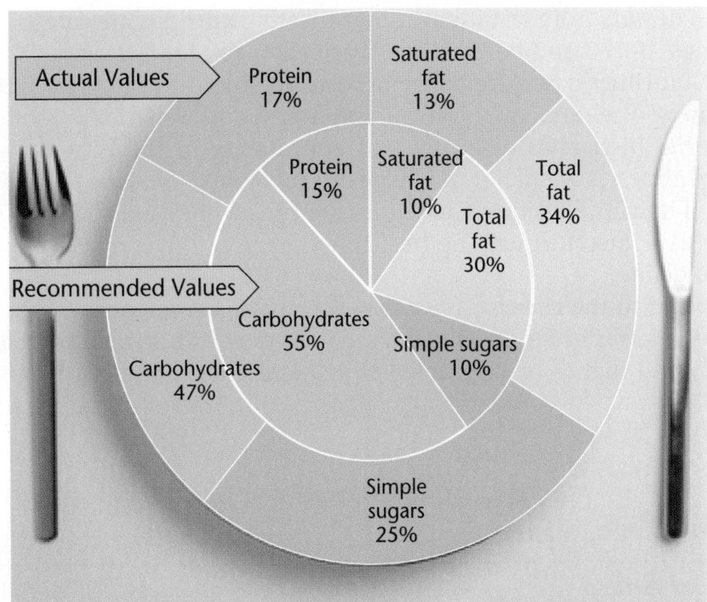

Illustration 6.4
Percentage of total calories in the average U.S. diet by food component: actual versus recommended.

Source: U.S. Department of Agriculture, 1997.

missing. The lack of variety has important health implications. Failure to consume foods from more than two of the five basic food groups daily is associated with a higher rate of death in adults than is consumption of foods from more than two of the food groups.[3] Low intakes of vegetables and fruits are related to increased body fat and the risk of cancer, stroke, and heart disease.[5–7] How do we strengthen the weak points of the American diet while maintaining its healthiest features? The answer is at hand.

Guides to Healthy Diets

Healthy diets for people in the United States are described in the "Dietary Guidelines for Americans" and the Food Guide Pyramid. The Dietary Guidelines primarily address the issue of dietary balance, and the Food Guide Pyramid focuses upon food choices for adequate diets. Both are updated periodically as knowledge about diet and health expands.

The Dietary Guidelines for Americans

The Dietary Guidelines were introduced in 1980 and are revised every five years. They address broad diet and health concerns and present advice on selecting balanced and, to a lesser degree, adequate diets. The 2000 edition of the Dietary

Food Group	People with No Intake on the Day Surveyed	Average Number of Servings Consumed Daily	Recommended Number of Servings
Dairy products	24%	1.1	2–3
Meats and alternates	6	2.4	2–3
Vegetables	18	2.2	3–5
Vegetables other than potatoes	50	—	—
Fruits	46	1.0	2–4
Breads and cereals	5	5.2	6–11

Table 6.1
Proportion of the U.S. population consuming zero servings of foods in the basic food groups in one day, average number of foods within each group consumed by adults daily, and recommended daily servings.[2,4]

Guidelines departs from tradition in that the 10 guidelines are categorized into three groups: fitness and body weight, basic diet, and food choices. It's the first time the Guidelines have given a major focus to physical fitness and weight and to food safety. The rationale for including fitness and weight and food safety is evidence that high rates of overweight and obesity in the United States and food-borne illnesses are major public health problems.[8] Because the recommendations made in the Dietary Guidelines are broad, they are easier to put into practice if explained further. The Health Action below gives a brief explanation of each guideline.

Compared to the current American diet, that recommended in the Dietary Guidelines is lower in fat, saturated fat, cholesterol, salt, sugars, and alcohol (if warranted) and higher in vegetables, fruits, and grain products. Foods recom-

HEALTH ACTION

The 2000 Dietary Guidelines for Americans Explained

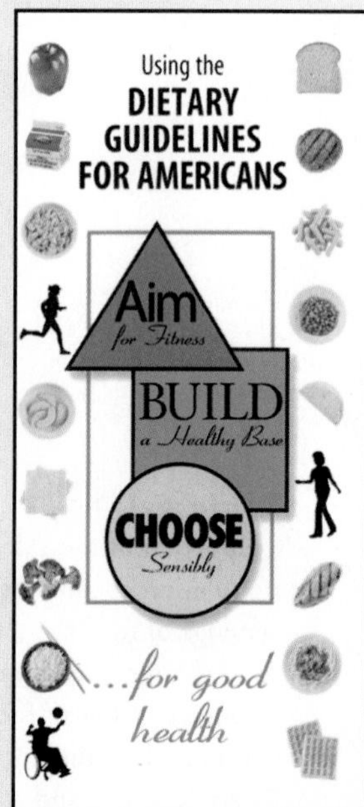

USDA

Aim, Build, Choose—for Good Health

Dietary Guideline	Explanation
Aim for fitness • Aim for a healthy weight. • Be physically active each day.	Following these two guidelines will help keep you and your family healthy and fit. Healthy eating and regular physical activity enable people of all ages to work productively, enjoy life, and feel their best. They also help children grow, develop, and do well in school.
Build a healthy base • Let the Pyramid guide your food choices. • Eat a variety of grains daily, especially whole grains. • Eat a variety of fruits and vegetables daily. • Keep food safe to eat.	Following these four guidelines builds a base for healthy eating. Let the Food Guide Pyramid guide you so that you get the nutrients your body needs each day. Make grains, fruits, and vegetables the foundation of your meals. This forms a base for good nutrition and good health and may reduce your risk of certain chronic diseases. Be flexible and adventurous—try new choices from these three groups in place of some less nutritious or higher calorie foods you usually eat. Whatever you eat, always take steps to keep your food safe to eat.
Choose sensibly • Choose a diet that is low in saturated fat and cholesterol and moderate in total fat. • Choose beverages and foods that limit your intake of sugars. • Choose and prepare foods with less salt. • If you drink alcoholic beverages, do so in moderation.	These four guidelines help you make sensible choices that promote health and reduce risk of certain chronic diseases. You can enjoy all foods as part of a healthy diet as long as you don't overdo on fat (especially saturated fat), sugars, salt, and alcohol. Read labels to identify foods that are high in saturated fats, sugars, and salt (sodium).

mended for inclusion in the diet reflect a "back to basics" theme: they include foods without added fat, salt, or sugars; fresh vegetables and fruits and whole-grain products; and fish, poultry, unprocessed meats, and dried beans. The Health Action below indicates types and serving sizes of foods within the Food Guide Pyramid groups that are consistent with the recommendations of the Dietary Guidelines.

The guideline "Choose a diet that is low in saturated fat and cholesterol and moderate in total fat" is perhaps the most complex. For individuals over the age of about five years, the guidelines recommend that fat intake not exceed 30% of total calories. You can get an idea of what a diet that provides less than 30% of total calories from fat looks like by examining the menus in Illustration 6.5. Most adults can obtain diets consistent with the guidelines by selecting low-fat or skim milk instead of whole milk; replacing high-fat meats such as regular hamburger and sausages with ground turkey, broiled fish, and other low-fat meats; and using low-fat alternatives to products such as regular mayonnaise and salad dressing.[5] Such diets also tend to be low in saturated fat and cholesterol. Excessive intakes of these two ingredients are linked to heart disease.

HEALTH ACTION

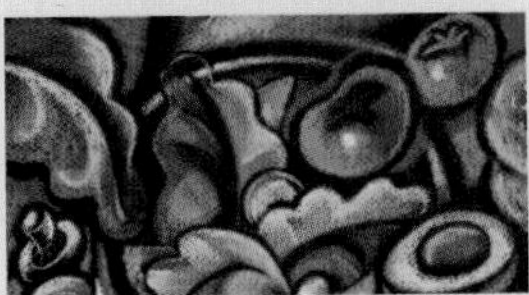

Food Group Choices à la the Dietary Guidelines

Food Group	Size of One Serving
A. Bread, cereal, rice, and pasta	
Whole-grain bread	1 slice
Rice, cooked	½ cup
Pasta	½ cup
Hot cereal	½ cup
Grits	½ cup
Cold cereal	1 cup
Granola	¼ cup
Tortilla	1
Chapati	1
Bagel	½
English muffin	½
Crackers	3–4 small
Muffin	½ (1 oz.)
Waffle	1 small
Pancake	1 small
Roll	1 small
B. Vegetables	
All types, raw or cooked	½ cup
Leafy vegetables	1 cup
Vegetable juices	¾ cup
C. Fruits	
Fruit, fresh (any type)	1
Cantaloupe	⅛ whole
Fruit, canned	½ cup
Applesauce	½ cup
Dried fruit	¼ cup
Berries, fresh	1 cup
Fruit juice	¾ cup
D. Milk	
Low-fat milk	1 cup
Fortified soy milk	1 cup
Low-fat yogurt	1 cup
Low-fat cottage cheese	½ cup
Low-fat cheese	1½ ounces
E. Meat, poultry, fish, dry beans, eggs, and nuts	
Lean meat, poultry, fish	3 ounces
Dried beans, cooked	½ cup
Tofu	½ cup
Seeds	½ cup
Egg whites	2
Peanut butter	4 tablespoons

Illustration 6.5
Comparison of the fat content of two menus.

BREAKFAST—HIGH-FAT MENU

Orange juice	6 ounces
Cornflakes	1 cup
Whole-wheat toast	2 slices
Margarine	2 teaspoons
Milk	1 cup

BREAKFAST—LOWER-FAT MENU

Orange juice	6 ounces
Cornflakes	1 cup
Whole-wheat toast	2 slices
Margarine	2 teaspoons
Skim milk	1 cup

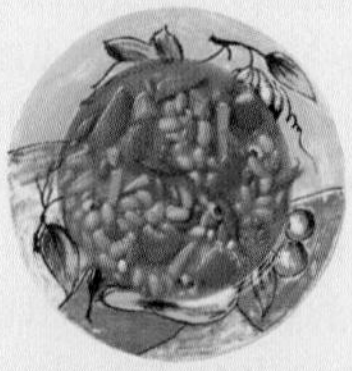

LUNCH—HIGH-FAT MENU

Cream of tomato soup	1 cup
Grilled cheese sandwich	1
Potato chips	1 ounce
Milk	1 cup

LUNCH—LOWER-FAT MENU

Vegetable soup	1 cup
Cheese sandwich	1
Tomatoes	½ cup
Skim milk	1 cup

SNACK—HIGH-FAT MENU

Peanut butter	2 tablespoons
Ritz crackers	4

SNACK—LOWER-FAT MENU

Orange	1
Whole-wheat crackers	4

DINNER—HIGH-FAT MENU

Tossed salad	1½ cups
Salad dressing	2 tablespoons
Fried chicken	4 ounces
French fries	1 cup
Green beans	½ cup
Margarine	½ teaspoon
Ice cream	1 cup
Milk	1 cup

DINNER—LOWER-FAT MENU

Tossed salad	1½ cups
Salad dressing	2 tablespoons
Baked chicken (no skin)	4 ounces
Boiled potatoes	1 cup
Green beans	½ cup
Margarine	½ teaspoon
Ice milk	1 cup
Skim milk	1 cup

FOOD CONSTITUENT ANALYSIS

High-fat		Lower-fat
2800	Calories	1800
38	Carbohydrates, % of calories	53
14	Protein, % of calories	21
48	Fat, % of calories	29
390	Cholesterol, milligrams	210

Photo Disc

Other Dietary Guidelines around the World Many other countries have developed guides for healthy eating. Examples of guides from Canada, Japan, and Costa Rica are shown in Table 6.2. The specific recommendations in each set of dietary guidelines address priority health problems that can be improved by dietary changes. Although the guidelines differ a bit, certain themes are evident in all of them: "eat a wide variety of foods," "maintain a healthy weight," and "consume foods low in total fat, saturated fat, and salt."

Table 6.2
Dietary guidelines from Canada, Japan, and Costa Rica.

Canadian Dietary Guidelines

A. Summary of Nutrition Recommendations

The Canadian diet should:

- provide energy consistent with the maintenance of body weight within the recommended range.
- include essential nutrients in amounts specified in the Recommended Nutrient Intakes.
- include no more than 30% of energy as fat (33 g/1000 kcal or 39 g/5000 kJ) and no more than 10% as saturated fat (11 g/1000 kcal or 13 g/5000 kJ).
- provide 55% of energy as carbohydrate (138 g/1000 kcal or 165 g/5000 kJ) from a variety of sources.
- be reduced in sodium content.
- include no more than 5% of total energy as alcohol, or 2 drinks daily, whichever is less.
- contain no more caffeine than the equivalent of 4 cups of regular coffee per day.
- Community water supplies containing less than 1 mg/litre should be fluoridated to that level.

B. Guidelines for Healthy Eating

- Enjoy a VARIETY of foods.
- Emphasize cereals, breads, other grain products, vegetables, and fruits.
- Choose low-fat dairy products, lean meats, and foods prepared with little or no fat.
- Achieve and maintain a healthy body weight by enjoying regular physical activity and healthy eating.
- Limit salt, alcohol, and caffeine.

Japan's Dietary Guidelines for Health Promotion

1. Eat a variety of foods to assure a well-balanced diet.
   (1) Eat 30 or more different kinds of food daily.
   (2) Balance main and side dishes around the staple food.
2. Match daily caloric intake with daily physical activity.
   (1) Avoid excess caloric intake to help prevent obesity.
   (2) Adjust physical activity to match daily caloric intake.
3. Be aware that both the quality and quantity of fats consumed are important.
   (1) Avoid too much fat.
   (2) Use vegetable oil, rather than animal fat.
4. Avoid too much salt.
   (1) Aim for a salt intake of less than 10g per day.
   (2) Resourceful cooking cuts down on excessive salt intake.
5. Make all activities pertaining to food and eating pleasurable ones.
   (1) Use the mealtime as an occasion for family communication.
   (2) Treasure family taste and home cooking.

Costa Rica Food Guide (Guías de Alimentación)

A. Energy

1. 60–65% of total calories from carbohydrates
   a. Less than 10% from simple sugars
   b. Consume 30 g fiber daily.

continued on next page

Table 6.2
Dietary guidelines from Canada, Japan, and Costa Rica, continued.

Costa Rica Food Guide (Guías de Alimentación), continued

2. 25% of total calories from fat
   a. Less than 10% from saturated fats
3. 10–15% of total calories from protein

B. Cholesterol
1. Consume no more than 300 mg cholesterol per day, or less than 100 mg per 1000 calories.

C. Increase consumption of antioxidant nutrients.

D. Vitamin A
1. Consume 750 µg of vitamin A daily.

E. Calcium
1. Consume 800 mg of calcium daily.

F. Iron
1. Consume 18 mg daily.

G. Sodium
1. Limit intake to 5–6 g daily if sweating is moderate.
2. Limit intake to 10 g daily if sweating is heavy.
3. Limit intake to less than 3 g daily in the absence of sweating.

H. Iodine
1. Use iodized salt.

I. Food Safety
Wash fruits and vegetables.
Cook foods to a high temperature.
Store perishable foods in a refrigerator at 10°C or less.
Keep countertops and cooking utensils clean.
Wash your hands regularly during food preparation.
Protect food from insects, rodents, and other animals.
Use clean water.

The Food Guide Pyramid: The Food Group Approach to an Adequate Diet

Food group guides for adequate diets have existed in the United States since 1916. In 1992 a substantially modified basic food guide, the Food Guide Pyramid, was released (Illustration 6.6 on p. 0–00).

The Food Guide Pyramid consists of five basic food groups and recommended ranges of daily servings for each group. It represents an outline of types of food to eat each day rather than a rigid prescription of what to eat. This food guide encourages decision making about which foods within each group fit best into an individual's diet.

Food groups are located in the pyramid based on their recommended number of servings and not on their relative importance to diet or health. Since each food group consists of foods that make particular contributions to nutrient intake, foods from one group cannot be substituted for those from another group. Healthy diets consist of foods from all of the basic groups.

Additional information provided with the Food Guide Pyramid (Table 6.3 on p. 6–11) defines serving sizes and differences in servings per day needed by women, men, children, and active people. This is important information to know because serving sizes and numbers of servings recommended are major sources of confusion about the Food Guide Pyramid.

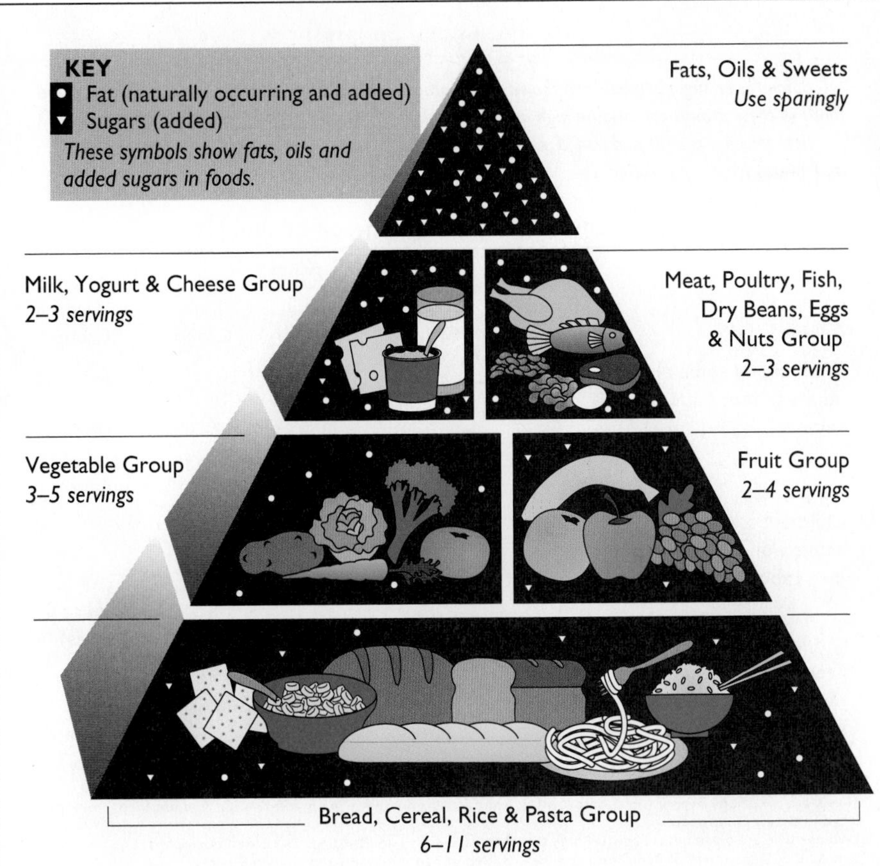

Illustration 6.6
The Food Guide Pyramid: A guide to daily food choices.

Table 6.3
Serving sizes and numbers of servings recommended in the Food Guide Pyramid.

What counts as one serving?

The amount you eat may be more than one serving. For example, a dinner portion of spaghetti would count as two or three servings.

Bread, Cereal, Rice, and Pasta Group	Vegetable Group	Fruit Group
1 slice of bread ½ cup of cooked rice or pasta ½ cup of cooked cereal 1 ounce of ready-to-eat cereal	½ cup of chopped raw or cooked vegetables 1 cup of leafy raw vegetables	1 piece of fruit or melon wedge ¾ cup of juice ½ cup of canned fruit ¼ cup of dried fruit

Milk, Yogurt, and Cheese Group	Meat, Poultry, Fish, Dry Beans, Eggs, and Nuts Group	Fats and Sweets
1 cup of milk or yogurt 1½ ounces of natural cheese 2 ounces of process cheese	2½ to 3 ounces of cooked lean meat, poultry, or fish Count ½ cup of cooked beans, or 1 egg, or 2 tablespoons of peanut butter as 1 ounce of lean meat	LIMIT CALORIES FROM THESE especially if you need to lose weight.

Fats and Added Sugars.

The small tip of the pyramid shows fats and sweets. These are foods such as salad dressings, cream, butter, margarine, sugars, soft drinks, candies, and sweet desserts. Alcoholic beverages are also part of this group. These foods provide calories but few vitamins and minerals. Most people should go easy on foods from this group.

continued on next page

Table 6.3
Serving sizes and numbers of servings recommended in the Food Guide Pyramid, continued.

Fats and Added Sugars.

Some fat or sugar symbols are shown in the other food groups. That's to remind you that some foods in these groups can also be high in fat and added sugars. When choosing foods for a healthful diet, consider the fat and added sugars in your choices from all the food groups, not just fats and sweets from the pyramid tip.

How many servings do you need each day?

	Servings				
Calorie Level[a]	Bread Group	Vegetable Group	Fruit Group	Milk Group	Meat Group
Women and some older adults (about 1600 calories)	6	3	2	2–3[b]	2 (for a total of 5 ounces)
Children, teenage girls, active women, and most men (about 2200 calories)	9	4	3	2–3[b]	2 (for a total of 6 ounces)
Teenage boys and active men (about 2800 calories)	11	5	4	2–3[b]	3 (for a total of 7 ounces)

[a]These are the calorie levels if you choose low-fat, lean foods from the five major food groups and use foods from the fats and sweets group sparingly.
[b]Women who are pregnant or breastfeeding, teenagers, and young adults to age 24 need three servings.
Source: U.S. Department of Agriculture and the U.S. Department of Health and Human Services.

Keep an Eye on Serving Size Serving sizes represent portions of food generally consumed by people studied in national surveys.[9] They tend to be smaller than people think, and much smaller than "super-sized" and "extra value meal" portions served at fast-food restaurants. According to the Food Guide Pyramid, a serving of pasta, rice, dried beans, or chopped broccoli equals ½ cup, for example. This may make the number of servings recommended for some of the food groups seem not *that* high. The intake range of 6 to 11 servings for grain products appears to some people to be high. It seems less high when one understands that the low end of the serving range applies to inactive women and the higher end to teenage boys and active men.

At the top of the pyramid is not a food group, but a message about fats and sweets. These foods are assigned the smallest area in the pyramid, sending the message that fat, oils, and sweets should be a small part of our diet.

In general, the more types of food an individual eats in a day, the wider the array of nutrients and other beneficial components of food, such as fiber and antioxidants, provided. People in the United States tend to consume 18 to 20 different foods in a day. In contrast, dietary guidelines developed for Japan recommend that people eat 30 different foods daily, and the Canadian food guide displays many more types of food in each group than does the pyramid.

Nutritional Recommendations for Canadians Canada has similarly chosen to stress grain products, fruits, and vegetables in its food guide. The Canadian Food Guide to Healthy Eating (Illustration 6.7 on p. 6–14) is depicted as a rainbow with grains, fruits, and vegetables located on the outer bands to emphasize these foods. The Canadian format was chosen to provide consumers with a visual display of a variety of food choices within each group.

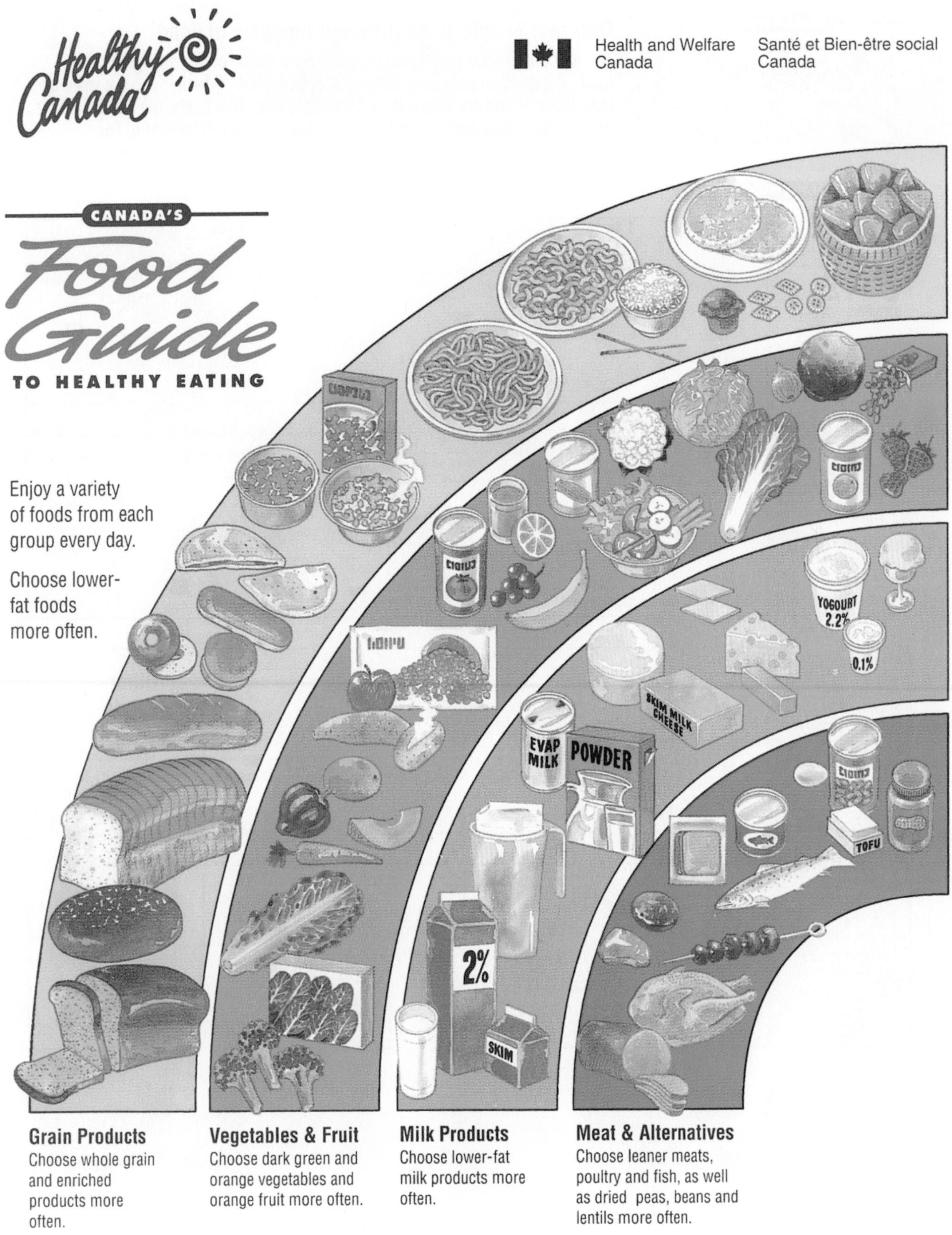

Health and Welfare Canada | Santé et Bien-être social Canada

CANADA'S

Food Guide

TO HEALTHY EATING

Enjoy a variety of foods from each group every day.

Choose lower-fat foods more often.

Grain Products
Choose whole grain and enriched products more often.

Vegetables & Fruit
Choose dark green and orange vegetables and orange fruit more often.

Milk Products
Choose lower-fat milk products more often.

Meat & Alternatives
Choose leaner meats, poultry and fish, as well as dried peas, beans and lentils more often.

Health and Welfare Canada

Illustration 6.7

Different People Need Different Amounts of Food

The amount of food you need every day from the 4 food groups and other foods depends on your age, body size, activity level, whether you are male or female and if you are pregnant or breast-feeding. That's why the Food Guide gives a lower and higher number of servings for each food group. For example, young children can choose the lower number of servings, while male teenagers can go to the higher number. Most other people can choose servings somewhere in between.

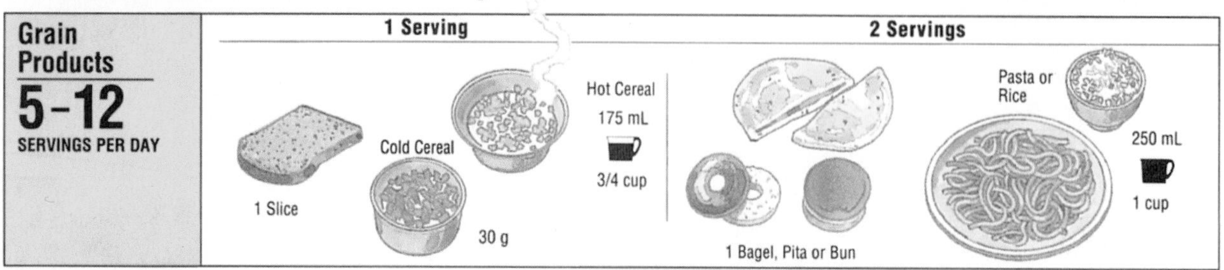

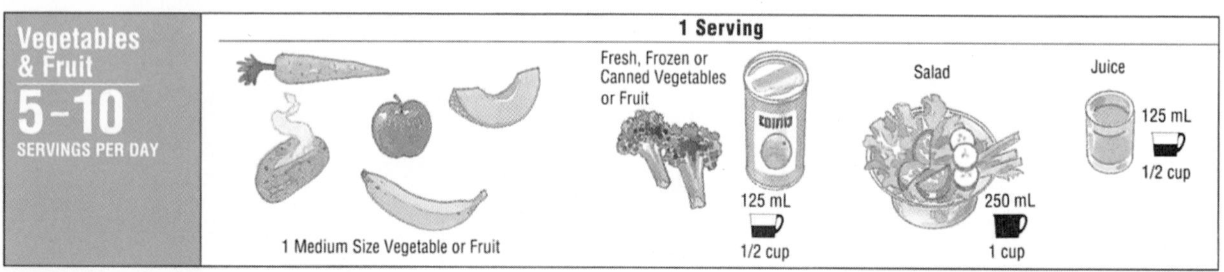

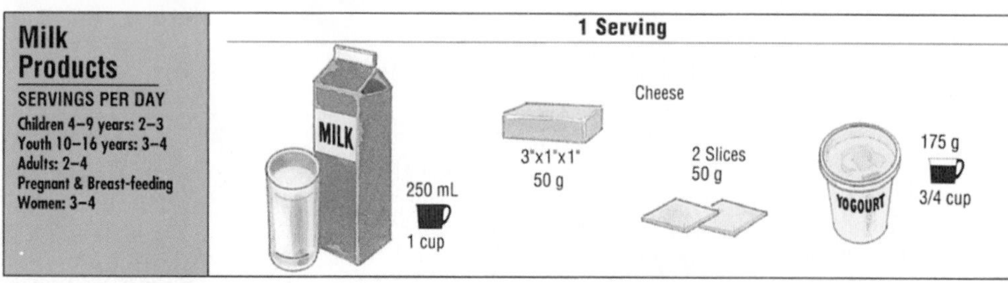

Other Foods

Taste and enjoyment can also come from other foods and beverages that are not part of the 4 food groups. Some of these foods are higher in fat or Calories, so use these foods in moderation.

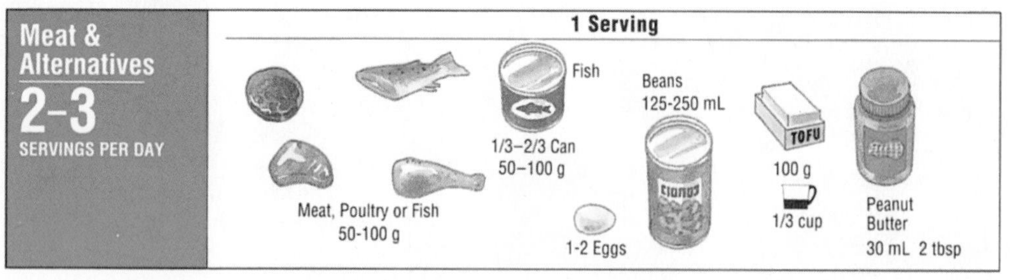

Enjoy eating well, being active and feeling good about yourself. That's VITALIT

Illustration 6.7, continued.

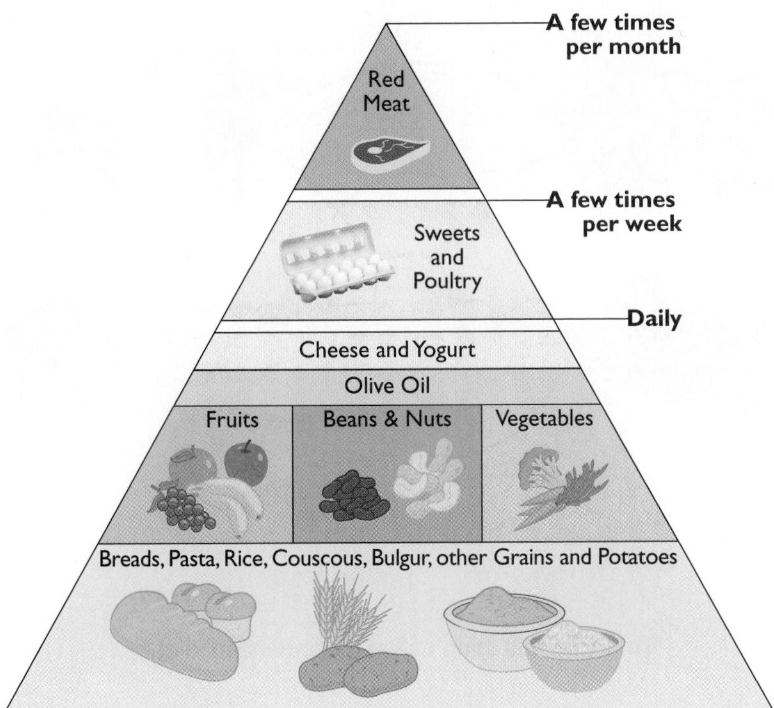

Illustration 6.8
The Mediterranean Diet Pyramid.

Specialized Food Guides Food groups have proved to be very useful in planning diets and are being adapted to meet the needs of people in various countries and ethnic groups. Of these, the most well known is the "Mediterranean Diet Pyramid" shown in Illustration 6.8. The World Health Organization developed this guide in 1994 to help popularize a type of diet associated with reduced risk of a number of chronic diseases.[10,11] It emphasizes olive oil, breads, pasta, rice, beans, vegetables, fruits, and wine in moderation. Intake of red meats is limited to monthly intake, and sweets and poultry to weekly intake.

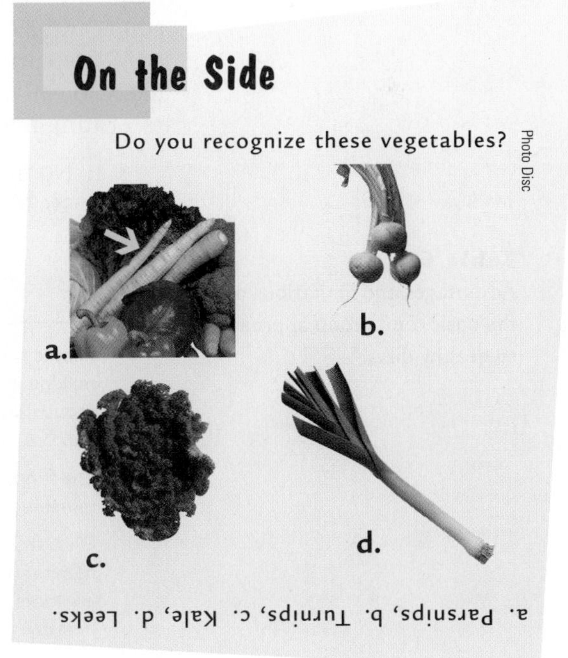

Advantages and Limitations of the Food Groups Whether a diet designed around the pyramid's groups is adequate and balanced depends on which foods are selected. For example, within the fruit group, you could choose avocados or apricots. Both are fruits, but their fat content varies substantially. Approximately 90% of the avocado's calories are from fat, compared to only 3.5% of the apricot's calories. Or you might choose to eat ice cream, fried chicken, and processed meats high in fat and sodium regularly. Such a diet may lack balance (Illustration 6.9 p. 6–16). Other limitations, as well as advantages, of the food group approach to healthy diets are presented in Table 6.4 (p. 6–16).

Dietary Guidelines, Food Groups, and Your Diet When you read the information on adequate and balanced diets, did you wonder how your diet rates? Illustration 6.10 (p. 6–17) encourages you to take a close look at your diet by running it through the Diet Analysis Plus Program.

Illustration 6.9

"I'm trying this thing called the 'balanced' diet."

Good Diets by Design

Some people find it easier to plan a healthy diet if they have a framework from which to work. Table 6.5 provides an example. This framework distributes foods from the various food groups into three meals and a snack.

Are You Better Off Eating Breakfast? People who eat breakfast *are* better off in a number of ways than people who don't.[12] Yet breakfast is the most frequently missed meal (Illustration 6.11 p. 6–18)). Approximately 15% of U.S. adults and 29% of people over the age of 15 in Canada skip it.[13,14] According to a U.S. Department of Agriculture (USDA) study, males aged 20 to 29 are the least likely to eat anything before lunch (72%).[15]

Eating breakfast pays off in these ways:

1. No midmorning hunger pangs—you don't regret *not* eating breakfast at 10:30 A.M.

Table 6.4
Advantages and limitations of the basic food group approach to healthy diets.[4]

Advantages	Limitations
· People eat foods and not nutrients, making it easier to relate to foods in groups rather than to recommended nutrient intake levels (i.e., the RDAs).	· Applicability of the food groups to vegetarians and people from a variety of ethnic groups may be unclear.
· The food groups are fairly easy to remember and teach.	· The food groups fail to directly address important nutritional concerns such as overweight or alcohol intake.
· Diets that include the recommended number of servings from the food groups are more likely to be adequate than diets that don't.	· The food groups do not include combination dishes such as pizza, soups, and stews that many people often eat. They do not include water (an essential nutrient).
· The food groups provide leeway for personal food choices within each group.	· People may end up with an inadequate diet if they select the same foods from each of the groups regularly.

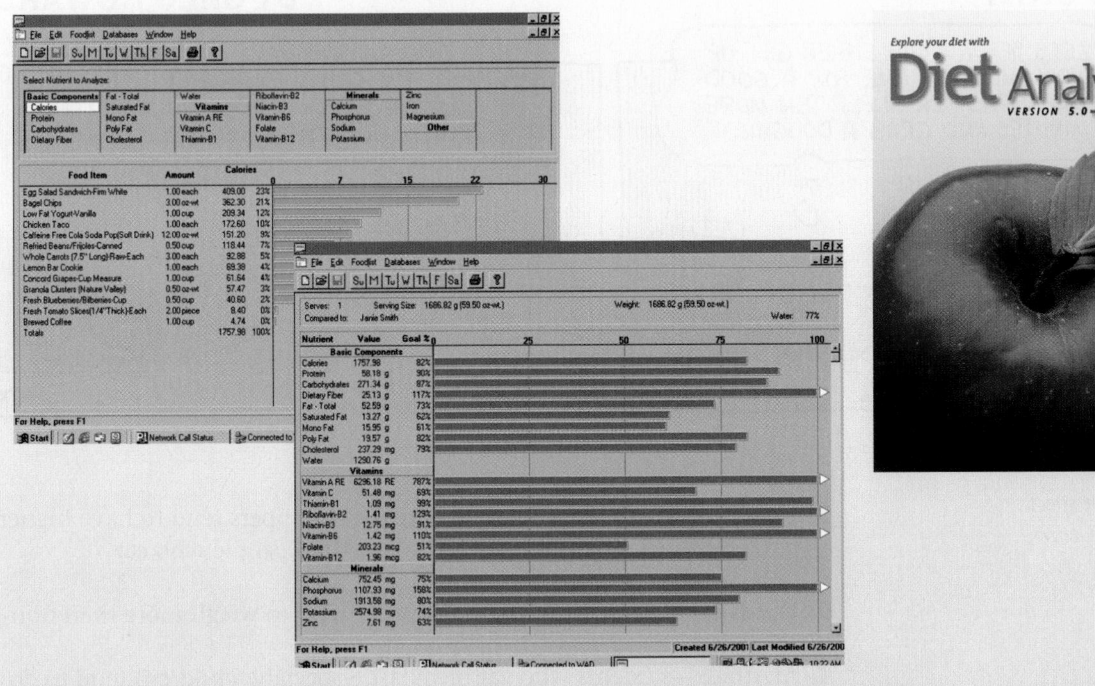

Illustration 6.10

Analyzing your diet. A convenient method is available for finding out how well your diet hits the targets recommended and where it misses. Keep track of your usual food and beverage intake for three days (include one weekend day) and enter the information into the Diet Analysis Plus Program.

The program works best when the dietary records prepared for input are complete and detailed. Record foods and beverages right after you eat them and write down the amounts. The quantities listed on food labels can often help you determine how much of a particular food you consumed.

Table 6.5
A framework for planning adequate diets based on the Food Guide Pyramid. Include one or more servings of foods from each group listed.

Adequate Diet Plan	Example
Meal one	
Fruits	Strawberries, ½ cup, with
Milk, yogurt, and cheese	low-fat yogurt, 1 cup
Bread, cereal, rice, and pasta	Whole-wheat toast, 2 slices
Meal two	
Vegetables Bread, cereal, rice, and pasta Meat, poultry, fish, dry beans, eggs, and nuts	Two bean tostadas with tomatoes, lettuce, and cheese
Fruits	Pear
Milk, yogurt, and cheese	Frozen yogurt, 1 cup
	Iced tea
Meal three	
Vegetables	Tossed salad, 2 cups
	Green beans, ½ cup
Bread, cereal, rice, and pasta	Rice, ½ cup
	Rolls, 2
Meat, poultry, fish, dry beans, eggs, and nuts	Salmon, 3 ounces
Milk, yogurt, and cheese	Skim milk, 1 cup
Snack	
Bread, cereal, rice, and pasta	Popcorn, 2 cups
Vegetables or fruits	Apple

SALLY FORTH BY GREG HOWARD

Illustration 6.11

Source: Reprinted with special permission of King Features Syndicate.

2. Lower blood cholesterol levels—breakfast skippers tend to have higher cholesterol levels and higher fat intake than people who eat breakfast.[16]
3. Lower body weight—breakfast skippers tend to weigh more than non-skippers.[17]
4. Alertness—students who eat breakfast (especially children) tend to do better on math tests.[12]
5. Higher-quality diets—skipping breakfast may lower the nutritional adequacy of your diet.[13]

Eating a full meal, like toast, cereal, juice, and milk, at breakfast is ideal, but when you don't have time for a complete meal, something nutritious is better than nothing. A carton of yogurt with fruit, a peanut butter sandwich, a raisin bagel, or an orange and crackers eaten on the way to class or work will give you the edge breakfast can provide.

Can You Still Eat Right When Eating Out? The question about what to eat often boils down to choosing a restaurant. According to recent USDA data, 50% of Americans eat out every day. In general, foods eaten away from home have lower nutrient content and are higher in fat than foods eaten at home.[18] In addition, children and teenagers who eat dinner with their families most days tend to have more healthful diets—including more vegetables, fruits, vitamins, minerals, and fiber, and less fat—than others who never or occasionally eat dinner with the family.[19]

The goal of eating right while eating out has become easier to reach in recent years because many restaurants have improved their menus. Restaurants frequently give customers the choice of fried or broiled fish, poultry, or beef and, upon request, will hold the sauces, salad dressings, and gravies or serve them on the side. Increasingly, restaurants are offering vegetarian entrées and "heart healthy" dishes low in fat, saturated fat, and sodium.

On the Side

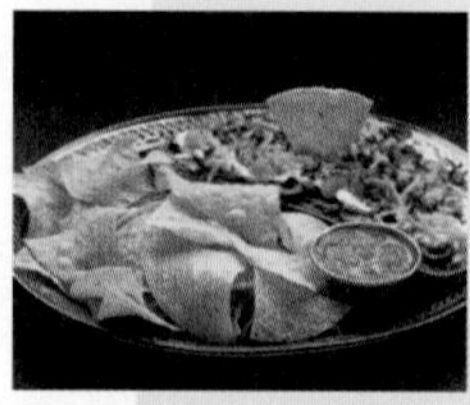

Photo Disc

Mexican is the most popular ethnic cuisine in the United States, followed by Italian and Chinese.[20]

Staying on Track While Eating Out You'll find it easier to stick to a healthy diet if you decide what to eat before you enter a restaurant and look over the menu. You could make the decision to order soup and a salad, broiled meat, a half-portion of the entrée, or no dessert *before* you enter the restaurant. "Impulse ordering" is a hazard that can throw diets out of balance. If you're going to a party or business event where food will be served, decide before you go what types of food you will eat and what you will drink. If only high-calorie foods are offered, plan on taking a small portion and stopping there.

Can Fast Foods Be Part of a Healthy Diet? For the most part, fast foods can find a place in a healthy diet. Although many fast foods deserve their reputation for being high in fat, you can enjoy fast foods and avoid the excess fat if you're selective.

Table 6.6 lists the calorie and fat contributions of a variety of foods sold in fast-food restaurants. Most of the choices are high in fat as indicated by the percentage of total calories provided by fat. Tostadas, bean burritos, grilled chicken

Table 6.6
The fat content of some fast foods.*

Food	Calories	Percentage of Calories from Fat
Sausage Breakfast Croissanwich®	538	69
Fried chicken thigh	278	62
Fried chicken wing	181	60
Bacon cheeseburger	610	58
Chicken McNuggets® (6)	323	56
Fried chicken breast	276	55
Fried chicken drumstick	147	55
Quarter Pounder with cheese®	525	55
Big Mac®	570	55
Sausage McMuffin®	427	55
Whopper with cheese®	711	54
Fried onion rings	274	53
Whopper®	628	52
Coleslaw	103	52
French fries, regular	227	52
Hot dog	280	51
Whaler Fish Sandwich®	488	50
Quarter Pounder®	427	48
Pan Supreme Pizza® (2 slices)	589	46
Cheeseburger	318	45
Ham and Cheese Specialty Sandwich®	471	44
Thin-n-Crispy Pizza with pepperoni® (2 slices)	413	44
Thin-n-Crispy Pizza Supreme® (2 slices)	459	43
Burrito Supreme®	457	43
Fillet-o-fish	373	43
Egg McMuffin®	340	42
Hamburger	350	41
Beef tostada	239	41
Beef n' Cheddar®	490	39
Taco	186	39
Thin-n-Crispy Cheese Pizza® (2 slices)	398	38
Beef burrito	357	38
Roast beef sandwich	353	38
Pan Pizza with pepperoni® (2 slices)	540	37
Pan Pizza with cheese® (2 slices)	492	33
Chili, regular	240	30
Tostada	179	30
Grilled chicken breast on bun	320	28
Pizza with pepperoni (2 slices)	380	28
Bean burrito	357	25
Pizza with cheese (2 slices)	376	24
Vanilla shake	280	19
Chocolate shake	447	16
Mashed potatoes, without gravy	62	15
Baked potato, plain	250	7

*Calories and fat content may vary somewhat depending on the fast-food restaurant or chain.

sandwiches, and pizza with cheese are among the lowest fat options available at fast-food restaurants. You can eat an occasional cheeseburger, fries, or specialty sandwich at your favorite fast-food restaurant and still maintain a healthy diet, however. It's the fat content of the total diet, and not of individual foods, that is important to health.

What if You Don't Know How to Cook? With so many convenience foods available and time at such a premium, there is growing concern that we're becoming a nation of cooking illiterates. Cooking at home gives you control over what you eat and how it's prepared. Some people enjoy cooking immensely and get a real thrill out of making their specialties for friends and family. It's becoming a popular leisure-time activity: 43% of U.S. adults have taken it up for enjoyment.[21]

If you don't know how to cook, there are several ways to learn. You could start on your own using the recipes on food packages like pasta, tomato sauce, or rice. Or you could buy a basic cookbook and make simple dishes like salads, tacos, shish kebab, and lentil soup. You could even take a community education course.

Illustration 6.12 shows some examples of good starter cookbooks. Read the sections on basic cooking skills, and learn what types of equipment and utensils you need to prepare basic dishes. Get the foods and other supplies you need. Select the recipes that look doable and pass your nutritional standards test. Voilà! You're cooking.

Illustration 6.12
Some good starter cookbooks and recipe sources. Most bookstores offer a wide variety of basic cookbooks.

Bon Appétit!

Japan's dietary guidelines contain one other rule of healthy eating that would serve Americans well: "Make all activities pertaining to food and eating pleasurable ones." Eating a healthy diet has to be enjoyable. If it's too much of a struggle, the healthy diet won't last. The best diets are those that keep us healthy and enhance our sense of enjoyment and well-being. The trick is to remember the broad array of nutritious foods we like that give us good taste and enjoyment when we eat them. And remember not to feel guilty when you occasionally eat hamburgers or ice cream. Enjoy them to the utmost—as much as ripe oranges, antipasto salads, homemade soups, roast turkey, hummus, and countless other nutritious delicacies.

WWW links

www.eatright.org
Web site of the American Dietetic Association and link to healthy eating advice.

www.foodtv.com
Offers recipes (among lots of ads and diversions).

www.nal.usda.gov/fnic
Your link to the Dietary Guidelines for Americans, ethnic and cultural resources, the Food Guide Pyramid, and dietary intake standards.

www.vrg.org
This excellent site covers dietary guidance for vegetarians.

www.nal.usda.gov/fnic/history
History buffs may like this site on food group recommendations through the years.

www.nutrition.gov
Sends you to a variety of ethnic and cultural food group plans.

www.usda.gov/cnpp
Select "Healthy Eating Index" to find a user-friendly, interactive dietary assessment program that allows you to enter one day's diet at a time.

www.dietitians.ca
Healthy eating information and resources are available from Dietitians of Canada.

www.cyberdiet.com
A fun and interactive site with a slew of tools for planning healthy diets.

www.fcs.uga.edu~selbon/apple/guides/choose.html
This long address leads you quickly to a listing of dietary guidelines for many countries.

www.hc-sc.gc.ca/nutrition
Check out Canada's Food Guide to Healthy Eating.

Nutrition UP CLOSE

The Food Guide Pyramid: A Balancing Act

FOCAL POINT: Discover how well-balanced your current diet is.

Think about what you typically eat each day for breakfast, lunch, dinner, and snacks. Next, think about the amount of food you normally consume at each meal. Now you are ready to compare your eating habits with the Food Guide Pyramid recommendations.

Check the appropriate boxes below, indicating the average number of servings you usually consume daily in each food group. For help in calculating serving sizes, refer to the Health Action on p. 6–8 and Table 6.3. Compare your responses with the Food Guide Pyramid recommendations in the Feedback section.

	Daily Servings						
Food Guide Pyramid Group	**0**	**1**	**2**	**3**	**4**	**5**	**6 or more**
How many daily servings of **bread, cereal, rice, and pasta?**	❍	❍	❍	❍	❍	❍	❍
How many daily servings of **fruit?**	❍	❍	❍	❍	❍	❍	❍
How many daily servings of **vegetables?**	❍	❍	❍	❍	❍	❍	❍
How many daily servings of **meat, poultry, fish, dry beans, eggs, and nuts?**	❍	❍	❍	❍	❍	❍	❍
How many daily servings of **milk, yogurt, and cheese?**	❍	❍	❍	❍	❍	❍	❍
How many daily servings of **fats, oils, and sweets?**	❍	❍	❍	❍	❍	❍	❍

Special note: You can use the Diet Analysis Plus software to determine how your diet matches the Food Guide Pyramid recommendations. Enter your food intake for one day; then, in the analysis/reports section, select "Pyramid." This report will show you whether you are getting enough, or too much, of each of the food groups.

FEEDBACK (answers to these questions) can be found at the end of Unit 6.

REFERENCES

1. Poor nutrition among adults. CDC Behavioral Risk Factor Survey, 1999, www.cdc.gov, accessed 11/00.
2. Krebs-Smith SM, et al. Characterizing food intake patterns of American adults. Am J Clin Nutr 1997;65(suppl):1264S–8S.
3. Kant AK, Schatzkin A, Harris TB, et al. Dietary diversity and subsequent mortality in the First National Health and Nutrition Examination Survey Epidemiological Follow-up Study. Am J Clin Nutr 1993;57:434–40.
4. Kant AK, Block G, Schatzkin A, et al. Dietary diversity in the U.S. population, NHANES II, 1976–1980. J Am Diet Assoc 1991;91:1526–31.
5. Weisburger JH. Approaches for chronic disease prevention based on current understanding of underlying mechanisms. Am J Clin Nutr 2000;71 (suppl):1710S–4S.
6. Liu S, et al. Fruit and vegetable intake and risk of cardiovascular disease: the Women's Health Study. Am J Clin Nutr 2000;72:922–8.
7. McCrory MA, et al. Dietary variety within food groups: association with energy intake and body fatness in men and women. Am J Clin Nutr 1999;69: 440–7.
8. Johnson RK, Kennedy E. The 2000 Dietary Guidelines for Americans: what are the changes and why were they made? J Am Diet Assoc 2000;100:769–74.
9. Food portions and servings: how do they differ? Nutrition Insights, USDA, 1999 March 11.
10. de Lorgeril M, et al. Mediterranean diet, traditional risk factors, and the rate of cardiovascular disease complications after myocardial infarction. Circulation 1999;99:779–85.
11. Lasheras C, et al. Mediterranean diet and age with respect to overall survival in institutionalized, nonsmoking elderly people. Am J Clin Nutr 2000:71:987–92.
12. Murphy JM, et al. The relationship of school breakfast to psychosocial and academic functioning. Arch Pediatr Adol Med 1998;152:899–907.
13. Eating breakfast greatly improves school children's diet quality. Nutrition Insights, USDA 1999 Dec.
14. Chao ESM, Vanderkooy PS. An overview of breakfast nutrition. J Can Diet Assoc 1989;50:225–8.
15. Consumption. CNI Nutrition Newsletter 1999; April 12:7.
16. Stanton JL Jr., Keast DR. Serum cholesterol, fat intake, and breakfast consumption in the United States adult population. J Am Coll Nutr 1989;8:567–72.
17. Heaton KW. Breakfast—do we need it? J Royal Society of Med 1989;82:770–1.
18. Committee on Diet and Health (Food and Nutrition Board; National Research Council). Diet and Health. Implications for reducing chronic disease risk. Washington, DC: National Academy Press; 1989; and Fat intake continues to drop; veggies still low in U.S. diet. USDA Research News 1996 Feb 6.
19. Gillman MW, et al. Family dinner and diet quality among older children and adolescents. Arch Fam Med 2000;9:235–40.
20. Economic trends in ethnic food restaurants. Results presented on Wall Street Week, 7/7/00.
21. Cooking. CNI Nutrition Newsletter 1999;April 12:8.

Nutrition

UP CLOSE

The Food Guide Pyramid: A Balancing Act

Feedback for Unit 6

You met the Food Guide Pyramid recommendations if your answers indicate that daily you eat 6–11 servings in the **bread, cereal, rice, and pasta** group; 2–4 servings in the **fruit** group; 3–5 servings in the **vegetable** group; 2–3 servings in the **meat, poultry, fish, dry beans, eggs, and nuts** group; and 2–3 servings in the **milk, yogurt, and cheese** group, while practicing moderation in the **fats, oils, and sweets** group.

If you did not meet the recommendations, balance your diet by choosing more foods from the group(s) in which you scored low, while cutting back on foods eaten in excess.

Unit 7

How the Body Uses Food

Photo Disc

Nutrition Scoreboard

	TRUE	FALSE
1. Almost all of the carbohydrate and fat you consume in foods is absorbed by the body, but only about half of the protein is.		
2. Disorders of the digestive system are the leading reason for surgical procedures and medical treatments in U.S. adults.		
3. Fewer than one in four African Americans is lactose intolerant.		

(Answers on next page)

Key Concepts and Facts

- Our bodies are in a continuous state of renewal. Materials used to renew body tissues come from the food we eat in the form of nutrients.
- The body uses nutrients for energy, body structures, and regulation of body processes.
- Digestion and absorption are processes that make nutrients in foods available for use by the body.
- Digestive disorders are common and often related to dietary intake.

Answers to Nutrition Scoreboard

	TRUE	FALSE
1. Over 90% of all the carbohydrates, fats, *and* proteins consumed in food are absorbed and become part of the body.		✓
2. Digestive disorders are the leading reason for surgery and medical treatment in adults in the United States. More than 5.3 million people are hospitalized each year due to hemorrhoids, ulcers, gallbladder troubles, chronic constipation, and other digestive system problems.[1]	✓	
3. Approximately 70% of African Americans are lactose intolerant, or unable to fully digest lactose in milk and milk products.		✓

Photo Disc

My Body, My Food

La vie est une fonction chimique. ("Life is a chemical process.")
—Antoine Lavoisier, late eighteenth century

You are not the same person you were a month ago. Although your body looks the same and you don't notice the change, the substances that make up the organs and tissues of your body are constantly changing. Tissues we generally think of as solid and permanent, such as bones, the heart and blood vessels, and nerves, are continually renewing themselves. The raw materials used in the body's renewal processes are the nutrients you consume in foods.

Each day, about 5% of our body weight is replaced by new tissue. Existing components of cells are renewed, the substances in our blood are replaced, and body fluids are recycled. Taste cells, for example, are replaced about every seven days, and cells lining the intestinal tract every one to three days. All of the cells of the skin are replaced every month. Red blood cells turn over every 120 days. If you thought it was hard to maintain a car, an apartment, or a house, just imagine what the maintenance is on a body! Maintenance is just one of the body's ongoing functions that require nutrients as raw material.

The Reasons the Body Needs Nutrients

The body needs nutrients for three basic reasons: for energy; for building and maintaining structural components of the body; and for regulation and control of body processes. These three roles are often referred to as the "energy, structural, and regulatory" functions of nutrients. The primary roles played by each of the six categories of nutrients are summarized in Table 7.1.

energy nutrients Carbohydrates, proteins, and fats are the "energy nutrients."

Which Nutrients Do What Carbohydrates, proteins, and fats are known as the ***energy nutrients.*** (Alcohol also provides energy but is not considered an

Table 7.1
Roles of nutrients in the body.

Nutrient Category	Primary Roles
1. Carbohydrates	Energy source
2. Fats	Energy source, structural components
3. Proteins	Structural components, energy source, regulation and control of body processes
4. Vitamins	Regulation and control of body processes
5. Minerals	Structural components, regulation and control of body processes
6. Water	Regulation and control of body processes, structural components

energy nutrient.) These nutrients contain energy that can be released and used by the body to fuel muscular work, thought processes, breathing, and other activities that require energy. In addition to supplying energy, protein and fat serve as structural components of body tissues. Fat is a structural component of every cell in the body and is found in particularly large amounts in cells of the nervous system. Protein is a part of bone, muscle, skin, intestines, hair, and many other body tissues.

Proteins, vitamins, minerals, and water are the body's regulators. They play major roles in regulating the rate at which the body forms energy, bone, fat, and other tissues. Minerals and water are a part of many body structures. Minerals, such as calcium and phosphorus, for example, are major components of bone. Water, which makes up approximately 60% of body weight in adults, is an important component of muscle, bone, ligaments, and other tissues.

How Do Nutrients in Food Become Available for the Body's Use?

> The human body is an amazing machine. Mine is, anyway. For example, I regularly feed my body truly absurd foods, such as cheez doodles, and somehow it turns them into useful body parts, such as glands. At least I assume it turns them into useful body parts.[2]
>
> —Dave Barry, 1987

The components of food that make "useful body parts" are nutrients. Through the processes of ***digestion*** and ***absorption,*** they are made available for use by every cell in the body.

➡ **digestion**
The breakdown of food in the digestive system.

➡ **absorption**
The process by which nutrients and other substances are transferred from the digestive system into the blood for transport throughout the body.

The Internal Travels of Food: An Overview The "food processor" of the body is the digestive system shown in Illustration 7.1. It consists of a 25- to 30-foot-long muscular tube and organs such as the liver and pancreas that secrete digestive juices. The digestive juices break foods down into very small particles that can be absorbed and used by the body. The absorbable forms of carbohydrates are monosaccharides, such as glucose and fructose. Proteins are largely absorbed as amino acids, and fats as fatty acids and glycerol. Vitamins and minerals are not broken down before they are absorbed; they are simply released from foods during digestion.

Much of the work of digestion is accomplished by ***enzymes*** manufactured by components of the digestive system such as the salivary glands, stomach, and pancreas. Enzymes are complex protein substances that speed up reactions that break down food. A remarkable feature of enzymes is that they are not changed by the chemical reactions they affect. This makes them reusable. Enzymes

➡ **enzymes**
Protein substances that speed up chemical reactions. Enzymes are found throughout the body but are present in particularly large amounts in the digestive system.

consumed in supplements, however, lose much of their activity during their journey through the gastrointestinal tract due to protein digestion. There is no evidence that enzyme supplements benefit digestion in healthy people.

Carbohydrates, proteins, and fat each have their own set of digestive enzymes. All together, over a hundred different enzymes participate in the digestion of food. Table 7.2 summarizes the major processes of the digestion of carbohydrates, fats, and proteins.

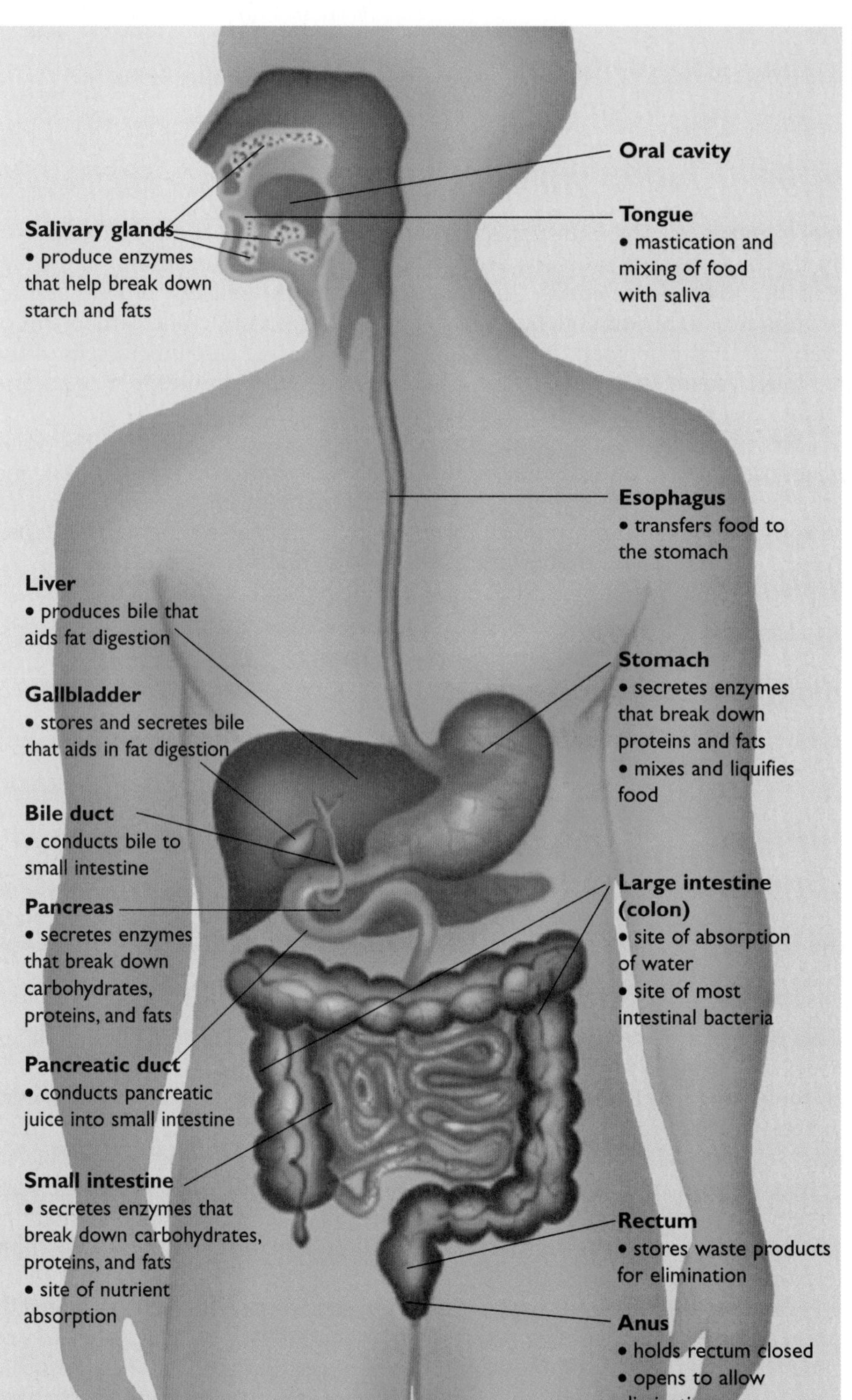

Illustration 7.1
The digestive system.

Table 7.2
Summary of the digestion of carbohydrates, fats, and proteins.

	Mouth	Stomach	Small Intestine, Pancreas, Liver, and Gallbladder	Large Intestine (Colon)
Carbohydrates (excluding fiber)	The salivary glands secrete saliva to moisten and lubricate food; chewing crushes and mixes it with a salivary enzyme that initiates starch digestion.	Digestion of starch continues while food remains in the stomach. Some alcohol (a carbohydrate-like substance) is absorbed. Acid produced in the stomach aids digestion and destroys bacteria in food.	The pancreas produces enzymes that digest starch and releases them into the small intestine. Intestinal cell walls possess enzymes that break sugars and starch fragments into simple sugars, which then are absorbed. Some alcohol is also absorbed here.	Undigested carbohydrates reach the colon and are partly broken down by intestinal bacteria.
Fiber	The teeth crush fiber and mix it with saliva to moisten it for swallowing.	No action.	Fiber binds cholesterol and some minerals.	Most fiber is excreted with feces; some fiber is digested by bacteria in the colon.
Fat	Fat-rich foods are mixed with saliva. The tongue produces traces of a fat-digesting enzyme that accomplishes some fat breakdown.	Fat tends to separate from the watery stomach fluid and foods and float on top of the mixture. Only a small amount of fat is digested. Fat is last to leave the stomach.	The liver secretes bile; the gallbladder stores it and releases it into the small intestine. Bile readies fat for enzyme action. The pancreas produces fat-digesting enzymes and releases them into the small intestine to split fats into their component parts (primarily fatty acids).	A small amount of fatty materials escapes absorption and is carried out of the body with other wastes.
Protein	In the mouth, chewing crushes and softens protein-rich foods and mixes them with saliva.	Stomach acid works to uncoil protein strands and to activate the stomach's protein-digesting enzyme. Then the enzyme breaks the protein strands into smaller fragments.	Enzymes from the small intestine and pancreas split protein fragments into peptides (compounds consisting of two or more amino acids) and free amino acids. Small peptides and amino acids are absorbed.	The large intestine carries undigested protein residue out of the body. Normally, almost all food protein is digested and absorbed.

A Closer Look As you chew food, glands under the tongue release saliva that lubricates the food so that it can be swallowed and pass easily along the intestinal tract. Saliva also gets food digestion started. It contains enzymes that begin to break down carbohydrates and fats.

After food is chewed, it is swallowed and passes down the esophagus to the stomach. Muscles that act as valves at the entrance and exit of the stomach

On the Side

Get Your Juices Flowing

You don't have to actually eat food to start your digestive juices flowing. You just have to think about food or see it.[3]

Put this information to the test. Clear your mind and take a close look at Illustration 7.2.

ensure that the food stays there until it's liquified, mixed with digestive juices, and ready for the digestive processes of the small intestine. Solid foods tend to stay in the stomach for over an hour, whereas most liquids pass through it in about 20 minutes. When the stomach has finished its work, it ejects one to two teaspoons of its liquified contents into the small intestine through the muscular valve at its end. Stomach contents continue to be ejected in this fashion until they are totally released into the small intestine. These small pulses of liquified food stimulate muscles in the intestinal walls to contract and relax; these movements churn and mix the food as it is digested by enzymes. When the diet contains a lot of fiber and sufficient fluids, the bulge of digesting food in the intestine tends to be larger. Larger food bulges stimulate a higher level of intestinal muscle activity than do smaller food bulges. Thus, high-fiber meals pass through the digestive system somewhat faster than low-fiber meals.

Digestion, as well as the absorption of nutrients, is greatly enhanced by the structure of the intestines (Illustration 7.3). Fingerlike projections called "villi" line the inside of the intestinal wall and increase its surface area tremendously. If laid flat, the surface area of the small intestine would be about the size of a baseball infield, or approximately 675 square feet. This large mass of tissue requires a high level of nutrients for maintenance. Much of this need (50% in the small intestine and 80% in the large intestine) is met by foods that are being digested.[4]

Digestion is completed when carbohydrates, proteins, and fats are reduced to substances that can be absorbed, and when vitamins and minerals are released from food. Most nutrients are absorbed in the small intestine. Water, sodium, and some of the end products of bacterial digestion, however, are absorbed from the large intestine. The large intestine is home to many strains of bacteria that consume undigested fiber and other types of complex carbohydrates that are not broken down by human digestive enzymes. These bacteria excrete gas as well as fatty acids that are partly absorbed in the large intestine. Substances in food that cannot be absorbed collect in the large intestine and are excreted in the stools.

Illustration 7.2
Testing, testing. This is a test of your salivary secretions. Does this food speak directly to your salivary glands?

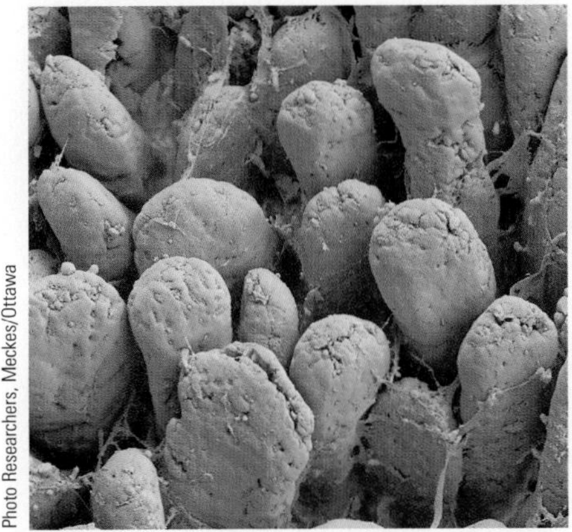

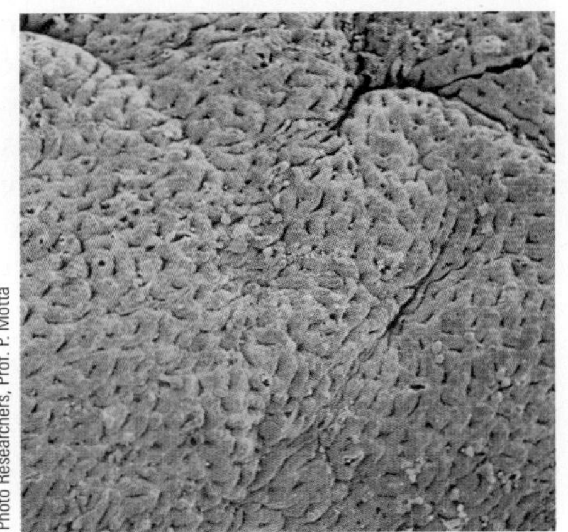

Illustration 7.3
Scanning electron micrographs of cross sections of the small intestine (left) and the large intestine (right). Note the high density of villi in the small intestine and the relative flatness of the lining of the large intestine.

lymphatic system
A network of vessels that absorb some of the products of digestion and transport them to the heart, where they are mixed with the substances contained in blood.

circulatory system
The heart, arteries, capillaries, and veins responsible for circulating blood throughout the body.

Absorption Absorption is the process by which the end products of digestion are taken up by the ***lymphatic system*** (Illustration 7.4) and the ***circulatory system*** (Illustration 7.5) for eventual distribution to cells of the body. Lymph

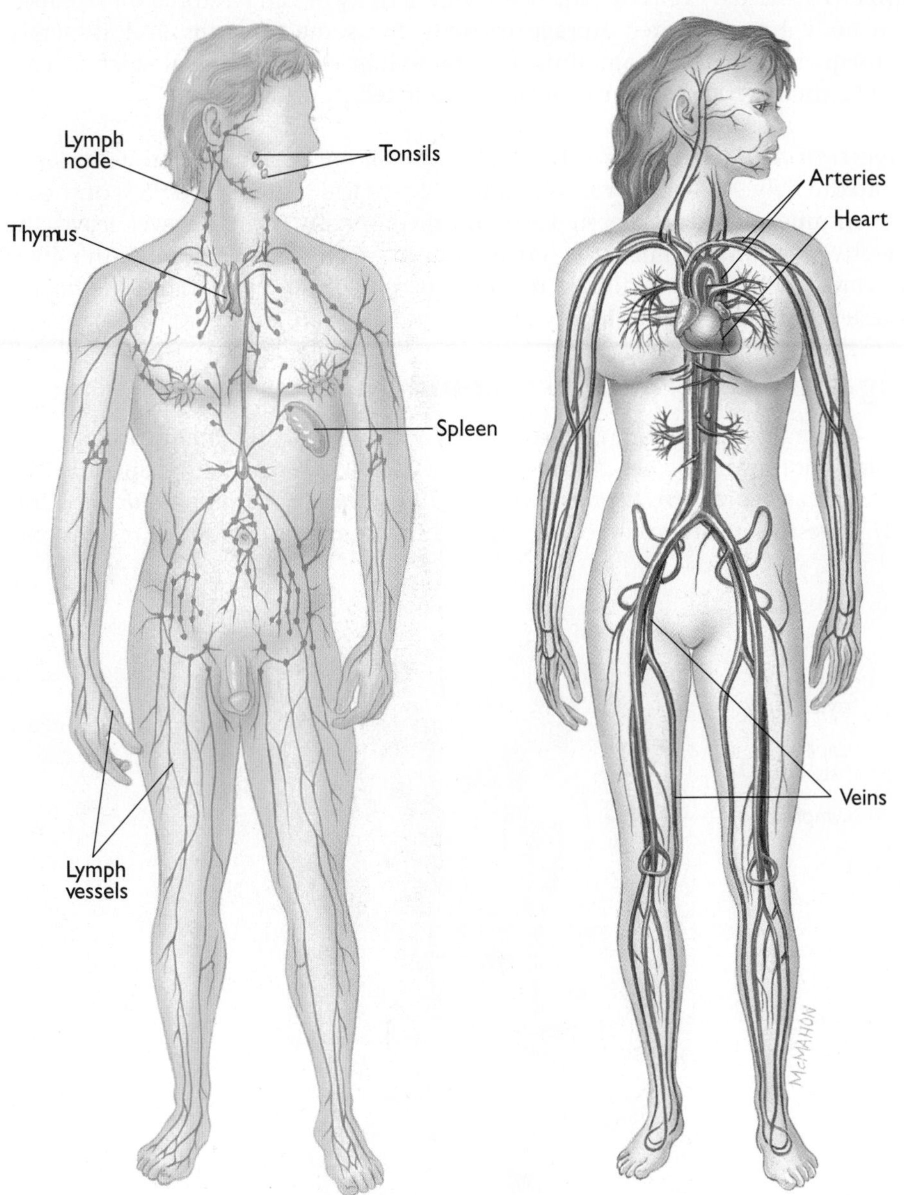

Illustration 7.4
(left) The lymphatic system.

Illustration 7.5
(right) The circulatory system includes the heart and blood vessels. This system serves as the nutrient transportation system of the body.

vessels and blood vessels infiltrate the villi that line the inside of the intestines (Illustration 7.6) and transport absorbed nutrients toward the major branches of the lymphatic and circulatory systems. The breakdown products of fat digestion are largely absorbed into lymph vessels, whereas carbohydrate and protein breakdown products enter the blood vessels.

The nutrient-rich contents of the lymphatic system are transferred to the bloodstream at a site near the heart where vessels from both systems merge into one vessel. From there the lymph and blood mixture is sent to the heart and subsequently throughout the body by way of the circulatory system. The circulatory system reaches every organ and tissue in the body, thereby supplying the cells with nutrients obtained from food.

Beyond Absorption Cells can use nutrients directly for energy, body structures, or the regulation of body processes, or convert them into other usable substances. For example, glucose delivered to cells can be used "as is" for energy formation or converted to glycogen and stored for later use. Fatty acids, an end product of fat digestion, can be incorporated into cell membranes or used in the synthesis of certain hormones. Vitamins and minerals freed from food by digestion can be used by cells to control enzyme activity or can be stored for later use. The body has a limited storage capacity for some vitamins and minerals. Consequently, excessive amounts of certain vitamins and minerals such as vitamin C, thiamin, and sodium are largely excreted in urine.

Digestion and Absorption Are Efficient Of our intake of energy nutrients, approximately 99% of the carbohydrate, 92% of the protein, and 95% of the fat we consume in food are digested and absorbed. Dietary fiber, however, leaves the digestive system in much the same form as it entered. Humans don't have enzymes that break down fiber. It should be noted, however, that some fiber is digested in the large intestine by bacteria.

Digestive Disorders Are Common

The leading cause of surgical procedures and medical treatments among U.S. adults is not heart disease, cancer, or accidents, but disorders of the digestive system such as ***heartburn, hemorrhoids,*** and ***duodenal and stomach ulcers*** (Table 7.3). Digestive disorders are common in children as well as adults. At least one-

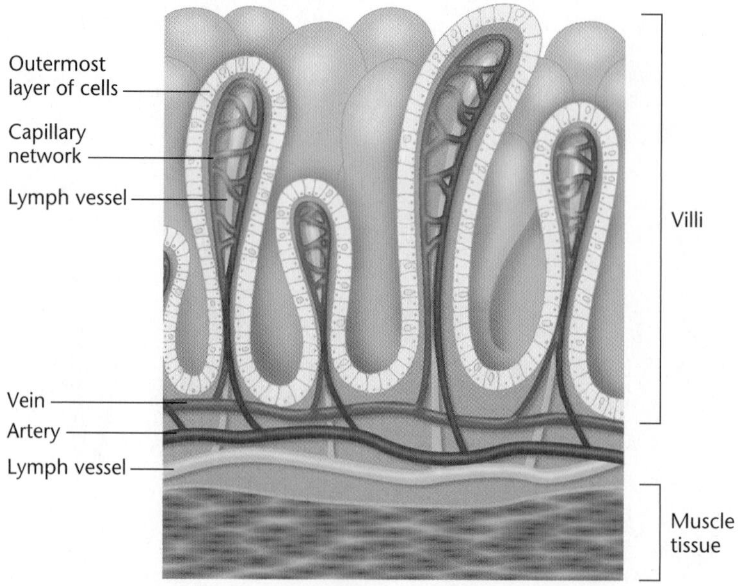

➡ **heartburn**
A condition that results when acidic stomach contents are released into the esophagus, usually causing a burning sensation.

➡ **hemorrhoids**
Swelling of veins in the anus or rectum.

➡ **duodenal and stomach ulcers**
Open sores in the lining of the duodenum (the uppermost part of the small intestine) or the stomach.

Illustration 7.6
Structure of villi, showing blood and lymph vessels.

third of U.S. adults experience heartburn, and 10% to 18% of schoolchildren experience constipation.[7]

Table 7.3
Common digestive disorders.[1,5–6]

	U.S. Adults Affected*
Heartburn	33.0%
Hemorrhoids	12.8
Ulcers	3.5
Chronic constipation	3.0
Chronic diarrhea	1.2

*Noninstitutionalized adults experiencing the disorder in the past 12 months.

Constipation and Hemorrhoids Both constipation and hemorrhoids are often due to diets that provide too little fiber. Fiber intakes by adults of 25 to 30 grams per day along with plenty of fluids (8 to 12 cups a day) can help prevent constipation and hemorrhoids in healthy people. Some good sources of dietary fiber are shown in Illustration 7.7.

Ulcers and Heartburn Ulcers develop when the protective barrier formed by cells lining the stomach and duodenum (the uppermost part of the small intestine) is damaged. This allows stomach acid and digestive enzymes to erode the lining of the stomach and duodenum and cause an "ulcer." Duodenal ulcers are 10 times more common than stomach ulcers and are closely associated with the presence of *Helicobacter pylori (H. pylori)* bacteria.[8] Excessive production of stomach acid can lead to ulcers, too, and may also cause heartburn. Heartburn is unrelated to heart conditions. It is called that because acid that escapes from the stomach causes a burning sensation in an area of the esophagus located near the heart.

Stress, anxiety, and frequent use of aspirin, ibuprofen, naproxen, and other medications appear to be related to the development of ulcers and heartburn in some people.[9] Fatty foods, coffee, alcohol, citrus fruits, soft drinks, and a variety of other foods have been implicated in the development of ulcers and heartburn. Rather than cause these disorders, certain foods appear to aggravate the symptoms of ulcers and heartburn.[10] High-fiber diets, on the other hand, appear to be protective against the development of ulcers and heartburn.[8]

A number of medications are available for the treatment of *H. pylori* and the excessive production of stomach acid. Reduction of stress, judicious elimination of offending foods and beverages from the diet, and small, frequent meals also help relieve ulcers and heartburn in some people.[9]

Illustration 7.7
Food sources of dietary fiber. Together, the foods shown provide 29 grams of dietary fiber, an amount that helps prevent constipation.

diarrhea
The presence of three or more liquid stools in a 24-hour period.

Diarrhea *Diarrhea* is a common problem in the United States and a leading public health problem in developing countries. Most cases of diarrhea are due to bacterial- or viral-contaminated food or water, lack of immunizations against infectious diseases, and vitamin A, zinc, and other nutrient deficiencies that make children particularly susceptible to diarrhea. Diarrhea can deplete the body of fluid and nutrients and produce malnutrition as well.[11] If it lasts more than two weeks or is severe, diarrhea can lead to dehydration, heart and kidney malfunction, and death. An estimated 3.5 million deaths from diarrhea diseases occur each year to the world's population of children five years of age or under.[11]

The vast majority of cases of diarrhea can be prevented through food and water sanitation programs, immunizations, and adequate diets.[11] The early use of oral rehydration fluids (for example, the formula provided by the World Health Organization and commercial formulas such as Pedialyte and Rehydralyte) shortens the duration of diarrhea.[12] Rehydration generally takes four to six hours after the fluids are begun.[13]

Rather than "resting the gut" during diarrhea as used to be recommended, children and adults, once rehydrated, should eat solid foods. Foods such as yogurt, lactose-free or regular milk, chicken, potatoes and other vegetables, dried beans, and rice and other cereals are generally well tolerated and provide nutrients needed for the repair of the intestinal tract.[13,14] It is best to avoid sugary fluids such as soft drinks. High-sugar beverages tend to draw fluid into the intestinal tract rather than increase the absorption of fluid.[12]

On the Side

Ancient Perspectives on Flatulence

Photo Disc

Passing gas is necessary to well-being.
—Hippocrates

All Roman citizens shall be allowed to pass gas whenever necessary.
—Claudius

Flatulence Young adults pass gas 10 times per day on average. The frequency with which people experience gas emissions, however, varies a good deal from person to person. Differences are largely due to diet and the types and amounts of bacteria that live in the gut. Poorly digested complex carbohydrates found in dried beans and some vegetables are primary causes of excess gas production.[15] Flatulence may be due to certain illnesses and medications and may cause a great deal of discomfort.

On the Side

Brain freeze" or that splitting headache you can get from eating ice cream too fast is caused by the quick drop in temperature in the back of your mouth. That causes vessels to constrict, and the result is an "ice cream headache."

Photo Disc

Lactose Maldigestion and Intolerance

I can't drink milk. It tears me up inside.

lactose maldigestion
A disorder characterized by reduced digestion of lactose due to the low availability of the enzyme lactase.

lactose intolerance
The term for gastrointestinal symptoms (flatulence, bloating, abdominal pain, diarrhea, and "rumbling in the bowel") resulting from the consumption of more lactose than can be digested with available lactase.

A very common digestive disorder is ***lactose maldigestion.*** The lactose found in milk and milk products presents a problem for the majority of the world's adult population who cannot digest it, either partially or completely (Table 7.4). The condition occurs more commonly in population groups that have no historical links to dairy farming and milk drinking.[18] Early humans in central and northwestern Europe and the regions of Africa and China highlighted in Illustration 7.8 tended to raise dairy animals and drink milk.

Lactose maldigestion is caused by a genetically determined low production of lactase, the enzyme that digests lactose.[20] People who lack this enzyme end up with free lactose in their large intestine after they consume milk or milk products that contain lactose. The presence of lactose in the large intestine produces the symptoms of ***lactose intolerance.*** These symptoms include a bloated feeling and diarrhea due to fluid accumulation, and gas and abdominal cramping caused by the excretion of gas by bacteria that digest lactose.[20]

Lactose maldigestion is rare in young children and affects adults to various degrees. Some adults produce little or no lactase and develop symptoms of lactose intolerance when they consume only small amounts of milk or milk products. Others produce some lactase and can tolerate limited amounts of lactose-containing milk and milk products, such as a cup of milk at a time or two cups of milk consumed in small portions throughout the day.[21,22]

Many people who are lactose maldigesters have no trouble eating yogurt and other fermented milk products such as cultured buttermilk and aged cheese.[23,24] The bacteria used to culture yogurt digest about half of the lactose. This reduction in lactose content is sufficient to prevent adverse effects in many people with lactose maldigestion.[23] Flavored and frozen yogurt and yogurt with added milk solids contain more lactose than plain yogurt and may not be tolerated as well, however.[23]

Milk solids, milk, and other lactose-containing components of milk may be added to foods you wouldn't expect. Consequently, it's best to examine food ingredient labels when in doubt. Milk, for instance, is a primary ingredient in some types of sherbet, and milk solids are added to many types of candy.

Table 7.4
Estimated incidence of lactose maldigestion among older children and adults in different population groups.[16,17]

Incidence of Lactose Maldigestion	
Asian Americans	90%
Africans	70
African Americans	70
Asians	65 or more
American Indians	62 or more
Mexican Americans	53 or more
U.S. adults (overall)	25
Northern Europeans	20
American Caucasians	15

Do You Have Lactose Maldigestion? The simple most reliable indicator of lactose maldigestion is the occurrence of lactose intolerance within hours after consuming lactose.[19] If you consistently experience these symptoms, visit your health care provider for a diagnosis. The symptoms could be due to lactose, other substances in milk, or another problem.

How Is Lactose Maldigestion Managed? Lactose maldigestion should *not* be managed by omitting milk and milk products from the diet! That would

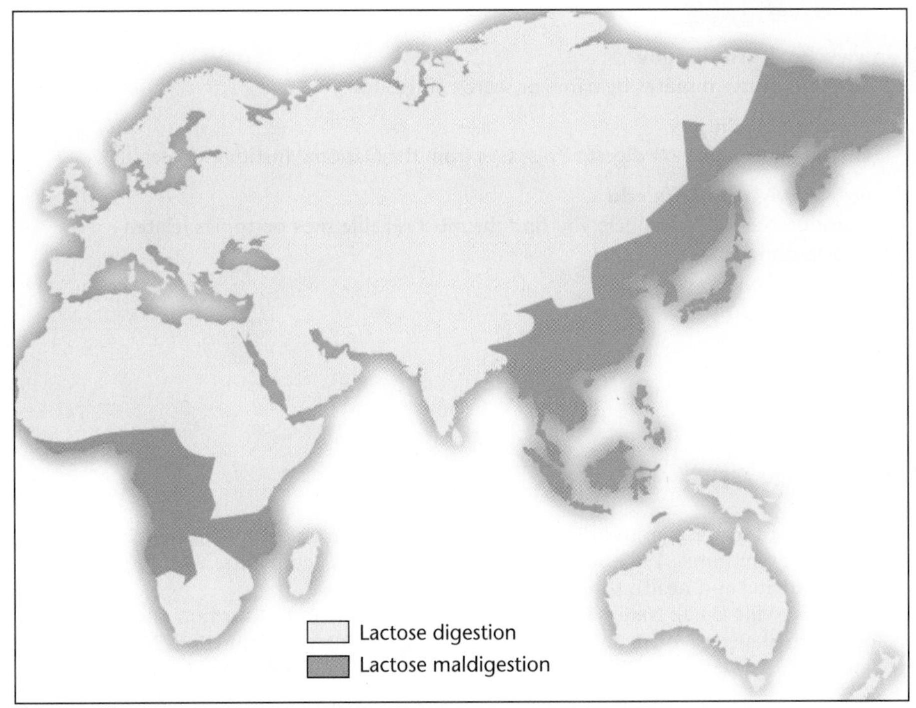

Illustration 7.8
Lactose maldigestion is less common among descendants of people who consumed milk from domesticated animals during prehistoric times (light areas) than among people whose early ancestors did not drink milk (dark areas).[19]

Illustration 7.9
Dairy products generally well tolerated by people with lactose maldigestion.

exclude a food group that contributes a variety of nutrients that cannot easily be replaced by other foods. The omission of milk and milk products from the diet of people with lactose intolerance appears to promote the development of osteoporosis.[25] Rather, fortified soy milk or cow's milk pretreated with lactase, or with lactase drops you add to milk prior to drinking it, and yogurt and other fermented milk products (if tolerated) should be consumed. Illustration 7.9 shows a variety of dairy products that are generally well tolerated by people with lactose maldigestion. Most people find that lactase-treated milk tastes a bit sweeter than regular milk.

Lactase tablets are also available, but they do not appear to work as well as lactase drops and pretreated milk.[26] The tablets may be less effective because some of the lactase is digested before it has a chance to act on the lactose. Lactase is an enzyme and enzymes are made of protein. Protein is partially digested in the stomach, so some of the lactase ingested in the tablets may never reach the small intestine where it would be used to break down lactose. Adding lactase to milk bypasses this problem because the lactase breaks down lactose before the milk is consumed.

Digestion is a remarkably complex and efficient process that could be covered in much more detail than has been presented here. Readers are encouraged to consult the web sites listed at the end of the unit for additional information.

www.healthfinder.gov
Search digestive diseases by name or search "digestion."

www.niddk.nih.gov
Obtain information on digestive diseases from the National Institute of Health.

www.navigator.tufts.edu
Nutrition Navigator will help you find the most reliable sites on topics related to digestion.

REFERENCES

1. Prevalence of major digestive disorders and bowel symptoms, 1989. Advance data from vital and health statistics; no. 212. Hyattsville (MD): National Center for Health Statistics; 1992; and Advance data no. 253. Hyattsville (MD): National Center for Health Statistics; 1994.
2. Barry D. Bad habits. New York: Henry Holt; 1987.
3. Schneeman B. Nutrition and gastrointestinal function. Nutr Today 1993;Jan/Feb:20–24.

Nutrition UP CLOSE

Personal History of Digestive Upsets

FOCAL POINT: Digestive disorders are common.

To bring you closer to your digestive system, review the following list and check the upsets you have experienced in the past month:

Digestive Upset	Experienced in the Past Month?
Heartburn or indigestion	______
Diarrhea	______
Constipation	______
Stomach cramps	______
Vomiting	______
Abdominal bloating	______
Hemorrhoids	______

FEEDBACK: Most bouts of digestive upsets are brief and only bothersome. Some will be related to diet (inadequate or excessive dietary fiber, bacterially contaminated food, for example) whereas others will be initiated by illness, stress, or another condition. Painful or prolonged episodes of digestive disorders should, of course, be brought to the attention of your health care provider.

4. Bengmark S. Econutrition and health maintenance: a new concept to prevent GI inflammation, ulceration, and sepsis. Clin Nutr 1996;15:1–10.
5. Isolauri J, Laippaia P. Prevalence of symptoms suggestive of gastrooesophageal reflux disease in an adult population. Ann Med 1995;27:67–70.
6. Prevent heartburn. Nutr Today 1999;34:213.
7. Feldman W, McGrath P, Hodgson C, Ritter H, Shipman RT. The use of dietary fiber in the management of simple, childhood, idiopathic, recurrent, abdominal pain. Results in a prospective, double-blind, randomized, controlled trial. Am J Dis Child 1985;139:1216–18.
8. Current medical diagnosis and treatment, 35th ed. Stamford (CT): Appleton & Lange; 1997.
9. Kurata AN, et al. Dyspepsia in primary care: perceived causes, reasons for improvement, and satisfaction with care. J Fam Practice 1977;44:281–8.
10. The Merck manual of medical information. Whitehouse (NJ): Merck Research Laboratory; 1997.
11. Kilgore PE, et al. Trends in diarrheal disease associated mortality in US children, 1968 through 1991. JAMA 1995;274:1143–8.
12. Goepp JG, Katz SA. Oral rehydration therapy. Am Fam Phys 1993;47:843–8.
13. Meyers A. Oral rehydration therapy: what are we waiting for? Am Fam Phys 1993;47:740–2.
14. Lima AAM, Guerrant RL. Persistent diarrhea in children: epidemiology, risk factors, pathophysiology, nutritional impact, and management. Epidemiol Rev 1992;34:222–42.
15. Furne JK, Levitt MD. Factors influencing frequency of flatus emission by healthy subjects. Dig Dis Sci 1996;41:1631–5.
16. Scrimshaw NS, Murray, EB. Prevalence of lactose maldigestion. Am J Clin Nutr 1988;48(suppl):1086–98.
17. Inman-Felton AE. Overview of lactose maldigestion (lactose nonpersistence). J Am Diet Assoc 1999;99:481–9.
18. Simmons FJ. Primary adult lactose intolerance and the milking habit: a problem in biological and cultural interrelationships. I. Review of the medical research. Am J Digestion 1981;14:819.

19. Simoons FJ. Primary lactose intolerance and the milking habit: a problem in biological and cultural interrelationships. Am J Digestive Diseases 1970;15:696–710.
20. Tolstoi LG. Adult-type lactase deficiency. Nutr Today 2000;35:134–42.
21. Suarez FL, et al. Tolerance to the daily ingestion of two cups of milk by individuals claiming lactose intolerance. Am J Clin Nutr 1997;65:1502–6.
22. Johnson AO, Semenya JG, Buchowski MS, et al. Adaptation of lactose maldigesters to continued milk intakes. Am J Clin Nutr 1993;58:879–81.
23. Martini MC, Smith DE, Savaiano DA. Lactose digestion from flavored and frozen yogurts, ice milk, and ice cream by lactase-deficient persons. Am J Clin Nutr 1987;46:636–40.
24. Sinden AA, Sufphen JL. Dietary treatment of lactose intolerance in infants and children. J Am Diet Assoc 1991;91:1567–71.
25. Pacifici R, Droke D, Avioli LV. Intestinal lactase activity and calcium absorption in the aging female with osteoporosis. Calcified Tissue International 1985;37:101–2.
26. Onwulata CI, Rao DR, Vankineni P. Relative efficiency of yogurt, sweet acidophilus milk, hydrolyzed-lactose milk, and a commercial lactase tablet in alleviating lactose maldigestion. Am J Clin Nutr 1989;49:1233–7.

Photo Disc

Unit 8

Calories! Food, Energy, and Energy Balance

Nutrition Scoreboard

	TRUE	FALSE
1. Gram for gram, carbohydrates provide the body with more energy than protein does.		
2. A teaspoon of butter has a higher caloric value than a teaspoon of margarine.		
3. Energy can be neither created nor destroyed. It can, however, change from one form to another.		

(Answers on next page)

Answers to

Nutrition Scoreboard

		TRUE	FALSE
1.	Carbohydrates and protein both provide four calories per gram. Fat, on the other hand, provides nine calories per gram, and alcohol seven calories per gram.		✓
2.	Butter and margarine contain the same amount of fat, so they provide the same number of calories (35 per teaspoon).		✓
3.	Stated is the conservation of energy principle, the first law of thermodynamics.	✓	

Key Concepts and Facts

- A calorie is a unit of measure of energy.
- The body's sources of energy are carbohydrates, proteins, and fats (the "energy nutrients").
- Fats provide over twice as many calories per unit weight as carbohydrates and proteins do.
- Most foods contain a mixture of the energy nutrients as well as other substances.
- Weight is gained when caloric intake exceeds the body's need for energy. Weight is lost when caloric intake is less than the body's need for energy.

Energy!

It's high in energy, you say? Sure, I'll eat one, thanks. I thought those snack bars were loaded with calories!

When you think of calories, do you think of energy? That is the "scientifically correct" way to think about them. Energy is what calories are all about.

Calories Are a Unit of Measure

calorie
(calor = heat). A unit of measure used to express the amount of energy produced by foods in the form of heat. The calorie used in nutrition is the large "Calorie," or the "kilocalorie" (kcal). It equals the amount of energy needed to raise the temperature of one kilogram of water (about four cups) from 15 to 16°C (59 to 61°F). The term kilocalorie, or "calorie" as used in this text, is gradually being replaced by the "kilojoule" (kJ) in the United States; 1 kcal = 4.2 kJ.

The ***calorie*** is like a centimeter or pound in that it is a unit of measure. Rather than serving as a measure of length or weight, the calorie is used as a measure of energy. Specifically, a calorie is the amount of energy needed to raise the temperature of one kilogram of water (about four cups) from 15°C to 16°C (59°F to 61°F) (Illustration 8.1). This amount of energy is used as a standard for assigning caloric values to foods.

Because calories are a unit of measure, they are not a component of food like vitamins or minerals. When we talk about the caloric content of a food, we're really talking about the caloric value of the food's energy content.

The caloric value of food is determined by burning it completely in a container surrounded by a specific amount of water (Illustration 8.2). The energy released by the food in the form of heat raises the temperature of the surrounding water. The rise in temperature indicates how many calories were released from the portion of food. Although the body doesn't literally burn food, the amount of heat released by food while burning is approximately the same as the amount of energy it supplies the body.

The Body's Need for Energy

The body uses energy from foods to fuel muscular activity, growth, and tissue repair and maintenance; to chemically process nutrients; and to maintain body temperature (to name a few examples). These needs for energy are subdivided

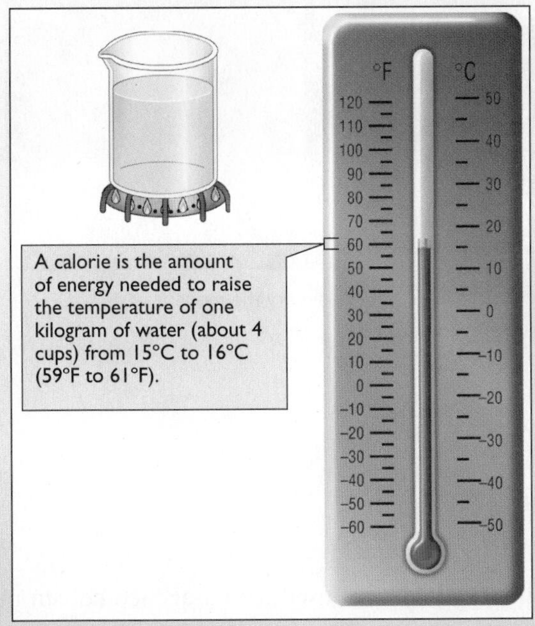

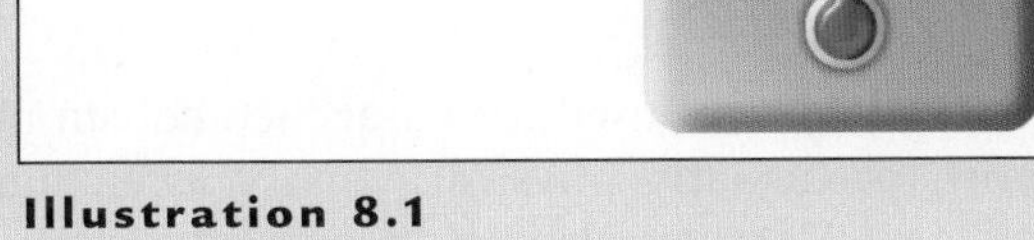

Illustration 8.1

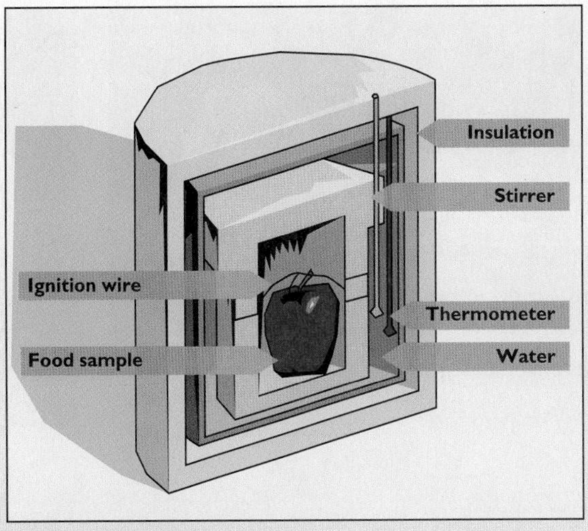

Illustration 8.2
How the energy content of food is measured. The caloric value of a food is determined by the amount of heat it releases and transfers to water when completely burned.

into three categories: basal metabolism, physical activity, and dietary thermogenesis (Table 8.1 and Illustration 8.3).

Basal Metabolism Ongoing activities that the body performs while at rest, work, or play make up the bulk of most people's need for calories—usually 60 to 80%. These activities sustain ***basal metabolism,*** or processes that are required to maintain health such as breathing, the beating of the heart, the maintenance of body temperature, growth, and the transport of nutrients to, and waste products out of, cells. Energy-using activities of basal metabolic processes require no conscious effort on our part; they are continuous activities that the body must perform to sustain life. The energy needed to carry out basal metabolic functions is assessed when the body is in a state of complete physical and emotional rest.

➡ **basal metabolism**
Energy used to support body processes such as growth, health, tissue repair and maintenance, and other functions. Assessed while at rest, basal metabolism includes energy the body expends for breathing, the pumping of the heart, the maintenance of body temperature, and other life-sustaining, ongoing functions.

How Much Energy Do I Expend for Basal Metabolism? You can quickly estimate the calories needed for basal metabolic processes:

- For men: Multiply body weight in pounds by 11.
- For women: Multiply body weight in pounds by 10.[1]

Thus, a man who weighs 170 pounds needs approximately 170 × 11, or 1870 calories per day for basal metabolic processes. A 135-pound woman needs 135 × 10, or 1350 calories.

Table 8.1
The three energy-requiring processes of the body.

Basal metabolism
- Energy required to maintain normal body functions while at rest

Physical activity
- Energy needed for muscular work

Dietary thermogenesis
- Energy used to chemically process foods. (The process gives off heat.)

basal metabolism

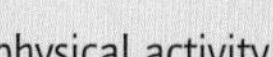

physical activity

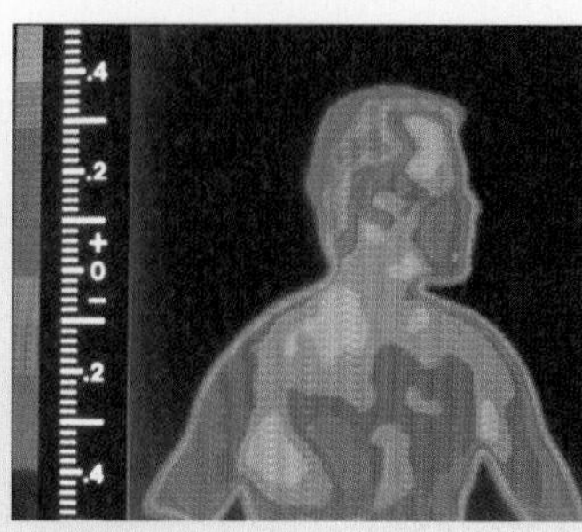

dietary thermogenesis

Illustration 8.3
Examples of the three types of energy-requiring processes in the body.

This formula gives an estimate of calories used for basal metabolism based on sex and weight. Other factors, particularly physical activity level, muscle mass, height, health status, and genetic traits, also affect how efficiently the body uses calories for basal metabolism.[2,3] Consequently, results obtained using this quick formula may be 10 to 20% lower or higher than the true number of calories required for basal metabolism.[4]

How Much Energy Do I Expend in Physical Activity? The caloric level needed for physical activity can vary a lot, depending on how active a person is. It usually accounts for the second highest amount of calories we expend.

The energy cost of supporting a physically inactive lifestyle (Table 8.2) is about 30% of the number of calories needed for basal metabolism. An "average" activity level requires roughly 50%, and an "active" level requires approximately 75% of the calories needed for basal metabolism.[5] A physically inactive person needing 1500 calories a day for basal metabolism, for example, would require about 450 calories (1500 calories × 0.30—for 30%) for physical activity.

A Common Source of Error People have a tendency to overestimate time spent in physical activity. This tends to overestimate calories needed for physical activity, and depending on the amount of overestimation, can lead to highly inaccurate estimates of total calorie need. Physical activity level should be based on time spent actually engaged in physical activity and not include time spent getting ready for the activity, on breaks, or between activities.

Table 8.2
Energy expenditure by usual level of activity.[5]

Activity Level	Percentage of Basal Metabolism Calories
Inactive. Sitting most of the day; less than two hours of moving about slowly or standing.	30%
Average. Sitting most of the day; walking or standing two to four hours, but no strenuous activity.	50
Active. Physically active four or more hours each day; little sitting or standing; some physically strenuous activities.	75

	Calories
1. Basal metabolism Multiply body weight in pounds by 10. (For men the figure is 11.)	130 × 10 = **1300**
2. Physical activity Multiply basal metabolism calories by 0.30 (for 30%) based on the usual energy expenditure level of "inactive" (see Table 8.2).	1300 × 0.30 = **390**
3. Dietary thermogenesis Add calories needed for basal metabolism and physical activity together: 1300 + 390 = 1690 Multiply the result by 0.10 (for 10%).	1690 × 0.10 = **169**
4. Total calorie need Add calories needed for basal metabolism, physical activity, and dietary thermogenesis together. 1300 + 390 + 169 =	**Total calorie need = 1859**

Table 8.3
Summary of calculations for estimating total calorie need of a 130-pound, inactive woman.

How Many Calories Does Dietary Thermogenesis Take? A portion of the body's energy expenditure is used for digesting foods, absorbing and utilizing nutrients, and transporting nutrients into cells. Some of the energy involved in such activities escapes as heat. These processes are referred to as the energy cost of ***dietary thermogenesis.*** Calories expended for dietary thermogenesis are estimated as 10% of the sum of basal metabolic and usual physical activity calories.

Say a person's basal metabolic need is 1500 calories and 450 calories are required for usual activity: 1500 calories + 450 calories = 1950 calories. Calories expended for dietary thermogenesis would equal approximately 10% of the 1950 calories, or 195 calories.

dietary thermogenesis Thermogenesis means the production of heat. Dietary thermogenesis is the energy expended during digestion of food and the absorption and utilization of nutrients. Some of the energy escapes as heat. It accounts for approximately 10% of the body's total energy need. Also called *specific dynamic action* of foods, *diet-induced thermogenesis,* and *thermic effect of foods.*

Adding It All Up Your estimated total daily need for calories is the sum of calories used for basal metabolism, physical activity, and dietary thermogenesis. In the preceding example, total calorie need would be 2145 calories (1500 + 450 + 195). A complete example of calculations involved in estimating a person's total calorie need is provided in Table 8.3. Although the caloric level calculated won't be exactly right, it should provide a good estimate of your total caloric need.

Where's the Energy in Foods?

Any food that contains carbohydrates, proteins, or fats (the "energy nutrients") supplies the body with energy. (That *includes* the foods listed in Table 8.4!)

1. A candy bar eaten with a diet soda
2. Celery and grapefruit
3. Hot chocolate, cheesecake, or soft drinks consumed to make you feel better
4. Cookie pieces
5. Foods "taste-tested" during cooking
6. Foods you eat while on the run
7. Foods you eat straight from their original containers (like ice cream from the carton, milk from a jug, or chips from a large bag)

Table 8.4
Foods that *do* have calories.

Illustration 8.4
If you have observed the flames produced by fat dripping from a steak or hamburger on a grill, you have seen the powerhouse of energy stored in fat. The carbohydrate and protein contents of grilled foods don't burn with nearly the same intensity. They have less energy to give.

Carbohydrates and proteins supply the body with four calories per gram, and fat provides nine calories per gram. Alcohol also serves as a source of energy. There are seven calories in each gram of alcohol in the diet (Table 8.5).

If you enjoy grilling foods on an outdoor barbecue, you have probably observed firsthand the high level of stored energy in fats (Illustration 8.4). Unlike the drippings of low-fat foods such as shrimp or vegetables, drips from high-fat foods cause bursts of flames to shoot up from the grill. The high-energy content of alcohol can be seen in the alcohol-fueled flames that adorn cherries jubilee and baked Alaska.

If you know the carbohydrate, protein, fat, and, if present, alcohol content of a food or beverage, you can calculate how many calories it contains. For example, say a cup of soup contains 15 grams of carbohydrate, 10 grams of protein, and 5 grams of fat. To calculate the caloric value of the soup, multiply the number of grams of carbohydrate and protein by 4 and the number of grams of fat by 9. Then add the results together:

$$\begin{aligned} \text{15 grams carbohydrate} \times \text{4 calories/gram} &= \text{60 calories} \\ \text{10 grams protein} \times \text{4 calories/gram} &= \text{40 calories} \\ \text{5 grams fat} \times \text{9 calories/gram} &= \underline{\text{45 calories}} \\ & \quad \text{145 calories} \end{aligned}$$

You can calculate the percentage of total calories from carbohydrate, protein, and fat in the soup by dividing the number of calories supplied by each nutrient by the total number of calories and then multiplying by 100:

- Carbohydrate: $\frac{\text{60 calories}}{\text{145 calories}} = 0.41 \times 100 = 41\%$
- Protein: $\frac{\text{40 calories}}{\text{145 calories}} = 0.28 \times 100 = 28\%$
- Fat: $\frac{\text{45 calories}}{\text{145 calories}} = 0.31 \times 100 = 31\%$

$$100\%$$

Reality Check Given this information about the caloric value of carbohydrates, proteins, and fats, which of the items in Illustration 8.5 would you expect to be highest in calories?

College students have been found to miss this question 47% of the time.[6] It's the margarine! Margarine contains the most fat; it is made primarily from oil. *High-fat foods provide more calories ounce for ounce than foods that contain primarily carbohydrate or protein.* That means bread and potatoes (rich in carbohydrates), catsup (rich in water and a low-calorie vegetable), sugar, and other low-fat or no-fat foods provide fewer calories than equal amounts of foods that contain primarily fat.

Table 8.5
Caloric value of the energy nutrients and alcohol.

	cals/gm
Carbohydrate	4
Protein	4
Fat	9
Alcohol	7

Most Foods Are a Mixture

Some foods (such as oil and table sugar) consist almost exclusively of one energy nutrient, but most foods contain many substances in varying amounts.

Bread is high in complex carbohydrates ("starch"), but it also contains many other substances such as protein, water, vitamins, minerals, and fiber. Likewise,

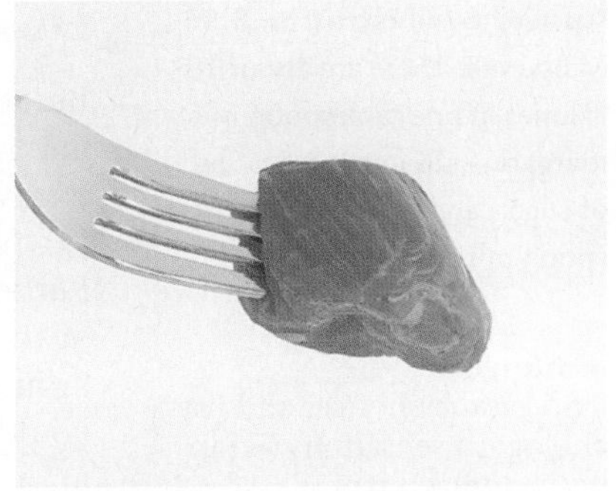

Illustration 8.5
Which contains the most calories—a tablespoon of margarine, sugar, or steak? (Turn the page to find the answer.)

steak is not all protein. Although protein constitutes about 32% of the total weight of lean sirloin steak, fats make up about 8%, and the largest single ingredient is water—59% of the weight. Yet we don't think of steak as a "high-water food"; instead, it's often thought of as pure protein.

So, even though some foods provide relatively more carbohydrate, protein, or fat than other foods, most foods contain a mixture of energy nutrients.

The Caloric Value of Foods: How Do You Know?

Some people say you can estimate the caloric value of food by how it tastes or by how appetizing it looks. Actually, it's not that simple. How many calories do you think are contained in a half-cup of peanuts, a boiled, medium-sized potato, and a cup of rice (Illustration 8.6)?

Taste, appearance, and reputation do not make good criteria for determining the caloric value of foods. Here's an example:

> I used to order the fish sandwich at fast-food restaurants because I was trying to avoid all the calories in hamburgers. Then I found out the fish sandwich had about the same number of calories as the quarter-pound hamburger! Where did I get the idea that fried fish loaded with tartar sauce has fewer calories than a hamburger?

This doesn't have to happen to you! By referring to Appendix A, you could determine that a typical fast-food fish sandwich and a quarter-pound hamburger weigh in at about 400 calories each. If you're interested in the caloric value of foods you frequently eat, check them out.

Illustration 8.6
What's the caloric value of these foods? One contains 420 calories, another 205, and the third 118 calories. Try matching the caloric values with the foods, and then check your answers on p. 8–10.

Answer to Illustration 8.5: Margarine. There are about 101 calories in one tablespoon of margarine, 46 in a tablespoon of sugar, and 40 in a tablespoon of relatively lean steak.

How Is Caloric Intake Regulated by the Body?

Because energy is critical to survival, the body has a number of mechanisms that encourage regular caloric intake. It has less effective means of discouraging an excessive intake of calories.[7]

Mechanisms that encourage food intake don't depend on weight status.[8] Rather, they are keyed to encouraging eating on a regular basis so that food, if available, will be consumed and carry the body through times when food isn't around. Humans developed on a schedule of "feast and famine." Those who could store enough fat to see them through the times when food was scarce had an advantage. So, no matter how thin or fat a person is, she or he experiences ***hunger*** if food is not consumed several times throughout the day. The "hungry" signal is thought to be sent by a series of complex mechanisms when cells run low on energy nutrients supplied by the last meal or snack.[8]

When we eat, a point comes when a fresh supply of energy nutrients enters the bloodstream; at this point we feel full and are no longer interested in eating. The signal is due to hormones and internal sensors in the brain, stomach, liver, and fat cells that indicate ***satiety***—the feeling that we've had enough to eat.[9]

For most people, hunger and satiety mechanisms adjust energy intake to match the body's need for energy. However, internal signals that urge us to eat or to stop eating can be overridden. People can resist eating, no matter how strong the hunger pains. Even after the "I'm full" siren has sounded, people can go on eating. And sometimes people eat because they have an ***appetite*** for specific foods and desire the pleasure foods can bring.

Appetite may or may not be related to being hungry. It can be triggered when we smell or see a tasty food right after a meal or when we're really hungry. Ever see those TV commercials for a juicy burger and fries that air about 11:00 P.M.? How many people who jump into their cars and head to the carry-out window are actually hungry? Or have you noticed how appealing the idea of eating is and how good food tastes when you're very hungry? That's appetite at work.

➡ **hunger**
Unpleasant physical and psychological sensations (weakness, stomach pains, irritability) that lead people to acquire and ingest food.

➡ **satiety**
A feeling of fullness or of having had enough to eat.

➡ **appetite**
The desire to eat; a pleasant sensation that is aroused by thoughts of the taste and enjoyment of food.

On the Side

People expend an average of 11 calories an hour chewing gum. If you chewed gum for an hour every day and didn't change any other component of energy balance, you'd lose about a pound a year.[10]

The Question of Energy Balance Unless you are currently losing or gaining weight, the number of calories you need is the number you usually consume in your diet. Adults who maintain their weight are in a state of energy balance (Illustration 8.7). Because they are not losing weight (using fat and other energy stores) or gaining it (storing energy), the body's expenditure of energy and its intake of energy are balanced. When energy intake

Illustration 8.7
The body's energy status.

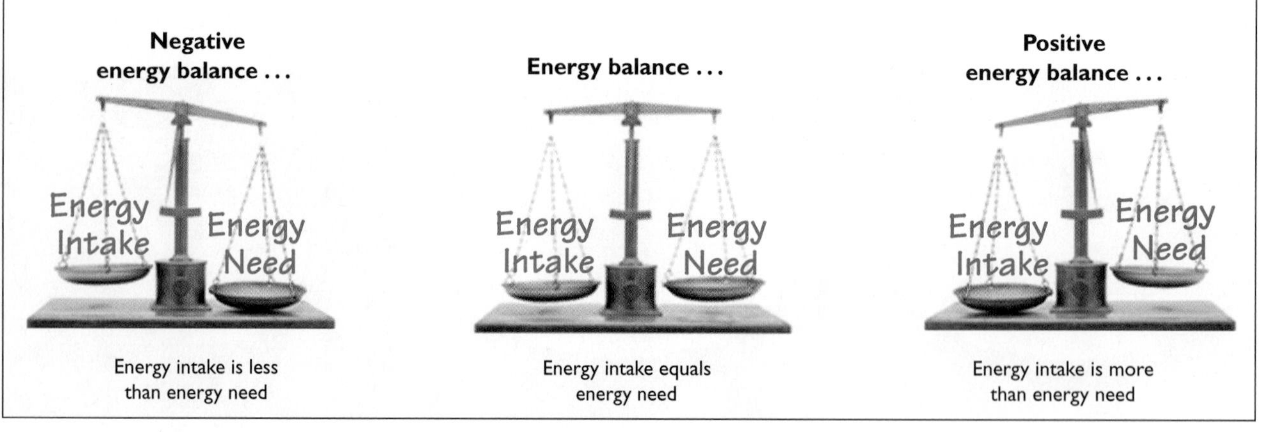

Nutrition
UP CLOSE

Food as a Source of Calories

FOCAL POINT:
Examine the distribution of calories in the foods you eat.

What are the sources of calories in food? Determine the caloric contribution of the fat, carbohydrate, and protein content of the following snack foods using the composition data and calorie conversion factors listed. Then answer the two questions that follow the caloric calculations.

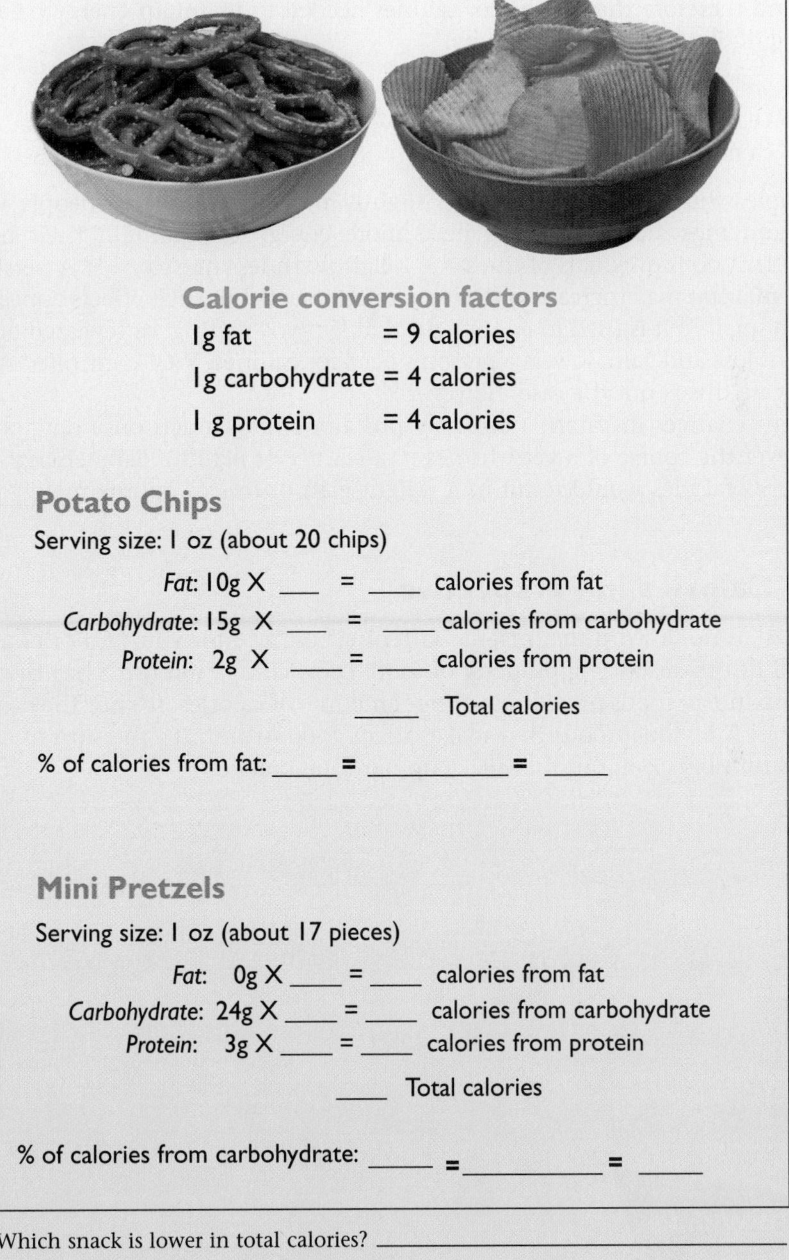

Calorie conversion factors

1g fat = 9 calories
1g carbohydrate = 4 calories
1 g protein = 4 calories

Potato Chips

Serving size: 1 oz (about 20 chips)

Fat: 10g X ___ = ____ calories from fat
Carbohydrate: 15g X ___ = ____ calories from carbohydrate
Protein: 2g X ____ = ____ calories from protein
_____ Total calories

% of calories from fat: _____ = __________ = _____

Mini Pretzels

Serving size: 1 oz (about 17 pieces)

Fat: 0g X ____ = ____ calories from fat
Carbohydrate: 24g X ____ = ____ calories from carbohydrate
Protein: 3g X ____ = ____ calories from protein
____ Total calories

% of calories from carbohydrate: _____ = __________ = _____

Which snack is lower in total calories? ______________________________

Which snack is lower in fat calories? ______________________________

FEEDBACK (answers to these questions) can be found at the end of Unit 8.

is less than the amount of energy expended, people are in negative energy balance. In this case, energy stores are used and people lose weight. When a positive energy balance exists, weight and fat stores are gained because more energy is available from foods than is needed by the body.

Sometimes, a positive energy balance is a healthy and normal circumstance. For example, a positive energy balance is normal when growth is occurring, as in childhood or pregnancy, or when a person is regaining weight lost during an illness.

Caloric values of the foods in Illustration 8.6:

	Calories
½ cup of peanuts	420
1 medium boiled potato	118
1 cup of white rice	205

A Few Final Notes about Energy Balance A person's total caloric need, and therefore the number of calories needed to maintain energy balance and weight, is affected somewhat by:

- Smoking
- Lean muscle mass
- Genetic makeup

People who smoke tend to burn slightly more calories than people who don't, and muscular individuals need more energy to maintain their body weight than do individuals of the same weight with less muscle.[11,12] A person's genetic makeup may increase or decrease caloric need, but such effects appear to be quite small.[13] It is used to be thought that "yo-yo" dieting, or repeated bouts of weight loss and gain, lowers a person's need for calories. Well-controlled studies indicate this is not the case, however.[14]

Slight changes in caloric need may not amount to much on a day-to-day basis. Over the course of a year, however, a *positive* or *negative* daily energy balance of 50 calories would result in a weight gain or loss of approximately five pounds.

Keep Calories in Perspective

"Calories" is not a word that means "fattening" or "bad for you." Calories are a life- and health-sustaining property of food. Diets compatible with health contain a mixture of foods providing various amounts of calories. It's not the caloric content of individual foods that makes them good or not. It's the sum of calories and nutrients in foods that make up our total diets.

www.fsci.umn.edu/nutrexp
Fun learning activities for children on how energy is stored in the body.

www.cyberdiet.com
A multifeatured, well-presented site that allows you to calculate energy expended in physical activity.

Nutrition UP CLOSE

Feedback for Unit 8

Food as a Source of Calories

Potato Chips

Serving size: 1 oz (about 20 chips)

Fat: 10 g × 9 = 90 calories from fat
Carbohydrate: 15 g × 4 = 60 calories from carbohydrate
Protein: 2 g × 4 = 8 calories from protein
158 Total calories

% of calories from fat: $\frac{90}{158}$ = 0.57 × 100 = 57%

Mini Pretzels

Serving size: 1 oz (about 17 pieces)

Fat: 0 g × 9 = 0 calories from fat
Carbohydrate: 24 g × 4 = 96 calories from carbohydrate
Protein: 3 g × 4 = 12 calories from protein
108 Total calories

% of calories from carbohydrate: $\frac{96}{108}$ = 0.89 × 100 = 89%

The pretzels are a better snack choice because they are lower in total calories and fat calories than the potato chips.

REFERENCES

1. Harris JA, Benedict FG. A biometric study of basal metabolism in men. Publication no. 279 of the Carnegie Institute of Washington, 1919.
2. Sjodin AM, et al. The influence of physical activity on BMR. Med Sci Sports Exercise 1996;28:85–91.
3. Klausen B, et al. Age and sex effects on energy expenditure. Am J Clin Nutr 1997;65:895–907.
4. Frankenfield DC, et al. The Harris-Benedict studies of human basal metabolism: history and limitations. J Am Diet Assoc 1998;98:439–45.
5. Boothby WM, Berksen J, Dunn HL. Studies of the energy of metabolism of normal individuals: a standard for basal metabolism, with a nomogram for clinical application. Am J Physiol 1936;116:468–84.
6. Melby CL, Femea PL, Sciacca JP. Reported dietary and exercise behaviors, beliefs and knowledge among university undergraduates. Nutr Res 1986;6:799–808.
7. Mattes RD. Hunger and satiety. Presentation at the American Dietetic Association annual meeting, Boston: 1997 Oct 27.
8. Schwartz MW, et al. Model for the regulation of energy balance and adiposity by the central nervous system. Am J Clin Nutr 1999;69:584–96.
9. McCrory MA, et al. Dietary determinants of energy intake and weight regulation in healthy adults. J Nutr 2000;130:276S–9S.
10. Levine J, et al. The energy expended in chewing gum. N Eng J Med 1999;341:2100.
11. Perkins KA. Effects of tobacco smoking on caloric intake. Br J Addiction 1992;87:193–205.
12. Poehlman ET, Horton ES. The impact of food intake and exercise on energy expenditure. Nutr Rev 1989;47:129–37.
13. Ravussin E, Bogardus C. Relationship of genetics, age, and physical fitness to daily energy expenditure and fuel utilization. Am J Clin Nutr 1989;49(suppl):968–75.
14. Bray GA. Physiology and consequences of obesity. Medscape Diabetes and Endocrinology Clinical Management Modules, http://womenshealth.medscape.com/medscape/endocrinology/ClinicalMgmt/CM.v03/pnt-CM.v03.html, 2001, accessed 1/01.

Unit 9

The Highs and Lows of Body Weight

Beth IsraelHospital

Nutrition Scoreboard

	TRUE	FALSE
1. There is usually little difference between the ideal body weights defined by cultural norms and those defined by science.		
2. Fat stores concentrated around the stomach present a greater hazard to health than do fat stores located around the hips and thighs.		
3. High-calorie diets are the cause of obesity.		
4. Healthy underweight people often find it nearly impossible to gain weight.		

(Answers on next page)

Answers to Nutrition Scoreboard

	TRUE	FALSE
1. Cultural norms of ideal body weights are often at odds with the "healthy" weights defined by science.		✓
2. Health is affected by the location of body fat stores as well as by obesity.[1] Adults who store body fat in the stomach, or central area of the body, are at higher risk for a number of health problems than are adults who store fat primarily in their hips and thighs.	✓	
3. Obesity appears to have many causes, and high-calorie diets are just one of them.		✓
4. Gaining weight is as difficult for many underweight people as losing weight is for many overweight people.	✓	

Key Concepts and Facts

→ Ideal body weight and shape are defined by culture and by health measures. The operating definition should be based on health.

→ The location of body fat stores is important to health. Fat stores concentrated around the stomach, or the central area of the body, present a greater hazard to health than do fat stores located lower, around the hips and thighs.

→ The basic causes of obesity remain unclear. It is likely that obesity has many types and causes, some of which are genetically determined.

→ Underweight in the United States usually results from a genetic tendency to be thin, poverty, illness, or the voluntary restriction of food intake.

Variations in Body Weight

Photo Disc

Wouldn't it be terrific if your body adjusted your food intake based on a healthy level of body fat stores? Then you wouldn't have to worry about being too thin or getting too fat. Why doesn't that happen?

Obesity appears to represent a weak link in the biological evolution of humans. Body processes that regulate food intake developed over 40,000 years ago when "feast and famine" cycles were common. Being underweight was a distinct disadvantage. The more body fat people could store after a feast, the better their chances of surviving the subsequent famines. Consequently, mechanisms that favored food intake and body fat storage developed.[2] These body mechanisms continue to encourage food intake even when body fat stores are ample and obesity poses a greater threat to survival than famine does. In the words of Theodore Van Itallie, a noted obesity researcher, in environments with an abundant supply of food and no requirement for vigorous physical activity, "perhaps thin people are the ones who are abnormal."[3]

Circumstances today are very different from those existing when the human genetic endowment and its food intake regulatory mechanisms were established. These changes may go a long way to explaining why obesity is or is becoming a major problem in many industrialized countries.

How Is Weight Status Defined?

Culture and science define the appropriateness of body size. In the fifteenth century, moon-faced and pear-shaped adults were considered beautiful.[4] In the 1930s, full-figured women such as Jean Harlow were box office stars, and Marilyn Monroe was an icon of the 1950s. Then came Twiggy, Michelle Pfeiffer,

Table 9.1
Criteria for the assessment of weight status in adults.[8]

A. Weight-for-height	
Underweight	less than 90% of average
Normal weight	90–110% of average
Overweight	115–120% of average
Obese	120% or more of average
B. Body mass index (BMI)	
Underweight	under 18.5 kg/m^2
Normal weight	18.5–25 kg/m^2
Overweight	25–30 kg/m^2
Obese	30 kg/m^2 or higher
C. Percent body fat	
Obese	women: 25% men: 20%

and Calista Flockhart. Thinness beyond health boundaries became the standard of beauty. With "too thin" as the cultural ideal, many people regard weights within the normal range as being too high. Men feel the pressure, too. For them, visible body fat is culturally taboo.[5]

Science defines standards for body weight based on disease risk and length of life.[6] ***Underweight,*** normal weight, and ***overweight*** are generally defined based on weight-for-height, whereas ***obesity*** is primarily based on body fat content.[7] Criteria for the assessment of weight status are overviewed in Table 9.1. You will note that some of the criteria for obesity use weight-for-height standards as a proxy for obesity rather than percent body fat.

Such criteria remain in use due to the closeness between weight-for-height and percent body fat in many individuals, and the ease of getting information on weight and height versus body fat. Body fat measures provide the most accurate assessment of obesity.[9]

➡ **underweight**
Usually defined as a low weight-for-height. May also represent a deficit of body fat.

➡ **overweight**
A high weight-for-height.

➡ **obesity**
A condition characterized by excess body fat.

Weight-for-Height Healthy weights-for-height are shown in Table 9.2. Individuals who weigh less than the weight range for their height are considered underweight. If weight is higher than the range, the person is considered overweight. The low ends of the weight ranges in the table are intended for women, and the higher numbers of the ranges are for men. Weights-for-height above these ranges are associated with an increased risk of developing health problems that shorten life expectancy.

A quick way to estimate standard weight-for-height is to use the Hamwi method,[11] which provides a fairly good estimate of appropriate weight-for-height. The Hamwi method can be applied to adults at least five feet tall and is calculated as follows:

- Women: Begin with 5 feet equals 100 pounds, and then add 5 pounds for each additional inch of height. The standard weight is ±10% of this amount. For example, the standard weight for a woman 5 feet 7 inches tall would be:

  5 feet = 100 pounds
  7 inches × 5 pounds = 35 pounds
  100 pounds + 35 pounds = 135 pounds
  135 pounds × 10% (or 0.10) = 13.5 pounds
  135 pounds − 13.5 pounds = 121.5 pounds
  135 pounds + 13.5 pounds = 148.5 pounds

  Thus, the range of standard weight-for-height for a woman 5 feet 7 inches tall is 121.5 to 148.5 pounds. Note that the healthy weight range for a person 5 feet 7 inches tall in Table 9.2 is 121 to 160 pounds. The higher figure applies primarily to men.

Table 9.2
Healthy weight ranges (BMI = 19–25 kg/m^2) for men and women.[10*]

Height	Weight in Pounds
4'10"	91–119
4'11"	94–124
5'0"	97–128
5'1"	101–132
5'2"	104–137
5'3"	107–141
5'4"	111–146
5'5"	114–150
5'6"	118–155
5'7"	121–160
5'8"	125–164
5'9"	129–169
5'10"	132–174
5"11"	136–179
6'0"	140–184
6'1"	144–189
6'2"	148–195
6'3"	152–200
6'4"	156–205
6'5"	160–211
6'6"	164–216

*Height without shoes, weight with light, indoor clothing.

- Men: Begin with 5 feet equals 106 pounds, and then add 6 pounds for each additional inch of height. The standard weight is ±10% of this amount. Calculating as before, the range of standard weight for a man 5 feet 7 inches tall is 133.2 to 162.8 pounds. As we noted, the healthy weight range for a person this height in Table 9.2 is 121 to 160 pounds. This calculation yields a weight-for-height at the higher end of that range, as is appropriate for men.

All methods of estimating weight status based on weight-for-height suffer the same limitation of not being based on percent body fat. Although often closely related, weight-for-height and percent body fat do not always correspond.

Body Mass Index Weight status is also assessed by a measure known as "body mass index," or "BMI." This measure is calculated by dividing body weight (in kilograms) by height (in meters) squared; it can also be determined from the nomogram shown in Illustration 9.1 or from the BMI table on the inside back cover of this book.

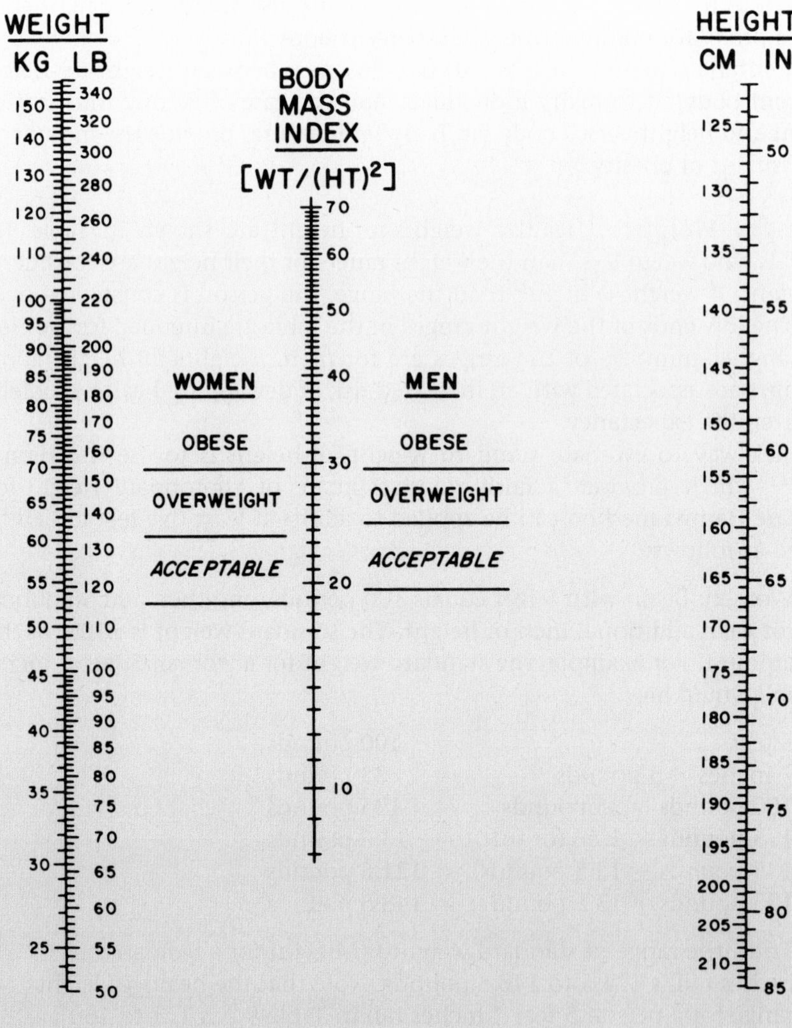

Illustration 9.1
A nomogram for determining body mass index. To use this nomogram, place a ruler or other straightedge between the body weight in kilograms (kg) or pounds (lb) (without clothes) on the left-hand line and the body height in centimeters (cm) or inches (in) (without shoes) on the right-hand line. The body mass index is read from the middle of the scale and is in metric units.

Source: Copyright 1978 George A. Bray.

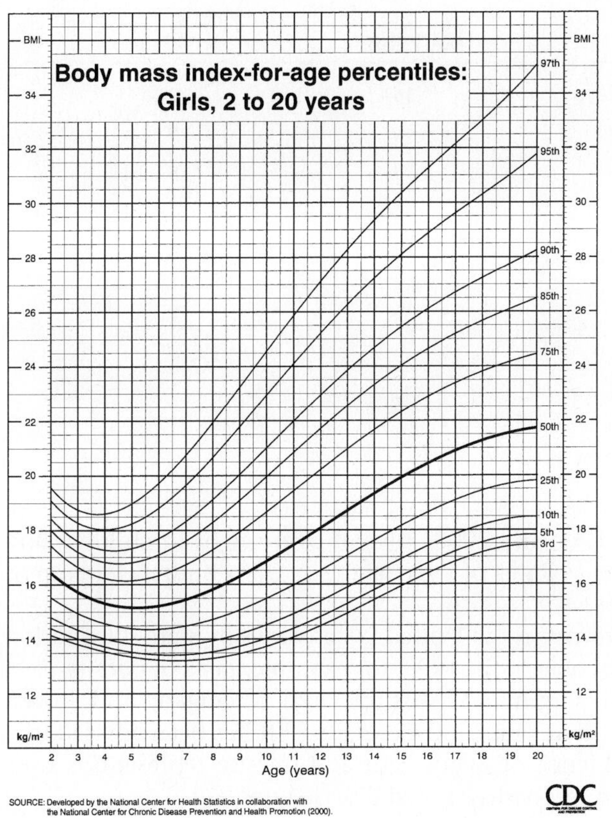

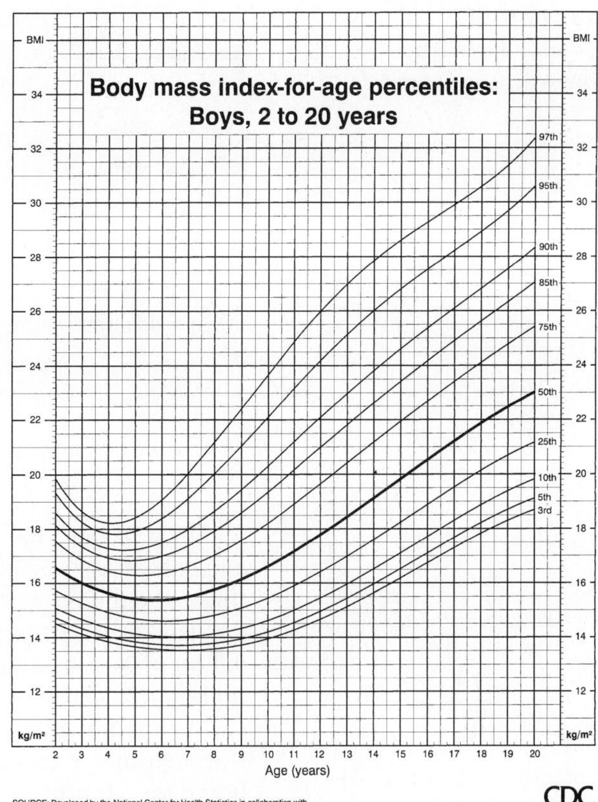

Illustration 9.2
An example of the CDC's 2000 Growth Charts: BMI-for-age. Copies of the growth charts can be obtained from www.cdc.gov/growthcharts. You can use the CDC's BMI calculator (www.cdc.gov/nccdphp/dnpa/obesity.htm) to determine BMI from weight and height.

Weight Status Assessment in Children and Adolescents It is recommended that the Centers for Disease Control's BMI-for-age charts for 2 to 20 year olds be used to assess weight status in children and adolescents.[12] These charts (shown in Illustration 9.2) provide percentile ranges for girls and boys based on age and BMI. The percentile ranges represent the proportion of boys or girls of a given age in the U.S. population with a similar BMI. For example, the BMI percentile ranges for a 14-year-old boy with a BMI of 20 kg/m^2 are the 50th and 75th percentiles. This means that the BMIs of 50 to 75% of other 14 year olds fall between those same percentile ranges.

Standards for identifying weight status based on BMI-for-age are provided in Table 9.3 (p. 9–5). Because the charts assess weight-for-height and not body fat content, they cannot be used to directly diagnose obesity.[12]

Most Adults in the United States Weigh Too Much

Being overweight or obese *is* the norm in the United States. The combined incidence of overweight and obesity among adults is 55% and is on its way up (Table

Table 9.3
Weight status standards for 2 to 20 year olds based on BMI-for-age.[12]

	Percentile Range(s)
Underweight	5th or less
At risk of underweight	5th–15th
Normal weight	15th–85th
At risk of overweight	85th–95th
Overweight	95th or higher

Table 9.4
Estimated incidence of overweight and obesity.[7,13–16]

Group	Overweight	Obese
U.S. adults	36%	19%
U.S. men	44	19
U.S. women	28	18
U.S. 6–11 year olds	14	–
U.S. 2–17 year olds	12	–
Canada, men*	35	–
Canada, women*	27	–
Canada, 7–13 year olds	26	13
Worldwide, children	3–5	–

*Assessed as BMI of 27 kg/m^2 or greater.

9.4 on p. 9–6).[13] Between 1990 and 2000, rates of overweight in children and adults increased nearly 50%.[17] Rates of overweight in U.S. children far surpass those of children worldwide.[15]

The incidence of overweight in the United States varies by state (Illustration 9.3 on p. 9–6), as do rates of obesity. In 1999, for example, rates of obesity were highest (22–24%) in West Virginia, Mississippi, and Michigan; and lowest (12–14%) in Arizona, New Hampshire, and Colorado.[13] It has been advised that the high and growing incidence of overweight and obesity represents a time bomb for a future explosion in obesity-related disease rates.[17]

The Influence of Obesity on Health True obesity is not a healthy state. It carries with it an increased risk of diabetes, hypertension, stroke, heart disease, elevated total cholesterol levels, low HDL-cholesterol levels (the "good cholesterol"), certain types of cancer, and other health problems (Table 9.5).[17] Life expectancy is also influenced by body weight. Contrary to the popular view that underweight people live the longest, adults with BMIs between 22 and 25 kg/m^2 live longer on average than do underweight adults (BMI <18.5 kg/m^2). Obese adults tend to have the shortest life expectancies.[18] The more obese a person is, the more common health problems related to obesity become.[19] In the category of good news are the facts that loss of 5–10% of body weight in obese

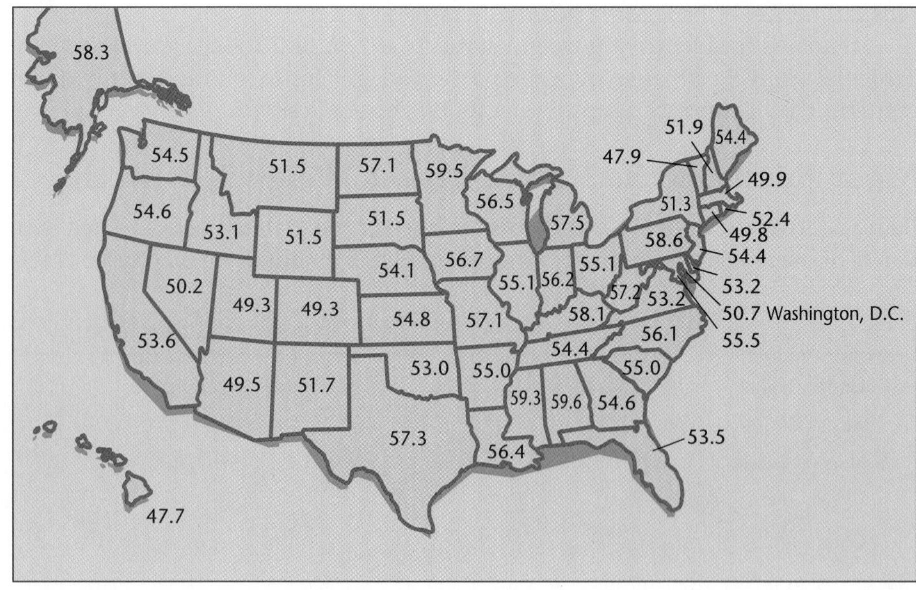

Illustration 9.3
Percent of overweight adults (BMI = 25 kg/m^2) by state.

Source: Generated from data from CDC, 1999.

adults decreases risks to health substantially. Weight loss lowers blood pressure, lowers LDL cholesterol, increases HDL cholesterol (the one you want to be high), and decreases the risk of diabetes or high glucose levels in people with diabetes.[20]

Another consequence of obesity is the ingrained cultural prejudice to which obese people are subjected. Children who are obese are more likely to suffer unfair or indifferent treatment from teachers than are nonobese children.[21] They experience more isolation, rejection, and feelings of inferiority than other children.[22] Obese adults are likely to be discriminated against in hiring and promotion decisions[23] and to be thought of as lazy or lacking in self-control, even by health professionals who care for them.[24] Society's prejudice against people who don't conform to the cultural ideal of body size may be the most injurious consequence of obesity.[25]

Table 9.5
The mismatch between obesity and health. All these problems are associated with obesity.[17]

- Diabetes
- Hypertension
- Stroke
- Elevated cholesterol level
- Low HDL-cholesterol level
- Heart disease
- Certain types of cancer
- Gallbladder disease
- Shortened life expectancy
- Discrimination
- Depression
- Infertility
- Accidents
- Skin disorders

Obesity Should Be Assessed by Body Fat Content As you can imagine, weight-for-height tables and BMI determinations do not identify obesity as well as tests of body fat content do. Weight doesn't distinguish between lean body mass (muscle, bone, and water) and body fat. Take a 165-pound, 30-year-old woman who is 5 feet 6 inches tall and trains heavily. Her body weight would indicate obesity, but she may have too little body fat to be truly obese. In that case, she may not be obese, no matter what the table indicates. But what if this woman does not exercise much? Chances are that she would qualify as obese based on body fat content.

People who weigh less than the range given on the weight-for-height and age table may actually be underweight, normal weight, or obese, depending on their body fat content.

Methods for Assessing Body Fat Content A number of methods for estimating body fat content are available and are reasonably accurate if undertaken correctly.[26] The most common methods are:

- skinfold thickness measures,
- bioelectrical impedance analysis (BIA),
- underwater weighing,
- magnetic resonance imaging (MRI),
- computerized axial tomography (CT or CAT scans), and
- dual-energy X-ray absorptiometry (DEXA)

The theory underlying each of these methods and its advantages and limitations are presented in Table 9.6 (p. 9–8). Body fat content can also be assessed by ultrasound, total body electrical conductivity, magnetic resonance spectroscopy, and neutron activation analysis. These methods are highly technical and expensive, however, and are used infrequently because of their cost or lack of availability.

Each body fat test has its advantages and drawbacks. The tests are most likely to provide accurate results when they are performed by skilled, experienced technicians using proper, well-maintained equipment, and when the measurements are converted into percent body fat by the appropriate formulas.

Everybody Needs Some Body Fat A certain amount of body fat—3 to 5% for men and 10 to 12% for women—is needed for survival.[27] Body fat serves essential roles in the manufacture of hormones; it's a required component of every cell in the body; and it provides a cushion for internal organs. Fat that serves these purposes is not available for energy formation no matter how low energy reserves become.[27] Low body fat levels are associated with delayed physical maturation during adolescence, infertility, accelerated bone loss, and problems that accompany starvation.[28]

Table 9.6
Commonly used methods for assessing body fat content.

Skinfold Measurement

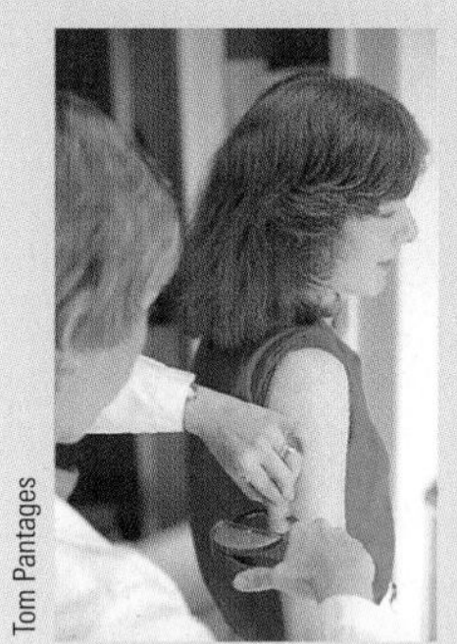

Tom Pantages

Body fat content can be estimated by measuring the thickness of fat folds that lie underneath the skin. Calipers are used to measure the thickness of fat folds, preferably over several sites on the body. Body fat content is estimated by "plugging" the thicknesses into the appropriate formula.

Advantages: Calipers are relatively inexpensive (they cost about $250–$350). The procedure is painless if done correctly and can yield a good estimate of percent body fat.

Limitations: This technique is often performed by untrained people. Skinfolds may be difficult to isolate in some individuals.

Bioelectrical Impedance Analysis (BIA)

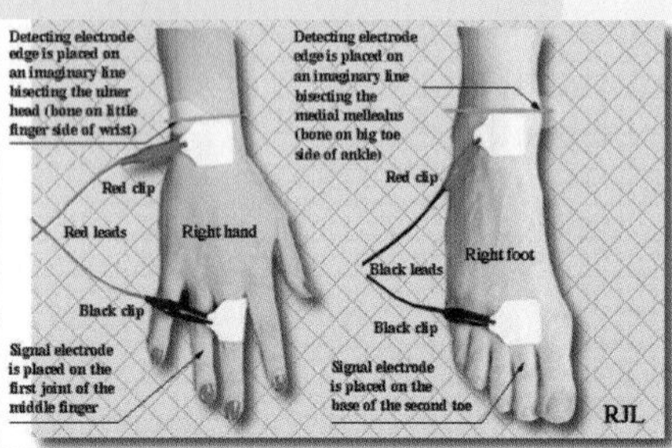

Because fat is a poor conductor of electricity, and water and muscles are good conductors, body fat content can be estimated by determining how quickly electrical current passes from the ankle to the wrist.

Advantages: The equipment required is portable, and the test is easy to do and painless. The results are fairly accurate for people who are not at the extremes of weight-for-height.

Limitations: Equipment is expensive; inferior equipment produces poor results. Hydration status may affect electrical conductivity and produce inaccurate results, as may inaccurate formulas used to calculate body composition from test results.

Underwater Weighing

Photo Take, Yoav Levy

Underwater weighing is sometimes called the "gold standard" of body fat assessment; results of this method are often used to judge the accuracy of other methods. The subject is first weighed on dry land; next he or she is submerged in water and exhales completely; then his or her weight is measured. The less the person weighs under water compared to the weight on dry land, the higher the percent body fat. (Fat, but not muscle or bone, floats in water.)

Advantages: If undertaken correctly and if appropriate formulas are used in calculations, this technique gives an accurate value of percent body fat.

Limitations: The equipment required is expensive and not easily moved. The test doesn't work well for people who don't swim or who are ill or disabled in some way.

Magnetic Resonance Imaging (MRI)

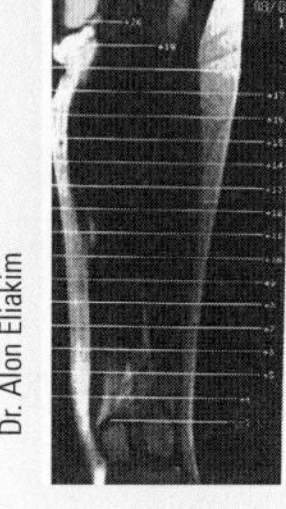

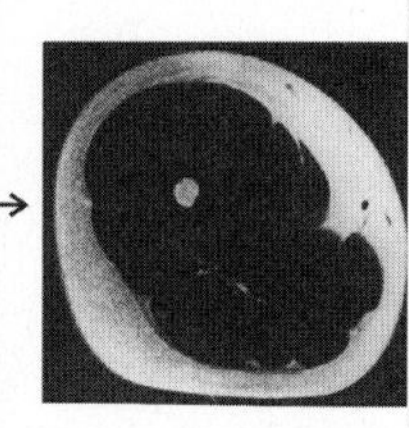

Dr. Alon Eliakim

MRI and CT or CAT scans provide similar results. Using this technology, a person's body fat and muscle mass can be photographed from cross-sectional images obtained when the body, or parts of it, is exposed to a magnetic field (or, in the case of CAT scans, to radiation). Based on the volume of fat and muscle observed, total fat and muscle content can be determined.

Advantages: Provides highly accurate assessment of fat and muscle mass.

Limitations: Expensive ($500–$1000 per assessment). Largely used for research purposes.

Dual-energy X-ray Absorpiotmetry (DEXA)

DEXA is a fairly new technique for quantifying body fat mass. The method is based on the principle that various body tissues can be differentiated by the level of X-ray absorption. The measure is made by scanning the body with a small dose of X rays (similar to the level of exposure from a transcontinental flight) and then calculating body fat content based on the level of X-ray absorption. DEXA may become the gold standard for body fat assessment.

Advantages: Provides highly accurate results when measurements are undertaken correctly. DEXA is safe and "user friendly" for people being measured, and can also be used to assess bone mineral content and lean tissue mass.

Limitations: DEXA machine is expensive and so is the cost of individual assessments ($150 per person). Machine must be operated by a trained and certified radiation technologist.

Table 9.6
Commonly used methods for assessing body fat content. (continued)

Body Fat Location Is Important to Health The location of body fat stores, as well as total body fat content, is important to health.[1] You may be better off health-wise if you're a "pear" rather than an "apple" (Illustrations 9.4 and 9.5).

Pear-shaped individuals store fat primarily in their hips and thighs whereas apples store fat centrally—around the waist. The apple shape is more common among obese men than the pear shape is. Obese women exhibit both shapes.[29]

Why a predominance of central body fat stores is riskier to health than high levels of fat stores in the lower body, arms, and legs isn't entirely clear. It appears, however, that fat cells that make up central fat deposits are larger and "resistant" to insulin. This decreases the ability of insulin to lower blood glucose levels. As a result of this state of ***insulin resistance,*** blood levels of insulin increase, as do levels of glucose over time. Increased blood levels of insulin are related to increased blood triglyceride levels and blood pressure, and reduced levels of beneficial, HDL cholesterol. These changes increase the risk of hypertension and

➡ **insulin resistance**
A condition in which cells "resist" the action of insulin in facilitating the passage of glucose into cells.

Illustration 9.4
Basic body shapes. The pear normally has narrow shoulders, a small chest, and an average-size waist. Fat is concentrated in the hips and thighs. Weight loss in these areas is usually difficult.

The apple looks round in the middle. Fat is concentrated in the waist and can be lost with diet and exercise. Apple-shaped people are at increased risk for diabetes, hypertension, high blood cholesterol, and heart disease.

Illustration 9.5
The man on the left is 6 feet tall and weighs 240 pounds; the man on the right is 6 feet 1 inch tall and weighs 230 pounds. Although similar in weight and height, the man on the left is at higher risk for diabetes, hypertension, and heart disease due to his "apple" shape and body fat content.

heart disease. Increasing levels of blood glucose promote the development of diabetes.[1]

Weight loss in people with high levels of body fat reduces disease risk by lowering both central and lower body fat stores.[17] Central body fat stores decline first with weight loss, making it easier for most people to lower central than lower body fat stores.[19]

Assessing Body Fat Distribution Body fat distribution can be determined by assessing a person's waist-to-hip ratio (Illustrations 9.6 and 9.7 on p. 9–11). People with the pear shape have lower waist-to-hip ratios than apples. Waist-to-hip ratios of more than 0.80 in women and 0.95 in men indicate central body fat distribution—or the apple shape.[25]

Waist circumference by itself predicts the risk of a number of chronic diseases. A waist circumference over 40 inches in men and over 35 inches in women increases the risk of diabetes, heart disease, and hypertension.[8]

What Causes Obesity?

We do not know with certainty what causes obesity. There are likely multiple causes and several different types of obesity.[25] Obesity should actually be thought of as "obesities."

Three major factors appear to contribute to the development of obesity:[31]

1. Genetic background
2. Dietary intake
3. Physical activity

Several diseases and medications are also associated with the development of obesity, but their contribution to the overall incidence of obesity is small (although meaningful to the affected people).

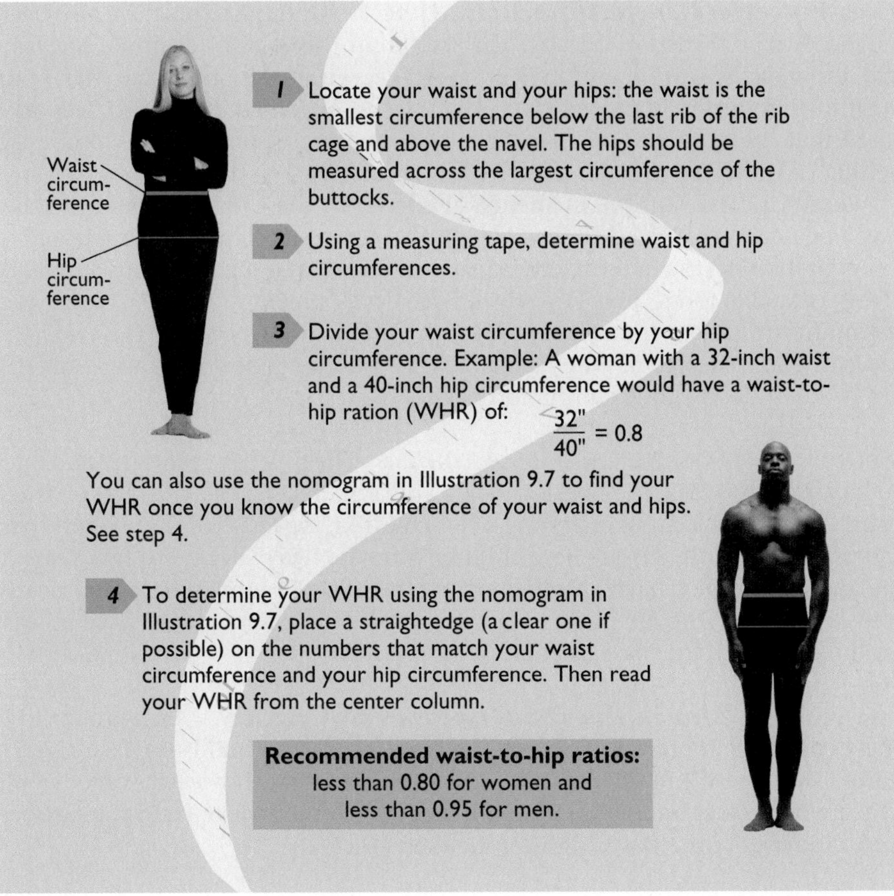

Photo Disc

Illustration 9.6
Determining your waist-to-hip ratio.

➡ **metabolism**
The chemical changes that take place in the body. The conversion of glucose to energy or to body fat is an example of a metabolic process.

➡ **environmental trigger**
An environmental factor, such as inactivity, a high-fat diet, or a high sodium intake, that causes a genetic tendency toward a disorder to be expressed.

Are Some People Born to Be Obese? Accumulating evidence indicates that genetics plays important roles in the development and maintenance of obesity. Heredity may account for approximately 25 to 40% of the incidence of obesity.[32]

Genetic influences on obesity take several paths. Some people are born with errors in ***metabolism*** that produce obesity (this appears to be rare). Others are born with genetic traits that predispose them to becoming obese (probably common). A predisposing genetic trait is expressed when the right ***environmental trigger*** exists. For example, a person with a genetic predisposition to becoming obese may maintain normal weight as long as she or he is physically active or consumes a low-fat diet. If activity level becomes low or the diet changes to one high in fat (two types of environmental triggers), the genetically susceptible person then gains weight.[33]

The propensity to gain weight when excessive levels of calories are consumed may be affected by genetic traits. Claude Bouchard and his colleagues at Lavel University in Canada fed adult identical twins 840 more calories a day than each twin needed to maintain his or her weight.[34] At the end of the 100-day feeding period, weight gain among the twins ranged from 9 to 29 pounds! (The average weight gain was 24 pounds.) Related twins tended to gain similar amounts of weight, but the weight gains varied widely even though all subjects consumed the same number of excess calories. Another study showed that identical twins reared apart are less likely to have similar body weights than identical twins who are reared together.[35]

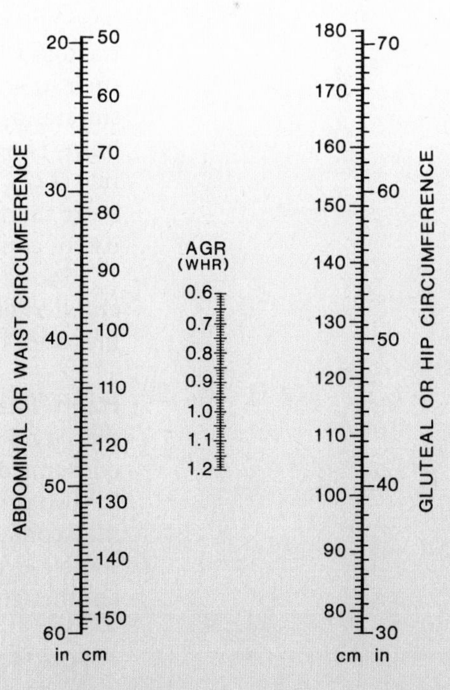

Illustration 9.7
Waist-to-hip ratio nomogram.[30]

Other Potential Genetic Influences Other hereditary factors have also been theorized to be related to obesity. The "set-point theory," for example, suggests that individuals are programmed to weigh a certain amount and that body weight will return to that level after weight is lost or gained. It has also been proposed that the number and size of fat cells predispose some people to obesity. Neither of these two popular theories has been shown to be true.[36]

Recent increases in the incidence of obesity in the United States and other developed countries indicate that genetic influences in combination with environmental factors, such as inactivity, high-fat diets, and a continuous supply of highly palatable foods, play key roles in the development of obesity. We are just beginning to understand the nature of the interactions between the two. The day will come, however, when people with inborn tendencies toward obesity can be identified by an examination of their genes.

Do Obese Children Become Obese Adults? The link between early and later obesity is very weak for obese infants and children under the age of three years.[17] It becomes stronger among older children, especially if one or both parents are obese. Only 8% of obese children who are heavy at one to two years of age and who do not have an obese parent are obese as adults. However, nearly 80% of children who are obese between the ages of 10 and 14 and have at least one obese parent are obese as adults.[37]

The Role of Diet in the Development of Obesity Everyone who is obese consumes more calories than the body needs and builds up excessive fat stores. That doesn't mean that all obese people overeat, however. Some people who are obese eat less than people who aren't.[38] A number of factors are associated with obesity besides a high-calorie diet.

Low Levels of Physical Activity Promote Obesity A widespread decline in physical activity is thought to be related to the high and increasing incidence of obesity among Americans.[39] To a large extent, physical activity has become voluntary. Many farmers now plow, sow, and reap in air-conditioned tractors with power steering. Lawn mowers propel themselves and sometimes their operators, too. Instead of walking or biking two and a half miles to the store and back, we drive. An average-size adult driving two and a half miles burns about 17 calories. Biking that distance would use seven times more calories (122). But walking two and a half miles would burn around 210 calories, more than 12 times the calories it takes to drive! We have traded physical activity for convenience, time, and, perhaps, personal fat stores.

Too much television watching, in particular, has been blamed for obesity in children and adults,[40] but it is not clear which comes first—obesity, sedentary lifestyles, or too much TV.[41,42]

High-Fat Diets May Promote Obesity Adults who consume high-fat diets tend to weigh more and are at increased risk of obesity than are adults who consume diets low in fat. It is speculated that high-fat diets promote weight gain for two primary reasons: (1) they tend to be higher in calories, and (2) the body may convert excess dietary fat into body fat stores with greater efficiency than similar excesses of dietary carbohydrate or protein.[43] This evidence provides another good reason to keep fat intake below 30% of total calories.

Obesity: The Future Lies in Its Prevention

Whether obesity is related to genetic predisposition, environmental factors, or a combination of both, certain steps can be taken to help prevent it.

Preventing Obesity in Children For children, the prevention of obesity includes the early development of good food and activity habits. Parents should offer a nutritious selection of food, but children themselves should be allowed to decide how much they eat.[44] Physical activities that are fun for every child—not just those who show athletic promise—should be routine in schools and summer programs.

Interactions between parents and children around eating and body weight can set the stage for the prevention or the promotion of overweight in children. Parents who overreact to a child's weight by focusing on it, restricting food access, and making negative comments to the child may increase the likelihood that eating and weight problems will develop or endure. Lifestyle changes for the whole family, such as incorporating fun physical activities into daily schedules, making a wide assortment of nutritious foods available in the home, and decreasing a focus on eating, foods, and weight, are some of the positive changes families can make to promote healthy eating and exercise habits, and normal weight in children.[45]

Illustration 9.8

Source: Reprinted with special permission of King Feature Syndicate.

Preventing Obesity in Adults Action needs to be taken to prevent weight gain during the adult years. Many adults gain weight at a slow pace (about three-fourths of a pound per year) as they age (Illustration 9.8).[46] Data from a national nutrition and health survey indicate that major gains in weight are most likely to occur in adults between the ages of 25 and 34 years.[47] Regular, vigorous exercise may prevent or lessen the amount of weight gain that occurs with age.

In situations where obesity cannot be prevented, society itself should take action. It would help if people acknowledged and accepted the fact that people come in all sizes. In reality, why *should* one size fit all (Illustration 9.9)?

Some People Are Underweight

Worldwide, underweight due to inadequate diets is much more common than obesity and a good deal more life-threatening.[15] There are many preventable reasons why people fail to get enough to eat. The lows of body weight discussed here relate to underweight as it occurs in many economically developed nations.

Illustration 9.9
People come in many different sizes and shapes.

Photo Disc

Illustration 9.10
Some underweight people are genetically thin and are healthy.

Photo Researchers, Richard Hucthings

In contrast to the desperate situations faced by many people in economically emerging countries, underweight in developed nations largely results from illnesses such as HIV/AIDS, pneumonia, and cancer; an eating disorder (anorexia nervosa); and the voluntary restriction of food. An important and preventable cause of underweight in the United States and some economically developed nations is poverty.[48]

Underweight Defined

People who are underweight have too little body fat, or less than 20% body fat in adult females and 12% body fat in males. They have weights that fall below the ranges in Table 9.2, or they have BMIs below the "acceptable" level in Illustration 9.1. A portion of people classified as underweight by the tables will not actually be underweight, just as a subset of people categorized as obese by the tables are not really obese.

Some people assessed as underweight for height are healthy and have a sufficient supply of body fat. Like the person in Illustration 9.10, they are probably genetically thin.[49] People who are naturally thin often have as much difficulty gaining weight as obese people have losing it.[32] Unless health is compromised, as indicated by fatigue, frequent illness, impaired concentration, apathy, or extreme intolerance to cold temperatures, there is little reason to be concerned about being underweight.

An Avoidable Cause of Underweight: Middle-Class Malnutrition

Underweight related to low-calorie, low fat diets has been observed in children of middle- to upper-income families in the United States.[50,51] Some of the parents say they implement this sort of diet to help their children form good lifelong eating habits and avoid obesity and heart disease. Parents' fear that their children will become obese appears to be the major motivation for restricting their children's food intake.

About 2% of adolescents from middle- and upper-income families limit their diets in order to prevent obesity.[50] While not having anorexia or bulimia, these adolescents moderately and continually restrict their weight gain. Unlike people with anorexia nervosa or bulimia, the goal of these adolescents is not to lose weight, but to avoid *gaining* weight.

The failure to gain weight during periods of growth has an undesired consequence: it may halt growth in height. Counseling has been effective in getting adolescents to eat enough to allow growth in height to proceed.[50,51]

Toward a Realistic View of Body Weight

A widespread belief among Americans is that individuals can achieve any body weight or shape they desire if they just diet and exercise enough. It's a myth. People naturally come in different weights and shapes, and these can only be modified so much.[44] Half of all women in the United States wear size 14 to 26, yet many clothing models are very underweight. Many men, no matter how hard they work out, will never have a washboard stomach or fit into "slim jeans."

Nutrition UP CLOSE

Are You an Apple or a Pear?

Follow the directions given in Illustration 9.6 for determining your waist-to-hip circumference ratio. If you don't have a measuring tape, use a string. Mark the string where the two ends intersect when placed around your waist and your hips. Then use a ruler to measure the distance between the end and each mark on the string.

FOCAL POINT: Determining your waist-to-hip ratio.

FEEDBACK can be found at the end of Unit 9.

The U.S. obsession with body weight and shape is spreading to other industrialized countries as part of popular culture. Ironically, however, strong societal bias against certain body sizes may contribute broadly to weight and health problems. Intolerance of overweight and obese children and adults tends to increase discrimination against them. This type of societal bias lowers the individual's feeling of self-worth, and may promote eating disorders, including the consumption of too much food.[52] Females are hardest hit by negative attitudes about body size. Although the incidence of overweight and obesity tends to be higher in males, obesity in females carries with it many more negative stereotypes.[53] Overreactive parents make things worse.

Of all the things societies can do to prevent obesity, acceptance of people of different sizes and a more realistic view of obtainable body weights and shape may be two of the most important.

WWW links

www.healthfinder.gov/
A gateway to information on obesity, body weight, underweight, and related topics.

www.nutribase.com
Interactive programs for estimating caloric requirement and weight status.

www.naafa.org
The National Association to Advance Fat Acceptance is dedicated to ending size discrimination and supporting high quality of life and health for fat people. Learn more about it at this site.

www.cdc.gov/growthcharts
The 2000 growth charts for 2 to 20 years old, including BMI-for-age graphs, are available at this site.

www.cdc.gov/nccdphp/dnpa/obesity.htm
Find the CDC's BMI calculator, obesity trend data, and much more at this site.

www.cyberdiet.com
Use their assessment tools to calculate waist-to-hip ratio, body mass index, and target heart rate.

REFERENCES

1. Bjorntorp P et al. "Portal" adipose tissue as a generator of risk factors for cardiovascular disease and diabetes. Atherosclerosis 1990;10:493–6.
2. Brown PJ, Konner M. An anthropological perspective on obesity. Annals of NY Acad Sci 1987;499:29–46.
3. Van Itallie TB. Obesity: the American disease. Food Technol 1979;(Dec):43–47.
4. Rossner S. Ideal body weight—for whom? (editorial). Acta Medica Scand 1984;216:241–2.
5. Fallon A, Rozin P. Sex differences in perception of desirable body shape. J Abnormal Psychology 1985;94:102–5.
6. Williamson DF. The prevention of obesity. N Engl J Med 1999;341:1140–1.
7. Sheerwood N. Epidemiology of obesity. Health Generations 2000;1:1–12.
8. Clinical guidelines on the identification, evaluation, and treatment of overweight and obesity in adults. National Heart, Lung, and

Blood Institute; 1998 June. http://www.nhlbi.nih.gov.
9. Willett WW et al. Guidelines for healthy weight. N Engl J Med 1999;342:427–34.
10. Dietary guidelines for Americans. Washington, DC: PHS; 1995.
11. Roth JH. What is optimum body weight? (letter) J Am Diet Assoc 1995;95:856–7.
12. CDC growth charts: United States. Advance Data, No. 314, 12/4/00 (Revised), www.cdc.gov/nchs.
13. www.cdc.gov/nccdphp/dnpa/obesity.htm, 1/15/01.
14. Chronic diseases and their risk factors: the nations leading causes of death, 1999. Atlanta:Centers for Disease Control;1999.
15. Onis M et al. Prevalence and trends of overweight among preschool children in developing countries. Am J Clin Nutr 2000;72:1032–9.
16. Macdonald SM et al. Obesity in Canada: a descriptive analysis. Canadian Heart Health Surveys Research Group. CMAJ 1997;157(suppl):S3–9.
17. Bray GA. Physiology and consequences of obesity. Medscape Diabetes and Endocrinology Clinical Management Modules, http://womenshealth.medscape.com/medscape/endocrinology/ClinicalMgmt/CM.v03/pnt-CM.v03.html, 2001.
18. Calle EE et al. Body mass index and mortality in a prospective cohort of U.S. adults. N Engl J Med 1999;341:1097–1105.
19. Pi-Sunyer FX. Health implications of obesity. Am J Clin Nutr 1991;53: 1595S–603S.
20. Harrard WR. Weight control and exercise. JAMA 1995;274:1964–5.
21. Mayer J. Overweight: causes, cost, and control. Englewood Cliffs (NJ): Prentice-Hall; 1968.
22. Patterson RE, et al. Factors related to obesity in preschool children. J Am Diet Assoc 1986;86:1376–81.
23. Kanarek RB, Orthen-Gambill N, Marks KR, Mayer J. Obesity: possible psychological and metabolic determinants. In: Galler JR, ed. Human nutrition: a comprehensive treatise, Vol. 5. New York: Plenum; 1984.
24. Price J, Desmond S, Krol R, Snyder F, O'Connell J. Family practice physicians' beliefs, attitudes and practices regarding obesity. Am J Prev Med 1987;3:339–45.
25. Solomon CG, Manson JE. Obesity and mortality: a review of the epidemiologic data. Am J Clin Nutr 1997; 66(suppl): 1044S–50S.
26. Houtkooper LB et al. Comparison of methods for assessing body composition changes over 1 y in postmenopausal women. Am J Clin Nutr 2000;72:401–6.
27. Oppliger RA, Nielsen DH, Vance CG. Wrestlers' minimal weight: anthropometry, bioimpedance, and hydrostatic weighing compared. Med Sci Sports 1991;23:247–53.
28. Gibson RS. Principles of nutritional assessment. New York: Oxford University Press; 1990.
29. Bouchard C, Bray GA, Hubbard VS. Basic and clinical aspects of regional fat distribution. Am J Clin Nutr 1990;52:946–50.
30. Bray GA. Cited in Committee on Diet and Health (Food and Nutrition Board; National Research Council). Diet and health. Implications for reducing chronic disease risk. Washington, DC: National Academy Press; 1989 p 575.
31. Bouchard C. Can obesity be prevented? Nutr Rev 1996;54:S125–S30.
32. Bouchard C. Heredity and the path to overweight and obesity. Med Sci Sports 1991;23:285–91.
33. Perusse L, Bouchard C. Role of genetic factors in childhood obesity and in susceptibility to dietary variations. Ann Med 1999;31(suppl):19–25.
34. Bouchard C, et al. The response to long-term overfeeding in identical twins. N Engl J Med 1990;322:1477–82.
35. Macdonald A, Stunkard AJ. Body-mass indexes of British separated twins (letter). N Engl J Med 1990;322:1530.
36. Weinsier R et al. Do adaptive changes in metabolic rate favor weight regain in weight-reduced individuals? An examination of the set-point theory. Am J Clin Nutr 2000;72:1088–94.
37. Whitaker JA, et al. Predicting obesity in young adulthood from childhood parental obesity. N Engl J Med 1997;337:869–73.
38. Miller WC. Diet composition, energy intake, and nutritional status in relation to obesity in men and women. Med Sci Sports 1991;23:280–4.
39. Heini AF, Weinsier RL. Divergent trends in obesity and fat intake patterns: the American paradox. Am J Med 1997;102:259–64.
40. Gortmaker SL, Dietz WH Jr., Cheung LWY. Inactivity, diet, and the fattening of America. J Am Diet Assoc 1990;90:1247–55.
41. Shannon B, Peacock J, Brown MJ. Body fatness, television viewing and calorie-intake of a sample of Pennsylvania sixth grade children. J Nutr Ed 1991;23:262–8.
42. Andersen RE, et al. Relationship of physical activity and television watching with body weight and level of fatness among children. JAMA 1998;279:938–42.
43. Hill JO et al. Dietary fat and regulation of energy balance: implications for obesity. J Nutr 2000;130:284S–88S.
44. Satter EM. Internal regulation and the evaluation of normal growth as the basis for prevention of obesity in children. J Am Diet Assoc 1996;96:860–4.
45. Birch LL, Davison KK. Effects on children of parent's reaction to overweight. Pediatrics 2001;107:46–53.
46. Brown JE, Kaye SA, Folsom AR. Parity-related weight change in women. Int J Obesity 1992;16:627–31.
47. Williamson DF, Madans J, Anda RF, Kleinman JC, Giovino GA, Byers T. Smoking cessation and severity of weight gain in a national cohort. N Engl J Med 1991;324:739–45.
48. Committee on Diet and Health (Food and Nutrition Board; National Research Council). Diet and health. Implications for reducing chronic disease risk. Washington, DC: National Academy Press; 1989.
49. Costanzo PR, Schiffman SS. Thinness—not obesity—has a genetic component. Neuroscience and Biobehavioral Reviews 1989;13:55–58.
50. Lifshitz F. Nutrition and growth. Clin Nutr 1985;4:40–47.
51. Pugliese MT, Weyman-Dunn M, Moses N, Lifshitz F. Parental health beliefs as a cause of nonorganic failure to thrive. Pediatrics 1987;80:175–82.
52. Schuster K. The dark side of nutrition. Food Mang 1999;34:34–9.
53. UK panel calls for media to end pressure on girls to be thin. Report of the UK's Cabinet's Body Image Summit. Reuter Health 2000;June 22.

Nutrition UP CLOSE

Are You an Apple or a Pear?

Feedback for Unit 9

If your waist-to-hip ratio is greater than 0.8, you're an apple. If it is less than 0.8, you're a pear. (You are all a peach for doing this exercise.)

Most men will turn out to be apples, and women will tend to be pears.

Unit 10

Weight Control: The Myths and Realities

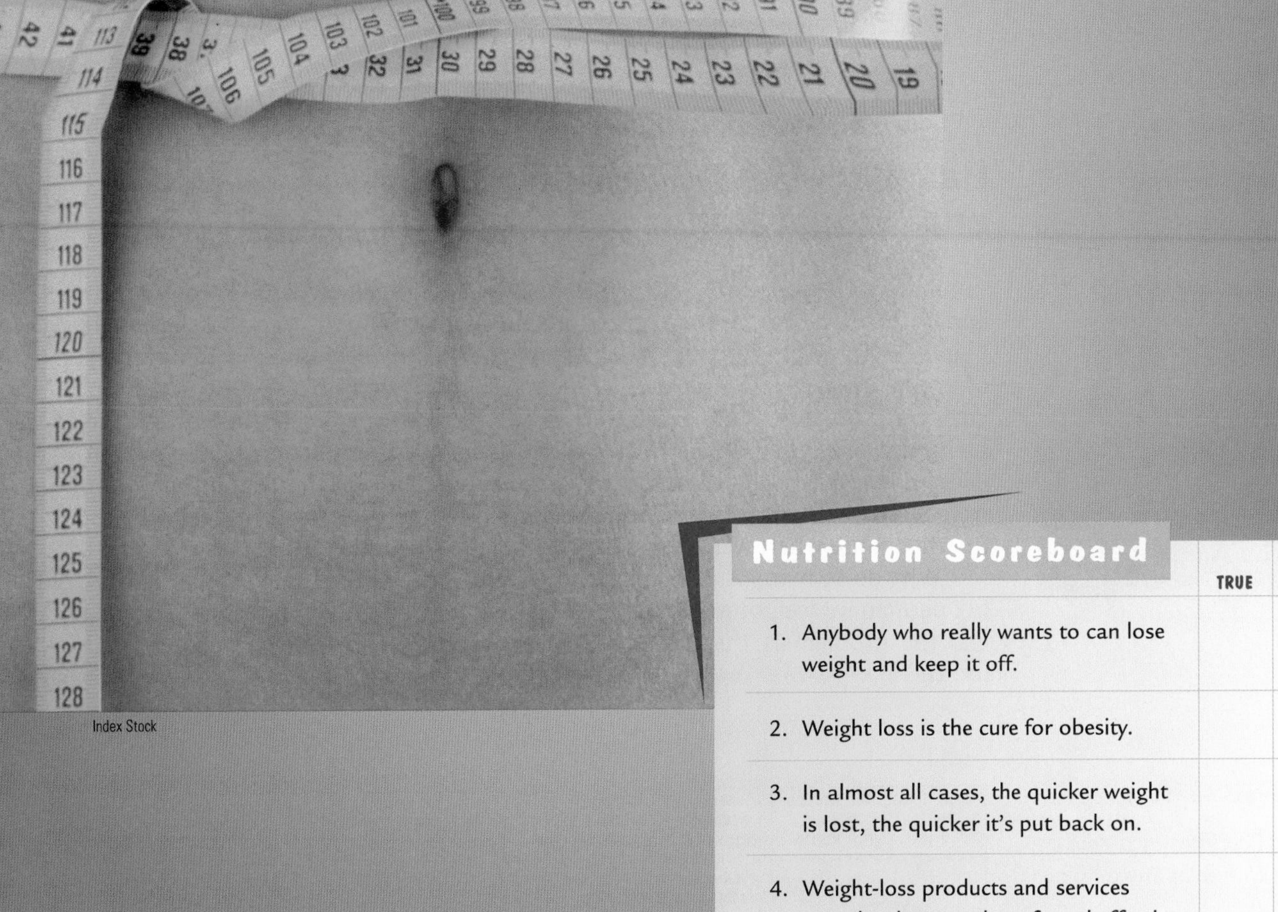

Index Stock

Nutrition Scoreboard

	TRUE	FALSE
1. Anybody who really wants to can lose weight and keep it off.		
2. Weight loss is the cure for obesity.		
3. In almost all cases, the quicker weight is lost, the quicker it's put back on.		
4. Weight-loss products and services must be shown to be safe and effective before they can be marketed.		

(Answers on next page)

Answers to

Nutrition Scoreboard

	TRUE	FALSE
1. If true, hardly anybody would be obese.		✔
2. Maintenance of weight loss is the cure for obesity.		✔
3. Quick weight-loss diets almost never work because weight lost is regained after the diet ends.	✔	
4. Unfortunately, many weight-loss products and services on the market have not been shown to be safe or effective. Laws and regulations do not fully protect the consumer from the introduction of bogus products and services.		✔

Key Concepts and Facts

- → The effectiveness of weight-control methods should be gauged by their ability to prevent weight regain.
- → That a weight-loss product or service is widely publicized and utilized doesn't mean it works.
- → Some people cannot lose weight and keep it off, no matter how hard they try.
- → Successful weight control is characterized by gradual weight loss from small, acceptable, and individualized changes in eating and activity.

Photo Disc

Baseball, Hot Dogs, Apple Pie, and Weight Control

Americans are preoccupied with their weight. On any given day, 44% of women and 30% of men are trying to lose weight—whether they need to or not.[1] To help get the weight off, Americans spend over *$33 billion* annually (an average of approximately $120 per person each year) on weight-loss products and services.[2] Yet Americans are gaining weight faster than they are losing it, and the incidence of obesity in the United States is on the rise.[3] Roughly 5 to 10% of people who lose weight keep it off.[4] Consumers are paying handsomely for weight loss without experiencing the desired cosmetic changes or health benefits that come when weight loss is maintained (Illustration 10.1).

Why do so many Americans fail at weight control? For some people, achieving permanent weight reduction on their own may be truly impossible.[5] For others, the problem is the ineffective methods employed, not the people who use them.

Illustration 10.1

Source: Reprinted by permission of Universal Press Syndicate.

Weight Loss versus Weight Control

"Jess—did I hear you say you wanted to lose some weight? It happens I know about a terrific diet! A couple of months ago I went on Dr. Quick's Amino Acid Diet and lost 15 pounds! You eat nothing but fish, papaya, and broccoli and the pounds just melt away!"

Jess was shocked. Selma looked heavier than she did last year when they were in class together.

"But Selma," Jess inquired, "what happened to the weight you lost?"

"Oh, I gained it back. I ran out of willpower and started eating everything in sight. I'm going to get back on track, though. Tomorrow I start the Slim Chance Diet. . . ."

Dietary torture and weight loss are not the cure for overweight. If losing weight was all it took to achieve the cultural ideal of thinness, nobody would be overweight. Any approach to weight loss (and some get pretty spectacular) that calls for a reduction in caloric intake can produce weight loss in the short run. These methods fail in the long run, however, because they become too unpleasant. Feelings of hunger, deprivation, and depression that often occur while on a weight-loss diet eventually lead to a breakdown of control, eating binges, a return to previous habits, and weight regain.[6] Humans are creatures of pleasure and not pain. Any painful approach to weight control is bound to fail. Improved and enjoyable eating and exercise habits are needed to keep excess weight off, and quick weight-loss approaches don't change habits.

Unfortunately, many dieters think weight-control methods are successful if they lead to a rapid loss of weight.[7] When the lost weight is regained, dieters tend to blame themselves and not the faulty method. Blaming themselves for the failure, many people are ready to try other quick weight-loss methods. But they usually fail, too.[8]

The Business of Weight Loss More than 29,000 weight-loss products and services are available. Some of these are shown in Illustration 10.2. Most of

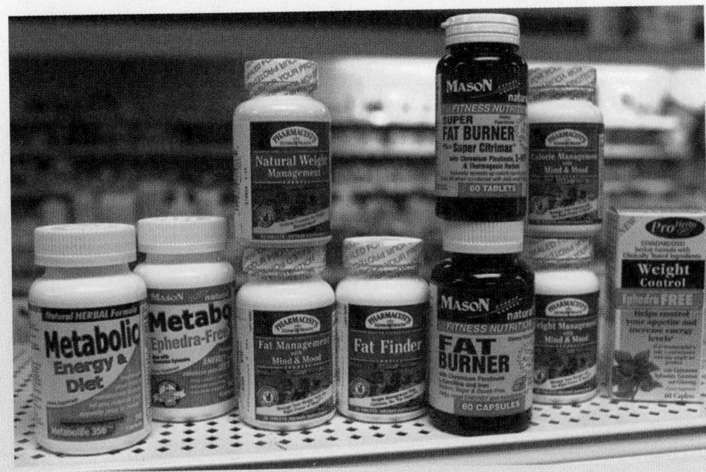

Illustration 10.2
There is no lack of weight-loss books and products. There is a lack of popular approaches that help people keep weight off.

them either don't work at all or don't prevent weight regain.[8] "Quick fix" weight-loss approaches that don't lead to long-term changes in behavior are the primary reason approximately 90 to 95% of people who lose weight gain it back. The demand for such products and services is so great, however, that many are successful—financially.

One reason so many weight-loss products and services are available is that almost none of them work. If any widely advertised approach helped people lose weight and keep if off, manufacturers of bogus methods would go out of business. The weight-loss industry also thrives because of the social pressure to be thin. Many people try new weight-loss methods even though they sound strange or too good to be true. Often people believe that a product or service must be effective, or it wouldn't be allowed on the market.[9] Although reasonable, this belief is incorrect.

The Lack of Consumer Protection The truth is that general societal standards for consumer protection do not apply to the weight-loss industry.[9] No laws require a product to be effective: in most cases, companies do not have to show that weight-loss products or services actually work before they can be sold. That's why products like herbal remedies, forks with stop and go lights, weight-loss skin patches, colored "weight-loss" glasses, electric cellulite dissolvers, mud and plastic wraps, and inflatable pressure pants are available for sale. These, like many other weight-loss products, do not work, and there is no reason why they should. Promotions for the products just have to make it *seem* as though they might work wonders on appetite or body fat.

Furthermore, weight-loss products and services are usually not tested for safety before they reach the market. The history of the weight-loss industry is littered with abject failures: fiber pills that can cause obstructions in the digestive tract, very low-calorie diets and intestinal bypass surgeries that lead to nutrient-deficiency disease, amphetamines that can produce physical addiction, diet pills that cause heart valve problems, and liquid protein diets that have led to heart problems and death from heart failure.[10] A brief history of weight-loss method failures is chronicled in Table 10.1.

Pulling the Rug Out Fraudulent weight-loss products and services may be investigated and taken off the market. Currently, the Federal Trade Commission

Table 10.1
A brief history of discontinued weight-loss methods.

Year	Method	Reason for Discontinuation
1940s–1960s	Amphetamines	Highly addictive, heart and blood pressure problems
	Vibrating machines	Did not work
1970s	Jejunoileal; bypass surgery	Often caused chronic diarrhea, vitamin and mineral deficiencies, kidney stones, liver failure, arthritis
	Liquid protein diet	Poor-quality protein caused heart failure, deaths
1990	Oprah Winfrey liquid diet success (she lost 67 pounds)	
1991	Oprah Winfrey gains 67 pounds and declares "No more diets!"	
1997	Phen-Fen (Redux)	Heart valve defects, hypertension in lung vessels

Fat magnet pills were purported to break into thousands of magnetic particles once swallowed. When loaded with fat, the particles simply flushed themselves out of the body. The FTC found the product too hard to swallow. The company took the product off the market and made $750,000 available for customer refunds.

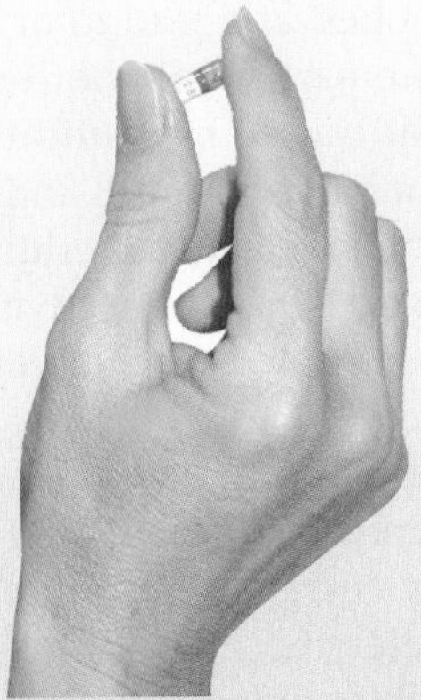

"Blast" away 49 pounds in less than a month with Slim Again, Absorbit-All, and Absorbit-AllPlus pills claimed ads in magazines, newspapers, and on the Internet. The company got blasted by the FTC to the tune of $8 million for making fraudulent claims.

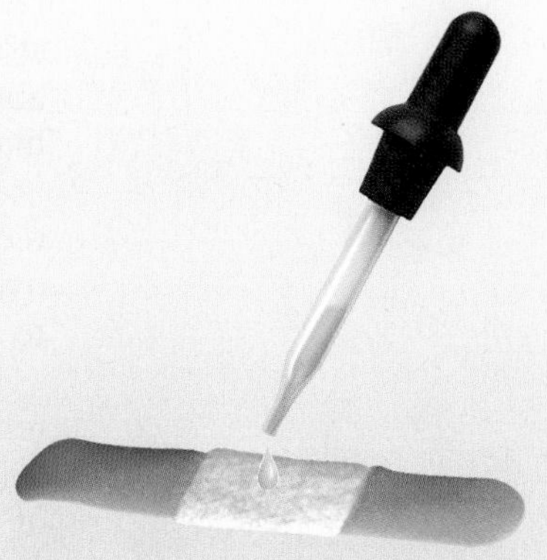

Take a few drops of herbal liquid, put them on a bandage patch, and voilà a new weight-loss product! The herbal liquid was supposed to reach the appetite center of the brain and turn the appetite off. Federal marshals weren't impressed. They seized $22 million worth of patch kits and banned the sale of others.

Illustration 10.3
A few examples of bogus weight-loss products that were removed from the market.

(FTC) monitors deceptive practices and weight-loss claims on a case-by-case basis. It has filed suits against companies making exaggerated claims.[11] Most investigations are performed in response to consumer complaints. Illustration 10.3 shows three examples of products that were taken off the market and explains why.

Requirements for truth in labeling keep many bogus products from including false or misleading information on their labels. These laws, however, do not keep outrageous claims from being made on television or printed in pamphlets, books, magazines, and advertisements. Few companies get caught if they include false or misleading information in weight-loss product ads.

What if the Truth Had to Be Told? Suppose the weight-loss industry had to inform consumers about the results of scientific tests of the effectiveness and safety of weight-loss products and services. What if this information had to be routinely included on weight-loss product labels (Illustration 10.4)? What do you think would happen to consumer choices and the weight-loss industry? Increased federal enforcement of truth-in-advertising laws may help put an element of honesty into promotions for many weight-loss products. In addition, the FTC has proposed that claims about long-term weight loss "must be based on the experience of patients followed for at least two years after they complete the [weight-loss] program."[12] If this proposal ever becomes law, it will change the weight-loss industry in the United States.

Illustration 10.4
Just imagine what would happen if weight-loss approaches were required to divulge their effectiveness and risks.

Diet Pills Two types of prescription diet pills are currently approved for use in obese people in the United States. One is Meridia (sibutramine), and the other is Xenical (orlistat).[3] Meridia works by enhancing satiety (which reduces food intake), and Xenical works by partially blocking fat absorption in the intestines. Both have side effects. Meridia may increase blood pressure, heart rate, and headaches, and lead to dry mouth. Xenical's primary side effect is oily stools, which are particularly bothersome if high fat meals are consumed. Malabsorption of fat caused by Xenical reduces absorption of a number of fat-soluble nutrients such as vitamin E and beta-carotene.[3]

It is recommended that Meridia and Xenical be used in conjunction with reduced calorie diets. Both effectively lower weight by an average of about 10% over a period of six months to a year. Weight regain after pill use stops appears to be common.[13]

Many other diet drugs are under development or testing. Much activity currently centers around leptin, a hormone produced in fat cells that may modify caloric intake or energy expenditure, and other drugs that modify the sensations of hunger or satiety.[3]

Even though people using prescribed diet drugs are carefully screened and monitored by medical professionals, serious side effects can develop. No diet drug is absolutely safe, and none are known to cure obesity forever.[2]

High Protein Diets Do high protein diets help people lose weight? Yes. Do they pose risks to health? Some. Do they help people maintain weight loss in the long run? No.

High protein, low carbohydrate diets can lead to weight loss, and may improve health status as a result (Illustration 10.5).[14] They promote weight loss by decreasing appetite. By depriving the body of carbohydrates, fats must be used as the body's primary source of energy. Use of fat to meet much of the body's energy needs leads to increased blood levels of by-products of fat utilization called "ketone bodies," and that decreases appetite. People often experience rapid weight loss when they first start high protein diets, and that can be very encouraging. The weight loss is primarily due to water excretion. High protein diets increase water need because the nitrogen component of much of the protein consumed is excreted from the body diluted in urine.[15]

All is not well with high protein diets, however. They are not recommended for people with liver or kidney problems, may promote heart disease and prostate cancer, may leave glycogen stores nearly empty, and may cause nausea,

Illustration 10.5
High protein diets don't work in the long run, and their long-term consequences on health are not clear.

headache, decreased stamina, and a reduced ability to concentrate. Long-term consequences of high protein diets on health are far from clear.[15,16]

What's the primary limitation of high protein diets as far as we know now? It's a big one: they don't work in the long run. As with other weight-loss diets requiring substantial changes in food intake, people revert to more pleasurable, usual diets. As soon as carbohydrates are returned to the diet in normal amounts, appetite returns and weight gain follows.

Organized Weight-Loss Programs The mark of successful weight-loss programs is maintenance of weight loss in the long term.[17] Whether organized weight-loss programs such as Weight Watchers, Jenny Craig, Overeaters Anonymous, or services provided by many medically supervised weight-loss clinics hit this mark is not known. Results of studies of the effectiveness of privately run weight-loss programs are rarely published. Programs requiring the largest changes in diet and lifestyle would be expected to have the lowest success rates in terms of weight-loss maintenance.[17] No weight-loss product, diet, program, or clinic that substantially alters usual eating and activity patterns has been shown to result in a long-term loss of excess body fat in more than 5 to 10% of people who use it.[6]

Drastic Measures

Fasting and surgery represent extreme approaches to body fat reduction. These approaches are riskier than weight-loss methods that produce a gradual loss in weight.

Fasting Sometimes people fast to lose weight, but this method is not recommended unless the fast is closely supervised by medical professionals. Abruptly stopping food intake not only does not foster long-term weight-loss maintenance, but it produces a high level of wear-and-tear on the gastrointestinal tract. Fasting causes "intestinal starvation" because it removes a primary source of nutrients from the intestines—foods that are being digested. Without sufficient nutrients to maintain the gastrointestinal tract, mucus and cells lining the tract are lost, constipation occurs, and infection is more likely. Fasting can also lead to a preoccupation with food and eating and a rapid rebound in weight once the fast ends.[18]

Surgical Techniques Excessive body fat can be treated by surgery, but this is a method of last resort. It is reserved for very obese persons whose weight jeopardizes their immediate health.[19] Surgical techniques are expensive and carry the risks associated with surgical interventions.

Stomach Surgery Several surgical techniques reduce the size of the stomach and thereby limit the amount of food people can eat. The technique that has the lowest failure rate (gastric bypass surgery) is shown in Illustration 10.6.[19] In gastric bypass surgery, most of the stomach is stapled shut, leaving a pouch at the top that can hold about two to four tablespoons of food. A section of the small intestine that connects to the bottom of the stomach is cut off and attached as a drain for the stomach pouch. The loose end of the small intestine is then reattached to the section of the small intestine that leads from the pouch. This way, digestive juices from the stomach and the upper part of the small intestine are made available for digesting food. People who have their stomachs stapled can eat only a small amount of food. If they eat too much, they feel nauseous and vomit.

Gastric bypass surgery typically costs more than $10,000 and leads to an average weight loss of about 100 pounds in very obese people. Most people

Illustration 10.6
Gastric bypass surgery.

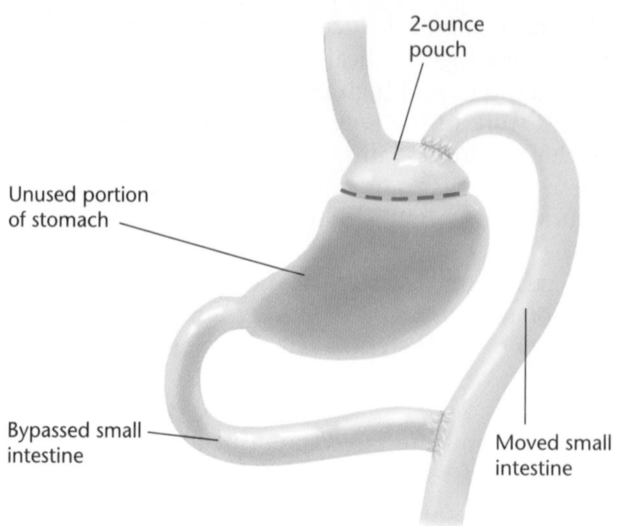

(70%) maintain a 50% loss in body weight five years after the surgery.[2] The procedure is generally reserved for people who weigh twice their normal weight and is not fail-proof. People can eat enough to separate the staples and end up back in surgery.[20] In addition, blocking off most of the stomach reduces vitamin B_{12} absorption and increases the risk of osteoporosis later in life.[19]

Liposuction At a cost of over $3000 per surgery (prices vary by fat deposit site, inflation, and surgeon), fat deposits in the thighs, hips, arms, back, or chin can be partially removed by liposuction. The procedure is considered cosmetic and is not intended for weight loss or to diminish health risks associated with obesity (Illustration 10.7). If the person gains a good deal of weight after liposuction, fat will be deposited to some extent in areas where fat was extracted and in other areas not operated on. These deposits can lead to a return of an undesired body shape.[21] In addition, surgery always carries a risk of infection and other complications, so it cannot be taken lightly.

Illustration 10.7
The effects of liposuction. The photograph on the left shows the thighs of a woman before liposuction, and the photo on the right shows the same woman's thighs after liposuction.

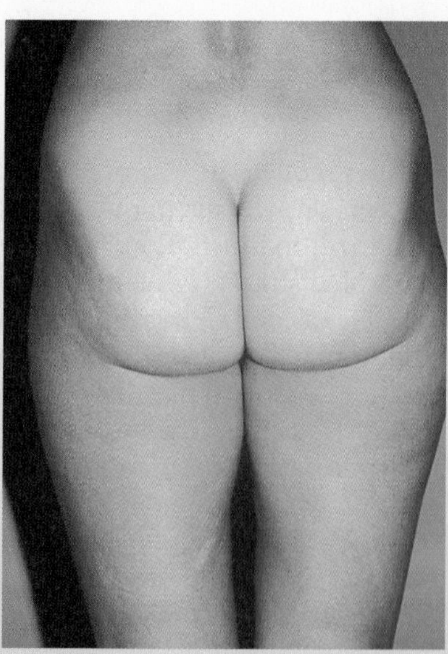

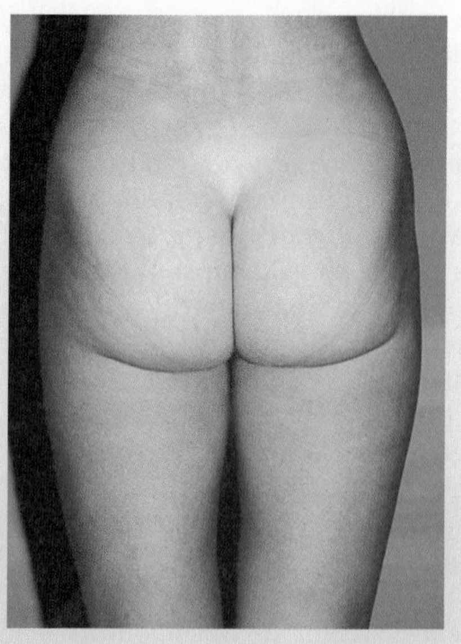

Custom Medical Stock, Dr. I. Richard Toranto

Table 10.2 Characteristics of weight-loss programs that promote maintenance of weight loss.[17,22]

1. Offer long-term approach to beneficial changes in diet and physical activity.
2. Employ reduced-calorie, nutritionally adequate diets.
3. Recommend foods from all food groups.
4. Do not exclude certain foods.
5. Employ relatively small and acceptable changes in usual lifestyle, dietary, and other behaviors.
6. Do not rely solely on drugs or herbal remedies.

Weight Loss: Making It Last

People come in all sizes, and not everyone can change his or her body shape or weight. Some can, however, and the most effective ways to lose weight and keep it off are known. Many weight-loss methods lead to a temporary reduction in body weight, but this is not the test of a successful method. An approach to weight loss can be considered successful only if it is safe, healthful, and prevents weight regain. Table 10.2 summarizes characteristics of weight-loss programs that promote maintenance of weight loss. An approach that fosters dietary and exercise patterns that prevent weight regain is best.[17]

Such methods take time, however, so they are not among the more popular approaches. Because they cannot be sold as a product, the weight-loss industry isn't very interested in them. Nevertheless, these methods work in the long run for many overweight people. The key ingredients of these methods are small, acceptable changes in diet and activity. These approaches focus on healthy eating and exercise for a lifetime, rather than "dieting." The goal is to achieve improvements in lifestyle, fitness, and health—not losing a certain number of pounds in a week. Simple though they sound, such methods are the difference between those who lose weight and keep it off and those who don't (see the Health Action below).

HEALTH ACTION

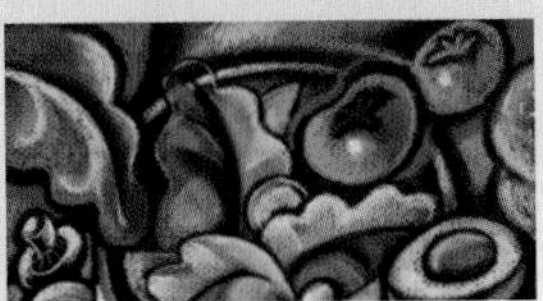

Weight-Loss Maintainers versus Weight Regainers[22–24]

Weight-Loss Maintainers	Weight Regainers
• Lose weight slowly by making small, acceptable changes in eating and activity.	• Are not committed to a gradual weight loss through behavioral changes.
• Consume regular meals.	• Change diet radically to lose weight.
• Exercise regularly.	• Exercise little.
• Make conscious efforts to avoid regaining weight.	• Eat unconsciously in response to stress.
• Avoid feeling deprived while changing habits (i.e., eat foods they like and undertake activities they enjoy).	• Take diet pills.
• Use available sources of social support.	• Do not seek out or have social support.
	• Cope with problems by escape and avoidance.

Photo Disc

Illustration 10.8
Small, acceptable changes are the key to a successful weight-loss/weight-maintenance program. Try to find activities that you enjoy; get out there and play!

The Importance of Small, Acceptable Changes

For most people, excess body fat accumulates slowly over time. Consuming just an extra 50 calories a day, for example, will lead to a gain of approximately five pounds in a year. (Approximately 3500 excess calories will produce a weight gain of a pound, and a 3500-calorie deficit will produce a loss of a pound in body weight in many people.) Excess weight is rarely all gained over the course of a few weeks or months. Fat is put on slowly and that's the best way to take it off.

Gradual losses in body fat do not require dramatic changes in diet or activity level. Only small changes in diet and activity are needed. By cutting back food intake by 100 calories per day, a person could lose 10 pounds in a year. Jogging 20 minutes three times a week in addition to usual activity could take off another 10 pounds in a year. Routinely engaging in more vigorous forms of exercise such as handball or rowing can contribute to body fat loss, weight-loss maintenance, and fitness.[23] The secret to success is including foods and activities that are enjoyed in the lifestyle change package. To be lasting, the changes have to be acceptable, even preferable, to existing patterns (Illustration 10.8). Often this means that the changes have to be small.[24]

It should be mentioned that a minority of people are able to make drastic changes in dietary intake (adopt a very low-fat diet, for example) or increase their exercise level substantially and maintain these changes and lower body weight over time.[23,25] Such people tend to be the exceptions, however.

Identifying Small, Acceptable Changes To identify changes in eating and activity that have staying power, first list the weak points in diet and activity. Dietary weak points might include the consumption of high-fat foods due to eating out often or relying on high-fat convenience foods that can be heated up in seconds. Another weak point might be skipping breakfast and overeating later in the day because of extreme hunger. Weak points in physical activity might include driving instead of walking, not engaging in sports, or spending too little time playing outside.

For each weak point, identify options that seem acceptable and enjoyable. A person who enjoys broiled chicken with barbecue sauce might not mind eating that at restaurants instead of fried chicken. That's a change people can make if they plan ahead. A person who gets too full from a large serving of fries might be happier ordering a small serving and not eating so many. A breakfast skipper might find grabbing a piece of fruit and a slice of cheese for breakfast acceptable and doable. People who enjoy walking may not mind leaving the car or bus behind and letting their feet carry them to class, the grocery store, or a friend's house.

Many acceptable options for making small improvements in diet and activity may be available (see the Health Action below). The easiest changes to accomplish are the ones that should be incorporated into the overall lifestyle improvement plan. Some people, for example, lose weight and keep it off simply by consciously cutting down on portion sizes. Others avoid eating too much at any meal and walk more. Simply adding breakfast helps some people lose weight and maintain the loss.[26] The easier the changes are to follow, the more likely they are to succeed.

Individualized plans that don't work out often include unacceptable or unenjoyable changes. The changes may be too large or too different from the usual. In that case, people need to go back to the drawing board and modify the plan so that it includes small changes that are acceptable in the long run. Perhaps the original plan included jogging, but it turned out that jogging wasn't any fun. In that case, jogging should be taken out of the plan! It should be replaced by any other physical activity that would be enjoyed. Midcourse corrections should be expected. Some experimentation may be required before the small changes that will last are identified.

What to Expect for Weight Loss If all goes well, weight loss will be gradual but lasting. The pattern of loss should be somewhat like that graphed in Illustration 10.9 where the person lost eighteen pounds over 30 weeks. The pattern will include peaks, valleys, and plateaus—not a straight downward curve. Sometimes a bit of weight will be gained, and other times more weight than expected will be lost. It's more important to enjoy and continue improved eating and activity patterns than to concentrate on the number of pounds lost.

HEALTH ACTION

Small Changes That Make a Big Difference

- Use less margarine or butter and mayonnaise on foods.
- Choose low-fat varieties of foods (cheese, milk, yogurt, fish, poultry) that you like.
- Trim the fat off meats or buy lean meats.
- Eat small portions of high calorie foods.
- Eat fried foods infrequently.
- Eat breakfast.
- Eat fruit and vegetables for snacks.
- Cook at home.
- Walk to school.
- Bike to work.
- Take the stairs.
- Go outside and play.

Illustration 10.9
People who lose weight gradually are more likely to keep it off than those who lose weight rapidly. The weight loss graphed here averages half a pound per week.

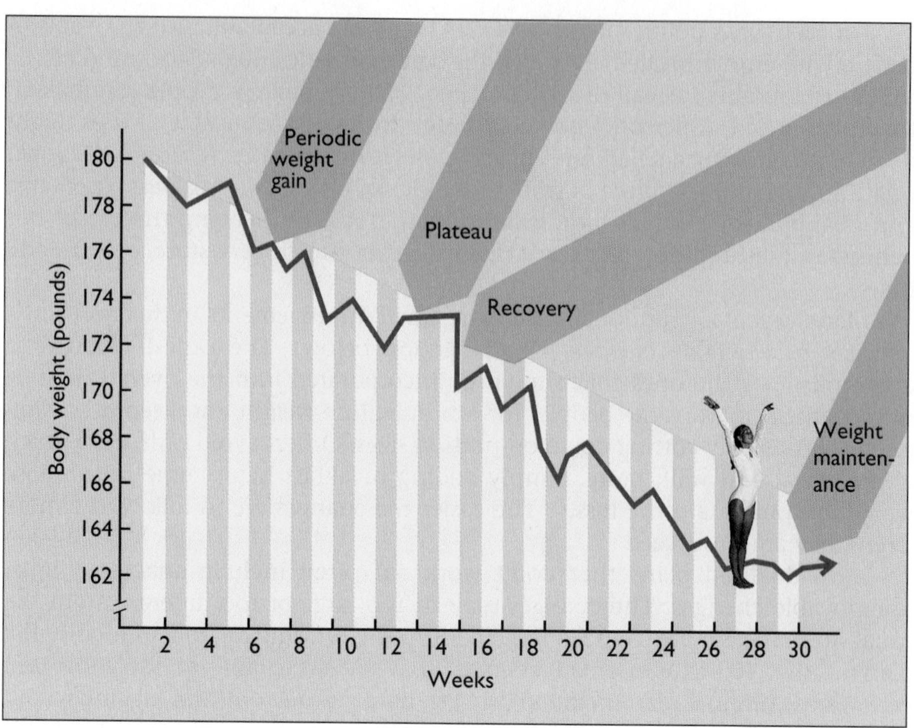

If diet and exercise behaviors are improved in acceptable ways, there is little need to become preoccupied with the number of calories consumed, the number of calories burned off in a bout of exercise, or the number of pounds lost last week. The goal is reached when the small changes become an enjoyable part of life on a day-to-day basis. Improved eating and activity patterns offer many benefits. Weight loss is only one of them.

www.cyberdiet.com
Construct a personal menu plan and get an estimate of calories burned in physical activity at this site.

www.eatright.org
The American Dietetic Association's web site provides a checklist for evaluating weight-loss diets and services as well as advice on safe and effective weight-loss methods.

www.ediets.com
A commercial site offering personalized diet plans and other resources developed by dietitians.

www.niddk.nih.gov/health/nutrit/win.htm
The Weight Control Information Network provides consumer information about weight control. Call toll free: 1-800-WIN-8098.

www.shapedown.com
This commercial site provides background information and methods for the tested Shapedown Program for weight management in adolescents.

www.quack.watch.com
Check out weight-loss promotions.

Nutrition UP CLOSE

Will You Maintain or Regain?

FOCAL POINT: People who lose weight and keep the pounds off share similar successful behaviors.

Have you recently lost weight? Answer the following questions to find out if your current eating and activity behaviors promote making the weight loss permanent.

		YES	NO
1.	Do you usually eat breakfast?	___	___
2.	Do you avoid skipping meals?	___	___
3.	Do you plan ahead for snack and mealtime foods?	___	___
4.	Do you prepare most of your meals at home?	___	___
5.	Do you limit consumption of high fat foods like butter, margarine, mayonnaise, and fried foods?	___	___
6.	Do you consume high calorie foods like desserts infrequently and in small portions?	___	___
7.	Do you walk instead of drive whenever possible?	___	___
8.	Do you engage in regular physical activities that you enjoy?	___	___
9.	Do you seek social support to help maintain your weight loss?	___	___

FEEDBACK (answers to these questions) can be found at the end of Unit 10.

REFERENCES

1. Serdula MK, et al. Prevalence of attempting weight loss and strategies for controlling weight. JAMA 1999;282:1353–58.
2. Cummings SM, et al. Position of the American Dietetic Association: weight management. J Am Diet Assoc 1997;97:71–74.
3. Bray GA. Physiology and consequences of obesity. Medscape Diabetes and Endocrinology Clinical Management Modules, http://womenshealth.medscape.com/medscape/endocrinology/ClinicalMgmt/CM.v03/pnt-CM.v03.html, 2001.
4. Sarlio-Lahteenkorva SS, et al. A descriptive study of weight loss maintenance: 6 and 15 year follow-up of initially overweight adults. Int J Obesity 2000;24:116–25.
5. Marcus MD, Wing RR, Lamparski DM. Binge eating and dietary restraint in obese patients. Addict Behav 1985;10:163–8.
6. Goodrick GK, Foreyth JP. Why treatments for obesity don't last. J Am Diet Assoc 1991;91:234–47.
7. Womble L, et al. Unrealistic weight loss goals and feelings of failure. Presentation at the North American Association for the Study of Obesity, Long Beach, CA, 2000 Nov.
8. Wooley SC, Garner DM. Obesity treatment: the high cost of false hope. J Am Diet Assoc 1991;91:1248–51.
9. Lustig A. Weight loss programs: failing to meet ethical standards? J Am Diet Assoc 1991;91:1252–4.
10. Wadden TA, Stunkard AJ, Brownell KD. Very low calorie diets: their efficacy, safety, and future. Ann Intern Med 1983;99:675–84.
11. Community Nutrition Institute newsletter. 1997 Apr and Oct 3.
12. FTC acts on diet claims. J Am Diet Assoc 1992;92:4.
13. Thomas TA et al. Sibutramine works best with lifestyle modification. Arch Intern Med 2001;1612:218–27; and Obesity: a ballooning disease with no quick fix. Drugs Ther Perspect 2000;15:10–13.
14. Heimowitz C. Presentation at the Southern Society of General Internal Medicine, New Orleans, 1999 Feb.
15. Stein K. High-protein, low carbohydrate diets: Do they work? J Am Diet Assoc 2000;100:760–1.
16. Metges CC, Barth CA. Metabolic consequences of a high dietary-protein intake in adulthood: assessment of the available evidence. J Nutr 2000;130:886–9.
17. USDA: Most popular weight loss diets flawed. USDA news release, 2001 Jan 10.

18. Polivy J. Psychological consequences of food restriction. J Am Diet Assoc 1996;96:589–92.
19. Atkinson RL. Obesity surgery as a model for understanding the regulation of food intake and body weight. Am J Clin Nutr 1997;66:184–5.
20. Graney AS, Smith LB, Hammer KA. Gastric partitioning for morbid obesity: postoperative weight loss, technical complications, and protein status. J Am Diet Assoc 1986;86:630.
21. Cohen S. Lipolysis: pitfalls and problems in a series of 1246 procedures. Aesthetic Plast Surg 1985;9:207.
22. St Jeor ST, et al. A classification system to evaluate weight maintainers, gainers, and losers. J Am Diet Assoc 1997;97:481–8.
23. Schoeller DA, et al. How much physical activity is needed to minimize weight gain in previously obese women? Am J Clin Nutr 1997;66:551–6.
24. National Institutes of Health. Methods for voluntary weight loss and control. Nutr Today 1992;27:27–33.
25. Shick SM, et al. Persons successful at long-term weight loss and maintenance continue to consume a low-energy, low-fat diet. J Am Diet Assoc 1998;98:408–14; and Cullinen K, Caldwell M. Weight training increases fat-free mass and strength in untrained young women. J Am Diet Assoc 1998;98:414–18.
26. Schlundt DG, Hill JO, Sbrocco T, Pope-Cordle J, Sharp T. The role of breakfast in the treatment of obesity: a randomized trial. Am J Clin Nutr 1992;55:645–51.

Nutrition UP CLOSE

Will You Maintain or Regain?

Feedback for Unit 10

The more checks you made in the Yes column, the greater the chances that you will maintain your weight loss. Reflect for a moment upon any checks you made in the No column. Enacting small, acceptable changes in your eating and activity plan now may make the difference between maintaining your weight loss or regaining the pounds.

Photo Disc

Unit 11

Disordered Eating: Anorexia Nervosa, Bulimia, and Pica

Nutrition Scoreboard

	TRUE	FALSE
1. The United States has one of the world's highest rates of anorexia nervosa.		
2. Eating disorders result from psychological, and not biological, causes.		
3. People in many different cultures may consume clay, dirt, and other nonfood substances.		

(Answers on next page)

Answers to

Nutrition Scoreboard

	TRUE	FALSE
1. Anorexia nervosa is most common in the United States and other Westernized countries. It is rarely observed in developing, non-Westernized nations.[1]	✓	
2. The cause (or causes) of eating disorders is not yet known. Both psychological and biological factors may play a role.		✓
3. Although not recommended for health reasons, people in many different cultures practice pica—the regular ingestion of nonfood items such as clay and dirt.	✓	

Key Concepts and Facts

- Anorexia nervosa, bulimia nervosa (bulimia), binge-eating disorder, and pica are the four major types of eating disorders. They may seriously threaten health.
- Eating disorders are much more common in females than males.
- The incidence of eating disorders in a society is related to the value placed on thinness by that society.
- An important route to the prevention of anorexia nervosa and bulimia is to change a society's cultural ideal of thinness and to eliminate biases against people (especially women) who are not thin.

The Eating Disorders

Three square meals a day, an occasional snack or missed meal, and caloric intakes that average out to match the body's need for calories—this set of practices is considered "orderly" eating. Self-imposed semi-starvation, feast and famine cycles, binge eating, ***purging,*** and the regular consumption of nonfood substances such as paint chips and clay—these behaviors are symptoms of disordered eating.

purging
The use of self-induced vomiting, laxatives, or diuretics (water pills) to prevent weight gain.

Four eating disorders have received official designation (that is, they have been established as scientifically recognized disorders). They are (1) anorexia nervosa, (2) bulimia nervosa, (3) binge-eating disorder, and (4) pica[2] (the regular ingestion of substances that aren't foods). Other forms of disordered eating, such as compulsive overeating, restrained eating, and chronic starvation, have been observed, but too little research exists to establish criteria for diagnosis.

Anorexia Nervosa

It's about 9:30 on a Tuesday night. You're at the grocery store picking up sandwich fixings and some milk. Although your grocery list contains only four items, you arrive at the checkout line with a half-filled cart. The woman in front of you has only five items: a bag with about 10 green beans, an apple, a bagel, a green pepper, and a four-ounce carton of nonfat yogurt. As she carefully places each item into her shopping bag, you notice that she is dreadfully thin.

The woman is Alison. She has just spent half an hour selecting the food she will eat tomorrow. Alison knows a lot about the caloric value of foods and makes only low-calorie choices. Otherwise, she will never get rid of her excess fat. To Alison, weight is everything—she cannot see the skeleton-like appearance others see when they look at her.

Illustration 11.1
A day's diet? For a person with anorexia nervosa, it was. The foods shown provide approximately 562 calories.

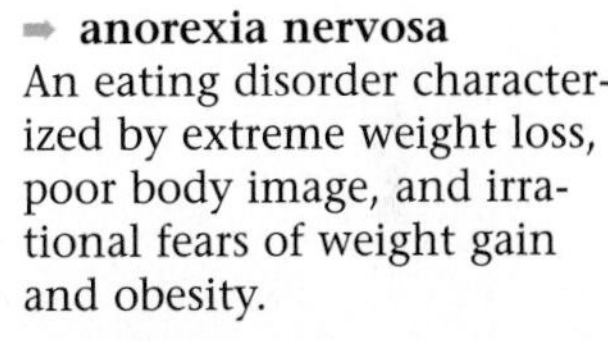
➡ **anorexia nervosa**
An eating disorder characterized by extreme weight loss, poor body image, and irrational fears of weight gain and obesity.

As a child, Alison enjoyed little independence. Decisions about her life were made for her, a situation that contributed to her low self-esteem. During her teen years, Alison was overweight and clearly remembers the painful teasing and ridicule she had to endure. Now Alison is on her own, away from home and in control in an out-of-control way.

You didn't know this about Alison when you saw her. There is much more to anorexia nervosa than meets the eye.

Individuals with ***anorexia nervosa*** starve themselves (Illustration 11.1). In one out of ten cases, they starve themselves to death.[2] For people with this disorder, thinness is everything. They can never be too thin—no matter how emaciated they may be. As shown in Illustration 11.2, people with anorexia nervosa look extraordinarily thin from the neck down. The face and the rest of the head usually look normal because the head is the last part of the body to be affected by starvation.

Instead of the normal amount of body fat (20–25% of body weight), people with anorexia nervosa have little fat (7–13% of body weight).[3] They become cold easily and have unusually low heart rates and sometimes an irregular heartbeat, dry skin, low blood pressure, absent or irregular menstrual cycles, infertility, and poor pregnancy outcomes (Table 11.1).[4] Approximately 9 in 10 women with anorexia nervosa have significant bone loss, and 38% have osteoporosis. The extent of bone loss correlates strongly with undernutrition: the lower the body weight, the lower the bone density. Improving calcium and vitamin D intakes and use of bone-density drugs have limited effectiveness in rebuilding bones in females with current or past anorexia nervosa.[6]

PhotoEdit, Felicia Martinez

Illustration 11.2
Eating disorders occur in males as well as females, but females make up approximately 95% of all cases.

The Female Athlete Triad Pediatricians, nutritionists, and coaches are beginning to be on the lookout for eating disorders, menstrual cycle dysfunction, and decreased bone mineral density in young, female athletes. Low caloric intakes and underweight related to eating disorders can lower estrogen levels and

Table 11.1
Features of anorexia nervosa.[2,5]

Reprinted with permission from the Diagnostic and Statistical Manual of Mental Disorders, Fourth Edition, Text Revision. Copyright 2000 American Psychiatric Association.

A. Essential Features:

1. Refusal to maintain body weight at or above 85% of normal weight for age and height.
2. Intense fear of gaining weight or becoming fat, despite being underweight.
3. Disturbance in the way in which body weight or shape is experienced, undue influence of body weight or shape on self-evaluation, or denial of the seriousness of current low body weight.
4. Lack of menstrual periods in teenage females and women (missing at least three consecutive periods).

Restricting type: Person does not regularly engage in binge-eating or purging behavior.

Binge-eating type: Person regularly engages in binge-eating or purging behavior (self-induced vomiting; laxative, diuretic, or enema use).

B. Common Features in Females

1. Low-calorie diet, extensive exercise, low body fat
2. Soft, thick facial hair, thinning scalp hair
3. Loss of heart muscle, irregular, slow heartbeat
4. Low blood pressure
5. Increased susceptibility to infection
6. Anemia
7. Constipation
8. Low body temperature (hypothermia)
9. Dry skin
10. Depression
11. History of physical or sexual abuse
12. Low estrogen levels
13. Low bone density
14. Infertility, poor pregnancy outcome

C. Common Features in Males

1. Most of the common features in females
2. Substance abuse
3. Mood and other mental disorders

disrupt menstrual cycles. The lack of estrogen decreases calcium deposition in bones and reduces bone density at a time when peak bone mass is accumulating.

Irregular or absent menstrual cycles used to be thought of as "no big deal." That attitude has changed, however, due to research results indicating that abnormal cycles in young females are related to delayed healing of bone and connective tissue injuries, and bone fractures and osteoporosis later in life.[7,8]

Motivations Underlying Anorexia Nervosa The overwhelming desire to become and remain thin drives people with anorexia nervosa to refuse to eat, even when ravenously hungry, and to exercise intensely. Half of the people with anorexia turn to ***binge eating*** and purging—features of bulimia nervosa—in their efforts to lose weight.[2] Preoccupied with food, people with anorexia may prepare wonderful meals for others, but eat very little of the food themselves. Family members and friends, distressed by their failure to persuade the person with anorexia to eat, report high levels of anxiety. Although adults often describe people with anorexia as "model students" or "ideal children,"

➡ **binge eating**
The consumption of a large amount of food in a small amount of time.

their personal lives are usually marred by low self-esteem, social isolation, and unhappiness.[9]

What Causes Anorexia Nervosa? The cause of anorexia nervosa isn't yet clear. It is likely that many different conditions, both psychological and biological, predispose an individual to become totally dedicated to extreme thinness.[2] The value that Western societies place on female thinness, the need to conform to society's expectations of acceptable body weight and shape, low self-esteem, and a need to control some aspect of one's life completely are commonly offered as potential causes for this disorder (Illustration 11.3).[4,10]

On the Side

The average size of female gymnasts on the U.S. Olympic team shrank from 5 feet 3 inches tall and 105 pounds in 1976 to 4 feet 9 inches tall and 88 pounds in 1992.

How Common Is Anorexia Nervosa? It is estimated that 1% of adolescent and young women in the Western world and less than 0.1% of young males have anorexia nervosa.[4] The disorder has been reported in girls as young as five and in women through their forties;[11] however, it usually begins during adolescence. It is estimated that one in ten females between the ages of 16 and 25 has "subclinical" anorexia nervosa, or exhibits some of the symptoms of the disorder.[4]

Certain groups of people are at higher risk of developing anorexia nervosa than others (Table 11.2). People at risk come from all segments of society, but tend to be overly concerned about their weight and food, and have attempted weight loss from an early age.[4]

Table 11.2
Risk groups for anorexia nervosa.[12,13]

- Dieters
- Ballet dancers
- Competitive athletes (gymnasts, figure skaters)
- Fitness instructors
- Dietetics majors

Treatment There is no "magic bullet" treatment that cures anorexia nervosa quickly and completely. In all but the least severe cases, the disorder generally takes a good deal of time and professional help to correct. Treating the disorder is often difficult because few people with anorexia believe their weight needs to be increased.[2]

Treatment programs for anorexia nervosa generally focus on restoring normal body weight, psychological counseling to improve self-esteem and attitudes

Illustration 11.3

There is a need for a more realistic body shape on television and in fashion magazines.
—Vivienne Nathanson, British physician, 2000

about body weight and shape, antidepressant or other medications, family therapy, and normalizing eating and exercise behaviors.[14] It is successful in 50% of people, and partially successful in almost all other cases.[4,10] One-third of people with full recovery from anorexia nervosa will relapse within seven years or less.[15] By eight years after diagnosis, 3% of people with anorexia nervosa will have died from the disorder, and it claims the lives of 18% 33 years later.[16] Results of treatment are often excellent when the disorder is treated early. Unfortunately, many people with the condition deny that problems exist and postpone treatment for years. Initiation of treatment is often prompted by a relative, coach, or friend.[17,18]

Bulimia Nervosa

Finally home alone, Lisa heads to the pantry and then to the freezer. She has carefully controlled her eating for the last day and a half and is ready to eat everything in sight.

It's a bittersweet time for her. Lisa knows the eating binge she is preparing will be pleasurable, but that she'll hate herself afterward. Her stomach will ache from the volume of food she'll consume, she'll feel enormous guilt from losing control, and she'll be horrified that she may gain weight and will have to starve herself all over again. Lisa is so preoccupied with her weight and body shape that she doesn't see the connection between her severe dieting and her bouts of uncontrolled eating. To get rid of all the food she is about to eat, she will do what she has done several times a week for the last year. Lisa avoids the horrible feelings that come after a binge by "tossing" everything she ate as soon as she can.

In just 10 minutes, Lisa devours 10 peanut butter cups (the regular size), a 12-ounce bag of chocolate chip cookies, and a quart of mint chocolate chip ice cream. Before five more minutes have passed Lisa will have emptied her stomach, taken a few deep breaths, thrown on her shorts, and started the five-mile route she jogs most days. As she jogs, she obsesses about getting her 138-pound, 5-foot 5-inch frame down to 115 pounds. She will fast tomorrow and see what news the bathroom scale brings.

➡ **bulimia nervosa**
An eating disorder characterized by recurrent episodes of rapid, uncontrolled eating of large amounts of food in a short period of time. Episodes of binge eating are often followed by purging.

Lisa is not alone. ***Bulimia nervosa*** occurs in 1 to 3% of young women and in about 0.5% of young males in the United States.[18] The disorder is characterized by regular episodes of dieting, binge eating, and attempts to prevent weight gain by purging; use of laxatives, diuretics, or enemas; dieting; and sometimes exercise. In most cases, bulimia nervosa starts with voluntary dieting to lose weight. At some point, voluntary control over dieting is lost, and people feel compelled to engage in binge eating and vomiting.[19] The behaviors become cyclic: food binges are followed by guilt, purging, and dieting. Dieting leads to a feeling of deprivation and intense hunger, which leads to binge eating and so on. Once a food binge starts, it is hard to stop.

Table 11.3 lists the features of bulimia nervosa. Approximately 86% of people with this condition vomit to prevent weight gain and avoid postbinge anguish. A smaller proportion use laxatives, diuretics (water pills), or enemas alone or in combination with vomiting.[2] Laxatives, enemas, and diuretics do not prevent weight gain, however, and their regular use can be harmful. The habitual use of laxatives and enemas causes "laxative dependency"—they become necessary for bowel movements. Diuretics can cause illnesses by depleting the body of water and certain minerals and disturbing its fluid balance.[14]

The lives of people with bulimia nervosa are usually dominated by conflicts about eating and weight. Some affected individuals are so preoccupied with food that they spend days securing food, bingeing, and purging. Others experience only occasional episodes of binge eating, purging, and fasting.[2]

Unlike those with anorexia nervosa, people with bulimia usually are not underweight or emaciated. They tend to be normal weight or overweight.[2] Like

Table 11.3
Features of bulimia nervosa.[2]

Reprinted with permission from the Diagnostic and Statistical Manual of Mental Disorders, Fourth Edition, Text Revision. Copyright 2000 American Psychiatric Association.

A. Essential Features:

1. Recurrent episodes of binge eating. An episode of binge eating is characterized by both of the following:
   a. Eating an amount of food within a two-hour period of time that is definitely larger than most people would eat in a similar amount of time and under similar circumstances.
   b. A sense of lack of control over eating during the episode; a feeling that one cannot stop eating or control what or how much one is eating.
2. Recurrent inappropriate compensatory behavior in order to prevent weight gain, such as self-induced vomiting; misue of laxatives, diuretics, enemas, or other medications; fasting; or excessive exercise.
3. The binge eating and inappropriate compensatory behaviors both occur, on average, at least twice a week for three months.
4. Self-evaluation is unduly influenced by body weight and shape.
5. The disturbance does not occur exclusively during episodes of anorexia nervosa.

Purging type: The person regularly engages in self-induced vomiting or the misuse of laxatives, diuretics, or enemas.

Nonpurging type: The person regularly engages in fasting or excessive exercise but does not regularly engage in self-induced vomiting or the misuse of laxatives, diuretics, or enemas.

B. Common Features

1. Weakness, irritability
2. Abdominal pain, constipation, bloating
3. Dental decay, tooth erosion
4. Swollen cheeks and neck
5. Binge on high-calorie foods
6. Eat in secret
7. Normal weight or overweight
8. Guilt and depression
9. Substance abuse
10. Dehydration
11. Impaired fertility
12. History of sexual abuse

anorexia nervosa, bulimia nervosa is more common among athletes (including gymnasts, weight lifters, wrestlers, jockeys, figure skaters, physical trainers, and distance runners) and ballet dancers than in other groups.[13,18]

Bulimia nervosa leads to major changes in metabolism. The body must constantly adjust to feast and famine cycles and mineral and fluid losses. Salivary glands become enlarged, and teeth may erode due to frequent vomiting of highly acidic foods from the stomach.[2]

Is the Cause of Bulimia Nervosa Known?

"Do you follow a special diet?" asks the dietitian at the eating disorder clinic. "Yes," answers the client with bulimia. "Feast or famine."

The cause of bulimia nervosa is not known with certainty, but the scientific finger is pointing at depression, abnormal mechanisms for regulating food intake, and feast and famine cycles as possible causes.[19,20] Fasts and ***restrained eating*** may prompt feelings of deprivation and hunger that may trigger binge eating.[21]

restrained eating The purposeful restriction of food intake below desired amounts in order to control body weight.

The ideal thinness may become more and more difficult to achieve as the feast and famine cycles continue.

Treatment The cornerstone of bulimia treatment is nutrition and psychological counseling to break the feast and famine cycles. Replacing the disordered pattern of eating with regular meals and snacks often reduces the urge to binge and the need to purge. Psychological counseling aimed at improving self-esteem and attitudes toward body weight and shape goes hand-in-hand with nutrition counseling. In many cases, antidepressants are a useful component of treatment.[14,22] The full recovery of women with bulimia nervosa is higher than that for anorexia nervosa. Nearly all women with bulimia achieve partial recovery, but one-third will relapse back into bingeing and purging within seven years.[15] Bulimia nervosa usually improves substantially during pregnancy; about 70% of women with the condition will improve their eating habits for the sake of their unborn baby.[23]

Binge-Eating Disorder

binge-eating disorder An eating disorder characterized by periodic binge eating, which normally is not followed by vomiting or the use of laxatives. People must experience eating binges twice a week on average over a period of six months to qualify for the diagnosis.

Psychiatrists now recognize an eating disorder called ***binge-eating disorder*** (Table 11.4). People with this condition tend to be overweight or obese, and one-third are male.[24] Like individuals with bulimia nervosa, people with binge-eating disorder eat several thousand calories' worth of food within a short period of time during a solitary binge, feel a lack of control over the binges, and experience distress or depression after the binges occur. People must experience eating binges twice a week on average over a period of six months to qualify for the diagnosis. Unlike individuals with bulimia nervosa, however, people with binge-eating disorder don't vomit, use laxatives, fast, or exercise excessively in an attempt to control weight gain.

It is estimated that 9 to 30% of people in weight-control programs have binge-eating disorder. The condition is far less common (perhaps 1 to 5%) in the general population.[23,25] Stress, depression, anger, anxiety, and other negative emotions appear to prompt binge-eating episodes.[22]

The Treatment Approach to Binge-Eating Disorder The treatment of binge-eating disorder focuses on both the disordered eating and the underlying psychological issues.[24] Persons with this condition will often be asked to record their food intake, indicate bingeing episodes, and note feelings, circumstances, and thoughts related to each eating event (Illustration 11.4). This information is used to identify circumstances that prompt binge eating and practical alternative behaviors that may prevent it. Individuals being treated for binge-eating disorder are usually given information about it, attend individual and group therapy sessions, and receive nutrition counseling on normal eating,

Table 11.4 Features of binge-eating disorder.[2]

Binge-Eating Disorder
1. Rapid consumption of extremely large amounts of food (several thousand calories) in a short period of time.
2. Two or more such episodes of binge eating per week over a period of six months.
3. Binge eating by oneself.
4. Lack of control over eating or an inability to stop eating during a binge.
5. Postbinge-eating feelings of self-hatred, guilt, and depression or disgust.
6. Purging, fasting, excessive exercise, or other compensation for high-calorie intakes not present.

Daily Food Record

Date________

Time	Type and amount of food and beverage	Meal, Snack, Binge?	Eating triggers (feelings, situation)
7:30 am	coffee, 2 cups	M	Hunger!
	sugar 2 tsp		
	cornflakes, 2 cups		
	skim milk, 1 cup		
11:30 am	tuna sandwich	M	Bored, hungry
	ice tea, 2 cups		
7:30 pm	3 hamburgers	B	Stressed out,
	2 large fries		angry at my coach
	24 oreos		
	1/2 gallon ice cream		

Illustration 11.4
Example of a food diary of a person with binge-eating disorder.

hunger cues, and meal planning. Antidepressants may be part of the treatment.[23,24] Treatment is successful in 85% of women treated for binge-eating disorder.[26]

Resources for Eating Disorders

Information and services related to eating disorders are available from a variety of sources. Services are best delivered by health care teams specializing and experienced in the treatment of eating disorders. Contact with a primary care physician, dietitian, or nurse practitioner is often a good start to the process of identifying qualified health care teams.

Undieting: The Clash between Culture and Biology

The pressure to conform to society's standard of beauty and acceptability is thought to be a primary force underlying the development of eating disorders.[1] Children acquire prevailing cultural values of beauty before adolescence. As early as age five, American children learn to associate negative characteristics with people who are overweight and positive characteristics with those who are thin.[11] Standards of beauty defined by models and movie and television stars often include thinness, but the body shape portrayed as best is often unhealthfully thin and unattainable by many. The disparity between this ideal and what people normally weigh has led to widespread discontent. Approximately 50% of normal-weight adult women are dissatisfied with their weight, and many diet, binge, and purge occasionally or fast in an attempt to reach the standard of beauty set for them.[27,28]

A movement toward acceptance of body size, fashionable attire for larger people, full-size models, and a more realistic view of individual differences in body shapes is emerging slowly in America (Illustration 11.5). Acceptance of a realistic standard of body weight and shape—one that corresponds to health and physical fitness—and respect for people of all body sizes may be the most effective measures that can be taken to prevent anorexia nervosa, bulimia nervosa, and binge-eating disorder.

Illustration 11.5
The trend toward size acceptance. Acceptance of a realistic standard of body weight and shape—one that corresponds to health and physical fitness—and respect for people of all body sizes may be the most effective measures that can be taken to prevent anorexia nervosa, bulimia nervosa, and binge-eating disorder.

Pica

When did I start eating clay? I know it might sound strange to you, but I started craving clay in the summer of '58. It was a beautiful spring morning—it had just rained. I smelled something really sweet in the breeze coming in my bedroom window. I went outside and knew instantly where the sweet smell was coming from. It was the wet clay that lies all around my house. I scooped some up and tasted it. That's when and how I started my craving for that sweet-smelling clay. I keep some in the fridge now because it tastes even better cold.

➡ **pica**
The regular consumption of nonfood substances such as clay or laundry starch.

➡ **geophagia**
Clay or dirt eating.

➡ **pagophagia**
Ice eating.

➡ **amylophagia**
Laundry starch or cornstarch eating.

A most intriguing type of eating disorder, ***pica*** has been observed in chimpanzees and in humans in many different cultures since ancient times.[29] The history and persistence of pica might suggest that the practice has its rewards. Nevertheless, important health risks are associated with eating many types of nonfood substances.

The characteristics of pica are summarized in Table 11.5. Young children and pregnant women are most likely to engage in the practice; for unknown reasons, it rarely occurs in men.[30] It most commonly takes the form of ***geophagia*** (clay or dirt eating), ***pagophagia*** (ice eating), ***amylophagia*** (laundry starch and corn-

Table 11.5
Characteristics of pica.

Pica
A. **Essential features:** Regular ingestion of nonfood substances such as clay, paint chips, laundry starch, paste, plaster, dirt, or hair.
B. **Other common features:** Occurs primarily in young children and pregnant women in the southern United States.

Table 11.6
A partial list of nonfood substances reported to be consumed by individuals with pica.

Animal droppings	Coal	Laundry starch	Paste
Baking soda	Coffee grounds	Leaves	Pebbles
Burnt matches	Cornstarch	Mothballs	Plaster
Cigarette butts	Crayons	Nylon stockings	Sand
Clay	Dirt	Paint chips	String
Cloth	Hair	Paper	Wool

Photo Disc

starch eating), or ***plumbism*** (lead eating). A potpourri of nonfood substances, listed in Table 11.6, may be consumed.

➡ **plumbism**
Lead (primarily from old paint flakes) eating.

Pica has a forceful calling card:

> Pica permits the mind no rest until it is satisfied.[31]

It is not clear why pica exists, although several theories have been proposed.

Geophagia Some people very much like to eat certain types of clay or dirt. Those who do often report that the clay or dirt tastes or smells good, quells a craving, or helps relieve nausea or an upset stomach. The belief that certain types of clay provide relief from stomach upsets may have some validity: a component of some clays is used in nausea and diarrhea medicines. There is no evidence that geophagia is motivated by a need for minerals found in clay or dirt, however.[32]

Although the reasons given for clay and dirt ingestion make the practice understandable, the consequences to health outweigh the benefits. Clay and dirt consumption can block the intestinal tract and cause parasitic and bacterial infections.[2] The practice is also associated with iron-deficiency anemia in some individuals.[33]

Pagophagia Have you ever known somebody who constantly crunches on ice? That person may have a nine-in-ten chance of being iron deficient.[34] Regular ice eating, to the extent of one or more trays of ice cubes a day, is closely associated with an iron-deficient state.[35] Ice eating usually stops completely when the iron deficiency is treated.[34]

Ice eating may be common during pregnancy. In one study of women from low-income households in Texas, 54% of pregnant women reported eating large amounts of ice regularly. Ice eaters had poorer iron status than other pregnant women who did not eat ice.[36]

PhotoEdit

Illustration 11.6
The regular consumption of lead-based paint chips from old houses is a major cause of lead poisoning in young children.

Amylophagia The sweet taste and crunchy texture of flaked laundry starch are attractive to a small number of women, especially during pregnancy. If the laundry starch preferred is not available, cornstarch may be used in its place.[30] Laundry starch is made from unrefined cornstarch. The taste for starch almost always disappears after pregnancy.[30]

Laundry starch and cornstarch have the same number of calories per gram as do other carbohydrates (four calories per gram). Consequently, starch eating provides calories and may reduce the intake of nutrient-dense foods. In addition, starch may contain contaminants because it is not intended for consumption. Starch eaters' diets are generally inferior to the diets of pregnant women who don't consume starch, and their infants are more likely to be born in poor health.[30]

Plumbism The consumption of lead-containing paint chips poses a major threat to the health of children in the United States and many other countries (Illustration 11.6). Many older homes and buildings, especially those found in substandard housing areas, are covered with lead-based paint and its dried-up flakes. Children may develop lead poisoning if they eat the sweet-tasting paint flakes or inhale lead from contaminated dust and soil near the buildings. An estimated one million young children in the United States have elevated blood lead levels.[37]

High levels of exposure to lead can cause profound mental retardation and death in young children. Low levels of exposure can lead to hearing problems, growth retardation, reduced intelligence, and poor classroom performance. Children with lead poisoning are more likely to fail or drop out of school than children not exposed to lead in their environment.[37]

Eating disorders affect the health and well-being of over a million people in the United States. Although there are treatment strategies, such as counseling and the removal of lead-based paints from old houses and apartments, the solution to eating disorders lies in their prevention. With the exception of certain types of pica discussed here, the most effective way to prevent eating disorders may be to adjust our expectations and cultural norms to reflect reality.

WWW links

www.laureate.com/aboutned.html
The National Eating Disorder Organization's web site provides information to people with eating disorders and their families.

www.edreferral.com
The Eating Disorders Referral Information Center provides information and treatment resources for all forms of eating disorders.

www.anred.com
Anorexia Nervosa and Related Eating Disorders, Inc., is a nonprofit organization providing information on recovery and prevention of a wide variety of eating disorders.

www.eatright.org/anorexiainter.html
The American Dietetic Association's location for information about nutrition intervention in the treatment of eating disorders.

www.ams.queeensu.ca/anab
This is the educational and services site for the Anorexia and Bulimia Association in Canada.

www.healthfinder.gov
A gateway for reliable consumer information about eating disorders.

www.win@info.niddk.nih.gov
The Weight Control Information Network's address at the National Institute of Health.

Nutrition UP CLOSE

Eating Attitudes Test

FOCAL POINT:
Discover if your eating attitudes and behaviors are within a normal range.

Date ________ Age ________ Gender ________

Height ________ Present weight ________ How long at present weight? ________

Highest past weight ________ How long ago? ________

Lowest past weight ________ How long ago? ________

Answer the following questions using these responses:

A = always **S** = sometimes

U = usually **R** = rarely

O = often **N** = never

_____ **1.** I am terrified of being overweight.

_____ **2.** I avoid eating when I am hungry.

_____ **3.** I find myself preoccupied with food.

_____ **4.** I have gone on eating binges where I feel that I may not be able to stop.

_____ **5.** I cut my food into very small pieces.

_____ **6.** I am aware of the calorie content of the foods I eat.

_____ **7.** I particularly avoid foods with a high-carbohydrate content.

_____ **8.** I feel that others would prefer that I ate more.

_____ **9.** I vomit after I have eaten.

_____ **10.** I feel extremely guilty after eating.

_____ **11.** I am preoccupied with a desire to be thinner.

_____ **12.** I think about burning up calories when I exercise.

_____ **13.** Other people think I am too thin.

_____ **14.** I am preoccupied with the thought of having fat on my body.

_____ **15.** I take longer than other people to eat my meals.

_____ **16.** I avoid foods with sugar in them.

_____ **17.** I eat diet foods.

_____ **18.** I feel that food controls my life.

_____ **19.** I display self-control around food.

_____ **20.** I feel that others pressure me to eat.

_____ **21.** I give too much time and thought to food.

_____ **22.** I feel uncomfortable after eating sweets.

_____ **23.** I engage in dieting behavior.

_____ **24.** I like my stomach to be empty.

_____ **25.** I enjoy trying new rich foods.

_____ **26.** I have the impulse to vomit after meals.

FEEDBACK (including scoring) can be found at the end of Unit 11.

Source: McSherry JA. Progress in the diagnosis of anorexia nervosa. Journal of the Royal Society of Health 1986;106:8–9. (Eating Attitudes Test developed by Dr. Paul Garfinkel.)

REFERENCES

1. Lake AJ et al. Effect of Western culture on women's attitudes to eating and preceptions of body shape. Int J Eat Disord 2000;27:83–9.
2. American Psychiatric Association. Diagnostic and statistical manual of mental disorders: DMS-IV, 4th ed, Text Revision. Washington, DC:2000.
3. Mazess RB, Barden HS, Ohlrich ES. Skeletal and body-composition effects of anorexia nervosa. Am J Clin Nutr 1990;52:438–41.
4. Mehler PS. Eating disorders: 1: Anorexia nervosa. Hosp Prac 1996;Jan 15:110–17.
5. Striegel-Moore RH et al. Psychiatric comorbidity of eating disorders in men: a national study of hospitalized veterans. Int J Eat Disord 1999;25:399–404.
6. Grinspoon S et al. Prevalence and predictive factors for regional osteopenia in women with anorexia nervosa. Ann Intern Med 2000;133:790–4.
7. Larkin M. Young female athletes at risk, say US experts. Lancet 2000;356:1007.
8. Lucas AR et al. Bone fracture risk in women with anorexia nervosa. Mayo Clinic Proc 1999;Oct 14.
9. Omizo SA, Oda EA. Anorexia nervosa: psychological considerations for nutrition counseling. J Am Diet Assoc 1988;88:49–51.
10. Patel DR et al. Eating disorders. Indian J Pediatr 1998;65:487–94.
11. Feldman W, Feldman E, Goodman JT. Culture versus biology: children's attitudes toward thinness and fatness. Pediatrics 1988;81:190–4.
12. Worobey J, Schoenfeld D. Eating disordered behavior in dietetics students and students in other majors. J Am Diet Assoc 1999;99:1100–2.
13. Edell D. Beware of personal trainers with their looks. IDEA Health and Fitness Source. www.healthcentral.com/drdean/DeanFullTextTopics.cfm?ID=41044&src=n2, Sept 2000.
14. Robinson PH. Recognition and treatment of eating disorders in primary and secondary care. Alimentary Pharmacol & Thera 2000;14:367–77.
15. Herzog DB et al. Recovery and relapse in anorexia and bulimia nervosa: a 7.5 year follow-up study. J Am Acad Child & Adol Psychiatr 1999;38:829–37.
16. Herzog DB et al. Mortality in eating disorders: a descriptive study. Int J Eat Disord 2000;28:20–6.
17. Treating eating disorders. Harvard Women's Health Watch 1996 May:4–5.
18. When eating goes awry: an update on eating disorders. Food Insight 1997 Jan/Feb:35; and Needleman HL, Schell A, Bellinger D, Leviton A, Allred EN. The long-term effects of exposure to low doses of lead in childhood. An 11-year follow-up report. N Engl J Med 1990;322:83–88.
19. Faris PL et al. Effect of decreasing afferent vagal activity with ondansetron on symptoms of bulimia nervosa: A randomized, double-blind trial. Lancet 2000;355:792–70.
20. Tolstoi LG. The role of pharmacotherapy in anorexia nervosa and bulimia. J Am Diet Assoc 1989;89:1640–6.
21. Herzog DB, Copeland PM. Bulimia nervosa—psyche and satiety (editorial). N Engl J Med 1988;319:716–8.
22. Eating disorders III. Disease definition, epidemiology, and natural history. www.mentalhealth.com, 2000 Dec.
23. Hohlstein LA. Eating disorders program. American Dietetic Association annual meeting, Boston, 1997 Oct 27.
24. Binge eating disorders. National Institute of Diabetes and Digestive and Kidney Diseases, 1998 Feb 9: http:www.info. niddk.nih.gov.
25. Basdevant A et al. Prevalence of binge eating disorder in different populations of French women. Int J Eat Disord 1999;18:309–15.
26. Fairburn CG et al. Distinctions between binge eating disorder and bulimia nervosa. Arch Gen Psychiatry 2000;57:659–65.
27. Moore DC. Body image and eating behavior in adolescent girls. Am J Dis Child 1988;142:1114–8.
28. Zuckerman DM, Colby A, Ware NC, Layerson JS. The prevalence of bulimia among college students. Am J Public Health 1986;76:1135–7.
29. Cooper M. Pica: a survey of the historical literature as well as reports from the fields of veterinary medicine and anthropology. Springfield (IL): Charles C. Thomas; 1957.
30. Edwards CH, McDonald S, Mitchell JR, et al. Clay- and cornstarch-eating women. J Am Diet Assoc 1959;35:810–15.
31. Craign FW. Observations on cachexia Africana or dirt-eating. Am J Med Sci 1935;17:365.
32. Johns T, Duquette M. Detoxification and mineral supplementation as functions of geophagy. Am J Clin Nutr 1991;53: 448–56.
33. Korman SH. Pica as a presenting symptom in childhood celiac disease. Am J Clin Nutr 1990;51:139.
34. Reynolds RD, Binder HJ, Miller MB, Chang WWY, Horan S. Pagophagia and iron deficiency anemia. Ann Intern Med 1968;69:435–40.
35. Coltman CA Jr. Pagophagia and iron lack. JAMA 1969;207:513–16.
36. Rainville AJ. Pica practices of pregnant women are associated with lower maternal hemoglobin level at delivery. J Am Diet Assoc 1998;98:293–6.
37. Child health USA. Washington, DC: Maternal and Child Health Bureau, U.S. PHS; 2000.

Nutrition UP CLOSE

Eating Attitudes Test

Feedback for Unit 11

Never = **3**
Rarely = **2**
Sometimes = **1**
Always, usually, and often = **0**

A total score under 20 points may indicate abnormal eating behavior. If you think you have an eating disorder, it is best to find out for sure. Careful evaluation by a qualified health professional is necessary to exclude any possible underlying medical reasons for your symptoms. Contacting a physician, nurse practitioner, dietitian, or the student health center is an important first step. You may wish to show your Eating Attitudes Test to the health professional.

Photo Disc

Unit 12

Useful Facts about Sugars, Starches, and Fiber

Nutrition Scoreboard

	TRUE	FALSE
1. Pasta, bread, and potatoes are good sources of complex carbohydrates.		
2. Ounce for ounce, presweetened breakfast cereals and unsweetened cereals provide about the same number of calories.		
3. A 12-ounce can of soft drink contains about three tablespoons (nine teaspoons) of sugar.		
4. Lettuce, onions, and celery are high in dietary fiber.		
5. Cooking vegetables destroys their fiber content.		

(Answers on next page)

Key Concepts and Facts

- Simple sugars, the "starchy" complex carbohydrates, and dietary fiber are members of the carbohydrate family.
- Ounce for ounce, sugars and the starchy complex carbohydrates supply fewer than half the calories of fat.
- The usual intake of complex carbohydrates in the United States is too low.
- The only established health problem directly related to sugar intake is tooth decay.
- Dietary fiber benefits health in a number of ways.
- Diabetes and hypoglycemia are related to the way the body processes carbohydrates.

Answers to

Nutrition Scoreboard

	TRUE	FALSE
1. If you get this right, you may be in the minority. In one study of college students,[1] only 38% could identify good sources of complex carbohydrates. Rice, crackers, grits, dried beans, corn, peas, tortillas, biscuits, and oatmeal are also good sources of complex carbohydrates.	✓	
2. That's true! Both sweetened and unsweetened cereals consist primarily of carbohydrate. A gram of carbohydrate provides four calories whether the source is sugar or flakes of corn.	✓	
3. That's a lot of sugar!	✓	
4. Although these are healthy food choices, they do not contain very much dietary fiber. (All that goes "crunch" is not a good source of fiber.)		✓
5. Cooking doesn't destroy dietary fiber.		✓

carbohydrates
Chemical substances in foods that consist of a simple sugar molecule or multiples of them in various forms.

simple sugars
Carbohydrates that consist of a glucose, fructose, or galactose molecule; or a combination of glucose and either fructose or galactose:
glucose + fructose = sucrose
glucose + glucose = maltose
glucose + galactose = lactose
High-fructose corn syrup and alcohol sugars are also considered simple sugars. Simple sugars are often referred to as "sugars."

complex carbohydrates
The form of carbohydrate found in starchy vegetables, grains, and dried beans and in many types of dietary fiber. The most common form of starch is made of long chains of interconnected glucose units.

The Carbohydrates

Carbohydrates are the major source of energy for people throughout the world. They are the primary ingredient of staple foods such as pasta, rice, cassava, beans, and bread. By and large, however, carbohydrates are underappreciated and underconsumed in the United States.

The carbohydrate family consists of three types of chemical substances:

1. ***Simple sugars***
2. ***Complex carbohydrates*** ("starch")
3. ***Dietary fiber***

Some food sources of these different types of carbohydrate are shown in Illustration 12.1. With the exception of dietary fiber, members of the carbohydrate family perform one role in common in the body: they provide energy. Simple sugars and starchy complex carbohydrates supply four calories per gram. Dietary fiber cannot be digested by enzymes produced by humans and is not considered a source of energy. Nevertheless, dietary fiber plays a number of important roles in maintaining health.

Carbohydrates have two sets of "chemical relatives" of significance to human nutrition. Both are derived from carbohydrates and have an alcohol group as part of their chemical structure. The first of these, the ***alcohol sugars,*** are presented in this unit. ***Alcohol*** (or, more precisely, ethanol), the other chemical relative of carbohydrates, supplies seven calories per gram. It is covered in a separate unit.

Illustration 12.1
The carbohydrate family. Some food sources of simple sugars (left), and some food sources of starch and dietary fiber (right) are shown.

Simple Sugar Facts

Simple sugars are considered "simple" because they are small molecules that require little or no digestion before they can be used by the body. They come in two types: ***monosaccharides*** and ***disaccharides.*** The monosaccharides consist of one molecule and include glucose ("blood sugar" or "dextrose"), fructose ("fruit sugar"), and galactose—which has no nickname. Disaccharides consist of two molecules. The combination of a glucose molecule and a fructose molecule makes sucrose (or "table sugar"); maltose ("malt sugar") is made from two glucose molecules; and lactose consists of a glucose molecule plus a galactose molecule. Honey, by the way, is a disaccharide. It is composed of glucose and fructose just as sucrose is, but it's a liquid rather than a solid because of the way the two molecules of sugar are chemically linked together. Disaccharides are broken down into their monosaccharide components during digestion; only glucose, fructose, and galactose are absorbed into the bloodstream.

High-fructose corn syrup, a liquid sweetener used in some soft drinks, fruit drinks, and other products, and alcohol sugars are also considered simple sugars.[2] Most of the simple sugars have a distinctively sweet taste.

The only simple sugar the body can use to form energy is glucose. That's not a problem because both fructose and galactose are readily converted by the body to glucose. When the body has more glucose than it needs for energy formation, it converts the excess to fat or ***glycogen,*** the body's storage form of glucose. Glycogen is a type of complex carbohydrate. It consists of chains of glucose units linked together in long strands. Glycogen is only produced by animals and is stored in the liver (Illustration 12.2) and muscles. When the body needs additional glucose, glycogen is broken down, making glucose available for energy formation.

Where's the Sugar? Sugars enter our diets primarily from foods that have added sugar.[3] Although used primarily because of their sweet taste, simple sugars improve the appearance, consistency, and cooking properties of many food products. Consequently, they are the most commonly used food additive.[3] On average, Americans consume 82 grams of added sugar per day, and that equals

➡ **dietary fiber**
Complex carbohydrates and carbohydrate-like substances in plants that cannot be broken down by human digestive enzymes.

➡ **alcohol sugars**
Simple sugars containing an alcohol group in their molecular structure. The most common are xylitol, mannitol, and sorbitol.

➡ **alcohol**
A substance formed by fermentation of carbohydrates.

➡ **monosaccharides**
(mono = one, saccharide = sugar): Simple sugars consisting of one sugar molecule. Glucose, fructose, and galactose are monosaccharides.

➡ **disaccharides**
(di = two, saccharide = sugar): Simple sugars consisting of two molecules of monosaccharides linked together. Sucrose, maltose, and lactose are disaccharides.

➡ **glycogen**
The body's storage form of glucose. Glycogen is stored in the liver and muscles.

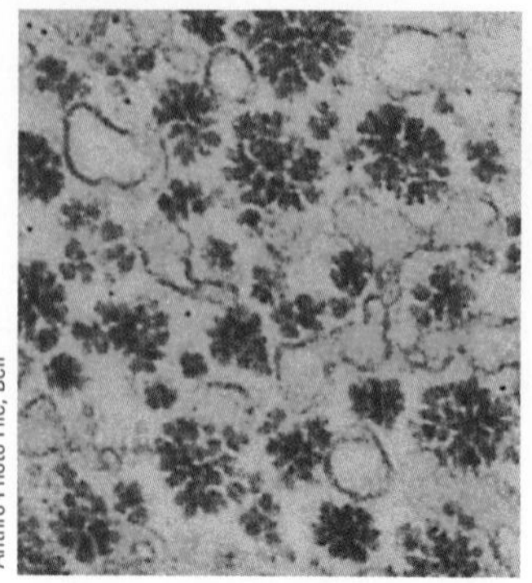

Illustration 12.2
Glycogen in a liver cell. The black "rosettes" are aggregates of glycogen molecules. This cell was photographed under an electron microscope at a magnification of 65,000×.

about 16% of the average caloric intake. The biggest source of sugar in many diets is soft drinks,[4] but sucrose is added to a lot of other foods, too. Sugar is a very common ingredient of breakfast cereals, for example (Table 12.1). Simple sugars are also present in fruits, and small amounts occur in some vegetables (Table 12.2). Unlike many foods with added sugar, both fruits and vegetables provide an array of nutrients. With the exception of milk, which contains the simple sugar lactose, animal products are nearly devoid of simple sugars.

The Name Game Sometimes the labels of food products hide the true identity of simple sugars they contain, confusing consumers who want to know whether a food product contains a good bit of sugar. Table 12.3 lists some of the various names given to simple sugars. Nutrition labels must list the total amount of sugar per serving under the heading "sugars" (Illustration 12.3). In addition, in the ingredient list, all simple sugars contained in the product must be lumped together under "sweeteners" followed by a parenthetical list itemizing each type in order of weight.[2]

What's So Bad about Sugar? Foods to which simple sugars have been added are often not among the top sources of nutrients. Simple sugars are among the few foods that provide only calories. Many foods high in simple sugars, such as cake, sweet rolls, cookies, pie, candy bars, and ice cream, are also high in fat. Furthermore, it is perfectly clear that the frequent consumption of sticky sweets causes tooth decay (Illustration 12.4).[3]

Simple sugars have been blamed for far more problems than tooth decay. Sugar has been accused of depleting the body of nutrients (now how would sugar do that?), causing violent and criminal behavior (this has been disproved by a number of studies),[5] and producing diabetes (also untrue).[6] Nor is there any

Table 12.1
The simple sugar content of a one-ounce serving of some breakfast cereals.

Cereal	Simple Sugars (Grams)	Percentage of Total Calories from Simple Sugars
Apple Jacks	13	52%
Corn Pops	14	47
Frosted Cheerios	13	43
Raisin Bran	19	40
Frosted Flakes	11	40
All Bran	5	29
Frosted MiniWheats	6	24
Bran Flakes	5	22
Life	6	20
Rice Krispies	3	13
Grape-Nuts	3	11
Product 19	3	11
Special K	3	11
Wheaties	3	11
Wheat Chex	2	8
Cornflakes	2	7
Cheerios	1	4
Shredded Wheat	0	0

Table 12.2
The simple sugar content of some common foods.

	Amount	Simple Sugars (Grams)*	Percentage of Total Calories from Simple Sugars
Sweeteners:			
Corn syrup	1 tsp	5	100%
Honey	1 tsp	6	100
Maple syrup	1 tsp	4	100
Table sugar	1 tsp	4	100
Fruits:			
Apple	1 medium	16	91
Peach	1 medium	8	91
Watermelon	1 wedge (4" × 8")	25	87
Orange	1 medium	14	86
Banana	1 medium	21	85
Vegetables:			
Broccoli	½ cup	2	40
Corn	½ cup	3	30
Potato	1 medium	1	4
Beverages:			
Fruit drinks	1 cup	29	100
Soft drinks	12 oz	38	100
Skim milk	1 cup	12	53
Whole milk	1 cup	11	28
Candy:			
Gumdrops	1 oz	25	100
Hard candy	1 oz	28	100
Caramels	1 oz	21	73
Fudge	1 oz	21	73
Milk chocolate	1 oz	16	44

*4 grams sucrose = 1 teaspoon.

evidence that sugar intake causes hyperactivity in children.[7,8] Actually, high sugar intakes may decrease children's activity level and tend to make adults sleepy.[7,9]

Advice on Sugar Intake What's the bottom line on eating sugary foods? Enjoy them in limited amounts. Brush your teeth after eating sticky sweets. And don't let sweet, high-fat foods like candy bars, pastries, and ice cream replace nutrient-dense foods such as vegetables, fruits, and whole-grain products in your diet!

The Alcohol Sugars—What Are They?

Nonalcoholic in the beverage sense, the alcohol sugars are like simple sugars except that they include a chemical component of alcohol. Like simple sugars, the alcohol sugars have a sweet taste. Xylitol is by far the sweetest—it's much sweeter than sucrose and the other two common alcohol sugars, mannitol and sorbitol.

Alcohol sugars are found naturally in very small amounts in some fruits. Their presence in foods is most often due to their use as sweetening agents in

Table 12.3
Which of the following are names for simple sugars? Check your answer below the table.*

- Brown sugar
- Confectioner's sugar
- Dextrose
- Fructose
- Glucose
- High-fructose corn syrup
- Honey
- Invert syrup
- Lactose
- Levulose
- Maltodextrin
- Maltose
- Mannitol
- Maple syrup
- Molasses
- Powdered sugar
- Raw sugar
- Sorbitol
- Sorghum
- Sorghum syrup
- Sucrose
- Sugar
- Turbinado sugar
- Xylitol

*All of them. (printed upside down)

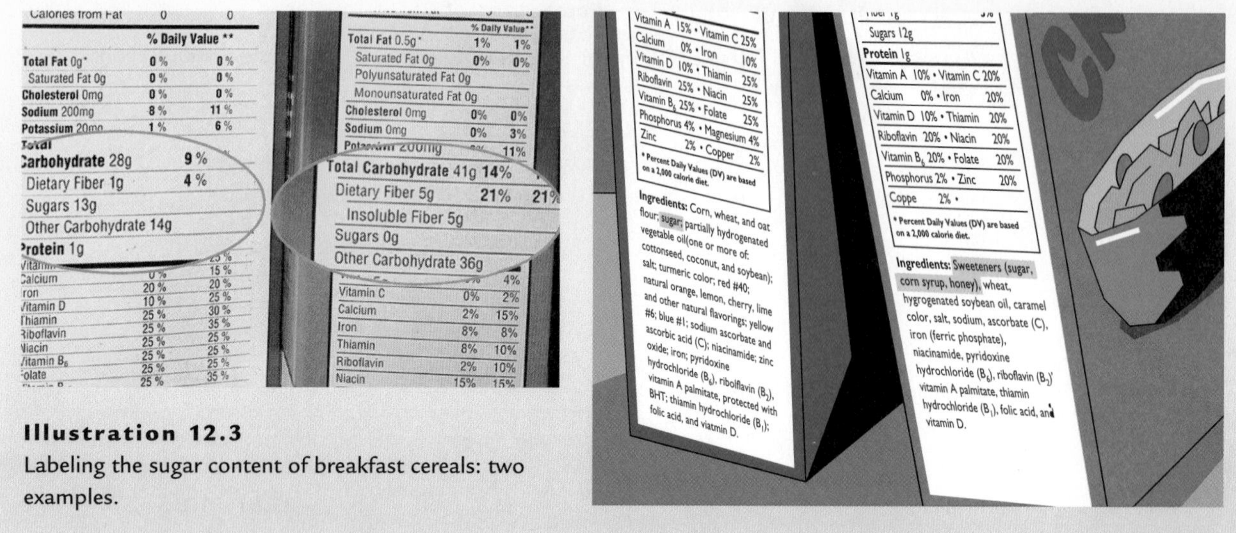

Illustration 12.3
Labeling the sugar content of breakfast cereals: two examples.

gums and candy (Illustration 12.5). Xylitol, mannitol, and sorbitol do not promote tooth decay as the simple sugars do.[10] The extensive use of alcohol sugars in food is limited by their tendency to cause diarrhea if consumed in large amounts.[11]

Artificial Sweetener Facts

Unwanted calories in simple sugars, the connection of sucrose with tooth decay, the need for a sugar substitute for people with diabetes, and sugar shortages such as occurred during the two world wars have all provided incentives for developing sugar substitutes. Four artificial sweeteners are currently on the market in the United States, and it appears a fifth one will join the list soon. Many more artificial sweeteners are being developed.

Artificial sweeteners, also known as "intense sweeteners" and "alternative sweeteners," are contained in a wide assortment of foods (Illustration 12.6). None of these sweeteners are signficant sources of calories, and they do not promote tooth decay or raise blood glucose levels. Artificial sweeteners do not appear to be effective for weight loss as their use is not related to reduced caloric intake.[12]

Illustration 12.4

Source: © 2001 H.L. Schwadron. Reprinted with permission.

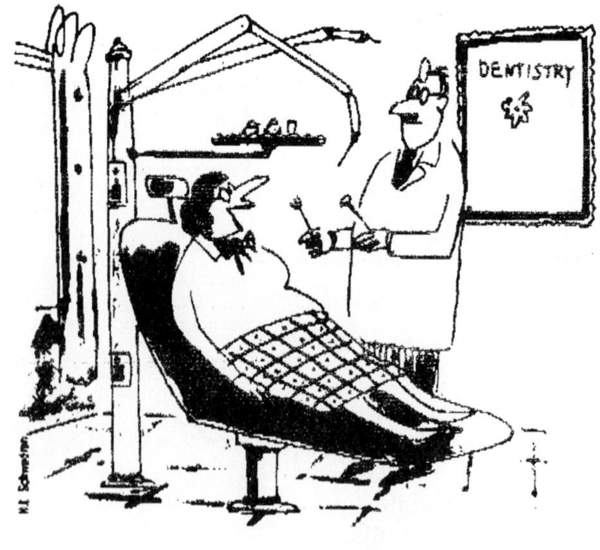

DENTISTRY

"Just pull my sweet tooth."

Saccharin Saccharin was the first artificial sweetener developed. Did you know that it was discovered in a laboratory in the late 1800s? That's right—saccharin is over a hundred years old.[3] The availability of this supersweet artificial sweetener helped relieve the sugar shortages that occurred during World Wars I and II.

In 1977 saccharin was taken off the market after very high doses were found to cause cancer in laboratory animals. At that time, however, saccharin was the only no-calorie, artificial sweetener available, and its removal sparked a public outcry. After many people complained to Congress, saccharin was returned to the market by congressional mandate. Products containing saccharin, however, had to be labeled with a warning that saccharin may be unsafe. Saccharin was deemed safe in 2000 after scientists concluded that there was

Illustration 12.5
Examples of products sweetened with xylitol, mannitol, and sorbitol. Alcohol sugars provide four calories per gram.

no clear evidence that it causes cancer in humans.[13] It is expected that the warning label will become a thing of the past.

Aspartame Early in the 1980s, the artificial sweetener aspartame was approved for use in the United States and more than 90 other countries. Known as Nutrasweet, this artificial sweetener is about 200 times as sweet as sugar. Aspartame is used in more than 4000 products worldwide, including soft drinks, whipped toppings, jellies, cereals, puddings, and some medicines.[12] Because high temperatures tend to break down aspartame, it is not used in products exposed to high temperatures.

Aspartame Was Developed by Chemists, Not Nature Aspartame is made from two amino acids. Both are found in nature, but it took chemists to arrange their chemical partnership. Because aspartame is made from amino acids (the building blocks of protein), it supplies four calories per gram. Aspartame is so sweet, however, that very little is needed to sweeten products. Illustration 12.7

Illustration 12.6
Some of the thousands of foods that contain artificial sweeteners.

SACCHARIN
SUCRALOSE
ASPARTAME
ACESULFAME K
FRUCTOSE
SUCROSE
XYLITOL
GLUCOSE
SORBITOL
GALACTOSE
MALTOSE
LACTOSE

Illustration 12.7
A ranking of various types of artificial sweeteners and naturally occurring sugars in order of sweetness.

shows the relative sweetening power of various artificial sweeteners and naturally occurring sugars.

Is Aspartame Safe? Scientists are satisfied that aspartame is a safe food additive when consumed below the limit established by the Food and Drug Administration (FDA).[12] A safe level of aspartame intake is defined as 50 milligrams per kilogram of body weight per day in the United States and as 40 milligrams per kilogram of body weight in Canada.[14] In food terms, the limit in the United States is equivalent to approximately 20 aspartame-sweetened soft drinks or 55 desserts per day (Illustration 12.8). The average intake of aspartame in the United States, Canada, Germany, and Finland, for example, ranges from 2 to 10 milligrams per day, well below the level of intake considered safe.[12,14] A small proportion of individuals, however, report that they are sensitive to aspartame and develop headaches, dizziness, or anxiety when they consume small amounts. Studies have failed to confirm these effects.[15]

Sucralose This intense sweetener is made from sucrose, is safe, is very sweet (600 times sweeter than sucrose), and does not leave a bitter aftertaste. Known as "Splenda" on product labels, it is used in both hot and cold food products, including soft drinks, baked goods, frosting, pudding, and chewing gum.

Acesulfame Potassium The fourth artificial sweetener currently available in the United States is acesulfame potassium, also known as acesulfame K and "Sunette." Approved by the FDA in 1988, acesulfame K is added to at least 4000 foods and is used in food production in about 90 countries. It is 200 times as sweet as sucrose.

Cyclamate Cyclamate was widely used in food products in the United States from the 1940s through most of the 1960s. It was banned in 1969 because large doses of a cyclamate-saccharin mixture caused cancer in laboratory mice. Cyclamates are available in over 50 countries, some of which (including Canada) have banned the use of saccharin!

Illustration 12.8
You would have to consume more than 20 cans of soft drinks sweetened with aspartame (Nutrasweet) a day to exceed the safe limit set for this artificial sweetener.

Table 12.4
The complex carbohydrate content of some common foods.

	Amount	Complex Carbohydrate (Grams)	Percentage of Total Calories from Complex Carbohydrates
Grain and grain products:			
Rice (white), cooked	½ cup	21	83%
Pasta, cooked	½ cup	15	81
Cornflakes	1 cup	11	76
Oatmeal, cooked	½ cup	12	74
Cheerios	1 cup	11	68
Whole wheat bread	1 slice	7	60
Dried beans (cooked):			
Lima beans	½ cup	11	64
White beans	½ cup	13	63
Kidney beans	½ cup	12	59
Vegetables:			
Potato	1 medium	30	85
Corn	½ cup	10	67
Broccoli	½ cup	2	40

Several national groups of experts have recently concluded that cyclamates are safe.[12] It is anticipated that the FDA will approve their use in the United States in the near future.

Complex Carbohydrate Facts

Starches, glycogen, and dietary fiber constitute the complex carbohydrates known as ***polysaccharides.*** Only plant foods such as grains, potatoes, dried beans, and corn that contain starch and dietary fiber are considered dietary sources of complex carbohydrates (Table 12.4). Very little glycogen is available from animal products.

➡ **polysaccharides** (poly = many, saccharide = sugar): Carbohydrates containing many molecules of sugar linked together. Starch, glycogen, and dietary fiber are the three major types of polysaccharides.

Are They Fattening? Starchy foods are caloric bargains (Illustration 12.9, next page). A medium baked potato weighs in at only 122 calories, a half-cup of corn at 85 calories, and a slice of bread at 70 calories. In contrast, a three-ounce lean hamburger patty supplies 239 calories, an ounce of American cheese 105 calories, a quarter-cup of beef gravy 150 calories, and a tablespoon of margarine over 100 calories. So, go ahead. Load up on starchy, complex carbohydrates.

Dietary Fiber Facts

What has no calories; prevents constipation; may lower the risk of heart disease, obesity, and diabetes; and is generally underconsumed by people selecting a Western-type diet?[16] The answer is dietary fiber.

Dietary fiber intake by U.S. children and adults is less than half the amount recommended.[16] People who consume the recommended 25 to 30 grams of dietary fiber select whole-grain breads, high-fiber cereal, and dried

On the Side

Photo Disc

Have you ever had buckwheat pancakes (3 grams of fiber per serving)? Did you know that buckwheat is a member of the rhubarb family—not a grain?

Illustration 12.9
Which has more calories? Check your answers below.*

One medium baked potato (four ounces) or three ounces of lean hamburger?

One slice of bread or a half-cup of low-fat cottage cheese?

One cup of spaghetti noodles or 17 french fries (three ounces)?

*Answers:
Potato = 122 calories;
Lean hamburger = 239 calories.
Bread = 70 calories;
Cottage cheese = 102 calories;
Spaghetti = 197 calories;
French fries = 265 calories.

beans most days and eat at least five servings of vegetables and fruits daily.[16] It is recommended that children consume 5 grams of fiber plus their age, so a five-year-old should consume 10 grams of fiber.[17]

Food sources of fiber are listed in Table 12.5. It doesn't matter whether the fiber foods are mashed, chopped, cooked, or raw. They retain their fiber value through it all. Note, though, that fast foods tend to be poor sources of fiber.

There Are Two Main Types of Dietary Fiber Dietary fibers are components of plants that cannot be broken down by digestive enzymes humans produce. They come in both fibrous and nonfibrous forms, and many foods contain both types of fiber.

➡ **insoluble fiber**
Dietary fiber that does not dissolve in water. It "holds" water and is found in plant cell walls. Some food sources of insoluble fiber are listed in Table 12.7.

➡ **hemorrhoids**
Swelling veins in the anus or rectum.

Insoluble Dietary Fibers—The "Fibrous" Form Fibrous sources of dietary fiber are found primarily in the cell walls of plants. They are called ***insoluble fibers*** because they do not dissolve in water. Insoluble fibers attract water, however, and speed up the elimination of waste products from the intestinal tract. That's why insoluble fibers help prevent ***hemorrhoids*** and constipation (Table 12.6 on p. 12–12).[16]

Table 12.5
Examples of good sources of dietary fiber. The recommended intake for adults is 25 to 30 grams per day.

	Amount	Dietary Fiber (Grams)
Grain and grain products:		
Bran Buds	½ cup	12.0
All Bran	⅓ cup	10.0
Raisin Bran	1 cup	7.0
Granola (homemade)	½ cup	6.0
Bran Flakes	¾ cup	5.0
Oatmeal	1 cup	4.0
Spaghetti noodles	1 cup	4.0
Shredded Wheat	1 biscuit	2.7
Whole wheat bread	1 slice	2.0
Bran (dry; wheat, oat)	2 tbs	2.0
Fruits:		
Avocado	½ medium	7.0
Raspberries	1 cup	5.0
Mango	1 medium	4.0
Pear (with skin)	1 medium	4.0
Orange (no peel)	1 medium	3.0
Apple (with skin)	1 medium	2.4
Peach (with skin)	1 medium	2.3
Strawberries	10 medium	2.1
Banana	6" long	1.8
Vegetables:		
Lima beans	½ cup	5.0
Green peas	½ cup	4.0
Potato (with skin)	1 medium	3.5
Broccoli	½ cup	3.4
Brussels sprouts	½ cup	3.0
Collard greens	½ cup	3.0
Corn	½ cup	3.0
Carrots	½ cup	2.8
Green beans	½ cup	2.7
Cauliflower	½ cup	2.0
Nuts:		
Almonds	¼ cup	4.5
Peanuts	¼ cup	3.3
Peanut butter	2 tbs	2.3
Dried beans (cooked):		
Pinto beans	½ cup	10.0
Black beans (turtle beans)	½ cup	8.0
Black-eyed peas	½ cup	8.0
Kidney or navy beans	½ cup	6.2
Peas	½ cup	5.7
Lentils	½ cup	5.0
Fast foods:		
Big Mac	1	3
French fries	1 regular serving	3
Whopper	1	3
Cheeseburger	1	2
Taco	1	2
Chicken sandwich	2	1
Egg McMuffin	1	1
Fried chicken, drumstick	1	1

Good sources of insoluble fibers are listed in Table 12.7 (p. 12–12); they include whole grains and vegetables such as green beans and green peas. The bran that naturally covers wheat, rice, and some other grains is a rich source of insoluble fiber (Illustration 12.10).

Table 12.6
Health benefits attributed to insoluble (fibrous) and soluble (nonfibrous) dietary fibers.

Type of Dietary Fiber	Health Benefits
Insoluble	Prevents constipation and hemorrhoids
Soluble	Lowers blood cholesterol Delays glucose absorption

soluble fiber
Dietary fiber that dissolves in water and produces a gel. Some food sources of soluble fiber are listed in Table 12.7.

Soluble Dietary Fibers—They Form Gels Dietary fiber that dissolves in water is referred to as ***soluble fiber.*** This type of fiber is found in dried beans, oats, apples, and other foods listed in Table 12.7.

Soluble fibers form a thick, gel-like substance when combined with water in the intestinal tract. They are credited with decreasing blood cholesterol levels by about 5% and slowing the absorption of glucose—a particular benefit to people with diabetes.[16] Of the soluble fibers, only oat and psyllium fiber are known to have laxative effects.[16]

Be Cautious When Adding More Fiber to Your Diet Newcomers to adequate fiber diets often experience diarrhea, bloating, and gas for the first week or so of increased fiber intake. These side effects can be avoided. They occur when too much fiber is added to the diet too quickly. Although humans do not produce enzymes that can break down fiber, bacteria that live in the lower intestines do. The bacteria consume dietary fiber as food and excrete gas and other products as wastes. The gas produced by the bacteria causes the bad feelings that can accompany unusually high fiber intakes.[18] Adding sources of dietary fiber to the diet gradually can prevent these side effects. In addition, dietary fiber can be constipating if consumed with too little fluid. Your fluid intake should increase along with your intake of dietary fiber. You know you've got the right amount of fiber in your diet when stools float and are soft and well formed.

The Health Action on the facing page suggests some food choices that will put fiber into your meals and snacks. Because most food sources contain 2 or 3 grams of dietary fiber per serving, consuming 25 to 30 grams daily may mean choosing seven or more servings of fiber-containing foods. Fiber should be added carefully to children's diets, because the large volume of some high-fiber diets can fill the children up before they consume enough calories.[17]

Illustration 12.10
Diagram of a grain of wheat showing the bran that is a rich source of dietary fiber. The germ contains protein, unsaturated fats, thiamin, niacin, riboflavin, iron, and other nutrients. (The bran and germ are removed in the refining process.) The endosperm primarily contains starch, the storage form of glucose in plants.

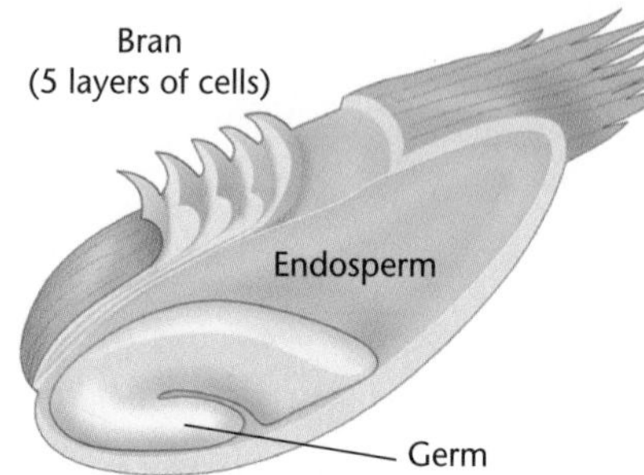

More to Come on Fiber As the science of nutrition advances, dietary fibers are being categorized into smaller groups based on their structure

Table 12.7
Some food sources of insoluble and soluble dietary fiber.

Insoluble Fiber	Soluble Fiber
Brown rice	Apples
Fruit and vegetable skins, peels	Barley
Green beans	Broccoli
Green peas	Carrots
Nuts	Corn
Rice bran	Dried beans
Seeds	Grapefruit
Wheat bran	Oat bran
Whole-grain products	Oats
	Oranges
	Potatoes

HEALTH ACTION

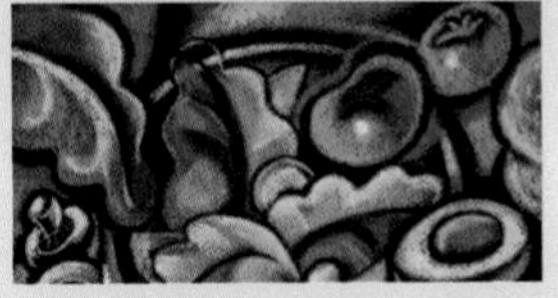

Putting the Fiber into Meals and Snacks

HIGH-FIBER OPTIONS FOR BREAKFAST

Food	Amount	Fiber
Whole-grain toast		2 g per slice
Bran cereal:		
Bran flakes	1 cup	7 g
All Bran	⅓ cup	10 g
Raisin bran	¾ cup	5 g
Oat bran	⅓ cup	5 g
Bran muffin, with fruit:	1 small	3 g
Strawberries	10	2 g
Raspberries	½ cup	3 g
Bananas	1 medium	2 g

LUNCHES THAT INCLUDE FIBER

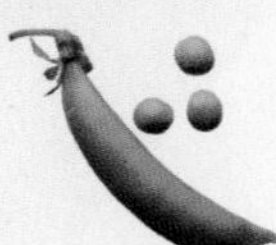

Food	Amount	Fiber
Whole-grain bread		2 g per slice
Baked beans	½ cup	10 g
Carrot	1 medium	2 g
Raisins	¼ cup	2 g
Peas	½ cup	4 g
Peanut butter	2 tablespoons	2 g

FIBER ON THE MENU FOR SUPPER

Food	Amount	Fiber
Brown rice	½ cup	2 g
Potato	1 medium	3 g
Dried cooked beans	½ cup	8 g
Broccoli	½ cup	3 g
Corn	½ cup	3 g
Tomato	1 medium	2 g
Green beans	½ cup	3 g

FIBER-FILLED SNACKS

Food	Amount	Fiber
Peanuts	¼ cup	3 g
Apple	1 medium	2 g
Pear	1 medium	4 g
Orange	1 medium	3 g
Prunes	3	2 g
Sunflower seeds	¼ cup	2 g
Popcorn	2 cups	2 g

*Prunes contain fiber, but their laxative effect is primarily due to a naturally occurring chemical substance that causes an uptake of fluid into the intestines and the contraction of muscles that line the intestines.

Photo Disc

and specific functions in the gut. These functions may not always be distinguished based on whether a fiber is soluble or insoluble.[19] In the years to come, much more will be known about the health benefits of other types of fiber and which foods are the best sources of them.

Carbohydrate-Related Disorders

A variety of health problems result when the body's utilization of carbohydrates goes awry. Two are discussed in this unit: diabetes and hypoglycemia. Lactose intolerance is presented in another unit.

Table 12.8
Characteristics of insulin-dependent and noninsulin-dependent diabetes.

Characteristic	Insulin-Dependent Diabetes (Type 1)	Noninsulin-Dependent Diabetes (Type 2)
Insulin deficiency	Yes	No
Proportion of total cases	10%	90%
Risk factors	Viral infection and/or early allergy to cow's milk that affects the pancreas, long-term use of certain medications, young age	Obesity (especially central-body fat stores), "Westernized" diet and lifestyle, family history, older age
Treatment	Insulin injections, diet, and exercise	Weight loss, diet, exercise, sometimes oral medications or insulin

➡ **diabetes**
A disease characterized by abnormally high blood glucose levels. There are three major types: insulin-dependent (type 1) diabetes mellitus, noninsulin-dependent (type 2) diabetes mellitus, and gestational diabetes.

➡ **noninsulin-dependent diabetes mellitus (type 2 or NIDDM)**
In this type of diabetes, insulin is produced by the body but is utilized poorly.

Diabetes

Diabetes is a complex, chronic disease that affects 21 million Americans, 6 million of whom do not know they have diabetes.[20] One in 14 Americans has diabetes, or will develop it eventually.[21] There are three major types of diabetes: insulin-dependent (type 1), noninsulin-dependent (type 2), and gestational diabetes. Table 12.8 lists the characteristics of the first two types. In all cases of diabetes, the central defect is elevated blood glucose levels caused by the body's inability to maintain blood glucose within a healthy range. Abnormally high levels of blood glucose (defined as fasting serum glucose levels of 126 mg/dL and higher[22]) cause short- and long-term problems. In the short run, poorly controlled diabetes produces blurred vision, susceptibility to infection, delayed wound healing, and extreme hunger and thirst. In the long run, diabetes may contribute to hypertension, blindness, heart disease, kidney failure, stroke, and the loss of limbs due to poor circulation. Many of the side effects of diabetes may be prevented or delayed if blood glucose levels are maintained within the normal range.[16,20]

Noninsulin-Dependent Diabetes (Type 2) Of the three types of diabetes, ***noninsulin-dependent diabetes mellitus*** is by far the most common. Excess body fat, especially around the stomach area (Illustration 12.11), is closely associated with the development of type 2 diabetes.[23] The disorder tends to run in families.[21]

People with type 2 diabetes produce lots of insulin (the hormone that lowers blood glucose levels after meals) for much of their lives but do not utilize it well due to insulin resistance. (Read about insulin resistance in Illustration 12.12.) Some people with type 2 diabetes run out of insulin after many years of producing an excess of it. When this happens, replacement insulin injections are needed.[22]

Managing Type 2 Diabetes with Diet and Exercise Diet and exercise are key components of the management of type 2 diabetes. Some Certified Diabetes Educators/Registered Dietitians promote incorporation of "low glycemic index" sources of carbohydrates into diet plans. The glycemic index value of high carbohydrate foods reflects their ability to raise blood glucose levels. (Illustration 12.13 on p. 12–16 gives interested readers additional information about the glycemic index.) In general, diets developed for people with diabetes emphasize:

- complex carbohydrates including whole-grain breads and cereals, vegetables, fruits, and high-fiber foods,
- limited intake of simple sugars and foods and beverages that contain them,

PhotoEdit, Alan Oddie

Illustration 12.11
Obesity characterized by central-body fat stores is a strong risk factor for type 2 diabetes.

Illustration 12.12

Insulin is a hormone released by the pancreas that increases glucose passage into cells by attaching to, and stimulating receptors on, cell membranes. When insulin is bound to these receptors, enzymes are activated that open cell membrane "doors" to the passage of glucose. With insulin resistance, cells "resist" this process and that lowers the amount of glucose that can be transported into the cells. In response to high levels of blood glucose, more insulin is released by the pancreas. If the capacity of the pancreas to produce increasingly large quantities of insulin needed for glucose absorption into cells is exceeded, blood glucose levels become and may remain high. About 25% of people with insulin resistance develop type 2 diabetes.[22]

It is estimated that about half of the cases of insulin resistance are related primarily to genetic factors, and half to behaviors. Most cases result from an interaction of genetic and behavioral factors. Obesity—especially large stores of central-body fat—high-fat diets, smoking, physical inactivity, and small size at birth have been related to insulin resistance. The condition is also related to the development of metabolic syndrome (or "Syndrome X") characterized by high blood glucose, triglycerides, and LDL cholesterol levels; high blood pressure; and increased risk of hypertension and heart disease.[24]

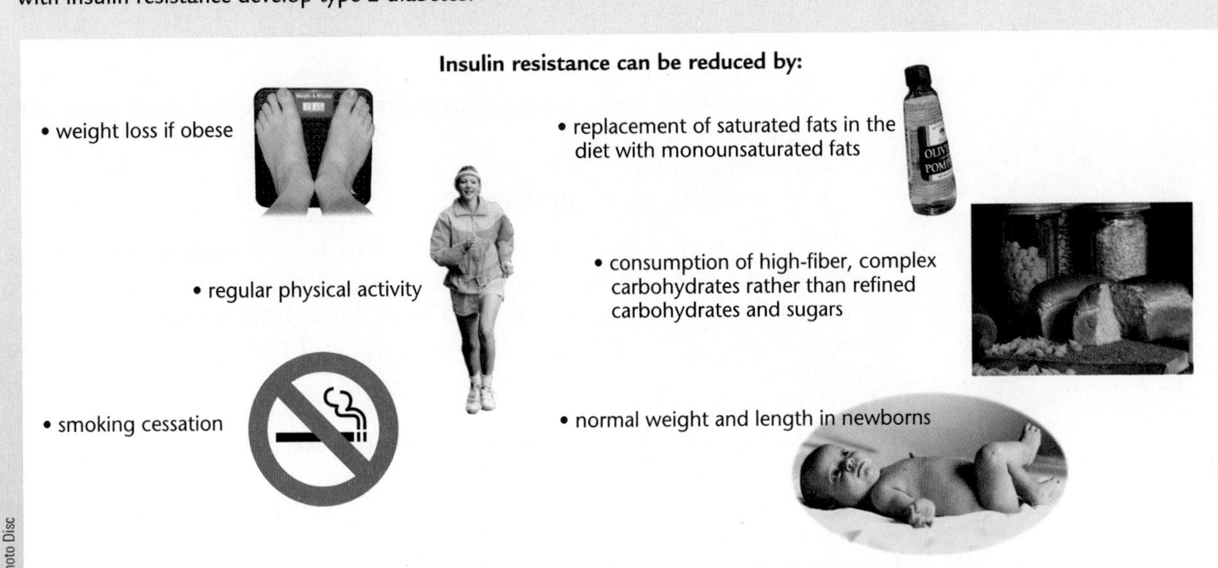

Photo Disc

Illustration 12.13
The glycemic index.

The glycemic index is a method of classifying foods by their blood glucose raising potential. It is determined by comparing levels of blood glucose after equal portions of a high carbohydrate food and a standard carbohydrate, such as white bread or glucose, are given. Food sources of carbohydrate are ranked by the level of blood glucose produced versus that from white bread or glucose. Examples of foods with high, moderate, and low glycemic index are given below.[25–27] (High glycemic index foods raise blood glucose levels more than low glycemic index foods.)

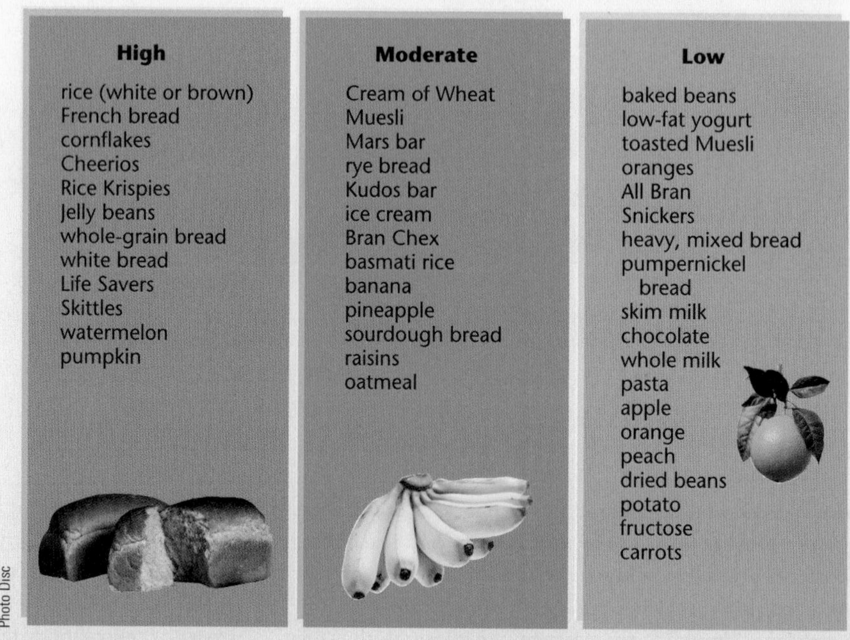

High	Moderate	Low
rice (white or brown)	Cream of Wheat	baked beans
French bread	Muesli	low-fat yogurt
cornflakes	Mars bar	toasted Muesli
Cheerios	rye bread	oranges
Rice Krispies	Kudos bar	All Bran
Jelly beans	ice cream	Snickers
whole-grain bread	Bran Chex	heavy, mixed bread
white bread	basmati rice	pumpernickel bread
Life Savers	banana	skim milk
Skittles	pineapple	chocolate
watermelon	sourdough bread	whole milk
pumpkin	raisins	pasta
	oatmeal	apple
		orange
		peach
		dried beans
		potato
		fructose
		carrots

Photo Disc

Individuals with diabetes or insulin resistance may benefit from consuming low glycemic index foods because they raise blood glucose a lesser extent than other carbohydrate foods, and therefore lower the need for insulin.[25] Whether the use of low glycemic foods should be encouraged for all people with high blood glucose or insulin levels is controversial. Selecting low glycemic index foods is a bit complicated to implement, and the glycemic value of foods varies somewhat depending on what else is eaten along with the carbohydrate foods.[28]

- low glycemic index foods, or carbohydrate foods that don't raise glucose levels greatly,
- unsaturated fats,
- three regular meals and snacks daily.

Relatively modest levels of weight loss (5 to 10% of body weight) in overweigh and obese people with type 2 diabetes can significantly improve blood glucose levels.[29] Exercising four times a week for at least 30 minutes is recommended because it improves insulin resistance and contributes to weight loss maintenance.[22]

The Incidence of Type 2 Diabetes Is Increasing The incidence of type 2 diabetes is increasing in the United States and other Westernized countries.[20] It appears that the Western diet and lifestyle are conducive to its development.[3,30] Further evidence comes from studies of populations who shift from a diet of basic foods and a physically active lifestyle to a "Western" diet, which is high in fat and protein and low in complex carbohydrates, and an inactive lifestyle. Part of the reason rates of type 2 diabetes increase in populations adopting a Western lifestyle is because that lifestyle promotes obesity. Increased rates of type 2 diabetes in countries such as Japan, Libya, and Bangladesh; in aboriginal children and adults in Canada and Australia; and in American Indians, are directly related to epidemics of obesity.[20,31]

Insulin-Dependent Diabetes Mellitus (Type 1) Diabetes that results from a deficiency of insulin accounts for about 10% of all cases. It usually occurs before the age of 40 and begins abruptly.[3] People with ***insulin-dependent diabetes (Type 1)*** require daily insulin shots, regular blood glucose self-monitoring, and an individualized and flexible diet and exercise program. Weight loss, even as little as 10 to 15 pounds, benefits blood glucose control in overweight and obese individuals with type 1 diabetes.[22]

➡ **insulin-dependent diabetes mellitus (type 1 or IDDM)** This type of diabetes is due to the destruction of insulin-producing cells in the pancreas, which leads to a lack of insulin.

Insulin deficiency appears to stem from a viral infection (such as mumps, rubella, or the flu), or an allergic reaction in genetically susceptible people that destroys the portion of the pancreas that produces insulin.[32] Medications used to treat high blood pressure, arthritis, and other disorders may also contribute to the development of type 1 diabetes.[3] Breastfeeding for the first four months of life or so may confer an element of protection against type 1 diabetes.[33]

Environmental factors appear to be more important than genetic background in the development of this type of diabetes. The incidence of type 1 diabetes varies 36-fold among countries (Illustration 12.14), suggesting that environmental factors play a key role in its development.[34]

Gestational Diabetes Approximately 3 to 6% of women develop ***gestational diabetes*** during pregnancy, but the incidence varies a good deal based on age, body weight, and ethnicity. Native Americans, African Americans, women over the age of 35 years, obese women, and those with habitually low levels of physical activity are at higher risk than other women.[36] Infants born to women with poorly controlled diabetes may be very fat at birth and have blood glucose control problems after delivery. They are at greater risk for developing diabetes later in life, and 6 to 20% will have a physical abnormality that may threaten survival or a high quality of life.[37]

➡ **gestational diabetes** Diabetes first discovered during pregnancy.

Most women with gestational diabetes are insulin resistant and can control their blood glucose levels with an individualized diet and exercise plan. Some women require daily insulin injections for blood glucose control.[22]

Gestational diabetes disappears after delivery, but noninsulin-dependent diabetes may appear later in life. Exercise, maintenance of normal weight, and consumption of a healthy diet reduce the risk that diabetes will return.[38]

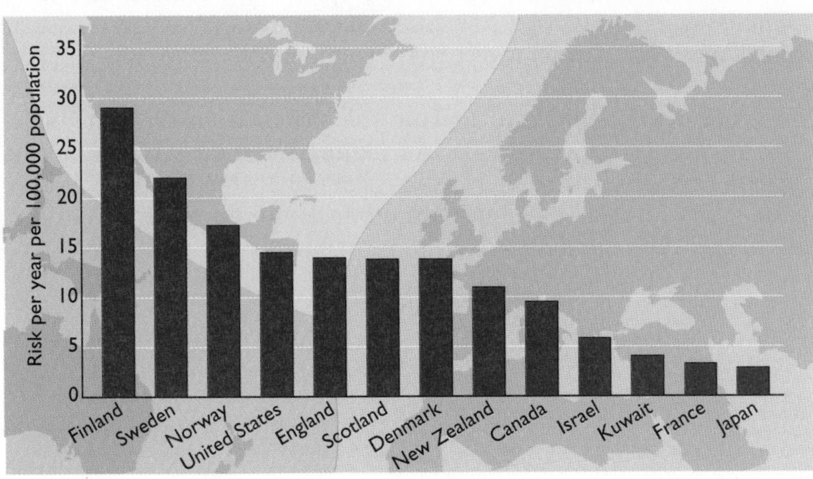

Illustration 12.14
Incidence of type 1 diabetes mellitus in various countries.[34,35]

Can People with Diabetes Eat Sugar? Small amounts of simple sugars consumed with meals or in a sweetened dessert do not raise blood glucose levels unduly. Therefore, people with diabetes do not have to exclude simple sugars from their diets altogether, but they should avoid the intake of concentrated sources of simple sugars between meals.[22]

Hypoglycemia: The Low Blood Sugar Blues?

➡ **hypoglycemia**
A disorder resulting from abnormally low blood glucose levels. Symptoms of hypoglycemia include irritability, nervousness, weakness, sweating, and hunger. These symptoms are relieved by eating foods that provide calories.

Hypoglycemia is due to abnormally low blood glucose levels. It is thought to be a rare disorder because it is not often diagnosed. The diagnosis is tricky—blood tests for glucose should be conducted when the symptoms of hypoglycemia are present rather than during a scheduled appointment.[39] The true incidence of hypoglycemia is not known.

Hypoglycemia is most often caused by an excessive availability of insulin in the blood. The oversupply of insulin may be caused by certain tumors that secrete insulin, by other health problems, by high alcohol intake on an empty stomach, or, in people with diabetes, by an insulin dose that is too high.[40] Hypoglycemia also occurs during prolonged starvation, but blood glucose levels become very low only when starvation threatens life.

Symptoms of hypoglycemia include weakness, sweating, nervousness, confusion, and irritability.[41] They may appear before blood glucose levels are low enough to be clinically considered abnormal. Apparently, people vary considerably in their response to low-normal glucose levels.[42] Symptoms of hypoglycemia disappear within 5 to 15 minutes after food is consumed.[39]

The notion that high sugar intakes cause hypoglycemia has not been verified by research.[43] People who suspect that they have this disorder should consult a physician and have their blood glucose level checked at a time when the symptoms normally occur.[41]

Standard diet therapy for hypoglycemia calls for people to consume five to six small meals that include a variety of complex carbohydrates and high-protein foods and to avoid alcohol and snacks high in simple sugars. High-sugar foods may cause a surge in insulin and a subsequent drop in glucose level in susceptible individuals.[41]

The Complex Carbohydrate Advantage

Recommendations for diets that promote health indicate that 55 to 60% of the total calories in a person's diet should come from carbohydrates, particularly complex sources such as whole-grain products, vegetables, fruits, and pasta. Foods rich in complex carbohydrates are often good sources of dietary fiber, vitamins, and minerals and are not high in calories. Most people in the United States get too few of their calories from complex carbohydrates and too many from simple sugars and fat. The low intake of complex carbohydrates also means that most Americans fail to ingest the recommended 25 to 30 grams of fiber daily.

There are many benefits to making the switch from high-fat, high-sugar foods to the complex carbohydrates. Consuming 5 or more servings of fruits and vegetables a day along with 6 to 11 servings of foods from the bread, cereal, rice, and pasta group (including whole-grain products) as recommended in the Food Guide Pyramid can help make the benefits happen.

Nutrition UP CLOSE

Does Your Fiber Intake Measure Up?

FOCAL POINT: Approximate the amount of fiber your diet contains.

Are you meeting your fiber quota, or do you consume the typical low-fiber American diet? To determine if your fiber intake is adequate, award yourself the allotted number of points for **each serving of the following foods that you eat in a typical day.** For example, if you normally eat 1 slice of whole-grain bread each day, give yourself 2 points. If you eat 2 slices daily, give yourself 4 points. After tallying your score, refer to the Feedback section for the results.

High-fiber food choices	
Fruits	**2 points for each serving**
1 whole fruit (e.g., apple, banana)	________
½ cup cooked fruit	________
¼ cup dried fruit	________
Grains:	
½ cup cooked brown rice	________
1 whole-grain slice of bread, roll, muffin, or tortilla	________
½ cup hot whole-grain cereal (e.g., oatmeal)	________
¾ cup cold whole-grain cereal (e.g., Cheerios)	________
2 cups popcorn	________
Nuts and seeds:	
¼ cup seeds (e.g., sunflower)	________
2 tablespoons peanut butter	________
Vegetables:	**3 points for each serving**
1 whole vegetable (e.g., potato)	________
½ cup cooked vegetable (e.g., green beans)	________
Bran cereals:	**7 points for each serving**
½ cup cooked oat bran cereal	________
1 cup cold bran cereal	________
Legumes:	**8 points for each serving**
½ cup cooked beans (e.g., baked beans, pinto beans)	________
	Total score ________

Special note: You can also calculate your fiber intake using the Diet Analysis Plus software. Input your food intake for one day. Then go to the Analyses/Reports section to view the total number of grams of fiber in your diet on that day.

FEEDBACK (including scoring) can be found at the end of Unit 12.

www.webmd.com
Provides a diabetes resource site for consumers.

www.diabetes.com
The "Healthy Diet and Exercise" page provides useful information for both newly diagnosed and experienced people with diabetes.

www.nal.usda.gov/fnic/foodcomp
Questions about the carbohydrate composition of food are answered through this site.

www.healthfinder.gov and www.mayohealth.
Check out advances in nutrition and diabetes.

www.diabetes.org (or 1-800-DIABETES)
The American Diabetes Association web site offers updates, practical advice, educational materials, and other resources.

www.diabetes.ca
The Canadian Diabetes Association web site provides a user-friendly format for delivering advice and information about diabetes.

www.ificinfo.health.org/
Provides information about artificial sweeteners.

REFERENCES

1. Melby CL, Femea PL, Sciacca JP. Reported dietary and exercise behaviors, beliefs and knowledge among university undergraduates. Nutr Res 1986;6:799–808.
2. Food and Drug Administration (Department of Health and Human Services). Nutrition labeling. Federal Register 1991 Nov 27.
3. Committee on Diet and Health (Food and Nutrition Board; National Research Council). Diet and Health. Implications for reducing chronic disease risk. Washington, DC: National Academy Press; 1989.
4. Guthrie JF, Morton JF. Food sources of added sweeteners in the diets of Americans. J Am Diet Assoc 2000;100:43–48, 51.
5. Bachorowski J-A, Newman JP, Nichols SL, Gans DA, Harper AE, Taylor SL. Sucrose and delinquency: behavioral assessment. Pediatrics 1990;86:244–53; and Gans DA, Harper AE, Bachorowski J-A, Newman JP, Shrago ES, Taylor SL. Sucrose and delinquency: oral sucrose tolerance test and nutritional assessment. Pediatrics 1990;86:254–62.
6. Colditz GA, et al. Diet and risk of clinical diabetes in women. Am J Clin Nutr 1992;55:1018–23.
7. Wolraich M, et al. The effect of sugar on behavior or cognition in children: a meta-analysis. JAMA 1995;274:1617–21.
8. Wolraich ML, Lindgren SD, Stumbo PJ, et al. Effects of diets high in sucrose or aspartame on the behavior and cognitive performance of children. N Engl J Med 1994;330:301–7; and Kinsbourne M. Sugar and the hyperactive child. N Engl J Med 1994;330:355–6.
9. Liberman HR. Sugars and behavior. Clin Nutr 1985;4:195–9; and Pivonka EEA, Grunewald KK. Aspartame- or sugar-sweetened beverages: effects on mood in young women. J Am Diet Assoc 1990;90:250–4.
10. Oberrieder HK, Fryer EB. College students' knowledge and consumption of sorbitol. J Am Diet Assoc 1991;91:715–7.
11. Badiga MS, Jain NK, Casanova C, Pitchumoni CS. Diarrhea in diabetics: the role of sorbitol. J Am Coll Nutr 1990;9:578–82.
12. Verni C. Artificial sweeteners: review and update for practitioner. SCAN'S Pulse, Summer 1999.
13. Saccharin deemed safe. Community Nutrition Institute, 2000; June 2: 8.
14. Butchko HH, Kotsonis FN. Acceptable daily intake vs. actual intake: the aspartame example. J Am Coll Nutr 1991;10:258–66.
15. Spiers PA, et al. Aspartame: neuropsychologic and neurophysiologic evaluation and chronic effects. Am J Clin Nutr 1998;68:531–7.
16. Marlett JA, Slavin JL. Position of the American Dietetic Association: health implications of dietary fiber. J Am Diet Assoc 1997;97:1157–9; and Burton-Freeman B. Dietary fiber and energy regulation. J Nutr 2000;130:272S–5S.
17. Williams CL. A summary of conference recommendations on dietary fiber in childhood. Pediatrics 1995;96:1023–8.
18. Topping DL. Soluble fiber polysaccharides: effects on plasma cholesterol and colonic fermentation. Nutr Rev 1991;49:195–206.
19. Schneeman BO. Building scientific consensus: the importance of dietary fiber. Am J Clin Nutr 1999;69:1.
20. Kahn R. Type 2 diabetes I children and adolescents. Pediatrics 2000;105:671–80.
21. Berdanier CD. Diabetes mellitus: a genetic disease. Nutr Today 1999;34:89–98.

22. Intensive diabetes self-management: 2000 update. American Association of Clinical Endocrinologists, Endocrine Practice 2000;6:42–67.
23. Steiner G. From an excess of fat, diabetics die. JAMA 1989;262:398; and Ohlson LO, Larsson B, Svardsudd K, et al. The influence of body fat distribution on the incidence of diabetes mellitus: 13.5 years of follow-up of the participants in the study of men born in 1913. Diabetes 1985;34:1055–8.
24. The ins and outs of insulin resistance. Food Insight 2000; May/June: 1,4–5.
25. Brand-Miller J, Foster-Powell K. Diets with low glycemic index: from theory to practice. Nutr Today 1999;34:64–72.
26. Soh NL, Brand-Miller J. The glycemic index of potatoes. Euro J Clin Nutr 1999;53:249–54.
27. Nishimune T, et al. Glycemic index and fiber content of some foods. Am J Clin Nutr 1991;54:414–19.
28. Franz MJ. In defense of the American Diabetes Association's recommendations on the glycemic index. Nutr Today 1999;34:78–81.
29. Heysmfield S, et al. Effects of weight loss with Orlistat on glucose tolerance and progression to type 2 diabetes in obese adults. Arch Inter Med 2000;160: 1321–6.
30. Salmeron J, et al. Dietary fiber, glycemic load, and risk of non-insulin dependent diabetes mellitus in women. JAMA 1997;277:472–7.
31. Young R, et al. Childhood obesity in a population at high risk for type 2 diabetes. J Pediatr 2000;136:365–69.
32. Winter WE, Obata M. Heritable origins of type 1 (insulin-dependent) diabetes mellitus: immunogenetic update. Growth, Genetics and Hormones 1991;7:1–6.;;
33. Virtanen SM, Rasanen L, Aro A, et al. Infant feeding in Finnish children over the age of 4 years with newly diagnosed IDDM. Diabetes Care 1991;14:415–7.
34. LaPorte RE, Taijima N, Akerblom HR, et al. Geographic differences in the risk of insulin-dependent diabetes mellitus. Updated in 2000. Diabetes Care 1985;8(suppl):101–7.
35. Muntoni S, et al. Nutritional factors and worldwide incidence of childhood type 1 diabetes. Am J Clin Nutr 2000;71:1525–9.
36. Solomon CG, et al. A prospective study of pregravid determinants of gestational diabetes mellitus. JAMA 1997;278: 1078–83.
37. Carr SR. Effect of maternal hyperglycemia on fetal development. Annual meeting of the American Dietetic Association, Boston, 1997 Oct 28.
38. Kjos SL, Buchanan TA. Gestational diabetes mellitus. N Engl J Med 1999;341:1749–56.
39. Hypoglycemia frequently "diagnosed" but rarely true. Envir Nutr 1998;Feb:7.
40. Amiel S. Reversal of unawareness of hypoglycemia. N Engl J Med 1993; 329:876–7.
41. Service FJ. Hypoglycemia and the postprandial syndrome. N Engl J Med 1989;321:1472–4.
42. Palard J, Havrankova J, LePage R, Matte R, Belander R, D'Amour P, Ste.-Marie L-G. Blood glucose measurements during symptomatic episodes in patients with suspected postprandial hypoglycemia. N Engl J Med 1989;321:1421.
43. Ryan CM, Atchison J, Puczynski S, Puczynski M, Arslanian S, Becker D. Mild hypoglycemia associated with deterioration of mental efficiency in children with insulin-dependent diabetes mellitus. J Pediatr 1990;117:32–38; Nelson RL. Hypoglycemia: fact or fiction? Mayo Clin Proc 1985;60:844–50.

Nutrition UP CLOSE

Does Your Fiber Intake Measure Up?

Feedback for Unit 12

The total number of points you scored approximates the **grams** of dietary fiber you typically consume daily.* Use this scale to find out if your fiber intake meets the recommended goal:

- **0–10 grams:** You consume less than the average American. Increase your fiber intake by including more fruits, vegetables, whole grains, and legumes in your diet overall.
- **11–15 grams:** Like other Americans, you consume too little fiber. Increase the number of servings of high-fiber foods you already enjoy, while substituting more high-fiber foods for refined food products. A quick way to add fiber to your diet is to consume more of the two fiber powerhouses: legumes and bran cereal.
- **15–20 grams:** You currently consume more fiber than the average American. Make sure you're including 5 or more servings of fruits and vegetables. Eat 6–11 servings of bread, cereal, rice, and pasta daily; choose whole-grain versions of these more often.
- **20+ grams:** Congratulations! Your dietary fiber intake meets the recommended goal. Keep up the good work.
- **Bonus:** High-fiber foods are also generally good sources of other complex carbohydrates. If you consume 25 to 30 grams of fiber daily, you likely have an ample intake of complex carbohydrates.

*Because different foods within a food group contribute varying amounts of dietary fiber, the point values have been averaged. Make sure to check the Nutrition Facts label on bran cereals because these cereals vary in the amount of dietary fiber they contain.

Unit 13

Alcohol: The Positives and Negatives

Photo Disc

Nutrition Scoreboard

	TRUE	FALSE
1. Alcohol is produced from carbohydrates.		
2. You can protect your body from the harmful effects of consuming excessive amounts of alcohol by eating a nutritious diet.		
3. Alcohol abuse plays a major role in injuries and deaths in the United States.		

(Answers on next page)

Answers to

Nutrition Scoreboard

	TRUE	FALSE
1. Alcohol (actually ethanol) is produced by the fermentation of carbohydrates in grains, fruits, and other foods.	✓	
2. High intakes of alcohol are harmful to the body, regardless of the quality of the diet.		✓
3. The statistics of alcohol overuse, injury, and death are startling. Alcohol abuse is a major personal, social, and public health problem in the United States.	✓	

Key Concepts and Facts

- → Alcohol is both a food and a drug and can have positive or negative effects on health.
- → Alcohol is produced from carbohydrates.
- → Alcohol abuse is harmful to the body and is associated with a high proportion of acts of violence and accidents.
- → Both genetic and environmental factors are associated with the development of alcoholism.

Alcohol Facts

Alcohol is both a food and a drug. It's a food because alcohol is made from carbohydrates and is used by the body as an energy source. Alcohol is a drug because it modifies various body functions.

The type of alcohol people consume in beverages is ethanol. (We refer to ethanol by the broader term "alcohol" in this unit.) Alcohol is produced from carbohydrates in grains, fruits, and other foods by the process of ***fermentation.*** Wines, brews made from grains, and other alcohol-containing beverages are a traditional part of the food supply of many cultural groups. In high doses, however, alcohol is harmful to the body and can cause a wide variety of nutritional, social, and physical health problems.

➡ **fermentation**
The process by which carbohydrates are converted to ethanol by the action of the enzymes in yeast.

The Positive

Whether alcohol has harmful effects on health depends on how much is consumed. The consumption of moderate amounts of alcohol by healthy adults who are not pregnant appears to cause no harm.[1] In fact, moderate alcohol consumption is associated with a significant level of protection against heart disease and stroke.[2] A moderate level of alcohol consumption is considered to be one or two standard-sized drinks per day (Illustration 13.1). Alcohol increases the body's production of high-density lipoprotein cholesterol (HDL).[3] HDL is known as the "good cholesterol" because it helps eliminate cholesterol from the body.

Although all types of alcohol-containing beverages reduce the risk of heart disease, drinking a glass or two of red wine daily appears to be particularly beneficial.[4] Red wine includes pigments that make grapes deep red, blue, or purple. A number of these pigments act as antioxidants and may decrease the tendency of blood to clot (which happens during a stroke or heart attack) and may decrease the ability of LDL cholesterol to stick to plaque in the arteries.[4] Purple grape juice contains these antioxidants, too, and may also help prevent heart disease.[5]

People don't have to consume alcohol to reduce their risk of heart disease. Diets low in saturated fat and cholesterol, liberal intakes of vegetables and fruits, and ample physical activity also reduce the risk of heart disease.

Illustration 13.1

The Negative

Heavy drinking, often defined as the consumption of five or more drinks per day, poses a number of threats to the health of individual drinkers and often to other people as well. Although health can be damaged by the regular consumption of high amounts of alcohol, the ill effects of alcohol are most obvious in people with ***alcoholism.***

alcoholism An illness characterized by a dependence on alcohol and by a level of alcohol intake that interferes with health, family and social relations, and job performance.

Habitually high alcohol intakes and alcoholism increase the risk of developing high blood pressure, stroke, cirrhosis of the liver, throat and stomach cancer, breast and bladder cancer, central nervous system disorders, and vitamin and mineral deficiency diseases.[6,7] Alcohol abuse is associated with a high proportion of deaths from homicide, drowning, fires, traffic accidents, and suicide (Illustration 13.2).[8] It is also involved in a large proportion of rapes and assaults. Alcohol poisoning from the consumption of a large amount of alcohol in a short period of time can cause death—and does to a number of college students each year.

Drinking during pregnancy may harm the fetus. Women who binge drink or drink regularly during pregnancy are at risk of delivering an infant with signs of Fetal Alcohol Syndrome (Illustration 13.3).[9] Children with Fetal Alcohol Syndrome suffer long-term growth and mental retardation.[10] The severity of the

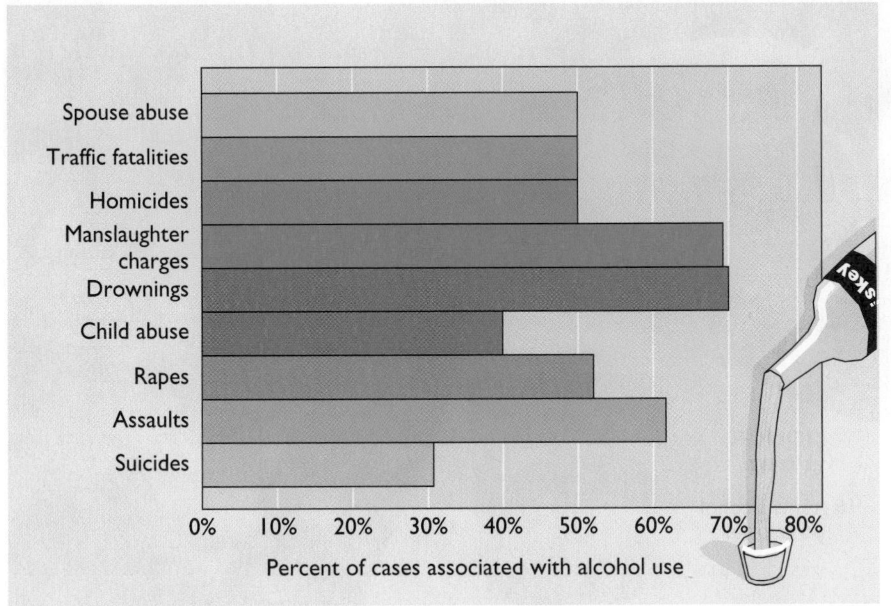

Illustration 13.2 Violence and injuries associated with alcohol.

Source: National Institute on Alcohol Abuse and Alcoholism, 2001.

PhotoEdit, Felicia Martinez

Illustration 13.3
This child was born to a woman who drank heavily during pregnancy. In addition to the unusual facial characteristics, children with Fetal Alcohol Syndrome experience growth and mental retardation.

condition depends on how much alcohol was consumed during pregnancy and whether excessive intake occurred early or late in pregnancy.[10] Although small amounts of alcohol do not appear to be hazardous,[11] it is recommended that women who are, or may become, pregnant refrain from drinking.[12]

Alcohol Intake and Diet Quality

Alcohol provides seven calories per gram, making alcohol-containing beverages rather high in caloric content (Illustration 13.4). Because many alcohol-containing beverages provide calories and few or no nutrients, they are considered "empty-calorie" foods. On average, alcohol accounts for 10% of the caloric intake of U.S. adults who drink.[13] The average goes up to around 50% among heavy drinkers.[14]

As caloric consumption from alcohol-containing beverages increases, the quality of the diet generally decreases. Diets of heavy drinkers frequently provide too little thiamin, vitamins A and C, calcium, and iron.[7] Deficiencies of nutrients, as well as direct, toxic effects of high levels of alcohol ingestion, produce most of the physical health problems associated with alcoholism. The lack of thiamin, for example, impairs the utilization of glucose by the brain. When people with alcoholism initially withdraw from alcohol, the thiamin deficiency may result in "delirium tremens," a condition called the "DTs" by people who staff detoxification centers. People with delirium tremens experience convulsions and hallucinations and are severely confused. Thiamin injections are a key component of treatment for delirium tremens.[14]

Because alcohol in excess is directly toxic to body tissues, consuming an adequate diet won't protect heavy drinkers from all of the harmful effects of alcohol.[15]

How the Body Handles Alcohol

Alcohol is easily and rapidly absorbed in the stomach and small intestine. Within minutes after it is consumed, alcohol enters the circulatory system and is on its way to the liver, brain, and other tissues throughout the body. Alcohol remains in blood and body tissues until it is broken down and used for energy or is converted into fat and stored. The process of converting alcohol into a source of energy takes several hours or more to complete, depending on the amount of

Photo Disc

Illustration 13.4
Alcohol content and caloric value of common alcohol-containing beverages.

Table 13.1
Alcohol doses and blood levels.

Number of Drinks*	Percent Blood Alcohol by Body Weight 100 lb	120 lb	150 lb	180 lb	200 lb
2	0.08	0.06	0.05	0.04	0.04
4	0.15	0.13	0.10	0.08	0.08
6	0.23	0.19	0.15	0.13	0.11
8	0.30	0.25	0.20	0.17	0.15
12	0.45	0.36	0.30	0.25	0.23
14	0.52	0.42	0.35	0.34	0.27

*Taken within an hour or so; each drink equal to ½ ounce pure ethanol.
Source: Sizer F, Whitney EN. Nutrition: current concepts and controversies, eighth edition. Belmont: Wadsworth; 2000, p. 172.

alcohol consumed. Because of the lag time between alcohol intake and utilization, blood levels of alcohol build up as drinking continues (Table 13.1).

The intoxicating effects of alcohol correspond to blood alcohol levels. After a drink or two, blood levels of alcohol reach approximately 0.03% in people who weigh about 150 pounds.[15] Blood alcohol levels of 0.03% correspond to mild intoxication. At this level, people lose some control over muscle movement and have slowed reaction times and impaired thought processes. A person's ability to drive or operate equipment in a safe manner is decreased at this level of blood alcohol content (Illustration 13.5). Blood alcohol levels of around 0.06% are associated with an increased involvement in traffic accidents. The legal limit for intoxication according to most states' highway safety ordinances is 0.08 to 0.1%—beyond the point where driving is impaired. When blood alcohol content increases to 0.13%, speech becomes slurred, "double vision" occurs, reflexes are dulled, and body movements become unsteady.[15]

If blood alcohol level continues to increase, drowsiness occurs and people may lose consciousness. Levels of blood alcohol above 0.6% can cause death, especially in individuals who have not developed a tolerance for alcohol.

Emerging evidence indicates that mixing alcohol with pain relievers may cause stomach problems. The combination of three plus drinks per day and the use of aspirin and nonaspirin pain relievers (acetaminophan, ibuprofen) may irritate the stomach lining and cause ulcers.[16]

A given amount of alcohol intake among women appears to produce higher blood levels of alcohol than for men of the same body weight.[17] Consequently, women may experience the intoxicating effects of alcohol on lower amounts of alcohol than men.

Illustration 13.5
The legal limit for intoxication in most states' highway safety ordinances is 0.08 to 0.1%—beyond the point where driving is impaired.

Acute Alcohol Poisoning Very high blood levels of alcohol can be extremely dangerous. Such levels can be reached by drinking large amounts of alcohol (usually in the form of liquor such as rum or gin) in a short period of time. People who have overdosed on alcohol become unconscious; their pulse is rapid, their blood pressure is low, and their pupils remain dilated.[15] They require emergency medical care.

Because of the dangers associated with high blood levels of alcohol, only moderate amounts of alcohol (if any) should be consumed. Drinks should be sipped slowly rather than gulped.

Illustration 13.6
The more severe type of alcoholism is characterized by early onset and signs of physical dependence.

Photo Researchers, Inc., Blair Seitz

How to Drink Safely if You Drink Many of the problems related to alcohol intake can be prevented by drinking responsibly. That means not drinking on an empty stomach, which can make a person intoxicated surprisingly fast. It means limiting alcohol intake to an amount that doesn't make you feel "tipsy"—without full control over mind and body—and *never* driving a car or boat, hunting, or operating heavy equipment while under the influence of alcohol.

What Causes Alcoholism?

Directly or indirectly, alcoholism touches the lives of one out of three people in the United States. It is a disease that appears to "run in families," but it can also be related to environmental influences.

Two major types of alcoholism are thought to exist. The first type generally develops after the age of 25 and is characterized by a psychological dependence on alcohol. This type of alcoholism appears to be strongly related to environmental factors. The other major type of alcoholism is more severe and usually develops prior to the age of 25 (Illustration 13.6). Early-onset alcoholism carries a high risk of physical dependence on alcohol and withdrawal symptoms when intake stops. Alcoholism that develops early is often related to history of alcoholism in the family.[18] In general, the younger the age when alcohol consumption begins, the greater the likelihood that an individual will develop a drinking problem at some point in life.[19]

Alcohol Use among Adolescents Alcohol consumption by young adolescents is a growing problem in the United States. In a national survey in 1999, binge drinking (defined as consumption of five or more drinks in a row in the previous two weeks) was reported by:

- 15% of 8th graders,
- 26% of 10th graders, and
- 31% of 12th graders.[20]

Reduction in alcohol intake by adolescents is a major public health initiative of the Year 2010 Health Objectives for the Nation.[21]

Until more is learned about the prevention of alcohol abuse and alcoholism, treatment programs such as Alcoholics Anonymous and educational programs that stress safe drinking will be relied upon to lessen the impact of alcohol abuse on personal and public health.

Nutrition UP CLOSE

Effects of Alcohol Intake

FOCAL POINT: Estimating blood alcohol levels and side effects.

Scenario: Ligia and Mark attend a wedding reception. Prior to the meal, they both drink a glass of champagne to toast the bride, and 15 minutes later, they drink another glass to toast the groom. Ligia weighs 150 pounds and Mark, 180.

Questions: Using the information in Table 13.1 and the information on "How the Body Handles Alcohol" (p. 13–4), answer the following questions:

A. After two glasses of champagne, what would be Ligia's estimated blood alcohol level? ______________ % blood alcohol

What would be Mark's? ______________ % blood alcohol

B. List three side effects of these blood alcohol levels:

1. ____________________________

2. ____________________________

3. ____________________________

FEEDBACK (answers to these questions) can be found at the end of Unit 13.

WWW links

www.gov.ab.ca/aadac
Provides information on alcohol-related services and information, and on how to get help in Canada.

www.niaaa.nih.gov
Additional information about alcohol abuse is available from the National Institute on Alcohol Abuse and Alcoholism.

www@potsdam.edu/alcohol-info
The Alcohol "Problems and Solution" site is designed to provide parents, students, and educators information on alcohol consumption and abuse.

www.mentalhealth.com
A comprehensive resource on alcohol abuse diagnostic information, treatment advice, and stories of recovery.

www.navigator.tufts.edu
Search alcohol to access reliable sites.

REFERENCES

1. Camargo CA Jr., et al. Prospective study of moderate alcohol consumption and mortality in US male physicians. Arch Int Med 1997;157:79–85.
2. Mullan Z. How much should we be drinking to minimise mortality? Lancet 2000;335:123.
3. Silva ERDE, et al. Alcohol consumption raises HDL cholesterol levels by increasing the transport rate of apolipoproteins A-I and A-II. Circ 2000;102:2347–52.
4. German JB, Walzem RL. The health benefits of wine. Ann Rev Nutr 2000;20:561–93.
5. Flavonoids in grape juice mirror benefits of red wine. Envir Nutr 1997;20:7.
6. Bunout D. Nutritional and metabolic effects of alcoholism: their relationship with alcoholic liver disease. Nutr 1999;15:583–89.
7. Levine JA, et al. Energy expenditure in chronic alcohol abuse. Euro J Clin Invest 2000;30:779–86.
8. Advance data from vital and health statistics; no. 205. Hyattsville (MD): National Center for Health Statistics (DHHS publication no. [PHS] 91-1250); 1991.
9. Spohr H-L, Willms J, Steinhausen H-C. Prenatal alcohol exposure and long-term developmental consequences. Lancet 1993;241:907–10.
10. Mattson SN, et al. Heavy prenatal alcohol exposure with or without physical features of fetal alcohol syndrome leads to IQ deficits. J Pediatrics 1997;131:718–21.
11. Knupfer G. Abstaining for fetal health: the fiction that even light drinking is dangerous. British J Addiction 1991;86:1063–73.
12. Polygenis D, et al. Moderate alcohol consumption during pregnancy and the incidence of fetal malformations: a meta-analysis. Neurotox Terat 1998;20:61–67.
13. Williamson DF, et al. Alcohol and body weight in United States adults. Am J Public Health 1987;77:1324–30.
14. Lieber CS. The influence of alcohol on nutritional status. Nutr Rev 1988;46:241–54.
15. Charness ME, Simon RP, Greenberg DA. Ethanol and the nervous system. N Engl J Med 1989;321:442–53.
16. The FDA warns against mixing nonprescription pain pills, booze. Associated Press interview with M. Friedman, acting commissioner, FDA, 1997 Nov 15.
17. Frezza M, et al. High blood alcohol levels in women. The role of decreased gastric alcohol dehydrogenase activity and first-pass metabolism. N Engl J Med 1990;322:95–99.
18. Ackerman SJ. Research on the genetics of alcoholism is still in ferment. J NIH Res 1992;4:61–65.
19. Grant BR, Dawson DA. Age of onset of alcohol use and its association with DSM-IV alcohol abuse and dependence: results from the National Longitudinal Alcohol Epidemiologic Survey. J Subst Abuse 1998;9:103–10.
20. Health, United States 2000, with Adolescent Health Chartbook. www.cdc.gov/nchs, 2000 Nov.
21. Health People 2010 Objectives for the Nation. web/health.gov/healthypeople, 2001 Jan.

Nutrition UP CLOSE

Effects of Alcohol Intake

Feedback for Unit 13

A. Ligia's estimated % blood alcohol = 0.05%.
   Mark's estimated % blood alcohol = 0.04%.

B. Three side effects of these blood alcohol levels:
   1. Loss of some control over muscle movements
   2. Slowed reaction time
   3. Impaired thought processes

Unit 14

Diet and Dental Health

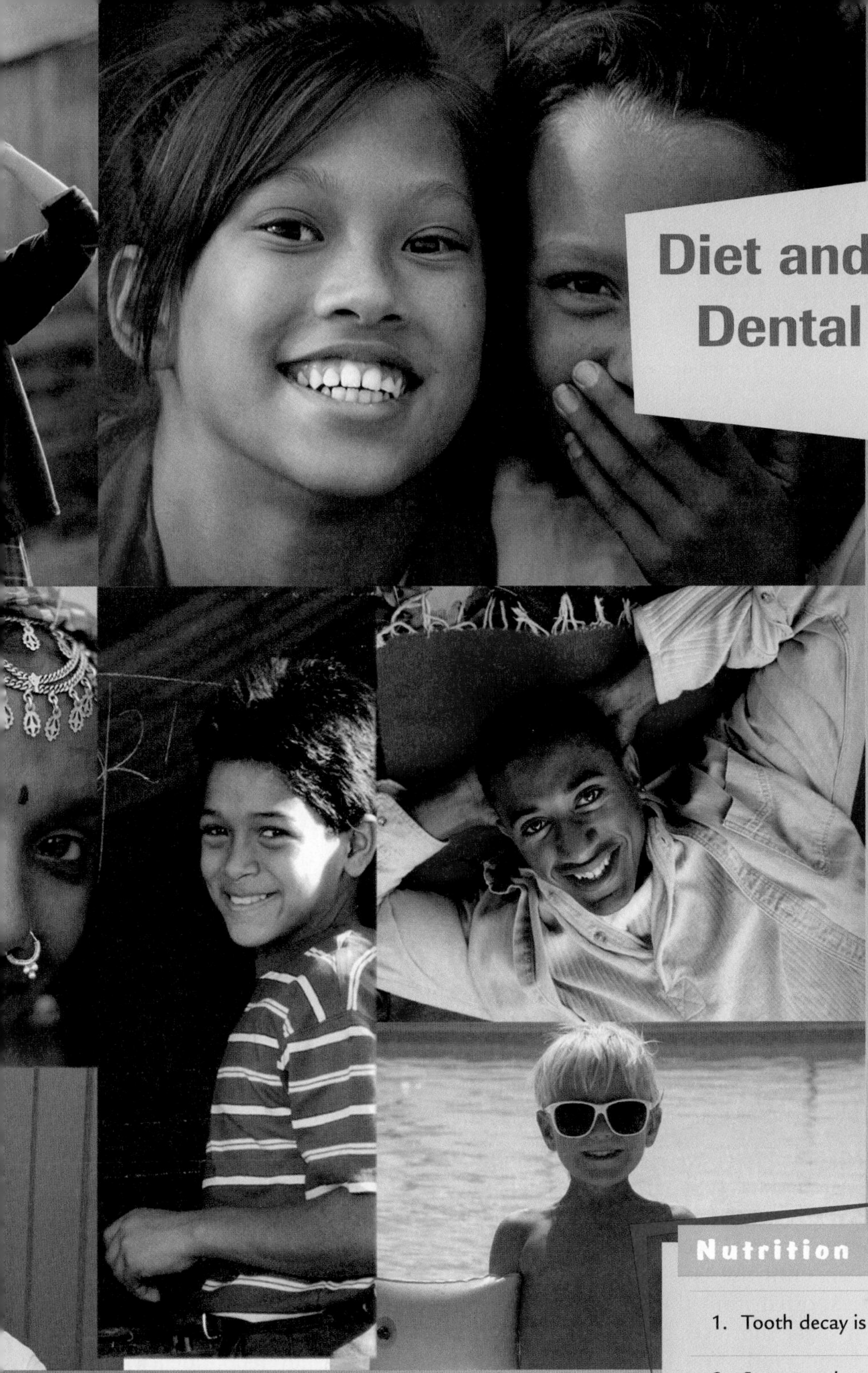

Photo Disc

Nutrition Scoreboard

	TRUE	FALSE
1. Tooth decay is an infectious disease.		
2. Sugary and starchy foods eaten between meals are more likely to cause tooth decay than are the same foods consumed with meals.		
3. A dramatic drop in sugar intake in the United States during the last 30 years is largely responsible for declining rates of tooth decay.		

(Answers on next page)

Key Concepts and Facts

- Simple sugars, especially sucrose, and starches promote tooth decay.
- Fluoridated water is the main reason for the widespread reduction in tooth decay in the United States.
- Tooth decay and tooth loss are largely preventable.

Answers to
Nutrition Scoreboard

	TRUE	FALSE
1. Cavities *are* caused by bacteria that live in the mouth.	✓	
2. Sugary and starchy foods (especially if they're sticky) undiluted by meals are more likely to cause tooth decay than are the same foods eaten with meals.[1]	✓	
3. The incidence of tooth decay in the United States has dropped by over 50% since the 1970s.[2] But the decline is largely attributed to increased intake of fluoride from drinking water and the use of fluoride-containing toothpastes, mouth rinses, and gels.[3]		✓

Your Diet and Your Teeth

tooth decay The disintegration of teeth due to acids produced by bacteria in the mouth that feed on sugar. Also called *dental caries* or *cavities*.

More than anything else, the health of your teeth depends on two components of your diet: sugar and fluoride. Sugar is the leading cause of ***tooth decay*** (the leading dental health problem in the world), and water fluoridation is the most effective means of preventing it.[4] This unit explores the realities of diet and dental health and presents practical ways to take care of your teeth.

How Close Is the Relationship between Sugar and Tooth Decay?

The relationship between sugar and tooth decay is very close, and the history of tooth decay closely parallels the availability of sugar. The incidence of tooth decay is estimated to have been very low (less than 5%) among hunter-gatherers who had minimal access to sugars.[5] Tooth decay did not become a widespread problem until the late seventeenth century, when great quantities of sucrose were exported from the New World to Europe and other parts of the world. When sugar shortages occurred in the United States and Europe during World War I and World War II, rates of tooth decay declined; they rebounded when sugar became available again.[6] Across 47 countries, rates of tooth decay in children have been found to be strongly related to the population's average intake of sugar.[7] Rates of tooth decay in children vary substantially among countries, but the highest rates are in countries where sugar is widely available in processed foods and beverages. Tooth decay is spreading rapidly in developing countries where sugar, candy, soft drinks, and fruit drinks are becoming widely available.[8]

Sweets are not the only culprit. Simple sugars that promote tooth decay can also come from starchy foods, especially pretzels, crackers, and breads that stick to your gums and teeth. Some of the starch is broken down to simple sugars by enzymes in the mouth.

To reduce the incidence of tooth decay, a number of countries have developed campaigns to help inform consumers about cavity-promoting foods. Switzerland, for example, labels foods safe for teeth with a "happy tooth" symbol (Illustration 14.1) and encourages the use of alcohol sugars (which don't promote tooth decay) in sweets. Other countries recommend that sweets be consumed with meals or that teeth be brushed after sweets are eaten.

Illustration 14.1
Switzerland's "happy tooth" symbol.

There's More to Tooth Decay Than Sugar Per Se How frequently sugary and starchy foods are consumed and how long they stick to gums and teeth make a difference in their tooth decay–promoting effects. People who make a habit of consuming sugary and starchy foods between meals risk tooth decay. Drinking coffee or tea with sugar throughout the day or consuming three or more regular soft drinks between meals hastens tooth decay (more than if these beverages are consumed with meals).[9] Candy, cookies, and crackers eaten between meals are much more likely to promote tooth decay than are the same foods consumed as part of a meal.[10] Chewing as few as two sticks of sugar-containing gum a day also significantly increases tooth decay.[11] Sugary foods eaten with meals rather than as snacks are more likely to be wiped off teeth or diluted by other foods that have less tooth decay–promoting effects.

Sugar-containing foods that stick to the teeth are more likely to cause tooth decay than those that don't (Table 14.1). Marshmallows, caramels, and taffy, for example, are much more likely to promote tooth decay than are apples and milk chocolate. Nevertheless, all of the foods listed in Table 14.1 can promote tooth decay if allowed to remain in contact with the gums and teeth.

Other Promoters of Tooth Decay Several other factors are also related to the development of tooth decay:

- Genetic traits
- Certain diseases and disorders
- Some medications
- Acidic beverages
- Excessive cleaning and polishing of teeth

Some people appear to be born with teeth that resist dental decay (others have teeth that appear to be prone to decay). Conditions such as HIV infection, poorly controlled diabetes, bulimia nervosa, and fasting are associated with high rates of tooth decay. Medications that make the mouth dry (antihistamines, for example), blood pressure pills, and medicines that contain a lot of sugar (cough drops, some liquid pain relievers, lozenges) can promote decay. Exposing the teeth to acidic medicines, such as chewable vitamin C tablets and aspirin, can erode enamel. Tooth erosion and cavities can also result from the frequent

Table 14.1
The "stickiness" value of some foods. The stickier the food, the worse it is for your teeth.

Very Sticky	Sticky	Somewhat Sticky	Barely Sticky
Caramels	Doughnuts	Bagels	Apples
Chewy cookies	Figs	Cake	Bananas
Crackers	Frosting	Cereal	Fruit drinks
Cream-filled cookies	Fudge	Dry cookies	Fruit juices
Granola bars	Hard candy	Milk chocolate	Ice cream
Marshmallows	Honey	Rolls	Oranges
Pretzels	Jelly beans	White bread	Peaches
Taffy	Pastries		Pears
	Raisins		
	Syrup		

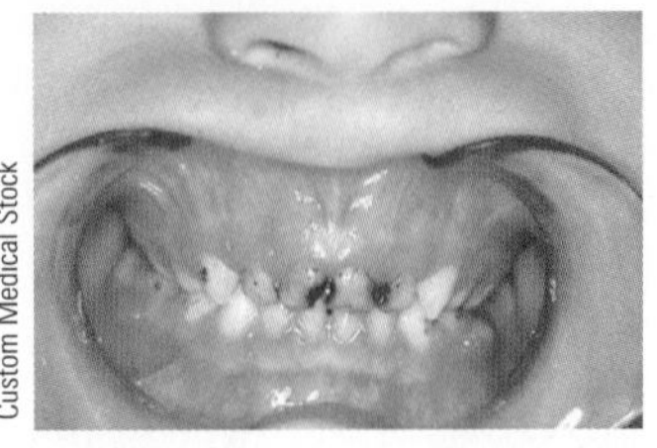

Custom Medical Stock

Illustration 14.2
"Baby bottle caries" (also called "nursing bottle syndrome") occurs in infants who habitually receive sweet fluids or milk in bottles when they go to sleep. Cavities occur first in the upper front teeth because that's where fluid pools when babies sleep.

consumption of acidic beverages, such as carbonated beverages, diet sodas, vinegar, certain herbal teas (rose hip, lemon, and mallow teas, for example), and lemon or lime juice. In addition, the frequent use of dental abrasives and bleaches for cleaning and polishing teeth removes enamel and makes teeth susceptible to decay.[12]

Baby Bottle Caries A startling example of the effect that frequent and prolonged exposure to sugary foods can have is "baby bottle caries" (Illustration 14.2). Infants and young children who routinely fall asleep while sucking a bottle of sugar water, fruit drink, milk, or formula, or while breastfeeding, may develop severe decay. After the child falls asleep, the fluid may continue to drip into the mouth. A pool of the fluid collects between the tongue and the front teeth, bathing the teeth in the sweet fluid for as long as the child sleeps. The upper front teeth become decayed first because the tongue protects the lower teeth. Baby bottle caries occur in 5 to 10% of infants and young children and can lead to the destruction of all baby teeth.[4]

Some Foods Protect Teeth against Tooth Decay Foods rich in fats such as cheese and peanut butter, protein (eggs, meats, yogurt), and dietary fiber (vegetables and whole-grain products, for example) inhibit tooth decay.[3] Artificial sweeteners and chewing gums sweetened with sorbitol, mannitol, or xylitol (three alcohol sugars) don't promote tooth decay either. Actually, these gums may help prevent tooth decay by increasing the flow of saliva.[11] A sampling of foods that don't promote tooth decay appears in Table 14.2.

Why Does Sugar Promote Tooth Decay?

Sugar promotes tooth decay because it is the sole food for certain bacteria that live in the mouth and excrete acid that dissolves teeth. In the presence of sugar, bacteria in the mouth multiply rapidly and form a sticky, white material called ***plaque.*** Tooth areas covered by plaque are prime locations for tooth decay because they are dense in acid-producing bacteria. Acid production by bacteria increases within five minutes of exposure to sugar. It continues for 20 to 30 minutes after the bacteria ingest the sugar.[13] If teeth are frequently exposed to sugar, the acid produced by bacteria may erode the enamel (Illustration 14.3), producing a cavity. If the erosion continues, the cavity can extend into the dentin and pulp and allow bacteria to enter the inside of the tooth. That can cause an infection and the loss of the tooth. It can be prevented if the plaque is removed before the acid erodes much of the enamel. Teeth are capable of replacing small amounts of minerals lost from enamel.[14]

On the Side

Bacteria on the tongue have recently been identified as the primary cause of bad breath. Mouth wash gets rid of it for about an hour, brushing teeth with toothpaste takes care of bad breath in 25% of people, but brushing the tongue eliminates it in 70 to 80% of people.[15]

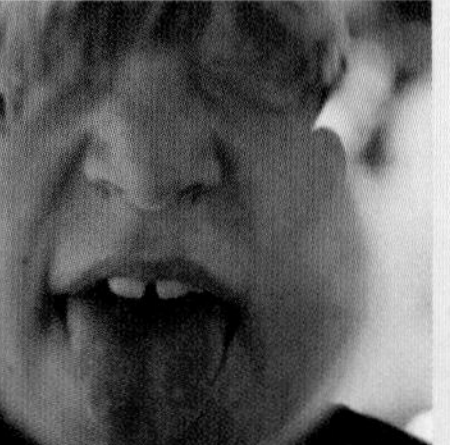

Photo Disc

Gum Disease Plaque that forms near the gums can cause tooth erosion and ***gingivitis.*** If the plaque is removed, the gingivitis and tooth erosion can be reversed.[16] If it isn't removed, the acid excreted by bacteria living in the plaque will continue to erode the tooth near the gums and eventually destroy the base of the tooth underneath the gums. The condition can lead to a loss of the tooth and the bony structures that support it.

It's important to limit the diet of bacteria in the mouth by reducing their access to sugars. Without sugar, bacteria don't multiply, form plaque or acid, or erode your teeth.

➡ **plaque**
A soft, sticky, white material on teeth formed by bacteria.

➡ **gingivitis**
A form of periodontal disease characterized by a minor inflammation of the gums.

Fluoride and Dental Health

> **Fluoridation is the most effective way to prevent dental caries in all children.**
>
> —National Maternal and Child Oral Health Resource Center

Table 14.2
Foods that don't promote tooth decay.[3]

Artificial sweeteners
Cheese
Coffee (no sugar)
Eggs
Fats and oils
Gum and candy sweetened with alcohol sugars
Meats
Milk
Nuts
Peanut butter
Tea
Water
Yogurt (plain)

Not all teeth are equally susceptible to tooth decay. Those exposed to fluoride during formation and regularly thereafter are more resistant to decay than teeth that aren't protected by fluoride. The protection offered by fluoride is available to teeth of all ages.

Fluoride in the blood is released into enamel during tooth formation and attaches to the outermost layer of enamel in exposed teeth. Fluoride helps protect teeth from acid produced by bacteria and promotes the remineralization of enamel damaged by erosion.[14] Although a few foods such as tea and seafood contain a small amount of fluoride, our major sources are fluoridated water, toothpastes, dental rinses, and beverages made from fluoridated water.

Recommended Fluoride Intake Recommended levels of daily fluoride intake were established for the first time in 1997. These recommendations are based on levels of fluoride intake that reduce tooth decay without causing a discoloration of teeth. (These recommendations are included in the DRI tables shown on the inside front covers of this book.)

The Remarkable Success of Water Fluoridation In the early 1930s, lower rates of tooth decay observed among children living in areas where water naturally contained fluoride provided the initial evidence that led to the fluoridation of many community water supplies. Fluoridated water has repeatedly been shown to reduce the incidence of tooth decay by 50% or more.[3]

There is no basis for the fears sometimes expressed about fluoridation of community water supplies. After intensive studies, water fluoridation has not

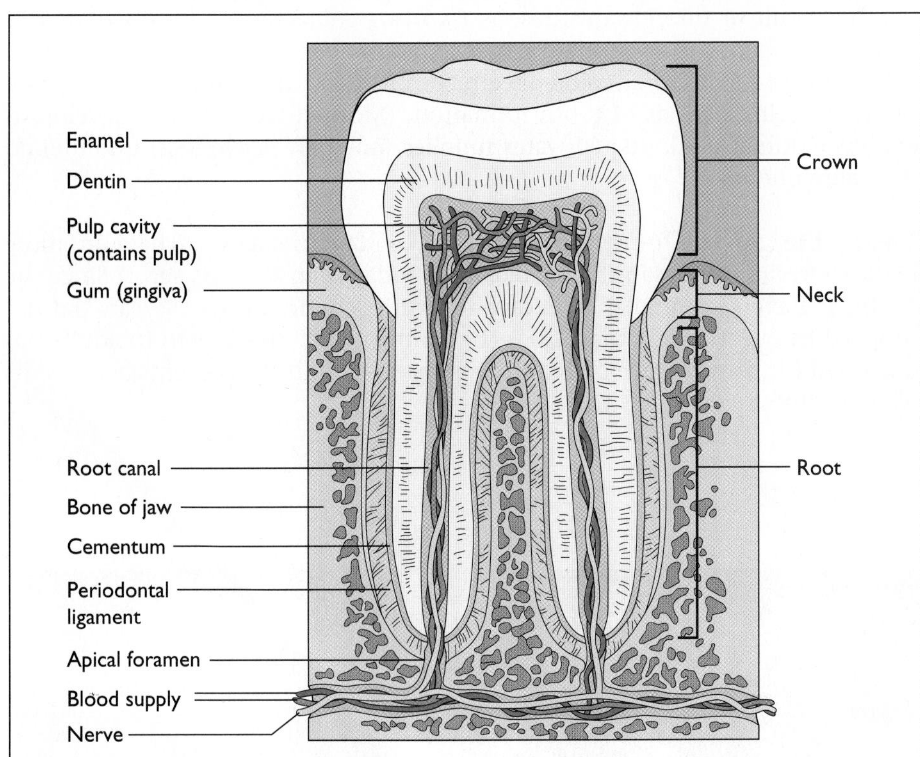

Illustration 14.3
Cross section of a healthy tooth.

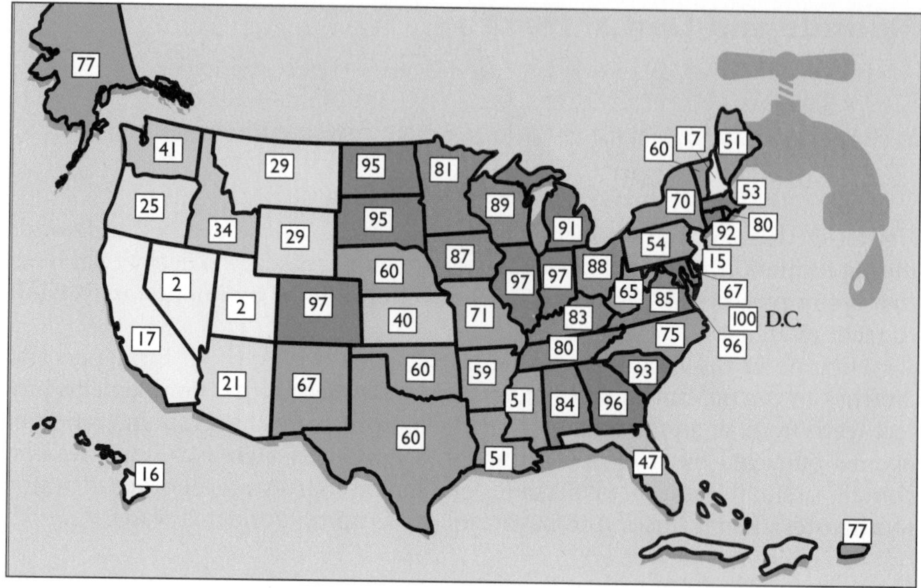

Illustration 14.4
Percentage of the population drinking fluoridated water in 1990 by state.[17]

been shown to promote cancer, heart disease, allergies, "brain lesions," mental retardation, or other health problems.[14] Fluoridation is a safe, effective, and cheap method of controlling dental disease. (Providing fluoridated community water supplies costs about 50 cents per person per year.[4]) Despite the advantages of fluoride, more than one million Americans consume water from a less than optimally fluoridated water supply.[4] The availability of fluoridated water varies a great deal from state to state (Illustration 14.4).

Can You Get Too Much Fluoride? It is possible to get too much fluoride. Fluoride overdose is most often due to naturally occurring, high levels of fluoride in water. It can also occur when excessive fluoride supplements are taken or when small children eat fluoridated toothpaste.[18]

The result of fluoride overdose is ***fluorosis*** characterized by discolored or "mottled" enamel that is unattractive but otherwise is not harmful to teeth (Illustration 14.5). How discolored teeth become depends on the dose of fluoride received during periods of tooth formation. (Mottled teeth do not develop in adults.) Optimally fluoridated water supplies (one part per million fluoride) do not cause fluorosis.[14]

➡ **fluorosis**
Fluoride overdose characterized by discolored ("mottled") teeth in children.

Tooth Decay Is Decreasing in the United States There are more children free of cavities in the United States today than 20 years ago.[19] Since the 1970s, the average number of decayed teeth in children 6 to 18 years old has dropped from 7.1 to less than 2.0.[2] The number of decayed teeth in adults has decreased from an average of 16.9 to less than 8.0, and the percentage of adults who are "toothless" is less than 4%.[19]

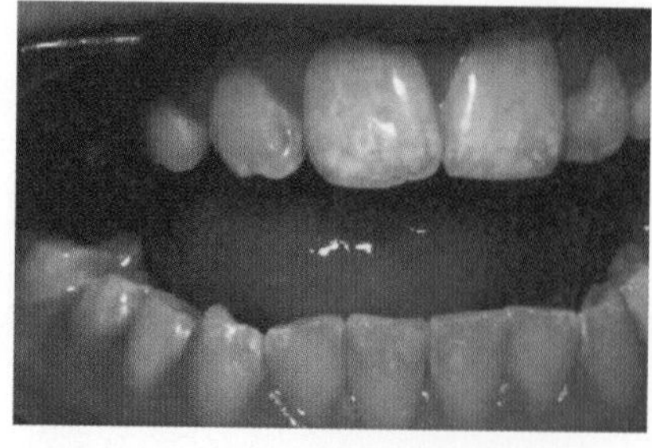

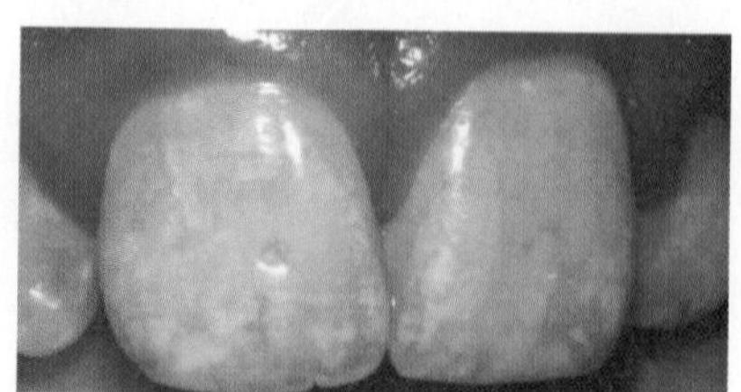

Illustration 14.5
The severity of fluorosis varies widely. With the mildest form (2 parts per million, or ppm, fluoride), the enamel shows occasional white flecks, whereas high fluoride levels in water (7 to 9 ppm) produce brown stains and pitted enamel.

Source: Hardwick F, Ketchem L. Pediatric basics. Gerber Co., 1990.

Age	Tooth Decay
6 years	50%
15 years	66%
17 years	80%

Table 14.3
Percentage of U.S. children and adolescents with tooth decay.[4]

Tooth decay remains an important health problem in the United States, however. Rates of tooth decay are still high and increase with age during childhood (Table 14.3). Approximately half of children do not have dental insurance.[4]

To maintain the downward trend in tooth decay, infants over six months of age and children living in areas where water contains little or no fluoride should receive fluoride supplements. The needed dose varies from 0.25 to 1.0 milligram per day, depending on age and the amount of fluoride naturally present in the water supply.[12] Giving fluoride supplements to pregnant women, however, does not help protect young children's teeth from decay.[20] Only minute amounts of fluoride are transferred to the fetus from the mother's blood.

The credit for the declines in tooth decay and tooth loss belongs primarily to the increased availability of fluoridated water supplies.[14] Fluoride supplements, toothpastes, rinses and gels, protective sealants, and improved dental hygiene and care deserve credit, too.[19] Decreased intakes of sucrose and increased use of high-fructose corn sweeteners, which are less caries promoting than sucrose, in soft drinks, cereals, and snacks may have contributed to the decline in tooth decay as well. Further declines in the spread of this infectious disease will come as more communities begin to fluoridate their water supplies, the use of dental sealants on the molars of children expands, and effective methods for controlling cavity-causing bacteria are discovered.

Nutrition UP CLOSE

Teeth Care Quiz

FOCAL POINT: Preventing dental problems.

	Yes	No
1. Is the water you drink fluoridated? (Note: Municipal water supplies are almost always fluoridated; many bottled waters do not contain fluoride.)	_____	_____
2. Do you brush your teeth at least twice a day with toothpaste?	_____	_____
3. Do you consume sticky sweets or starchy foods between meals without brushing your teeth afterward?	_____	_____
4. Do you floss your teeth daily?	_____	_____
5. Do you get regular dental checkups?	_____	_____

FEEDBACK (answers to these questions) can be found at the end of Unit 14.

WWW links

www.ncemch.org
Link to the National Maternal and Child Oral Health Resources Center for data on trends in oral health and information on preventive and educational programs.

www.nidr.nih.gov
The National Institute for Dental Research, National Institutes of Health site provides research updates and information for the public on dental health.

www.ada.org
Consumer information on dental health is available from the American Dental Association.

www.adha.org
Consumer information from the American Dental Hygienists' Association is also available.

www.nin.ca
Get information on dental health from Canada's National Institute of Nutrition.

REFERENCES

1. Fitzsimons D, et al. Nutrition and oral health guidelines for pregnant women, infants, and children. J Am Diet Assoc 1998;98:182–9.
2. Brown LJ, et al. Trends in untreated caries in permanent teeth of children 6 to 18 years old. J Am Dent Assoc 2000;131:1637–441.
3. Position of the American Dietetic Association: the impact of fluoride on health. J Am Diet Assoc 2001;101:126–32.
4. Trends in children's oral health. National Maternal and Child Oral Health Resource Center, www.ncemch.org, 1999.
5. Cornero S, Puche RC. Diet and nutrition of prehistoric populations at the alluvial banks of the Parana River. Medicina 2000;60:109–14.
6. Shaw JH. Diet and dental health. Am J Clin Nutr 1985;41:1117–31.
7. Committee on Diet and Health (Food and Nutrition Board; National Research Council). Diet and health. Implications for reducing chronic disease risk. Washington, DC: National Academy Press; 1989.
8. Parajas IL. Sugar content of commonly eaten snack foods of school children in relation to their dental health status. J Phillipine Dent Assoc 1999;51:4–21.
9. Ismail A, Burt BA, Eklund SA. The cariogenicity of soft drinks in the United States. J Am Dental Assoc 1984;109:241–5.
10. Ismail A. Food cariogenicity in Americans aged from 9 to 29 years accessed in a national cross-sectional study, 1971–1974. J Dental Res 1986;65:1435–40.
11. Edgar WM. Sugar substitutes, chewing gum, and dental caries—a review. British Dental J 1998;184:29–32; and Rugg-Gunn AJ. The benefits of using sugar-free chewing gum: a proven anti-caries effect. British Dental J 1998;184:26.
12. Touger-Decker R, Mobley C. Position of the American Dietetic Association: oral health and nutrition. J Am Diet Assoc 1996;96:184–9.
13. Schachtele CF, Harlander SK. Will the diets of the future be less cariogenic? J Canadian Dental Assoc 1984;3:213–9.
14. Palmer CA, Watson LM. Position of the American Dietetic Association: the impact of fluoride on dental health. J Am Diet Assoc 1994;94:1428–31.
15. Bad breath. Nutr Today 2000;35:6.
16. The surgeon general's report on nutrition and health. Washington, DC: Public Health Service (DHHS publication no. PHS 88-50210); 1988.
17. Fluoridation Census 1990 Summary. Atlanta: Public Health Service, Centers for Disease Control; Apr 1993.
18. Riordan PJ. The place of fluoride supplements in caries prevention today. Aust Dental J 1996;41:335–42.
19. Brown LJ, et al. Most U.S. schoolchildren are caries-free in their permanent teeth. Public Health Reports 1995;110:531–3.
20. Leverett DH, et al. Randomized clinical trial of the effect of prenatal fluoride supplements in preventing dental caries. Caries Res 1997;31:174–9.

Nutrition
UP CLOSE

Teeth Care Quiz

Feedback for Unit 14

1. If your water supply is not fluoridated, make sure you use a fluoridated toothpaste regularly. You may want to ask a dentist or dental hygienist about using fluoridated rinses and other preventive measures.
2. Regular brushing is a must for healthy teeth!
3. Carbohydrates that stick to your teeth and gums should be rinsed or brushed away after eating.
4. Flossing regularly helps prevent plaque.
5. Dental problems are easier and less expensive to treat if identified early.

Unit 15

Proteins and Amino Acids

Food Stock America

Nutrition Scoreboard

	TRUE	FALSE
1. Protein is the most important component of our diet.		
2. "Nonessential amino acids" are not required for normal body processes. Only "essential amino acids" are.		
3. High-protein diets enhance muscle development in individuals who work out a lot.		

(Answers on next page)

Answers to

Nutrition Scoreboard

	TRUE	FALSE
1. The problem here is not that protein isn't important, but that its importance is overrated. Most Americans get plenty of protein—probably too much.		✓
2. "Nonessential amino acids" are required by the body, but they are not required components of our diet. (Yes, it is confusing.)		✓
3. Muscles contain protein, but you can't increase muscle mass by consuming protein, protein powders, or amino acid supplements—whether you're a couch potato or a nationally ranked athlete.		✓

Key Concepts and Facts

- Proteins are made of amino acids. Some amino acids are "essential" (required in the diet), and some are "nonessential" (not a required part of diets).
- Although protein can be used for energy, its major function in the body is the construction and maintenance of protein tissues.
- Protein tissue construction in the body proceeds only when all nine essential amino acids are available.
- Appropriate combinations of plant foods can supply sufficient quantities of all the essential amino acids.
- Americans are being urged to select more foods high in complex carbohydrates and fewer foods high in protein.

Protein's Image versus Reality

protein Chemical substance in foods made up of chains of amino acids.

The term ***protein*** is derived from the Greek word *protos* for "first." The derivation indicates the importance ascribed to this substance when it was first recognized. An essential structural component of all living matter, protein is involved in almost every biological process in the human body. As important as protein is for growth and health, however, many Americans tend to overestimate the amount of protein needed in daily diets.

Protein has a very positive image (Illustration 15.1). It's so positive that you don't have to talk about the importance of protein—people are already convinced of it. In fact, many would say protein is *the* most important of all the essential nutrients. What *does* need to be talked about, though, is the relative importance of protein in the Western diet.

Illustration 15.1
The protein perception.

Rich or poor, nearly all people in the United States get enough protein in their diets. Actually, most people consume far more protein than they need. Average intakes of protein exceed the Recommended Dietary Allowance (RDA) level for all age and sex groups. Approximately 17% of total calories in the average U.S. adult diet are supplied by protein compared to the recommended 10 to 12%.[1]

The current level of protein intake in the United States is not beneficial to health.[2] Instead, high protein intakes generally correspond to the high-fat, low-fiber diet of many Americans. Foods high in protein such as hamburger, cheese, nuts, and eggs are high in fat, too. Many lean meats available in the market provide a considerable proportion of their total calories as fat (Illustration 15.2).

To bring the American diet into better balance, consumers are being urged to choose foods high in complex carbohydrates, such as grains, pastas, fruits, and vegetables, and to select fewer high-protein and high-fat foods or to consume smaller portions of them.[3] This does not mean Americans should omit

(a) Hamburger (90% lean): 45%

(b) Tenderloin: 43%

(c) Top loin: 41%

(d) Round tip: 34%

(e) Sirloin: 33%

(f) Eye of round: 26%

(g) Top round: 25%

(h) Pork chop, lean: 48%

(i) Pork loin roast: 36%

(j) Pork tenderloin: 28%

(k) Chicken thigh, no skin: 47%

(l) Baked chicken breast, no skin: 19%

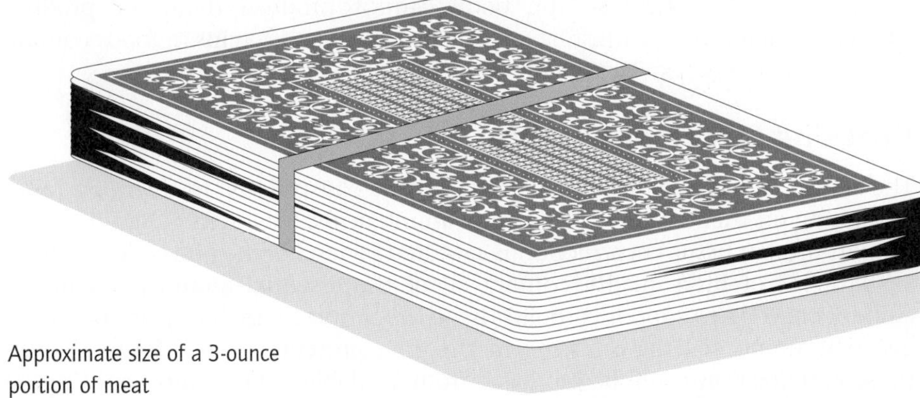

Approximate size of a 3-ounce portion of meat

Illustration 15.2
The fat content of 3-ounce portions of "lean" meats. The percentage of calories from fat is indicated for each portion. (A 3-ounce portion of meat is about the size of a deck of cards.) Each portion of meat provides approximately 21 grams of protein.

lean meats from their diets, but they are being advised to balance their intake of meats with plant foods low in fat so they can realize the goal of consuming less than 30% of total calories from fat.

Protein Facts

The primary role of protein is to provide the building material for body tissues. Muscles, enzymes, hormones, connective tissues (skin, cartilage, ligaments), ***antibodies,*** and red blood cells, for example, are built mainly from protein. Healthy adults need about 45 to 60 grams of protein daily to maintain and replace tissues.[4] When protein intake exceeds that level, the body uses the protein to meet immediate energy needs or converts it to fat and stores it. The average intake of protein by adults in the United States is 74 grams, an amount well

➡ **antibodies**
Blood proteins that help the body fight particular diseases. They help the body develop an immunity, or resistance, to many diseases.

above the RDA of 46 grams per day for women aged 19 to 24 years and 58 grams for men aged 19 to 24 years.[4]

Illustration 15.3

Source: © 2001 by Sidney Harris.

Amino Acids

The "building blocks" of protein are amino acids (Illustration 15.3). Protein consumed in food is broken down by digestive enzymes and absorbed into the bloodstream as amino acids. There are 20 common amino acids (Table 15.1) that form proteins when linked together. Every protein in the body is composed of a unique combination of amino acids linked together in chains (Illustration 15.4). The organization of amino acids into the chains is orchestrated by ***DNA,*** the genetic material within each cell that directs protein synthesis. Once formed, the chains of amino acids may fold up into a complex shape. Some proteins are made of only a few amino acids, while other proteins contain hundreds. Whatever the number of amino acids, the specific amino acids involved and their arrangement determine whether the protein is an enzyme, a component of red blood cells, a muscle fiber, or another tissue made from protein. Illustration 15.5 shows two examples of protein substances in the body.

Nine of the 20 common amino acids are considered ***essential,*** and 11 are ***nonessential.*** Despite the labels, all 20 amino acids are required to build and maintain protein tissues. The essential amino acids are called "essential" because the body cannot produce them, or produce enough of them, so they must be provided by the diet. Proteins in foods contain both essential and nonessential amino acids.

➡ **DNA (deoxyribonucleic acid)**
Genetic material contained in cells that initiates and directs the production of proteins in the body.

➡ **essential amino acids**
Amino acids that cannot be synthesized in adequate amounts by humans and therefore must be obtained from the diet. They are sometimes referred to as "indispensable amino acids."

➡ **nonessential amino acids**
Amino acids that can be readily produced by humans from components of the diet. Also referred to as "dispensable amino acids."

Proteins Differ in Quality

The ability of proteins to support protein tissue construction in the body varies depending on their content of essential amino acids. How well dietary proteins support protein tissue construction is captured by tests of the protein's "quality."

Proteins of high quality contain all of the essential amino acids in the amounts needed to support protein tissue formation by the body. If any of the essential amino acids are missing in the diet, proteins are not formed—even those proteins that could be produced from available amino acids. Shutting off all protein formation for want of an amino acid or two may appear inefficient, but if the body did not cease all protein formation, cells would end up with an imbalanced assortment of proteins. This would seriously affect cell functions. When the required level of an essential amino acid is lacking, the remaining amino acids are used as a source of energy or are converted to fat and stored.

Table 15.1
Essential and nonessential amino acids.

Essential		Nonessential	
Histidine	Tryptophan	Alanine	Glutamine
Isoleucine	Valine	Arginine	Glycine
Leucine		Asparagine	Proline
Lysine		Aspartic acid	Serine
Methionine		Cysteine	Tyrosine
Phenylalanine		Glutamic acid	
Threonine			

Amino acids cannot be stored very long in the body, so we need a fresh supply of essential amino acids daily. This means we need to consume foods that provide a sufficient amount of all essential amino acids every day.

Complete Proteins Food sources of high-quality protein (meaning they contain all of the essential amino acids in the amount needed to support protein formation) are called ***complete proteins.*** Proteins in this category include those found in animal products such as meat, milk, and eggs (Illustration 15.6). ***Incomplete proteins*** are deficient in one or more essential amino acids. Proteins in plants are "incomplete," although soybeans are considered a complete source of protein for adults. (Soybeans may not meet the essential amino acid requirements of young infants.)[5] You can "complement" the essential amino acid composition of plant sources of protein by combining them to form a "complete" source of protein. Illustration 15.7 shows a few plant combinations that produce complete proteins.

Vegetarian Diets Diets that consist only of plant foods can provide an adequate amount of complete proteins. The key to success is the consumption of a variety of complementary sources of protein each day. Illustration 15.7 shows some of the combinations of plant foods that provide a complete source of protein. Vegetarian diets have been practiced for centuries by some religious and cultural groups, bearing testimony to their general adequacy and safety. (The unit on vegetarian diets expands on this topic.)

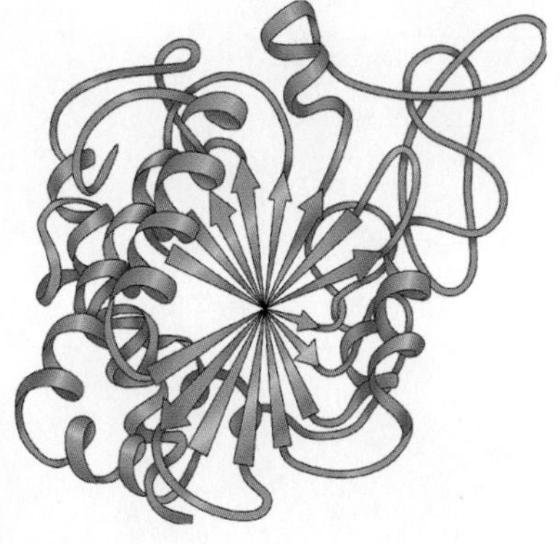

Illustration 15.4
Amino acid chains in the protein insulin (shown on the left), and the structure of insulin (shown above).

Amino Acid Supplements

Because amino acids occur naturally in foods, people often assume they are harmless, no matter how much is taken (Illustration 15.8). Researchers have known for decades, however, that high intakes of individual amino acids can harm health.[6] Nevertheless, this issue didn't receive much attention until a few years ago, because very few people took amino acids in supplement form in the past. Furthermore, manufacturers are not required to demonstrate the safety of amino acids before selling them as supplements, so the public becomes aware of problems only when something goes wrong—like the disaster of the contaminated tryptophan supplements.

➡ **complete proteins**
Proteins that contain all of the essential amino acids in amounts needed to support growth and tissue maintenance.

➡ **incomplete proteins**
Proteins that are deficient in one or more essential amino acids.

The Tryptophan Supplement Scare

Nature's original sleeping pill? Safe as a cup of warm milk? Supplements of the amino acid tryptophan (pronounced "trip-toe-phane") turned into heartbreaking stories of a serious illness that terrified many people.

Late in the summer of 1989, people from coast to coast developed a mysterious illness characterized by painful muscles and joints, weakness, fever, cough, a rash, and swelling. Some people who had the disease described the muscle pain as "like having a charley horse" in every muscle. Diagnosing the cause of the

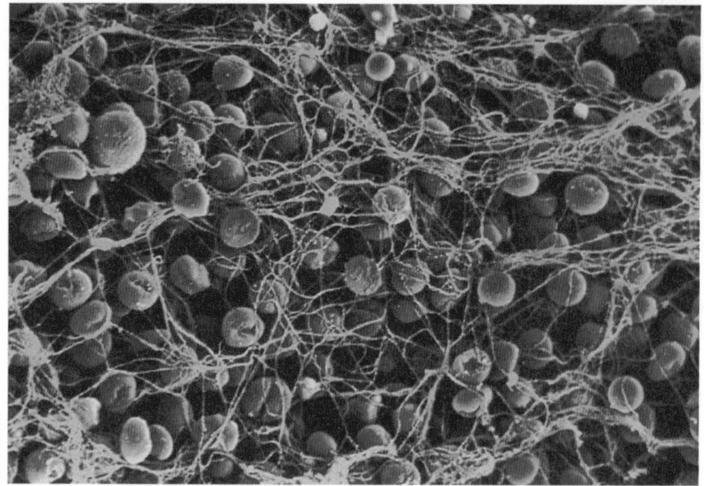

Illustration 15.5
Red blood cells enmeshed in fibrin in a color-enhanced microphotograph. Red blood cells and fibrin (which helps stop bleeding by causing blood to clot) are made primarily from protein.

symptoms turned out to be difficult. Within a few months, however, the disease was diagnosed as eosinophilia-myalgia syndrome (EMS). It has no cure. The outbreak was quickly related to tryptophan supplement use when researchers confirmed that 98% of the patients with EMS used tryptophan.[7] In March 1990, tryptophan was taken off the market, and people were urged to stop taking the supplements. By April 1990, when tryptophan was banned altogether, 1477 cases of EMS had been reported and at least 28 people had died from the disease. Contaminants in certain batches of the supplement and other factors appear to have been related to the development of EMS.[8] Although history now, the tryptophan scare taught us an important lesson about the potential hazards of amino acid supplements.

Melatonin: The Tryptophan Replacement? A derivative of tryptophan, the hormone melatonin (Illustration 15.9), entered the marketplace shortly after tryptophan supplements were banned. In addition to promoting drowsiness and sleep, melatonin has been touted as a cure for jet lag and a way to help your body switch to a night shift. Studies have produced conflicting results on the ability of melatonin to facilitate sleep in young adults, but doses of about 1 milligram per day may improve sleep in older adults lacking normal levels of brain melatonin.[9] It does not appear to be effective against jet lag or helpful in the switch to night shifts.[10]

Illustration 15.6
Animal sources of protein supply "complete proteins." Each food shown is a source of complete protein.

Illustration 15.7
Each of these combinations of plant foods is also a source of complete protein.

Low doses of melatonin (1 milligram per day) do not appear to be harmful, but the safety of melatonin supplements is still unclear. As was the case for tryptophan, there is little oversight of the purity or dose levels of melatonin in supplements.

Illustration 15.8
A wide array of amino acid supplements are available over-the-counter, but the safety of these supplements is unknown.

Can Amino Acid Supplements and Protein Powders Build Muscle?

You can't just consume amino acids or protein powders and watch your muscles grow (no matter how convincing the ads that sell such products are).[11,12] If that happened, everyone who wanted a rippled stomach and bulging triceps could have them. No amino acids or protein powders are delivered directly to our muscles and cause muscles to enlarge. Only exercise and a good diet build muscles (Illustration 15.10).

Illustration 15.9
Tryptophan supplements were banned in the United States on April 23, 1990, a month after their recall. A close chemical relative, melatonin, is available, but its safety is unclear.

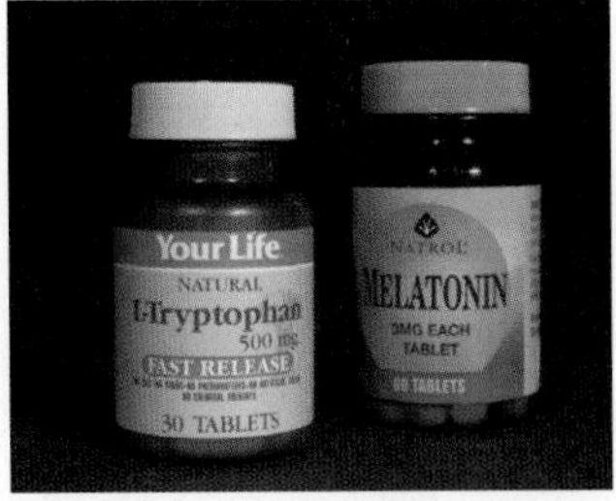

Amino acid supplements may be harmful. High amounts may disrupt normal protein production by overwhelming cells with a surplus of some amino acids and a relative deficit of others. Amino acid supplements have been known to cause loss of appetite, diarrhea, and other gastrointestinal upsets. They increase the work load of the liver and kidneys and the likelihood of dehydration.[11,12] Amino acids in protein powders are largely converted to glucose and fat by the liver. The conversion of amino acids to glucose or fat entails clipping off the nitrogen component of amino acids in the liver and the excretion of the nitrogen by the kidneys. (The conversion of high amounts of amino acids to energy sources by the body is made evident by dark-colored urine.) Converting excess amino acids into glucose and fat increases the body's need for water because water is excreted along with the excess nitrogen.[11,12] In addition, the purity of and the amino acid dose levels in amino acid supplements do not have to be guaranteed by manufacturers.

Most athletes get more than enough protein from their usual diet to support their need for it—which is not extraordinarily high.[13]

Illustration 15.10
No amount of protein powders or amino acid supplements will build muscles such as these. Exercise and a good diet are needed to build muscles.

Photo Researchers, Inc. Jean MacBarey

Food as a Source of Protein

Approximately 70% of the protein consumed by Americans comes from meats, milk, and other animal products.[14] Dried beans and grains are not as well known for their protein but are nevertheless good sources (Table 15.2). Plant sources of protein are generally low in fat, making them a wise choice for consumers who are trying to limit their intake of fat. Nearly all food sources of protein provide an assortment of vitamins and minerals as well. Beef and pork are particularly good sources of iron, a mineral often lacking in the diets of women (Table 15.3).

What Happens When a Diet Contains Too Little Protein?

Protein deficiency can occur by itself or in combination with a deficiency of calories and nutrients. Because food sources of protein generally contain essential nutrients such as iron, zinc, vitamin B_{12}, and niacin, diets that produce protein deficiency usually cause a variety of other deficiencies, too. Protein deficiency

Table 15.2
Food sources of protein. The adult RDA is 46 grams for women aged 19 to 24 years and 58 grams for men aged 19 to 24 years.

Food	Amount	Protein Content: Grams	Protein Content: Percentage of Total Calories
Animal products			
Tuna (water packed)	3 oz	24	89%
Shrimp	3 oz	11	84
Cottage cheese (low-fat)	½ cup	14	69
Beef steak (lean)	3 oz	26	60
Chicken (no skin)	3 oz	24	60
Pork chop (lean)	3 oz	20	59
Beef roast (lean)	3 oz	23	45
Skim milk	1 cup	9	40
Fish (haddock)	3 oz	19	38
Leg of lamb	3 oz	22	37
Yogurt (low-fat)	1 cup	13	34
Hamburger (lean)	3 oz	24	34
Egg	1 medium	6	32
Swiss cheese	1 oz	8	30
Sausage (pork links)	3 oz	17	28
2% milk	1 cup	8	26
Cheddar cheese	1 oz	7	25
Whole milk	1 cup	8	23
Dried beans and nuts			
Tofu	½ cup	14	38
Soybeans (cooked)	½ cup	10	33
Split peas (cooked)	½ cup	5	31
Lima beans (cooked)	½ cup	6	27
Dried beans (cooked)	½ cup	8	26
Peanuts	¼ cup	9	17
Peanut butter	1 tbs	4	17
Grains			
Corn	½ cup	3	29
Egg noodles (cooked)	½ cup	4	25
Oatmeal (cooked)	½ cup	3	15
Whole-wheat bread	1 slice	2	15
Macaroni (cooked)	½ cup	3	13
White bread	1 slice	2	13
White rice (cooked)	½ cup	2	11
Brown rice (cooked)	½ cup	2	10

Table 15.3
Iron content in a 3-ounce serving of various meats. The RDA for women aged 19 to 24 years is 15 milligrams. The RDA for men aged 19 to 24 years is 10 milligrams.

Meat	Iron Content (mg)
Pork chop (lean)	3.4
Round steak (lean)	3.1
Hamburger (lean)	3.0
Shrimp	2.6
Tuna	1.6
Baked chicken (no skin)	1.4
Lamb (lean)	1.3

leads to a loss of muscle tissue, growth retardation, reduced resistance to disease, weakness, edema, and kidney and heart problems.

Kwashiorkor is a severe form of protein deficiency in children. It usually develops after a child has been weaned from breast milk and given high-carbohydrate, low-protein foods such as cassava or a watery gruel made from oats, rice, or corn. A protein-deficient body cannot grow, maintain a normal level of blood volume, produce digestive enzymes, fight infections adequately, or maintain existing tissues in proper working order. As seen in Illustration 15.11, children with kwashiorkor may look fat due to ***edema*** (swelling), but they are actually very skinny. Children with protein deficiency are apathetic, irritable, small, and highly vulnerable to infection.

Protein deficiency may be accompanied by a lack of calories. In this case, a condition called ***marasmus,*** or protein-energy malnutrition exists. Unfortunately, this condition is all too frequent in areas of the world where hunger and famine are common. It is also the type of malnutrition that occurs among people with severe anorexia nervosa and certain diseases such as cancer and AIDS. Individuals with marasmus look starved and wasted (see Illustration

➡ **kwashiorkor (kwa-she-or-kor)**
A deficiency disease primarily caused by a lack of complete protein in the diet. It usually occurs after children are taken off breast milk and given solid foods that contain protein of low quality.

➡ **edema**
Swelling due to an accumulation of fluid in body tissues.

➡ **marasmus**
A condition of severe body wasting due to deficiencies of both protein and calories. Also called *protein-energy malnutrition* and *protein-calorie malnutrition.*

15.12). They have precious little body fat and must utilize protein from muscles, the liver, and other tissues as an energy source.

In addition to having a "skin and bones" appearance, children with marasmus are apathetic, highly susceptible to infection, and have dry skin and brittle hair. Use of the fat pads in the cheeks for energy (which are among the last fat deposits to be used for energy during starvation) gives children with advanced cases of marasmus the look of a very old person.

Both kwashiorkor and marasmus are generally accompanied by multiple vitamin and mineral deficiencies due to limited food intake. The conditions are frequently complicated by infection, diarrhea, and dehydration.

Trends

In the early 1900s about half of the protein consumed in the United States came from plant sources and half from animal sources. Now, approximately two-thirds of the protein intake comes from animal products.[1,15] As countries develop economically, the proportion of dietary protein obtained from meats tends to increase. The increased intake of protein from meats is accompanied by an increased consumption of fat and by elevated rates of some of the "diseases of Western civilization" such as heart disease and certain cancers. Adequate intakes of protein without excess levels of fat can easily be obtained from diets that include dried beans, cereals, and other grains as sources of protein as well as lean meats and low-fat dairy products.

Illustration 15.11
Children suffering from a severe form of protein deficiency called kwashiorkor experience swelling in the arms, legs, and stomach area. The swelling hides the devastating wasting that is taking place within their bodies. The child at left has the characteristic "moon face" (edema), swollen belly, and patchy dermatitis (from zinc deficiency) often seen with kwashiorkor. At right, the same child after nutritional therapy.

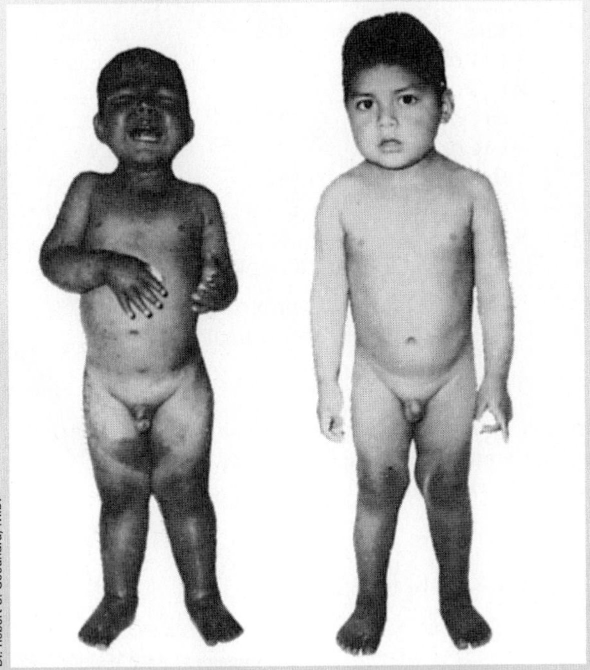

Illustration 15.12
Individuals with marasmus look as starved as they are. People with this condition lack both protein and calories. The child at left is suffering from the extreme emaciation of marasmus. At right is the same child after nutritional therapy.

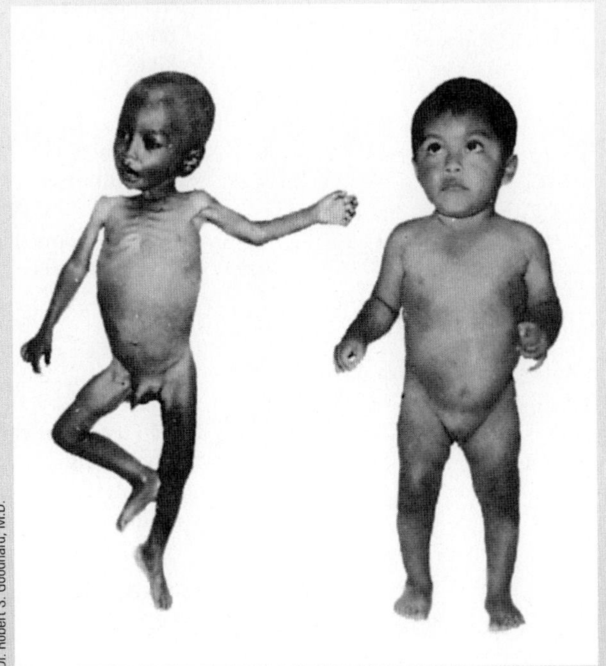

Nutrition UP CLOSE

Are You Overconsuming Protein?

FOCAL POINT: Determine the amount of protein in your diet yesterday.

If you eat a typical American diet, chances are you consume too much protein. To assess your protein intake, consult the list of foods below. For *each serving* of a food item you ate yesterday, write the grams of protein the food contains in the corresponding blank. For example, a standard serving of meat is 3 ounces (about the size of the palm of your hand or a deck of cards). If you had *one* 3-ounce pork chop yesterday, write *20 grams* in the corresponding blank. If you had *two* 3-ounce pork chops, write *40 grams*. If a protein food you ate yesterday is not included, choose the item on the list closest to it. Then, total the grams of protein you ate yesterday from both plant and animal sources.

Food	One Serving	Protein in One Serving (grams)	Protein You Ate (grams)
Animal products			
Milk (whole)	1 c (8 oz)	8	______
Yogurt	1 c (8 oz)	13	______
Cottage cheese	½ c (4 oz)	14	______
Hard cheese	1 oz	7	______
Hamburger (lean)	3 oz	24	______
Beef steak (lean)	3 oz	26	______
Chicken (no skin)	3 oz	24	______
Pork chop (lean)	3 oz	20	______
Fish	3 oz	19	______
Hot dog	1	6	______
Sausage	3 oz	17	______
		Subtotal from animal foods:	______
Plant products			
Bread	1 slice	2	______
Rice	½ c (4 oz)	2	______
Pasta	½ c (4 oz)	3	______
Cereals	½ c (4 oz)	3	______
Vegetables	½ c (4 oz)	2	______
Peanut butter	1 tbs	4	______
Nuts	¼ c (2 oz)	7	______
Cooked beans (legumes)	½ c (4 oz)	8	______
		Subtotal from plant foods:	______
		Total grams of protein from plant and animal foods:	______

Special note: You can also calculate your protein intake using Wadsworth Diet Analysis Plus software. Input your food intake for one day. Then go to the Analyses/Reports section to view the total number of grams of protein in your diet.

FEEDBACK can be found at the end of Unit 15.

www.healthfinder.gov
Search topics related to protein.

www.nemsn.org
The National EMS Network provides answers to questions about EMS, tryptophan, and melatonin.

REFERENCES

1. McDowell MA, et al. Energy and macronutrient consumption of persons 2 months and over in the United States: third national Health and Nutrition Examination Survey, 1988–91. Hyattsville, MD: National Center for Health Statistics, 1994.
2. Food and Drug Administration. Nutrition labeling. Federal Register 1991 Nov 27.
3. Dietary guidelines for Americans. Washington, DC: U.S. Public Health Service; 1995.
4. National Academy of Sciences (Institute of Medicine). Recommended dietary allowances, 10th ed. Washington, DC: National Academy Press; 1989.
5. Young VR. Soy protein in relation to human protein and amino acid nutrition. J Am Diet Assoc 1991; 91:828–35.
6. Harper AE, et al. Effects of ingestion of disproportionate amounts of amino acids. Physiol Rev 1970;50:428–558.
7. Centers for Disease Control. Update: eosinophilia-myalgia syndrome associated with ingestion of L-tryptophan—United States, as of January 9, 1990. JAMA 1990;263:633.
8. Fernstrom JD. Can nutrient supplements modify brain function? Am J Clin Nutr 2000;71(suppl):1669S–73S.
9. Brzezinski A. Melatonin in humans. New Engl J Med 1997;336:186–95.
10. Arendt J. Jet-lag and shift work. Therapeutic uses of melatonin. J Royal Soc Med 1999;92:402–5.
11. Lund BC, Perry PJ. Nonsteroid performance-enhancing agents in athletic competition: an overview for clinicians. Medscape Pharmacotherapy: www.medscape.com/Medscape/pharmacology/journal/2000/v02,n05; Sept. 25, 2000.
12. Ergogenic aids: reported facts and claims. SCAN's Pulse 1999;winter:12–21.
13. Hultman E. Nutritional effects on work performance. Am J Clin Nutr 1989;49(suppl):949–57.
14. Scrimshaw NS. Nutrition and preventive medicine. In: Last JM, ed. Public health and preventive medicine, 12th ed. Norwalk (CT): Appleton-Century-Crofts; 1986:1515–42.
15. The surgeon general's report on nutrition and health. Washington, DC: Public Health Service (DHHS publication no. PHS 88-50210); 1988.

Nutrition UP CLOSE

Are You Overconsuming Protein?

Feedback for Unit 15

The recommended amount of daily protein for adults between the ages of 19 and 24 is 46 *grams* for *women* and 58 *grams* for *men*. If you exceeded these amounts, you are consuming unnecessary protein. The average American intake is 74 grams daily—far in excess of the required amounts.

Now, compare your subtotals to find out which protein source you prefer, plant or animal. Protein from animal products is often accompanied by fat. If you are concerned about calories and fat in your diet, choose plant protein sources more often. When choosing animal products, make sure to opt for low-fat dairy products and lean meats.

Unit 16

Vegetarian Diets

Nutrition Scoreboard

	TRUE	FALSE
1. The human body developed to function best on a vegetarian diet.		
2. People who don't eat meat have more health problems than people who do.		
3. Macrobiotic diets cure some types of cancer.		
4. In order to consume enough high-quality protein, vegetarians need to consume combinations of plant foods that provide a complete source of protein at every meal.		

(Answers on next page)

Key Concepts and Facts

- Vegetarianism is more than a diet. It is often part of a value system that influences a variety of attitudes and behaviors.
- Appropriately planned vegetarian diets are health promoting.
- Vegetarian diets that lead to caloric and nutrient deficiencies generally include too narrow a range of foods.

Answers to **Nutrition Scoreboard**

	TRUE	FALSE
1. Early humans developed on an omnivorous diet (meat and plant diet). However, health can be fostered on an omnivore or a vegetarian diet.		✔
2. Persons who consume appropriately planned vegetarian diets do not have more health problems than other people. Furthermore, such diets may reduce the risk of developing several chronic diseases.		✔
3. No type of vegetarian or quasi-vegetarian diet, including macrobiotic and vegan diets, has been found to cure cancer.		✔
4. Vegetarians do need to plan their diets to ensure that they obtain enough high-quality protein, but they need to consume complete sources of protein daily, not at every meal. Appropriately planned vegetarian diets provide sufficient amounts of high-quality protein.		✔

Perspectives on Vegetarianism

. . . unless vegetarianism comes from the soul, it will just be a passing fad.

—Janis Barkas, 1978

It wouldn't be any fun to live without meat. . . . I do not believe in this meatless, dairyless, butterless society.

—Julia Child, 1991

Vegetarianism in the United States, Canada, and other economically developed countries is moving from the realm of counterculture to the mainstream. Yet even with the increased acceptance of vegetarian diets, people tend to be for vegetarianism or against it, often without knowing much about it. Few of those opposed to vegetarianism have tried to learn about the vegetarian way of life or understand that vegetarian diets can be healthful. A small percentage of health professionals are vegetarians, and those who are not are often skeptical about how healthy a vegetarian diet can be. The possibility that something so different from the customary diet can be nourishing to the body may be rejected out of hand.

An objective look at vegetarianism reveals that both appropriately planned vegetarian diets and the lifestyle often followed by vegetarians can be very good for health.[1] Vegetarians tend to be health conscious, often avoid using alcohol, tobacco, and illicit drugs, and engage in regular physical activity.[2] Diet is usually one of several characteristics shared by people practicing particular types of vegetarianism.

Why Do People Become Vegetarians?

For many people, vegetarianism is not a choice (Table 16.1). Most of the world's population subsists on vegetarian diets because meat and other animal products are scarce or too expensive.[3] In other societies, people have the luxury of choosing a healthy assortment of food from an abundant and affordable food supply that includes a wide variety of items acceptable to vegetarians. When food availability is not an issue, people tend to adopt vegetarian diets because of a desire to cause no harm to animals, a desire to preserve the environment and the world's food supply by "eating low on the food chain," or a belief that animal products are unhealthful or unsafe. They may avoid animal products as part of a value or religious belief system. Others follow vegetarian diets to keep their weight down or to lower the risk of developing specific diseases such as cancer or heart disease.[3]

Table 16.1
Reasons for vegetarianism.

- Lack of availability or affordability of animal products
- Desire not to cause harm to animals
- Religious beliefs
- Desire to "eat low on the food chain"
- Desire to preserve the world's food supply
- Health

Vegetarian Diets Come in Many Types

> Yeah, I guess I'm kind of a vegetarian, too. I try not to eat meat—well, I love chicken but I don't eat red meat. I mean, I eat hamburgers. I can't remember the last time I had a steak.
>
> —A 17-year-old "vegetarian"

There is no one vegetarian diet. People who consider themselves to be vegetarians range from those who eat all foods except red meat (mainly beef) to those who exclude all foods from animal sources, including honey. According to the American Vegetarian Society, a person is a vegetarian only if she or he eats *no* meat.

Vegetarian Diet Options The types of foods included in common vegetarian diets are summarized in Table 16.2. The least restrictive form of vegetarian diet has been unofficially labeled "far vegetarian" because it excludes only red meats. This diet is very much like that consumed by omnivores, or meat and plant eaters. Quasi-vegetarian diets (also called semivegetarian diets) vary somewhat, but generally exclude beef, pork, and poultry, while including fish, eggs, dairy products, and plant foods. The lacto-ovovegetarian diet excludes additional foods. Individuals practicing this type of diet exclude all meats, but include milk and milk products, eggs, and plant foods. The lactovegetarian diet, as the name implies, includes only milk and milk products and plant foods.

Table 16.2
Vegetarian diets come in many types.

	Foods Included					
Type of Diet	Beef, Lamb, Pork ("Red Meat")	Poultry	Fish	Eggs	Milk and Milk Products	Plant Foods
"Far" vegetarian		x	x	x	x	x
Quasi-vegetarian		*	x	x	x	x
Lacto-ovovegetarian				x	x	x
Lactovegetarian					x	x
Macrobiotic			**			x
Vegan						x

*Quasi-vegetarian diets (also called semivegetarian) may include poultry; they tend to vary in the type of animal products consumed.
**Macrobiotic diets may include fish and other animal products.

Vegan diets are a more restrictive type of vegetarian diet. Vegans eat only plant foods. In addition, they may avoid honey and clothes made from wool, leather, or silk.

Macrobiotic diets fall somewhere between quasi-vegetarian and vegan diets. The formulation of macrobiotic diets has changed dramatically over the past decade. In addition to brown rice, other grains, and vegetables, these diets may now include fish, dried beans, spices, fruits, and many other types of foods. No specific foods are prohibited,[4] and locally grown and whole foods are emphasized.[5]

Each of these vegetarian diets has philosophical underpinnings. Persons adhering to the macrobiotic philosophy, for example, place value on consuming organic foods grown locally and balancing the intake of foods nonjudgmentally assigned as "yin" and "yang." Foods are classed as yin or yang based on beliefs about the food's relationship to the emotions and the physical condition of the body. Yin foods such as corn, seeds, nuts, fruits, and leafy vegetables are considered negative, dark, cold, and feminine. Yang foods represent opposing positive forces of light, warmth, and masculinity. Poultry, fish, and eggs and cereal grains such as buckwheat are yang foods.[4]

For a number of vegetarian regimes, the spiritual or emotional importance assigned to certain foods supersedes consideration of their contribution to an adequate diet.[6] Vegetarians adhering to the "living foods diet" consume uncooked and fermented plant foods only. This diet is inadequate in a number of nutrients, including vitamin B_{12}.[5] ***Fruitarians*** consume only fruit and olive oil. People adopting this type of diet rarely stick with it for long—it does not sustain health.

➡ **fruitarian**
A form of vegetarian diet in which fruits are the major ingredient. Such diets provide inadequate amounts of a variety of nutrients.

Vegetarian Diets and Health

> Appropriately planned vegetarian diets are healthful, are nutritionally adequate, and provide health benefits in the prevention and treatment of certain diseases.
>
> —Position of the American Dietetic Association, 1997

Long-standing vegetarian dietary regimes that promote health have been tested by time (if not by science) and found to be adequate in essential nutrients.[1] All of the types of vegetarian diets listed in Table 16.2 are health promoting if implemented correctly. On the other hand, highly restrictive vegetarian diets, such as fruitarian regimes and newly popular vegetarian diets that stray from established regimes, can be harmful to health. In general, the more restrictive the diet, the more likely it is to be inadequate and lead to health problems. This is especially true for pregnant women, children, and persons who are ill, all of whom have relatively high needs for nutrients. Reports of caloric and nutrient deficiencies in young children of vegan parents are more common than is the case with less restrictive vegetarian diets.[7,8] Vegan diets may supply too little vitamin B_{12}, vitamin D, calcium, and zinc—a mineral that is poorly absorbed from plant sources.[9] Vegetarians in economically developed countries are not at greater risk for iron deficiency than nonvegetarians.[9,10] Sufficient iron intake from plants and ample intake of vitamin C, which enhances absorption of iron from plants, meet their need for iron. In contrast, the diets of vegetarians in developing countries often include too few iron-rich plants, making iron deficiency common. Both the quality and the quantity of protein in vegetarian diets can also be a source of concern. However, vegetarians in developed countries generally have adequate protein intakes.[9,10]

What Are the Benefits to Health of Vegetarian Diets?

> I am a great eater of beef, and I believe that does great harm to my wit.
>
> —William Shakespeare, 1589

Compared to the usual American diet and lifestyle, vegetarianism is associated with a lower risk of developing heart disease, stroke, hypertension, diabetes, chronic bronchitis, gallstones and kidney stones, and colon cancer.[11–13] People who follow a vegetarian diet rarely become obese or develop high blood cholesterol levels.[14] For children and adults at risk of early heart disease, the low-fat and high-fiber, vegetable and fruit content of vegetarian diets may help reduce blood cholesterol levels and the risk of heart disease.[15] Appropriately implemented vegetarian diets that meet a person's need for essential nutrients clearly can be health promoting.[16]

Dietary Recommendations for Vegetarians

Because vegetarian diets exclude one or more types of foods, it's important that the foods included provide sufficient calories and the assortment and quantity of nutrients needed for health. No matter what the motivation underlying the assortment of foods included, vegetarian diets that fail to provide all the nutrients humans need in the required amounts will not sustain health.

How to Eat Vegetarian

The watchword for vegetarians striving for a nutritionally complete diet is *variety.* At the minimum, vegetarian diets should include an assortment of fruits, vegetables, grains and grain products (such as cereals, pasta, and breads), nuts, seeds, dried beans, and dairy products or fortified soy substitutes.

Illustration 16.1 shows the Food Guide Pyramid adapted by the USDA for use by vegetarians. Although a number of vegetarian food guides have been developed, this one is considered the "official" guide. The specific foods selected

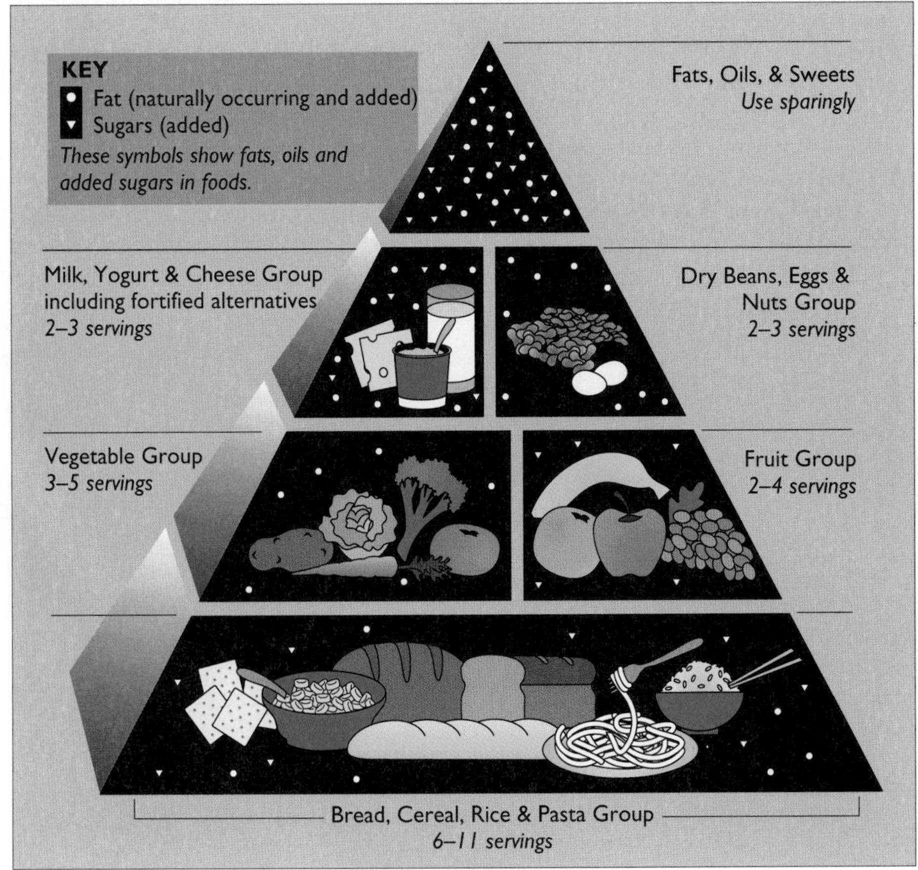

Illustration 16.1
The "official" vegetarian Food Guide Pyramid.

from each group will vary depending on the type of vegetarian diet consumed. For example, lacto-ovovegetarians could include eggs from the "dry beans, eggs, and nuts group." Lactovegetarians would exclude eggs but could include food from the "milk, yogurt, and cheese group."

The American Dietetic Association has developed a set of guidelines for vegetarian meal planning. These guidelines are summarized in Table 16.3.

Special Considerations for Vegetarian Diets Diets that include few or no animal products may be low in sources of ***complete protein,*** vitamin B_{12}, vitamin D, calcium, and zinc. With appropriate food selection, these potential nutrient inadequacies can be prevented.

complete proteins Proteins that contain all of the nine essential amino acids in amounts sufficient to support protein tissue construction by the body.

essential amino acids Amino acids that cannot be synthesized in adequate amounts by the human body and must therefore be obtained from the diet.

Complementing Plant Protein Sources Animal products such as meat, eggs, and milk provide all of the nine ***essential amino acids*** in sufficient quantity to qualify as a complete source of protein. In addition, tests have shown soy proteins to be complete protein sources for children and adults.[17] The diet must include complete proteins because the body needs a sufficient supply of each essential amino acid if it is to build and replace protein substances such as red blood cells and enzymes. If any of the essential amino acids are missing in the diet, protein tissue construction stops, and the available amino acids will be used for energy instead. Essential amino acids consumed in foods are not stored, so the body needs a fresh supply each day.

Vegetarians who don't consume animal products can meet their need for essential amino acids by combining plant foods to yield complete proteins. This is done by consuming plant foods that *together* provide all the essential amino acids although each individual food is missing some of these essential nutrients. The goal is to "complement" plant sources of essential amino acids, or to consume ***complementary protein sources*** from plant foods regularly.[3]

complementary protein sources Plant sources of protein that together provide sufficient quantities of the nine essential amino acids.

Many different combinations of plant foods yield complete proteins. Basically, complete sources of protein can be obtained by combining grains such as rice, bulgur (whole wheat), millet, or barley with dried beans, tofu, or green peas, or corn with lima beans or dried beans, or seeds with dried beans. Some examples of complementary sources of plant proteins are shown in Illustration 16.2. Milk, meat, and eggs contain complete proteins and will complement the essential amino acids profile of any plant source of protein.

It used to be thought that you had to consume complementary sources of protein in the same meal to meet the body's need for essential amino acids. It is now concluded, however, that complementary protein sources should be consumed approximately daily.[3]

Table 16.3
The American Dietetic Association's guidelines for vegetarian meal planning.[16]

1. Choose a variety of foods, including whole grains, vegetables, fruits, legumes, nuts, seeds, and, if desired, dairy products and eggs.
2. Choose whole, unrefined foods often and minimize intake of highly sweetened, fatty, and highly refined foods.
3. Choose a variety of fruits and vegetables.
4. If animal foods such as dairy products and eggs are used, choose lower-fat versions of these foods. Cheeses and other high-fat dairy foods and eggs should be limited in the diet because of their saturated fat content and because their frequent use displaces plant foods in some vegetarian diets.
5. Vegans should include a regular source of vitamin B_{12} in their diets along with a source of vitamin D if sun exposure is limited.
6. Do not restrict dietary fat in children younger than two years. For older children, include some foods higher in unsaturated fats (e.g., nuts, seeds, nut and seed butters, avocado, and vegetable oils) to help meet nutrient and energy needs.

Getting Enough Vitamin B_{12} Vitamin B_{12} is of particular concern to vegetarians because it is present only in animal products. It was once believed that sea vegetables, algae, and fermented soybean products such as tamari, tempeh, and miso provided vitamin B_{12}. We now know that the vitamin B_{12} in these foods is in a form that humans utilize poorly.[5] Vegetarians who don't consume animal products can obtain vitamin B_{12} from fortified products such as soy milk and breakfast cereals and from vitamin B_{12} supplements.

Where's the Vitamin D? People get vitamin D from two primary sources: the sun and milk. Exposure of the skin to sunlight causes vitamin D to be produced from a substance that circulates in the blood. People can meet their need for vitamin D this way, but not everyone lives in a climate where the skin can be exposed to the sun regularly enough to produce all the vitamin D needed.

Vitamin D is found in only a few foods, and most of our intake comes from the consumption of vitamin D–fortified milk. (Milk is fortified with vitamin D, but other dairy products such as yogurt and cheese generally are *not.*) If milk is not part of the vegetarian's diet plan, and exposure of the skin to sunlight is limited, vitamin D–fortified soy milk and breakfast cereals or a vitamin D supplement should be consumed. Because vitamin D is toxic at high levels, supplements should provide around 200 to 400 IU (or 5 to 10 micrograms) per day.

On the Side

There are more than 70 varieties of dried beans. They are an important part of the cuisine of almost all cultures. Bean cuisine includes chili, baked beans, beans and rice, hummus, and refried beans.

Illustration 16.2

Some combinations of plant foods that provide complete protein:

- Rice and dried beans
- Rice and green peas
- Bulgur (wheat) and dried beans
- Barley and dried beans
- Corn and dried beans
- Corn and green peas
- Corn and lima beans (succotash)
- Soybeans and seeds
- Peanuts, rice, and dried beans
- Seeds and green peas

Food combinations that supply complete protein are shown below:

Rice and black beans.

Hummus and bread.

Corn and black-eyed peas.

Bulgur (whole wheat) and lentils.

Tofu and rice.

Corn and lima beans (succotash).

Tortilla with refried beans (e.g., a bean burrito).

Pea soup and bread.

Contributors of Calcium to the Vegetarian Diet Vegetarians who exclude milk and milk products need to rely on good plant sources of calcium (kale, broccoli, and dried beans, for example) and calcium-fortified products such as soy milk and orange juice. High intakes of soy products such as fermented soybeans, miso, and tofu appear to promote bone mass and protect against the development of osteoporosis.[18] Although plant foods contribute significantly to calcium intake, it is difficult to obtain the 1000 milligrams or more needed each day without milk or milk products.[16] Many plant sources of calcium contain substances that block the body's ability to absorb calcium. For example, four and one-half servings of broccoli contain about the same amount of calcium as one cup of milk. One would have to consume 20 servings of broccoli, however, to meet daily calcium needs after adjusting for the availability of calcium in broccoli.[19]

Solving the Zinc Problem Many plant foods are good sources of zinc. The problem is that zinc from plants is poorly absorbed, so only a small portion of the zinc consumed becomes available for use by the body. To obtain enough zinc, vegans should consume whole grains, dried beans, nuts, zinc-fortified breakfast cereals, and other sources of zinc regularly.[20]

Where to Go for More Information on Vegetarian Diets You can find information about vegetarian diets in a variety of sources. World Wide Web resources appear at the end of this unit, and several cookbooks that offer information about vegetarian nutrition and delicious recipes are pictured in Illustration 16.3.

Unfortunately, some of the information available on vegetarianism is wrong or misconstrued. Some vegetarian organizations are more committed to selling particular beliefs, memberships, or books and magazines than to promoting healthy vegetarian diets. Beware of vegetarian diets that claim to cure cancer, AIDS, or other serious illnesses or promise you'll experience inner peace or spiritual renewal. Appropriately planned vegetarian diets are health promoting, but they are not magic bullets that will cure all of the ills of body and soul.

Illustration 16.3
Many excellent vegetarian cookbooks are available; most provide information on diet as well as recipes. The books pictured here are *Moosewood Restaurant Low-Fat Favorites: Flavorful Recipes for Healthful Meals* edited by Pam Krauss; *Vegetarian Times Complete Cookbook* by the editors of *Vegetarian Times; Vegetarian Cooking for Everyone* by Deborah Madison; *The Whole Soy Cookbook* by Patricia Greenberg; *Vegetable Heaven* by Mollie Katzen.

Nutrition UP CLOSE

Vegetarian Main Dish Options

FOCAL POINT: Serving up vegetarian alternatives to meat dishes.

Assume you are a member of the "vegetarian option" planning committee for your college. You are asked to identify three meatless main dishes that could be served in the dining halls for lunch and another three that could be served for dinner. The one stipulation is that they should be main dishes you would enjoy eating.

What meatless dishes would you identify as options for lunch and for dinner?

Lunch Dishes	Dinner Dishes
1. ______	1. ______
2. ______	2. ______
3. ______	3. ______

FEEDBACK (answers to these questions) can be found at the end of Unit 16.

WWW links

www.nal.usda.gov/fnic/Fpyr/pyramid.html
Check this site for the Vegetarian Food Guide Pyramid, food guides from other countries, and links to other sites.

www.vrg.org
Contact the Vegetarian Resource Group.

www.veg.org/veg
Puts you in contact with the vegetarian pages.

www.macrobiotics.org
Learn more about the macrobiotic philosophy.

www.goodkarmacafe.com
Funky graphics; purchase vegetarian books, print recipes, or brush up on cooking skills.

www.vegweb.com
Provides guide to vegetarianism, recipes, and books.

REFERENCES

1. Segasothy M, Phillips PA. Vegetarian diet: panacea for modern lifestyle diseases: QJM 1999;92:531–44.
2. Buffonge I. The vegetarian/vegan lifestyle. West Indian Med J 2000;49:17–19.
3. Havala S, Dwyer J. Position of the American Dietetic Association: vegetarian diets—technical support paper. J Am Diet Assoc 1988;88:352–5.
4. Kushi M. The book of macrobiotics. Tokyo: Japan Publications, Inc.; 1977.
5. Rauma A-L, et al. Vitamin B-12 status of long-term adherents of a strict, uncooked vegan diet ("living foods diet") is compromised. J Nutr 1995;125:2511–15.
6. Mutch PB. Food guides for the vegetarian. Am J Clin Nutr 1988;48:913–19.
7. Inadequate vegan diets at weaning. Nutr Rev 1990;48:323–6.
8. Van Dusselforp M, et al. Risk of persistent cobalamin deficiency in adolescents fed a macrobiotic diet early in life. Am J Clin Nutr 1999;69:664–71.
9. Haddad EH, et al. Dietary intake and biochemical, hematologic, and immune status of vegans compared with nonvegetarians. Am J Clin Nutr 1999;70(suppl):586S–93S.
10. Ball MJ, Bartlett MA. Dietary iron status of Australian vegetarian women. Am J Clin Nutr 1999;70:353–8.
11. Key TJA, et al. Dietary habits and mortality in 11,000 vegetarians and health conscious people: results of a 17 year follow up. Brit Med J 1996;313:775–9.
12. Vecchia CL, et al. Vegetable consumption and risk of chronic disease. Epidemiol 1998;9:208–10.
13. Key TJ, et al. Health benefits of a vegetarian diet. Proc Nutr Soc 1999;58:271–5.
14. National Academy of Sciences web audiocast: New Dietary Recommendations, www.national-academies.org, 1/09/01.
15. Resnicow K, et al. Diet and serum lipids in vegan vegetarians: a model for risk reduction. J Am Diet Assoc 1991;91: 447–53.
16. Position of the American Dietetic Association: vegetarian diets. J Am Diet Assoc 1997;97:1317–21.
17. Young VR. Soy protein in relation to human protein and amino acid nutrition. J Am Diet Assoc 1991;91:828–35.
18. Somekawa Y, et al. Soy intake and bone mass in postmenopausal women, Obstet Gynecol 2000;97:109–15.
19. Smith CF, et al. Vegetarian and weight-loss diets among young adults. Obesity Res 2000;8:123–9.
20. Hunt JR, et al. Zinc absorption, mineral balance, and blood lipids in women consuming controlled lacto-ovo-vegetarian and omnivorous diets for 8 weeks. Am J Clin Nutr 1998;67:421–30.

Nutrition

UP CLOSE

Vegetarian Main Dish Options

Feedback for Unit 16

Vegetarians definitely had the advantage in completing this exercise! For the meat eaters, were the main dishes you identified meat-free?

Here are six vegetarian options for lunch and dinner main dishes:

- Eggplant parmesan
- Falafel
- Red beans and rice
- Pasta with mixed vegetables
- Melted cheese, tomato, and sprout sandwiches
- Bean burritos
- Vegieburgers

Unit 17

Food Allergies and Intolerances

Food Stock America, Eising

Nutrition Scoreboard

	TRUE	FALSE
1. About two in every five Americans are allergic to at least one food.		
2. Food labels must list allergenic ingredients.		
3. Skin prick tests are an accurate way to diagnose food allergies.		
4. Food intolerances cause less severe reactions than food allergies.		

(Answers on next page)

Answers to

Nutrition Scoreboard

	TRUE	FALSE
1. About two out of every five Americans *believe* they are allergic to at least one food. Probably one to two in a hundred adults actually are.[1]		✔
2. Listing potential allergens on food labels is voluntary. Congratulations are due to those companies that do list them.		✔
3. Skin prick tests for food allergies are often inaccurate.[2]		✔
4. (Yikes! Four false answers in a row! There are a lot of myths about food allergies.) Food allergies and food intolerances have different causes, but both can produce life-threatening reactions.		✔

Key Concepts and Facts

- True food allergies involve a response by the body's immune system to a particular substance in food.
- Food allergies may be caused by hundreds of different foods. However, most food allergies are due to cow's milk, eggs, peanuts, nuts, and wheat.[1]
- Food intolerances cover an array of adverse reactions to foods that do *not* involve the body's immune system.
- The most accurate method of identifying food allergies and intolerances is the double-blind, placebo-controlled food challenge.

Food Allergy Mania

Talk about your controversial subjects. This one is very hot. People use the term "food allergy" to refer to virtually any type of problem they have with food. At one extreme are people who believe allergies to milk, wheat, and sugar are to blame for hyperactivity and a host of other behavioral problems in children. At the other extreme are some health professionals who think people who complain of food allergies need to have their heads examined rather than their bodies' reactions to foods.

Food allergies are real and can be very serious. At the minimum, true food allergies can cause a rash or an upset stomach. At the maximum, they can result in death. Unreal food allergies can cause problems, too. They can lead people to eliminate nutritious foods from their diet unnecessarily, resulting in inadequate diets and eventually in health problems. One of the most intriguing aspects of food allergies is the frequency and ease with which foods are falsely blamed for a variety of mental and physical health problems.

food allergy
Adverse reaction to a normally harmless substance in food that involves the body's immune system. (Also called *food hypersensitivity*.)

immune system
Body tissues that provide protection against bacteria, viruses, and other substances identified by cells as harmful.

food intolerance
Adverse reaction to a normally harmless substance in food that does *not* involve the body's immune system.

Adverse Reactions to Foods

People experience adverse reactions to food for three primary reasons. One is food poisoning. The other two involve the body's reaction to substances in food that are normally harmless. With ***food allergies,*** the body's ***immune system*** reacts to a substance in food (almost always a protein) that it identifies as harmful (Illustration 17.1). The immune system is not involved in ***food intolerances,*** which encompass other adverse reactions to normally harmless substances in food.

Why Do Food Allergies Involve the Immune System? It may seem odd that a system that helps the body conquer bacteria and viruses is involved

PhotoEdit, Felicia Martinez

Illustration 17.1
What do these foods have in common? Each food shown, plus many more not shown, may cause adverse reactions in some individuals.

in "protecting" us from normal constituents of food. In people with allergies, however, the cells recognize some food components as harmful—just as they recognize bacteria and viruses. Components of food that trigger the immune system are called ***food allergens.***

In a small proportion of susceptible people, exposure to trace amounts of an allergen triggers an allergic reaction. Highly sensitive people can develop an allergic reaction simply by being in the room when the food to which they are allergic is prepared or by touching the food. Allergic reactions to peanuts have been triggered by kissing someone who has recently consumed them[3] and by breathing peanut vapors on airplanes.[4]

In response to the allergen, the immune system forms ***antibodies*** (Illustration 17.2). The antibodies attach to cells located in the nose, throat, lungs, skin, eyes, and other areas of the body. When the allergen appears again, the body is ready. The previously formed antibodies recognize the allergen, attach to it, and signal the body to secrete ***histamine*** and other substances that cause the physical signs of allergic reactions. Most commonly, allergic reactions cause a rash, diarrhea, congestion, or wheezing (Table 17.1), but the symptoms may be much more serious.

Exposure to trace amounts of an allergen in peanuts, nuts (peanuts are actually a legume, not a nut), fish, and shellfish can cause ***anaphylactic shock*** (Illustration 17.3). This massive reaction of the immune system can result in death, caused by the cutoff of the blood supply to tissues throughout the body. People experiencing anaphylactic shock can be revived with an injection of epinephrine.[5]

Symptoms of food allergies may occur within seconds after the body is exposed to the food allergen, or they may take up to two hours to occur.[6] The symptoms often disappear within two hours after they begin.[7]

➡ **food allergen**
A substance in food (almost always a protein) that is identified as harmful by the body and elicits an allergic reaction from the immune system.

➡ **antibodies**
Proteins the body makes to combat allergens.

➡ **histamine**
A substance released in allergic reactions. It causes dilation of blood vessels, itching, hives, and a drop in blood pressure and stimulates the release of stomach acids and other fluids. Antihistamines neutralize the effects of histamine and are used in the treatment of some cases of allergies.

➡ **anaphylactic shock**
(an-ah-fa-lac-tic) Reduced oxygen supply to the heart and other tissues due to the body's reaction to an allergen in food. Symptoms of anaphylactic shock include paleness, a flush discoloration of the lips and fingertips, a "faceless" expression, weak and rapid pulse, and difficulty in breathing.

What Foods Are Most Likely to Cause Allergic Reactions?

Many foods may cause allergic reactions in susceptible individuals (Table 17.2 on p. 17–5). Nevertheless, approximately 90% of all food allergies are caused by cow's milk, eggs, peanuts, nuts, and wheat. Soy, fish, and shellfish are also

Table 17.1
The three most common types of symptoms caused by food allergies. Some people experience more than one type of reaction.[2]

Symptom	Percentage of People with Food Allergies Who Develop the Symptom
· Skin eruptions: rash, hives	84%
· Gastrointestinal upsets: diarrhea, vomiting, cramps, nausea	52
· Respiratory problems: congestion, runny nose, cough, wheezing, asthma	32

➡ **celiac disease**
A disease characterized by inflammation of the small intestine lining resulting from a genetically based intolerance to gluten. The inflammation produces diarrhea, fatty stools, weight loss, and vitamin and mineral deficiencies. (Also called *celiac sprue* and *gluten-sensitive enteropathy.*)

Illustration 17.2
The development of allergic reactions to foods.

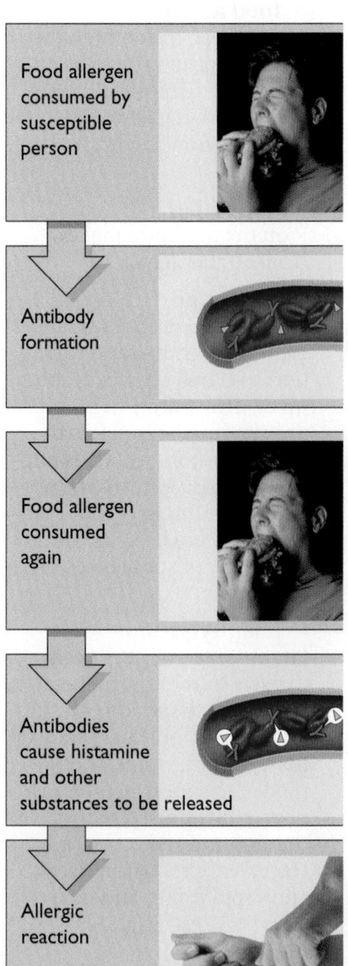

relatively common sources of food allergies.[1] Together, these are considered "the big eight" food allergies.[1]

A Closer Look at Wheat Allergy

Minnie is 46 years old and has three sisters and two brothers, all of whom are taller than she is. (She considers herself the "runt of the litter.") Since childhood, Minnie has had problems with diarrhea and cramps. Over the past two years, these problems have become worse, and she has been chronically tired. Assuming that she was lactose intolerant, Minnie stopped eating all dairy products. Her problems persisted, however. When she almost lost her job as a receptionist due to her frequent bathroom breaks and her obvious fatigue, she decided to see a doctor again.

Diagnosing anemia, the doctor sent Minnie to a registered dietitian. Hearing Minnie's story, the dietitian suspected celiac disease caused by an allergy to a component of wheat and other grains. After two weeks on an allergen-free diet, Minnie started to regain her strength and spent much less time in the bathroom. Diagnostic tests subsequently confirmed that Minnie had celiac disease, a condition that likely had existed since childhood.

An allergy to wheat, or more specifically to a component of gluten found in wheat and other grains such as barley, rye, triticale, and possibly oats, has been under-diagnosed in the past. Called ***celiac disease*** (and also celiac sprue and gluten-sensitive enteropathy), it affects one in 250–500 people in most countries, including the United States.[8] It differs somewhat from traditional immunoglobulin E-mediated allergies in that the immune system reaction to gluten is localized in the lining of the small intestine. Celiac disease can be difficult to diagnose because the symptoms (diarrhea, weight loss, cramps, anemia) are similar to those of other diseases.[9] With new awareness about the disorder, diagnostic tests are being ordered more frequently, and the rate of confirmed cases of celiac disease is increasing.[8]

Celiac disease is treated with a life-long gluten-free diet. Because gluten is included in many food additives as well as grain products, it takes the help of an experienced registered dietitian to design and monitor gluten-free diets.[8]

Do Genetically Modified Foods Contain Allergens? It's possible but unlikely.

Genetically modified foods contain proteins from different types of plants or animals. Consequently, it is possible that a protein that causes allergic reactions in some people would end up in a genetically modified food. This possibility is recognized by manufacturers, and genetically modified foods that may contain allergens are tested for safety before they are sold. Pretests of soybeans with added Brazil nut proteins, for example, found that people allergic to Brazil nuts were allergic to the genetically modified soybeans. Plans to make the soybeans commercially available were dropped, and the manufacturer halted further production.[1] Due to manufacturers' testing procedures and FDA regulations related to these foods, the possibility that genetically modified foods contain hidden allergens is limited.[1]

PhotoEdit, Felicia Martinez

Illustration 17.3
These foods can cause anaphylactic shock, a severe allergic reaction, in highly sensitive people. (Most food allergies cause milder symptoms.)

How Common Are Food Allergies?

Infants and young children experience food allergies more often than adults do. Up to 7% of infants develop a food allergy, which most often is due to a protein in cow's milk.[10] To prevent allergic reactions, it is recommended that cow's milk and other potentially allergenic foods such as peanuts, nuts, fish, seafood, eggs, and wheat be excluded from the infant's diet during the first year of life.[11] The majority of infants outgrow their allergy to cow's milk or to other foods such as egg white, wheat, and soy by the time they are two or three years old.[6]

The incidence of food allergies in adults is estimated to be 1–2% and appears to be increasing.[1] Nevertheless, 40% of Americans believe they are allergic to one or more foods.[1] About 60% of complaints of food allergy fail to be confirmed by testing.[2] And only 1 out of 10 people with food allergies is allergic to more than one food.[12]

Why the Difference between Reality and Perception? One reason food allergies seem more widespread than they actually are is that popular wisdom endows various foods with certain properties. The following are some of the more widely held beliefs:

- Allergies to sugar or food colors lead to hyperactivity in children.
- Yeast causes chronic fatigue syndrome.
- Dairy products, especially milk, cause phlegm.

Table 17.2
The top five foods that cause allergies and examples of some other foods that cause food allergies in susceptible people.[1]

The Top Five Foods That Cause Allergies*

- Cow's milk
- Eggs
- Nuts
- Peanuts
- Wheat

Examples of Other Foods Known to Cause Allergies

· Fish and shellfish (shrimp, lobster, oysters, clams, crab) · Soybeans · Fruits, (oranges, pineapple, kiwi, tomatoes)	· Vegetables (Chinese cabbage, corn, green beans, mushrooms, celery) · Dried beans	· Meat (beef, chicken, pork) · Yeast · Seeds

*To help remember these, think of "NEW Milk" for nuts, eggs, wheat, and milk.

None of these associations has been documented by scientific studies.

Bad timing and mental attitudes about certain foods can also lead to erroneous diagnoses of food allergies. Here are two examples:

1. Isaiah, a lover of blueberries and blueberry pie, hasn't touched a blueberry since 1985. That year, Isaiah had a piece of blueberry pie in a restaurant and later became violently sick to his stomach. Bingo! Isaiah thought he must be allergic to blueberries.

Actually, Isaiah gave up blueberries for no good reason. He was coming down with the flu when he ate the pie. Nevertheless, to this day when he thinks of blueberries, he gets a bit queasy.

2. For 11 years, Emilia rigidly avoided even small amounts of cow's milk. She was convinced that just a few drops of cow's milk would cause pressure in her head, blurred vision, dizziness, cramps, and nausea.

Finally, Emilia's presumed allergy to cow's milk was put to the test. She was given liquid through a dark tube inserted into her stomach from her mouth. When she was told the fluid was cow's milk—even though it was actually water—the familiar symptoms appeared within 10 minutes. When she was given milk but told it was water, no symptoms appeared.[13]

Emilia was shocked. For 11 years she had scrutinized nearly everything she ate and had wasted hundreds of dollars on special food products and supplements. Finishing a frosted brownie and a glass of milk, Emilia contemplated the practical realities of the power of suggestion.

Diagnosis: Is It a Food Allergy?

> It is essential to realize that the [allergic] reactions occur in human beings and are therefore enmeshed in the interplay between body, mind and soul. . . . Only through experience with blind food challenges can the amazing power of self-deception be appreciated.
>
> —Charles May, 1980

A variety of tests are used to diagnose food allergies, but there is only one "gold standard." That test is the ***double-blind, placebo-controlled food challenge.***[14]

➡ **double-blind, placebo-controlled food challenge** A test used to determine the presence of a food allergy or other adverse reaction to a food. In this test, neither the patient nor the care provider knows whether a suspected offending food or a placebo is being tested.

Suppose you suspect you are allergic to grapes because you got an annoying rash twice after eating them. Your doctor arranges an appointment for you at an allergy clinic. At the clinic, you will be given grapes, either in a concentrated pill form or blended into a liquid, and placebo pills (or a liquid) without being told which is the grapes and which the placebo. After an appropriate amount of time, your reaction to the grapes and to the placebo will be noted. If the rash appears after the grapes but not after the placebo, you have an adverse reaction to grapes that may be an allergy.

Food challenges are best undertaken under medical supervision. True food allergies can cause serious reactions, and immediate help may be needed. Foods suspected of causing anaphylactic shock in the past should not be given in food challenge tests. They should be totally omitted from the diet.[14]

The Other Tests for Food Allergy

Other tests for food allergy are reliable to varying degrees. None of them is always accurate, and the results of bogus tests are not to be trusted at all.

Skin Prick Tests

Skin prick tests (also called prick skin tests) for food allergy are commonly employed and are useful for identifying *the absence* of a food allergy. A positive test result isn't proof positive that an allergy exists, however, because positive test results are inaccurate about 50% of the time.[2]

For this test, a few drops of food extract are placed on the skin, and the skin is then pricked with a needle. At approximately the same time, another area of skin is pricked using only water. If the area around the food skin prick becomes redder and more swollen than the area pricked with water (Illustration 17.4), it is concluded that the person *might* be allergic to the food tested.

Skin prick tests are useful for ruling out allergies and for reducing the number of foods that may have to be tested by double-blind, placebo-controlled food challenges.

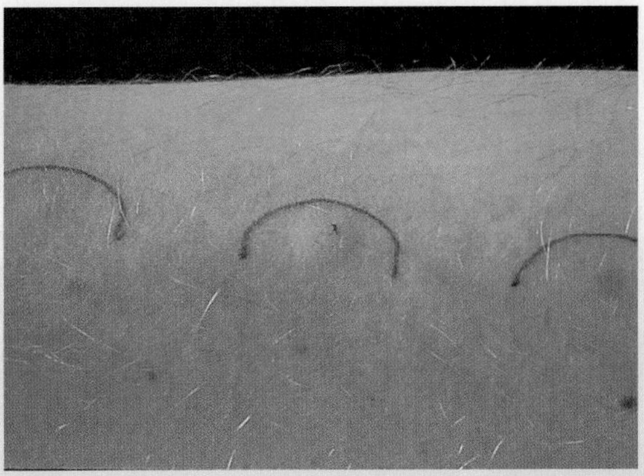

Illustration 17.4
A positive reaction to a skin prick test for a food allergy usually looks like this. The result indicates that an allergy to the food tested *may* be present, but doesn't guarantee it.

The RAST Test The acronym RAST stands for "radioallergosorbent test." In this test, a sample of blood is examined for immune substances formed in response to an allergen. The test is not specific for individual food allergens, nor does it reliably indicate the presence of a food allergy.[2]

Bogus Tests A number of companies offer food allergy tests through the mail. In one study, researchers sent five of these companies the required samples, nine from adults who were allergic to fish and nine from adults who had no allergies. All five companies missed the fish allergy. When duplicate samples from the same adults were sent separately to each company, reports came back with different results for each of the pairs of samples. In addition, the companies reported their laboratories had identified various food allergies that none of the adults actually had.[13] Mail-order allergy tests are a waste of money; they do not use reliable assessment techniques.

Other unreliable methods for identifying food allergies include hair analysis, cytotoxic blood tests, iridology, and sublingual (under the tongue) and food injection provocation tests.[15]

What's the Best Way to Treat Food Allergies?

After a food allergy is confirmed, the food is eliminated from the diet. This is the only treatment available for food allergies. "Allergy" shots for food allergens are not yet available but may be in the future.[16] If the eliminated food is an important source of nutrients or is found in many food products (such as milk or wheat), consultation with a registered dietitian is recommended. As noted earlier, for infants and young children, an allergy-producing food such as cow's milk, eggs, soy, or wheat can usually be reintroduced into the diet in a year or two. Most (but not all) infants outgrow food allergies by the age of two or three years.[10] Children and adults with severe allergies to peanuts, nuts, fish, or seafood may have to eliminate the food from their diets for a lifetime.[1]

Table 17.3
Foods and substances in them linked to food intolerance reactions.

Aged cheese
Anchovies
Aspartame
Beer
Catsup
Chocolate
Dried beans
Lactose
Pickled cabbage
Red wine
Sardines
Sausage (hard, cured)
Soy sauce
Spinach
Tomatoes
Yeast extracts

Food Intolerances

Food intolerances produce some of the same reactions as food allergies, but the reactions develop by different mechanisms. Food intolerance reactions do not involve the immune system; they are due to a missing enzyme or other cause.[11,17]

It is clear that a small proportion of people are intolerant of lactose, sulfite, components of red wine, aged cheese, and other foods (Table 17.3). It is also clear that not all food intolerances are real. Just as with food allergies, the best way to separate the real from the unreal is the double-blind, placebo-controlled food challenge. True food intolerances produce predictable reactions. Problems such as headache, diarrhea, swelling, or stomach pain will occur every time a person consumes a sufficient amount of the suspected food.

➡ **lactose maldigestion** A disorder characterized by reduced digestion of lactose due to the low availability of the enzyme lactase.

➡ **lactose intolerance** The term for gastrointestinal symptoms (flatulence, bloating, abdominal pain, diarrhea, and "rumbling in the bowel") resulting from the consumption of more lactose than can be digested with available lactase.

Lactose Maldigestion and Intolerance

Lactose maldigestion, or the inability to break down lactose in dairy products due to the lack of the enzyme lactase, results in the condition known as ***"lactose intolerance."*** Symptoms of lactose intolerance, such as flatulence, bloating, abdominal pain, diarrhea, and "rumbling in the bowel," occur in lactose maldigesters within several hours of consuming more lactose than can be broken down by the available lactase. These symptoms are due to the breakdown of undigested lactose by bacteria in the lower intestines (which produce gas as a by-product of lactose ingestion) and by fluid accumulation.[18]

Reduced intake of lactose-containing dairy products (milk, ice cream, and cottage cheese, for example) reduces lactose intolerance. Foods such as hard cheese, low or no lactose milk, buttermilk, and yogurt without added milk solids contain low amounts of lactose and can generally be consumed, however. Small amounts of milk or other lactose-containing dairy products are generally well tolerated by people with lactose maldigestion. Care should be taken to ensure adequate intake of calcium and vitamin D if milk and dairy product intake is restricted.[18]

Lactose maldigestion is a very common disorder. It occurs in about 25% of U.S. adults, but the incidence varies dramatically by population group. By one set of estimates, 62 to 100% of Native American adults, 90% of Asian Americans, 80% of African Americans, 53% of Mexican Americans, and 15% of Caucasian Americans are lactose maldigesters.[19]

Sulfite Sensitivity

Sulfite is a food additive used by food manufacturers to keep vegetables and fruits looking fresh. It is also added to beer, wine, some processed foods, and medications as a preservative (see Table 17.4).

Very small amounts of sulfite can cause anaphylactic shock and bring on an asthma attack in sensitive people.[20] Increasingly, other preservatives are replacing sulfite in food products.

Table 17.4
Some foods that may contain sulfite.

Wine
Beer
Hard cider
Tea
Fruit juices
Vegetable juices
Guacamole
Dried fruit
Potato products
Canned vegetables
Baked goods
Spices
Gravy
Soup mixes
Jam
Trail mix
Fish and seafood

Because many foods have added sulfite, people with asthma who are sensitive to sulfite and others who are sensitive should read food ingredient labels carefully. The FDA requires that processed foods containing sulfites list sulfites on the ingredient label. It also prohibits the use of sulfite on fresh vegetables and fruits sold or served raw to consumers. (FDA Consumer 1993 Food Allergies)

Red Wine, Aged Cheese, and Migraines

Some people develop migraine headaches when they consume red wine. The presence of histamine in wine has been blamed for the headaches. People with this intolerance are unable to break down histamine during digestion, so it accumulates in the blood and causes headaches.[21] Histamine in also found in beer, sardines, anchovies, hard cured sausage, pickled cabbage, spinach, and catsup. Tyramine, a compound closely related to histamine and found in aged cheese, soy sauce, and other fermented products, causes migraine headaches in sensitive people.[22] Chocolate, too, may bring on migraine headaches in sensitive individuals, although this finding is controversial.[11,12]

MSG

Monosodium glutamate (MSG) has been blamed for flushing, headache, and tightness of the chest that may follow the ingestion of foods, soups, and other food products that contain this flavor enhancer. The reaction has been dubbed the "Chinese restaurant syndrome" because Chinese foods often contain MSG. Nevertheless, controlled clinical trials have not demonstrated that MSG causes these problems.[4] It has been suggested that histamines, soy, or wheat proteins that are often present in the same foods as MSG may be responsible, but this is not yet clear.

Precautions

People with food allergies or intolerances have to be very careful about what they eat and should have a plan of action ready in case they develop a serious reaction. Becoming highly knowledgeable about which foods and food products contain ingredients that cause an adverse reaction is essential. When eating out, people with allergies and intolerances should ask a lot of questions to make sure they know what is being served.

Increasingly, food product ingredients such as cow's milk protein, egg white, sulfites, peanuts, and peanut oil are listed on labels, and these should be checked carefully. Products that contain sulfites must be labeled. Genetically engineered foods (which contain proteins from another plant or animal) must list potential allergens such as soy or nut protein on the label.

Importantly, people who develop anaphylactic shock should always carry a preloaded syringe of epinephrine and know how to use it. They should also know where to call for emergency help.[14] People whose reactions to foods lessen with the use of asthma inhalers and antihistamines should keep them readily available.

One final thought about food allergies and intolerances—they are never caused by studying about them.

On the Side

Foods may be cross-contaminated with potential allergens during production or preparation. Yogurt-covered peanuts have been accidentally mixed with yogurt-covered raisins, and chocolate-covered peanuts with chocolate-coated raisins.

Nutrition UP CLOSE

Prevention of Food Allergies

FOCAL POINT: Foods that may cause allergies in the first year of life.

Scenario: Darla and Mitch have a five-month-old son who has been breastfed exclusively. The time has come to add semisolid food to their son's diet. Mitch had food allergies as a child, so he and Darla are concerned that their son may develop allergies, too.

Question: If you were Darla or Mitch, what five foods would you omit from your son's diet during the first year of life?

1. ____________________
2. ____________________
3. ____________________
4. ____________________
5. ____________________

FEEDBACK (answers to these questions) can be found at the end of Unit 17.

www.nim.nih.gov/search.html
Select Food Allergies to get reliable information from the National Institutes of Health.

www.specialtyfoodshop.com
The Toronto Hospital for Sick Children's website offers access to registered dietitians, reliable nutrition information, and food products for people with allergies.

www.foodallergy.org
Connects to the Food Allergy Network. Or call the network (in Fairfax, VA) at 1-800-929-4040 for advice on how to deal with schools and support.

REFERENCES

1. Taylor SL, et al. Food allergies and avoidance diets. Nutr Today 1999;34:15–22.
2. Sampson HA. State of the art theories on food allergies. Annual meeting of the American Dietetic Association, Boston, 1997 Oct 28.
3. Scadding GK, Brostoff J. The dietetic treatment of food allergy. In: Schmidt E, editor. Food allergy. New York: Vevey/Raven Press; 1988; and Carroll P, Caplinger KJ, France GL. Guidelines for counseling parents of young children with food sensitivities. J Am Diet Assoc 1992;92:602–3.
4. Koerner CB. Food allergies. Annual meeting of the American Dietetic Association, Boston, 1997 Oct 28.
5. Bock SA. Natural history of severe reactions to foods in young children. J Pediatr 1985;107:676–80.
6. Burks AW. Food hypersensitivity. Annual meeting of the American Dietetic Association, Boston, 1977 Oct 28.
7. Leinhas JL, et al. Food allergy challenges: guidelines and implications. J Am Diet Assoc 1987;87:604–8.
8. Murray JA. The widening spectrum of celiac disease. Am J Clin Nutr 1999;69:354–65.
9. Collin P. New diagnostic findings in celiac disease. Ann Med 1999;31:399–405.
10. Arvola T, et al. Concerns and expectations of parents with atopic infants. Pediatr Allergy Immunol 2000;11:183–88.
11. Chandra RK. Food hypersensitivity and allergic disease: a selective review. Am J Clin Nutr 1997;66:526S–9S.
12. Bock SA, Atkins FM. Patterns of food hypersensitivity during sixteen years of double-blind, placebo-controlled food challenges. J Pediatr 1990;117:561–7.
13. May CD. Food sensitivity: facts and fancies. Nutr Rev 1984;42:72–8.
14. Dutau G. Food allergies and diagnostic alternatives: labial food challenge, oral food challenge. Revue Franscaise d'Allergologie et d'Immunologie Clinique. 2000;40:728–41.
15. Ferguson A. Food sensitivity or self-deception? N Engl J Med 1990;323:476–8.
16. MacLennan A. Recombinant allergens offer new hope for allergy treatment, diagnosis. J Allergy Clin Immunol 2000:106:409–18.
17. Fouchard T. Development of food allergies with special reference to cow's milk allergy. Pediatrics 1985;75(suppl): 177–81.
18. Tolstoi LG. Adult-type lactase deficiency. Nutr Today 2000;35:134–42.
19. Inman-Felton AE. Overview of lactose maldigestion (lactose nonpersistence). J Am Diet Assoc 1999;99:481–89.
20. Yang WH, Purchase EC. Adverse reactions to sulfites. Can Med Assoc J 1985;133:865–7.
21. Jarisch R, Wantke F. Wine and headache. Int Arch Allergy Immunol 1996;110:7–12.
22. Hannington E. Preliminary report on tyramine headache. Br Med J 1967;2:550–1.

Nutrition

UP CLOSE

Prevention of Food Allergies

Feedback for Unit 17

Although it is not known whether the son would develop food allergies, Darla and Mitch would be wise to omit foods most likely to cause allergic reactions during the first year of life. These foods include:

1. Cow's milk
2. Wheat
3. Nuts, peanuts
4. Soy products
5. Eggs
6. Fish, seafood

Unit 18

Fats and Cholesterol in Health

Photo Disc

Nutrition Scoreboard

	TRUE	FALSE
1. It is currently recommended that all foods consumed by Americans should contain less than 30% of calories from fat.		
2. The food ingredient that Americans report they are trying hardest to cut down on is fat.		
3. Cholesterol is present in every food.		
4. Saturated fat intake has a stronger influence on blood cholesterol levels than does cholesterol intake.		

(Answers on next page)

Answers to

Nutrition Scoreboard

	TRUE	FALSE
1. Diets, not individual foods, should contain less than 30% of calories from fat.		✓
2. Most Americans (88%) are trying to cut down on fat—more than are trying to cut down on any other ingredient in their diet.[1]	✓	
3. Cholesterol is present only in animal products.		✓
4. Saturated fat, which is found primarily in animal products, raises blood cholesterol levels more than does dietary cholesterol.	✓	

Key Concepts and Facts

- Fats are our most concentrated source of food energy. They supply nine calories per gram.
- Dietary fats "carry" the essential fatty acids, fat-soluble vitamins, and healthful phytochemicals along with them in foods.
- Dietary fat intake should be kept below 30% of total calories. Cholesterol intake should not exceed 300 milligrams per day. (Note: These recommendations are *not* for children younger than two years.)
- Saturated fats raise blood cholesterol levels more than does dietary cholesterol.

A Basic Understanding of Fats and Cholesterol

Ask Americans to name the top two dietary villains and they are likely to say "fats and cholesterol." We hear a lot about these two components of our diet, and there's good reason for that. Diets high in animal fat and cholesterol are related to high rates of heart disease.[2] The national fondness for fats also appears to be related to prostate and colon cancer, gallbladder disease, and obesity.[1,3] This unit presents facts about fats and cholesterol, including information on how the body uses them, their food sources, and ways to reduce fat intake.

Facts about Fats

Fats are a group of substances found in food that have one major property in common. They dissolve in fat and not in water. If you have ever tried to mix vinegar and oil when making salad dressing, you have observed the principle of water and fat solubility firsthand.

Fats are actually a subcategory of the fat-soluble substances known as ***lipids.*** Lipids include all types of fats and oils. Fats such as butter, margarine, and shortening are often distinguished from oils by their property of being solid at room temperature. This physical difference between fats and oils is due to their chemical structures.

lipids
Compounds that are insoluble in water and soluble in fat. Triglycerides, saturated and unsaturated fats, and essential fatty acids are examples of lipids, or "fats."

essential fatty acids
Components of fats (linoleic acid and alpha-linolenic acid) required in the diet. Linoleic acid is needed for growth and reproduction, and alpha-linolenic acid for normal vision and brain development. Deficiency symptoms, such as poor growth, reproductive failure, and impaired vision and learning ability, occur with deficient intake.

The Contributions of Dietary Fats Fats in foods supply energy and fat-soluble nutrients. Fats include the ***essential fatty acids*** (linoleic acid and alpha-linolenic acid) and "carry" the fat-soluble vitamins D, E, K, and A (the "deka" vitamins). So, part of the reason we need fats in our diet is to get a supply of the essential nutrients they contain (Table 18.1).

Fats increase the flavor and palatability of foods. Although "pure" fats by themselves are tasteless, they absorb and retain the flavor of substances that surround them. Thus, fats in meats and other foods pick up flavors from their environment and give those flavors to the food. This characteristic of fat is why butter, if placed next to the garlic in the refrigerator, tastes like garlic.

Fats contribute to the sensation of feeling full (as they should at nine calories per gram!).[1] They tend to stay in the stomach longer than carbohydrates or proteins and are absorbed over a longer period of time. That's why foods with fat "stick to your ribs."

A crucial role of fat is to serve as a component of cell membranes. Some types of fats give cell membranes flexibility and help regulate the transfer of nutrients into and out of cells.[4]

Table 18.1
Roles of dietary fat.

- Provides a concentrated source of energy
- Carries the essential fatty acids, the fat-soluble vitamins, and certain phytochemicals
- Increases the flavor and palatability of foods
- Provides sustained relief from hunger
- Serves as a component of cell membranes

Excess Dietary Fat, Carbohydrates, and Protein Are Stored as Fat The body has two sources of fats: those consumed in the diet and those produced from carbohydrates and proteins. Humans eat only a few times a day, but energy is needed throughout the day. To ensure a constant supply of energy, the body converts carbohydrates and proteins, which have been supplied from foods and are not used to meet immediate needs, to storage forms of energy. Some of the excess carbohydrate and protein is converted to glycogen, the storage form of glucose under normal circumstances, but most of the excess is changed to fat and stored in fat cells (Illustration 18.1).

Body fat is not just skin deep. Fat is also located around organs such as the kidneys and heart. It's there to cushion and protect the organs and keep them insulated. Cold-water swimmers can attest to the effectiveness of fat as an insulation material. They purposefully build up body fat stores because they need the extra layer of insulation (Illustration 18.2).

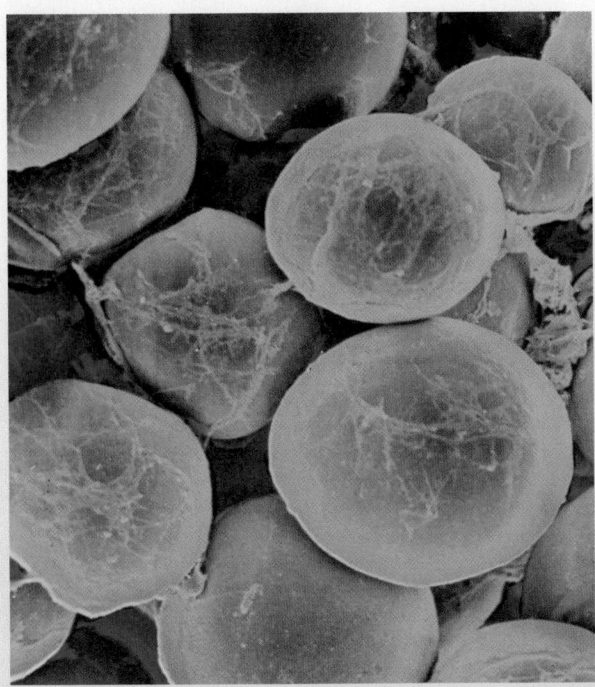

Illustration 18.1
A close look at fat cells (color-enhanced microphotograph).

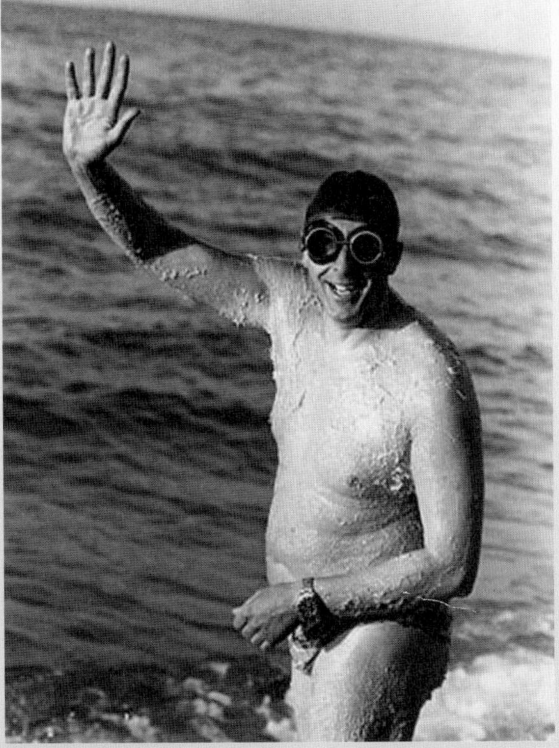

Illustration 18.2
Although their body fat stores don't fit the image of the superathlete, cold-water swimmers need the fat to help stay warm. Pictured here is the English swimmer Mike Read, who had swum the English Channel 20 times by age 39.

Table 18.2
Basic facts about the types of fat.

- Fats can be:
  - Monoglycerides
  - Diglycerides
  - Triglycerides
- Fats can be:
  - Saturated
  - Monounsaturated
  - Polyunsaturated
- Unsaturated fats come in:
  - "Cis" forms
  - "Trans" forms

Fats Come in a Variety of Types

There are many types of fat and fatlike substances in food and our bodies (Table 18.2). Of primary importance are ***triglycerides*** (or "triacylglycerols"), ***saturated*** and ***unsaturated fats,*** and ***cholesterol*** (for definitions of these and other types of fat, see Table 18.3). The different types of fats have different effects on health.

Triglycerides, which consist of one ***glycerol*** unit (a glucoselike substance) and three fatty acids (Illustration 18.3), make up 98% of our dietary fat intake and the vast majority of our body's fat stores. Triglycerides are transported in blood attached to protein carriers and are used by cells for energy formation and tissue maintenance. A minority of fats take the form of ***diglycerides*** (glycerol plus two fatty acids) and ***monoglycerides*** (glycerol and one fatty acid).

As far as health is concerned, the glycerol component of fat is relatively unimportant. It's the fatty acids that influence what the body does with the fat we eat and are responsible, in part, for how fat affects health. There are many different types of fatty acids in fats. You've heard of the major ones: those that make fat "saturated" or "unsaturated."

Saturated and Unsaturated Fats Fatty acids found in fats consist primarily of hydrogen atoms attached to carbon atoms (Illustration 18.4). When the carbons are attached to as many hydrogens as possible, the fatty acid is "saturated"—that is, saturated with hydrogen. Saturated fats tend to be solid at room temperature. With the exception of palm and coconut oil, only animal products are rich in saturated fats (Illustration 18.5 on p. 18–6).[5] Fatty acids that contain fewer hydrogens than the maximum are "unsaturated." They tend to be liquid at room temperature. By and large, plant foods are the best sources of unsaturated fats.

Table 18.3
A glossary of fats.

Triglycerides: Fats in which the glycerol molecule has three fatty acids attached to it; also called *triacylglycerol.* Triglycerides are the most common type of fat in foods and in body fat stores.

Saturated fats: Molecules of fat in which adjacent carbons within fatty acids are linked only by single bonds. The carbons are "saturated" with hydrogens; that is, they are attached to the maximum possible number of hydrogens. Saturated fats tend to be solid at room temperature. Animal products and palm and coconut oil are sources of saturated fats.

Unsaturated fats: Molecules of fat in which adjacent carbons are linked by one or more double bonds. The carbons are not saturated with hydrogens; that is, they are attached to fewer than the maximum possible number of hydrogens. Unsaturated fats tend to be liquid at room temperature and are found in plants, vegetable oils, meats, and dairy products.

Glycerol: A syrupy, colorless liquid component of fats tht is soluble in water. It is similar to glucose in chemical structure.

Cholesterol: A fat-soluble, colorless liquid found in animals but not in plants. Cholesterol is used by the body to form hormones such as testosterone and estrogen and is a component of animal cell membranes.

Diglyceride: A fat in which the glycerol molecule has two fatty acids attached to it; also called *diacylglycerol.*

Monoglyceride: A fat in which the glycerol molecule has one fatty acid attached to it; also called *monoacylglycerol.*

Monounsaturated fats: Fats that contain a fatty acid in which one carbon-carbon bond is not saturated with hydrogen.

Polyunsaturated fats: Fats that contain a fatty acid in which two or more carbon-carbon bonds are not saturated with hydrogen.

Illustration 18.3
A triglyceride.

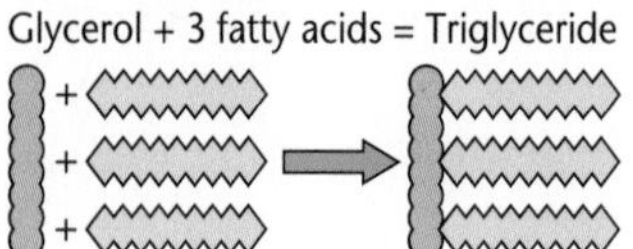

Hydrogen
Oxygen
Oxygen
Carbon
CH_3
Saturated fatty acid

CH CH
Monounsaturated fatty acid

CH CH CH CH
Polyunsaturated fatty acid

Two hydrogens are missing from each of these carbon-carbon links, making the fatty acid polyunsaturated. With fewer hydrogens to attach to, these carbons are doubly bonded to each other. Monounsaturated fatty acids have only one carbon-carbon bond that is "unsaturated" with hydrogen atoms.

Illustration 18.4
A look at the difference between a saturated and an unsaturated fatty acid.

Unsaturated fats are classified by their degree of unsaturation. If only one carbon-carbon bond in the fatty acid is unsaturated, the fat is called "monounsaturated." If two or more carbon-carbon bonds are unsaturated with hydrogen, the fat qualifies as "polyunsaturated."

Fish Oils Contain Highly Unsaturated Fatty Acids Many unsaturated fats contain fatty acids with one or two unsaturated carbon-to-carbon bonds. The fatty acids in fish oils, however, stand out in that they contain "omega-3 fatty acids" with four-to-six double bonds. This high level of unsaturation in fish oils makes them highly fluid even at cold temperatures. This property allows the fatty acids to act as antifreeze, keeping cell membranes and other fat-containing tissues in fish flexible in cold water. Fatty fish and fish that live in cold water contain the highest amounts of omega-3 fatty acids (Table 18.4 on p. 18–7).

Regular consumption of omega-3 fatty acid from fish (two or more fish-containing meals per week) appears to confer some protection against heart disease, stroke, certain cancers, and arthritis.[4] But large amounts of omega-3 fatty acid, such as those obtained when fish oil supplements are consumed on a regular basis, may cause more harm than good. Fish oil capsules can increase bleeding and cause easy bruising; they can also lead to gastrointestinal upsets, and some types increase the risk of vitamin A and vitamin D toxicity.[8] Some fish, particularly swordfish, shark, king mackerel, and tilefish, and large fish (over 20 inches) may contain unacceptably high levels of mercury due to industrial contamination of fresh and salt waters. It is advised that pregnant and breastfeeding women, and children under the age of six, not eat these fish and limit consumption to 12 ounces per week of a variety of allowed fish.[9]

On the Side

Fishy eggs? A new source of fish oils may become part of North Americans' diets in the future. Hens given feed with added cod liver oil produce eggs with yolks containing 100 to 200 milligrams of highly unsaturated omega-3 fatty acids.[7]

Illustration 18.5
Fat profiles of selected foods.

Percentage Fatty Acid Composition

Food	Saturated	Monounsaturated	Polyunsaturated
DAIRY PRODUCTS			
Brick cheese	67	30	3
Cheddar cheese	67	30	3
Cottage cheese	67	30	3
American cheese	67	30	3
Whole milk	65	31	4
Butter	65	31	4
HUMAN MILK	48	40	12
EGG (medium)	37	48	15
OILS AND MARGARINES			
Canola oil	6	58	36
Coconut oil	92	6	2
Corn oil	13	25	62
Corn oil margarine	17	61	22
Olive oil	14	77	9
Palm oil	52	38	10
Peanut oil	18	48	34
Soybean oil	15	25	60
Soybean oil margarine	22	50	28
Sunflower oil	11	21	68
MEATS			
Hamburger, regular (21% fat)	53	43	4
Chicken, roasted (no skin)	26	46	28
Pork chop	39	49	12
Turkey, roasted	40	25	35
NUTS AND SEEDS			
Cashews	21	62	17
Macadamia nuts	16	46	2
Peanuts, dry–roasted	15	52	33
Peanut butter	18	51	32
Sunflower seeds (dried)	11	20	69

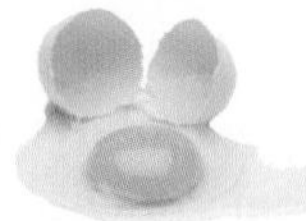

Table 18.4
Omega-3 fatty acid content of selected seafoods.[5,6]

Seafood (3½-ounce serving)	Omega-3 Fatty Acids (grams)	Seafood (3½-ounce serving)	Omega-3 Fatty Acids (grams)
Mackerel	2.6	Shrimp	0.5
Lake trout	2.0	Trout, rainbow	0.5
Herring	1.8	Tuna	0.5
Salmon, sockeye	1.5	Carp	0.3
Whitefish, lake	1.5	Crab	0.3
Anchovies	1.4	Pike, walleye	0.3
Salmon, Chinook	1.4	Flounder	0.2
Bluefish	1.2	Haddock	0.2
Halibut	1.2	Lobster	0.2
Salmon, pink	1.0	Salmon, red	0.2
Bass, striped	0.8	Scallops	0.2
Oysters	0.6	Snapper, red	0.2
Catfish	0.5	Swordfish	0.2
Pollock	0.5		

Modifying Fats Unsaturated fats aren't as stable as saturated fats. They are more likely to turn rancid with time and exposure to air and heat than are saturated fats. Additionally, solid fats are preferable to oils for some cooking applications. These problems with unsaturated fats have a solution. It's called "hydrogenation."

What's Hydrogenation? *Hydrogenation* is a process that adds hydrogen to liquid unsaturated fats, thereby making them more saturated and solid. The shelf life, cooking properties, and taste of vegetable oils are improved in the process. Hydrogenation has two drawbacks, however. Hydrogenated vegetable oils contain more saturated fat than the original oil. Corn oil, for example, contains only 6% saturated fats, but corn oil margarine has 17%. The other negative is that hydrogenation causes a change in the structure of the unsaturated fatty acids. Specifically, hydrogenation converts some unsaturated fats into ***trans fats.***

➡ **hydrogenation**
The addition of hydrogen to unsaturated fatty acids.

➡ **trans fats**
Unsaturated fatty acids in fats that contain atoms of hydrogen attached to opposite sides of carbons joined by a double bond:

```
H
–C = C–
     H          H  H
              –C = C–
Trans fatty   Cis fatty
   acid          acid
```

Fats containing fatty acids in the trans form are generally referred to as "trans fats."

Cis fatty acids are the most common, naturally occurring form of unsaturated fatty acids. They contain hydrogens located on the same side of doubly bonded carbons.

Trans Fatty Acids Up to 30% of the fatty acids in unsaturated fat molecules may be converted from their naturally occurring "cis" to the "trans" form as a result of hydrogenation.[10] Overall, fats containing trans fatty acids make up about 7% of the total fat intake of North Americans.[11] Trans fatty acids raise blood cholesterol levels more than cis fatty acids do and appear to be related to the development of heart disease.[12]

It is recommended that people limit their consumption of trans fatty acids.[13] Determining the trans fatty acid content of foods is a bit difficult, however, because Nutrition Facts panels are not required to list this information. (New labeling requirements for the trans fatty acid content of foods are in the works, however.) Until food products are labeled with trans fats content, you can get a guesstimate of whether a food contains them by looking at ingredient labels for the terms "hydrogenated fat," "hydrogenated oil," and "partially hydrogenated oils." Table 18.5 lists the trans fat content of a sampling of foods.

Food manufacturers have developed methods for hydrogenating oils that don't cause the formation of trans fats. You may see products labeled "trans fat free" and "no trans fat" (Illustration 18.6). Proposed labeling requirements

Table 18.5
Trans fatty acid content of foods.

Food	Trans Fatty Acids (grams)
Kentucky Fried Chicken pot pie	8
French fries, large serving	4–7
Cake doughnut	6
Breaded fish sticks, 3	5
Margarine, 1 tbs	4
Shortening, 1 tbs	1
Tub margarine, 1 tbs	1
Butter, 1 tbs	0
"No trans fat" margarine	0
Olive oil	0

would allow these announcements to be used only on foods that are very low in saturated fats, too.[14] Small amounts of trans fatty acids occur naturally in dairy products, beef, and lamb.[15]

Illustration 18.6
Products that feature "no trans fats" labels.

Checking Out Cholesterol

Cholesterol is a fatlike substance found *only* in animal products. It is a tasteless, odorless, clear liquid that feels oily and is contained in both the lean and fat parts of animal products. Table 18.6 lists some sources of cholesterol. Plants don't contain cholesterol because they can't produce it and don't need it to function and grow normally.

Table 18.6
Food sources of cholesterol. Note that all the sources are animal products. The recommended daily intake for adults is 300 milligrams or less per day.

Cholesterol is a clear, oily liquid found in the fat and lean portions of many animal products. It looks like this:

Animal Product	Amount	Cholesterol (milligrams)
Brain	3 oz	1746
Liver	3 oz	470
Egg	1	212
Veal	3 oz	128
Shrimp	3 oz	107
Prime rib	3 oz	80
Chicken (no skin)	3 oz	75
Turkey (no skin)	3 oz	65
Hamburger, regular	3 oz	64
Pork chop, lean	3 oz	60
Fish, baked (haddock, flounder)	3 oz	58
Ice cream	1 cup	56
Sausage	3 oz	55
Hamburger, lean	3 oz	50
Milk, whole	1 cup	34
Crab, boiled	3 oz	33
Lobster	3 oz	29
Cheese (cheddar)	1 oz	26
Milk, 2%	1 cup	22
Yogurt, low-fat	1 cup	17
Milk, 1%	1 cup	14
Butter	1 tsp	10
Milk, skim	1 cup	7

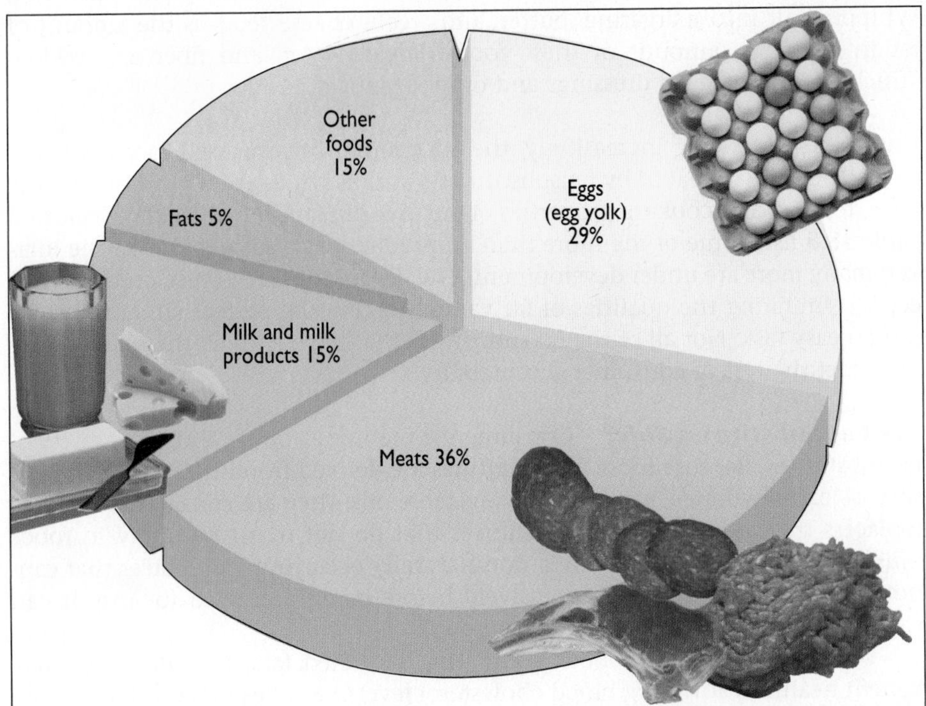

Illustration 18.7
Food sources of cholesterol in the U.S. diet.[16] The percentages indicate the proportion of cholesterol each type of food contributes to the diet.

The cholesterol used by the body comes from two sources. Most (about two-thirds) of the cholesterol available to the body is produced by the liver. The rest comes from the diet (Illustration 18.7). Because the liver produces cholesterol from other substances in our diet, cholesterol does not qualify as an essential nutrient.

The Contributions of Cholesterol Would you be surprised to learn that cholesterol:

- is found in every cell in your body?
- serves as the building block for estrogen, testosterone, and the vitamin D that is produced in your skin upon exposure to sunlight?
- is a major component of nerves and the brain?
- cannot be used for energy (so it provides no calories)?

The body has many uses for cholesterol (Table 18.7 on p. 18–9). It doesn't just accumulate in arteries!

Table 18.7
How the body uses cholesterol.

- Cholesterol is a component of all cell membranes, the brain, and nerves.
- It is needed to produce estrogen, testosterone, and vitamin D.

Reducing the Fat Content of Foods

In response to strong consumer demand for lower-fat food choices, the food industry is investing heavily in the development of low-fat foods and fat substitutes.[1] More than 5000 foods with reduced-fat content are currently available, and 9 out of 10 Americans consume at least some of these products.[1]

Taking the Fat Out of Foods: How It's Done The food industry is reducing the fat content of foods by adding less fat to processed foods, trimming the fat from meats, selling poultry products without the skin, and adding less fat to ground meats. Fat content is also being lowered by adding water to ham, processed meats, and margarine. When the weight of a product is increased by adding water, the percentage of fat in the product automatically decreases. Whipping air into margarine, butter, and cream cheese reduces the amount of fat in a specific amount of these foods. Sugar, water, and fiber are used to "thicken" low-fat salad dressings and other products.

Fat Substitutes Increasingly, the fat content of processed foods is being partially or fully replaced by fat substitutes. Fat substitutes attempt to imitate the taste, texture, and cooking properties of our favorite fats, but with fewer calories. Table 18.8 lists some of the more than 60 products that are already on the market; many more are under development. The degree to which these products succeed in imitating the qualities of fat varies. Making the perfect fat substitute is not an easy task. Not all of the fat substitutes that have entered the marketplace have met the test of consumer acceptability.[1]

Are Fat Substitutes Safe? Consumers sometimes wonder about the safety of fat substitutes. Because most fat substitutes are derived from ingredients of food such as carbohydrates, protein, and vegetable oils, they are considered safe. Fat replacers that are made from substances that do not occur naturally in foods must be tested for safety. Olean, a non-naturally occurring fat replacer that cannot be digested, was tested extensively before being approved for use. It can cause diarrhea if consumed in large amounts.[18]

When substituted for foods containing saturated fats, fat substitutes may benefit health by lowering blood cholesterol levels.[1] Studies suggest that the use of foods containing fat substitutes reduces total fat consumption without affecting total caloric intake.[19] So, although they may not be the answer to weight control, fat substitutes may provide part of the solution to high fat intakes and the health problems related to them.

One of the primary concerns about fat substitutes is their effect on food choices. Will people eat more foods low in nutrients if they contain low-calorie fat substitutes? Many high-fat foods such as potato chips and rich desserts aren't that nutrient dense to begin with—do we need to eat more of them? And will children meet their need for calories if their diets are loaded with low-calorie foods made with fat substitutes? These are legitimate concerns that are yet to be addressed by scientific studies.

Finding Out about the Fat Content of Food

Not all of the fat in food is visible. To avoid being fooled, it helps to use a reference on the fat composition of foods. Table 18.9 lists the fat content of common

Table 18.8
Some currently available fat substitutes.[1,17]

Carbohydrate-Based (0–4 cal/g)	Protein-Based (1–2 cal/g)	Fat-Based (0–5 cal/g)	Combinations
Amalean I and II	K-Blazer	Benefat	Fitesse
CrystaLean	Proplus	Caprenin	Nutrifat
Instant Stellar	Simplesse	Olean	Prolestra
Juguar			
Litesse			
Maltrin			
N-Lite			
Oat fiber			
Oatrim			
Opta Grade			
Pure-gel			
Sta-Slim			
Sta-Lite			

Table 18.9
The fat content of some foods.

Food	Amount	Grams	Percentage of Total Calories from Fat
Fats and oils			
Butter	1 tsp	4.0	100%
Margarine	1 tsp	4.0	100
Oil	1 tsp	4.7	100
Mayonnaise	1 tbs	11.0	99
Heavy cream	1 tbs	5.5	93
Salad dressing	1 tbs	6.0	83
Meats and fast foods			
Hot dog	1 (2 oz)	17.0	83
Bologna	1 oz	8.0	80
Sausage	4 links	18.0	77
Bacon	3 pieces	9.0	74
Salami	2 oz	11.0	68
Pork steak	3 oz	18.0	62
Hamburger, regular (20% fat)	3 oz	16.5	62
Chicken, fried with skin	3 oz	14.0	53
Big Mac	6.6 oz	31.4	52
Quarter Pounder with cheese	6.8 oz	28.6	50
Whopper	8.9 oz	32.0	48
Steak (rib eye)	3 oz	9.9	47
Hamburger, lean (10% fat)	3 oz	9.5	45
Steak (T-bone), lean	3 oz	8.9	44
Rabbit	3 oz	7.0	38
Veggie pita	1	17.0	38
Ranch chicken pita	1	18.0	34
Steak (round), lean only	3 oz	5.2	29
Chicken, baked without skin	3 oz	4.0	25
Turkey wrap	1	9.0	24
Hamburger, extra lean (4% fat)	3 oz	2.3	23
Venison	3 oz	2.7	18
Subway, club	6"	5.0	14
Flounder, baked	3 oz	1.0	13
Subway, veggie	6"	3.0	11
Shrimp, boiled	3 oz	1.0	10
Milk and milk products			
Cheddar cheese	1 oz	9.5	74
American cheese	1 oz	6.0	66
Milk, whole	1 cup	8.5	49
Cottage cheese, regular	½ cup	5.1	39
Milk, 2%	1 cup	5.0	32
Milk, 1%	1 cup	2.7	24
Cottage cheese, 1% fat	½ cup	1.2	13
Milk, skim	1 cup	0.4	4
Yogurt, frozen	¾ cup	0.0–6.6	0–3
Other			
Olives	4 medium	1.5	90
Avocado	½	15.0	84
Almonds	1 oz	15.0	80
Sunflower seeds	¼ cup	17.0	77
Peanut butter	1 tbs	8.0	76
Peanuts	¼ cup	17.5	75
Cashews	1 oz	13.2	73
Egg	1	6.0	61
Potato chips	1 oz (13 chips)	11.0	61
Chocolate chip cookies	4	11.0	54
French fries	20 fries	20.0	49
Taco chips	1 oz (10 chips)	6.2	41

continued

Table 18.9
Continued

Food	Amount	Grams	Percentage of Total Calories from Fat
Candy			
Mr. Goodbar	1.7 oz	15.0	56
Peanut butter cups, 2 regular	1.6 oz	15.0	54
Milk chocolate	1.6 oz	14.0	53
Almond Joy	1.8 oz	14.0	50
Kit Kat	1.5 oz	12.0	47
M and M's, peanut	1.7 oz	13.0	47
Nestlé's Crunch	1.6 oz	11.0	45
Twix	2.0 oz	14.0	45
Baby Ruth	2.1 oz	14.0	43
Pay Day	1.9 oz	12.0	43
Snickers	2.1 oz	13.0	42
Butterfinger	2.1 oz	12.0	39
M and M's, plain	1.7 oz	10.0	39
Milky Way	2.2 oz	11.0	35
3 Musketeers	2.1 oz	9.0	31
Tootsie Roll	2.3 oz	6.0	21

food sources of fat, including candy. Vegetables and fruits (except avocado) and grains are not listed because they contain relatively little fat. Other references can also be used, such as the food composition tables in Appendix A, the Diet Analysis Plus Program software, and the nutrition labels on food products (Illustration 18.8).

Knowledge of the caloric and fat content of a food can be used to calculate the percentage of calories provided by fat. For example, suppose that a slice of cherry pie provides 350 calories and 15 grams of fat. To calculate the percentage

Illustration 18.8
Labeling foods for fat content: an example.

Table 18.10
What claims about the cholesterol content of foods that normally contain cholesterol must mean.

- **No cholesterol** or **cholesterol-free:** Contains less than 3 milligrams of cholesterol per serving.
- **Low cholesterol:** Contains 20 milligrams or less of cholesterol per serving.
- **Reduced cholesterol:** Contains at least 75% less cholesterol than normal.
- **Less cholesterol:** Contains at least 25% less cholesterol than normal. The percentage less must be stated on the label.

of fat calories, multiply 15 grams by 9 (the number of calories in each gram of fat), divide the result by 350 calories, and then multiply by 100:

$$\begin{aligned} 15 \text{ grams fat} \times 9 \text{ calories/gram} &= 135 \text{ calories} \\ 135 \text{ calories} \div 350 &= 0.39 \\ 0.39 \times 100 &= 39\% \text{ of total calories from fat} \end{aligned}$$

Fat Labeling Remember when bananas and peanut butter were labeled "cholesterol-free" and one-ounce servings of cheesecake were declared "low fat"? The days of labeling foods that are naturally devoid of cholesterol as "cholesterol-free" and reducing serving sizes of rich desserts to lower their fat content are over.

Nutrition labeling regulations for fat and cholesterol that took effect in 1994 represent substantial improvements over prior standards. These regulations require food manufacturers to adhere to standard definitions of "low fat," "fat-free," and related terms used on food labels. Similarly, claims made about the cholesterol content of food products must comply with standard definitions (Table 18.10). If a claim is made about the fat content of a food, the Nutrition Facts panel must specify the food's fat, saturated fat, and cholesterol content. If a claim is made about cholesterol content (and claims can be made only for products that normally contain cholesterol), the nutrition panel must also reveal the product's fat and saturated fat content. To prevent the use of unrealistically small serving sizes as a way to appear to cut down on a product's fat content, standard serving sizes must be used on food labels.

The Fat American Diet

Americans' collective consciousness about the hazards of excessive fat has been raised. Nearly 90% of Americans are trying to cut down on fat.[1] Their efforts have paid off—the average intake of fat in the United States has dropped from 42% of total calories in 1965 to 34% in recent years.[20] Reducing fat intake to the recommended level of less than 30% of total calories, however, turns out to be harder than was previously imagined (Illustration 18.9). Most people find it difficult to get fat intake down and keep it there.[10,21] Why are we so attached to high-fat foods?

Illustration 18.9
The U.S. diet is heavy in fats and cholesterol.

> The average American diet is high in fat because fat tastes good. It doesn't take a genius to figure that out.
>
> —Max Brown

Given access to high-fat foods, most people readily develop a preference for them. People generally like the texture and flavor of high-fat foods and miss those foods if they are excluded from the diet.[10,21] How, then, are Americans supposed to reach the healthful goal of cutting back on fat?

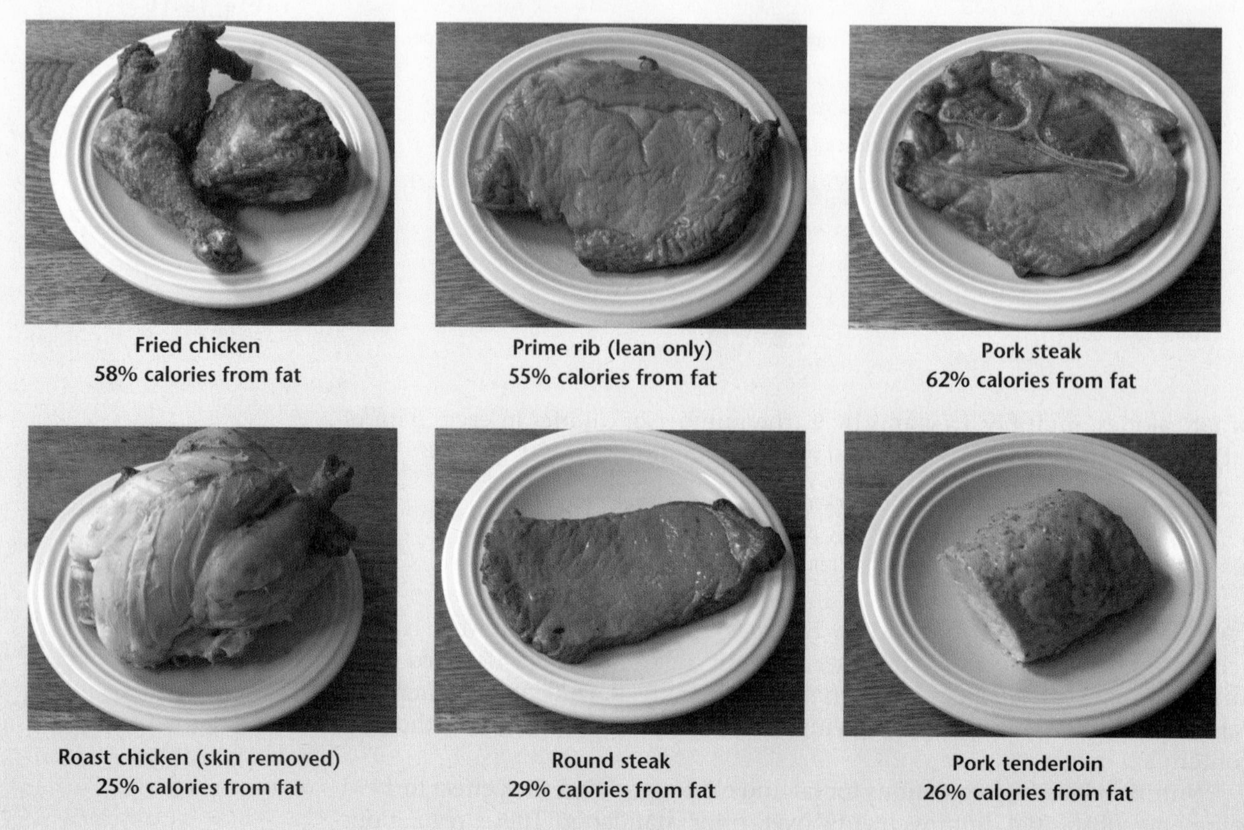

Illustration 18.10 Which meats are lean? Take a look.

Taking the Pain Out of Reducing Fat Intake Fat-reduction strategies that require people to give up their favorite high-fat foods almost never work.[22] The change is too drastic and unpleasant for most people to accept. Other approaches to fat reduction have been shown to work for many people, however. A basic element of these approaches is practical knowledge about the fat content of foods.[22] Without that knowledge, it's unlikely that much progress in reducing fat intake will be made.

Topping the list of effective strategies for reducing fat intake is the substitution of low- and no-fat foods for their higher-fat counterparts.[23] Switching from whole milk to reduced-fat or skim milk, baking instead of frying, trimming the fat off meats, eating meats low in fat (such as skinless chicken and fish), reducing intake of luncheon meats (bologna, salami), and choosing plant sources of protein are examples of changes some people find easy to make. Examples of the reduction in fat intake that can be obtained by selecting specific types of meat are shown in Illustration 18.10.

Table 18.11
The number of fat grams allowed per day for diets that provide less than 30% of total calories as fat.*

Usual Caloric Intake	Fat Grams Allowed
1500	49
1750	57
2000	66
2250	74
2500	82
2750	91
3000	99
3250	107
3500	116
3750	124

*To calculate fat grams for a caloric intake level different from those shown, multiply caloric intake by 0.3, then divide the result by 9. For example: 1800 calories × 0.3 = 540 calories; 540 calories ÷ 9 = <60 grams of fat allowed.

Counting Fat Grams People who want to guarantee a low fat intake might use another method for reducing fat intake. It involves keeping track of fat grams to make sure fat intake remains less than 30% of total calories. The individual identifies the "allowable" number of grams of fat per day (Table 18.11) and then keeps a tally of grams of fat consumed, adjusting food choices as necessary to keep to the limit.

This method isn't recommended for everyone, but some people do find that counting fat grams is the key to a lower-fat diet. The approach that works best will vary from individual to individual. Clearly, however, the most successful method will be the one that an individual finds acceptable and easiest to do.

Nutrition UP CLOSE

Score Your Diet for Fat

FOCAL POINT: Identify the high-fat food choices in your diet.

Do the foods you eat provide more fat than is good for you? Answer the questions below to see how your diet rates.

How Often Do You Eat:	Seldom or Never	1–2 Times per Week	3–5 Times per Week	Almost Daily
1. Bacon, sausage, hot dogs, ribs, and luncheon meats?	☐	☐	☐	☐
2. Heavily marbled steaks or roasts and chicken with the skin?	☐	☐	☐	☐
3. High-fat protein alternatives such as refried beans, nuts, and sunflower seeds?	☐	☐	☐	☐
4. Whole milk, cheese, and ice cream?	☐	☐	☐	☐
5. Whipped cream, table cream, sour cream, and cream cheese?	☐	☐	☐	☐
6. Butter or margarine on vegetables, dinner rolls, and toast?	☐	☐	☐	☐
7. French fries, egg rolls, taco shells, and other deep-fat-fried, breaded, or battered foods?	☐	☐	☐	☐
8. Salad dressings, guacamole, and mayonnaise?	☐	☐	☐	☐
9. Rich sauces and gravies?	☐	☐	☐	☐
10. Croissants, muffins, and coffee cakes?	☐	☐	☐	☐
11. Snacks such as potato chips, tortilla chips, buttered popcorn, and candy bars?	☐	☐	☐	☐
12. Desserts such as pies, pastries, cookies, and cakes.	☐	☐	☐	☐

FEEDBACK (including scoring) can be found at the end of Unit 18.

Source: Adapted from Reduce fat, saturated fat, and cholesterol intake. Washington, DC: USDA Home and Garden Publication 232-2; 1986.

WWW links

www.nal.usda.gov
Find out more about fats and fat substitutes.

www.healthfinder.gov
Good source of reliable information on fats and cholesterol.

www.ificinfo.health.org/
This food industry-sponsored site provides information on fat, trans fatty acids, fat substitutes, and cholesterol.

REFERENCES

1. Mattes RD. Position of the American Dietetic Association: fat replacers. J Am Diet Assoc 1998;98:463–8.
2. Hu FB, et al. Dietary fat and the risk of coronary heart disease in women. N Engl J Med 1997;337:1491–9.
3. Fair WR, et al. Cancer of the prostate: a nutritional disease? Urology 1997;50:840–8.
4. Connor WE. Importance of n-3 fatty acids in health and disease. Am J Clin Nutr 2000;71(suppl):171S–5S.
5. Hepburn FN, et al. Provisional tables on the content of omega-3-fatty acids and other fat components of selected foods. J Am Diet Assoc 1986;86:788–93.
6. National Fisheries Institute. In: Catch of the day: omega-3-fats, but from fish or capsules? Envir Nutr 1997 July:1, 6.
7. Meluzzi A, et al. Effects of dietary vitamin E on the quality of table eggs enriched with n-3 long-chain fatty acids. Poultry Sci 2000;79:539–45.
8. Stone NJ. Fish consumption, fish oil, lipids, and coronary heart disease. Am J Clin Nutr 1997;65:1083–6.
9. An expectant mother's guide to eating Minnesota fish. Minnesota Health Department, 2000; and FDA Press Release 2001 Jan 13.
10. Walsh J. Low fat, no fat, some fat . . . high fat? Envir Nutr 1998;21(April):1, 6.
11. Allison DB, et al. Estimated intakes of trans fatty and other fatty acids on the US population. J Am Diet Assoc 1999;99:166–74.
12. Krauss RM, et al. Revision 2000: A statement for health professionals from the Nutrition Committee of the American Heart Association. J Nutr 2001;131:132–46.
13. Stampfer MJ, et al. Primary prevention of coronary heart disease in women through diet and lifestyle. N Engl J Med 2000;343:16–22.
14. FDA proposes labeling of trans fatty acids. Federal Register 1999; Nov 12:1.
15. Ascherio A, Willett WC. Health effects of trans fatty acids. Am J Clin Nutr 1997;66:1006S–10S.
16. Sabar AF, et al. Dietary sources of nutrients among U.S. adults. J Am Diet Assoc 1998;98:537–47.
17. Gershoff SN. Nutrition evaluation of dietary fat substitutes. Nutr Rev 1995;53:305–13.
18. Olestra and steatorrhea. Lancet 2000;355:300.
19. Rodriquez L, et al. Use of low-fat foods by people with diabetes decreases fat, saturated fat and cholesterol intake. J Am Diet Assoc 2000;100:531–6.
20. Ernst ND, et al. Consistency between US dietary fat intake and serum total cholesterol concentrations: the National Health and Nutrition Examination Surveys. Am J Clin Nutr 1997;66(suppl): 965S–72S.
21. Drewnowski A. Sensory properties of fats and fat replacements. Nutr Rev 1992;50:17–20.
22. Kristal AR, et al. Nutrition knowledge, attitudes and perceived norms as correlates of selecting low-fat diets. Health Education Research 1990;5:467–77.
23. Gorbach SL, et al. Changes in food patterns during a low-fat dietary intervention in women. J Am Diet Assoc 1990;90:802–9.

Nutrition

UP CLOSE

Score Your Diet for Fat

Feedback for Unit 18

The more responses in the last two columns, the higher your fat intake. Perhaps it's time you cut back. If so, the following strategies can help you reduce the amount of fat in your diet:

1. Substitute lower-fat versions of high-fat foods you enjoy. For example, instead of having a blueberry muffin topped with butter for breakfast, try a piece of whole-grain toast spread with reduced-fat margarine along with a bowl of blueberries.
2. Watch how foods are prepared. Try baking, steaming, grilling, and roasting instead of deep-fat frying and sautéing.
3. Read the Nutrition Facts label to compare the amount of fat in similar products before buying. Many lower-fat food products are now available.
4. When you do choose a high-fat food, try eating a smaller portion of it.

Nutrition and Heart Disease

Photo Disc

Nutrition Scoreboard

	TRUE	FALSE
1. Heart disease is the leading cause of death among men and women in the United States.		
2. The most effective way to lower blood cholesterol is to exercise regularly.		
3. Inadequate intake of folate (a B vitamin) is a risk factor for heart disease.		
4. The risk factors for heart disease are the same for women as for men.		

(Answers on next page)

Answers to

Nutrition Scoreboard

	TRUE	FALSE
1. About 32% of the deaths of men *and* women in the United States each year are related to heart disease.	✓	
2. The most effective way to lower blood cholesterol levels is to reduce saturated fat intake.		✓
3. Increasing folate intake may substantially reduce the risk of heart disease in many adults.	✓	
4. Risk factors for heart disease differ between women and men.		✓

Key Concepts and Facts

- Heart disease is the leading cause of death in men and women in the United States.
- Of the multiple contributors to heart disease, genetic and dietary factors appear to be the most important.
- Throughout the world, populations with high saturated fat intakes have high rates of heart disease.
- Lowering high blood cholesterol levels reduces the risk of heart disease.

The Diet–Heart Disease Connection

Suspicions that dietary fat may be related to heart disease were first raised over 200 years ago. During the late eighteenth century, physicians noted that people who died of heart attacks had fatty streaks and deposits in the arteries that led to the heart. Unlike many early theories about the causes of diseases, this one has basically held up over time. It is clearly established that diets high in saturated fat (rather than total fat as originally suspected) are the number one risk factor for heart disease in populations worldwide.[1] Other risk factors for heart disease have been identified and, except for genetic tendencies and age, can be modified. That makes heart disease a highly preventable cause of death and disability.

A Primer on Heart Disease

There is no bigger health problem in the United States and other developed countries than heart disease (Illustration 19.1). Worldwide, heart disease accounts for one out of every four deaths.[2] It is an "equal opportunity" disease, striking as many women as men, although women on average die 10 years later from heart disease than do men.[3] It is an *un*equal opportunity disease in that African Americans are more likely to die from it than are Caucasian Americans.

What Is Heart Disease? ***Heart disease,*** or more correctly "coronary" heart disease, is a term that covers several disorders that result from inadequate blood circulation to parts of the heart. It is almost always due to a narrowing of the arteries leading to the heart. Arteries become narrow due to a buildup of ***plaque*** (Illustration 19.2). People with narrowed arteries have ***atherosclerosis,*** or "hardening of the arteries" as it is often called. Heart disease develops silently over time (usually decades).

When arteries are narrowed by 50% or more, the shortage of blood to the heart can produce chest pain (called "angina"). If the blockage is extensive

heart disease
One of a number of disorders that result when circulation of blood to parts of the heart is inadequate. Also called *coronary heart disease.* ("Coronary" refers to the blood vessels at the top of the heart. They look somewhat like a crown.)

plaque
Deposits of cholesterol, other fats, calcium, and cell materials in the lining of the inner wall of arteries.

atherosclerosis
"Hardening of the arteries" due to a buildup of plaque.

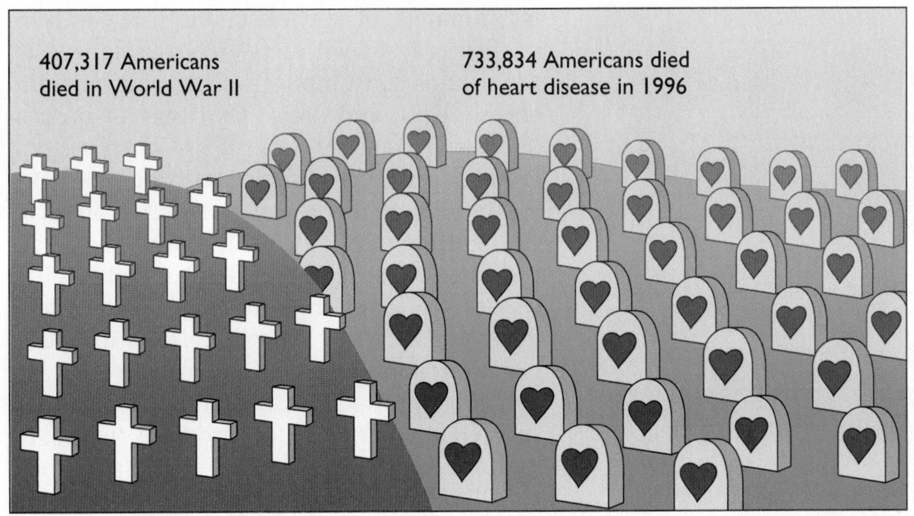

Illustration 19.1
The impact of heart disease. If deaths from stroke and other diseases related to atherosclerosis are included, the annual death toll reaches 951,302.

Source: National Center for Health Statistics, 1997.

enough, a heart attack occurs due to the lack of blood supply to a particular part of the heart (Illustration 19.3). Although heart disease primarily affects individuals over the age of 55, it's a progressive disease that may begin in childhood.[4]

Arteries leading to the heart aren't the only ones affected by atherosclerosis. Plaque can also build up in the arteries in the legs, neck, brain, and other body parts. If the blood supply to the legs is reduced, pain and muscle cramps may result after brief periods of exercise. Clogged arteries in the brain can contribute to stroke—an event that occurs when the blood supply to a part of the brain is inadequate. Health problems due to atherosclerosis in arteries of the heart, brain, neck, and legs are collectively referred to as ***cardiovascular disease.***

➡ **cardiovascular disease** Disorders related to plaque buildup in arteries of the heart, brain, and other organs and tissues.

What Causes Atherosclerosis? A number of conditions are known to increase plaque formation in arteries; the most influential is blood cholesterol level. (Some people with atherosclerosis don't have elevated blood cholesterol, however.) In general, the higher the blood cholesterol, the more likely it is that plaque will build up in arteries and that heart disease will occur (Illustration 19.4).[5]

What Raises Blood Cholesterol?

A person's cholesterol level is determined by a number of factors, including dietary intake, behaviors such as smoking and exercise, and heredity. Diets high

Illustration 19.2
The progression of atherosclerosis. As plaque builds up, arteries narrow, reducing or stopping the supply of blood to the heart, brain, muscle, or other affected parts of the body.

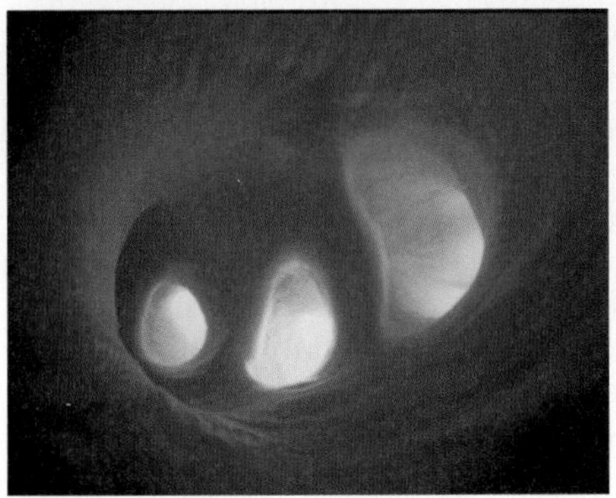

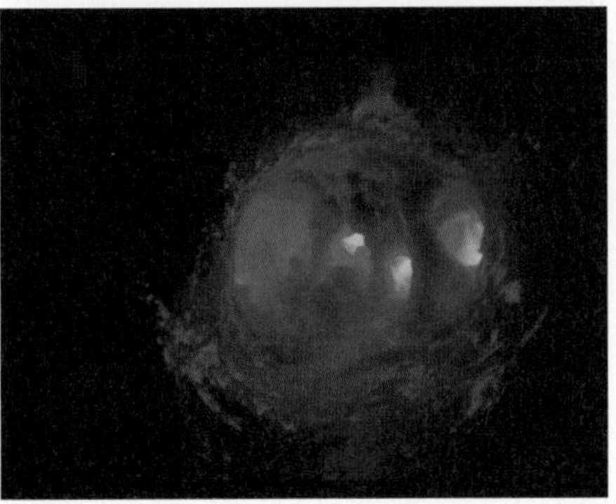

ICI Pharmaceutical Division, Cheshire, England

Illustration 19.3
The heart after a heart attack.

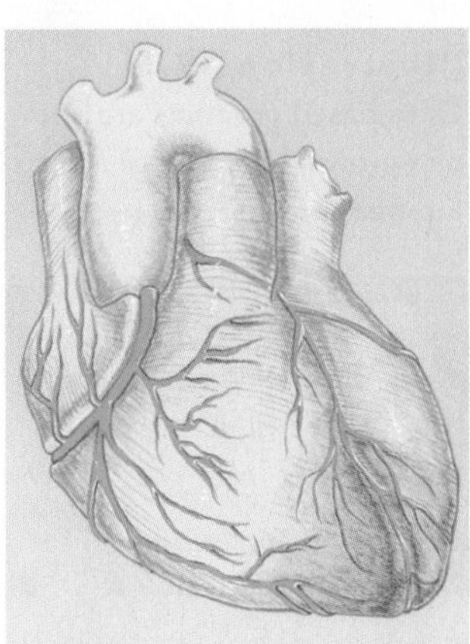

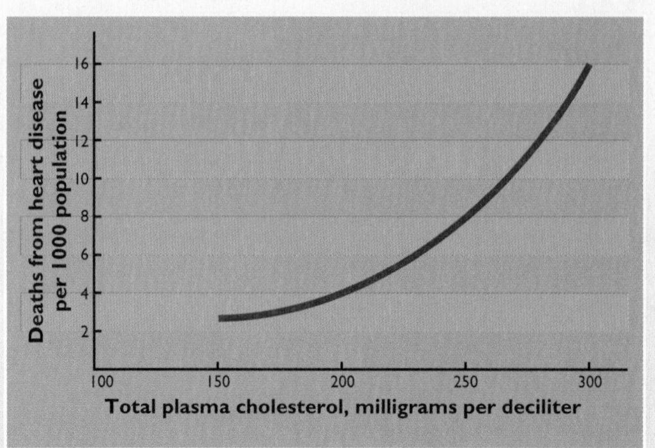

Illustration 19.4
The relationship between blood cholesterol level and death from heart disease.

Source: Grundy SM. Cholesterol and coronary heart disease: a new era. JAMA 1986;256:2849–58. Copyright 1986, American Medical Association.

in saturated fat elevate cholesterol levels in most people.[6] Such diets are characterized by the regular consumption of high-fat milk and cheese, eggs, and beef.[7] One type of unsaturated fat, "trans fat," raises blood cholesterol levels.[6,8] This type of fat is produced when vegetable oils are hydrogenated—made solid by the addition of hydrogen. The process changes the chemical structure of unsaturated fats in a way that enables them to raise blood cholesterol levels. Technological advances in hydrogenation and increased use of olive oil are reducing the trans fat content of margarines, shortenings, and other products made by hydrogenating oils (Illustration 19.5).

Blood cholesterol levels can also be raised by high cholesterol intakes. But blood cholesterol responds less to dietary cholesterol intake than to saturated fat intake.[9] Depending on genetic background, some individuals "hyper-respond"; that is, they experience a sizable rise in blood cholesterol level when they consume large amounts of saturated fat or cholesterol. Other people are "hypo-responders" and experience little change in blood cholesterol level when they consume saturated fat or cholesterol.[10] For many individuals, the addition of an egg or two a day to a low-fat diet has little or no effect on blood cholesterol levels.[11]

All Blood Cholesterol Is Not Equal Cholesterol is soluble in fat, but blood is mostly water. Therefore, cholesterol must be bound to compounds that mix with water, or it would float in blood. For this reason, cholesterol present in blood is bound to protein, which is soluble in water. The resulting combination is called a "lipoprotein," and there are a number of different types. Two lipoproteins have gained notoriety by virtue of their coverage in the popular press and their role in the development of heart disease. One is HDL-cholesterol (for "high-density lipoprotein" cholesterol), and it's the "good" type of cholesterol. LDL-cholesterol (low-density lipoprotein cholesterol) is the villain. (See Illustration 19.6 for a lesson and a chuckle.) The composition and roles of both types of lipoproteins are summarized in Illustration 19.7.

HDL and LDL Highlights HDL gets its reputation as the good cholesterol because it helps remove cholesterol from the blood. An avid cholesterol acceptor, HDL escorts cholesterol to the liver for its eventual excretion from the body. High HDL-cholesterol levels (over 60 milligrams per deciliter) are protective against heart disease.[12]

LDL-cholesterol carries more cholesterol than does HDL, and its cholesterol can be incorporated into plaque. The higher the LDL-cholesterol level, the more likely it is that atherosclerosis will develop and progress into heart disease.[1]

Illustration 19.5
Olive-based products are one of a growing number of trans fat–free spreads.

Modification of Blood Levels of HDL and LDL Table 19.1 (p. 19–6) summarizes factors that influence blood HDL and LDL levels. Blood levels of HDL-cholesterol are lowered by diets high in polyunsaturated fats and those low in saturated fats.[9,13] This is unfortunate since diets rich in polyunsaturated fats lower LDL-cholesterol, too. Monounsaturated fats, found in olive and other oils, lower LDL-cholesterol and not HDL-cholesterol, making them a preferred source

Close to Home: *By John McPherson*

"To help you better understand this good cholesterol / bad cholesterol thing, Nurse Bowman and Nurse Strickling are going to do a little skit for you."

Illustration 19.6

Source: © 2001 John McPherson. Distributed by Universal Press Syndicate.

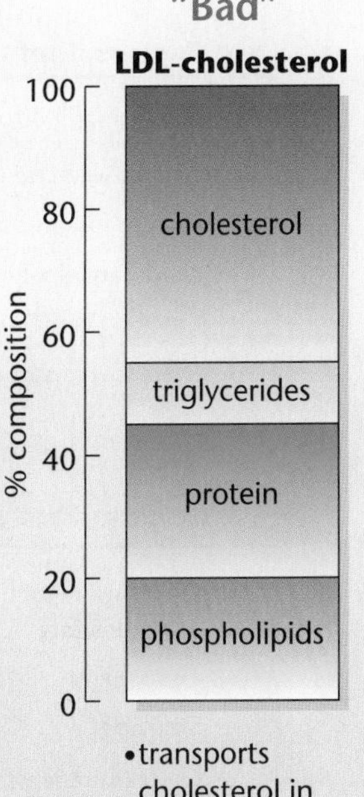

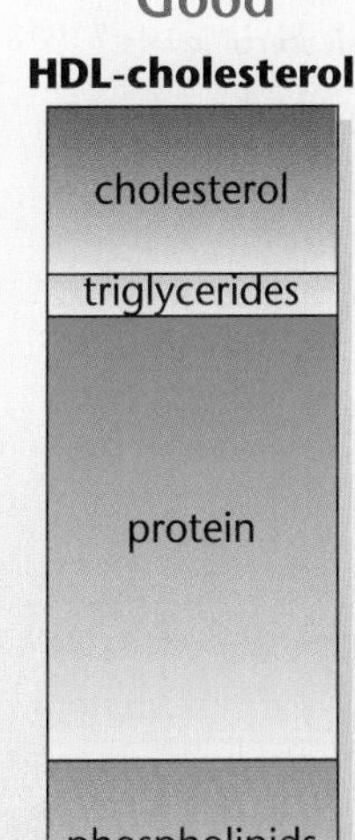

Illustration 19.7

The "bad" cholesterol (LDL) and the "good" cholesterol (HDL) found in blood.

of fat for improving blood lipid levels. Physical inactivity, obesity (especially high levels of central body fat), and smoking also tend to lower HDL-cholesterol levels.[9,10] HDL-cholesterol levels are often increased by regular exercise, reduction in body fat (if overweight), smoking cessation, saturated fats, and moderate alcohol intake (one to two drinks per day).[9,10]

Diets high in saturated fats from animal products and in trans fats are associated with elevated LDL-cholesterol levels.[14] High cholesterol diets raise blood levels of LDL-cholesterol, but to a lesser extent than do saturated fats. Regular consumption of boiled coffee, and coffee made without filters, has been found to raise LDL-cholesterol.[15] Other factors, including inactivity, obesity, and smoking, also increase LDL-cholesterol levels.[10] Personality type and stress have not been shown to have a consistent effect on blood cholesterol levels.[5]

Blood levels of LDL-cholesterol are generally reduced by diets providing primarily poly- and monounsaturated fats. Weight loss (if overweight) and ample consumption of soybean products, dried beans, vegetables, fruits, oats, and other high fiber and whole grain foods tend to lower LDL-cholesterol levels as well.[6,16,17]

Table 19.1
Factors affecting HDL- and LDL-cholesterol levels.[9,10,12]

HDL-Cholesterol (High levels are protective against heart disease.)*	
Factors That Increase HDL	**Factors That Decrease HDL**
· Genetic tendency to produce high levels	· Genetic tendency to produce low levels
· Regular exercise	· Diets high in polyunsaturated fats
· Weight loss in overweight people	· Physical inactivity
· Moderate alcohol intake	· Overweight
· Smoking cessation	· Smoking
· Diets low in saturated fats	
LDL-Cholesterol (High levels promote heart disease.)	
Factors That Increase LDL	**Factors That Decrease LDL**
· Diets *high* in saturated fats, cholesterol, and trans fats	· Diets *low* in saturated fats, cholesterol, and trans fats
· Overweight	· Diets high in mono- and polyunsaturated fats
· Smoking	· Diets high in dried beans, vegetables, and whole grains
· Coffee (made without a filter)	· Weight loss in overweight people
	· Physical activity

*Memory note: You want your *HDL* cholesterol to be *high* and your *LDL* cholesterol to be *low.*

Other Dietary Factors Related to Atherosclerosis

Plaque formation in arteries has been linked to vitamin E; antioxidants in vegetables, fruits, and grains; and iron and folate status. Of these, evidence for a role of iron in the development of atherosclerosis is the weakest.

Supplementary intakes of vitamin E in the range of 200 to 500 IU per day decrease plaque formation.[18] Vitamin E attaches to LDL-cholesterol and protects it from becoming oxidized and reactive. Oxidized, reactive LDL-cholesterol is much more likely to become incorporated into plaque than unoxidized LDL is. Other antioxidants that occur naturally in vegetables, fruits, and grains also help reduce plaque formation.[16]

On the Side

Seven out of 10 Americans over the age of 18 do not know the difference between "good" cholesterol (HDL) and "bad" cholesterol (LDL).

Excessive levels of iron stores in men and the elderly have been reported to increase plaque formation, but other studies have found either no effects or protective effects of high iron stores.[19] The jury is still out on the guilt or innocence of high iron intakes and stores on the development of atherosclerosis.[20,21]

Folate and Atherosclerosis A truly exciting development related to folate (a B vitamin) and atherosclerosis has emerged. Numerous studies have confirmed that inadequate folate is strongly related to the development of plaque-lined arteries.[14] This news is exciting because a practical solution is at hand—folate intakes can easily be increased.

Folate helps to prevent atherosclerosis by reducing blood levels of homocysteine, an amino acid that increases plaque formation if present in high amounts in the blood. Folate occurs primarily in two forms in our diets: naturally occurring folate in vegetables and fruits, and synthetic folic acid, the form

Folate or Folic Acid Intake/Day	Decline in Blood Homocysteine Level
A. 418 micrograms (mcg) folate from vegetables and fruits (about 8 servings/day)	A. 9%
B. 70–120 mcg folic acid from fortified, refined grain products	B. 7%
C. 300 mcg folic acid from fortified cereals	C. 24%
D. 250 mcg folic acid from supplements	D. 21%
E. 400 mcg folic acid from supplements	E. 24%

Table 19.2
Declines in blood homocysteine levels associated with intake of folate from vegetables and citrus fruits, folic acid–fortified cereals, and folic acid supplements.[22–24]

of folate added to fortified foods and used in supplements. Naturally occurring folates are only about 50% absorbed, whereas folic acid is approximately 90% absorbed. Consequently, it takes more naturally occurring folates to reduce blood homocysteine levels than it takes folic acid. As shown in Table 19.2, very high intakes of green vegetables (spinach, green peas, asparagus, broccoli, and collard greens, for example) and citrus fruit decrease blood levels of homocysteine about 9%, but consumption of 250–400 micrograms of folic acid in fortified foods or supplements reduces it by over 20%. Intakes of about 100 micrograms of folic acid each day reduce the percentage of adults with high homocysteine levels by 50%.[23] High homocysteine levels are rare among people with folic acid intakes of 400 micrograms per day.[23]

Since 1998 refined grain products such as bread, crackers, pasta, and rice have been fortified with folic acid. Most breakfast cereals are fortified with 100 micrograms of folic acid per serving, and some are highly fortified (see Table 19.3). The recommended intake of folate of 400 micrograms can be more than met by consuming five servings of vegetables and fruits and six servings of folic acid–fortified bread and cereal products daily.

Are you getting enough folate? You can estimate your intake by following the steps in the Health Action on p. 19–8.

Table 19.3
Breakfast cereals fortified with 400 micrograms of folic acid.

All Bran
Complete
Healthy Choice
Just Right
Life
Multi Grain Cheerios
Product 19
Smart Start
Special K
Total

What about Fish Oils? Are fish good food for the heart? Would Americans benefit if they ate more fish? The answer to both questions appears to be "yes."

Fish contain polyunsaturated fats, which tend to lower LDL-cholesterol levels. But fish is protective against heart disease to a greater extent than can be accounted for by its polyunsaturated fat content. As few as two fish meals a week are associated with a reduced risk of heart disease.[25]

The beneficial effect of fish on heart disease rates appears to be related to a particular type of polyunsaturated fat found in certain fish. Many types of cold-water fish such as tuna and salmon contain omega-3 fatty acids, which are highly unsaturated. These fatty acids appear to help protect the body against heart disease by decreasing the tendency of blood to clot and to block arteries and by lowering blood triglyceride levels.[26]

Although Americans are urged to consume more fish, this recommendation does not extend to fish oil supplements because concerns remain about their safety.[26] Large fish, such as swordfish, shark, king mackerel, and tilefish, may

HEALTH ACTION

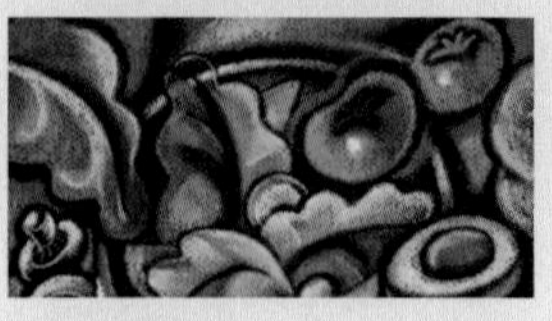

Are You Getting Enough Folate from Your Diet?

To estimate whether you are consuming approximately 400 micrograms of absorbable folate per day, follow these steps:

	Number of Servings		Folate Content	
1. Write down the number of servings of vegetables you usually consume in a day. Multiply that number by 20 to estimate the amount of folate that will be absorbed from vegetables.	______	× 20 =	______	mcg
2. Write down the number of servings of cold breakfast cereal you eat each day. (A serving is generally an ounce or a cup.) Multiply that number by 100. If you consume a highly fortified cereal (see Table 19.3), multiply the number of daily servings by 400.	______	× 100 =	______	mcg
3. Write down the number of servings of bread, pasta, rice, bagels, crackers, biscuits, or other refined grain products you tend to eat in one day. Multiply this figure by 40.	______	× 40 =	______	mcg
4. Write down the number of half-cup servings of cooked dried beans you eat in a usual day. Multiply that number by 70.	______	× 70 =	______	mcg
5. Add the figures in the Folate Content column.		Total =	______	mcg

If the result is approximately 400 micrograms of folate, you're right on target! If less—well, you know what to do.

P.S. The Upper Limit for folic acid intake is 1000 micrograms per day.

contain levels of mercury that might interfere with brain development. Consequently, it is advised that pregnant and breastfeeding women, and children under the age of six, limit their intake to small fish (less than 20 inches long).[27]

Special Spreads and Blood Cholesterol Reduction Margarine spreads containing ***plant stanols,*** such as Take Control and Benecol, are widely available in grocery stores in many countries. By competing with cholesterol for absorption in the small intestine, plant stanols lower cholesterol absorption.[28] Daily consumption of two tablespoons of a spread containing plant stanols is related to a 10% drop in total blood cholesterol level, and a 14% decline in LDL-cholesterol concentration. Blood levels of HDL-cholesterol are not affected by consumption of plant stanols.[29] Although expensive, these spreads represent one more tool for lowering high blood cholesterol levels.

➡ **plant stanols**
Substances in corn, wheat, oats, rye, olives, wood, and some other plants that are similar in structure to cholesterol but that are not absorbed by the body. They decrease cholesterol absorption.

Who's at Risk for Heart Disease?

Frankly, you may be. Take a look at Table 19.4 and see. Most Americans are at risk for heart disease because they have high blood cholesterol levels, hypertension, or a family history of heart attack before the age of 55, or because they smoke, are physically inactive, are obese, or have diabetes or high blood pressure (Illustration 19.8).[30] High blood cholesterol levels are common. If 100 adults representing a cross section of Americans were to file past you, 20 of them would be at risk of heart disease due to high cholesterol levels.[31]

A portion of the risk for heart disease may arise very early in life. Recent evidence suggests that infants who weigh less than 5.5 pounds at birth due to the mother's poor malnutrition during pregnancy may be at increased risk for heart disease later in life. According to this "fetal origins hypothesis" of chronic disease development, it is theorized that adaptations made by a fetus in response to an inadequate supply of energy or nutrients may compromise long-term health.[32] Small, malnourished infants may be at increased risk for heart disease as adults due to less than optimal development and growth of the liver.[33] Although researchers have long believed that heart disease begins in childhood, the risk for heart problems may begin even earlier than previously imagined.

Are the Risks the Same for Women as for Men? It is generally assumed that risk factors for heart disease identified in men apply to women as well. This assumption, along with the notion that heart disease represents a major health problem only for men, led to the exclusion of women from research studies.[34] Both of these assumptions have been shown to be incorrect. Heart disease is a major health problem of women. Indeed, it is the leading cause of death among women in many countries (Illustration 19.9). Furthermore, risk factors for heart disease identified in studies of men do not always apply to women.

Differences in Risk Factors for Heart Disease between Women and Men Results of recent studies involving women reveal that high blood cholesterol levels are a weak predictor of heart disease in older women and that low HDL-cholesterol levels are a stronger predictor than in men. Diabetes, obesity, high blood triglyceride levels, and increasing age are particularly strong risk factors for heart disease in women.[34–36]

The risk for heart disease in women increases substantially after menopause, which occurs around the age of 50 years. Menopause is characterized by declines in the levels of the hormone estrogen and HDL-cholesterol and increases in the levels of LDL-cholesterol. It is not clear, however, whether these increases in

Table 19.4
Leading risk factors for heart disease.[12]

- High blood cholesterol level (particularly LDL-cholesterol level)
- Low HDL-cholesterol level
- Diet high in saturated fat, cholesterol, and trans fats
- Family history of early heart attack
- Inadequate folate intake
- Diet low in fruits, vegetables, and whole grains
- High blood triglyceride level
- Hypertension
- Smoking
- Physical inactivity
- Obesity (especially central fat)
- Diabetes

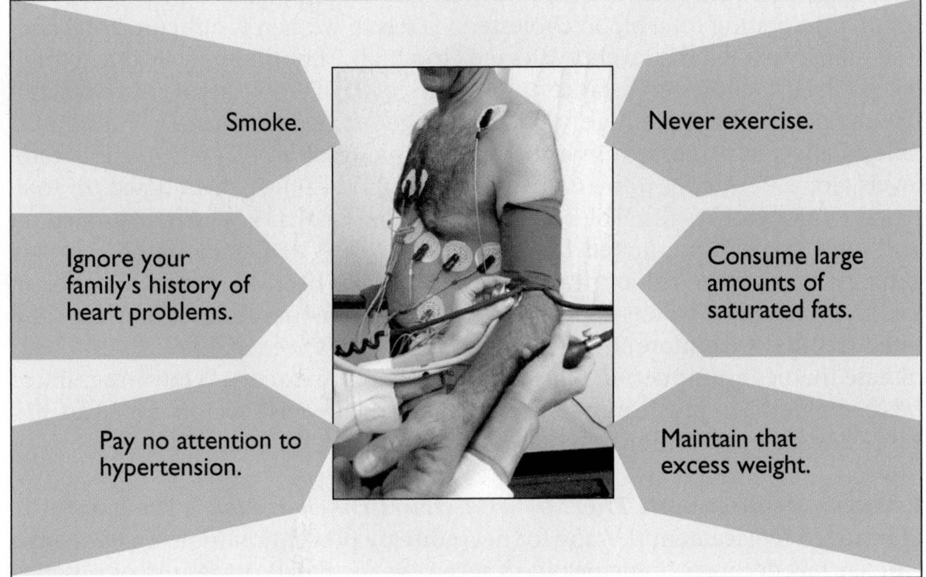

Photo Disc

Illustration 19.8
How to have a heart attack.

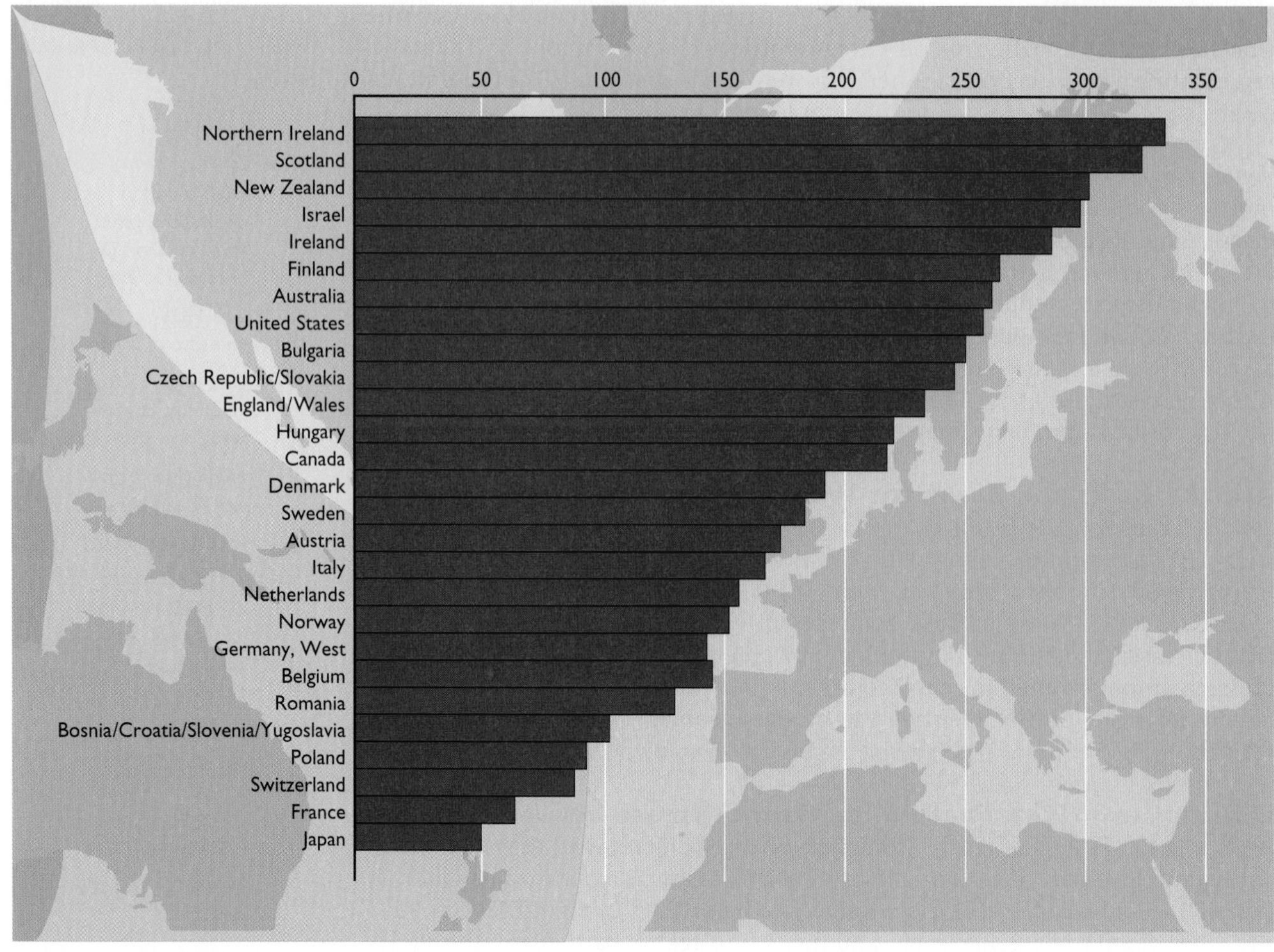

Illustration 19.9
Death rates for heart disease by country per 1000 women 35–74 years of age.

Source: Adapted from material from the National Heart, Lung, and Blood Institute; data are from the World Health Organization, 1985.

LDL-cholesterol or other factors increase heart disease risk and whether lowering LDL-cholesterol levels reduces risk in women.[37] Compared to men, it appears that LDL-cholesterol levels are a much weaker predictor of heart disease in women.[38]

Women normally have higher levels of HDL-cholesterol throughout their lives than men, and their total blood cholesterol level is a bit higher, too.[39] As a result, interpreting total blood cholesterol levels in women is difficult. Total cholesterol may exceed the level considered "too high," but that may be due to high levels of HDL-cholesterol. Since high levels of HDL-cholesterol (over 60 milligrams per deciliter) are highly protective against heart disease,[12,40] it's a mistake to allocate risk based on either total cholesterol levels or LDL-cholesterol levels alone. Yet many women are advised to modify their diets based on total blood cholesterol results. The recommended diet, which is low in saturated fat and high in polyunsaturated fat, tends to decrease HDL-cholesterol and may actually increase the risk of heart disease.[9,37] It has been suggested that weight loss by overweight women and substitution of monounsaturated fats for saturated fats may be a more healthful way for women to reduce their risk of heart disease than consumption of diets high in polyunsaturated and low in saturated fats.[38,41] We have much more to learn about dietary risk factors for heart disease—and other conditions such as cancer—in women.

Estrogen Replacement Therapy and Heart Disease Risk The popularity of estrogen replacement therapy for heart disease prevention in postmenopausal women has declined dramatically in recent years. Based on results of observation studies, it was assumed that raising estrogen levels in older women would reduce LDL-cholesterol, raise HDL-cholesterol, and reduce the risk of heart disease. Clinical trials have not, however, found that estrogen replacement therapy reduces the risk of heart attack in women with a history of heart disease. Additionally, estrogen replacement therapy increases the risk of blood clots and of breast and other types of reproductive cancer. Other preventive approaches, including diet and lifestyle changes, and the use of new, effective drugs for heart disease are recommended.[34,42] (The new generation of heart disease drugs is discussed later in the unit.)

Managing Blood Cholesterol

> No disease that can be treated by diet should be treated with any other means.
>
> —Maimonides, 1135–1204

As Americans have become more aware of the relationship between diet and heart disease, food choices have changed, and the average cholesterol level in adults has dropped from 240 milligrams in the early 1970s to 203 in recent years.[43] As cholesterol levels have declined, so has the rate of heart disease in the United States (Illustration 19.10). The 50% drop in deaths from heart disease since 1972 appears to be primarily related to declines in blood cholesterol levels, although reduced rates of smoking, improved blood pressure control, and advances in medical care have also contributed.[39]

Despite the general decline in cholesterol levels and rates of heart disease, the average cholesterol level of 203 milligrams among U.S. adults is still too high. Lowering the average cholesterol level to 180 to 200 milligrams would substantially reduce the rate of heart disease. Every 1% drop in average cholesterol level in countries with high rates of heart disease produces a 2% decline in heart disease rate. Raising HDL-cholesterol levels by 1 milligram per deciliter reduces rates by 2–3%.[12]

Risky Blood Lipid Levels "Desirable," "borderline high," and "high" risk categories for levels of cholesterol and triglycerides have been identified and are listed in Table 19.5.[12,44] In addition, a desirable ratio of total cholesterol to HDL-cholesterol has been established. It is considered to be a ratio of less than 4.5, or total cholesterol levels that are no higher than four and one-half times the level of HDL-cholesterol.

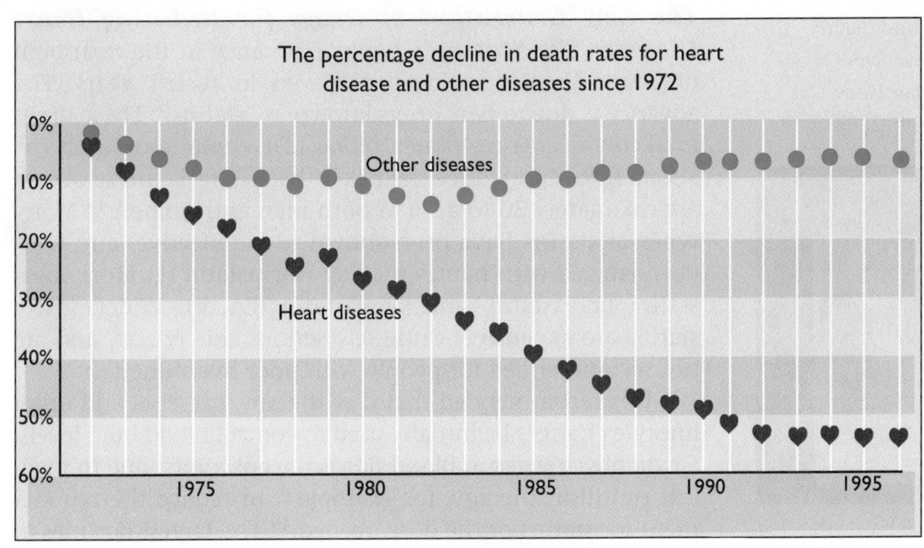

Illustration 19.10
The percentage decline in death rates for heart disease and other diseases since 1972.

Source: National Center for Health Statistics, 1997.

Table 19.5
Risk categories for total, LDL-, and HDL-cholesterol levels and triglyceride levels in adults and youth.*[12,44]

	Total Cholesterol (mg/dl)	LDL-Cholesterol (mg/dl)	HDL-Cholesterol (mg/dl)	Triglyceride (mg/dl)
Adults				
Desirable	<200	<130	>60	<150
Borderline high	200–239	130–159	(low) 35–45	150–200
High	>240	>160	(too low) <35	>200
Youth (<20 Years)				
Desirable	<180	—	—	—

*High cholesterol levels among women and people over the age of 70 may be a poor predictor of heart disease risk.

Individuals with a parent, sister, or brother who experiences a heart attack before the age of 55 years should monitor their blood cholesterol and maintain it at an acceptable level. Most cases of heart disease occur among people without a family history of early heart attack, however. That means every adult should be aware of his or her cholesterol levels and risk profile for heart disease.

Blood cholesterol levels vary from day to day. Consequently, more than one determination is needed to assess a person's usual cholesterol level.[45]

Keeping Blood Cholesterol from Becoming Too High Diets low in saturated fats are the cornerstone of heart disease prevention (Illustration 19.11).[5] Practically speaking, that means consuming diets low in fat from dairy products and meats. Consuming low- or no-fat dairy products and lean meats such as fish, chicken, and very low fat beef and pork can often reduce dietary intakes of saturated fat to the recommended level of less than 10% of total calories.[46] It can also lower elevated cholesterol levels by 8 to 16% within three weeks.[47] Limiting intakes of foods high in cholesterol and trans fat while increasing consumption of vegetables, fruits, and whole grain products also helps maintain healthy levels of blood cholesterol in many people.[9,17] Substituting olive oil and other plant sources of monounsaturated fats for animal fat is increasingly being recommended. Monounsaturated fats do not lower HDL-cholesterol levels as polyunsaturated fats do.[9]

Illustration 19.11
Excellent resources are available for anyone who wants to eat in a heart-healthy way. These cookbooks include recipes and sound dietary guidance: *The American Heart Association's Quick and Easy Cookbook* and *Don't Eat Your Heart Out Cookbook*, Second Edition, by Joseph C. Piscatella.

The New Generation of Drugs for Reducing Heart Disease: The Statins A major advance in the treatment of heart disease has been achieved in recent years. The advance is due to new drugs known as "statins." These drugs markedly decrease elevated blood LDL-cholesterol levels (by about 40%) and reduce the risk of heart attack and stroke by approximately 30 to 40% in both men and women.[48] Many types of statins have been shown to be effective, and have been labeled with names such as atorvastatin (Lipitor), lovastatin, pravastatin, and simvastatin (Zocor). Widely used, statins are expensive, cause few serious side effects, and are not recommended for people with poor liver function.

It is recommended that diet therapy, exercise, and other lifestyle changes be initially used to control blood lipid levels. Favorable changes in blood lipid concentrations due to medical nutrition therapy, for example, can reduce the requirement for statins or the dose needed.[49] The largest declines in blood cholesterol levels occur when diets low in saturated fat and cholesterol are combined with drug therapy.[12]

Medications that increase HDL-cholesterol are also available. They are principally used only after behavioral attempts to raise HDL-cholesterol have not succeeded. People with a genetic tendency toward low HDL-cholesterol levels may benefit greatly from such medications.

Are Very Low Cholesterol Levels Risky? Do low cholesterol levels cause cancer, depression, aggressive behavior, or other serious problems? The answer appears to be "no."[50,51] Unusually low cholesterol levels are more likely to be due to the overly ambitious use of cholesterol-lowering drugs and certain diseases such as cancer that lower blood cholesterol than to diets low in saturated fat and cholesterol.[52]

Cholesterol and Kids Recommendations for diets that help prevent heart disease are *not* intended for children under the age of two years. Very young children have a high need for calories, which may not be met if dietary fat intake is severely restricted. Growth failure has been documented in children placed on low-fat diets by overzealous parents (Illustration 19.12).[53]

For children over the age of two, the use of 1% milk instead of whole milk has been found to reduce average saturated fat intake by over 50%. This single change may bring the dietary intake of saturated fat in older children down to the desired intake of less than 10% of total calories.[54]

Illustration 19.12

Source: © Tribune Media Services. All rights reserved. Reprinted with permission.

Looking toward the Future Approaches to preventing heart disease and other disorders related to atherosclerosis are changing. It is becoming clear that "one size does *not* fit all" in terms of preventive dietary guidance. In the future, advice will be individualized based on a person's susceptibility to diets high in saturated fat and cholesterol. Increased emphasis will likely be given to encouraging consumption of olive oil and other monounsaturated fats, and less emphasis will be placed on polyunsaturated fats. The type of fat consumed is being viewed as more important than the amount of fat. Adequate intakes of folate, vegetables and fruits, and whole grain products are moving toward center stage in the fight against heart disease. Risk factors for women are being disentangled from heart disease risks for men. Although remarkable progress has been made, many opportunities for further reductions in heart disease and atherosclerosis are emerging.

WWW links

www.healthfinder.gov
A search engine for consumer information on heart disease and other problems.

www.nhlbi.nih.gov/chd
This site from the National Heart, Lung, and Blood Institute aims to educate consumers on reducing heart disease.

www.deliciousdecisions.org
The American Heart Association nutrition web site provides information on nutrition, heart health, and heart disease.

www.americanheart.org
The American Heart Association's home page that leads to information on heart disease. Or call 1-800-242-8721.

www.eatright.org
Information on heart-healthy diets from the American Dietetic Association.

Nutrition UP CLOSE

Heart Health Quiz

FOCAL POINT: Assess your chances for developing heart disease.

To estimate your risk for developing heart disease, circle the one number in each column that best describes you. Then, total the numbers from all categories to determine if your lifestyle encourages heart health or heart disease.

Heredity	Exercise	Age	Weight	Tobacco	Habits of Eating Fat
1 No known familial history of heart disease	1 **Intensive** exercise at work and during recreation	1 15–24	0 More than 5 lb below standard weight	0 Nonuser	1 No animal fat
2 **One** immediate family member who developed heart disease **over** age 55	2 **Moderate** exercise at work and during recreation	2 24–34	1 ±5 lb standard weight	1 Cigar or pipe	2 Very little animal fat (<10% of total calories)
3 **Two** immediate family members who developed heart disease **over** age 55	3 Sedentary work with **intensive** exercise during recreation	3 35–44	2 6–20 lb overweight	2 10 cigarettes or fewer per day	3 Some animal fat (11–20% of total calories)
4 **One** immediate family member who developed heart disease **under** age 55	5 Sedentary work with **moderate** exercise during recreation	4 45–54	4 21–35 lb overweight	4 20 cigarettes or more per day	4 Moderate animal fat (21–30% of total calories)
6 **Two** immediate family members who developed heart disease **under** age 55	6 Sedentary work with **light** exercise during recreation	6 55+ years	6 36–50 lb overweight	6 30 cigarettes or more per day	6 Excessive animal fat (>30% of total calories)

Total of all categories ______________

FEEDBACK (including scoring) can be found at the end of Unit 19.

Source: Adapted from *Heart Health Quiz* (Loma Linda University).

REFERENCES

1. Committee on Diet and Health (Food and Nutrition Board; National Research Council). Diet and health. Implications for reducing chronic disease risk. Washington, DC: National Academy Press; 1989.
2. Nissinen A, Stanley K. Unbalanced diets as a cause of chronic diseases. Am J Clin Nutr 1989;49:993–8.
3. Maynard C, et al. Association of gender and survival in patients with acute myocardial infarction. Arch Int Med 1997;157:1379–84.
4. Dwyer T, et al. Differences in HDL cholesterol concentrations in Japanese, American, and Australian children. Circ 1997;76:2830–6.
5. Conner WE. The decisive influence of diet on the progression of coronary heart disease. Am J Clin Nutr 1996;64:253–4.
6. Krauss RM, et al. Revision 2000: a statement for health professionals from the Nutrition Committee of the American Heart Association. J Nutr 2001;131:132–46.
7. Shea S, et al. Age, sex, educational attainment, and race/ethnicity in relation to consumption of specific foods contributing to the atherogenic potential of diet. Prev Med 1993;22:203–18.
8. Ascherio A, Willett WC. Health effects of trans fatty acids. Am J Clin Nutr 1997;66:1006S–10S.
9. Hegsted M, Kritchevsky D. Diet and serum lipid concentrations: where are we? Am J Clin Nutr 1997;65:1893–6.
10. Krauss RM. Understanding the basis for variation in response to cholesterol-lowering diets. Am J Clin Nutr 1997; 65:885–6.
11. Lewis NM, et al. Enriched eggs as a source of n-3 polyunsaturated fatty acids for humans. Poultry Sci 2000;79:971–4.
12. Jellinger PS, et al. The American Association of Clinical Endocrinologists medical guidelines for the diagnosis and treatment of dyslipidemia and prevention of atherogenesis. Endocrin Prac 2000;6:1–213.
13. Feldman EB, et al. Monounsaturated fatty acids promote healthy hearts. Am J Clin Nutr 1999;70:953–4.
14. Stampfer MJ, et al. Primary prevention of coronary heart disease in women through diet and lifestyle. N Engl J Med 2000;343:16–22.
15. Grubben MJ, et al. Unfiltered coffee consumption and cardiovascular disease risk. Am J Clin Nutr 2000;71:403,480–84.
16. Liu S, et al. Fruit and vegetable intake and the risk of cardiovascular disease: the Women's Health Study. Am J Clin Nutr 2000;72:922–8.
17. Liu S, et al. Whole grain consumption and the risk of coronary heart disease: results from the Nurse's Health Study. Am J Clin Nutr 1999;70:412–19.
18. Kushi LH. Vitamin E and heart disease: a case study. Am J Clin Nutr 1999;69(suppl):1322S–9S.
19. Klipstein-Grobusch K, et al. Serum ferritin and risk of myocardial infarction in the elderly: the Rotterdam Study. Am J Clin Nutr 1999;69:1231–36.
20. Kiechl S, et al. Body iron stores and the risk for carotid artery atherosclerosis. Circ 1997;96:3300–7.
21. Corti M-C, et al. Iron status and risk of cardiovascular disease. Ann Epidemiol 1997;7:62–68.
22. Riddell LJ, et al. Dietary strategies for lowering homocysteine concentrations. Am J Clin Nutr 2000;71:1448–54.
23. Jacques PF, et al. The effect of folic acid fortification on plasma folate and total homocysteine concentrations. N Engl J Med 1999;340:1449–54.
24. Brouwer IA, et al. Dietary folate from vegetables and citrus fruits decrease plasma homocysteine concentrations in humans in a dietary controlled trial. J Nutr 1999;129:1135–9.
25. Iso H, et al. Intake of fish and omega-3 fatty acids and risk of stroke in women. JAMA 2001;285:304–12.
26. AHA: revision 2000. Dietary guidelines for healthcare professionals from the nutrition committee of the American Heart Association. Circ 2000;102:2296–2311.
27. An expectant mother's guide to eating Minnesota fish. Minnesota Health Department, 2000; and FDA Press Release 2001 Jan 13.
28. Levine BS. Plant stanol esters. Nutr Today 2000;35:61–6.
29. Hendriks HFJ, et al. Spreads enriched with three different levels of vegetable oil sterols and the degree of cholesterol lowering in normocholesterolaemic and mildly hypercholesterolaemic subjects. Euro J Clin Nutr 1999;53:319–27.
30. Berg JE, Høstmark AT. Coronary artery disease and lipid fractions. Epidemiol 1995;6:91.
31. Forman A. Heart disease handbook—part 2: deciphering blood cholesterol. Envir Nutr 1997;20:1, 4.
32. Yajnik C. Interactions of perturbations in intrauterine growth and growth during childhood on the risk of adult-onset disease. Proc Nutr Soc 2000;59:257–65.
33. Brown JE, Kahn ESB. Maternal nutrition and the outcome of pregnancy. Clin Perinatol 1997;24:433–49.
34. Kereiakes DJ, et al. Managing cardiovascular disease in women. Medscape Women's Health, http://womenshealth.medscape.com/CMECircleWomensHealth/2000/CME02/pnt-CME02.html, accessed 1/01.
35. Sprecher DL, et al. Predictors of high mortality after coronary bypass surgery. J Am Coll Cardiol 2000;36:1159–65.
36. Millen BE, et al. Diet and plasma lipids in women. II. Macronutrients and plasma triglycerides, high density lipoproteins, and the ratio of total to high density lipoprotein cholesterol in women: the Framingham Nutrition Studies. J Clin Epidemiol 1996;49:665–72.
37. Jeppesen J, et al. Effects of low-fat, high-carbohydrate diet on risk factors for ischemic heart disease in postmenopausal women. Am J Clin Nutr 1997;65:1027–33.
38. Hu FB, et al. Dietary fat and the risk of coronary heart disease in women. N Engl J Med 1997;337:1491–9.
39. Ernst ND, et al. Consistency between US dietary fat intake and serum total cholesterol concentrations: the National Health and Nutrition Examination Surveys. Am J Clin Nutr 1997; 66(suppl):965S–72S.
40. Bass KM, et al. Plasma lipoprotein levels as predictors of cardiovascular death in women. Arch Intern Med 1993;53:2209–16.
41. Nicklas BJ, et al. Effects of an American Health Association diet and weight loss on lipoprotein lipids in obese, postmenopausal women. Am J Clin Nutr 1997;66:853–9.
42. Barbour MM. Hormone replacement therapy should not be used as secondary prevention of coronary heart disease. Pharmacotherapy 2000;20:1021–27.
43. Health, United States, 2000, p 246, www.cdc.gov/nchs, 1/01.
44. National Cholesterol Education Project Expert Panel. Report of the National Cholesterol Education Project Expert Panel on detection,

Nutrition UP CLOSE

Heart Health Quiz

Feedback for Unit 19

Find your score below to learn your estimated risk of developing heart disease:

4–9 very remote
10–15 less than average
16–20 average
21–25 moderately increased
26–30 excessive
31–35 too high—reduce score!

You have no control over certain risk factors for heart disease, such as heredity and age. Other conditions that detract from heart health, such as high blood pressure, diabetes, and high blood cholesterol, should be evaluated by your physician. Some risk factors are within your control, however. These lifestyle changes can help reduce your risk of developing heart disease:

- Keep your weight within a healthy range.
- Exercise regularly.
- Don't smoke.
- Reduce the amount of saturated fat in your diet if it is high.
- Eat at least five servings of vegetables and fruits each day.

Source: Adapted from *Heart Health Quiz* (Loma Linda University).

evaluation, and treatment of high blood cholesterol in adults. Arch Intern Med 1988;148:36–69.
45. Davis CE, et al. A single cholesterol measurement underestimates the risk of coronary heart disease. An empirical example from the Lipid Research Clinics Mortality Follow-up Study. JAMA 1990;264:3044–6.
46. Kromhout D. Dietary fats: long-term implications for health. Nutr Rev 1992;50:49–53.
47. Lowering cholesterol with diet and lovastatin. Nutr Rev 1991;49:315–7.
48. Chong PH, Bachenheimer BS. Current, new and future treatments in dyslipidaemia and atherosclerosis. Drugs 2000;60:55–93.
49. Sikand G, et al. Dietitian intervention improves lipid values and saves medication costs in men with combined hyperlipidemia, and a history of niacin noncompliance. J Am Diet Assoc 2000;100:218–24.
50. Eichholzer M, et al. Association of low plasma cholesterol with mortality for cancer at various sites in men: 17-y follow-up of the prospective basal study. Am J Clin Nutr 2000;72:569–74.
51. Wardle J, et al. Randomized trial of the effects of cholesterol lowering dietary treatment on psychological function. Am J Med 2000;108:547–53.
52. Fagot-Campagna A. Serum cholesterol and mortality rates in a Native American population with low cholesterol levels. Circ 1997;95:1408–15.
53. Lifshitz F, Moses N. Growth failure. A complication of dietary treatment of hypercholesterolemia. Am J Dis Child 1989;143:537–42.
54. Basch CE, Shea S, Zybert P. Food sources, dietary behavior, and the saturated fat intake of Latino children. Am J Public Health 1992;82:810–5.

Stock Food America - Rieck

Unit 20

Vitamins and Your Health

Nutrition Scoreboard

	TRUE	FALSE
1. The only documented benefit of consuming sufficient amounts of vitamins is protection against deficiency diseases.		
2. Vitamins provide energy.		
3. Vitamin C is found only in citrus fruits.		
4. Nearly all cases of illness due to excessive intake of vitamins result from the overuse of vitamin supplements.		

(Answers on next page)

Answers to

Nutrition Scoreboard

	TRUE	FALSE
1. For many vitamins, intake levels above those known to prevent deficiency diseases help protect humans from certain cancers, heart disease, and other disorders.		✔
2. Nope—only carbohydrates, proteins, and fats provide energy to the body. Vitamins are needed, however, to convert the energy in food into energy the body can use.		✔
3. Citrus fruits are good sources of vitamin C, but so are green peppers, collards, broccoli, strawberries, and a number of other fruits and vegetables.		✔
4. True. Nearly all cases of illness due to vitamin overdoses result from the excessive intake of vitamin supplements.	✔	

Key Concepts and Facts

→ Vitamins are chemical substances found in food that are required for normal growth and health.

→ Adequate intakes of vitamins protect people against deficiency diseases and help prevent a number of chronic diseases.

→ Each vitamin has a range of intake in which it functions optimally. Intakes below and above the range impair health.

→ Eating five or more servings of fruits and vegetables is a good way to get enough vitamins in your diet each day.

Vitamins: They're on Center Stage

On the Side

Why aren't vitamins named A through F? Vitamins as we know them today don't follow alphabetical order because errors were made when vitamins were discovered. They were named in alphabetical sequence, but some substances turned out not really to be vitamins. The letters assigned to these substances were dropped, so the sequence is interrupted.

These are exciting times for people interested in vitamins. Long relegated to the role of preventing deficiency diseases, vitamins are now being viewed from a very different vantage point. These essential nutrients clearly do more than the important job of protecting us from vitamin deficiency diseases. Vitamins are taking a preeminent position as protectors against a host of ills, ranging from certain birth defects and cataracts to heart disease and cancer.[1]

But first, what are vitamins? How much of them do we need? Where do we get them? And what's behind this new interest in them? Background information essential to understanding the current interest in the benefits of vitamins is next.

Vitamin Facts

Vitamins are chemical substances that perform specific functions in the body. They are essential nutrients because the body cannot produce them. If we fail to consume enough of any of the vitamins, specific deficiency diseases develop.

Thirteen vitamins have been discovered so far, and they are listed in Table 20.1. It is possible that a few more substances will be added to the list of vitamins in years to come.

➡ **vitamins**
Chemical substances that perform specific functions in the body.

Bogus Vitamins Some substances that don't belong on the list of vitamins and won't end up there in years to come are listed in Table 20.2. This list

The Water-Soluble Vitamins*	The Fat-Soluble Vitamins
B-complex vitamins	Vitamin A (retinol) (provitamin is beta-carotene)
Thiamin (B_1)	Vitamin D (1,25 dihydroxy-cholecalciferol)
Riboflavin (B_2)	Vitamin E (tocopherol)
Niacin (B_3)	Vitamin K (phylloquinone, menaquinone)
Vitamin B_6 (pyridoxine)	
Folate (folacin, folic acid)	
Vitamin B_{12} (cyanocobalamin)	
Biotin	
Pantothenic acid (pantothenate)	
Vitamin C (ascorbic acid)	

*Choline has been assigned Adequate Intake levels in the Dietary Reference Intakes. It is not included here as it has not been shown to be required in the diet during all stages of the life cycle.

Table 20.1
Thirteen vitamins are known to be essential for health. Nine are water soluble—that is, they dissolve completely in water—and four dissolve only in fats.

➡ **coenzymes**
Chemical substances, including many vitamins, that activate specific enzymes. Activated enzymes increase the rate at which reactions take place in the body, such as the breakdown of fats or carbohydrates in the small intestine, and the conversion of glucose and fatty acids into energy within cells.

includes some of the most common substances claimed to be vitamins by enterprising quacks, misdirected manufacturers of supplements, and some weight-loss and cosmetic product producers. Although these substances may help sales, they are not a required part of our diets. Because they are not required in our diet, they aren't essential and therefore cannot be considered vitamins. People do not develop deficiency diseases when they consume too little of the bogus "vitamins."

Table 20.2
The "real" vitamins are listed in Table 20.1. But a number of bogus vitamins are also on the market. They turn up in supplements, skin care creams, weight-loss and hair care products, and other items. These are some of the more popular nonvitamins.

Nonvitamins
Bioflavonoids (vitamin P)
Coenzyme Q_{10}
Gerovital H-3
Hesperidin
Inositol
Laetrile (vitamin B_{17})
Lecithin
Lipoic acid
Nucleic acids
Pangamic acid (vitamin B_{15})
Para-amino benzoic acid (PABA)
Provitamin B_5 complex
Rutin

Water- and Fat-Soluble Vitamins Vitamins come in two basic types—those soluble in water (the B-complex vitamins and vitamin C) and those that dissolve in fat (vitamins D, E, K, and A, or the "deka" vitamins). Their key features are summarized in Table 20.3, which provides an "intensive course" on vitamins. With the exception of vitamin B_{12}, the water-soluble vitamins can be stored in the body only in small amounts. Consequently, deficiency symptoms generally develop within a few weeks to several months after the diet becomes deficient in water-soluble vitamins. Vitamin B_{12} is unique in that the body can build up stores that may last for a year or more after intake of the vitamin stops. Of the water-soluble vitamins, niacin, vitamin B_6, and vitamin C are known to produce ill effects if consumed in excessive amounts.

The fat-soluble vitamins are stored in body fat, the liver, and other parts of the body. Because the body is better able to store these vitamins, deficiencies of fat-soluble vitamins generally take longer to develop than deficiencies of water-soluble vitamins when intake from food is too low.

What Do Vitamins Do?

For starters, vitamins *don't* provide energy or serve as components of body tissues such as muscle and bone. A number of vitamins do play critical roles as ***coenzymes*** in the conversion of proteins, carbohydrates, and fats into energy. Coenzymes are also involved in reactions that build and maintain body tissues such as bone, muscle, and red blood cells. Thiamin, for example, is needed for reactions that convert glucose into energy. People who are thiamin deficient tire easily and feel weak (among other things). Folate, another B-complex vitamin, is required for reactions that build body proteins. Without enough folate, proteins such as those found in red blood cells form abnormally and function poorly. Vitamin A is needed for reactions that replace worn-out cells that line the mouth, esophagus, intestines, and eyes with new ones. Without enough vitamin

Table 20.3
An intensive course on vitamins.

The Water-Soluble Vitamins

	Primary Functions	Consequences of Deficiency
Thiamin (vitamin B_1) AI[a] women: 1.1 mg men: 1.2 mg	· Helps body release energy from carbohydrates ingested · Facilitates growth and maintenance of nerve and muscle tissues · Promotes normal appetite	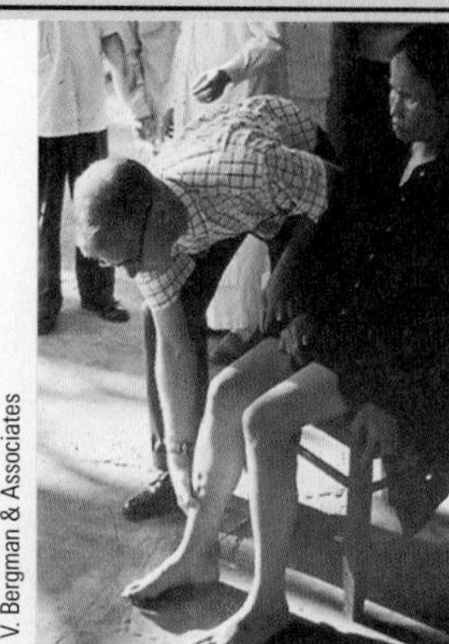 L. V. Bergman & Associates · Fatigue, weakness · Nerve disorders, mental confusion, apathy · Impaired growth · Swelling · Heart irregularity and failure A look at beriberi, a thiamin-deficiency disease.
Riboflavin (vitamin B_2) AI women: 1.1 mg men: 1.3 mg	· Helps body capture and use energy released from carbohydrates, proteins, and fats · Aids in cell division · Promotes growth and tissue repair · Promotes normal vision	· Reddened lips, cracks at both corners of the mouth. · Fatigue
Niacin (vitamin B_3) RDA women: 14 mg men: 16 mg UL: 35 mg (from supplements and fortified foods)	· Helps body capture and use energy released from carbohydrates, proteins, and fats · Assists in the manufacture of body fats · Helps maintain normal nervous system functions	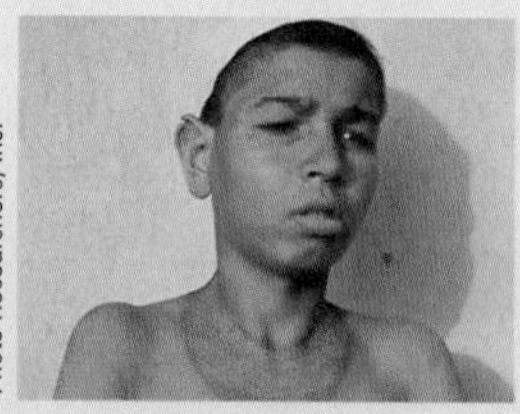 Photo Researchers, Inc. Pellagra: the niacin-deficiency disease. · Skin disorders · Nervous and mental disorders · Diarrhea, indigestion · Fatigue
Vitamin B_6 (pyridoxine) AI women: 1.3 mg men: 1.3 mg UL: 100 mg	· Needed for reactions that build proteins and protein tissues · Assists in the conversion of tryptophan to niacin · Needed for normal red blood cell formation · Promotes normal functioning of the nervous system	· Irritability, depression · Convulsions, twitching · Muscular weakness · Dermatitis near the eyes · Anemia · Kidney stones

[a]AI (Adequate Intakes) and RDAs (Recommended Dietary Allowances) are for 19–30 year olds; UL (Upper Limits) are for 19–70 year olds, 1997–2001.

Table 20.3
An intensive course on vitamins, continued.

The Water-Soluble Vitamins

Consequences of Overdose	Primary Food Sources	Highlights and Comments
· High intakes of thiamin are rapidly excreted by the kidneys. Oral doses of 500 mg/day or less are considered safe.	· Grains and grain products (cereals, rice, pasta, bread) · Ready-to-eat cereals · Pork and ham, liver · Milk, cheese, yogurt · Dried beans and nuts	· Need increases with carbohydrate intake. · There is no "e" on the end of thiamin! · Deficiency rare in the U.S.; may occur in people with alcoholism. · Enriched grains and cereals prevent thiamin deficiency.
· None known. High doses are rapidly excreted by the kidneys.	· Milk, yogurt, cheese · Grains and grain products (cereals, rice, pasta, bread) · Liver, poultry, fish, beef · Eggs	· Destroyed by exposure to light
· Flushing, headache, cramps, rapid heartbeat, nausea, diarrhea, decreased liver function with doses above 0.5 g per day	· Meats (all types) · Grains and grain products (cereals, rice, pasta, bread) · Dried beans and nuts · Milk, cheese, yogurt · Ready-to-eat cereals · Coffee · Potatoes	· Niacin has a precursor—tryptophan. Tryptophan, an amino acid, is converted to niacin by the body. Much of our niacin intake comes from tryptophan.
· Bone pain, loss of feeling in fingers and toes, muscular weakness, numbness, loss of balance (mimicking multiple sclerosis)	· Oatmeal, bread, breakfast cereals · Bananas, avocados, prunes, tomatoes, potatoes · Chicken, liver · Dried beans · Meats (all types), milk · Green and leafy vegetables	· Vitamins go from B_3 to B_6 because B_4 and B_5 were found to be duplicates of vitamins already identified.

continued

Table 20.3
An intensive course on vitamins, continued.

The Water-Soluble Vitamins

	Primary Functions	Consequences of Deficiency	
Folate (folacin, folic acid) RDA: women: 400 mcg men: 400 mcg UL: 1000 mcg (from supplements and fortified foods)	· Needed for reactions that utilize amino acids (the building blocks of protein) for protein tissue formation · Promotes the normal formation of red blood cells	· Anemia · Diarrhea · Red, sore tongue 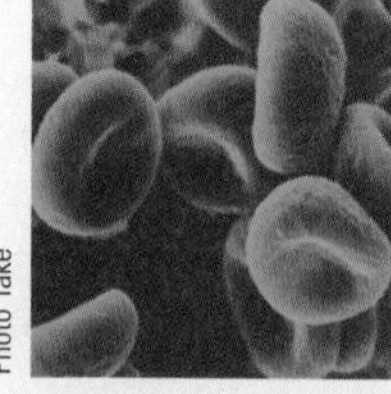 Normal red blood cells.	· Neural tube defects, low birthweight (in pregnancy); increased risk of heart disease and stroke 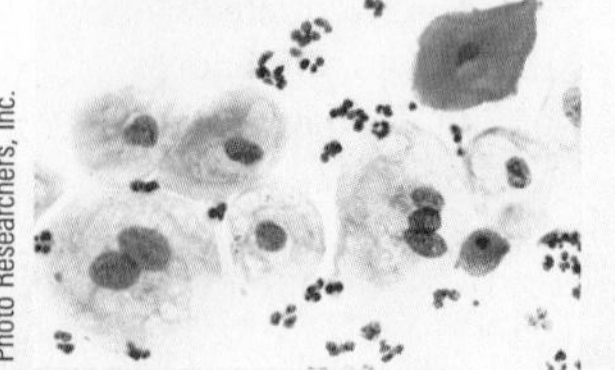 Red blood cells in folate deficiency.
Vitamin B_{12} (cyanocobalamin) AI women: 2.4 mcg men: 2.4 mcg	· Helps maintain nerve tissues · Aids in reactions that build up protein tissues · Needed for normal red blood cell development	· Neurological disorders (nervousness, tingling sensations, brain degeneration) · Pernicious anemia · Fatigue · Deficiency reported in 39% of adults in one study	
Biotin AI women: 30 mcg men: 30 mcg	· Needed for the body's manufacture of fats, proteins, and glycogen	· Depression, fatigue, nausea · Hair loss, dry and scaly skin · Muscular pain	
Pantothenic acid (pantothenate) AI women: 5 mg men: 5 mg	· Needed for the release of energy from fat and carbohydrates	· Fatigue, sleep disturbances, impaired coordination · Vomiting, nausea	
Vitamin C (ascorbic acid) RDA women: 75 mg men: 90 mg UL: 2000 mg	· Needed for the manufacture of collagen · Helps the body fight infections, repair wounds · Acts as an antioxidant · Enhances iron absorption	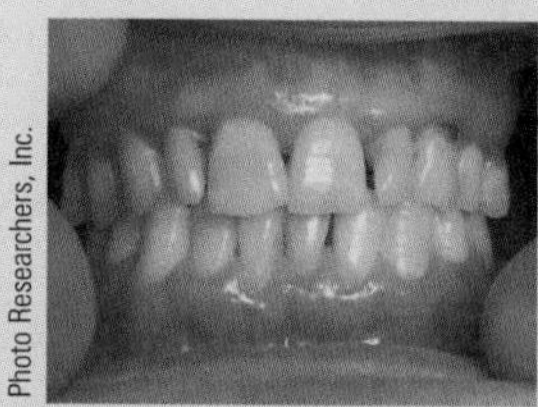 Gums that are swollen and bleed easily are signs of scurvy, the vitamin C–deficiency disease.	· Bleeding and bruising easily due to weakened blood vessels, cartilage, and other tissues containing collagen · Slow recovery from infections and poor wound healing · Fatigue, depression · Deficiency reported in 9–24% of adults in one study

Photo Take; Photo Researchers, Inc.; Photo Researchers, Inc.

Table 20.3
An intensive course on vitamins, continued.

Consequences of Overdose	The Water-Soluble Vitamins Primary Food Sources	Highlights and Comments
· May cover up signs of vitamin B_{12} deficiency (pernicious anemia)	· Fortified, refined grain products (bread, flour, pasta) · Ready-to-eat cereals · Dark green, leafy vegetables (spinach, collards, romaine) · Broccoli, brussels sprouts · Oranges, bananas, grapefruit · Milk, cheese, yogurt · Dried beans	· Folate means "foliage." It was first discovered in leafy green vegetables. · This vitamin is easily destroyed by heat. · Synthetic form added to fortified grain products is better absorbed than naturally occurring folates.
· None known. Excess vitamin B_{12} is rapidly excreted by the kidneys or is not absorbed into the bloodstream. · Vitamin B_{12} injections may cause a temporary feeling of heightened energy.	· Animal products: beef, lamb, liver, clams, crab, fish, poultry, eggs · Milk and milk products · Ready-to-eat cereals	· Older people and vegans are at risk for vitamin B_{12} deficiency. · Some people become vitamin B_{12} deficient because they are genetically unable to absorb it. · Vitamin B_{12} is found in animal products and microorganisms only.
· None known. Excesses are rapidly excreted.	· Grain and cereal products · Meats, dried beans, cooked eggs · Vegetables	· Deficiency is extremely rare. May be induced by the overconsumption of raw eggs.
· None known. Excesses are rapidly excreted.	· Many foods, including meats, grains, vegetables, fruits, and milk	· Deficiency is very rare.
· Intakes of 1 g or more per day can cause nausea, cramps, and diarrhea and may increase the risk of kidney stones.	· Fruits: oranges, lemons, limes, strawberries, cantaloupe, honeydew melon, grapefruit, kiwi fruit, mango, papaya · Vegetables: broccoli, green and red peppers, collards, cabbage, tomato, asparagus, potatoes · Ready-to-eat cereals	· Need increases among smokers (to 110–125 mg per day). · Is fragile; easily destroyed by heat and exposure to air · Supplements may decrease duration and symptoms of colds. · Deficiency may develop within 3 weeks of very low intake.

continued

Table 20.3
An intensive course on vitamins, continued.

The Fat-Soluble Vitamins	Primary Functions	Consequences of Deficiency	
Vitamin A **I. Retinol** RDA women: 700 mcg men: 900 mcg UL: 3000 mcg	· Needed for the formation and maintenance of mucous membranes, skin, bone · Needed for vision in dim light	· Increased susceptibility to infection, increased incidence and severity of infection (including measles and HIV) · Impaired vision, blindness · Inability to see in dim light	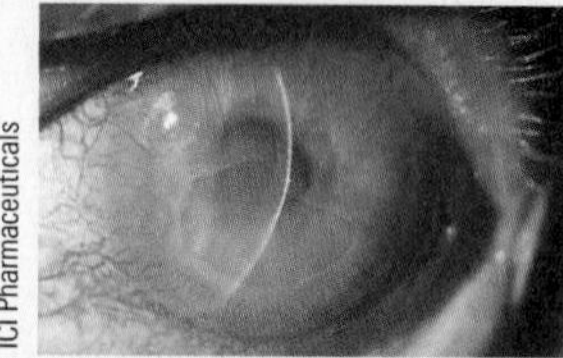 ICI Pharmaceuticals Xerophthalmia. Vitamin A deficiency is the leading cause of blindness in developing countries.
2. Beta-carotene (a vitamin A precursor or "provitamin") No RDA; suggested intake: 6 mg	· Acts as an antioxidant; prevents damage to cell membranes and the contents of cells by repairing damage caused by free radicals	· Deficiency disease related only to lack of vitamin A	
Vitamin E (alpha-tocopherol) RDA women: 15 mg men: 15 mg UL: 1000 mg	· Acts as an antioxidant, prevents damage to cell membranes in blood cells, lungs, and other tissues by repairing damage caused by free radicals · Reduces the ability of LDL-cholesterol (the "bad" cholesterol) to form plaque in arteries	· Muscle loss, nerve damage · Anemia · Weakness · Many adults may have non-optimal blood levels.	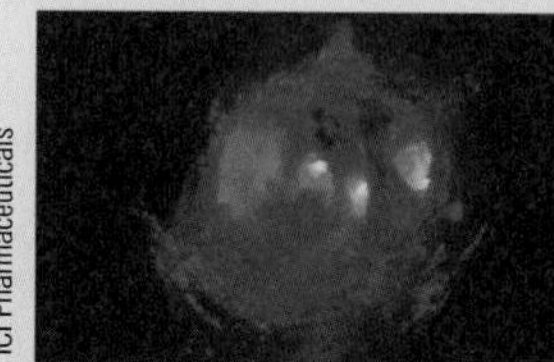 ICI Pharmaceuticals Vitamin E helps prevent plaque formation in arteries, may improve circulation in people with diabetes, and may decrease symptoms of Alzheimer's disease.

Table 20.3
An intensive course on vitamins, continued.

Consequences of Overdose	The Fat-Soluble Vitamins Primary Food Sources	Highlights and Comments
· Vitamin A toxicity (hypervitamosis A) with acute doses of 500,000 IU, or long-term intake of 50,000 IU per day. Limit retinol use in pregnancy to 5000 IU daily. · Nausea, irritability, blurred vision, weakness · Increased pressure in the skull, headache · Liver damage · Hair loss, dry skin · Birth defects	· Vitamin A is found in animal products only. · Liver, butter, margarine, milk, cheese, eggs · Ready-to-eat cereals	· Symptoms of vitamin A toxicity may mimic those of brain tumors and liver disease. Vitamin A toxicity is sometimes misdiagnosed because of the similarities in symptoms. · 1 mcg retinol equivalent = 5 IU vitamin A or 6 mcg beta-carotene
· High intakes from supplements may increase lung damage. · With high intakes and supplemental doses (over 12 mg/day for months), skin may turn yellow-orange. · Possibly related to reversible loss of fertility in women	· Deep orange, yellow, and green vegetables and fruits are often good sources. · Carrots, sweet potatoes, pumpkin, spinach, collards, red peppers, broccoli, cantaloupe, apricots, tomatoes, vegetable juice	· The body converts beta-carotene to vitamin A. Other carotenes are also present in food, and some are converted to vitamin A. Beta-carotene and vitamin A perform different roles in the body, however. · May decrease sunburn.
· Intakes of up to 800 IU per day are unrelated to toxic side effects; over 800 IU per day may increase bleeding (blood-clotting time). · Avoid supplement use if aspirin, anticoagulants, or fish oil supplements are taken regularly.	· Oils and fats · Salad dressings, mayonnaise, margarine, shortening, butter · Whole grains, wheat germ · Leafy, green vegetables, tomatoes · Nuts and seeds · Eggs	· Vitamin E is destroyed by exposure to oxygen and heat. · Oils naturally contain vitamin E. It's there to protect the fat from breakdown due to free radicals. · Eight forms of vitamin E exist, and each has different antioxidant strength. · Natural form is better absorbed than synthetic form: 15 IU alpha-tocopherol = 22 IU d-alpha tocopherol (natural form) and 33 IU synthetic vitamin E.

continued

Table 20.3
An intensive course on vitamins, continued.

The Fat-Soluble Vitamins	Primary Functions	Consequences of Deficiency
Vitamin D (1,25 dihydroxy-cholecalciferol) AI women: 5 mcg (200 IU) men: 5 mcg (200 IU) UL: 50 mcg (2000 IU)	· Needed for the absorption of calcium and phosphorus, and for their utilization in bone formation, nerve and muscle activity.	· Weak, deformed bones (children) · Loss of calcium from bones (adults), osteoporosis 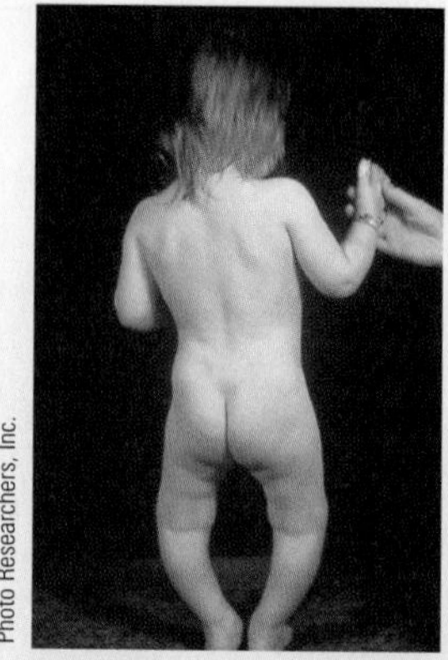 The vitamin D–deficiency disease: rickets
Vitamin K (phylloquinone, menaquinone) AI women: 90 mcg men: 120 mcg	· Is an essential component of mechanisms that cause blood to clot when bleeding occurs · Aids in the incorporation of calcium into bones 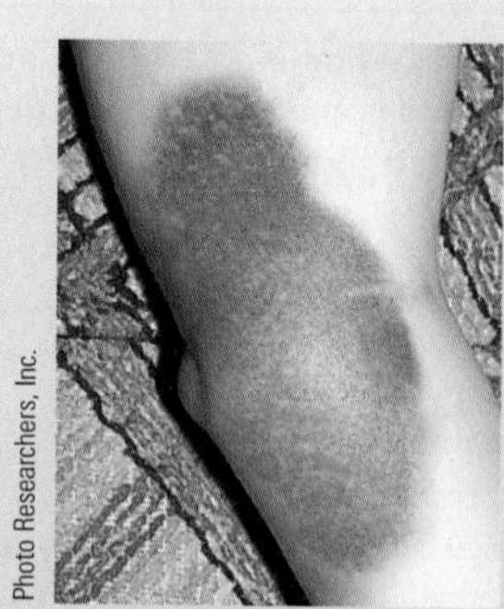 The long-term use of antibiotics can cause vitamin K deficiency. People with vitamin K deficiency bruise easily.	· Bleeding, bruises · Decreased calcium in bones · Deficiency is rare. May be induced by the long-term use (months or more) of antibiotics.

A, old cells aren't replaced and the affected tissues are damaged. And vitamin C is required for reactions that build and maintain collagen, a protein found in skin, bones, blood vessels, gums, ligaments, and cartilage. (Approximately 30% of the total amount of protein in the body is collagen.) With vitamin C deficiency, collagen becomes weak, causing tissues that contain collagen to weaken and bleed easily.

The examples just given all relate to the physical effects of vitamins. Vitamins participate in reactions that affect behaviors, too. Alterations in behaviors such as reduced attention span, poor appetite, irritability, depression, or paranoia often precede the physical signs of vitamin deficiency.[2] Vitamins are truly "vital" for health.

Protection from Vitamin Deficiencies and More Current research on vitamins centers around their effects on disease prevention and treatment. It

Table 20.3
An intensive course on vitamins, continued.

Consequences of Overdose	The Fat-Soluble Vitamins Primary Food Sources	Highlights and Comments
· Mental retardation in young children · Abnormal bone growth and formation · Nausea, diarrhea, irritability, weight loss · Deposition of calcium in organs such as the kidneys, liver, and heart	· Vitamin D–fortified milk and margarine · Butter · Fish · Eggs · Mushrooms · Milk products such as cheese, yogurt, and ice cream are generally not fortified with vitamin D.	· Vitamin D is manufactured from cholesterol in cells beneath the surface of the skin upon exposure of the skin to sunlight. · Deficiency may be common in ill, homebound, and elderly and hospitalized adults. · Breastfed infants with little sun exposure may benefit from vitamin D supplements.
· Toxicity is only a problem when synthetic forms of vitamin K are taken in excessive amounts. That may cause liver disease.	· Leafy, green vegetables · Grain products	· Vitamin K is produced by bacteria in the gut. Part of our vitamin K supply comes from these bacteria. · Newborns are given a vitamin K injection because they have "sterile" guts and consequently no vitamin K-producing bacteria.

is a very active area of research and, no doubt, new and important results will be announced after this text goes to press. (Nutrition textbooks should really be updated every few weeks. Stay tuned to your instructor for the latest developments.) Here are a few examples of recent developments in research on vitamins and disease prevention and treatment:

- *Folate, neural tube defects, and heart disease.* Daily consumption of 400 micrograms of folic acid (the synthetic form of folate added to refined grain products) before and early in pregnancy prevents about two-thirds of cases of neural tube defects in newborns.[3] Neural tube defects are abnormalities of the spinal cord and brain (Illustration 20.1). They are the most common type of malformation of newborns in the United States. Adequate intake of folate also reduces the risk of developing heart disease and other diseases associated with hardening of the arteries such as stroke.[4]

Illustration 20.1
A baby with spina bifida, a form of neural tube defect associated with poor folate status early in pregnancy.

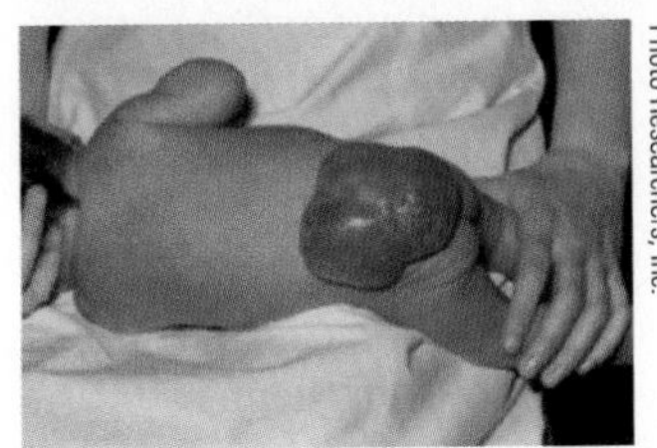

Photo Researchers, Inc.

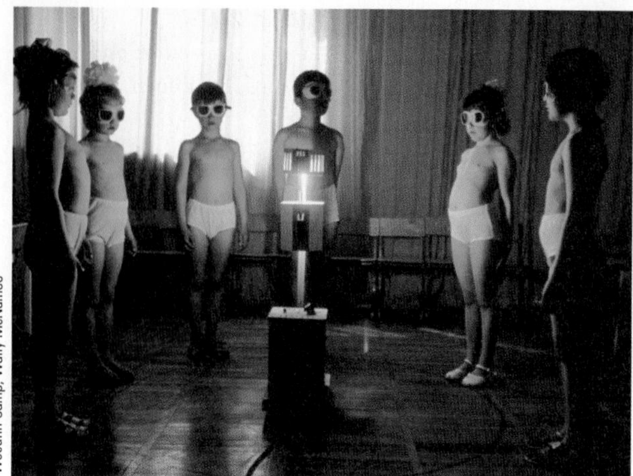

Woodfin Camp, Wally McNamee

Illustration 20.2
Russian children are exposed to a quartz lamp to prevent vitamin D deficiency during the long winter.

In 1998 manufacturers started fortifying refined grain products such as bread, pasta, and rice with folic acid. The addition of folic acid to these foods has produced substantial gains in people's folate status. Blood levels of folate have increased 177%, and the prevalence of poor folate status has declined by 92%.[5] Rates of neural tube defects appear to be declining substantially[6] as are blood levels of ***homocysteine,*** a risk factor for heart disease and stroke.[5]

- *Vitamin A—from measles to "liver spots."* Studies undertaken in both developing countries and the United States indicate that adequate vitamin A status decreases the severity of measles and other infectious diseases.[7] (It has been known for decades that adequate vitamin A intake also prevents blindness, an all-too-common consequence of vitamin A deficiency in developing nations.) This vitamin is showing promise as a treatment for skin and cervi-cal cancer,[8] and a vitamin A derivative (retinoic acid) is being used successfully in the treatment of serious cases of acne, skin wrinkles, and "liver" or "aging" spots on the skin due to overexposure to the sun.[9]
- *Vitamin D, osteoporosis, and rickets.* Vitamin D intakes of 800 to 1000 IU per day are associated with decreased bone loss and a reduced risk of osteoporosis.[10]

  Increased rates of breastfeeding in the United States have had the unintended side effect of increased vitamin D deficiency (rickets) in infants.[11] Breast milk contains a very low amount of vitamin D, so infants receiving human milk need to get it from exposure of their skin to sunlight or from supplements. Infants who develop vitamin D deficiency fail to get enough sunlight exposure due to winters in cold climates, or being fully clothed or covered with sun screen to protect their skin from the sun. Modest exposure to sunshine (about 30 minutes a week with diapers only) is sufficient to supply the required amount of vitamin D (Illustration 20.2).[12] In cold climates, a vitamin D supplement is needed.[11]
- *Vitamin C and the common cold revisited.* Recent research confirms that vitamin C, the popular cold remedy, reduces the symptoms and duration of the common cold, but does *not* affect how often colds occur.[13]

Some of the most exciting recent findings on vitamins and disease concern vitamins that function as ***antioxidants.***

The Antioxidant Vitamins Beta-carotene (a ***precursor*** to vitamin A), vitamin E, and vitamin C function as antioxidants. This means they prevent or repair damage to components of cells caused by exposure to ***free radicals.*** A free radical is formed when an atom of hydrogen or oxygen loses an electron. Without the electron, there is an imbalance between the atom's positive and negative charges. This makes the atom reactive—it needs to steal an electron from a nearby atom or molecule in order to reestablish a balance between its positive and negative charges. Atoms and molecules that have lost electrons to free radicals are said to be "oxidized." These oxidized substances are reactive and can damage cell membranes, DNA, and other cell components. Antioxidants, such as beta-carotene, vitamin E, and vitamin C, donate electrons to stabilize oxidized molecules or repair them in other ways.

➡ **homocysteine**
A compound produced when the amino acid methionine is converted to another amino acid, cysteine. High blood levels of homocysteine increase the risk of hardening of the arteries, heart attack, and stroke.

➡ **antioxidants**
Chemical substances that prevent or repair damage to cells caused by exposure to free radicals. Beta-carotene, vitamin E, and vitamin C function as antioxidants.

➡ **precursor**
In nutrition, a nutrient that can be converted into another nutrient. (Also called *provitamin.*) Beta-carotene is a precursor of vitamin A.

➡ **free radicals**
Chemical substances (usually oxygen or hydrogen) that are missing an electron. The absence of the electron makes the chemical substances reactive and prone to oxidizing nearby atoms or molecules by stealing an electron from them.

Why do free radicals exist in the body? Free radicals play a number of roles in the body, so they are always present. They are produced during energy formation, by breathing, and by the immune system to help destroy bacteria and viruses that enter the body. They can also be formed when the body is exposed to alcohol, radiation emitted by the sun, smoke, ozone, smog, and other environmental pollutants.

Cellular damage due to free radicals appears to play a role in the development of some types of cancer, bronchitis, emphysema, heart disease, cataracts, and premature aging.[14] Vitamin E specifically is associated with slowed progression of Alzheimer's disease,[15] reduced plaque formation in arteries,[16] and improved circulation in people with diabetes.[17] The antioxidant vitamins, along with the mineral selenium, substances found in plants, and other substances produced by the body that function as antioxidants, may help prevent or postpone these disorders by protecting cell components against free radical damage.

Vitamins: Getting Enough without Getting Too Much

Vitamins are widely present in basic foods (Table 20.4). Adequate amounts of them can be obtained from diets that include the variety of foods recommended in the Food Guide Pyramid. Most fruits and vegetables are good sources of vitamins, and eating five or more servings a day is one way for Americans to get their vitamins. In the United States, 80% of adults consume two or fewer servings of vegetables and fruits daily, and 10% consume none.[18]

Table 20.4
Food sources of vitamins.

Food	Thiamin Amount	Thiamin (Milligrams)
Meats:		
Pork roast	3 oz	0.8
Beef	3 oz	0.4
Ham	3 oz	0.4
Liver	3 oz	0.2
Nuts and seeds:		
Sunflower seeds	¼ cup	0.7
Peanuts	¼ cup	0.1
Almonds	¼ cup	0.1
Grains:		
Bran flakes	1 cup (1 oz)	0.6
Macaroni	½ cup	0.1
Rice	½ cup	0.1
Bread	1 slice	0.1
Vegetables:		
Peas	½ cup	0.3
Lima beans	½ cup	0.2
Corn	½ cup	0.1
Broccoli	½ cup	0.1
Potato	1 medium	0.1
Fruits:		
Orange juice	1 cup	0.2
Orange	1	0.1
Avocado	½	0.1

continued

Photo Disc

Beef, broccoli, and potatoes all contain thiamin.

Table 20.4
Food sources of vitamins, continued.

Milk and milk products are good sources of riboflavin.

Photo Disc

Salmon is a good source of niacin.

Riboflavin

Food	Amount	Riboflavin (Milligrams)
Milk and milk products:		
Milk	1 cup	0.5
2% milk	1 cup	0.5
Yogurt, low-fat	1 cup	0.5
Skim milk	1 cup	0.4
Yogurt	1 cup	0.1
American cheese	1 oz	0.1
Cheddar cheese	1 oz	0.1
Meats:		
Liver	3 oz	3.6
Pork chop	3 oz	0.3
Beef	3 oz	0.2
Tuna	3 oz	0.1
Vegetables:		
Collard greens	½ cup	0.3
Broccoli	½ cup	0.2
Spinach, cooked	½ cup	0.1
Eggs:		
Egg	1	0.2
Grains:		
Macaroni	½ cup	0.1
Bread	1 slice	0.1

Niacin

Food	Amount	Niacin (Milligrams)
Meats:		
Liver	3 oz	14.0
Tuna	3 oz	10.3
Turkey	3 oz	9.5
Chicken	3 oz	7.9
Salmon	3 oz	6.9
Veal	3 oz	5.2
Beef (round steak)	3 oz	5.1
Pork	3 oz	4.5
Haddock	3 oz	2.7
Scallops	3 oz	1.1
Nuts and seeds:		
Peanuts	1 oz	4.9
Vegetables:		
Asparagus	½ cup	1.5
Grains:		
Wheat germ	1 oz	1.5
Brown rice	½ cup	1.2
Noodles, enriched	½ cup	1.0
Rice, white, enriched	½ cup	1.0
Bread, enriched	1 slice	0.7
Milk and milk products:		
Cottage cheese	½ cup	2.6
Milk	1 cup	1.9

Table 20.4
Food sources of vitamins, continued.

Food	Vitamin B6 Amount	Vitamin B_6 (Milligrams)
Meats:		
Liver	3 oz	0.8
Salmon	3 oz	0.7
Other fish	3 oz	0.6
Chicken	3 oz	0.4
Ham	3 oz	0.4
Hamburger	3 oz	0.4
Veal	3 oz	0.4
Pork	3 oz	0.3
Beef	3 oz	0.2
Eggs:		
Egg	1	0.3
Legumes:		
Split peas	½ cup	0.6
Dried beans, cooked	½ cup	0.4
Fruits:		
Banana	1	0.6
Avocado	½	0.4
Watermelon	1 cup	0.3
Vegetables:		
Turnip greens	½ cup	0.7
Brussels sprouts	½ cup	0.4
Potato	1	0.2
Sweet potato	½ cup	0.2
Carrots	½ cup	0.2
Peas	½ cup	0.1

Photo Disc

Bananas are a good source of vitamin B_6.

Food	Folate Amount	Folate (Micrograms)
Vegetables:		
Garbanzo beans	½ cup	141
Navy beans	½ cup	128
Asparagus	½ cup	120
Brussels sprouts	½ cup	116
Black-eyed peas	½ cup	102
Spinach, cooked	½ cup	99
Romaine lettuce	1 cup	86
Lima beans	½ cup	71
Peas	½ cup	70
Collard greens, cooked	½ cup	56
Sweet potato	½ cup	43
Broccoli	½ cup	43
Fruits:		
Cantaloupe	¼ whole	100
Orange juice	1 cup	87
Orange	1	59
Grains:[a]		
Ready-to-eat cereals	1 cup/1 oz	100–400
Oatmeal	½ cup	97
Noodles	½ cup	45
Wheat germ	2 tbs	40
Wild rice	½ cup	37

[a]Fortified, refined grain products such as bread, rice, pasta, and crackers provide approximately 40 to 60 micrograms of folic acid per standard serving.

Photo Disc

Beans are an especially rich source of folate.

continued

Table 20.4
Food sources of vitamins, continued.

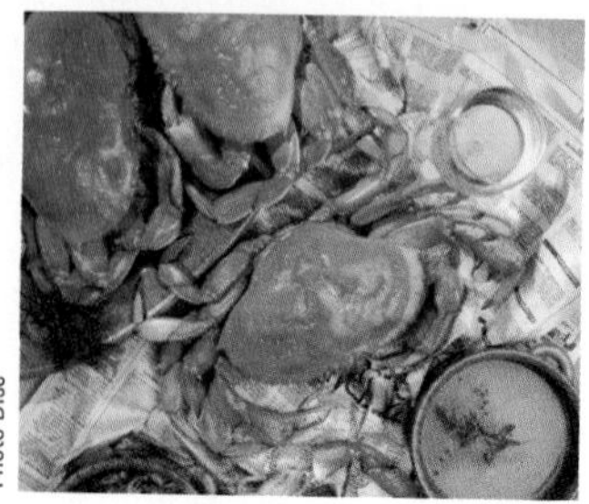

Photo Disc

Crab is one food source of vitamin B_{12}.

Vitamin B_{12}

Food	Amount	Vitamin B_{12} (Micrograms)
Meats:		
Liver	3 oz	6.8
Trout	3 oz	3.6
Beef	3 oz	2.2
Clams	3 oz	2.0
Crab	3 oz	1.8
Lamb	3 oz	1.8
Tuna	3 oz	1.8
Veal	3 oz	1.7
Hamburger, regular	3 oz	1.5
Milk and milk products:		
Skim milk	1 cup	1.0
Milk	1 cup	0.9
Yogurt	1 cup	0.8
Cottage cheese	½ cup	0.7
American cheese	1 oz	0.2
Cheddar cheese	1 oz	0.2
Eggs:		
Egg	1	0.6

Vitamin C

Food	Amount	Vitamin C (Milligrams)
Fruits:		
Orange juice, vitamin C–fortified	1 cup	108
Kiwi fruit	1 or ½ cup	108
Grapefruit juice, fresh	1 cup	94
Cranberry juice cocktail	1 cup	90
Orange	1	85
Strawberries, fresh	1 cup	84
Orange juice, fresh	1 cup	82
Cantaloupe	¼ whole	63
Grapefruit	1 medium	51
Raspberries, fresh	1 cup	31
Watermelon	1 cup	15
Vegetables:		
Green peppers	½ cup	95
Cauliflower, raw	½ cup	75
Broccoli	½ cup	70
Brussels sprouts	½ cup	65
Collard greens	½ cup	48
Vegetable (V-8) juice	¾ cup	45
Tomato juice	¾ cup	33
Cauliflower, cooked	½ cup	30
Potato	1 medium	29
Tomato	1 medium	23

Photo Disc

Citrus fruits are an excellent source of vitamin C.

Updated recommendations for vitamin intakes associated with the prevention of deficiency and chronic diseases are represented in the new intake standards called Dietary Reference Intakes, or DRIs. DRIs include Recommended Dietary Allowances (RDAs) for vitamins for which convincing scientific data exist for establishing intake standards. Adequate Intakes, or AIs, are assigned to

Table 20.4
Food sources of vitamins, continued.

Vitamin A (Retinol)

Food Sources of Vitamin A (Retinol)	Amount	Vitamin A (Micrograms RE)[b]
Meats:		
Liver	3 oz	9124
Salmon	3 oz	53
Tuna	3 oz	14
Eggs:		
Egg	1 medium	84
Milk and milk products:		
Skim milk, fortified	1 cup	149
2% milk	1 cup	139
American cheese	1 oz	82
Whole milk	1 cup	76
Swiss cheese	1 oz	65
Fats:		
Margarine, fortified	1 tsp	46
Butter	1 tsp	38

Photo Disc

Eggs are one food source of vitamin A (retinol).

Vitamin A and Beta-Carotene

Food Sources of Beta-Carotene	Amount	Vitamin A Value (Micrograms RE)[b]
Vegetables:		
Pumpkin, canned	½ cup	2712
Sweet potato, canned	½ cup	1935
Carrots, raw	½ cup	1913
Spinach, cooked	½ cup	739
Collard greens, cooked	½ cup	175
Broccoli, cooked	½ cup	109
Winter squash	½ cup	53
Green peppers	½ cup	40
Fruits:		
Cantaloupe	¼ whole	430
Apricots, canned	½ cup	210
Nectarine	1 medium	101
Watermelon	1 cup	59
Peaches, canned	½ cup	47
Papaya	½ cup	20

[b]RE (retinol equivalent) = 3.33 IU.

continued

Photo Disc

Spinach and winter squash are two sources of vitamin A and beta-carotene.

vitamins for which scientific information about levels of intake associated with chronic disease prevention is less convincing. Tolerable Upper Levels of Intake, abbreviated ULs, are also assigned to vitamins and indicate levels of vitamin intake from foods, fortified foods, and supplements that should not be exceeded. The RDAs or AIs, and ULs (where available) for the vitamins are given in Table 20.3. (World Wide Web addresses listed at the end of this unit can lead you to the latest developments on the DRIs.)

Although people can get all the vitamins they need from supplements, it makes more sense to get them from basic foods. Foods offer fiber, minerals, and other healthful ingredients that don't come in supplements, and they certainly taste better on the way down!

Table 20.4
Food sources of vitamins, continued.

Photo Disc

Seafood and asparagus both provide vitamin E.

Photo Disc

Shrimp and other seafoods are good sources of vitamin D.

Vitamin E

Food	Amount	Vitamin E (IU)[c]
Oils:		
Oil	1 tbs	6.7
Mayonnaise	1 tbs	3.4
Margarine	1 tbs	2.7
Salad dressing	1 tbs	2.2
Nuts and seeds:		
Sunflower seeds	¼ cup	27.1
Almonds	¼ cup	12.7
Peanuts	¼ cup	4.9
Cashews	¼ cup	0.7
Vegetables:		
Sweet potato	½ cup	6.9
Collard greens	½ cup	3.1
Asparagus	½ cup	2.1
Spinach, raw	1 cup	1.5
Grains:		
Wheat germ	2 tbs	4.2
Bread, whole wheat	1 slice	2.5
Bread, white	1 slice	1.2
Seafood:		
Crab	3 oz	4.5
Shrimp	3 oz	3.7
Fish	3 oz	2.4

Vitamin D

Food	Amount	Vitamin D (IU)[d]
Milk:		
Milk, whole, low-fat, or skim	1 cup	100
Fish and seafoods:		
Salmon	3 oz	340
Tuna	3 oz	150
Shrimp	3 oz	127
Organ meats:		
Beef liver	3 oz	42
Chicken liver	3 oz	40
Eggs:		
Egg yolk	1	27

[c] 15 milligrams alpha-tocopherol = 22 IU d-alpha tocopherol (natural form) and 33 IU synthetic vitamin E.
[d] 40 IU = 1 microgram.

WWW links

www.healthfinder.gov
Search vitamins.

www.nal.usda.gov/fnic/foodcomp
Look up the vitamin and other nutrient content of thousands of foods on this site.

www.iom.edu/fnb
Provides updated information on the DRIs.

www.merckhomeedition.com
The Merck Manual of Medical Information is free and searchable. It can be used to find out more about health problems related to vitamins.

Nutrition UP CLOSE

Antioxidant Vitamins—How Adequate Is Your Diet?

FOCAL POINT: Determine if you eat enough antioxidant-rich foods.

Vitamin C, beta-carotene, and vitamin E, the antioxidant vitamins, help to maintain cellular integrity in the body. They may act to reduce the risk of heart disease, certain cancers, and other ailments. Check below to find out how frequently you consume foods containing these important, health-promoting nutrients.

How Often Do You Eat:	Seldom or Never	1–2 Times per Week	3–5 Times per Week	Almost Daily
Vitamin C–rich foods:				
1. Grapefruit, lemons, oranges, or pineapple?	☐	☐	☐	☐
2. Strawberries, kiwi, or honeydew melon?	☐	☐	☐	☐
3. Orange juice, cranberry juice cocktail, or tomato juice?	☐	☐	☐	☐
4. Green, red, or chili peppers?	☐	☐	☐	☐
5. Broccoli, Chinese cabbage, or cauliflower?	☐	☐	☐	☐
6. Asparagus, tomatoes, or potatoes?	☐	☐	☐	☐
Beta-carotene–rich foods:				
7. Carrots, sweet potatoes, pumpkin, or winter squash?	☐	☐	☐	☐
8. Spinach, collard greens, or chard?	☐	☐	☐	☐
9. Cantaloupe, papayas or mangoes?	☐	☐	☐	☐
10. Nectarines, peaches, or apricots?	☐	☐	☐	☐
Vitamin E–rich foods:				
11. Whole-grain breads, whole-grain cereals, or wheat germ?	☐	☐	☐	☐
12. Crab, shrimp, or fish?	☐	☐	☐	☐
13. Peanuts, almonds, or sunflower seeds?	☐	☐	☐	☐
14. Oils, margarine, butter, mayonnaise, or salad dressing?	☐	☐	☐	☐

FEEDBACK (including scoring) can be found at the end of Unit 20.

REFERENCES

1. Hathcock JN. Vitamins and minerals: efficacy and safety. Am J Clin Nutr 1997;66:427–37.
2. Buzina R, et al. Workshop on functional significance of mild-to-moderate malnutrition. Am J Clin Nutr 1989;50:172–6.
3. American Academy of Pediatrics. Folic acid and the prevention of neural tube defects. Pediatrics 1999;104:325–27.
4. Boushey CJ, et al. A quantitative assessment of plasma homocysteine as a risk factor for vascular disease: probable benefits of increasing folic acid intakes. JAMA 1995;274:1049–57.
5. Jacques PE, et al. The effect of folic acid fortification on 4 total homocysteine concentrations. N Engl J Med 1999;340:1449–54.
6. Neural tube defect surveillance and folic acid intervention: Texas-Mexico border. MMWR 2000;49:1–4.
7. Frieden TR, et al. Vitamin A levels and severity of measles. Am J Dis Child 1992;146:182–6.
8. Lippman SM, et al. 13-cis-Retinoic acid and interferon alpha-2a: effective combination therapy for advanced squamous cell carcinoma of the skin. J National Cancer Institute 1992;84:235–41.
9. Rafal ES, et al. Topical tretinoin (retinoic acid) treatment for liver spots associated with photodamage. N Engl J Med 1992;326:368–74.
10. Utiger RD. The need for more vitamin D. N Engl J Med 1998;338:828–9.
11. Welch TR, et al. Vitamin D–deficient rickets: the reemergence of a once-conquered disease (editorial). J Pediatr 2000;1137:152–7.
12. Specker BL, et al. Sunshine exposure and serum 25-hdyroxyvitamin D concentrations in exclusively breast-fed infants. J Pediatr 1985;107:372–6.
13. Hemilä H, Herman ZS. A retrospective analysis of Chalmer's review. J Am Coll Nutr 1995;14:116–23.
14. Hu GZ, Cassano P. Antioxidant nutrients and pulmonary function. Am J Epidemiol 2000;151:975–81.
15. Grundman M. Vitamin E and Alzheimer disease: the basis for additional clinical trials. Am J Clin Nutr 2000;71(suppl): 630S–6S.
16. Stephens NG, et al. Randomized control trial of vitamin E in patients with coronary disease. Lancet 1996;347:781–6.
17. Bursell S-E, et al. High doses of vitamin E supplementation normalize retinal blood flow and creatinine clearance in patients with type 1 diabetes. Diabetes Care 1999;22:1245–51.
18. Johnston CS, Corte C. People with marginal vitamin C status are at high risk of developing vitamin C deficiency. J Am Diet Assoc 1999;99:854–56.

Nutrition

UP CLOSE

Antioxidant Vitamins—How Adequate Is Your Diet?

Feedback for Unit 20

Several responses in the last two columns indicate adequate antioxidant vitamin consumption. If you need to boost your intake, increase the overall amount of fruits, vegetables, and whole grains in your diet. Although nuts, seeds, oils, margarine, butter, mayonnaise, and salad dressing all contribute vitamin E, they are high-fat choices and should be consumed in moderation.

Unit 21

Phytochemicals and Genetically Modified Food

Photo Disc

Nutrition Scoreboard

	TRUE	FALSE
1. Phytochemicals are found only in plants.		
2. Phytochemicals taken as supplements provide the same health benefits as do phytochemicals consumed in plant foods.		
3. Humans eat lots of DNA every day.		
4. Some chemical substances that occur naturally in food may be harmful to health.		

(Answers on next page)

Answers to

Nutrition Scoreboard

	TRUE	FALSE
1. "Phytochemicals" mean "plant chemicals."	✔	
2. Beneficial effects of phytochemicals on health appear to result primarily from complex interactions among them in foods. Supplements do not provide these benefits.		✔
3. The average meal contains approximately 250,000 miles (150,000 kilometers) of uncoiled DNA.[1] It's digested.	✔	
4. Some foods contain "naturally occurring toxins" that can be harmful if consumed in excess.	✔	

Key Concepts and Facts

- Plants contain thousands of substances in addition to essential nutrients that affect body processes and health.
- Diets containing lots of vegetables, fruits, whole grains, and other plant foods are strongly associated with the prevention of chronic diseases such as heart disease and cancer.
- Biotechnology is rapidly changing the characteristics and types of foods available to consumers.
- Not all substances that occur naturally in foods are safe to eat.

Photo Disc

Phytochemicals: Nutrition Superstars

Things don't happen by accident in nature. If you observe it, it has a reason for being there.

—Norman Krinsky, Tufts University Medical Center

As recently as 20 years ago, the science of nutrition was primarily concerned with the study of the actions and health effects of protein, vitamins, minerals, and the other essential nutrients. Those days are gone. Now nutrition scientists are investigating the effects of thousands of other substances in food. Although we have barely scratched the surface of knowledge about these substances, current research shows that impressive benefits can be gained by diets high in plant foods. People who habitually consume lots of vegetables, fruits, whole grains, and other plant foods are less likely to develop heart disease, cancer, adult-onset diabetes, infections, eye disease, premature aging, and a number of other health problems than people who do not.[2–5] Research results are sufficiently positive to change the way nutritionists and other scientists are thinking about food and health relationships.

At center stage in this new era in nutrition are the ***phytochemicals,*** or chemical substances found in plants (Illustration 21.1). The number of potentially beneficial substances in plants is mind-boggling. There are thousands of individual phytochemicals in plants, and some foods contain hundreds of them. A sampling of the phytochemicals and other substances in two foods is shown in Illustration 21.2. Meats, eggs, dairy products, and other foods of animal origin also contain chemical substances that are not considered essential nutrients but nevertheless affect body processes. Much less is known about these ***zoochemicals*** because their effects on health are not as clear as those of phytochemicals.

phytochemicals (*phyto* = plant) Chemical substances in plants, some of which perform important functions in the body.

zoochemicals Chemical substances in animal foods, some of which perform important functions in the body.

Phytochemicals are not considered to be essential nutrients for humans because we do not develop a deficiency disease if we consume too little of them. They are similar to essential nutrients in that the body cannot make them.

Illustration 21.1
The foods shown here have star qualities. They have been featured in recent research reports on the benefits of phytochemicals in food. Foods shown are grape juice, green tea, tomatoes, cabbage, carrots, tofu, crushed garlic; green, red, and yellow peppers; brown rice, whole-grain pita bread, and peanut butter. (Food preparation courtesy of The Blue Moon Café, Menomonie, WI.)

Consequently, they must be obtained from the diet.[6] The chemical properties of most phytochemicals tend to make them heat and light stable, so they are not easily destroyed by cooking or storage.[7]

Many phytochemicals are excreted within a day or two after ingestion, so intake of vegetables, legumes, nuts, fruits, and other food sources should be maintained.[6] Cooking and chopping up vegetables, or consuming them with a small amount of fat, increases the body's absorption of some phytochemicals.

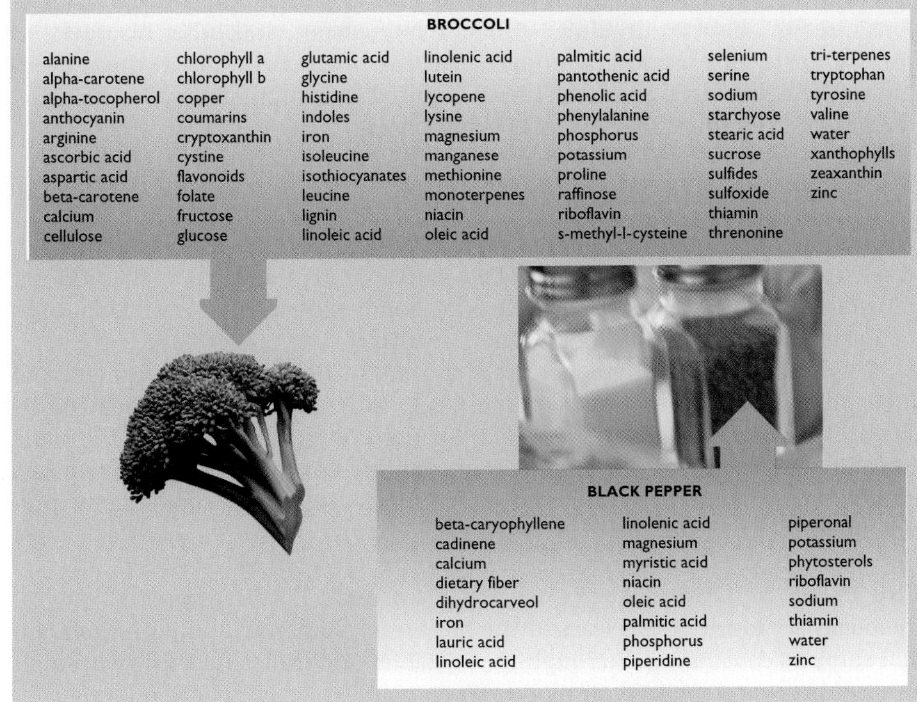

Photo Disc

Illustration 21.2
A sampling of the chemical substances in two foods.

Photo Disc

Characteristics of Phytochemicals

Phytochemicals serve a wide variety of functions in plants. They provide color and flavor and protect plants from insects, microbes, and oxidation due to exposure to sunlight and oxygen. Some phytochemicals are components of a plant's energy-making processes, and others act as plant hormones. More than 2000 types of phytochemicals that act as pigments have been identified. There are at least 600 types of carotenoids, for example. These pigments give plants yellow, orange, and red color; they primarily function as antioxidants.

The amount and type of phytochemicals present in plants vary a good deal depending on the plant. Some plant foods, such as citrus fruits and dried beans, which are considered rich sources of vitamins and minerals, contain an abundance of phytochemicals. Apples, celery, green tea, and garlic, foods considered by nutrient composition tables to be relative "nutrient weaklings," also provide ample amounts of phytochemicals (see Illustration 21.1). Many of the richest sources of beneficial phytochemicals are foods that are infrequently consumed.

Not all of the phytochemicals in plants are beneficial to health, however. Some are "naturally occurring toxins" and can be harmful. This type of phytochemical is discussed later in the unit.

Phytochemicals and Health

It has been known for more than 30 years that diets rich in vegetables and fruits are protective against heart disease and certain types of cancer.[8] For most of this time, the benefits of such diets were attributed to the vitamin, mineral, or dietary fiber content of vegetables and fruits. More recently, scientists discovered that the essential nutrient content of the diet didn't explain all of the differences in disease incidence between people consuming diets low in plants and those eating diets rich in plants. As researchers looked for the reason for the difference, they discovered that other chemical substances in food might be responsible. Many chemically active substances that could contribute to disease prevention were identified and their actions in the body described.

Phytochemicals are associated with a reduced risk of developing heart disease, certain types of cancer (lung, breast, cervical, esophageal, stomach, and colon cancer, for example), ***age-related macular degeneration, cataracts,*** infectious diseases, diabetes in adults, stroke, hypertension, and other disorders.[1–9] Although evidence supports a primary role for phytochemicals in the prevention of these disorders, research studies required to demonstrate *cause-and-effect relationships* in humans have yet to be completed.

➡ age-related macular degeneration
Eye damage caused by oxidation of the macula, the central portion of the eye that allows you to see details clearly. It is the leading cause of blindness in U.S. adults over the age of 65. Antioxidants provided by the carotenoids in dark green, leafy vegetables such as kale, collard greens, spinach, and Swiss chard may help prevent macular degeneration.[9]

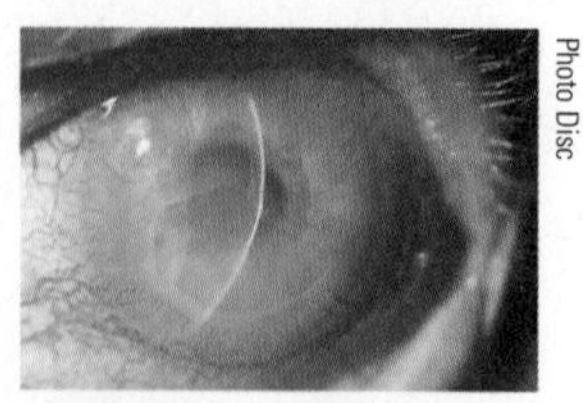
Photo Disc

➡ cataracts
Complete or partial clouding-over of the lens of the eye (shown in illustration).

Phytochemicals Work in Groups

News about the potential benefits of phytochemicals has sent consumers in search of pills that contain them. The market is replete with such products (Illustration 21.3), even though there is no solid evidence that individual phytochemicals extracted from foods benefit health.[10]

The absorption of phytochemicals appears to depend on the presence of other phytochemicals and nutrients in foods. Most (if not all) phytochemicals act together, producing a desired effect in the body if consumed at the same time. Since the optimal combinations of the different types of phytochemicals are not yet known, it is recommended that foods rather than supplements provide them.[11]

The Case of Beta-Carotene Supplements The lesson that phytochemicals act together was learned the hard way. Numerous studies had shown that smokers consuming diets high in vegetables and fruits and those with high levels of beta-carotene in their blood were much less likely to develop lung can-

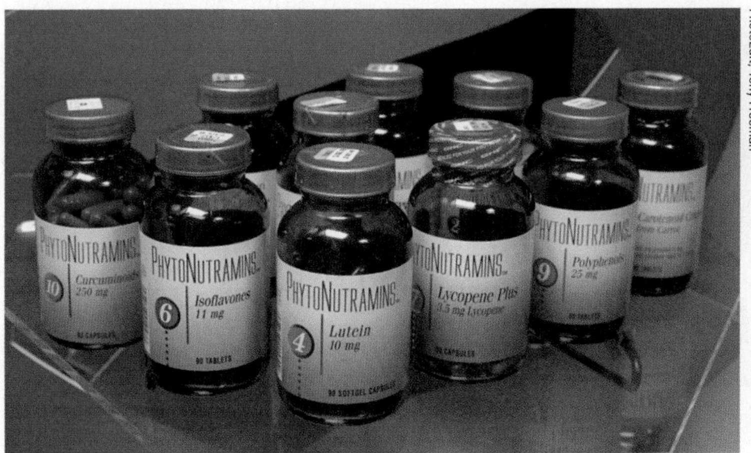

PhotoEdit, Tony Freeman

Photo Disc

Illustration 21.3
Consumers have their choice of phytochemicals in pills. The problem is that the supplements have not been shown to benefit health. Phytochemicals taken individually in supplements may actually harm health. They are best consumed in their natural packaging: vegetables, fruits, whole grains, and other plant foods.

cer than smokers who consumed few vegetables and fruits. Beta-carotene is a powerful antioxidant present in many vegetables and fruits, and researchers theorized that the beta-carotene provided by these plant foods might prevent lung cancer.

Three large and expensive clinical trials of beta-carotene supplementation among male smokers were conducted to test this theory. The results? None of the trials showed a benefit from the beta-carotene supplements, and two studies found a higher rate of lung cancer among the groups using the supplements.[12]

Scientists drew two major conclusions from these studies:

1. High blood levels of beta-carotene found in the group of smokers who did not develop cancer were likely a marker for a high intake of vegetables and fruits, and not the cause of the reduced risk.
2. Other phytochemicals in plants, or in a combination of plant foods, are likely responsible for the reduced risk of cancer.[3]

Vegetable Extracts and Essences Dehydrated, powdered extracts of vegetables high in certain phytochemicals and of the parts of vegetables richest in phytochemicals are also widely available. As is the case for supplements of individual phytochemicals, there is no solid evidence that these extracts benefit health. One such product called "Vegetable Essence" claimed to have "200 pounds of vegetables in a bottle." That is, of course, impossible. "Broccoli Concentrate," another vegetable extract, was found to contain primarily sulforaphane, one of the many phytochemicals in vegetables of the ***cruciferous family*** (Illustration 21.4). The problem with this and other vegetable concentrates is that you can fit only so much of a vegetable into a capsule. You would have to consume approximately 100 of the pills to get the amount of sulforaphane present in one serving of broccoli.[13] Even if you did that, you would still be missing out on the other phytochemicals and essential nutrients that are part of those little green trees.

Until scientific studies demonstrate that supplements of phytochemicals, individually or in combination, are safe and effective, it is best to forget about using them.

cruciferous vegetables Sulfur-containing vegetables whose outer leaves form a cross (or crucifix). Vegetables in this family include broccoli, cabbage, cauliflower, brussels sprouts, mustard and collard greens, kale, bok choy, kohlrabi, rutabaga, turnips, broccoflower, and watercress.

Illustration 21.4
Some examples of cruciferous vegetables.

How Do Phytochemicals Work?

A variety of body processes involved in disease development are affected by phytochemicals. Some of these processes, and the phytochemicals responsible, are listed in Table 21.1. In general, phytochemicals can

1. act as hormone-inhibiting substances that prevent the initiation of cancer,
2. serve as antioxidants that prevent and repair damage to cells due to oxidation,
3. block or neutralize enzymes that promote the development of cancer and other diseases,
4. modify the absorption, production, or utilization of cholesterol, or
5. decrease formation of blood clots.

Some plant pigments are particularly powerful antioxidants. Zeaxanthin, which give plants a corn yellow color, anthocyanin (the blue in blueberries and grapes), and lycopene, which helps make tomatoes, strawberries, and guava red, are all strong antioxidants. Some of the phytochemicals that act as antioxidants help reduce plaque formation in arteries by preventing the oxidation of LDL-cholesterol. Oxidized LDL is reactive and easily becomes a component of plaque.[1]

Diets High in Plant Foods

On the Side

Plants have been on this planet for about a billion years longer than humans. Although the total number of edible plants hasn't changed over the centuries, most modern humans consume only 150–200 different types of plants, or about 0.2% of the total possibilities.[14]

Photo Disc

The discovery of a wide array of health-promoting phytochemicals in plants is one more very important reason why vegetables, fruits, whole grains, and other plant foods should be a major part of the diet. They are a minor part now for many Americans. In the United States, 80% of adults consume two or fewer servings of vegetables and fruits daily, and 10% consume none.[13] In reality, people would probably be healthier if they consumed more than the five servings of vegetables and fruits recommended. It's hard to increase the recommendation when so many Americans aren't even close to eating five a day. Perhaps broadening the options for plant food choices beyond those commonly consumed would help. Did you know that we consume only a small fraction of the total supply of edible plants? If you are taken by surprise, discover the amazing facts presented in On the Side.

Due to the explosion of information about the benefits of some of the phytochemicals and consumer demand, new food products are being developed with specific applications to health in mind. Increasingly, these products are being designed in research laboratories using ***biotechnology.***

➡ **biotechnology**
As applied to food products, the process of modifying the composition of foods by biologically altering their genetic makeup. Also called *genetic engineering* of foods. The food products produced are sometimes referred to as "transgenic," "GM," and GMOs (genetically modified organisms).

Genetically Modified Foods

By using biotechnology, researchers can modify the phytochemical and nutrient makeup of foods by altering the genetic makeup of plants. Resulting plants are called "genetically modified" or "GM" plants (Illustration 21.5). The process of biotechnology basically entails identifying a favorable genetic trait in one plant and transplanting that gene into another plant that lacks the characteristic.

Table 21.1
Examples of phytochemicals, their food sources, and potential mechanisms of action in disease prevention.[1–12]

Phytochemical	Comments	Food Sources	Proposed Action
Indoles, isothiocynates (sulforaphanes)	These sulfur-containing compounds may be particularly protective against breast cancer.	Cruciferous vegetables (broccoli, brussels sprouts, cabbage, cauliflower)	Interfere with the cancer growth-promoting effects of genes on cells; increase the body's ability to neutralize cancer-causing substances.
Allicin	Health benefits are associated with as little as a half clove of garlic per day on average.	Garlic, onions, leeks, shallots, chives, scallions	Interferes with the replication of cancer cells; decreases the production of cholesterol by the liver, reduces blood clotting.
Terpenes (monoterpenes, limonene)	These substances give citrus fruits a slightly bitter taste. Limonene is from the same family of compounds as tamoxifen, a drug used to treat breast cancer.	Oranges, lemons, grapefruit, and their juices	Facilitate the excretion of cancer-causing substances; decrease tumor growth.
Phytoestrogens (plant estrogens; isoflavones, genistein, daidzein, lignans)	May decrease risk of some cancers, heart disease, and osteoporosis. Estrogen effects, if any, are weak. High doses of some individual phytoestrogens may have adverse effects on health.	Soybeans, soy food, chickpeas, other dried beans, peas, peanuts	Interfere with cancer growth-promoting effects of estrogen; block the action of cancer-causing substances; lower blood cholesterol; decrease menopausal symptoms and bone loss, act as antioxidants.
Lignans (phytoestrogens)	Flaxseed is an extremely rich source of lignans. In the gut, lignans are converted to substances that may help prevent breast cancer.	Flaxseed, flaxseed oil, seaweed, soybeans and other dried beans, bran	Interfere with the action of estrogen.
Saponins	Nearly every type of dried beans is an excellent source of saponins.	Dried beans, whole grains, apples, celery, strawberries, grapes, onions, green and black tea, red wine	Neutralize certain potentially cancer-causing enzymes in the gut.
Flavonoids (tannins, phenols)	There are over 4000 flavonoids, some of which are plant pigments. Originally called "vitamin P." Gives red wines and dark teas astringent or bitter taste.	Apples, celery, strawberries, grapes, onions, green and black tea, red wine, soy, purple grape juice, broccoli	Protect cells from oxidation; decrease plaque formation and blood clotting; increase HDL-cholesterol; decrease DNA damage related to cancer development.
Carotenoids (alpha-carotene, beta-carotene, lutein, zeaxanthin, beta-cryptoxanthin, lycopene)	There are more than 600 types of colorful carotenoids in plants. An orange contains at least 20 types. Dark green vegetables are often good sources; the green chlorophyll obscures the colors of carotenoids in these plants. Fat intake increases absorption.	Dark green vegetables, orange, yellow, and red vegetables and fruits	Neutralize oxidation reactions that can damage eyes and promote macular degeneration and cataracts; increase LDL-cholesterol and cancer risk.
Isoprenoids	Refining wheat eliminates the germ and bran and causes a 200- to 300-fold loss in phytochemical content.	Fruits, vegetables, whole grains, flaxseed	Decrease cholesterol production by the liver; lower blood cholesterol.

Illustration 21.5
Genetic modification of plants.

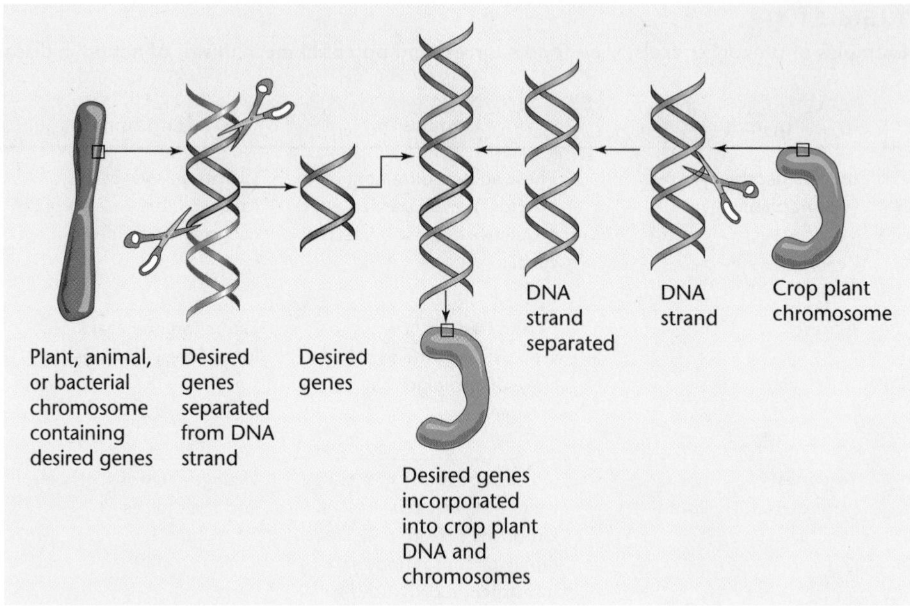

Thus, disease-resistant genes in some plants can be shared with another plant, conferring upon it an improved ability to resist disease. Flavor genes can be transported from one plant to another, creating new taste sensations like garbanzo beans with a nutty, peppery taste and basil preseasoned with cinnamon. Colors of vegetables and fruits can be modified by transferring the appropriate genes from one plant to another. Carrots have been modified to be dark red so they would match Texas A&M's school color. Tomatoes have been genetically altered to stay firm during shipment, and rice to be rich in beta-carotene, the precursor of vitamin A. High beta-carotene rice was produced to make a good source of vitamin A available to people in countries where vitamin A deficiency is widespead. Called "golden rice," it contains two genes extracted from daffodils and one bacterial gene that together lead to the production of beta-carotene within rice seeds.[15] Table 21.2 summarizes reasons for genetically modifying plant crops.

GM Foods: Are They Safe?

> **People have a fear of eating DNA. We eat 150,000 kilometers [250,000 miles] of DNA in an average meal.**
>
> —David Cove, PhD, Leeds University's Genetics Department

Table 21.2
Desired traits incorporated into crop plants.[16]

- Herbicide tolerance
- Insect resistance
- Viral, fungal, and bacteria resistance
- Food quality enhancement
- Nutritional improvement

People generally don't think of DNA as a normal component of food, but since we eat cells when we consume foods, we get hefty helpings of DNA and chromosomes in every meal. People have been consuming foods with novel DNA in hybrid plants for many years without ill effects. Obviously, consuming DNA is not harmful by itself. Concerns about the safety of GM foods have less to do with consuming them and more to do with their potential impact on the environment.

Controversies surrounding GM foods include both unscientific, emotionally driven speculation and legitimate concerns (Table 21.3). Although strong, anti-GM foods sentiment exists in Europe, most U.S. adults expect to benefit from them.[19] The line separating sides in the debate about the safety of these foods blurred recently when uncritical news reports announced that GM corn kills Monarch butterfly larvae and canola oil modified with Brazil nut genes produces

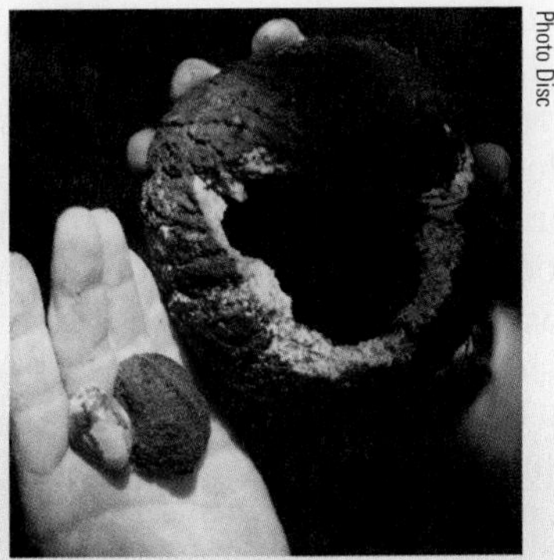

Illustration 21.6
Monarch butterfly larvae and Brazil nuts stirred the pot of controversy over GM foods.

Due to highly publicized research reports, Monarch butterflies have become a poster organism for anti-GM food organizations. Researchers coated milkweed leaves with the pollen of corn genetically modified to fend off corn borers and corn earworms. They then placed Monarch butterfly larvae on leaves to feast, and 45% of them died. The study was highly criticized for its unrealistic dose of pollen.[17] It did increase awareness that insect-resistant crops could potentially harm desired insects such as butterflies.

Adding genes from Brazil nuts to rapeseed and beans improved the methionine (an essential amino acid) content of canola oil (made from rapeseeds) and beans. Food scientists working for the manufacturer realized that genes from the nut could potentially cause an allergic reaction in people sensitive to Brazil nuts, so University of Nebraska researchers were hired to test the products. The foods were found to produce allergic symptoms in 7 of 9 people with an allergy to the nuts. The company halted production of the GM foods, but the concerns that GM foods would prompt allergic reactions in unsuspecting people increased. Because potentially allergic foods are tested, this is a minor concern related to GM food products.

allergic reactions (see Illustration 21.6). Companies that produce GM foods test them for allergy potential and grow the foods in test greenhouses before they are planted outside to assess their potential to cross-pollinate.[16] These tests are beneficial, but do not cover all the possible consequences of GM crops.

We All Eat GM Foods But May Not Know It

> **There is no current evidence that GM technologies used to produce foods are inherently harmful.**
>
> —United Kingdom's government on health implications of genetically modified foods, 2000

Table 21.3
Legitimate concerns about biotechnology and genetically modified foods.[1,17–18]

- GM crops may pollinate nearby plants, making them resistant to herbicides.
- GM crops may monopolize the market, and that may lead to higher food prices and lower biodiversity of crops and other plants.
- Insect- and herbicide-resistant crops are not reason enough to justify GM foods; too little focus is being placed on improving nutritional quality and availability of food to malnourished populations.
- Interested consumers cannot know which foods contain GM ingredients and which do not. (Labeling GM foods is being implemented in Korea and Japan.)

The U.S. Department of Agriculture has approved dozens of types of GM plant crops since 1995. About 57% of soybeans and 33% of the corn grown in the United States are from genetically modified seeds, and GM tomatoes, squash, cantaloupe, and potatoes are also grown and distributed.[16] Companies that produce these products must inform the FDA that they intend to market a new product and must provide results of safety tests. It is the FDA's position that GM foods, rather than the process used to develop them, are the objects of safety evaluations. Because foods containing genetically modified ingredients have not been demonstrated to harm health, there is no requirement for labeling products that contain GM ingredients.[16] Required labeling is, however, being considered due to consumer concerns about the safety of "frankenfoods."

The Future of GM Foods Genetically modified foods hold much promise for increasing food availability, decreasing dependence on pesticides and herbicides, and increasing the nutritional quality of plants. Whether the promise is fulfilled will depend on consumer acceptance of GM foods. In turn, that will depend on research that convincingly demonstrates that they are safe for people and the environment, beneficial to health, and reduce rather than raise food costs.

Naturally Occurring Toxins in Food

All that occurs naturally in foods is not necessarily good. There are many examples of foods containing phytochemicals that can have negative side effects. Spinach, collard greens, rhubarb, and other dark green, leafy vegetables contain oxalic acid. Eating too much of these foods can make your teeth feel as though they are covered with sand and give you a stomachache. Have you ever seen a potato that was partly colored green? (If not, see Illustration 21.7.) The green area contains solanine, a bitter-tasting, insect-repelling phytochemical that is normally found only in the leaves and stalks of potato plants. Small amounts of solanine are harmless, but large quantities (an ounce or so) can interfere with the transmission of nerve impulses.

Phytates are another example of a naturally occurring substance that can have harmful effects. Phytates, which are found in whole grains, tightly bind zinc, iron, and other minerals, making them unavailable for absorption. Although not toxic, phytates do reduce the availability of some minerals in whole-grain products. As another example, cassava, a root consumed daily in many parts of tropical Africa, can be very toxic if not prepared properly because it contains cyanide. Soaking cassava roots in water for three nights gets rid of the cyanide, but shorter periods of soaking do not. When the soaking time is cut to one or two nights, as sometimes happens during periods of food shortage, enough of the toxin remains in the root to cause konzo, a disease caused by the cyanide overdose.[20] Konzo is characterized by permanent, spastic paralysis.

Caffeine is another phytochemical in plants (primarily, coffee beans) that some people consider to be toxic. To others, it's a gift from the gods every morning. Does caffeine belong in the list of naturally occurring, harmful substances in food?

PhotoEdit, Tony Freeman

Illustration 21.7
Potatoes grown partly above ground develop a green color in the part exposed to the sun. The green section contains solanine, a naturally occurring, toxic phytochemical.

What's the Scoop on Caffeine? Caffeine is one of many phytochemicals in coffee, cocoa, and some teas. Coffee, however, is far-and-away the leading source of caffeine in the diet. (See Table 21.4 for the caffeine content of various beverages, chocolate, and nonprescription drugs.) Although the effects of caffeine on health are widely debated, it has received a clean bill of health with a couple of exceptions. Too much caffeine causes sleeplessness and "caffeine jitters" in many people.[21] Large amounts of coffee (more than four cups a day) may somewhat delay the time it takes women to become pregnant and may slightly increase the risk of miscarriage.[22] Drinking unfiltered, boiled coffee (as is the custom in Turkey and some Scandinavian countries) raises LDL-cholesterol levels and lowers HDL-cholesterol. That increases the risk of heart disease. However, two other phytochemicals in coffee, kahweol and catestol, and not caffeine, may be responsible for this effect.[23,24] Suspicions that filtered coffee or caffeine is related to heart disease, cancer, or osteoporosis have not been confirmed.[25,26]

Regular coffee drinking can be habit forming. Cessation of coffee intakes that average two-and-a-half cups a day leads to headaches in about 50% of people.[27]

Table 21.4
Caffeine content of foods, beverages, and some drugs.

Source	Caffeine (mg)
Coffee, one cup	
Drip	137–153
Decaffeinated (ground or instant)	0.5–4.0
Instant	61–70
Percolated	97–125
Espresso (2 ounces)	100
Tea, one cup	
Black, brewed 5 minutes, major U.S. brands	32–144
Black, brewed 5 minutes, major imported brands	40–176
Green, brewed 5 minutes	25
Instant	40–80
Soft drinks/bottled water	
Coca-Cola (8 ounces)	31
Cherry Coke (8 ounces)	31
Diet Coca-Cola (8 ounces)	31
Dr. Pepper (12 ounces)	40
Ginger ale (12 ounces)	0
Jolt (12 ounces)	108*
Kick (12 ounces)	56
Mountain Dew (12 ounces)	54
Pepsi-Cola (12 ounces)	38
Diet Pepsi-Cola (12 ounces)	37
7-Up (12 ounces)	0
Surge (8 ounces)	35
Water Joe (8 ounces)	125
Chocolate	
Cocoa, chocolate milk, one cup	10–17
Milk chocolate candy, one ounce	1–15
Chocolate syrup, one ounce (two tablespoons)	4
Nonprescription drugs, one tablet	
Nodoz	200
Vibrin	200
Dexatrim	200

*Estimated from company Web site description of "twice the caffeine."

Nutrition UP CLOSE

Have You Had Your Phytochemicals Today?

FOCAL POINT: Consuming good sources of the "other" beneficial components of food.

Good food sources of beneficial phytochemicals are listed below. Indicate foods you consumed at least twice last week, foods you like, and those you have never tried.

	Foods Eaten at Least Twice Last Week	Foods You Like	Foods You've Never Tried
Broccoli	______	______	______
Cabbage	______	______	______
Carrots	______	______	______
Cauliflower	______	______	______
Celery	______	______	______
Collard greens	______	______	______
Garlic	______	______	______
Onions	______	______	______
Tomatoes	______	______	______
Turnips	______	______	______
Apple/juice	______	______	______
Orange/juice	______	______	______
Grapefruit/juice	______	______	______
Grapes/juice	______	______	______
Strawberries	______	______	______
Brown rice	______	______	______
Dried beans	______	______	______
Tofu	______	______	______
Whole-grain bread, cereal	______	______	______

FEEDBACK (answers to these questions) can be found at the end of Unit 21.

WWW links

www.aphis.usda.gov/biotechnolog
Provides scientific information on GM foods.

www.citizens.org
Presents opponents' view of GM foods.

www.nas.edu
Select biotechnology to get reports on results from the National Academy of Science's subcommittee on genetically modified plants.

www.easynet.co.uk/ifst
Check for food biotechnology and related information from Britain's Institute for Food Science and Technology.

www.navigator.tufts.edu
Search engine for information on phytochemicals and biotechnology.

REFERENCES

1. Bunk S. Researchers feel threatened by disease genes. The Scientist 1999;13 (Oct. 11):7–8.
2. Arab L, Steck S. Lycopene and cardiovascular disease. Am J Clin Nutr 2000;71(suppl):1691S–5S
3. Michaud DS et al. Intake of specific carotenoids and risk of lung cancer in 2 prospective US cohorts. Am J Clin Nutr 2000;72:990–7.
4. Kelloff GJ et al. Progress in cancer chemoprevention: development of diet-derived chemopreventive agents. J Nutr 2000;130:467S–71S.
5. Potter SM et al. Soy protein and isoflavones: their effect on blood lipids and bone density on postmenopausal women. Am J Clin Nutr 1998; 68(suppl):1375S–9S.
6. Gross M. Flavonoids, platelets, and the risk of cardiovascular disease. Epidemiology Seminar, Minneapolis, 1999 Nov 10.
7. Forman A. As beta-carotene promise fades, focus turns to other carotenoids. Environmental Nutrition Newsletter 1997 Aug:1, 4.
8. Craig WT. Phytochemicals: guardians of our health. J Am Diet Assoc 1997;97(suppl 2):S199–S204.
9. Bloch A, Thomson CA. Position of the American Dietetic Association: phytochemicals and functional foods. J Am Diet Assoc 1995;95:493–6.
10. Gingsburg J, Prelevic GM. Lack of significant hormonal effects and controlled trials of phytoestrogens. N Engl J Med 2000;355:163–4.
11. Hollon T. Isoflavonoids: always healthy? The Scientist, 2000;14(Sept 4):21.
12. Knekt P, Jarvinen R, et al. Dietary flavonoids and the risk of lung cancer and other malignant neoplasms. Am J Epidemiol 1997;146:223–30.
13. Johnston CS, Corte C. People with marginal vitamin C status are at high risk of developing vitamin C deficiency. J Am Diet Assoc 1999;99:854–6.
14. Herber D. The stinking rose: organosulfur compounds and cancer. M J Clin Nutr 1997;66:425–6.
15. Palevita BA, Lewis R. In Bt football, it's Cornell vs. Cornell. The Scientist 1999;13 (Oct 11):1,8.
16. Nettleton JA. The promise and politics of agricultural biotechnology. Nutr Today 2000;35:118–28.
17. Palevitz B. With GM crops, who needs vitamin pills? The Scientist 2000;14 (Feb 7):14–15.
18. Biotechnology, food and fiber production and the environment. Food and Nutrition Board Update, National Academy of Sciences, Winter 1999–2000, www.nas.edu.
19. Dwyer S. The growing pains of GMOs. Prepared Foods 2000; Jan:14,15,16. Ohr LM. A defining time for GMOs. Prepared Foods 2000; June:41–42,44,46.
20. Tylleskar T, Banea M, et al. Dietary determinants of a non-progressive spastic paparesis (konzo). Int'l J Epidemiol 1995;24:949–56.
21. Shirlow MJ, Mathers CD. A study of caffeine consumption and symptoms: indigestion, palpitations, tremor, headache, and insomnia. Int'l J Epidemiol 1985;14:239–48.
22. Wei C, Shu X, et al. Caffeine consumption and reproductive outcomes. (in press) 1998.
23. Tol AV, Urgert R, et al. The cholesterol-raising diterpenes from coffee beans. Atherosclerosis 1997;132:251–4.
24. Grubben MJ et al. Unfiltered coffee increases plasma homocysteine concentrations in healthy volunteers: a randomized trial. Am J Clin Nutr 2000;71:480–4.
25. Lloyd T, Rollings N, et al. Dietary caffeine intake and bone status of postmenopausal women. Am J Clin Nutr 1997;65:1826–30.
26. Lloyd T et al. Caffeine not related to bone loss in postmenopausal women. Am Coll Nutr 2000;19:256–61.
27. Silverman K, Evans SM, et al. Withdrawal syndrome after the double-blind cessation of caffeine consumption. N Engl J Med 1992;327:1109–14.

Nutrition
UP CLOSE

Have You Had Your Phytochemicals Today?

Feedback for Unit 21

Did you eat four or more of the foods listed at least twice last week?If yes, listen carefully and you'll hear your cells say "thank you."

If you didn't eat four or more twice last week, go for the foods you didn't eat but like. Be adventurous! Try some of the phytochemical-rich foods you have never eaten before.

Unit 22

Diet and Cancer

PhotoEdit, Felicia Martinez

Nutrition Scoreboard

	TRUE	FALSE
1. Some types of cancer are contagious.		
2. Cancer is primarily an inherited disease.		
3. People who eat plenty of fruits and vegetables are less likely to get cancer than people who don't.		
4. High-fat diets contribute to the development of certain types of cancer.		
5. Two out of three Americans never get cancer.		

(Answers on next page)

Answers to Nutrition Scoreboard

	TRUE	FALSE
1. Cancer doesn't spread from person to person.		✔
2. Cancer is primarily related to environmental factors such as diet, smoking, and overexposure to the sun. Genetic background plays a role by placing some individuals at increased risk for developing cancer.		✔
3. Among the arsenal of cancer-protective measures, the consumption of five or more servings of fruits and vegetables a day stands out.	✔	
4. High-fat diets (and high levels of body fat) are related to the development of several types of cancer.	✔	
5. Statistics from the National Cancer Institute demonstrate that most people don't get cancer.	✔	

Key Concepts and Facts

- Cancer has many different causes. Diet is a major factor that influences the development of most types of cancer.
- Some components of food increase the risk of cancer whereas others decrease the risk.
- Consuming fruits and vegetables regularly, eating a low-fat diet, and maintaining normal levels of body fat appreciably reduce the risk of developing many types of cancer.
- Cancer is largely preventable, but there are no absolute guarantees that an individual will not develop cancer.

What Is Cancer?

Cancer, the second leading cause of death in the United States, is a group of conditions that result from the uncontrolled growth of abnormal cells. Although these cells can begin to grow in any tissue in the body, the respiratory and digestive systems, ***prostate,*** and breasts are the most common sites for cancer development (Illustration 22.1).[1] Some forms of cancer are highly curable.

cancer A group of diseases in which abnormal cells grow out of control and can spread throughout the body. Cancer is not contagious and has many causes.

prostate A gland located above the testicles in males. The prostate secretes a fluid that surrounds sperm.

How Does Cancer Develop?

Cancer develops by complex processes that are not yet fully understood. Adding to the complexity is the fact that cancer development often does not proceed in a straight line—cancer can progress two steps forward and then take a step or two back.

Illustration 22.2 summarizes the processes involved in the development of cancer, as we currently understand them. Cancer begins when something goes wrong within cells that modifies cell division. Every minute, 10 million cells in the body divide. Usually, the cells divide the right way and on schedule due to a set of regulatory mechanisms that control the replication (duplication) of DNA, the genetic material that becomes part of new cells.

During the ***initiation*** phase of cancer development, something alters the DNA in certain cells. During the ***promotion*** phase of cancer, the cells with altered DNA divide, eventually producing large numbers of abnormal cells. This phase of cancer development commonly takes place over a span of 10 to 30 years. Unless they are hindered by the body or corrected by some other means,

initiation The start of the cancer process; it begins with the alteration of DNA within cells.

promotion The period in cancer development when the number of cells with altered DNA increases.

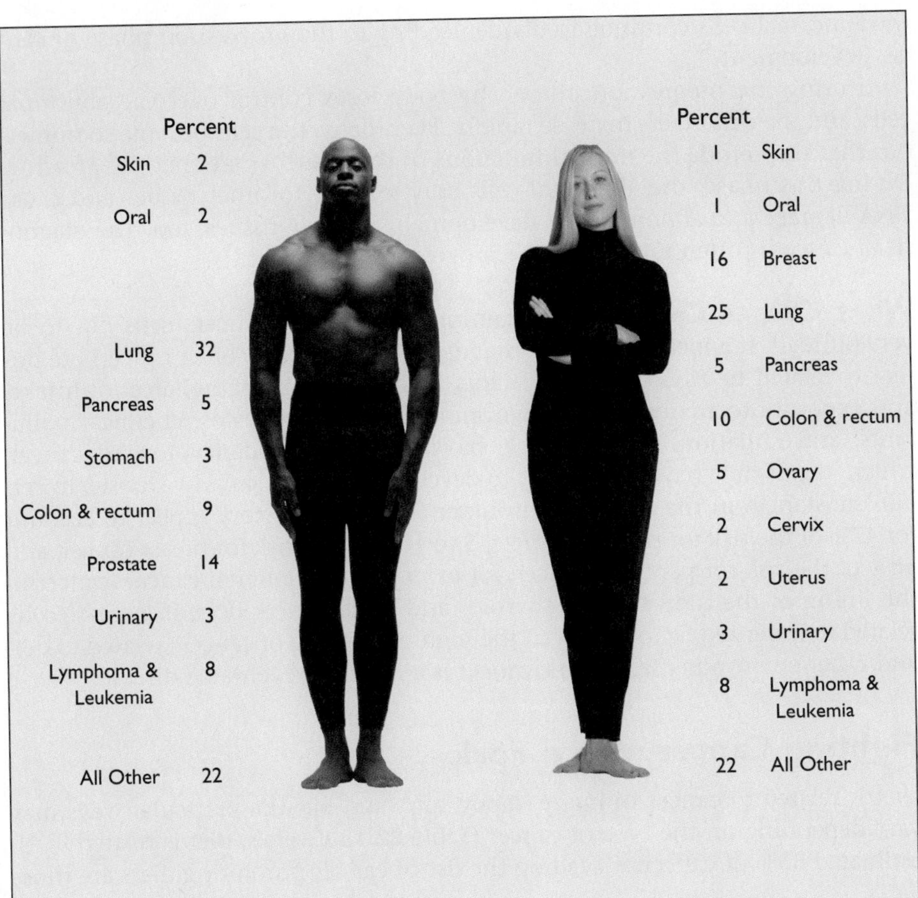

Illustration 22.1
Percentage of cancer deaths by selected sites and sex, 1998 estimates. In 1998, 23% of deaths in the United States were due to cancer.

Source: American Cancer Society, Surveillance Research, 1998 and Health, United States 2000, www.cdc.gov/nchs.

Illustration 22.2
Steps in the development of cancer.[2,3] Antioxidants, other substances in food, and the body's protective mechanisms may repair damage to cells and halt the progression of cancer.

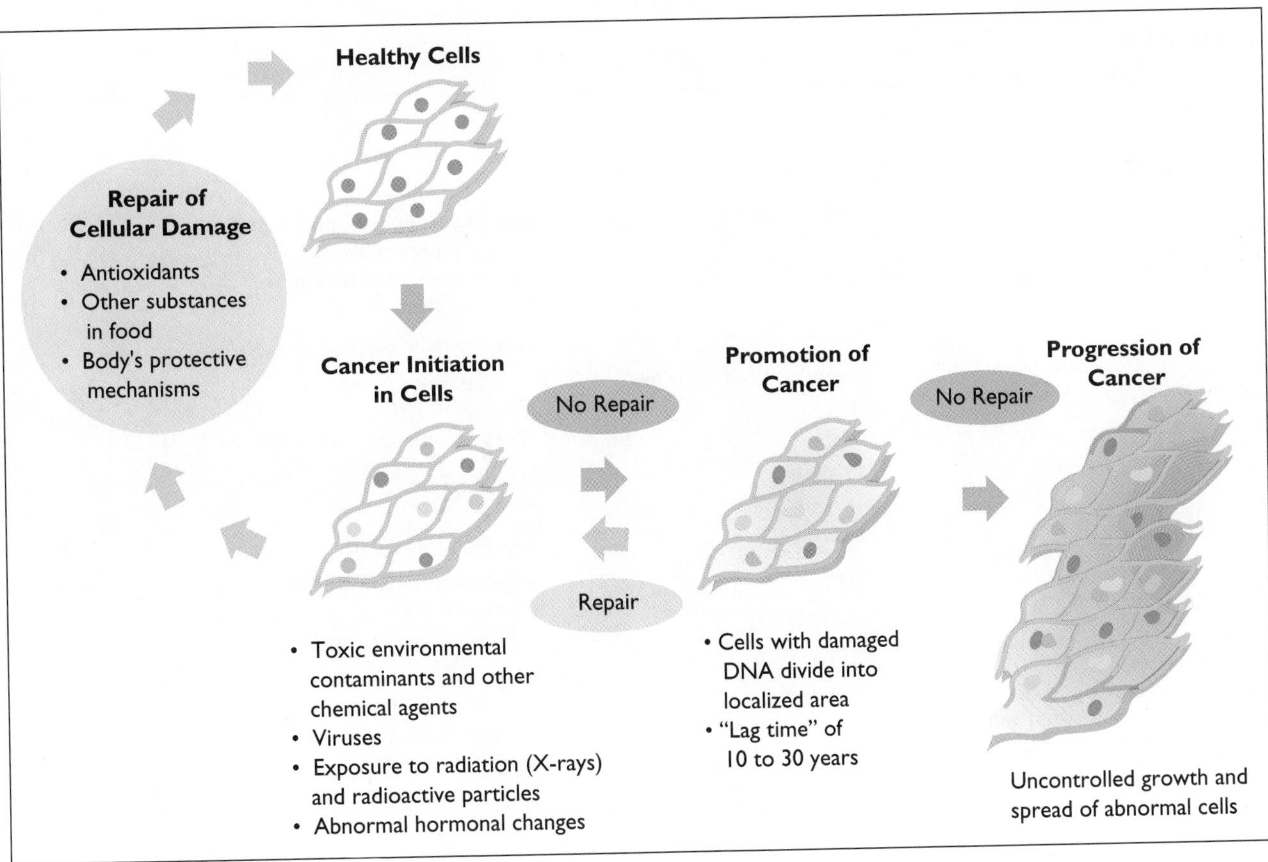

progression
The uncontrolled growth of abnormal cells.

the abnormal cells continue to divide, leading to the ***progression*** phase of cancer development.

During the progression phase, the body loses control over the abnormal cells, and their numbers increase rapidly. Eventually, the cells become so numerous that they erode the normal functions of the tissue where they are growing. During this phase, the abnormal cells may migrate to other tissues and cause DNA damage and abnormal cell development in these tissues, too. The diagnosis of cancer is often made after the progression phase occurs.[2]

What Causes Cancer? Explaining what causes cancer turns out to be very difficult. It appears, however, that approximately 90% of all cancers are primarily related to environmental factors such as diet, smoking, alcohol intake, and exposure to toxins in the environment such as asbestos, chemical pollutants, and radiation.[3] Some people have a genetic tendency toward cancer, which means they have a tendency to develop cancer if regularly exposed to certain substances in the diet or environment.[3] Genetic factors appear to account for 42% of the risk for prostate cancer, 5 to 10% of the risk for breast cancer, and 36% of the risk for pancreas cancer, for example. Endometrial cancer (cancer of the lining of the uterus), oral, thyroid, and bone cancer do not appear to be related to inherited factors.[3] Given the high percentage of cancers related to diet and other environmental factors, cancer is a largely preventable disease.

Fighting Cancer with a Fork

Diet is related to cancer in many important ways, but the particular ways may vary depending on the type of cancer (Table 22.1). Overall, diet is related to an estimated 35% of cancers.[4] Leading the list of cancer-promoting diets are those

Table 22.1
The connections between diet and cancer.[5]

Cancer Site	Dietary Risk Factors
Breast	Excessive alcohol intake Low vegetable and fruit intake High body fat
Colon or rectum	Low vitamin D and calcium intake High red meat intake Low vegetable and fruit intake Excessive alcohol intake
Lung	Low vegetable and fruit intake
Ovary and endometrium	High-fat diets Obesity Low vegetable and fruit intake
Prostate	High-fat diets High body fat
Pancreas	Low vegetable and fruit intake
Stomach	Regular consumption of smoked foods and foods cured with salt or nitrite compounds Low vegetable and fruit intake
Liver	Ingestion of *aflatoxin*-contaminated grains Regular consumption of smoked foods and foods cured with salt or nitrite compounds
Mouth and throat	Excessive alcohol intake Low vegetable and fruit intake

aflatoxins
Naturally occurring toxins produced by molds that may form on nuts, grains, seeds, and other plant products during storage.

low in vegetables and fruits.[2] High intakes of fat—especially when associated with excessive levels of body fat, excessive alcohol consumption, and the regular intake of smoked foods or foods cured with nitrite compounds—are also associated with the development of cancer.[6]

Body Fat, Dietary Fat, and Cancer Both excessive levels of body fat and the tendency to store fat around the stomach or central part of the body are related to an increased risk of developing breast and endometrial cancer (or cancer of the lining of the uterus).[7] High levels of body fat appear to promote the excessive production of certain hormones that may foster the growth of abnormal cells.

High-fat diets are related to the development of a number of cancers. The effect of high-fat diets may be due in part to their association with excessive caloric intake and high levels of body fat.[2,6]

Fruits, Vegetables, Whole Grains, and Cancer People who regularly consume plenty of fruits and vegetables have a lower risk of developing a number of types of cancer than people who eat few of those foods.[8] It is not yet clear why fruits and vegetables exert this effect, but a number of possibilities exist.

One possibility is the antioxidant effect of vitamin C, beta-carotene, and vitamin E found in fruits and vegetables. Much of the damage to DNA is thought to be caused by exposure to oxygen and other oxidizing substances that disrupt molecules within DNA; antioxidants may be able to repair damaged DNA.

Vegetables, fruits, and whole grain products contain "phytochemicals"—nutrient-like substances that may perform specific functions in the body. Phytochemicals in plants appear to protect cells from damage due to oxidation and inhibit the multiplication of abnormal cells.[9]

Garlic, onions, broccoli, cabbage, cauliflower, carrots, soybeans, sweet potatoes, and dark green vegetables are a few of the many plant foods associated with a decreased risk of developing cancer (Illustration 22.3).[9]

Alcohol Intake and Cancer Consumption of excessive amounts of alcoholic beverages has been linked to cancers of the breast, mouth, throat, and liver.[6,8] The risk of developing cancers of the mouth and throat increases for people who smoke cigarettes or chew tobacco as well as drink.[3]

Smoked, Cured, and Pickled Foods and Cancer Cancer of the stomach and liver appears to be related to the regular consumption of foods preserved by smoking, salting, nitrites, and pickling. Foods in these categories include ham, smoked sausage and pork chops, bacon, pickled eggs, and pickled vegetables such as cucumbers and radishes. (You can identify foods that contain nitrites by examining food ingredient labels.) Most cases of cancer associated with smoked, salted, and pickled foods now occur in China, parts of the former Soviet Union, and Central and South America where such foods are commonly consumed.[6]

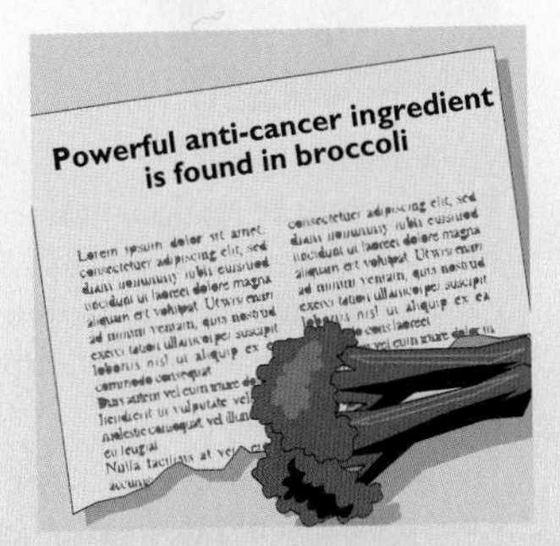

Illustration 22.3
An example of headline coverage of the cancer-protective effects of vegetables.

Grilled Foods and Cancer Certain nitrogen-containing and other substances in beef, chicken, fish, and other meats may become cancer promoting if heated to a high temperature. Such high temperatures can be

PhotoEdit, David Young-Wolff

Illustration 22.4
The charred and black, oily coating on grilled or broiled meats is the part you shouldn't eat.

reached by broiling and grilling food. The potentially cancer-promoting substances formed by meats are found in the charred portions of the meat and the fatty coating that accumulates on meat when fat drips into the heat source and smokes (Illustration 22.4).[10]

The Department of Agriculture concludes that eating broiled or grilled foods several times a week poses no threat to health. People should be careful not to eat the charred portions of meat and to wipe or rinse off any coating on meat that results from fatty flare-ups on the grill.

Diet and Cancer Guidelines To help people lower their risk of cancer, the American Cancer Society, the World Cancer Research Fund,[6] and the National Cancer Institute have developed dietary guidelines for cancer prevention. These recommendations are condensed in Table 22.2. In general, these guidelines are similar to, and compatible with, dietary recommendations to reduce the risk of heart disease.

Bogus Cancer Treatments

Unorthodox, purported cancer cures such as macrobiotic diets, hydrogen peroxide ingestion, laetrile tablets, vitamin, mineral, and herbal supplements, and animal gland therapy have *not* been shown to be effective treatments for cancer. Such remedies have been promoted since the early 1900s. They still exist because, although not proven to work, they offer some cancer patients their last ray of hope. They should *not* be used as a substitute for conventional cancer treatments.

Table 22.2
A summary of guidelines on diet, nutrition, and cancer.

1. Maintain a desirable body weight.
2. Eat a varied diet.
3. Include five or more servings of vegetables and fruits in the daily diet.
4. Eat more high-fiber foods, such as whole grain cereals, legumes, vegetables, and fruits.
5. Cut down on total fat intake.
6. Limit consumption of alcoholic beverages, if you drink at all.
7. Limit consumption of salt-cured, smoked, and nitrite-preserved foods.

Eating to Beat the Odds

Two out of every three people in the United States do *not* develop cancer. You can likely improve your odds of being among the two by not smoking or drinking excessively, by regularly consuming five or more servings of fruits and vegetables each day, by sticking to a low-fat diet, and by achieving or maintaining a normal level of body fat. Although there are no guarantees, people can help themselves prevent cancer by adhering to a good diet and a healthy lifestyle.

WWW links

www.nci.nih.gov
Website for the National Cancer Institute; provides evidence-based information about cancer prevention and treatment, and research updates.

www.cancer.org
The American Cancer Society's site provides a wealth of information on cancer clinics, treatments, prevention, and statistics.

www.altmed.od.nih.gov
Gets you to the National Center for Complementary and Alternative Medicine and state-of-the-science information on alternative cancer treatments.

cancer.med.upenn.edu
Provides sections on "Diet and its Relationship to Cancer Prevention" and "Nutrition during Cancer Treatment."

www.healthfinder.gov
Search engine for reliable sources of information on cancer and other health concerns.

www.nal.usda.gov/fnic
Click on cancer.

Nutrition
UP CLOSE

A Cancer Risk Checkup

FOCAL POINT:
Reducing cancer risk.

A number of behaviors that help protect people from developing cancer are listed below. Check those that apply to you.

	Yes	No	Don't Know
1. I eat five or more servings of vegetables and fruits daily.	_____	_____	_____
2. I consume whole grain products.	_____	_____	_____
3. I eat a high-fat diet.	_____	_____	_____
4. I smoke or chew tobacco.	_____	_____	_____
5. I exercise regularly.	_____	_____	_____
6. I have too much body fat.	_____	_____	_____

FEEDBACK (including scoring) can be found at the end of Unit 22.

REFERENCES

1. Guyer B, et al. Annual summary of vital statistics. Pediatrics 1997;100:905–18.
2. Archer MC. Cancer and diet. In: Ziegler EE, Filer LJ, editors. Present knowledge in nutrition. Washington, DC: ILSI Press; pp 482–7.
3. Lichtenstein P, et al. Environmental and heritable factors in the causation of cancer. N Engl J Med 2000;343:78–85.
4. National Center for Nutrition and Dietetics. Food choices for reducing cancer risk. Nutrition Fact Sheet. Chicago: American Dietetic Association; 1997.
5. Willett W. Diet and cancer. Presentation at the Experimental Biology Annual Meeting, San Diego, 2000 April 17.
6. Food, nutrition and the prevention of cancer: a global perspective. Washington, DC: World Cancer Research Fund, American Institute for Cancer Research; 1997.
7. Schapira DV, et al. Upper-body fat distribution and endometrial cancer risk. JAMA 1991;266:1808–11.
8. Steinmetz KA, Potter JD. Vegetables, fruit, and cancer prevention: a review. J Am Diet Assoc 1996;96:1027–39.
9. Kelloff GJ, et al. Progress in cancer chemoprevention: development of diet-derived chemopreventive agents. J Nutr 2000;130:467S–71S.
10. de Verdier MG, et al. Meat, cooking methods and colorectal cancer. Int'l J Cancer 1991;49:520–5.

Nutrition

UP CLOSE

A Cancer Risk Checkup

Feedback for Unit 22

Best answers:

1. Yes 4. No
2. Yes 5. Yes
3. No 6. No

If you responded "Don't Know" to statements 1, 2, or 3, analyze your diet using the Diet Analysis Plus program). "Regular exercise" in statement 5 means exercising 20 to 30 minutes three times per week. If you didn't know the answer to statement 6, can you pinch an inch of fat above your ribs while you are standing? If yes, you probably have too much body fat.

Unit 23

Good Things to Know about Minerals

Index Stock

Nutrition Scoreboard

	TRUE	FALSE
1. The sole function of minerals is to serve as a component of body structures such as bone, teeth, and hair.		
2. Bones continue to grow and mineralize through the first 30 years of life.		
3. Ounce for ounce, spinach provides more iron than beef does.		
4. Worldwide the most common nutritional deficiency is iron deficiency.		
5. One in four Americans has hypertension.		

(Answers on next page)

Key Concepts and Facts

- Minerals are single atoms that cannot be created or destroyed by the human body or by any other ordinary means.
- Minerals serve as components of body structures and play key roles in the regulation of body processes.
- Deficiency diseases occur when too little of any of the 15 essential minerals is provided to the body, and overdose reactions occur when too much is provided.
- Inadequate intakes of certain minerals are associated with the development of chronic disorders, including osteoporosis, iron deficiency, and hypertension.

Answers to Nutrition Scoreboard

	TRUE	FALSE
1. Minerals serve as structural components of the body, but they also play important roles in stimulating muscle and nerve activity and in other functions.		✓
2. Bones continue to grow and mineralize well after we reach adult height. Bone growth and development can continue to about age 30.	✓	
3. Spinach is a nutritious food, providing 0.6 milligrams iron per ounce. It contains less iron than beef (1 milligram iron per ounce). The iron in spinach is poorly absorbed, but the iron in meat is well absorbed by the body. (When several hundred college students were asked this question, 82% voted for spinach.)[1]		✓
4. Iron deficiency is the most common nutritional deficiency in both developed and developing countries. Approximately one-third of the world's population is iron deficient.[2]	✓	
5. An estimated 24% of Americans have hypertension (high blood pressure).[3]	✓	

Mineral Facts

minerals
In the context of nutrition, minerals are specific, single atoms that perform particular functions in the body. There are 15 essential minerals—or minerals required in the diet.

What are neither animal nor vegetable in origin, cannot be created or destroyed by living organisms (or by any other ordinary means), and consist of the substances from which all things on earth are made? The answer is the ***mineral*** elements, and they are displayed in full in the periodic table presented in Illustration 23.1. Minerals considered "essential," or required in the diet, are highlighted.

The body contains 40 or more minerals. Only 15 are an essential part of our diets; we obtain the others through the air we breathe or from other essential nutrients in the diet such as protein and vitamins.

Minerals are unlike the other essential nutrients in that they consist of single atoms. A single atom of a mineral typically does not have an equal number of protons (particles that carry a positive charge) and electrons (particles that carry a negative charge) and therefore carries a charge. The charge makes minerals reactive. Many of the functions of minerals in the body are related to this property.

Getting a Charge out of Minerals

The charge carried by minerals allows them to combine with other minerals of the opposite charge and form fairly stable compounds that become part of bones, teeth, cartilage, and other tissues. In body fluids, charged minerals serve as a source of electrical power that stimulates muscles to contract and nerves to react. The electrical current generated by charged minerals when performing

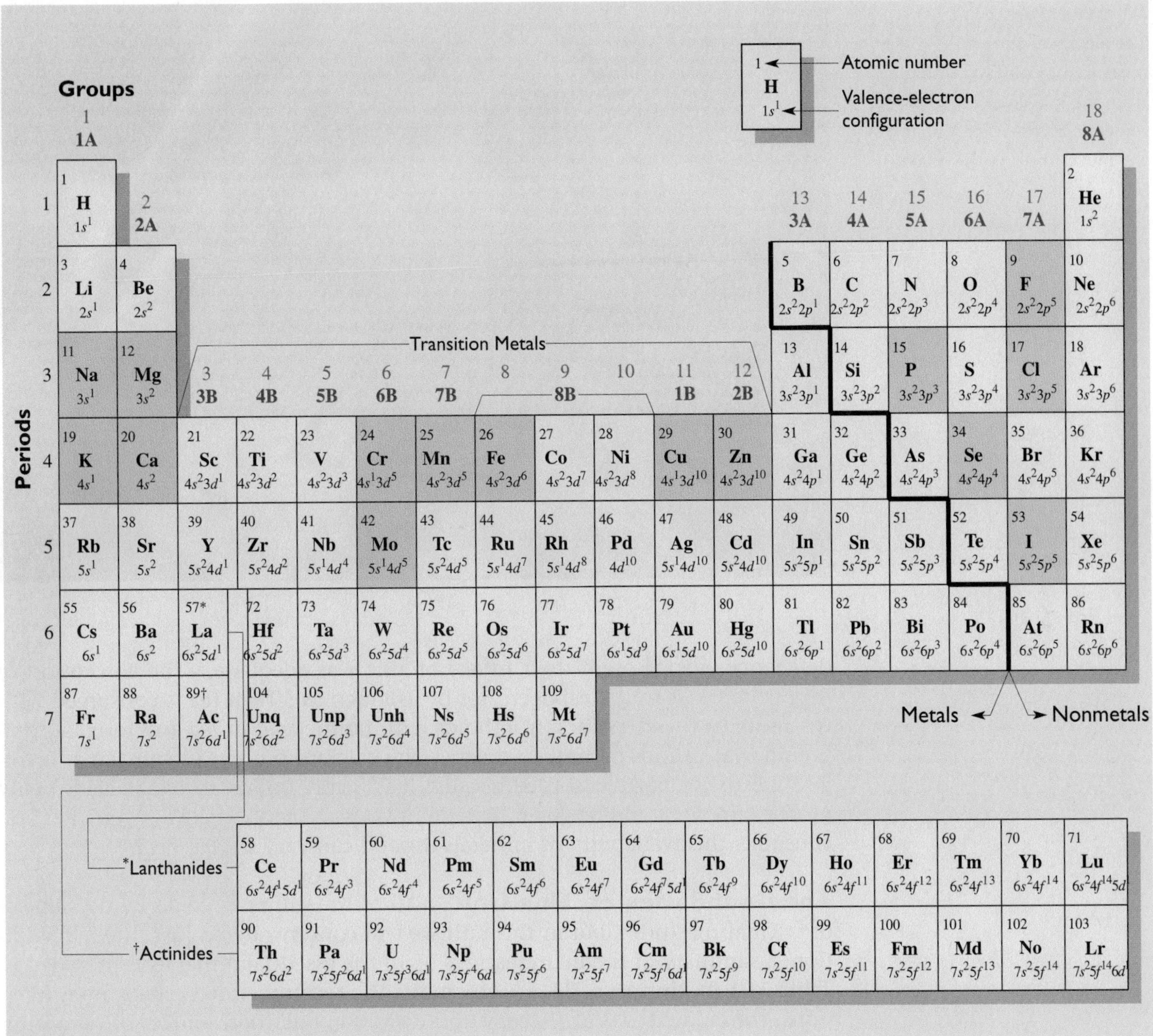

Illustration 23.1
The periodic table lists all known minerals. The highlighted minerals are required in the human diet.

these functions can be recorded by an electrocardiogram (abbreviated EKG or ECG) and an electroencephalogram (EEG). Abnormalities in the pattern of electrical activity in EKGs signal pending or past problems in the heart muscle. An EKG recording is shown in Illustration 23.2. Electroencephalograms similarly record electrical activity in the brain.

The charge minerals carry is related to many other functions. It helps maintain an adequate amount of water in the body and assists in neutralizing body fluids when they become too acidic or basic. Minerals that perform the roles of ***cofactors*** are components of proteins and enzymes, and provide the "spark" that initiates enzyme activity.

cofactors
Individual minerals required for the activity of certain proteins. For example:
- iron is needed for hemoglobin's function in oxygen and carbon dioxide transport,
- zinc is needed to activate or is a structural component of over 200 enzymes, and
- magnesium activates over 300 enzymes involved in the formation of energy and proteins.

Charge Problems Because minerals tend to be reactive, they may combine with other substances in food and form highly stable compounds that are not easily absorbed. Absorption of zinc from foods, for example, can vary from zero to 100%, depending on what is attached to it. Zinc in whole-grain products is very poorly absorbed because it is bound tightly to a substance called phytate. In contrast, zinc in meats is readily available because it is bound to protein. People whose sole source of zinc was whole grains have developed zinc

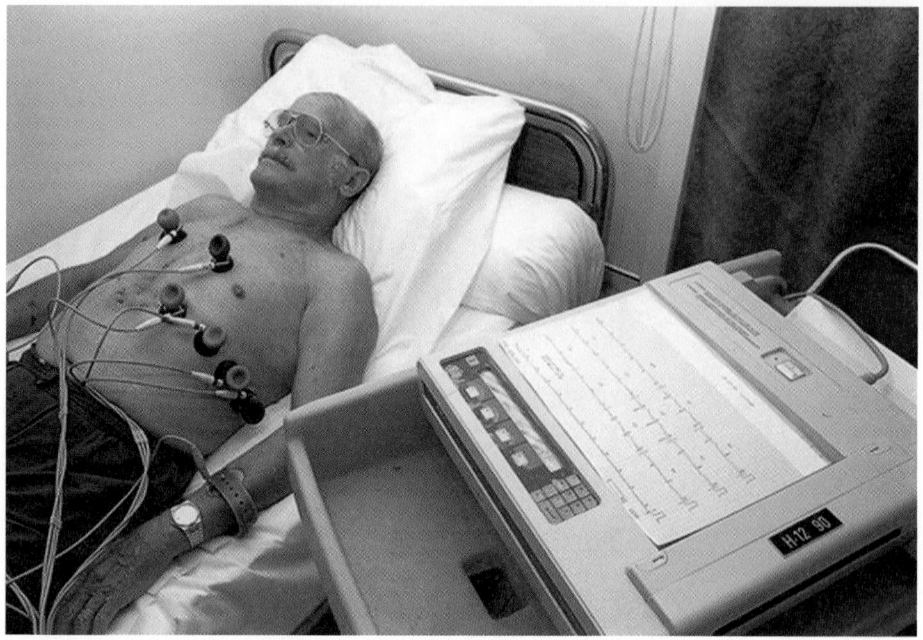

Illustration 23.2
The electrical current measured by an EKG results from the movement of charged minerals across membranes of the muscle cells in the heart.

deficiency even though their intake of zinc was adequate.[4] The absorption of iron from foods in a meal decreases by as much as 50% if tea is consumed with the meal. In the intestines, iron binds with tannic acid in tea and forms a compound that cannot be broken down.[5] The calcium present in spinach and collard greens is poorly absorbed because it is firmly bound to oxalic acid. Many more examples could be given. The point is that you don't always get what you consume; the availability of minerals in food can vary a great deal.

The Boundaries of This Unit All of the minerals could be the subject of fascinating stories, but in this unit we will concentrate on just three. In addition, a summary of the main features of all the essential minerals is provided in Table 23.1 (begins on p. 23–6), which lists the recommended intake level, functions, consequences of deficiency and overdose, and food sources of each mineral. Table 23.12 at the end of the unit lists food sources for many of the essential minerals.

Minerals highlighted are calcium, iron, and sodium. They have been selected primarily because they play important roles in the development of osteoporosis, iron-deficiency anemia, and hypertension, respectively. These disorders are widespread in the United States and many other countries, and improved diets offer a key to their prevention and treatment.

Selected Minerals: Calcium

What you've heard about calcium is true: it's good for bones and teeth. About 99% of the three pounds of calcium in the body is located in bones and teeth. The remaining 1% is found in blood and other body fluids. We don't hear so much about this 1%, but it's very active. Every time a muscle contracts, a nerve sends out a signal, or blood clots to stop a bleeding wound, calcium in body fluids is involved. Calcium's most publicized function, however, is its role in bone formation and the prevention of osteoporosis.

A Short Primer on Bones Most of the bones we see or study are hard and dead. As a result, people often have the impression that bones in living bodies are that way. Nothing could be further from the truth. The 206 bones in our bodies are slightly flexible living tissues infiltrated by blood vessels, nerves, and cells.

The solid parts of bones consist of networks of strong protein fibers (called the "protein matrix") embedded with mineral crystals (Illustration 23.3). Calcium is the most abundant mineral found in bone, but many other minerals such as phosphorus, magnesium, and carbon are also incorporated into the protein matrix. The combination of water, the tough protein matrix, and the mineral crystals makes bone very strong yet slightly flexible and capable of absorbing shocks.

Teeth are a type of bone. They have the same properties as other bone plus a hard outer covering called enamel, which is not infiltrated by blood vessels or nerves. Enamel serves to protect the teeth from destruction by bacteria and mechanical wear and tear.

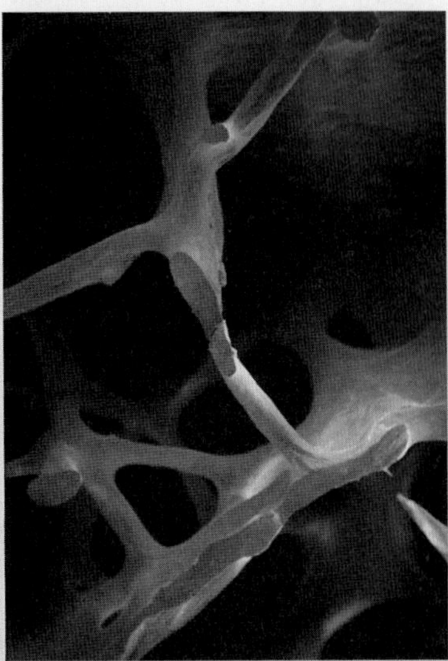

Illustration 23.3
(left) Electron micrograph of healthy bone. (right) Electron micrograph of bone affected by osteoporosis. *Reproduced with permission from Dempster et al, J Bone Min. Res 1, 15–21, 1986*

Remodeling Your Bones Bones slowly and continually go through a repair and replacement process known, appropriately enough, as ***remodeling.*** During remodeling, the old protein matrix is replaced and remineralized. If insufficient calcium is available to complete the remineralization, or if other conditions prevent calcium from being incorporated into the protein matrix, ***osteoporosis*** results.

➡ **remodeling**
The breakdown and buildup of bone tissue.

➡ **osteoporosis**
(*osteo* = bones, *poro* = porous, *osis* = abnormal condition) A condition characterized by porous bones; it is due to the loss of minerals from the bones.

Osteoporosis If you are female, you have a one in four chance of developing osteoporosis in your lifetime. If you are a Caucasian or Asian female, you have a higher risk of developing osteoporosis than if you are an African American woman or a male. If you are male, your chance of developing osteoporosis is one in eight. Approximately 24 million adults in the United States have osteoporosis, and one million have suffered a broken wrist or hip or crushed spinal vertebrae because of the disease.[6]

Osteoporosis is a disabling disease that reduces the quality of life and dramatically increases the need for health care.[6] Because the incidence of osteoporosis increases with age, its importance as a personal and public health problem is intensifying as the U.S. population ages. It currently appears, however, that a large percentage of the cases of osteoporosis can be prevented. The key to prevention is to build dense bones during childhood and the early adult years and then keep bones dense as you age.[6]

The Timing of Bone Formation Bones develop and mineralize throughout the first three decades of life.[7] Even after the growth spurt occurs during adolescence and people think they are as tall as they will ever be, bones continue to increase in width and mineral content for 10 to 15 more years. Peak bone density, or the maximal level of mineral content in bones, is reached somewhere after the age of 30 years. After that, bone mineral content no longer increases. The higher the peak bone mass, the less likely it is that osteoporosis will develop. People with higher peak bone mass simply have more calcium to lose before bones become weak and fracture easily. That also is the reason males

Table 23.1
An intensive course on essential minerals.

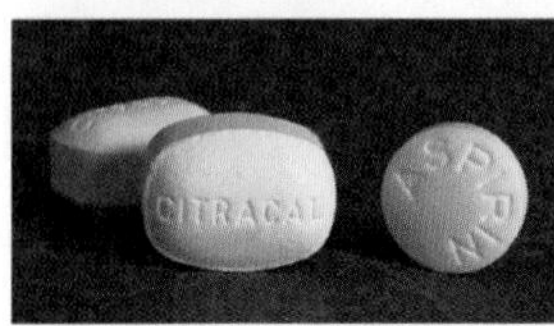

Calcium supplements are large because our daily need for calcium is high (1000 milligrams/day). Four pills provide 800 milligrams calcium, the amount in two and two-third cups of milk. An aspirin is shown for comparison.

	Primary Functions	Consequences of Deficiency
Calcium AI* women: 1000 mg men: 1000 mg UL: 2500 mg	· Component of bones and teeth · Needed for muscle and nerve activity, blood clotting	· Poorly mineralized, weak bones (osteoporosis) · Rickets in children · Osteomalacia (rickets in adults) · Stunted growth in children · Convulsions, muscle spasms
Phosphorus RDA women: 700 mg men: 700 mg UL: 4000 mg	· Component of bones and teeth · Component of certain enzymes and other substances involved in energy formation · Needed to maintain the right acid-base balance of body fluids	· Loss of appetite · Nausea, vomiting · Weakness · Confusion · Loss of calcium from bones
Magnesium RDA women: 310 mg men: 400 mg UL: 350 mg (from supplements only)	· Component of bones and teeth · Needed for nerve activity · Activates enzymes involved in energy and protein formation	· Stunted growth in children · Weakness · Muscle spasms · Personality changes
Iron RDA women: 18 mg men: 8 mg UL: 45 mg	· Transports oxygen as a component of hemoglobin in red blood cells · Component of myoglobin (a muscle protein) · Needed for certain reactions involving energy formation	· Iron deficiency · Iron-deficiency anemia · Weakness, fatigue · Pale appearance · Reduced attention span and resistance to infection · Mental retardation, developmental delay in children
Zinc RDA women: 8 mg men: 11 mg UL: 40 mg	· Required for the activation of many enzymes involved in the reproduction of proteins · Component of insulin, many enzymes	· Growth failure · Delayed sexual maturation · Slow wound healing · Loss of taste and appetite · In pregnancy, low-birth-weight infants and preterm delivery

*AIs and RDA are for women and men 19–30 years of age, ULs are for males and females 19 to 70 years of age, 1997–2001.

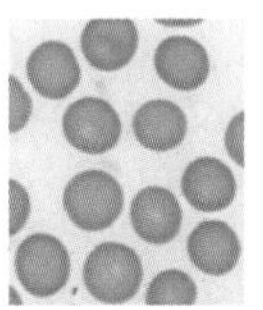 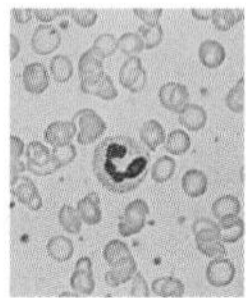

Iron-deficiency anemia is characterized by small, pale red blood cells (right photo). Normal red blood cells are shown on the left.
CNRI - Martin M. Rotker

Table 23.1
An intensive course on essential minerals, continued.

Consequences of Overdose	Primary Food Sources	Highlights and Comments
· Drowsiness · Calcium deposits in kidneys, liver, and other tissues · Suppression of bone remodeling · Decreased zinc absorption	· Milk and milk products (cheese, yogurt) · Broccoli · Dried beans · Calcium-fortified foods (some juices, breakfast cereals, bread, for example)	· The average intake of calcium among U.S. women is approximately 60% of the DRI. · One in four women and one in eight men in the U.S. develop osteoporosis. · Adequate calcium and vitamin D status must be maintained to prevent bone loss.
· Muscle spasms	· Milk and milk products (cheese, yogurt) · Meats · Seeds, nuts · Phosphates added to foods	· Deficiency is generally related to disease processes.
· Diarrhea · Dehydration · Impaired nerve activity due to disrupted utilization of calcium	· Plant foods (dried beans, tofu, peanuts, potatoes, green vegetables) · Milk · Bread · Ready-to-eat cereals · Coffee	· Magnesium is primarily found in plant foods where it is attached to chlorophyll. · Average intake among U.S. adults is below the RDA.
· Hemochromatosis ("iron poisoning") · Vomiting, abdominal pain · Blue coloration of skin · Liver and heart damage, diabetes · Decreased zinc absorption · Atherosclerosis (plaque buildup) in older adults	· Liver, beef, pork · Dried beans · Iron-fortified cereals · Prunes, apricots, raisins · Spinach · Bread · Pasta	· Cooking foods in iron and stainless steel pans increases the iron content of the foods. · Vitamin C, meat, and alcohol increase iron absorption. · Iron deficiency is the most common nutritional deficiency in the world. · Average iron intake of young children and women in the U.S. is low.
· Over 25 mg/day is associated with nausea, vomiting, weakness, fatigue, susceptibility to infection, copper deficiency, and metallic taste in mouth. · Increased blood lipids	· Meats (all kinds) · Grains · Nuts · Milk and milk products (cheese, yogurt) · Ready-to-eat cereals · Bread	· Like iron, zinc is better absorbed from meats than from plants. · Marginal zinc deficiency may be common, especially in children. · Zinc supplements may decrease duration and severity of the common cold.

continued

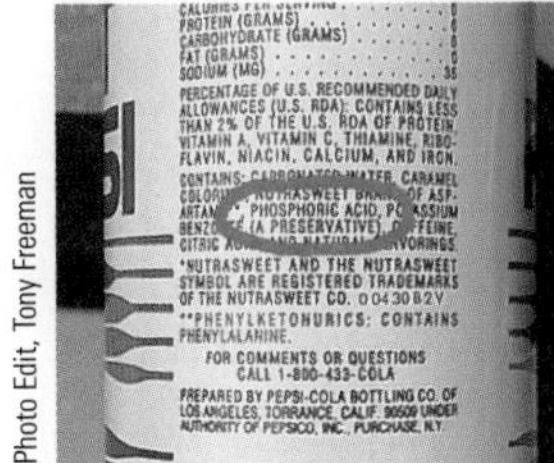
Photo Edit, Tony Freeman

Phosphates are a common, multipurpose food additive.

Photo Disc

Magnesium is to chlorophyll in plants as iron is to hemoglobin in humans.

Table 23.1
An intensive course on essential minerals, continued

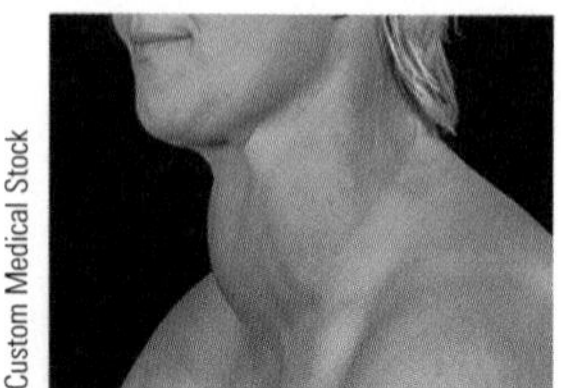
Custom Medical Stock

Goiter is a highly visible sign of iodine deficiency.

	Primary Functions	Consequences of Deficiency
Fluoride AI women: 3 mg men: 4 mg UL: 10 mg	· Component of bones and teeth (enamel)	· Tooth decay and other dental diseases
Iodine RDA women: 150 mcg men: 150 mcg UL: 1100 mcg	· Component of thyroid hormones that help regulate energy production and growth	· Goiter · Cretinism (mental retardation, hearing loss, growth failure)
Selenium RDA women: 55 mcg men: 55 mcg UL: 400 mcg	· Acts as an antioxidant in conjunction with vitamin E (protects cells from damage due to exposure to oxygen) · Needed for thyroid hormone production	· Anemia · Muscle pain and tenderness · Keshan disease (heart failure), Kashin-Beck disease (joint disease)
Copper RDA women: 900 mcg men: 900 mcg UL: 10,000 mcg	· Component of enzymes involved in the body's utilization of iron and oxygen · Functions in growth, immunity, cholesterol and glucose utilization, brain development	· Anemia · Seizures · Nerve and bone abnormalities in children · Growth retardation
Manganese AI women: 2.3 mg men: 1.8 mg	· Needed for the formation of body fat and bone	· Weight loss · Rash · Nausea and vomiting
Chromium AI women: 35 mcg men: 25 mcg	· Required for the normal utilization of glucose and fat	· Elevated blood glucose and triglyceride levels · Weight loss

Table 23.1
An intensive course on essential minerals, continued

Consequences of Overdose	Primary Food Sources	Highlights and Comments
· Fluorosis · Brittle bones · Mottled teeth · Nerve abnormalities	· Fluoridated water and foods and beverages made with it · Tea · Shrimp, crab	· Toothpastes, mouth rinses, and other dental care products may provide fluoride. · Fluoride overdose has been caused by ingestion of fluoridated toothpaste.
· Over 1 mg/day may produce pimples, goiter, and decreased thyroid function.	· Iodized salt · Milk and milk products · Seaweed, seafoods · Bread from commercial bakeries	· Iodine deficiency was a major problem in the U.S. in the 1920s and 1930s. Now intakes are marginally excessive. Deficiency remains a major health problem in some developing countries. · Amount of iodine in plants depends on iodine content of soil. · Most of the iodine in our diet comes from the incidental addition of iodine to foods from cleaning compounds used by food manufacturers.
· "Selenosis." Symptoms of selenosis are hair and fingernail loss, weakness, liver damage, irritability, and "garlic" or "metallic" breath.	· Meats and seafoods · Eggs · Whole grains	· Content of foods depends on amount of selenium in soil, water, and animal feeds. · May play a role in the prevention of some types of cancer.
· Wilson's disease (excessive accumulation of copper in the liver and kidneys) · Vomiting, diarrhea · Tremors · Liver disease	· Bread · Potatoes · Grains · Dried beans · Nuts and seeds · Seafood · Ready-to-eat cereals	· Toxicity can result from copper pipes and cooking pans. · Average intake in the U.S. is below the RDA.
· Infertility in men · Disruptions in the nervous system (psychotic symptoms) · Muscle spasms	· Whole grains · Coffee, tea · Dried beans · Nuts	· Toxicity is related to overexposure to manganese dust in miners.
· Kidney and skin damage	· Whole grains · Wheat germ · Liver, meat · Beer, wine · Oysters	· Toxicity usually results from exposure in chrome-making industries or overuse of supplements. · Supplements do not build muscle mass or increase endurance.

continued

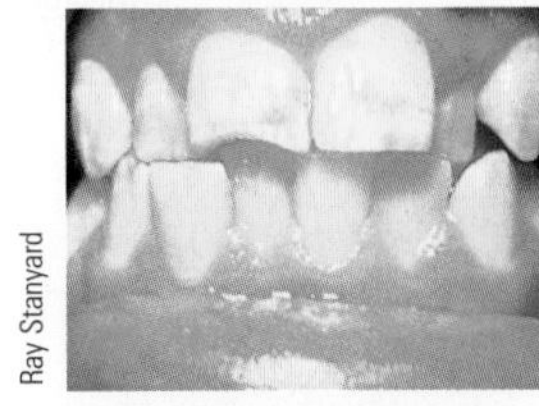
Ray Stanyard

"Mottled teeth" result from an excessive intake of fluoride in children.

Ray Stanyard

Iodine deficiency during pregnancy produces cretinism in the offspring.

Table 23.1
An intensive course on essential minerals, continued

	Primary Functions	Consequences of Deficiency
Molybdenum RDA women: 45 mcg men: 45 mcg UL: 2000 mcg	· Component of enzymes involved in the transfer of oxygen from one molecule to another	· Rapid heartbeat and breathing · Nausea, vomiting · Coma
Sodium No DRI. Recommended intake: 2400 mg	· Needed to maintain the right acid-base balance in body fluids · Helps maintain an appropriate amount of water in blood and body tissues · Needed for muscle and nerve activity	· Weakness · Apathy · Poor appetite · Muscle cramps · Headache · Swelling
Potassium No DRI	· Same as for sodium	· Weakness · Irritability, mental confusion · Irregular heartbeat · Paralysis
Chloride No DRI	· Component of hydrochloric acid secreted by the stomach (used in digestion) · Needed to maintain the right acid-base balance of body fluids · Helps maintain an appropriate water balance in the body	· Muscle cramps · Apathy · Poor appetite · Long-term mental retardation in infants

PhotoEdit - Robert Brenner

Illustration 23.4
This woman's stooped appearance is due to osteoporosis.

experience osteoporosis less often than females do: they have more bone mass to lose.

Bone size and density often remain fairly stable from age 30 to the mid-40s, but then bones tend to demineralize with increasing age. By the time women are 70, for example, their bones are 30 to 40% less dense on average than they once were.[8] A woman may lose an inch or more in height with age and develop the "dowager's hump" that is characteristic of osteoporosis in the spine (Illustration 23.4).

What has just been described is the way things are in the United States, but this situation is neither normal nor inevitable. The incidence of osteoporosis is far less in many other countries than in the United States.[9] Although genetic and other factors play a role, the causes of osteoporosis are mainly related to diet and other lifestyle behaviors. That means that osteoporosis can largely be prevented.

How Do You Build and Maintain Dense Bones? Since the late 1980s, a rash of studies have examined the relationships among dietary calcium intake, vitamin D status, and bone density. Results have been quite consistent:

Consequences of Overdose	Primary Food Sources	Highlights and Comments
· Loss of copper from the body · Joint pain · Growth failure · Anemia · Gout	· Dried beans · Grains · Dark green vegetables · Liver · Milk and milk products	· Deficiency is extraordinarily rare.
· High blood pressure in susceptible people · Kidney disease · Heart problems	· Foods processed with salt · Cured foods (corned beef, ham, bacon, pickles, sauerkraut) · Table and sea salt · Bread · Milk, cheese · Salad dressing	· Very few foods naturally contain much sodium. · Processed foods are the leading source of dietary sodium. · High-sodium diets are associated with the development of hypertension in "salt-sensitive" people.
· Irregular heartbeat, heart attack	· Plant foods (potatoes, squash, lima beans, tomatoes, plantains, bananas, oranges, avocados) · Meats · Milk and milk products · Coffee	· Content of vegetables is often reduced in processed foods. · Diuretics (water pills) and other antihypertension drugs may deplete potassium. · Salt substitutes often contain potassium.
· Vomiting	· Same as for sodium. (Most of the chloride in our diets comes from salt.)	· Excessive vomiting and diarrhea may cause chloride deficiency. · Legislation regulating the composition of infant formulas was enacted in response to formula-related chloride deficiency and subsequent mental retardation in infants.

Table 23.1
An intensive course on essential minerals, continued

bone mass before the age of 30 is increased by calcium intakes of approximately 1200 milligrams per day along with adequate vitamin D status.[10,11] In adults over the age of 50 years, bone density tends to be preserved with calcium intakes around 1200 to 1500 milligrams per day and 800 IU of vitamin D.[6] In addition, calcium intakes of 1200 milligrams or more per day with adequate vitamin D status decrease the incidence of stress fractures in athletes.[12] Low bone mineral density can be increased somewhat by switching from a low to an adequate intake of calcium and improved vitamin D status. The earlier in life this occurs, the more improvement in bone density is noted.[13] High levels of bone density once achieved do not last forever; bone must be continually renewed by an adequate supply of calcium and vitamin D.[12]

Vitamin D is important for building dense bones because it increases calcium absorption and the deposition of calcium into bone. Our need for vitamin D can be met from vitamin D–fortified milk, vitamin D–fortified breakfast cereals, and seafoods or by exposing the skin to the sun. A precursor of vitamin D lies beneath our skin. When skin is exposed to ultraviolet rays from sunlight, this precursor is converted to vitamin D. Exposing the hands and face to sunshine for 5 to 15 minutes produces a day's supply of vitamin D. (Windows and sunscreen block ultraviolet rays and prevent the production of vitamin D in the skin.) The Health Action summarizes behaviors that help build dense bones and prevent osteoporosis.

Women Consume Too Little Calcium Unfortunately, about half of U.S. women belong to the "600 Club," or consume 600 milligrams of calcium per day or less.[14] Only 10% of girls and 25% of boys aged 9 to 17 years consume the recommended amount of calcium during these critical bone density building years.[15] Men tend to consume more calcium than women, but still fall short of the recommended intake on average.[16]

Why Do Women Consume Too Little Calcium? Women consume too little calcium for several reasons. Although dairy products such as milk, cheese, and yogurt are our richest sources of calcium, some women consider them "fattening" and give them up during the teen years. Skipping breakfast and eating out are common among teens and tend to reduce calcium intake.[17] Many think that milk is a food that only children need, while others stop drinking milk because it "gives them gas." (Such people may be lactose intolerant.)[17]

Increasing intakes of soft drinks, combined with reduced consumption of milk, also appear to be contributing to bone fractures and future osteoporosis in today's youth.[18] The wide availability of soft drink machines in schools (Illustration 23.5) is thought to be a factor related to increased consumption by teens. According to some college students, one reason their intake of calcium is low is that they frequently eat at fast-food restaurants that feature soft drinks rather than milk on the menu. But when asked if they would choose specials that included fresh-tasting milk (not the kind that tastes like the milk container) as part of the meal, some students reported they would select the specials (Illustration 23.6).[19]

What Else Influences Osteoporosis? Table 23.2 (p. 23–14) summarizes the risk factors identified for osteoporosis. If you were to find all of the various risk factors combined in one person, that individual would be a thin woman with light skin who has consumed too little calcium and vitamin D and

Illustration 23.5
The wide availability of soft drink vending machines in some high schools is thought to contribute to excessive consumption of these beverages at the expense of milk. This concern has prompted some soft drink manufacturers and school districts to add milk, fruit juice, and water to vending machines.

HEALTH ACTION

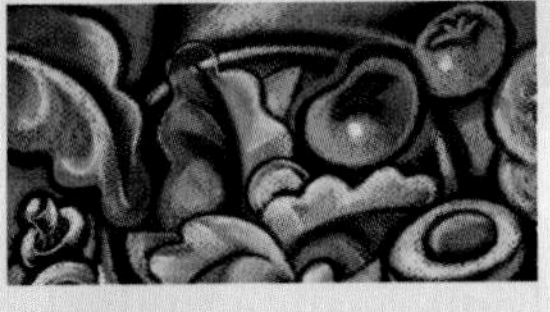

Behaviors That Help Prevent Osteoporosis

- Consuming enough dietary calcium (1000 to 1200 milligrams daily; see Table 23.4 on p. 23–15)
- Getting regular physical activity
- Not smoking
- Consuming adequate amounts of Vitamin D (400 IU) and/or exposing the hands and face to sunlight for 5 to 15 minutes daily

is physically inactive, an excessive alcohol drinker, a smoker, and genetically "small-boned." Additionally, she would have had her ovaries surgically removed for medical reasons before the age of 45.[6] Few people meet every aspect of that description, but people who have several of these characteristics are more likely to develop osteoporosis than those who don't.

At one time, researchers thought that a relatively high intake of phosphorus or protein might intensify the effects of low intakes of calcium on the development of osteoporosis. Studies in humans, however, have failed to demonstrate harmful effects of high phosphorus or protein intake on the development of osteoporosis.[20]

Can Osteoporosis Be Treated? Calcium supplements or adequate calcium intakes (around 1000–1500 milligrams per day), vitamin D supplements (800 IU), and weight-bearing exercise like walking and tennis decrease the progression of osteoporosis in many people.[6] Table 23.3 (p. 23–14) summarizes current recommendations for the treatment of osteoporosis.

Calcium: Where to Find It About three-fourths of the calcium supplied by the diets of Americans comes from milk and milk products. Milk (including chocolate milk), cheese, and yogurt are all good sources of calcium (see Table 23.12). Some plants such as kale, broccoli, and bok choy provide appreciable amounts of calcium, too. On average, 32% of the calcium content of milk and milk products is absorbed, compared to approximately 5 to 60% of the calcium from different plants (Table 23.4 on p. 23–15).[20] The best way to meet the recommended level of intake (1000 milligrams per day for most adults) is to choose rich sources of absorbable calcium.

Illustration 23.6
Would you be more likely to drink milk at fast-food restaurants if it was included as part of a bargain special? Some college women think this is a good idea—one they would go for.

Table 23.2
Risk factors for osteoporosis.[6]

- Female
- Menopause
- Deficient calcium intake
- Caucasian or Asian heritage
- Thinness ("small bones")
- Cigarette smoking
- Excessive alcohol intake
- Ovarectomy (ovaries removed) before age 45
- Physical inactivity
- Deficient vitamin D status
- Genetic factors

Many foods rich in calcium aren't loaded with calories or fat. As Table 23.4 shows, low-fat yogurt, skim and soy milk, kale, and broccoli, for example, provide good amounts of calcium at a low cost in calories.

Calcium is increasingly appearing in unexpected places such as in candy, snack bars, waffles, and bread (Illustration 23.7). The array of calcium-fortified foods is increasing dramatically in the United States in response to publicity about our need for more of it. Although development of such products is making it easier to consume more calcium, the availability of calcium in many of these foods has not been established.[21]

Watch for Calcium on the Label Health claims relating to the benefits of calcium in reducing the risk of osteoporosis may appear on the labels of food products that qualify as good sources of calcium such as milk, yogurt, and calcium-fortified foods such as orange juice, breakfast cereals, and grain products (Illustration 23.8 on p. 23–16).

Can You Consume Too Much Calcium? Yes, indeed, you can consume too much calcium. Supplemental calcium in doses exceeding 2.5 grams per day may produce drowsiness, lead to constipation, and cause calcium to deposit in tissues such as the liver and kidneys.[6]

Selected Minerals: Iron

➡ hemoglobin
The iron-containing protein in red blood cells.

➡ myoglobin
The iron-containing protein in muscle cells.

Most of the body's iron supply is found in ***hemoglobin.*** Small amounts are present in ***myoglobin,*** and free iron is involved in processes that capture energy released during the breakdown of proteins and fats.

The Role of Iron in Hemoglobin and Myoglobin What happens to a car when its paint gets scratched? After a while, the exposed metal rusts. The iron in the metal combines with oxygen in the air, and the result is iron oxide or "rust." Iron readily combines with oxygen, and that property is put to good use in the body. From its location in hemoglobin in red blood cells, iron loosely attaches to oxygen when blood passes near the inner surface of the lungs. The bright red, oxygenated blood is then delivered to cells throughout the body. When the oxygenated blood passes near cells that need oxygen for energy formation or other reasons, oxygen is released from the iron and diffuses into cells (Illustration 23.9). The free iron in hemoglobin then picks up carbon dioxide, a waste product of energy formation. When carbon dioxide attaches to iron, blood turns from bright red to dark bluish red. Blood then circulates back to the lungs, where carbon dioxide is released from the iron and exhaled into the air. The free iron attaches again to oxygen that enters the lungs, and the cycle continues.

Iron in myoglobin traps oxygen delivered by hemoglobin, stores it, and releases it as needed for energy formation for muscle activity. In effect, myoglobin boosts the supply of oxygen available to muscles.

The functions of iron just described operate smoothly when the body's supply of iron is sufficient. Unfortunately, that is often not the case.

Table 23.3
Approaches to the treatment of osteoporosis.[6]

- Daily intake of 1000–1500 milligrams of calcium
- Regular physical activity
- Adequate vitamin D intake (800 IU daily)
- Medication

➡ iron deficiency
A disorder that results from a depletion of iron stores in the body. It is characterized by weakness, fatigue, short attention span, poor appetite, increased susceptibility to infection, and irritability.

Iron Deficiency Is a Big Problem The most widespread nutritional deficiency in both developing and developed countries is ***iron deficiency*** (Table 23.5 on p. 23–17).[2] It is estimated that one out of every three people in the world is iron deficient.[2] For the most part, iron deficiency affects children and women of childbearing age who have a high need for iron and frequently consume too little of it (Table 23.6 on p. 23–17). Iron deficiency may develop in people who have lost blood due to injury, surgery, or ulcers. Donating blood more than three times a year can also precipitate iron deficiency.[22]

Table 23.4
Caloric content, calcium level, and percentage of available calcium from different foods.

Food	Amount	Calories	Calcium (mg)	Calcium Absorbed (%)	Available Calcium (mg)
Yogurt, low fat	1 cup	143	413	32	132
Skim milk	1 cup	85	301	32	96
Soy milk (fortified)	1 cup	79	300	31	93
1% milk	1 cup	163	274	32	88
Tofu	1 cup	188	260	31	81
Cheese	1 ounce	114	204	32	65
Kale, cooked	1 cup	42	94	49	46
Broccoli, cooked	1 cup	44	72	61	44
Bok choy, raw	1 cup	9	73	54	39
Dried beans	1 cup	209	120	24	29
Spinach, cooked	1 cup	42	244	5	12

Consequences of Iron Deficiency Many body processes sputter without sufficient oxygen. People with iron deficiency usually feel weak and tired. They have a shortened attention span and a poor appetite, are susceptible to infection, and become irritable easily.[22] If the deficiency is serious enough, ***iron-deficiency anemia*** develops and additional symptoms occur. People with iron-deficiency anemia look pale, are easily exhausted, and have rapid heart rates.[22] Iron-deficiency anemia is a particular problem for infants and young children as it is related to lasting retardation in mental development.[23]

iron-deficiency anemia A condition that results when the content of hemoglobin in red blood cells is reduced due to a lack of iron. It is characterized by the signs of iron deficiency plus paleness, exhaustion, and a rapid heart rate.

Getting Enough Iron in Your Diet "Enough" iron according to the RDAs is 8 milligrams for men and 18 milligrams per day for women aged 19 to 50 years. Consuming that much iron can be difficult for women. On average, 1000 calories' worth of food provides about 6 milligrams of iron. Women would have to consume around 2500 calories per day to obtain even 15 milligrams of iron on average. The caloric intake of most women in the United States is well below 2500.[16]

Iron is found in small amounts in many foods, but only a few foods such as liver, beef, and prune juice are rich sources. (A list of food sources of iron is given in Table 23.12.) Foods cooked in iron and stainless steel pans, however, can be a significant source of iron. Some of the iron in the pan leaches out during cooking. On average, approximately one milligram of iron is added to each three-ounce serving of food cooked in these pans.[24]

Most of the iron in plants and eggs is tightly bound to substances such as phytates or oxalic acid that limit iron absorption, making these foods relatively poor sources of iron even though they contain a fair amount of it. (Differences in the proportions of iron absorbed from various food sources are shown in Illustration 23.10.) A three-ounce hamburger and a cup of asparagus both contain approximately three milligrams of iron, for example. But 20 times more iron can be absorbed from the hamburger than from the asparagus.[25] Absorption of iron from plants can be increased if foods containing vitamin C are included in the same meal.

Illustration 23.7
The number of calcium-fortified foods is increasing.

"Regular exercise and a healthy diet with enough calcium help maintain good bone health and may reduce the risk of osteoporosis later in life."

Nutrition Facts

Serving Size: 1 cup (240ml)
Servings per Container: 16

Amount per Serving	
Calories 110	Calories from Fat 20
	% Daily Value*
Total Fat 2.5g	**4%**
Saturated Fat 1.5g	**8%**
Cholesterol 15g	**4%**
Sodium 135mg	**6%**
Total Carbohydrate 13g	**4%**
Dietary Fiber 0g	**0%**
Sugars 12g	
Protein 8g	

Vitamin A 10% • Vitamin C 4%
Calcium 30% Iron 0% Vitamin D 25%
Phosphorus 10%

*Percent Daily Values are based on a 2,000 calorie diet.

Illustration 23.8
Food products that are good sources of calcium can be labeled with a health claim.

Iron absorption is also increased by low levels of iron stores in the body (Illustration 23.11 on p. 23–18). In other words, if you're in need of iron, your body sets off mechanisms that allow more of it to be absorbed from foods or supplements. When iron stores are high, less iron is absorbed.

The body's ability to regulate iron absorption provides considerable protection against iron deficiency and overdose. The protection is not complete, however, as evidenced by widespread iron deficiency and the occurrence of iron toxicity.

Overdosing on Iron Excess iron absorbed into the body cannot be easily excreted. Consequently, it is deposited in various tissues such as the liver, pancreas, and heart. There, the iron reacts with cells, causing damage that can result in liver disease, diabetes, and heart failure. One in 200 people in the United States has an inherited tendency to absorb too much iron (a disorder called hemochromatosis). Other people develop iron toxicity from consuming large amounts of iron with alcohol (alcohol increases iron absorption) or from very high iron intakes, usually due to overdoses of iron supplements.

Each year in the United States, more than 10,000 people accidentally overdose on iron supplements.[27] Commonly, victims are young children who mistakenly think iron pills are candy (Table 23.7 and Illustration 23.12 on p. 23–18). The lethal dose of iron for a two-year-old child is about three grams,[28] the amount of iron present in 25 pills containing 120 milligrams of iron each. Iron supplements that are not being used should be thrown away or stored in a place where toddlers cannot get to them.

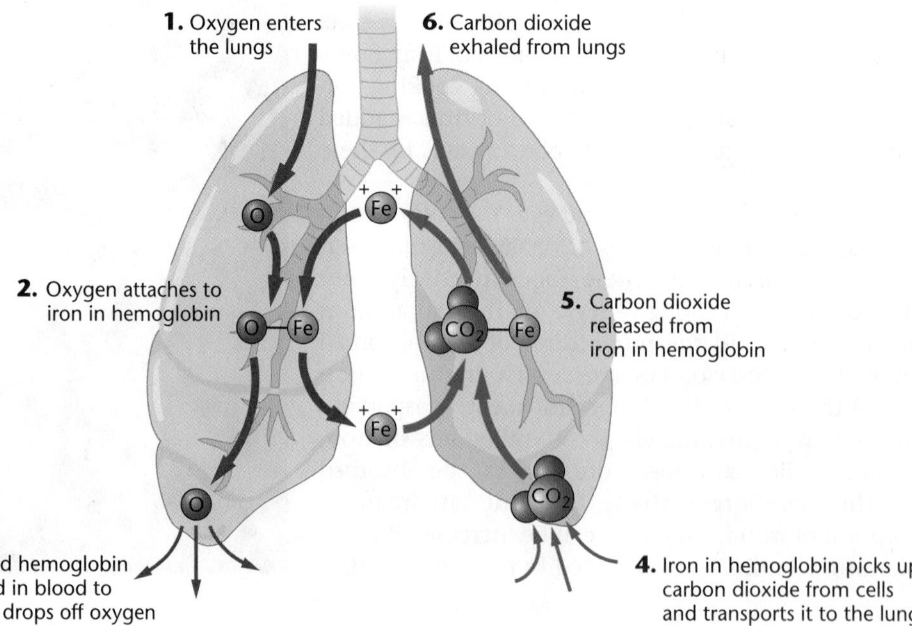

Illustration 23.9
Iron's role in carrying oxygen to cells and carbon dioxide away from them by way of the bloodstream. Fe^{++} represents charged particles of iron. The charge is neutralized when iron combines with oxygen or carbon dioxide.

Table 23.5
Incidence of iron deficiency.[2,9]

	Population with Iron Deficiency (%)
Worldwide (children under 5 years):	
Developing countries	51
Developed countries	12
United States:	
Children, 1 to 2 years	9
Pregnant women	12
Females, 16 to 19 years	11
Females, 20 to 49 years	11
Males, 12 to 49 years	1 or less

Selected Minerals: Sodium

On the one hand, sodium is a gift from the sea (Illustration 23.13); on the other, it is a hazard to health. Sodium's life- and health-sustaining functions are frequently overshadowed by its effects on blood pressure when consumed in excess.

An extremely reactive mineral, sodium occurs in nature combined with other elements. The most common chemical partner of sodium is chloride, and much of the sodium present on this planet is in the form of sodium chloride—table salt. Table salt is 40% sodium by weight; one teaspoon of salt contains about 2300 milligrams of sodium.

Although it is found in abundance in the oceans, humans have not always had access to enough salt to satisfy their taste for it. During times when salt has

Table 23.6
Iron intake of children, women, and men in the United States.[2]

	Percentage of Recommended Intake
Children	67%
Women	41
Men	84

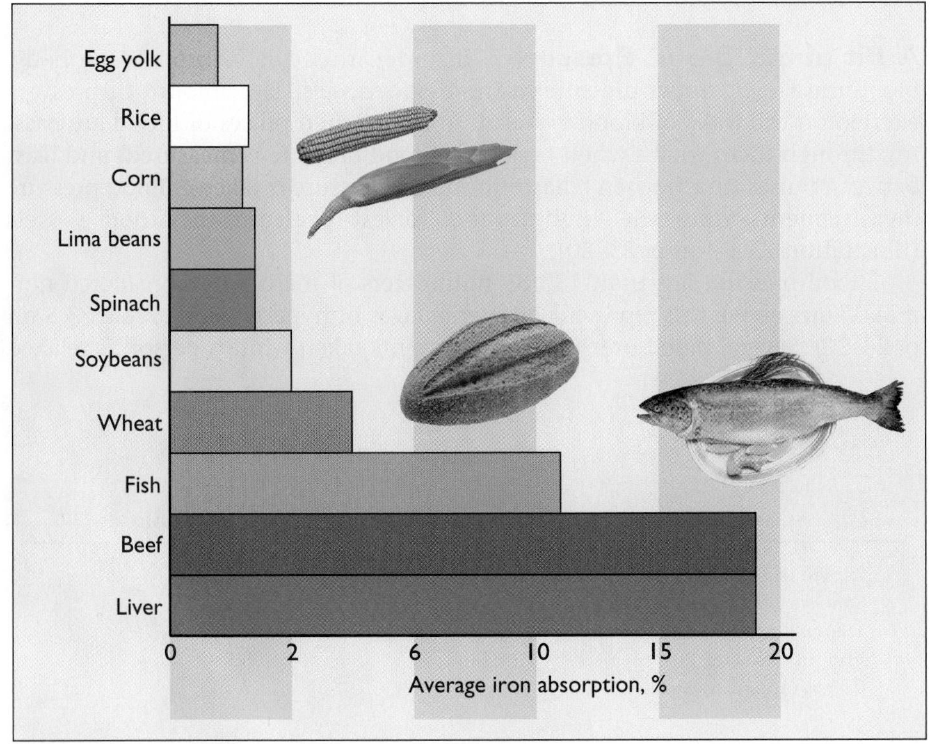

Illustration 23.10
Average percentage of iron absorbed from selected foods by healthy adults.[25]

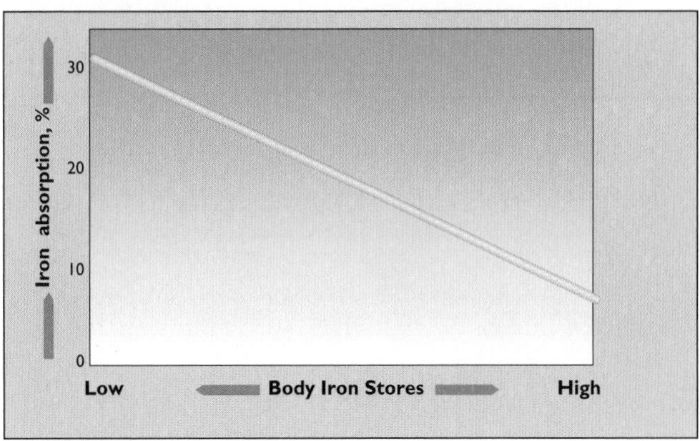

Illustration 23.11
People absorb more iron from foods and supplements when body stores of iron are low than when stores of iron are high.

been scarce, wars have been fought over it. In such periods, a pocketful of salt was as good as a pocketful of cash. (The word *salary* is derived from the Latin word *salarium,* "salt money.") People were said to be "worth their salt" if they put in a full day's work. Today, salt is widely available in most parts of the world, and the problem is limiting human intakes to levels that do not interfere with health.

What Does Sodium Do in the Body?

Sodium appears directly above potassium in the periodic table, and the two work closely together in the body to maintain normal ***water balance.*** Both sodium and potassium chemically attract water, and under normal circumstances, each draws sufficient water to the outside or inside of cells to maintain an optimal level of water in both places.

water balance
The ratio of the amount of water outside cells to the amount inside cells; this balance is needed for normal cell functioning.

Water balance and cell functions are upset when there's an imbalance in the body's supplies of sodium and potassium. You have probably noticed that you become thirsty when you eat a large amount of salted potato chips or popcorn. Salty foods make you thirsty because your body loses water when the high load of dietary sodium is excreted. The thirst signal indicates that you need water to replace what you have lost.

The loss of body water that accompanies ingestion of large amounts of salt explains why seawater neither quenches thirst nor satisfies the body's need for water. When a person drinks seawater, its high concentration of sodium causes the body to excrete more water than it retains. Rather than increasing the body's supply of water, the ingestion of seawater increases the need for water.

In healthy people, the body's adaptive mechanisms provide a buffer against upsets in water balance due to high sodium intakes. It appears that many people are overwhelming the body's ability to cope with high sodium loads, however. High dietary intakes of sodium appear to play an important role in the development of ***hypertension*** in many people.[29]

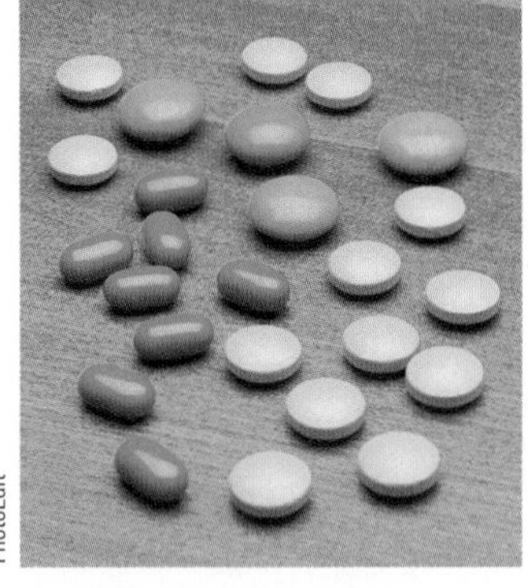
PhotoEdit

Illustration 23.12
If you were three years old, could you tell which "pills" are candy and which are the iron supplements?* Overdoses of iron supplements are a leading cause of accidental poisoning in young children.

**The iron supplements are the lightest green in color.*

A Bit about Blood Pressure In order to circulate through the body, blood must exist under pressure in the blood vessels. The amount of pressure exerted on the walls of blood vessels is greatest when pulses of blood are passing through them (that's when "systolic" blood pressure is measured) and least between pulses (that's when "diastolic" blood pressure is taken). Blood pressure measurements note the highest and lowest pressure in blood vessels (Illustration 23.14 on p. 23–20).

Blood pressure less than 130/85 millimeters of mercury is considered normal. Values above this represent different stages of hypertension (Table 23.8 on p. 23–20). Several blood pressure measurements taken while a person is relaxed

Table 23.7
Accidental overdoses of chemical substances by children up to age four in the United States.[26]

Substance	Cases per 100,000 Children
Aspirin and substitutes	12.7
Solvents/petroleum products	11.2
Tranquilizers	9.9
Iron supplements	8.7
Corrosives and caustics	7.4

are needed to obtain an accurate result. (Even going to a clinic or doctor's office to have blood pressure measured can raise it for some people!)

Hypertension is a major public health problem in the United States and other countries. Although not considered a disease by itself, the presence of hypertension substantially increases the risk that a person will develop heart disease or kidney failure or will experience a heart attack or stroke.[29] Hypertension occurs in 24% of U.S. adults[3] and in 10 to 20% of adults worldwide.[30] It is more likely to develop as people become older. About 40% of all cases of hypertension are classified as mild.[9]

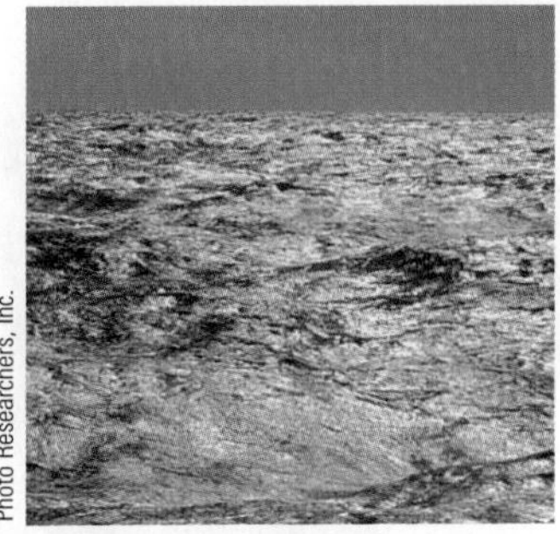
Photo Researchers, Inc.

Illustration 23.13
The oceans contain enough salt to cover the surface of the earth to a depth of more than 400 feet.

What Causes Hypertension? Only 5% of all cases of hypertension can be directly linked to a cause. People who have hypertension with no identifiable cause (95% of all cases) are said to have ***essential hypertension.***

A number of risk factors for hypertension have been identified, and dietary factors are among the most important (Table 23.9 on p. 23–20). Foremost among the evidence that links diet to hypertension are population studies that show a relationship between average salt intake and hypertension. As average salt intake rises, so do rates of hypertension.[32] When people with hypertension reduce their salt intake, their blood pressure tends to drop.[33] Further scrutiny of the data on salt and blood pressure, however, reveals that not everyone is equally susceptible to high-salt diets. Some people are genetically susceptible to salt, and their blood pressure rises in response to high-sodium diets. This condition is known as ***salt sensitivity.***

➡ **hypertension**
High blood pressure. It is usually defined as blood pressure that typically exceeds 140/90 millimeters of mercury.

➡ **essential hypertension**
Hypertension of no known cause; also called *primary* or *idiopathic hypertension,* it accounts for 95% of all cases of hypertension.

➡ **salt sensitivity**
A genetically influenced condition in which a person's blood pressure rises when high amounts of salt or sodium are consumed. Such individuals are sometimes identified by blood pressure increases of 10% or more when switched from a low-salt to a high-salt diet.

Salt Sensitivity Approximately 51% of people with hypertension and 26% of people with normal blood pressure are salt sensitive.[34] Reduction in salt intake by people who are salt sensitive, along with weight loss if overweight, substantially improves blood pressure in most cases.[35] Since it is currently difficult to identify who in the population is salt sensitive and who is not, because most Americans consume much more sodium than needed, and since high-sodium foods are often high in fat, official advice is that Americans limit their sodium intake to 2400 milligrams per day (the equivalent of approximately one level teaspoon of salt from all sources).[36]

Other Conditions Related to Hypertension Obesity is a major risk factor for hypertension. For obese people with hypertension, the most effective treatment is weight loss.[37] Excessive alcohol intake can prompt the development of hypertension, and moderation of intake (to two or fewer alcoholic drinks per day) can bring blood pressure back down.[9] Physically inactive lifestyles also appear to play a role in the development of hypertension.[38]

A number of other dietary factors have been reported to contribute to the development of hypertension. These include low intakes of calcium, vegetables and fruits, and potassium.[38] Diets that limit fat intake and are high in vegetables and fruits have been found to lower blood pressure to the same extent as an often-prescribed antihypertension drug.[39]

The primary example of the type of diet associated with substantial improvement in blood pressure in many people, the "DASH" diet, is presented in greater detail in Illustration 23.15 (p. 23–21).

What Is the Best Way to Prevent Hypertension? The Health Action on the next page highlights recommendations for decreasing the chances that hypertension will develop. These recommendations contribute to an overall healthy lifestyle whose benefits extend beyond the prevention of hypertension.

How Is Hypertension Treated? Once hypertension is diagnosed, there are several approaches to its treatment (Table 23.10 on p. 23–22). The preferred

HEALTH ACTION

Recommendations for Preventing Hypertension*

- Consume a diet rich in vegetables and fruits.
- Use low-fat dairy products.
- Reduce saturated and total fat intake.
- Reduce sodium intake.
- Limit alcohol intake.
- Exercise regularly.
- Lose weight if overweight.

*From the National Heart, Lung, and Blood Institute, 1997.

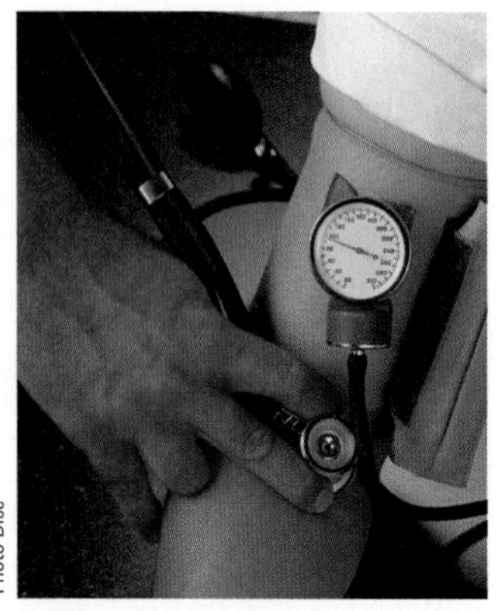

Photo Disc

Illustration 23.14
A blood pressure test.
Blood pressure is expressed as two numbers:

110 — systolic pressure measures the force of the blood when the heart contracts
70 — diastolic pressure measures the force of the blood when the heart is at rest

approach for cases of mild hypertension is weight loss (if needed), a low-sodium diet, regular exercise, and moderate alcohol consumption (if any).[35] These same recommendations apply for people with moderate or severe hypertension, but they may require antihypertension drugs as well. People with mild or moderate hypertension are urged to try first to reduce blood pressure by improving behaviors. If blood pressure remains too high, antihypertension drugs, which can have a number of side effects, are then prescribed along with needed changes in weight and diet.[37,38]

Cutting Back on Salt The leading sources of salt (and therefore of sodium) in the U.S. diet are processed foods (Table 23.11 on p. 23-22). Restricting the use of foods to which salt has been added during processing is the most effective way to lower salt intake and achieve a moderate- to low-salt diet.[32] High-salt processed foods include salad dressings, pickles, canned soups, corned beef, sausages and other luncheon meats, and snack foods such as potato chips and cheese twists.

Only a small proportion of our total sodium intake enters our diet from fresh foods. Very few foods naturally contain much sodium—at least not until they are processed (Illustration 23.16 on p. 23-22)).

Using spices and lemon juice to flavor foods, consuming fresh vegetables (rather than processed ones) and fresh fruits, and checking out the sodium content of foods and then selecting low-sodium ones can also help reduce sodium intake.

Table 23.9
Risk factors for hypertension.[31]

- Age
- Family history
- High-sodium diet
- Obesity
- Physical inactivity
- Excessive alcohol consumption

Table 23.8
Hypertension categories.

Category	Systolic (mm Hg)[a]	Diastolic (mm Hg)
Normal	<130	<85
High normal	130–139	85–89
Hypertension[b]		
Stage 1 (mild)	140–159	90–99
Stage 2-3 (moderate to severe)	160+	100+

[a]mm Hg = millimeters of mercury.
[b]Based on readings taken at two or more visits following an initial screening.
Source: NIH, NHLB Institute, 1997.

Illustration 23.15

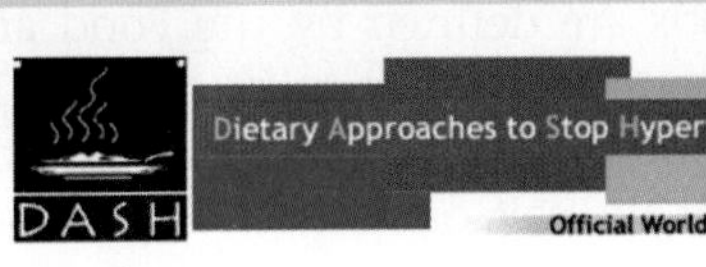

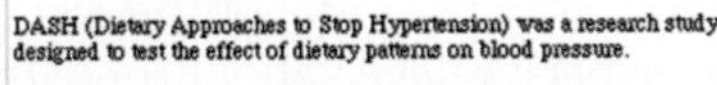

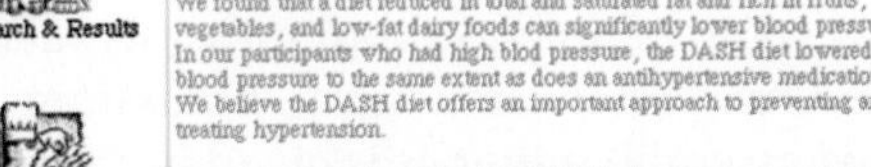

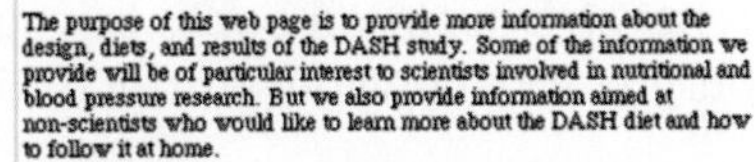

The DASH diet

In the mid-1990s a revolutionary approach to the control of mild and moderate hypertension was tested, and the results have changed health professionals' thinking about high blood pressure prevention and management. Called the DASH diet (Dietary Approaches to Stop Hypertension), it didn't focus on salt restriction and was related to significant reductions in blood pressure within two weeks in a majority of people tested. In some people, reductions in blood pressure were sufficient to erase the need for antihypertension medications, and for others the diet reduced the amount or variety of medications needed.[39] A subsequent study showed that a low-sodium diet boosts the blood pressure lowering effects of the DASH diet, especially in African Americans.[40]

The DASH diet consists of eating patterns made up of the following food groups:

	Daily Servings
Vegetables	4–5
Fruits	4–5
Grain products	7–8
Low-fat milk and dairy products	2–3
Meats, fish, poultry	2 or less
Nuts, seeds, dried beans	about 1

Although it does not work for all individuals with hypertension, the DASH diet is quickly becoming a mainstay in clinical efforts to control hypertension. Benefits of the diet last as long as the diet does.[41] A very useful website is available that describes refinements in the diet based on research studies, and that gives practical tips on how to follow the diet.

DASH diet sample meals, snacks, and complete guidelines are available on the website: dash.bwh.harvard.edu

Table 23.10
Approaches to the treatment of hypertension.[37]

- Weight loss (if needed)
- Moderate salt intake (2 grams of sodium per day or less)
- Moderate alcohol consumption (if any)
- Regular physical activity
- Increase vegetable and fruit intake
- Decrease fat intake
- Antihypertension drugs (if hypertension not controlled by above measures)

Label Watch Not all processed foods with added sodium taste salty. To find out which processed foods are high in sodium, you have to examine the label. Increasingly, low-salt processed foods are entering the market and can be easily identified by the "low-salt" message on the label (Illustration 23.17). Terms used to identify low-salt (or low-sodium) foods are defined by the Food and Drug Administration. To be considered "low-sodium," foods must contain 140 milligrams or less of sodium per serving. Food manufacturers must adhere to the definitions when they make claims about the salt or sodium content of a food on the label.

This unit ends having presented the story behind 3 of the 15 essential minerals. Much more information could have been relayed about these three, not to mention the wealth of knowledge that could be shared about minerals such as phosphorus, magnesium, zinc, iodine, selenium, and potassium. Critical information about these and other essential minerals is given in Table 23.1, and Table 23.12 lists food sources of many of the essential minerals. For additional information, look at the comprehensive nutrition books in your college library or take notes on the information presented by your instructor in class.

Table 23.11
Major sources of sodium in the U.S. diet.[42]

Sodium Source	Contribution to Sodium Intake (%)
Processed foods	77
Fresh foods	12
Salt added at the table	6
Salt added during cooking	5

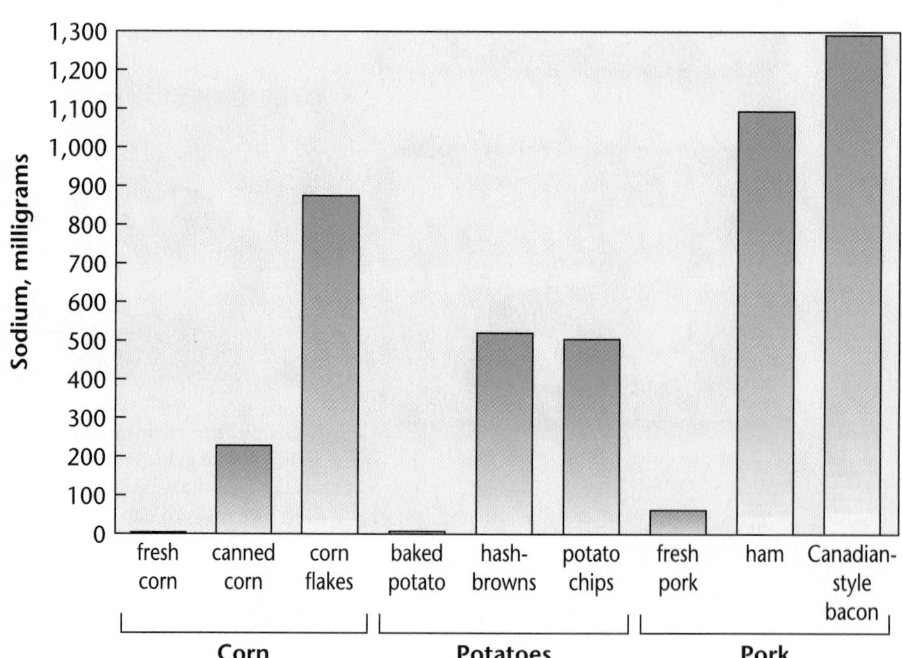

Illustration 23.16
Examples of how processing increases the sodium content of foods. Sodium values are for a 3-ounce serving of each food shown.

Term used on label	Definition
LOW SODIUM	Product must contain 140 milligrams or less of sodium per serving.
Reduced Sodium	Product must contain 75% less sodium than the customary amount (comparison data must be given on the label).
Sodium FREE	Product must contain less than 5 milligrams sodium per serving.
LIGHTLY SALTED	Product must contain at least 25% less sodium than the customary amount (comparison data must be given on the label).
UNSALTED No Salt Added	Salt must not have been added to foods customarily salted such as canned green beans, popcorn, and peanuts.

Illustration 23.17
When a label makes a claim about the sodium content of a food product, the label must list the amount of sodium in a serving and adhere to these definitions.

Table 23.12
Food sources of minerals, continues on next page.

Food	Amount	Magnesium (mg)	Food	Amount	Magnesium (mg)
Legumes:			**Vegetables:**		
Lentils, cooked	½ cup	134	Bean sprouts	½ cup	98
Split peas, cooked	½ cup	134	Black-eyed peas	½ cup	58
Tofu	½ cup	130	Spinach, cooked	½ cup	48
			Lima beans	½ cup	32
Nuts:			**Milk and milk products:**		
Peanuts	¼ cup	247	Milk	1 cup	30
Cashews	¼ cup	93	Cheddar cheese	1 oz	8
Almonds	¼ cup	80	American cheese	1 oz	6
Grains:			**Meats:**		
Bran buds	1 cup	240			
Wild rice, cooked	½ cup	119	Chicken	3 oz	25
Breakfast cereal, fortified	1 cup	85	Beef	3 oz	20
Wheat germ	2 tbs	45	Pork	3 oz	20

continued

Calcium*

Food	Amount	Calcium (mg)
Milk and milk products:		
Yogurt, low-fat	1 cup	413
Milk shake (low-fat frozen yogurt)	1¼ cup	352
Yogurt with fruit, low-fat	1 cup	315
Skim milk	1 cup	301
1% milk	1 cup	300
2% milk	1 cup	298
3.25% milk (whole)	1 cup	288
Swiss cheese	1 oz	270
Milk shake (whole milk)	1¼ cup	250
Frozen yogurt, low-fat	1 cup	248
Frappuccino	1 cup	220
Cheddar cheese	1 oz	204
Frozen yogurt	1 cup	200
Cream soup	1 cup	186
Pudding	½ cup	185
Ice cream	1 cup	180
Ice milk	1 cup	180
American cheese	1 oz	175
Custard	½ cup	150
Cottage cheese	½ cup	70
Cottage cheese, low-fat	½ cup	69
Vegetables:		
Spinach, cooked	½ cup	122
Kale	½ cup	47
Broccoli	½ cup	36
Legumes:		
Tofu	½ cup	260
Dried beans, cooked	½ cup	60
Foods fortified with calcium:		
Orange juice	1 cup	350
Frozen waffles	2	300
Soy milk	1 cup	200–400
Breakfast cereals	1 cup	150–1000

*Actually, the richest source of calcium is alligator meat; 3½ ounces contain about 1231 milligrams of calcium; but just try to find it on your grocer's shelf!

Selenium

Food	Amount	Selenium (mcg)
Seafood:		
Lobster	3 oz	66
Tuna	3 oz	60
Shrimp	3 oz	54
Oysters	3 oz	48
Fish	3 oz	40
Meats:		
Liver	3 oz	56
Ham	3 oz	29
Beef	3 oz	22
Bacon	3 oz	21
Chicken	3 oz	18
Lamb	3 oz	14
Veal	3 oz	10
Eggs:		
Egg	1 medium	37

Sodium*

Food	Amount	Sodium (mg)
Miscellaneous:		
Salt	1 tsp	2132
Dill pickle	1 (4½ oz)	1930
Sea salt	1 tsp	1716
Chicken broth	1 cup	1571
Ravioli, canned	1 cup	1065
Spaghetti with sauce, canned	1 cup	955
Baking soda	1 tsp	821
Beef broth	1 cup	782
Gravy	¼ cup	720
Italian dressing	2 tbs	720
Pretzels	5 (1 oz)	500
Green olives	5	465
Pizza with cheese	1 wedge	455
Soy sauce	1 tsp	444
Cheese twists	1 cup	329
Bacon	3 slices	303
French dressing	2 tbs	220
Potato chips	1 oz (10 pieces)	200
Catsup	1 tbs	155
Meats:		
Corned beef	3 oz	808
Ham	3 oz	800
Fish, canned	3 oz	735
Meat loaf	3 oz	555
Sausage	3 oz	483
Hot dog	1	477
Fish, smoked	3 oz	444
Bologna	1 oz	370
Milk and milk products:		
Cream soup	1 cup	1070
Cottage cheese	½ cup	455
American cheese	1 oz	405
Cheese spread	1 oz	274
Parmesan cheese	1 oz	247
Gouda cheese	1 oz	232
Cheddar cheese	1 oz	175
Skim milk	1 cup	125
Whole milk	1 cup	120
Grains:		
Bran flakes	1 cup	363
Cornflakes	1 cup	325
Croissant	1 medium	270
Bagel	1	260
English muffin	1	203
White bread	1 slice	130
Whole-wheat bread	1 slice	130
Saltine crackers	4 squares	125

Iron			Zinc		
Food	**Amount**	**Iron (mg)**	**Food**	**Amount**	**Zinc (mg)**
Meat and meat alternates:			Meats:		
Liver	3 oz	7.5	Liver	3 oz	4.6
Round steak	3 oz	3.0	Beef	3 oz	4.0
Hamburger, lean	3 oz	3.0	Crab	½ cup	3.5
Baked beans	½ cup	3.0	Lamb	3 oz	3.5
Pork	3 oz	2.7	Turkey ham	3 oz	2.5
White beans	½ cup	2.7	Pork	3 oz	2.4
Soybeans	½ cup	2.5	Chicken	3 oz	2.0
Pork and beans	½ cup	2.3	**Legumes:**		
Fish	3 oz	1.0	Dried beans, cooked	½ cup	1.0
Chicken	3 oz	1.0	Split peas, cooked	½ cup	0.9
Grains:			**Grains:**		
Breakfast cereal, iron-fortified	1 cup	8.0 (4–18)	Breakfast cereal, fortified	1 cup	1.5–4.0
Oatmeal, fortified, cooked	1 cup	8.0	Wheat germ	2 tbs	2.4
Bagel	1	1.7	Oatmeal, cooked	1 cup	1.2
English muffin	1	1.6	Bran flakes	1 cup	1.0
Rye bread	1 slice	1.0	Brown rice, cooked	½ cup	0.6
Whole-wheat bread	1 slice	0.8	White rice	½ cup	0.4
White bread	1 slice	0.6	**Nuts and seeds:**		
Fruits:			Pecans	¼ cup	2.0
Prune juice	1 cup	9.0	Cashews	¼ cup	1.8
Apricots, dried	½ cup	2.5	Sunflower seeds	¼ cup	1.7
Prunes	5 medium	2.0	Peanut butter	2 tbs	0.9
Raisins	¼ cup	1.3	**Milk and milk products:**		
Plums	3 medium	1.1	Cheddar cheese	1 oz	1.1
Vegetables:			Whole milk	1 cup	0.9
Spinach, cooked	½ cup	2.3	American cheese	1 oz	0.8
Lima beans	½ cup	2.2			
Black-eyed peas	½ cup	1.7			
Peas	½ cup	1.6			
Asparagus	½ cup	1.5			

Phosphorus					
Food	**Amount**	**Phosphorus (mg)**	**Food**	**Amount**	**Phosphorus (mg)**
Milk and milk products:			**Grains:**		
Yogurt	1 cup	327	Bran flakes	1 cup	180
Skim milk	1 cup	250	Shredded wheat	2 large biscuits	81
Whole milk	1 cup	250	Whole-wheat bread	1 slice	52
Cottage cheese	½ cup	150	Noodles, cooked	½ cup	47
American cheese	1 oz	130	Rice, cooked	½ cup	29
Meats:			White bread	1 slice	24
Pork	3 oz	275	**Vegetables:**		
Hamburger	3 oz	165	Potato	1 medium	101
Tuna	3 oz	162	Corn	½ cup	73
Lobster	3 oz	125	Peas	½ cup	70
Chicken	3 oz	120	French fries	½ cup	61
Nuts and seeds:			Broccoli	½ cup	54
Sunflower seeds	¼ cup	319	**Other:**		
Peanuts	¼ cup	141	Milk chocolate	1 oz	66
Pine nuts	¼ cup	106	Cola	12 oz	51
Peanut butter	1 tbs	61	Diet cola	12 oz	45

continued

Food	Amount	Potassium (mg)	Food	Amount	Potassium (mg)
Vegetables:			Hamburger	3 oz	480
Potato	1 medium	780	Lamb	3 oz	382
Winter squash	½ cup	327	Pork	3 oz	335
Tomato	1 medium	300	Chicken	3 oz	208
Celery	1 stalk	270	**Grains:**		
Carrots	1 medium	245	Bran buds	1 cup	1080
Broccoli	½ cup	205	Bran flakes	1 cup	248
Fruits:			Raisin bran	1 cup	242
Avocado	½ medium	680	Wheat flakes	1 cup	96
Orange juice	1 cup	469	**Milk and milk products:**		
Banana	1 medium	440	Yogurt	1 cup	531
Raisins	¼ cup	370	Skim milk	1 cup	400
Prunes	4 large	300	Whole milk	1 cup	370
Watermelon	1 cup	158	**Other:**		
Meats:			Salt substitutes	1 tsp	1300–2378
Fish	3 oz	500			

Nutrition UP CLOSE

The Salt of Your Diet

FOCAL POINT: Estimate your sodium intake.

Salt added to foods during processing, during home cooking, and at the table is the major source of sodium in the American diet. To determine if you consume too much sodium, answer the following questions.

How Often Do You Usually:	Daily	4–6 Times per Week	1–3 Times per Week	Less than Weekly
1. Eat processed foods such as luncheon meats, bacon, hot dogs, sausage, canned soups, broths, gravy, TV dinners, or smoked fish?	☐	☐	☐	☐
2. Eat pickles, green olives, potato or tortilla chips, pretzels, or salted crackers?	☐	☐	☐	☐
3. Use soy sauce, garlic salt, steak sauce, or catsup in recipes or on foods?	☐	☐	☐	☐
4. Salt foods at the table?	☐	☐	☐	☐
5. Add salt to cooking water or to foods you are preparing?	☐	☐	☐	☐

FEEDBACK (including scoring) can be found at the end of Unit 23.

Source: Adapted from Limit use of sodium. Washington, DC: USDA Home and Garden Publication

WWW links

www.nhlbi.nih.gov/nhlbi
NIH's website on hypertension includes current research and recommendations on prevention and treatment.

dash.bwh.harvard.edu
Visit this site for detailed information on the DASH diet for lowering blood pressure.

www.bloodpressure.com
Provides weekly research updates on hypertension.

www.dairycouncil.ofca.org
The National Dairy Council provides information on calcium and osteoporosis.

www.nof.org
Web address for the National Osteoporosis Foundation

www.navigator.tufts.edu
An excellent search engine for nutrition and hypertension, anemia, and osteoporosis.

REFERENCES

1. Mitchell SJ. Changes after taking a college basic nutrition course. J Am Diet Assoc 1990;90:955–61.
2. Parvanta I. The hidden hunger: micronutrient malnutrition. Centers for Disease Control, www.cdc.gov/epo/mmwr/preview/mmwrhtml/00051880.htm, March 2000.
3. Health, United States 2000 www.cdc.gov/nchs, Nov 2000.
4. Prasad AS. Discovery and importance of zinc in human nutrition. Federation Proc 1984;43:2829–34.
5. Monsen ER, Balintfy JL. Calculating dietary iron bioavailability: refinement and computerization. J Am Diet Assoc 1982;80:307–11.
6. Iqbal MM. Osteoporosis: epidemiology, diagnosis, and treatment. South Med J 2000;93:2–18.
7. Guthrie JR, et al. Risk factors for osteoporosis: a review. Medscape Women's Health 2000;5:1–6, www.medscape.com/Medscape/WomensHealth/journal/2000/v05.n04/wh7276.guth/wh7276.guth-01.htm, Nov 2000.
8. Maximizing peak bone mass: calcium supplementation increases bone mineral density in children. Nutr Rev 1992;50:335–7.
9. Committee on Diet and Health (Food and Nutrition Board; National Research Council). Diet and health. Implications for reducing chronic disease risk. Washington, DC: National Academy Press; 1989.
10. Heaney RP. Bone mass, nutrition, and other lifestyle factors. Nutr Rev 1996;54:S3–S10.
11. Holick MF. Vitamin D and bone health. J Am Coll Nutr (abstract) 1997;16:496.
12. Harris S. Calcium trials. Experimental Biology Annual Meeting, San Diego, Calif. 2000 Apr 4.
13. Heaney RP, et al. Dietary changes favorably affect bone remodeling in older women. J Am Diet Assoc 1999:99:1228–33.
14. Calcium Summit 2000, Executive Summary. National Dairy Council, 2000.
15. NIH Consensus Panel on Osteoporosis prevention, diagnosis, and therapy, www.nih.gov/news/pr/mar2000/od-29.htm, March 2000.
16. What we eat in America. Nutr Today 1997;Jan/Feb:37–40.
17. Bone builders. FDA Consumer 1997 Sept/Oct.
18. Wyshak G. Consumption of carbonated drinks by teenage girls associated with bone fractures. Arch Pediatr Adol Med 2000;154:542–3,610–13.
19. Lewis NM, Hollingsworth M. Food choices of young college women consuming low- or moderate-calcium diets. Nutr Res 1992;12:843–8.
20. Heaney R. Dietary protein and phosphorus intake do not affect calcium absorption. Am J Clin Nutr 2000;72:758–61.
21. Klausner A. EN's guide to calcium in unexpected places. Envir Nutr 1999;May:5.

22. Fairbanks VF. Iron in medicine and nutrition. In: Modern nutrition in health and disease, Shils ME et al. eds. 9th ed. Philadelphia: Lippincott Williams & Wilkins, 1999: pp 193–221.
23. Hurtado EK, et al. Early childhood anemia and mild or moderate mental retardation. Am J Clin Nutr 1999; 69:115–9.
24. Park J, Brittin HC. Increased iron content of food due to stainless steel cookware. J Am Diet Assoc 1997;97:659–61.
25. Monsen ER, et al. Estimation of available dietary iron. Am J Clin Nutr 1978;31:134–41.
26. Trinkoff AM, Baker SP. Poisoning hospitalizations and deaths from solids and liquids among children and teenagers. Am J Public Health 1986;76:657–60.
27. Shannon M. Ingestion of toxic substances by children. N Engl J Med 2000;342:186–89.
28. National Research Council Subcommittee on Iron (Committee on Medical and Biological Effects of Environmental Pollutants). Iron. Washington, DC: Division of Medical Sciences, Assembly of Life Sciences, National Academy of Sciences; 1979.
29. Messerli RE, et al. Salt: a perpetrator of hypertensive target organ disease? Arch Int Med 1997;157:2449–52.
30. Nissinen A, Stanley K. Unbalanced diets as a cause of chronic diseases. Am J Clin Nutr 1989;49:993–8.
31. Wong ZY, et al. Genetic linkage of beta and gamma subunits of epithelial sodium channel to systolic blood pressure. Lancet 1999;353:1222–5.
32. Law MR, et al. By how much does dietary salt reduction lower blood pressure? I. Analysis of observational data among populations. Br Med J 1991;302:811–15.
33. Korhonen MH, et al. Effects of a salt restricted diet on the intake of other nutrients. Am J Clin Nutr 2000;72:414–20.
34. Kaplan NM. The dietary guideline for sodium: should we shake it up? No. Am J Clin Nutr 2000;71:1020–6.
35. NHLBI Statement on sodium intake and high blood pressure. National Heart, Lung, and Blood Institute, NIH, www.nhlbi.nih.gov/nhlbi, Aug 1998.
36. Salt review. Nutr Today 1999;34:1–2.
37. The Trials of Hypertension Prevention Collaborative Research Group. The effects of nonpharmacologic intervention on blood pressure of persons with high normal levels. Results of the trials of hypertension prevention, phase 1. JAMA 1992;267:1213–20.
38. Caulin-Glaser T. Primary prevention of hypertension in women. J Clin Hypertens 2000;2:204–9,214.
39. Conlin PR, et al. Dash diet can control stage 1 hypertension. Am J Hypertens 2000;13:949–55.
40. Sacks FM, et al. Effects on blood pressure of reduced dietary sodium and the Dietary Approaches to Stop Hypertension (DASH) diet. N Engl J Med 2001;344:3–10.
41. Karanja NM, et al. Descriptive characteristics of the dietary patterns used in the Dietary Approaches to Stop Hypertension trial. J Am Diet Assoc 1999;99(suppl):S19–S27.
42. Mattes RD, Donnelley D. Relative contribution of dietary sodium sources. J Am Coll Nutr 1991;10:383–93.

Nutrition

UP CLOSE

The Salt of Your Diet

Feedback for Unit 23

People in the United States generally consume two to three times more sodium than they need, and you may be doing so too, if you answered "daily" or "4–6 times per week" to several of the questions. If you wish to cut back on sodium, here are some suggestions:

- Choose fresh instead of processed foods more often.
- Read labels to check for sodium content.
- Reduce salt in cooking and at the table.
- When you do choose processed foods, try the lower-sodium versions available.
- To enhance the flavor of foods, try herbs and spices instead of salt.

Unit 24

Herbal and Other Dietary Supplements

Nutrition Scoreboard

	TRUE	FALSE
1. Products classified as "dietary supplements" consist of herbs and vitamins and mineral supplements.		
2. Dietary supplements must be tested for safety and effectiveness before they can be sold.		
3. Herbal remedies have been used for over a hundred years in Germany so they must be safe and effective.		

(Answers on next page)

Answers to

Nutrition Scoreboard

	TRUE	FALSE
1. Dietary supplements include herbs, vitamin and mineral supplements, protein powders, amino acid and enzyme pills, fish oils and fatty acids, hormone extracts, and other products.		✔
2. Dietary supplements can be sold without proof of their safety or effectiveness.		✔
3. Results of clinical trials, and not historical use, are the gold standard for determining the safety and effectiveness of herbal remedies and other dietary supplements.		✔

Key Concepts and Facts

- → Dietary supplements include vitamin and mineral pills, herbal remedies, proteins and amino acids, fish oils, and other products.
- → Dietary supplements do not have to be shown to be safe or effective prior to being sold.
- → Although food is the preferred source of vitamins and minerals, certain people benefit from judiciously selected vitamin or mineral supplements.

What do vitamin E supplements, amino acid pills, and herbal remedies all have in common? They are members of the increasingly popular group of products called ***dietary supplements***—and they are all discussed in this unit! Types of dietary supplements currently on sale are presented in Table 24.1, and this list will grow longer with time. Dietary supplements are supposed to "supplement the diet." In truth, however, many such products are intended to prevent or treat health problems rather than improve the quality of one's dietary intake. These disparate products are grouped together because they are regulated by a common set of rules.

You will learn from this unit that taking many of the dietary supplements available on the market is similar to gambling. Little is known about the usefulness and safety of many dietary supplements, and claims for them range from the realistic to the absurd. For the most part, it is not clear which supplements benefit, and which harm health. This "buyer beware" situation exists because of the loose rules that govern dietary supplements.

Photo Disc

Regulation of Dietary Supplements

> You can call anything a dietary supplement, even something you grow in your back yard.
>
> —Donna Porter, RD, PhD, Congressional Research Service

Dietary supplements are loosely regulated by the Food and Drug Administration (FDA); they do not have to be tested prior to marketing or shown to be safe or effective.[1] Although often advertised to relieve certain ailments, they are not considered to be drugs. Consequently, dietary supplements are not subjected to vigorous testing to prove safety and effectiveness as drugs must be. Supplements are considered safe until demonstrated to be hazardous by the FDA. Since few

Type	Examples
1. Vitamins and minerals	Vitamins C and E, selenium
2. Herbs (botanicals)	Dong quai, ginseng, saw palmetto
3. Proteins and amino acids	Shark cartilage, chondroitin, creatine
4. Hormones, hormone precursors	DHEA, "Andro"
5. Fats	Fish oils, DHA, lecithin
6. Other plant extracts	Garlic capsules, fiber, cranberry concentrate, bee pollen

Vitamin, mineral, protein, and amino acid supplements.

Botanical supplements, such as dong quai, often come in gel caps.

Plant extracts can also be taken in liquid form, as tinctures.

Herbal teas are a familiar source of dietary supplements.

Photo Disc

Table 24.1
Types of dietary supplements.

dietary supplements have been adequately tested,[2] and because results of studies showing negative effects may never see the light of day,[3] it is difficult to prove them to be unsafe. The FDA largely relies on reports of ill effects from health professionals and consumers to assess supplement safety.[1] Between 1993 and 2000, the FDA received over 2800 reports of adverse effects of supplements (primarily for herbs), including 105 deaths.[4]

According to FDA regulations (Table 24.2), dietary supplements must be labeled with a "Supplemental Facts" panel that lists serving size, ingredients, and percent Daily Value (%DV) of essential nutrient ingredients. Products can be labeled with a health claim, such as "high in calcium" or "low fat," if the product qualifies according to the nutrition labeling regulations. Supplements can also be labeled with "structure/function" claims. These are not subject to regulation, with the exception that they cannot refer to disease prevention or treatment effects. Claims such as "improves circulation," "supports the immune system," and "helps maintain mental health" can be used, whereas "prevents heart disease" or "cures depression" cannot be. If a function claim is made on the label or package inserts, the label or insert must also acknowledge that the FDA does not support the claim.[1] (This is done to reduce the FDA's liability for problems that may be caused by supplements.) Nonetheless, many people wholeheartedly believe health claims made for supplements.[1]

dietary supplements Any products intended to supplement the diet, including vitamin and mineral supplements; proteins, enzymes, and amino acids; fish oils and fatty acids; hormones and hormone precursors; and herbs and other plant extracts. Such products must be labeled "Dietary Supplement."

The Federal Trade Commission (FTC) regulates claims for dietary supplements made in print and broadcast advertisements, including direct marketing, Internet sites, and infomercials. Claims made for dietary supplements in advertisements are supposed to be truthful but often are not.[5] Although some companies have been prosecuted for making false and misleading claims,[6] neither the FDA nor the FTC has sufficient resources to fully monitor products and enforce laws related to dietary supplements.[1]

Vitamin and Mineral Supplements

Multivitamin and mineral supplements, such as vitamins E or C, calcium, or magnesium, are among the wide variety of vitamin and mineral supplements used by

Table 24.2

FDA regulations for dietary supplement labeling.

Nutrition Facts
Serving size 1 Tablet

Amount Per Serving	% DV
Melatonin 3 mg	*

*Daily Value (DV) not established

Other Ingredients: Dicalcium Phosphate, Cellulose (Plant Origin), Vegetable Stearic Acid, Vegetable Magnesium Stearate, Silica, Croscarmellose.

GUARANTEED FREE OF: wheat, yeast, soy, corn, sugar, starch, milk, eggs. No artificial colors, flavors. No chemical additives. No preservatives. No animal derivatives.

Directions: As a dietary supplement for adults, take one (1) tablet, under the direction of a physician, only at bedtime as Melatonin may produce drowsiness. **DO NOT EXCEED 3 MG IN A 24 HOUR PERIOD.**

Warning: For Adults. Use only at bedtime. This product is not to be taken by pregnant or lactating women. If you are taking medication or have a medical condition such as an auto-immune condition or a depressive disorder, consult your physician before using this product. **NOT FOR USE BY CHILDREN 16 YEARS OF AGE OR YOUNGER.** Do not take this product when driving a motor vehicle, operating machinery or consuming alcoholic beverages.

In case of accidental overdose, seek professional assistance or contact a Poison Control Center immediately.

KEEP OUT OF REACH OF CHILDREN

1. Product must be labeled "Dietary Supplement."
2. Product must have a "Supplemental Facts" label that includes serving size, amount of the product per serving, % Daily Value of essential nutrients, a list of other ingredients, and the manufacturer's name and address.
3. Health claims made on the label (e.g., "high fiber" or "low salt") must be justified based on nutrition labeling regulations established for foods.
4. Structure claims (such as those listing the parts of a plant used or a particular form of a vitamin) and function claims (e.g., "promotes normal bowel function") can be made on product labels. If a function claim is made, this FDA disclaimer must appear:

This statement has not been evaluated by the FDA. This product is not intended to diagnose, treat, cure, or prevent any disease.

consumers. They represent the most popular type of dietary supplement, with approximately 49% of U.S. adults using them daily.[7] Intake levels of vitamins and minerals below the "Tolerable Upper Limits" of the Dietary Reference Intake values (given on the inside front cover of this book) are safe for the vast majority of people, although lower amounts are related to their optimal functioning in the body. Certain hazards related to vitamin and mineral supplements exist, however. Some calcium supplements, for example, have lead contents similar to what they were 10 years ago when the problem was highly publicized,[8] and the ***bioavailability*** of nutrients contained in supplements remains uncertain. In general, the wider the assortment of vitamins and minerals in a supplement, the lower the absorption of each.[9] Minerals are particularly prone to forming unabsorbable complexes with each other, reducing the bioavailability of multiple minerals in the supplement.

➡ **bioavailability**
The amount of a nutrient consumed that is available for absorption and use by the body.

USP Designation

Few reports of the bioavailability of vitamins and minerals in supplements are available. Consumers can, however, protect themselves to some extent from supplements that are poorly absorbed by looking for "USP" on labels. The letters USP indicate that the U.S. Pharmacopeia has tested the supplement for ingredient availability for absorption, purity, and strength. The USP label does not indicate whether a product is effective. You can also find this label on some herbal supplements.[1]

Vitamin and Mineral Supplements: Who Benefits?

Supplements can indeed have a positive effect on health (Table 24.3). Vitamin and mineral supplements are the standard treatment for deficiency diseases. (The long-term cure for deficiency diseases is generally dietary improvements, however.) Supplemental fluoride helps prevent tooth decay in people whose water supply doesn't contain fluoride. Vegans who don't consume fortified soy milk or other fortified products may benefit from vitamin B_{12} and vitamin D supplements. The use of iron supplements prevents iron-deficiency anemia in pregnancy, and folate supplements before and early in pregnancy prevent a majority of cases of spinal cord malformation in newborns. Folate supplements may also reduce the risk of heart disease and stroke. Iron supplements are often required by women with heavy menstrual periods and by other people who experience significant blood loss.[10,11]

Vitamin and mineral supplements are also showing promise as "preventive medicine" for certain disorders and as an important part of the treatment of others. Supplemental vitamin C and E are associated with a decreased risk of cataracts in the elderly.[11] Calcium and vitamin D supplements play a role in the prevention and treatment of osteoporosis in some people.[10] Doses of vitamin C of about one gram daily appear to decrease the severity and duration of colds.[11] Vitamin E may help protect the lungs against the harmful effects of air pollution[12] and may reduce the risk of heart disease.[13] Multivitamin and mineral supplements that contain approximately 100% of the DRI (Dietary Reference Intake) of nutrients—shown as 100% under the "% Daily Value" column on supplement labels—may benefit some elderly persons, dieters, and others with limited food intakes.[9,14]

Table 24.3

Who may benefit from vitamin and mineral supplements? Here are some examples:

- People with vitamin and mineral deficiency diseases
- Newborns (vitamin K)
- People living in areas without a fluoridated water supply (fluoride)
- Vegans (vitamins B_{12} and D)
- Pregnant women (iron and folate)
- People experiencing blood loss (iron)
- Elderly persons on limited diets (multiple vitamins and minerals)
- People on restricted diets (multiple vitamins and minerals)
- Adults at risk for heart disease (folate, vitamin E)
- People at risk for osteoporosis due to low calcium intake and poor vitamin D status (calcium, vitamin D)

Using Vitamin and Mineral Supplements for the Wrong Reason

People often take supplements as a sort of insurance policy against problems caused by poor diets. Although multivitamin and mineral supplements may help fill in some of the nutrient gaps caused by poor food habits, they can't make a bad diet good. Whether a diet is good or bad is determined by more than its vitamin and mineral content. The "goodness" of a diet also depends on its caloric, fat, cholesterol, fiber, water, complex carbohydrate, and protein content. Additionally, supplements do not provide substances found in food such as flavones, phenols, lycopene, thiocynates, and other compounds that may have a positive influence on health.

One of the most serious consequences of supplements results when they are used as a remedy for health problems that can be treated, but not by vitamins or minerals. Vitamin and mineral supplements have not been found to be an effective treatment for behavioral problems, diabetes, chronic fatigue syndrome, obesity, or stress, for example.

The Rational Use of Vitamin and Mineral Supplements Like all medications, vitamin and mineral supplements should be taken only if there is a need for them. If they are taken, dosages should not be excessive. Guidelines for the selection and use of vitamin and mineral supplements can be found in the Health Action on the next page.

Herbal Remedies

> **A weed is what we call a plant whose virtues have not yet been discovered.**
>
> —Ralph Waldo Emerson

Herbal remedies have a long history, are used worldwide, and are moving from alternative to mainstream health care in North America. In 2000, 12% of U.S. adults reported daily use of herbal supplements, and the percentage is on its way up.[7]

HEALTH ACTION

Guidelines for Choosing and Using Vitamin and Mineral Supplements

1. Purchase supplements labeled "USP."
   - Terms such as "release assured," "laboratory tested," "quality tested," and "scientifically blended" on supplement labels guarantee nothing.
2. Check the expiration date on supplements. Use unexpired supplements.
3. Choose supplements containing 100% of the Daily Value or less.
4. Take supplements with meals.
5. If you have a diagnosed need for a specific vitamin or mineral, take that individual vitamin or mineral and not a multiple supplement.
   - Avoid calcium supplements made from oyster shells or bone. They may contain lead or aluminum.
6. Store supplements where small children cannot get at them.

REMEMBER! Consult your health care provider about health problems before you start taking supplements to try to treat the problems.

The herb pharmacopeia includes over 550 primary herbs known by at least 1800 names. Approximately 30% of all modern drugs are derived from plants,[2] and there's no doubt that many additional plants and plant ingredients benefit health. Plant products known to treat disease are considered drugs, however. Those that have not passed the scientific tests needed to demonstrate safety and effectiveness in disease treatment are considered herbs. Yet, the truth is that some products sold as herbs in the United States have drug-like effects on body functions. An example of an herb that acts powerfully like a drug is given in illustration 24.1.

Effects of Herbal Remedies

Purported effects and side effects of herbal remedies are listed in Table 24.4. Herbal remedies, like drugs, have biologically active ingredients that can have positive, negative, and neutral effects on body processes. Basically, an herbal

Illustration 24.1
Herb or drug?

The line between food, supplements, and drugs is so blurry these days that it's hard to know what any product is.
—Donna Porter, Congressional Research Service

Statins are a popular type of prescription drug that effectively lower LDL cholesterol and raise HDL cholesterol (the good one), and reduce the risk of heart disease. It turns out that statins are also available in herbal form.

Cholestin is an "herb" made from rice grown with red yeast. If subjected to specific growing conditions, the concentration of naturally occurring statins in the rice-yeast mixture can be increased greatly. Cholestin works like prescription statins, costs less, and is widely available to the public. Because cholestin is marketed as an herb rather than a drug, there is more room for the product to be mislabeled, contaminated, or unsafe, and to contain less active ingredients than claimed on the label.[12]

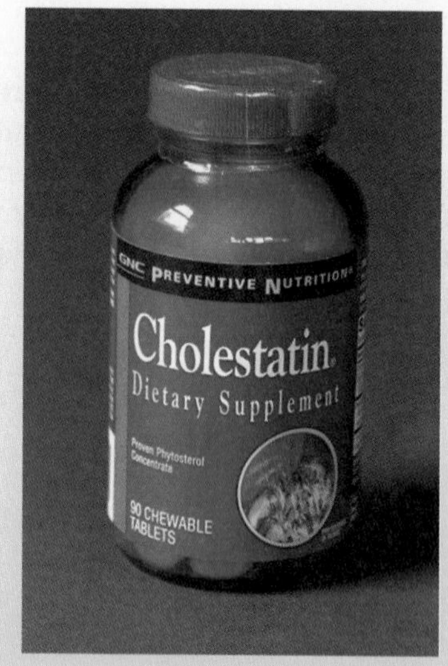

Herb/Other Remedy	Proposed Effects	Potential Side Effects
Glucosamine-chondroitin	Slows progression of osteoarthritis and its pain	Gastrointestinal upset, fatigue, headache
Ginseng	Increases energy, normalizes blood glucose, stimulates immune function, relieves impotence in males	Insomnia, hypertension, low blood glucose, menstrual dysfunction. Interacts with estrogen and blood thinners.*
SAMe	Relieves mild depression, pain relief for arthritis	May trigger manic excitement, nausea
Garlic	Lowers blood cholesterol, relieves colds and other infections, asthma, other	Heartburn, gas, blood thinner
Cholestin	Maintains desirable blood cholesterol levels	Safety of some ingredients unknown
Echinacea	Prevents and treats colds and sore throat	Allergies to plant components
DHEA	Improves memory, mood, physical well-being	Increases risk of breast cancer
Creatine	Sport supplement (increased performance in short, high intensity events)	Kidney disease
Saw palmetto	Improves urine flow, reduces urgency of urination in men with prostate enlargement	Nausea, abdominal pain
Ginkgo biloba	Increases mental skills, delays progression of Alzheimer's disease, increases blood flow, decreases depression	Nervousness, headache, stomachache. Interacts with blood thinners.
Shark cartilage	Treats lung cancer	Safety unknown
Saint John's wort	Relieves depression	Dry mouth, dizziness. Interacts with many drugs.
Ephedra (Ma Huang)	Promotes weight loss	Insomnia, headaches, nervousness, seizures, death

*Blood thinners include aspirin, warfarin, coumarin.

Table 24.4
Proposed effects and potential side effects of herbal and other dietary supplement remedies.[5,13,15,16,19]

remedy (or a drug) is considered valuable if it has beneficial effects on body processes, and if the benefits are not outweighed by the risks. Knowledge of the risks and benefits of many herbal supplements remains incomplete. However, available evidence suggests that some herbal remedies are safe and effective while others appear to be neither.

Which herbal remedies are likely ineffective or unsafe? Human experimentation with various botanicals over the centuries to the present time has helped to identify herbs that lack beneficial effects or have negative side effects. Table 24.5 lists some of these herbs. The extent to which the herbs included in the table pose a risk to health depends on the amount taken and the duration of use, the age and health status of the user, and other factors.

Which Herbal Supplements Are Potentially Beneficial? Results of many studies on the safety and benefits of specific herbal remedies have been mixed. The bulk of existing evidence, however, indicates that some may be beneficial:

- ginkgo biloba may decrease symptoms in patients with Alzheimer's disease and increase blood flow to the brain,
- echinacea may decrease the severity of colds,
- cranberry may relieve urinary tract infection,
- garlic can reduce elevated blood cholesterol levels and blood pressure,
- Saint John's wort may relieve mild and moderate depression,

Table 24.5
Examples of dietary supplements that may not be effective or safe.

Apricot pits (laetrile)
Belladonna
Blue cohosh
Borage
Broom
Chaparral
Comfrey
Dong quai
Eyebright
Licorice root
Liferoot
Lily of the valley
Ma Huang/ephedra
Mandrake
Mistletoe
Pennyroyal oil
Sassafras
Vinca
Wild yam
Wormwood

- rose hips have mild laxative and diuretic effects (Illustration 24.2),
- SAMe may relieve mild depression and arthritis pain,
- saw palmetto may reduce frequent bouts of a need to urinate and improve urine flow in older men,
- senna may relieve constipation, and
- ginger may prevent motion sickness.[4,15–18]

The list of effective herbal remedies may grow, or shrink, with time as additional studies are completed.

Use of Herbal Remedies

Countries vary widely in the use of herbal remedies in medical care. In some European countries, herbs are widely used and accepted as safe and effective if taken as prescribed. The safety of herbs in these countries is largely assumed based on historical use of the herbs. Other countries lack the resources to purchase drugs, and may use herbs because they are often substantially cheaper. In the United States and Canada, herbal remedies are only considered safe and effective if demonstrated to be so by results of controlled clinical trials.

Approval of herbal remedies for disease treatment is proceeding very slowly in the United States and Canada because too few clinical trials have reported results on their safety and effectiveness. (Many such trials are underway, so stay tuned.) Rather than ban or strictly regulate them, herbs are available to consumers on a "buyer beware" basis. There is no guarantee that the herbs we purchase are safe, effective, uncontaminated, or even present in the product.

A Measure of Quality Assurance for Herbal Products Tests by independent laboratories confirm that some herbal supplements contain less than the amount of the herb claimed on the label, and are contaminated with pesticides or heavy metals.[13] There is no government body that monitors the contents of herbal supplements. Private groups, such as the U.S. Pharmacopeia, the National Formulary, and Consumer Laboratories, offer testing services to ensure that dietary supplements meet standards for disintegration, purity, potency, and labeling. Products that pass these tests can display "USP," "NP," or the CL symbol boldly on product labels. These letters represent quality ingredients and labeling, but do not address product safety or effectiveness. Yet, they are the best standards currently available to judge the content of herbal remedies.

➡ **functional foods**
A marketing term for foods that contain added ingredients intended to benefit health. They do not have to be demonstrated to benefit health, however.

Health-related claims stated on herbal supplement labels may or may not represent the truth, and unfortunately there is no remedy for that aside from looking at the results of scientific studies. (To obtain this information, search the websites and resources listed at the end of the unit.) Due to the lack of studies and potential dangers, the FDA has advised dietary supplement manufacturers not to make claims related to pregnancy for herbs and other products.[14] Considerations for the use of herbal supplements are summarized in the Health Action on the next page.

Illustration 24.2
Rose hips.

Functional Foods

> Leave no aisle unfortified.
>
> —Headline for article in a food industry magazine

Also known as "neutraceuticals" and "designer foods," ***functional foods*** include a variety of products that have theoretically been modified to enhance their contribution to a healthy diet (Table 24.6 on p. 24–10).[17,20] Functional foods such as yogurt with active cultures, buttermilk, and Postum and Bosco beverage mixes were on the grocery shelves decades ago. Some have been taken off the market due to safety concerns, and others, including one popular beverage, have had to

HEALTH ACTION

Considerations for the Use of Herbal Remedies

1. Don't use herbal remedies for serious, self-diagnosed conditions such as depression, persistent headaches, and memory loss. (You might benefit more from a different treatment.)
2. Let your doctor know what herbal remedies you take.
3. If you take prescription medications, clear the use of herbal remedies with your doctor.
4. Don't use herbal remedies without medical advice if you are attempting to become pregnant or if you are pregnant or breastfeeding.
5. Don't mix herbal remedies.
6. If you are allergic to certain plants, make sure herbal remedies are not going to be a problem before you use them.
7. If you have a bad reaction to a herbal remedy, stop using it.
8. Buy herbs labeled "USP," "NP," or the CL in a beaker symbol.
9. Investigate brands, herb safety, and effectiveness by checking into one or more of the websites and resources listed at the end of this unit.

remove ingredients. (Read more about it in On the Side.) Because there is no statutory definition for what constitutes functional foods, there are no regulations that specifically apply to them.[8] Functional foods containing food additives not on the Generally Accepted As Safe (GRAS) list must be tested and approved by the FDA before being sold. Products such as Benecol and Nutrasweet are examples of functional food ingredients that have received FDA approval. Increasingly, the list of functional foods is becoming infiltrated with soups, beverages, and pastas spiked with herbs.

Some functional food ingredients have been tested for safety and effectiveness, but most have not. Functional food ingredients that appear to benefit health include fortified orange juice, breakfast cereals, and grain products; fiber-fortified, therapeutic liquid meals; and plant extracts added to margarine that lower blood cholesterol level.[2,3]

On the Side

Taking the coke out of coke Coca cola used to be fortified with cocaine. During the early part of the 1900s, the "coke" ingredient was removed.

Prebiotics and Probiotics: From "Pharm" to Table

The terms ***prebiotics*** and ***probiotics*** were derived from "antibiotics" due to their purported effects on increasing resistance to various diseases. Prebiotics may decrease disease risk by supporting the growth of beneficial bacterial colonies in the intestine. Probiotics is the term for bacteria added to foods, especially to dairy products.

Availability of foods containing prebiotics and probiotics is much more common in European countries than in Canada or the United States. However, availability of such products is increasing in these countries as research results shed light on their safety and effectiveness. It is known that specific carbohydrates found in breast milk enhance the growth of bacteria that protect infants against *E. coli* and other harmful bacteria. In addition, it appears that certain types of intestinal bacteria provide an effective barrier against diarrhea, irritable

prebiotics
"Intestinal fertilizer." Certain types of nondigestible carbohydrates added to some dairy products that support the growth of beneficial bacteria in the gut.

probiotics
Strains of *lactobacillus acidophilus* and *lactobacillus johnsnii* added to some dairy products that limit the growth of harmful bacteria.

Table 24.6
How are foods made to be "functional"?

Foods are made to be "functional" by:

1. taking out potentially harmful components (e.g., cholesterol in egg yolk and lactose in milk),
2. increasing the amount of nutrients and beneficial nonnutrients (e.g., fiber-fortified liquid meals, calcium-, and vitamin C–fortified orange juice), or
3. adding new beneficial compounds to foods (e.g., "friendly" bacteria to yogurt and other milk products).

bowel syndrome, and food allergies. In the future, strains of bacteria may be introduced that decrease fat absorption, help digest lactose in milk, and add an element of protection against colon cancer.[21,22] Growth of beneficial bacterial strains would be supported by their introduction into the diet and by ingestion of specific types of nondigestible carbohydrates that serve as the bacteria's favorite foods. Strains of these "friendly" bacteria that better survive exposure to stomach acid and digestive enzymes are being developed.

Prebiotics and probiotics appear to be safe and potentially beneficial.[21,22] But it is too soon to tell whether they are a magnet for corporate profit or represent a breakthrough in disease prevention and treatment.

Although availability and use of functional foods are increasing, it's not a household term yet. Many adults consume functional foods, but only 79% have heard the term.[8] (If the term functional foods is new to you, you're certainly not alone.)

Final Thoughts

You're leafing through the Sunday magazine insert in the newspaper when your attention becomes engaged by an article declaring that almond extract can be used to fight early morning fatigue. Finding it hard to keep awake through early morning classes, you're quietly thrilled that there's such a simple and inexpensive remedy.

The next article was selling the Brooklyn Bridge. . . . It's another "gotcha" and it happens all the time. Insist on scientific proof that herbal remedies and other dietary supplements work as advertised. Otherwise, there will be no end to the proliferation of useless and possibly unsafe ideas and products for what ails us.

Nutrition UP CLOSE

Supplement Use and Misuse

FOCAL POINT: Decide if a dietary supplement is warranted in these situations.

People take dietary supplements for many reasons, but is their use justified? Apply the information from this chapter to determine if you agree with the decisions made in each of the following scenarios.

1. Martha works part-time and takes a full load of classes. Like many college students, she is always on the go, often grabbing something quick to eat at fast-food restaurants or skipping meals altogether. Nevertheless, Martha feels confident her health will not suffer because she takes a daily vitamin and mineral supplement.

Is a supplement warranted in this case? ______________

Why or why not? ____

____

2. Sylvia is a 23-year-old student diagnosed with iron-deficiency anemia. She has learned in her nutrition class that it is preferable to get vitamins and minerals from food instead of supplements. Therefore, instead of taking the iron pills her doctor has prescribed, Sylvia has decided to counteract the anemia by increasing her consumption of iron-rich foods.

Is a supplement warranted in this case? ______________

Why or why not? ____

____

3. John is a 21-year-old physical education major involved in collegiate sports. He is very aware that nutrition plays an important role in the way he feels, so he is careful to eat well-balanced meals. In addition, John takes megadoses of vitamins and minerals daily. He is convinced they enhance his physical performance.

Is a supplement warranted in this case? ______________

Why or why not? ____

____

4. Roberto, a native Californian, is backpacking through Europe when he is slowed down by constipation. He visits a pharmacy where English is spoken and is given senna by the clerk rather than a fast-acting medicine he expected. Roberto has never taken an herb before and is not sure how his body will react to it, or if it will work.

Should Roberto try the senna, ask for a nonherbal drug, or take another action? (Assume they cost the same.) ______________

What's the rationale for this decision? ______________________________

____

5. While shopping at the mall, Yuen notices a kiosk selling "Hypermetabolite," a weight-loss product that guarantees you'll lose 5 pounds a week without dieting. Having gained 10 pounds since she started working full time, Yuen decides to try it. Her examination of the product's label reveals that ephedra and Asian ginseng are major ingredients.

Should Yuen take Hypermetabolite for weight loss? ______________

Why or why not? ____

____

FEEDBACK (including answers) can be found at the end of Unit 24.

WWW links

www.altmed.od.nih.gov/
Lots of excellent documents from NIH's Office of Alternative Medicine are available from this site.

www.alternativemedicine.com
Directs consumers to informative, credible websites on alternative therapies.

www.ncahf.org
Use to get into contact with the National Council Against Health Fraud.

www.opd.od.nih.gov/ods
The National Institute of Health's Office of Dietary Supplements provides a wealth of information on many vitamin, mineral, and herbal supplements.

www.consumerlab.com
Evaluate brands and effects of dietary supplements free of charge.

www.uic.edu/pharmacy/research/diet/ce.html
The University of Illinois provides handouts on some herbal supplements.

www.fda.gov/medwatch
Report adverse reactions to supplements.

www.amfoundation.org/herbalmed.htm
The Alternative Medicine Foundation's herbal medicine website.

REFERENCES

1. ADA/APhA Special Report from the Joint Working Group on Dietary Supplements. A healthcare professional's guide to evaluating dietary supplements. American Dietetic Association, American Pharmaceutical Association, 2000:47; 1–40.
2. Belew C. Herbs and the childbearing woman. Guidelines for midwives. J Nurse Midwifery 1999;44:231–52.
3. Ernst E, Pittler MH.*Ginkgo biloba* for dementia: a systematic review of double-blind, placebo-controlled trials. Clin Drug Invest 1999;17:301–8.
4. Community Nutrition Institute Newsletter. 2000 July 21:2.
5. Radimer KL, Subar AF, Thompson FE. Nonvitamin, nonmineral dietary supplements: issues and findings from NHANES III [see comments]. J Am Diet Assoc 2000;100:447–54.
6. Shark cartilage, Environ Nutr Newsletter, Aug 2000, p. 8.
7. Nutrition and you: trends 2000 survey results released. ADA Courier 2000;39:1,3.
8. Ross S. Functional foods: the Food and Drug Administration perspective. Am J Clin Nutr 2000;71:1735S–8S; discussion 1739S–42S.
9. Park YK, Kim I, Yetley EA. Characteristics of vitamin and mineral supplement products in the United States. Am J Clin Nutr 1991;54:750–9.
10. Horwitt MK. Relative biological values of d-alpha-tocopheryl acetate and all-rac-alpha-tocopheryl acetate in man. Am J Clin Nutr 1980;33:1856–60.
11. Brown JE, et al. Predictors of red cell folate level in women attempting pregnancy. JAMA 1997;277:548–52.
12. Havel RJ. Dietary supplement or drug? The case of cholestin [editorial]. Am J Clin Nutr 1999;69:175–6.
13. www.consumerlab.com, 2000.
14. HHS Statement: FDA Statement concerning structure/function rule and pregnancy claims. 2000 Jan 6.
15. Ramos L. SAMe as a supplement: can it really help treat depression and arthritis? J Am Diet Assoc 2000;100:414.
16. Foster S, Tyler VE. Tyler's honest herbal. 4th ed. NY: Haworth Herbal Press, 1999.
17. Monsen ER. The question of herbal supplements: are they here to stay? [editorial]. J Am Diet Assoc 2000;100:409.
18. Wilt TJ, et al. Saw palmetto extracts for treatment of benign prostatic hyperplasia: a systematic review. JAMA 1998;280:1604–9.
19. Stoll BA. Dietary supplements of dehydroepiandrosterone in relation to breast cancer risk. Eur J Clin Nutr 1999;53:771–5.
20. Roberfroid MB. Concepts and strategy of functional food science: the European perspective. Am J Clin Nutr 2000;71:1660S–4S; discussion 1674S–5S.
21. Kopp-Hoolihan L. Prophylactic and therapeutic uses of probiotics: a review. J Am Diet Asso 2001;101:229–38,241.
22. Sanders ME. Considerations for use of probiotic bacteria to modulate human health. J Nutr 2000;130:384S–90S.

Nutrition

UP CLOSE

Supplement Use and Misuse

Feedback for Unit 24

1. *Is a supplement warranted in this case?* No
   Why or why not? Martha is deceiving herself! She may be getting the vitamins and minerals she needs by taking a supplement, but this cannot make up for her poor food habits. She needs to improve the overall quality of her diet to ensure optimal health. If Martha follows the Food Guide Pyramid recommendations, she should be able to get all the nutrients she needs from what she eats.
2. *Is a supplement warranted in this case?* Yes
   Why or why not? This is one time when a supplement is in order. Sylvia needs to follow her doctor's advice and take the prescribed iron preparation to increase her hemoglobin and replete her iron stores. However, she is correct in consuming more iron-rich foods, too, so that after the anemia has been treated, she will not experience a relapse.
3. *Is a supplement warranted in this case?* No
   Why or why not? There is no scientific evidence that vitamin and mineral megadoses enhance physical performance. In fact, John may be setting himself up for toxicity reactions with prolonged intake of supplements at extremely high dosages.
4. *Should Roberto try the senna, ask for a nonherbal drug, or take another action?* He should take another action—get more information from the pharmacist so he can make a better informed decision.
5. *Should Yuen take Hypermetabolite for weight loss?* No
   Why or why not? She shouldn't take the product because ephedra can have serious side effects and might interact with ginseng, and because quick weight-loss strategies don't work in the longer run. Plus, there's no guarantee that the product will work, or that it isn't mislabeled.

Unit 25

Water Is an Essential Nutrient

Photo Disc

Nutrition Scoreboard

	TRUE	FALSE
1. Bottled water is better for health than municipal water from your faucet.		
2. Drinking lots of water helps moisturize the skin.		
3. Adequate water intake reduces the risk of some types of cancer and kidney stones.		
4. Physically inactive adults living in moderate climates require about 8 cups of water each day.		

(Answers on next page)

Answers to

Nutrition Scoreboard

	TRUE	FALSE
1. Bottled waters have not been shown to be better for health or to be "purer" than tap water obtained from city water supplies.		✓
2. Excess water intake is excreted—it doesn't go to the skin.		✓
3. Adequate intake of water from fluids and food (9 to 12 cups a day) reduces the risk of breast, colon, and bladder cancer and of kidney stone formation.[1]	✓	
4. Make that 9 cups of water a day for women and 12 cups a day for men.[1]		✓

Key Concepts and Facts

- Water is an essential nutrient. It is a required part of the diet. Deficiency symptoms develop when too little is consumed, and toxicity symptoms occur when too much is ingested.
- Functions of water include maintenance of body hydration and temperature, removal of waste products, and participation in energy formation. It is our major source of fluoride.
- Most adults consume too little water.
- Water is a precious resource whose availability and quality are threatened by wasteful use and pollution.

Photo Disc

Water: Where Would We Be without It?

Ask any three people you know to name as many essential nutrients as they can. If they mention water, give them a prize. Our need for water is so obvious that it is often taken for granted. Well, that's not going to happen in this text!

Water differs from other essential nutrients in that it is liquid, and our need for it is measured in cups rather than grams or milligrams. Furthermore, without it, our days are limited to about six. Water is the largest single component of our diet and body. It is a basic requirement of all living things. Now, how could something this important be so easily forgotten?

Water's Roles as an Essential Nutrient

Water qualifies in all respects as an essential nutrient. It is a required part of our diet; it performs specific functions in the body; and deficiency and toxicity signs develop when too little or too much is consumed.

Water is also our body's main source of fluoride, an essential mineral needed for the formation and maintenance of enamel resistant to tooth decay. Fluoride is naturally present in some water supplies, and is added to virtually all municipal water supplies in the United States and Canada. Because it is a gas that dissolves in water, fluoride is not removed by water pitcher or faucet filters. According to the American Dental Association, it may be removed by reverse-osmosis filtration systems that may be installed under sinks.

Water is the medium in which many chemical reactions take place within our body. Water plays key roles in energy formation and is needed to "carry" nutrients to cells and waste products away from them. Additionally, water acts as the body's cooling system. When our internal temperature gets too high, water transfers heat to the skin and releases it in perspiration. When we're too cool, less water—and heat—is released through the skin. Water's functions are summarized in Table 25.1.

Table 25.1
Key functions of water in the body.

- Provides a medium for chemical reactions
- Participates in energy formation
- Transports nutrients and waste products
- Helps regulate body temperature

Water has been given credit for other functions in the body, but undeservedly. Drinking more water than is normally needed does not prevent dry, wrinkled skin or weight gain. Nor will it cure constipation, fatigue, arthritis, migraines, or hypertension.

Water, Water Everywhere . . . The body of a 160-pound person contains about 12 gallons of water in blood and other tissue (Illustration 25.1). Adults are approximately 60% water by weight. The proportion of water in body tissues varies: blood is 83% water, muscle 75%, and bone 22%. Even fat cells are 10% water.[2]

Illustration 25.1
On average, the body of a 160-pound person contains 12 gallons of water. That's 96 pounds of water!

Most Foods Contain Lots of Water, Too Most beverages are more than 85% water, and fruits and vegetables are 75 to 90% (Illustration 25.2). Meats, depending upon their type and how well-done they are, contain between 50 and 70% water. Although it is nearly impossible to meet your need for water from solid foods alone, the water content of foods makes an important contribution to our daily intakes. On average, adults consume about six cups of water per day from fluids and four cups of water from foods.

Health Benefits of Water Adequate water consumption may benefit long-term as well as day-to-day health. Consumption of over 10 cups of fluid each day is associated with a decreased risk of bladder, breast, and colon cancer and of kidney stone formation.[1,3] Most people consume too little water, however. Average intake of water in the United States is a little over seven cups a day.[1]

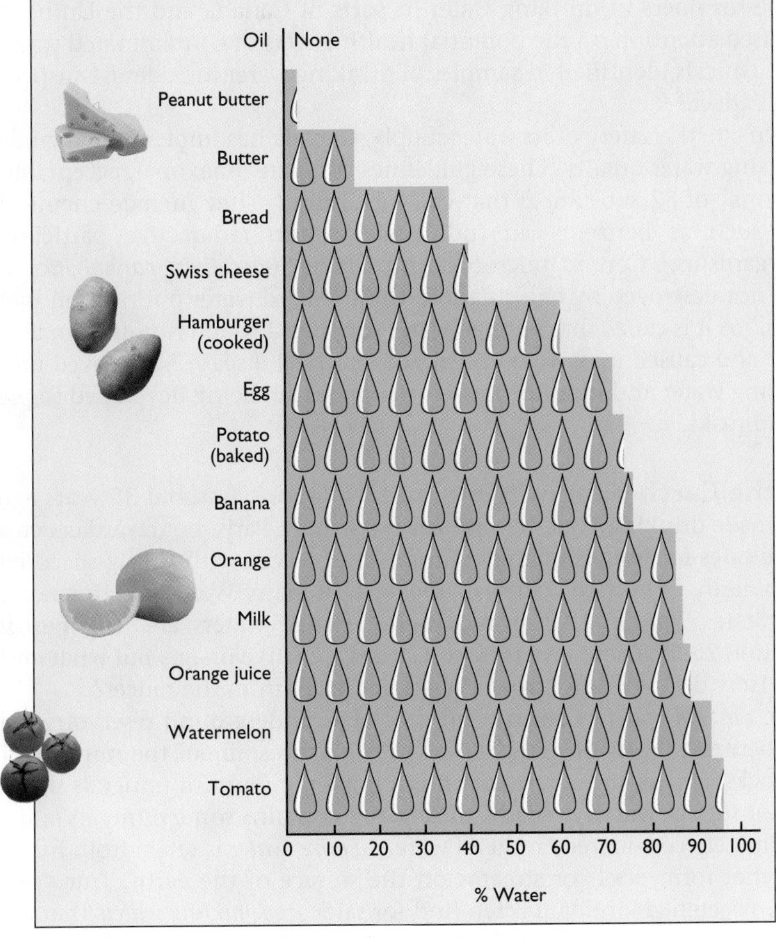

Illustration 25.2
The water content of some foods.

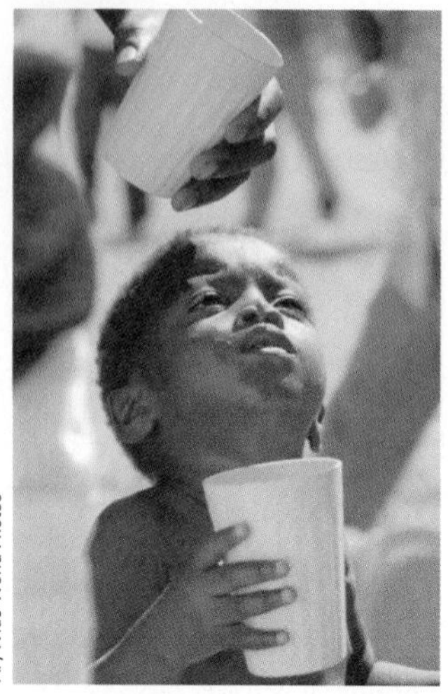

AP/Wide World Photos

Illustration 25.3
A child waits for water at a refugee camp.

The Nature of Our Water Supply

Water covers about three-quarters of the earth's surface, yet very little of it is drinkable. Nearly 97% of the total supply is salt water, and only 3% is fresh. Of the freshwater supply, only a fourth of the total amount is available for use. The rest is located in polar and glacier ice. Although fresh water is readily available in most locations in the United States, this is not the case in a number of other countries.

Water is sufficiently scarce in parts of Russia that drinking water dispensers are coin operated. Fresh water is so highly prized in sections of the arid Middle East that having a decorative fountain in one's home is considered a sign of affluence. Drinking water is sampled, judged, and celebrated in Middle Eastern countries as ceremoniously as fine wines are in France. Although water is not a subject of envy or a cause for celebration in most parts of America, it is being increasingly viewed as a precious resource that must be protected.

Water scarcity may become a primary concern of many more nations in the twenty-first century (Illustration 25.3). The demand for water is increasing by 100 to 500% per year in countries in Central and South America, Africa, and Europe, but the supply is dwindling. Wasteful use of water, groundwater depletion, pollution, and leaky public water systems are contributing to water shortages. Without more effective protection of the world's water, the quantity of clean water available for agriculture and consumption may be insufficient, limiting food production and increasing the incidence of water-borne illnesses.[4]

Water Safety Concerns Currently, concerns are being voiced about the quality and safety of the water supply in America (Illustration 25.4). The detection of hydrocarbons, chloroform, radioisotopes, lead, antibiotics, hormones, and asbestos fibers in drinking water in parts of Canada and the United States has focused attention on the potential health effects of contaminated water. The list of chemicals identified in samples of drinking water includes 64 suspected of causing cancer.[5,6]

To ensure the safety of its water supply, Canada has implemented guidelines for drinking water quality. These guidelines stipulate "maximal, acceptable concentrations" of 82 substances that can be harmful. They include chemical pollutants such as benzene, an industrial solvent; radioactive particles; and microorganisms.[7] Certain microorganisms (*Cryptosporidium parvum,* for example) are not destroyed by chlorination, the standard water purification method. "Crypto," as it is called, has entered water supplies through runoff from livestock pastures and caused massive outbreaks of diarrheal disease.[8] Improved methods for treating water and monitoring water supplies are being developed to prevent such outbreaks.

Illustration 25.4
Contaminated water supplies are a threat to the public's health.

Source: Greenpeace.

Does the Earth Supply "Gourmet" Water? About 35 years ago, the French made drinking "fine waters" very stylish. In Paris, boulevardiers crowded sidewalk cafes for hours sipping chilled Perrier served with thinly sliced lemon. The popularity of bottled waters skyrocketed in many Western countries. In the United States, "mineral," "spring," and "seltzer" waters are now best-sellers (Illustration 25.5). These waters have a strong, positive image. But what *are* these waters? How do they differ from the water we get from the faucet?

True *mineral water* is taken from protected underground reservoirs that are lodged between layers of rock. The water dissolves some of the minerals found in the rocks, and as a result, it contains a higher amount of minerals than most sources of surface water. Actually, most water contains some minerals and could be legitimately considered "mineral water." *Spring water* is taken from freshwater springs that form pools or streams on the surface of the earth. True "seltzers" (not the sweetened kind you often find for sale) are *sparkling waters* that are nat-

Illustration 25.5
Is bottled water better for you than tap water? Consumers often perceive that bottled waters have fewer impurities and are better for health than tap water. Such beliefs are unfounded.

urally carbonated. Most seltzers, however, become bubbly by the commercial addition of pressurized carbon dioxide.

Bottled waters have their advantages. They are calorie-free, are generally sodium-free or low in sodium, and quench a thirst better than their major competitor, soft drinks. But they are no "purer" or better for you than tap water.[7] As a matter of fact, some bottled waters contain tap water. The Food and Drug Administration (FDA) estimates that 25% of the bottled water sold in the United States is bottled tap water.

Is Bottled Water Safe? Every now and then, a story about contaminated bottled water makes the news, raising concerns about the safety of bottled water and the way the industry is monitored. The FDA regulates the bottled water industry. Domestic bottlers must conform to specified standards of water safety (such as allowable levels of chemical contaminants) and labeling requirements. "Mineral" or "spring" water must be just that under FDA regulations. Bottled tap water must be labeled as such.

Bottlers are also subject to unannounced inspections by the FDA. Bottled water is classified as a very low risk product, however, so domestic bottling plants are inspected, on average, every five years.[9] The FDA supplements its monitoring role by requiring that bottlers periodically test their water. Foreign bottling plants are not under the FDA's jurisdiction, but imported waters may be tested when they enter the United States.

Bottling plants that adhere to FDA regulations produce safe water. If you have questions about your favorite brand or want more information, call the Environmental Protection Agency's Safe Drinking Water Hotline (1-800-426-4791) between 8:30 A.M. and 5:00 P.M. Eastern Standard Time.

Water Gimmicks New waves of bottled water products are finding their way to grocery shelves throughout America. Tapping into consumer interest in vitamins, minerals, herbs, and fitness, producers are marketing "enhanced" bottled water fortified with everything from soy isolates to ginseng (Illustration 25.6). Although it does appear that some fortified water products, such as calcium-fortified water, can improve nutrient status and potentially health,[10] the rationale for many of the products (other than their commercial appeal) remains elusive. It makes little sense to fortify water with herbs, or with nutrients better obtained from foods. Safety of some of the combinations of herbs put into bottled water is unclear, and they do not appear to supplement the diet in meaningful ways.[11]

Meeting Our Need for Water

In general, individuals require enough water each day to replace the water lost in urine, perspiration, stools, and exhaled air. Physically inactive males living in moderate climates need about 12 cups of water each day, and females 9 cups.[1] That amount can easily be obtained from the fluid and food content of a normal diet. The best sources, however, are tap and bottled water (if preferred), brothy soups, and nonalcoholic beverages such as milk and fruit and vegetable juices. Alcohol tends to increase water loss, so beverages such as beer and wine are not "hydrating." Caffeinated beverages, including coffee and tea, provide some hydration, but not as much as water.[12] People who are physically active, live in hot climates, or have illnesses that involve vomiting, diarrhea, or fever need more water to replace their losses than other people.

Water Deficiency A deficiency of water can lead to dehydration. Dehydrated people feel very sick. They are generally nauseated, have a fast heart rate and increased body temperature, feel dizzy, and may find it hard to move. The ingestion of fluids produces quick recovery in all but the most serious cases of dehydration. If it is not resolved, however, dehydration can lead to kidney failure and death.

Built-in mechanisms that trigger thirst generally protect people from consuming too little water and becoming dehydrated. Some people, however, lose more water through physical activity and sweating than can be compensated by thirst mechanisms. People who do strenuous work, athletic or otherwise, in hot weather have to drink beyond their thirst to make up for water lost in sweat. One runner in the grueling "Death Valley" race—a 55-mile, 17-hour event—had

Illustration 25.6
"Specialty" bottled water products are of uncertain value to health or fitness.

to drink 30 pounds of water (the equivalent of almost four gallons) to keep up with his water losses. It is recommended that people performing strenuous physical activity drink more water than is indicated by thirst. You know you are drinking enough water if your mouth isn't dry and your urine is light yellow and produced in normal volume.

Prolonged bouts of vomiting, diarrhea, and fever can produce dehydration, and that is why you should drink plenty of fluids when you experience these conditions. High-protein and high-fiber diets and alcohol increase your need for water. Losses in body water that accompany high levels of protein consumption are the reason people on high-protein weight-loss diets are encouraged to "drink a lot of water." Adding fiber to your diet augments water need because fiber increases water loss in stools. The increased loss of water that goes along with alcohol intake explains why people get very thirsty after overindulging in spirits.

Water Toxicity It may sound fantastic, but people can overdose on water. Water intoxication develops when an individual consumes so much water (approximately three to four gallons) that it dilutes the blood and upsets the body's normal water balance. Water intoxication may occur in psychotic patients who have an inordinate appetite for water caused by medication.[13] In such cases, the drive for water can be so strong that access to water (even when showering) has to be limited. Water intoxication has been reported in infants from low-income families who receive water for days at a time after money for formula runs out or are routinely given overdiluted formula (Illustration 25.7).[14]

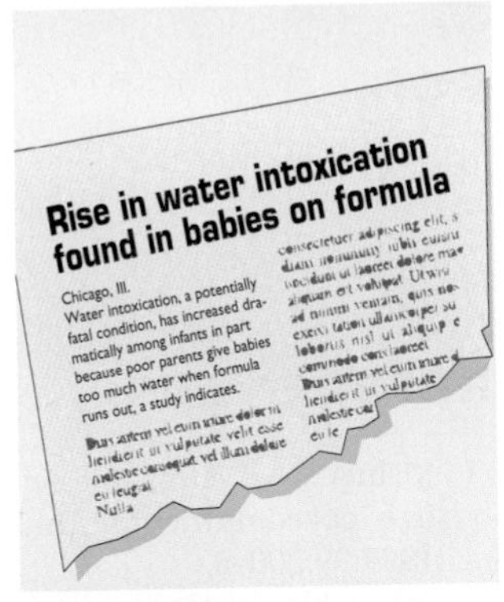

Rise in water intoxication found in babies on formula

Chicago, Ill.
Water intoxication, a potentially fatal condition, has increased dramatically among infants in part because poor parents give babies too much water when formula runs out, a study indicates.

Illustration 25.7
This headline is accurate—overdilution of infant formula can lead to water intoxication in infants.

Nutrition UP CLOSE

Foods as a Source of Water

FOCAL POINT: Water is a primary component of many foods.

Our requirement for water is met by fluids and solid foods. But how much water is in foods? You may be surprised. Circle the food in each set that you think contains the highest percentage of water by weight.

1. avocado potato, boiled corn, cooked
2. egg almonds ripe (black) olives
3. watermelon celery pineapple, fresh
4. 2% milk coke cranberry juice, low calorie
5. cheddar cheese banana refried beans
6. hot dog pork sausage, cooked ham, extra lean
7. apple mushrooms, raw orange
8. onions, cooked lettuce okra, cooked
9. peanut butter butter margarine
10. cake with frosting bagel Italian bread

FEEDBACK (answers to questions) can be found at the end of Unit 25.

WWW links

www.epa.gov/enviro/html/sdwis/sdwis_ov.html
Concerned about the safety of your water supply? Check it out on this Environmental Protection Agency web site.

www.navigator.tufts.edu
Search "water and health."

www.healthfinder.gov
A gateway to reliable information on water.

www.cdc.gov
The address for the Centers for Disease Control; provides access to water safety information and advice for travelers.

REFERENCES

1. Kleiner SM. Water: an essential but overlooked nutrient. J Am Diet Assoc 1999;99:200–6.
2. Askew EW. Water. In: Ziegler EE, Filer LJ Jr., editors. Present knowledge in nutrition. Washington, DC: ILSI Press; 1996: pp 98–108.
3. Michaud DS, et al. Fluid intake and the risk of bladder cancer in men. N Engl J Med 1999;340:1390–7.
4. Water resources in the twenty-first century. Washington, DC: Food Policy Research Institute; 1997.
5. Wigle DT. Contaminants in drinking water and cancer risks in Canadian cities. Can J Public Health 1986;77:335–42.
6. Morris RD, et al. Chlorination, chlorination by-products, and cancer: a meta-analysis. Am J Public Health 1992;82:955–63.
7. The undiluted truth about drinking water. Health Canada, Dept. of National Health and Welfare's World Wide Web site, 1997 Apr 10.
8. Messer AE. Climate change may impact waterborne diseases. Penn State World Wide Web site, 1998 May 26.
9. How safe is bottled water? American Institute for Cancer Research Newsletter 1992;36:8.
10. Guillemant J, et al. Mineral water as a source of dietary calcium: acute effects on parathyroid function and bone resorption in young men. Am J Clin Nutr 2000;71:999–1002.
11. Whatever happened to plain old H_2O? Environ Nutr 2000;23(Aug):3.
12. Welland D. Drink to good health, especially water: Here's why and how much. Environ Nutr 1999;22(Oct):1,6.
13. Goldman MB, Luchins DJ, Robertson GL. Mechanisms of altered water metabolism in psychotic patients with polydipsia and hyponatremia. N Engl J Med 1988;318:397–403.
14. Keating JP, Schears GJ, Dodge PR. Oral water intoxication in infants: an American epidemic. Am J Dis Child 1991;145:985–90.

Nutrition
UP CLOSE

Foods as a Source of Water

Feedback for Unit 25

The percentage of water is listed after each food.

1. **avocado (80%)** potato, boiled (77%) corn, cooked (73%)
2. egg (75%) almonds (4%) **ripe (black) olives (80%)**
3. watermelon (91%) **celery (95%)** pineapple, fresh (86%)
4. 2% milk (89%) coke (89%) **cranberry juice, low calorie (95%)**
5. cheddar cheese (37%) **banana (74%)** refried beans (72%)
6. hot dog (53%) pork sausage, cooked (45%) **ham, extra lean (74%)**
7. apple (84%) **mushrooms, raw (92%)** orange (87%)
8. onions, cooked (88%) **lettuce (96%)** okra, cooked (90%)
9. peanut butter (1%) **butter (16%)** margarine (16%)
10. cake with frosting (22%) bagel (33%) **Italian bread (36%)**

Photo Disc

Unit 26

Nutrition and Physical Fitness for Everyone

Nutrition Scoreboard

	TRUE	FALSE
1. Physical fitness means being very muscular.		
2. Overweight people can be physically fit.		
3. Exercise is more effective than diet in preventing heart attacks.		
4. Physical fitness can be achieved only by exercising intensively for at least an hour five times a week.		

(Answers on next page)

Key Concepts and Facts

- Physical fitness, along with a good diet, confers a number of physical and mental health benefits.
- You don't have to be an athlete or be lean to be physically fit. Fitness depends primarily on muscular strength, endurance, and flexibility.
- People who are physically fit have respiratory and circulatory systems capable of delivering large amounts of oxygen to muscles, and muscular systems that can utilize large amounts of oxygen for prolonged periods of time.
- Physical fitness can be achieved by resistance training, aerobic exercises, and stretching.
- The fitness level of the majority of people in the United States is poor.

Answers to

Nutrition Scoreboard

	TRUE	FALSE
1. Muscle strength is one part of physical fitness. The other primary components are endurance and flexibility.		✓
2. You can be fit and fat. Physical fitness is not measured by body fat content.	✓	
3. Regular exercise does help reduce the risk of heart attack. However, eating a diet low in animal fat and cholesterol is more effective at reducing the risk of heart attacks than is exercise alone.		✓
4. You don't have to exercise to that extent to become physically fit.		✓

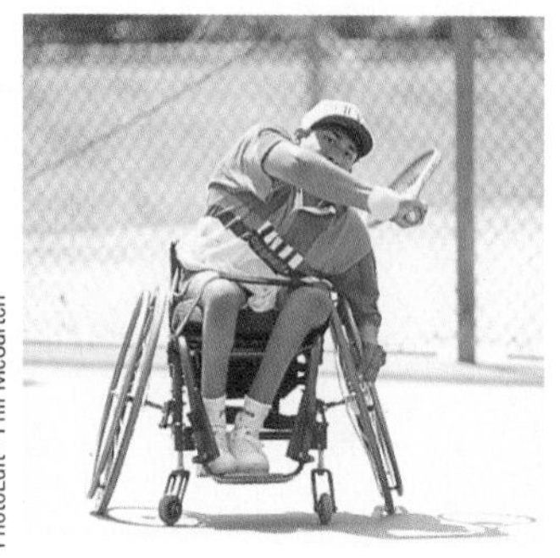
PhotoEdit - Phil McCarten

Physical Fitness: It Offers Something for Everyone

> I would not wish to imagine a world in which there were no games to play and no chance to satisfy the natural human impulse to run, to jump, to throw, to swim, to dance. Sport and recreation are, in themselves, eminently worthwhile and desirable. . . . Our aim should be, I suggest, to inspire everyone to become involved in sport and recreation by making the choice irresistible in its scope and variety. . . . I see health as a happy consequence of sporting activity.
>
> —Sir Roger Bannister, 1989

Physical activity has much to offer the athlete and nonathlete alike. It provides recreation good for the body and soul, it doesn't have to cost anything, and it benefits almost everyone. At its best, physical activity is play with happy consequences for health and well-being. As long as there are activities people enjoy doing, there's an "athlete" in everyone.

The "Happy Consequences" of Physical Activity

Ask people who are in good physical condition what they get from exercise, and you're likely to hear a variety of responses. Someone who is 20 years old may say he wants to stay in shape and improve his stamina. A 40-year-old individual may say that exercise helps keep her cholesterol and weight from getting too high. People who are 80 might tell you they exercise so that they won't have to use a walker and will be able to maintain their independence. Ask children why they exercise, however, and they may not understand the question. Children engage in active play; they don't exercise. For them, fitness is truly an unintended consequence of play.

Regular physical activity benefits both physical and psychological health in people of all ages (Table 26.1). By improving a person's physical health, regular exercise may help ward off heart disease, some types of cancer, hypertension and stroke, osteoporosis, back injury, and diabetes. It tends to increase a person's feeling of well-being and helps relieve depression, anxiety, and stress.[1]

The Bonus Pack: Exercise Plus a Good Diet Exercise benefits health most when combined with a good diet and other healthy behaviors. The risk of developing heart disease, for example, is meaningfully lower when people combine a diet low in animal fat and cholesterol with regular exercise than when exercise alone is relied upon for protection against heart disease.[2] Regular physical activity helps build bone mass and reduces the risk of osteoporosis. But the risk is lowered to a greater extent if exercise is combined with a diet that supplies adequate amounts of calcium and vitamin D. Some of the other benefits of exercise, such as a reduced chance of developing colon and breast cancer, may be related to the effect of exercise on body fat content. Regular exercise is one of the few factors yet identified that helps people achieve long-term weight control.[1]

Exercise and Body Weight Combined with a moderate decrease in usual caloric intake, on the order of 200 calories per day, exercise helps people lose fat, build muscle mass, and become physically fit.[3,4] Because the body needs more calories to maintain muscle than fat, exercise that results in an increase in muscle mass leads to an increase in caloric requirements. For some people, this increase in caloric requirement makes it easier to maintain weight.[5]

Exercise combined with a stable caloric intake can lead to weight loss, but the loss is generally smaller than can be achieved by reducing caloric intake.[6] Even if weight loss is not intended or is achieved at a slow pace, the happy consequences of exercise come to all basically healthy individuals who undertake it on a regular basis.

Table 26.1 Benefits of regular physical activity.

Reduced Risk of Certain Diseases and Disorders

- Heart disease
- Colon and breast cancer
- Hypertension
- Stroke
- Osteoporosis
- Back and other injuries
- Obesity
- Diabetes (non-insulin dependent)
- Bone and joint diseases

Photo Disc

Improved Sense of Well-Being

- Increases feeling of well-being
- Decreases depression and anxiety
- Helps relieve stress

Physical Activity and Fitness

➡ **physical fitness**
The health of the body as measured by muscular strength, endurance, and flexibility in the conduct of physical activity.

Many of the benefits of physical activity are related to the ***physical fitness*** it can produce. Physical fitness is not defined by bulging muscles, thin waistlines, or amount of physical activity. Obese as well as thin people can be physically fit or not.[7] According to the American College of Sports Medicine, it is a state of health measured by strength, endurance, and flexibility.[8]

The strength component of physical fitness relates to the level of maximum force that muscles can produce. Endurance refers to the length of time muscles can perform physical activities, and flexibility refers to a person's range of motion. Physical fitness results when all three exist.

Muscle Strength In adults, muscles grow in size and strength when muscle cells increase in size. Muscle cell size—and correspondingly the strength of muscles—increases in response to weight-bearing or resistance exercise. Activities that require muscles to work harder than usual increase muscular strength.[9]

To increase muscle strength, people should lift or push against a heavy weight; to increase muscle endurance, they should lift lighter weights and push against them repeatedly. Resistance training should include exercises that build both muscle strength and endurance.

Endurance: A Measure of Aerobic Fitness How long a person can perform an activity depends on inherited traits and conditioning.[10] Both contribute to a person's stamina—or the extent of one's ability to deliver oxygen to muscles and the ability of muscles to use the oxygen for work. The amount of oxygen an individual is able to deliver to muscles that can be used by muscles for physical activity corresponds to the level of ***aerobic fitness.*** The more aerobically fit a person is, the longer and harder she or he is able to exercise.

➡ **aerobic fitness**
A state of respiratory and circulatory health as measured by the ability to deliver oxygen to muscles, and the capacity of muscles to use the oxygen for physical activity.

➡ **maximal oxygen consumption**
The highest amount of oxygen that can be delivered to, and utilized by, muscles for physical activity. Also called *VO_2 max* and *maximal volume of oxygen.*

Aerobic activities include jogging, basketball, swimming, soccer, and other low- and moderate-intensity activities. They give the whole body, or most of it, a continuous workout.

How Is Aerobic Fitness Determined? Aerobic fitness is classically assessed by measuring ***maximal oxygen consumption*** (abbreviated as VO_2 max) in a specially equipped laboratory (Illustration 26.1). In the lab, individuals are exercised at increasingly higher intensities, for example, by elevating the grade or speed of a treadmill. The individual performing the exercise breathes through a tube that delivers room air. Monitoring equipment measures the amount of oxygen from the air that is breathed and utilized during the exercise. The maximal amount of oxygen a person delivers to working muscles is the amount used when the intensity of exercise can no longer be increased. The higher the level of oxygen utilized at the peak level of activity, the higher the level of aerobic fitness and the longer physical activity can be performed.

People can perform physical activity at 100% of VO_2 max for only a few minutes. Consequently, aerobic fitness goals are set below that level. In general, it is recommended that beginners start an aerobic fitness program with a goal of exercising at 40 to 60% of VO_2 max and working up to a higher level.[11] Aerobically fit people may train at 70 to 85% of VO_2 max.

How Do You Know Your VO_2 Max? There are several ways to estimate VO_2 max that don't require a supervised laboratory test. One commonly employed method uses the percentage of maximal heart rate. This formula uses heart rate as a proxy measure of oxygen consumption and provides a fairly good estimate of VO_2 max because the amount of blood circulated to muscles is roughly equivalent to the number of times the heart sends out a pulse of blood each minute.

Custom Medical Stock

Illustration 26.1
Determining VO_2 max. This man is running on a treadmill. His nose is plugged, so he must breath through a tube in his mouth. The tube is attached to an apparatus that measures the total amount of air breathed in and out, and the difference between the amount of oxygen inhaled and exhaled. Every few minutes the slope or speed of the treadmill is increased, making the exercise more intense. The amount of oxygen he utilizes at the point when his exercise intensity can go no higher is considered "maximal oxgen consumption," or VO_2 max.

The more oxygen muscles need to perform an activity, the more oxygen the lungs have to deliver and the faster the heart has to beat (that is, up to the point where the lungs and heart can no longer send an additional supply of oxygen to muscles). The heart rate at which the highest level of oxygen consumption by the body occurs is considered "maximal heart rate." It roughly corresponds to VO_2 max.

The formula for estimating maximal heart rate is quite simple: subtract your age from 220. To obtain a target heart rate for exercise, multiply the result by the desired percentage of maximal heart rate. (Table 26.2 shows an example.) The resulting figure is your "target heart rate" for aerobic exercise. Table 26.3 lists maximal heart rate, 60% of the maximum, and 75% of the maximum rate for different ages.

How Do You Know Your Heart Rate? You can determine if you are exercising at the appropriate level for aerobic fitness by measuring your pulse rate during a break or as soon as you're done exercising. To take your pulse, place a fingertip gently on an artery in your wrist or neck (Illustration 26.2). Count the number of pulses you feel in a 10-second period and multiply that number by 6. That's your heart rate, measured as heartbeats (or pulses) per minute.

Flexibility Stretching major muscle groups to the point of mild discomfort increases the range of motion of joints and muscles. It helps protect muscles and their connective tissues from injury and you from resulting stiffness and pain. Stretching also helps your muscles and joints warm up prior to exercise, and to cool down after exercise. Appropriate stretching involves slow stretches of 15 to 20 seconds, with no bouncing, for about 10 minutes three to seven days a week. Breathe rhythmically during stretches and other exercises. Your body needs the oxygen.

Table 26.2
Applying the percentage of maximal heart rate (MHR) formula to a 22-year-old individual who will exercise at 60% of MHR.

- Target heart rate = (220 − age) × %MHR
- Target heart rate = (220 − 22) × 0.60

  = 198 × 0.60

  = 119

Table 26.3
Target heart rates estimated at 60% and 75% of maximum for people at different ages using the percentage of maximal heart rate formula.

Age (Years)	Average Maximal Heart Rate (MHR)	60% of MHR	75% of MHR
20	200	120	150
25	195	117	146
30	190	114	142
35	185	111	138
40	180	108	135
45	175	105	131
50	170	102	127
55	165	99	123
60	160	96	120
65	155	93	116
70	150	90	113

PhotoEdit - Michael Newman

Illustration 26.2
To find out if you're exercising at the right level, gently take your pulse during a break or right after you finish exercising.

Nutrition and Fitness

Your diet can make a strong contribution to your level of physical fitness. That's because good diets contribute to health, and it is easier to become physically fit if you are in good health. Diet also makes a difference to physical fitness because the foods you eat serve as the source of energy for physical activity. For some activities, the body needs glucose as its primary energy source. For others, the fuel of choice is fat.

Muscle Fuel Muscles use fat, glucose, and amino acids for energy. The proportion of each that is used, as well as the amount, depends on the intensity of activity (Illustration 26.3). When we're inactive, fat supplies between 85 and 90% of the total amount of energy needed by muscles. The rest is provided by glucose (about 10%) and amino acids (5% at most).[12] Fat is also the primary source of fuel for activities of low-to-moderate intensity such as walking, running, and swimming. As can be seen from Illustration 26.3, amino acids make a relatively small contribution to meeting energy needs for physical activity.

Oxygen is required to convert fat into energy. That makes activities of low-to-moderate intensity "aerobic"—or "oxygen requiring." Consequently, fat-burning, oxygen-requiring exercises are the type used to increase aerobic fitness. Jogging, long-distance running, swimming, and aerobic dance are all aerobic activities.

High-intensity, short-duration activities like spiking a volleyball, running down a deep fly ball, or sprinting down the block to catch a bus are fueled primarily by glucose. Our supply of glucose for intense activities comes principally from glycogen, the storage form of glucose. Glycogen is stored in muscles and the liver and can be rapidly converted to glucose when needed by working muscles. The conversion of glucose to energy for intense activity doesn't require oxygen (it's *an*aerobic). People can undertake very intensive activity only as long as their stores of glucose last.

Activities such as basketball, hockey, tennis, and soccer that involve walking, running, sprinting, and the occasional high-intensity, quick move use mainly fat or glucose as the energy source depending on the intensity of the physical effort.

Stiff Muscles in the Out of Shape People who are out of shape and exercise intensely, or who change the type of exercise performed, may experience stiff muscles 24 to 48 hours after exercise (Illustration 26.4).[13] Although the cause is not clear, the stiffness usually lasts a day or two. The probability this will happen decreases as people get into shape.

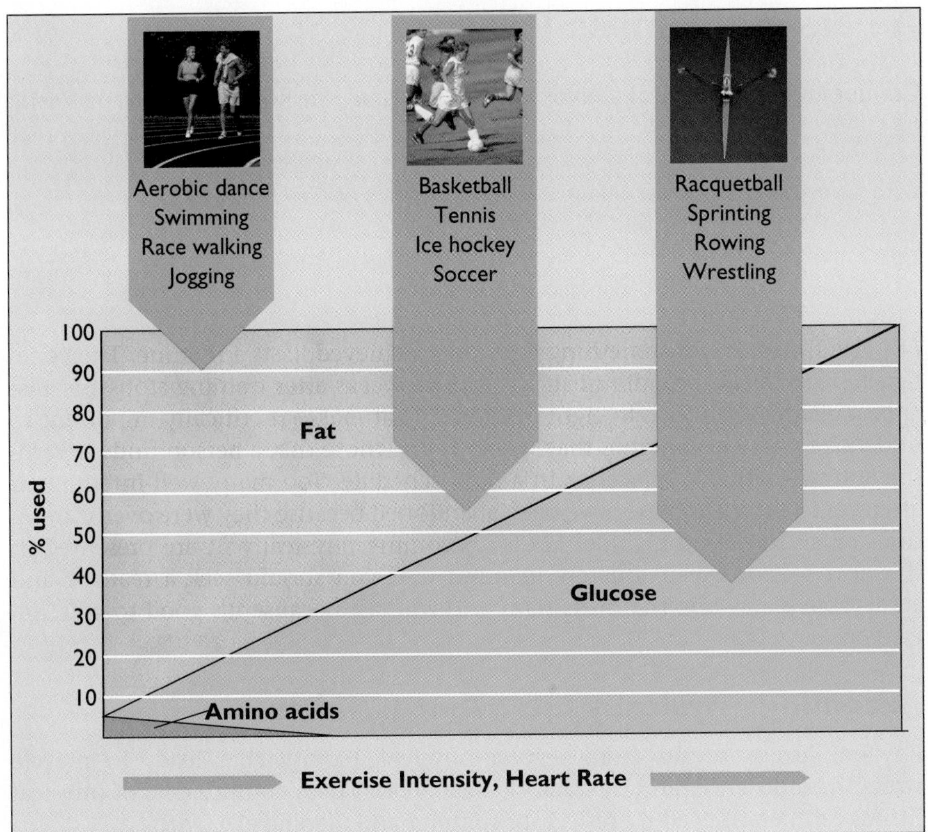

Illustration 26.3
Schematic representation showing the proportionate use of sources of energy for physical activities of various intensities. Most activities are fueled by both fat and glucose.

Diet and Aerobic Fitness The best type of diet for physical fitness is the same one that is recommended for people in general. The recommendations for healthy eating described in the Dietary Guidelines for Americans and the Food Guide Pyramid (presented elsewhere in this book) promote physical fitness because they facilitate maintenance of normal weight and adequate intake of carbohydrates, dietary fiber, protein, vitamins, and minerals. Diets that promote physical fitness most certainly include well-planned vegetarian diets. There are no supplements, protein powders, or herbal mixes that enhance physical fitness or strength beyond the levels obtained from a healthy diet and regular exercise.[14]

Actually, people who exercise tend to have better diets than people who don't. They are often health-conscious, select foods with care, and, because they generally eat more than sedentary people, have a better chance of getting all the nutrients they need.[15]

Photo Disc

Illustration 26.4
She's going to feel this tomorrow. People may get stiff muscles if they exercise too hard when they're out of shape.

A Reminder about Water Physical activity increases the body's need for water, and if the climate is hot and humid, this need increases even more. In general, people should consume two cups of water two hours prior to exercise and 4 to 8 ounces (½ to 1 cup) of water every 15 to 20 minutes during the exercise. For exercise that lasts over an hour, alternating water with a sports drink or diluted fruit juice may improve hydration.[16] It's important to keep up fluid intake when exercising in cold weather. Cold air holds less water vapor than hot air, so you lose more water through breathing in cold weather.

A Personal Fitness Program

Commenting on physical activity for the long run, Sir Roger Bannister has said:

> The best advice to those who are unfit is to take exercise unobtrusively. I am not an enthusiast for fitness and exercise schemes that are boring because they tend to fizzle out. . . . The vast majority of people, I think, can only be attracted for any length of time towards recreational activities if these are rewarding, enjoyable, and satisfying in themselves.

Physical fitness is not something that, once achieved, lasts a lifetime. The beneficial effects of training diminish within two weeks after training stops and disappear altogether in two to eight months.[1] That makes it critically important to undertake physical activities that "wear well"—those that a person finds rewarding and will reserve a place for in a busy schedule. Too many well-intentioned attempts to get into shape have been abandoned because they were overly ambitious or no fun. As the guidelines for becoming physically fit are presented in the next section, keep squarely in mind that you should seek a realistic and achievable plan—one that feels right and will last because it's good to you and for you.

Becoming Physically Fit: What It Takes

Physical fitness results from regular physical activity that leads to muscle strength, endurance, and flexibility (Table 26.4). These components of physical fitness can be achieved by resistance training and aerobic exercise.

Table 26.4
An overview of the components of physical fitness training.

A Resistance Training Plan The resistance training program for physical fitness doesn't prepare people for bodybuilding contests. It increases muscle

(Top) Moving muscles against resistance for strength.
(Middle) Aerobic exercise for oxygen delivery.
(Bottom) S-T-R-E-T-C-H-I-N-G for flexibility.

A. **Resistance training:**
- Goal: To improve muscle strength, flexibility, and endurance.
- Components: Two to three resistance training sessions per week.
  - → Each session includes 10 exercises repeated in sets of 8 to 12.
  - → Exercises should involve all major muscle groups.

B. **Aerobic training:**
- Goal: To increase the ability of the respiratory and circulatory systems to deliver oxygen to working muscles (to build "cardiorespiratory" fitness).
- Components: Three exercise sessions per week (lasting 20 to 30 minutes each) at 60 to 75% of maximal heart rate.
  - → Aerobic exercise should be a continuous activity (not stop-and-go) and involve all or most major muscle groups.

C. **Flexibility:**
- Goal: Increase range of motion of muscles and joints.
- Components: Stretching of all major muscle groups.
  - → Stretching about 10 minutes three to seven times a week.
  - → Stretch arm, legs, back, and stomach area slowly in multiple directions to the point of mild discomfort. Hold each stretch 15 to 20 seconds. Don't forget to breathe during the stretches, and don't bounce.
  - → Warm up prior to exercise, and cool down afterwards, by stretching your muscles and joints.

strength, flexibility, and endurance and improves the performance of aerobic exercises.

The American College of Sports Medicine recommends two to three resistance training sessions per week. Two sessions are enough for those who are not interested in "bulking up." Strength training twice weekly will give you 70 to 80% of the benefit of training three times a week.

Each workout session includes about 10 different exercises repeated 8 to 12 times each. Each set of exercises should leave the muscles exercised feeling a bit tired. Exercises selected should involve each of the major muscle groups in the legs and buttocks, arms and shoulders, and abdomen and back. Table 26.5 lists examples of exercises for the different muscle groups.

Resistance training can also be used for additional gains in muscular strength. Muscle strength is increased by the use of progressively heavier weights in workouts.

On the Side

Photo Disc

What's the largest muscle in your body? If your body is like most, its largest muscle is the gluteus maximus (buttocks).

The Aerobic Fitness Plan

Aerobic fitness can be achieved by exercising at 60 to 75% of maximal heart rate for 20 to 30 minutes, three times or more each week (Illustration 26.5). The best kind of aerobic exercise is one that involves all, or most, of the body's major muscle groups in the workout. The activity should be continuous (like jogging) and not stop-and-go (like tennis, for example). Beginners should start at 40 to 60% of maximum with low-intensity activities, such as alternating walking with jogging, and increase to a higher level as exercise gets easier. Exercises such as swimming, jogging, running, aerobic dance, and cycling are good choices for subsequent aerobic exercise. Activities such as mowing the lawn, shoveling dirt or snow, and household tasks that elevate heart rate to the target level for 20 to 30 minutes are also aerobic exercise. Actually, the most common physical activities of American adults are walking, gardening, and yard work.[1]

The U.S. Surgeon General's recommended levels of physical activity would benefit health and perhaps contribute to a decline in the rising incidence of obesity in the United States.[1] Only 15% of Americans exercise three or more times per week for at least 20 minutes each time,[1] so the surgeon general set modest goals that have a better chance of being adopted by sedentary people than more beneficial but also more rigorous goals. The recommendations call for people to engage in the following types of activities on most, if not all, days of the week:

- 30 minutes of brisk walking, bicycling, leaf raking, housecleaning, calisthenics, in-line skating, and/or similar activities, or
- 15 minutes of running, rope skipping, vigorous swimming, or similar activities.

To complement these aerobic activities, the surgeon general also suggests that people engage in strength-building exercises twice a week.

Table 26.5
The match between resistance exercises and the body's major muscle groups.

Muscle Group Exercised	Activity
Legs and buttocks	Stair climbing, resistance leg pushing, cycling
Arms and shoulders	Free weight lifts, pull-ups, push-ups
Abdomen and back	Abdominal curls, sit-ups, leg lifts (front and side)

Illustration 26.5

Source: © 2001 United Feature Syndicate, Inc. Reprinted by permission.

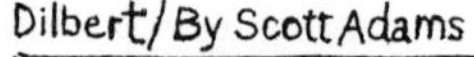

Some Exercise Is Much Better Than None Sedentary people who take up walking, golfing, biking, gardening, or other similar activities improve their fitness levels. Duration of low- and moderate-intensity exercises counts toward improved blood cholesterol concentrations, decreased body fat, and aerobic fitness.[17]

The Caloric Value of Exercise Some people prefer to base their aerobic exercise workout on a particular level of calories (such as 200), while others simply like to know how many calories they're burning off. Table 26.6 lists the caloric value of various exercises in case you would like to know.

A Cautionary Note People who have a chronic disease such as diabetes or hypertension, or are out of shape and over 40, should consult a physician before starting an exercise program. (It may well turn out that exercise will be just what the doctor orders.) Moderate exercise programs (those undertaken at 40 to 60% of VO_2 max) are beneficial and safe for almost everyone.[1]

The Paradox of Death During Exercise "Famous athlete dies while jogging" reads the headline. Why would this happen if exercise is good for you? That's the exercise paradox.

Most people who die while shoveling snow, playing basketball, or jogging are those with existing heart disease either due to hardened arteries (atherosclerosis) or an inborn heart problem. It is a very rare event (because exercise really is good for the heart) and the risk can be reduced by regular exercise versus sporadic and intense physical activity.[18]

U.S. Fitness: America Needs to Shape Up

The benefits of regular exercise are sufficient to make it a matter of public health policy in the United States. Americans have much to gain by getting in better shape.

Studies are undertaken periodically to see how well Americans are doing at increasing their physical activity level. The results are not encouraging (Illustration 26.6). Approximately 75% of adults do not achieve recommended levels of physical activity.[19] Only one in three students in U.S. schools has physical education every school day (yet it's the favorite subject of most young schoolchildren).[20] Many students engage in regular physical activity at school only if they're involved in competitive sports. In addition, less than one in seven adults engages in any type of physical exercise for 30 minutes or more most days of the week.[1]

On the Side

Can you trust your calorie counter? Perhaps not. Fitness equipment calculators often overestimate individuals' caloric expenditures by more than 30%.

A Focus on Fitness in Children Physical fitness advocates are especially concerned about the lack of opportunity for physical

Table 26.6
Average energy output per pound of body weight for selected types of exercise.

Exercise	Intensity	Calories (Pound/Hour)	Exercise	Intensity	Calories (Pound/Hour)
Walking	3 mph (20 min/mi)	1.6	Weight lifting		2.9
	3½ mph (17 min/mi)	1.8	Wrestling		6.2
	4 mph (15 min/mi)	2.7	Handball	Moderate	4.8
	4½ mph (13 min/mi)	2.9		Vigorous	6.2
Jogging	5 mph (12 min/mi)	4.1	Swimming	Resting strokes	1.4
	5½ mph (11 min/mi)	4.5		20 yd/min (mod.)	2.9
	6 mph (10 min/mi)	4.9		40 yd/min (vig.)	4.8
	6½ mph (9 min/mi)	5.2	Rowing	(sculling or machine)	4.8
	7 mph (8½ min/mi)	5.6	Downhill skiing		3.8
	7½ mph (8 min/mi)	6.0	Cross-country skiing (level)	4 mph (15 min/mi)	4.3
Running	8 mph (7½ min/mi)	6.3		6 mph (10 min/mi)	5.7
	8½ mph (7 min/mi)	6.7		8 mph (7½ min/mi)	6.7
	9 mph (6⅔ min/mi)	7.1		10 mph (6 min/mi)	7.6
	9½ mph (6⅓ min/mi)	7.4	Aerobic dancing	Moderate	3.4
	10 mph (6 min/mi)	7.8		Vigorous	4.3
	11 mph (5½ min/mi)	8.5	Racquetball/squash	Moderate	4.3
	12 mph (5 min/mi)	9.5		Vigorous	4.8
Rebound trampoline	50–60 steps/min	4.1	Tennis	Moderate	3.4
Cycling (stationary)	Mild effort	2.9		Vigorous	4.3
	Moderate effort	3.4	Volleyball	Moderate	3.4
	Vigorous effort	4.3		Vigorous	3.8
Cycling (level)	6 mph (10 min/mi)	1.5	Basketball	Moderate	3.8
	8 mph (7½ min/mi)	1.8		Vigorous	4.8
	10 mph (6 min/mi)	2.0	Football	Moderate	3.8
	12 mph (5 min/mi)	2.8		Vigorous	4.3
	15 mph (4 min/mi)	3.9	Baseball/golf/woodcutting/horseback riding/badminton/canoeing		2.4
	20 mph (3 min/mi)	5.7	Soccer/hill climbing/fencing/judo/snowshoeing		5.3
Skating		2.9	Bowling/archery/pool		1.2
Calisthenics	Moderate	2.4			
	Vigorous	2.9			
Rope skipping	Moderate	4.8			
Bench stepping	12" high, 24 steps/min	3.2			

Source: Values calculated from Guidelines for graded exercise testing and exercise prescription, 2d ed. Philadelphia: American College of Sports Medicine, Lea & Febiger; 1984.

activity in schools. They point out that children who are not physically active are more likely to be overweight and that a lifelong love for physical activity and a physically active lifestyle often develop during childhood.[21] Regular exercise benefits obese children by lowering body fat, increasing bone density, and improving aerobic fitness.[22] Yet, one-third of U.S. children are physically unfit, and most do not exercise enough to develop desirable levels of cardiovascular endurance.[1]

Schools are being encouraged to include physical activity as a daily part of the curriculum for all students in all grades. In addition, physical fitness goals should be stressed above competitive or sport performance goals.[23] Tests of

Illustration 26.6
The proportion of people aged six and older who engage in no leisure-time physical activity.

Sources: National Center for Health Statistics; Centers for Disease Control and Prevention.

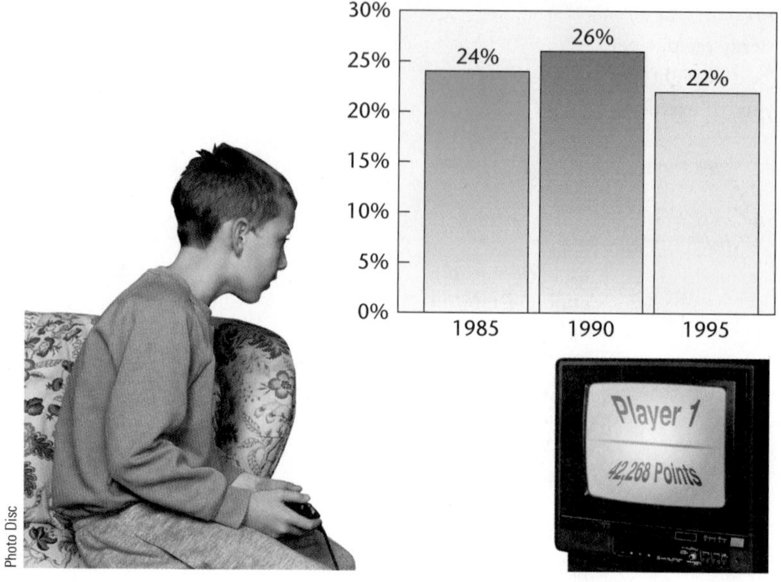

physical fitness, such as the muscle strength and endurance assessment shown in Illustration 26.7, should be performed periodically on all children and the results used to modify instruction and activities.

Physical activity programs for schoolchildren are most likely to be successful when they offer recreational and sporting activities that are irresistible and fulfill the human desire to run, jump, and play. The happy consequences of physical activity come easiest to those who do it for play.

Illustration 26.7
This child is performing modified pull-ups as part of a fitness assessment.

Greg Merhar, reprinted with permission of the Journal of Physical Education, Recreation and Dance, 59 (9), 1987.

Nutrition UP CLOSE

Exercise: Your Options

FOCAL POINT:
Assess your level of physical activity.

More than one-quarter of U.S. adults are physically inactive, and most Americans do not exercise regularly. How much physical activity are you getting?

Physical Activity

Check the category that best describes your usual overall daily activity level. *Do not* consider activities that are part of an exercise program.

- *Inactive:* Sitting most of the day, with less than two hours moving about slowly or standing. ☐
- *Average:* Sitting most of the day, walking or standing two to four hours each day, but not engaging in strenuous activity. ☐
- *Active:* Physically active four or more hours each day. Little sitting or standing, engaging in some physically strenuous activities. ☐

Exercise

Do you exercise three times a week or more for at least 20–30 minutes? ☐ **Yes** ☐ **No**

FEEDBACK (including scoring) can be found at the end of Unit 26.

WWW links

www.cyberdiet.com
Assessment tools provided to determine target heart rate, calories used in exercises, and more.

www.foundation.sdsu.edu/projects/spark/index.html
The "Sparks" site offers effective physical education materials and services for children K–6 grade to schools, recreation departments, after-school programs, and other nonprofit organizations.

www.activeliving.csep.ca (Canada), www.acefitness.org (U.S.)
Canadian/American Council on Exercise offers policy information, research results, and other data relevant to improving levels of exercise.

www.healthfinder.gov
A government link to fitness and health information.

REFERENCES

1. Physical activity and health: a report of the surgeon general. Washington, DC: U.S. Department of Health and Human Services; 1996.
2. Johnson C, et al. Effects of exercise, dietary cholesterol, and dietary fat on blood lipids. Arch Int Med 1990;150:137–43.
3. Broeder CE, et al. The effects of either high-intensity resistance or endurance training on resting metabolic rate. Am J Clin Nutr 1992;55:802–10.
4. Broeder C, et al. The effects of aerobic fitness on resting metabolic rate. Am J Clin Nutr 1992;55:795–801.
5. Foreyt JP, Goodrick GK. Factors common to successful therapy for the obese patient. Med Sci Sports Exer 1991;23:292–7.
6. Saris WHM. Physiological aspects of exercise in weight cycling. Am J Clin Nutr 1989;49(suppl):1099–1104.
7. Wei M, et al. Relationship between low cardiorespiratory fitness and mortality in normal weight, overweight, and obese men. JAMA 1999;278:1547–53.
8. ACSM Fitness Book, 2nd edition. American College of Sports Medicine, Human Kinetics Publishers, 1998.
9. Coleman E. Eating for endurance. Palo Alto (CA): Bull Publishing Co.; 1988.
10. Bouchard C, et al. Heredity and trainability of aerobic and anaerobic performances. Sports Med 1988;5:69–73.
11. Position of the American Dietetic Association and the Canadian Dietetic Association: nutrition for physical fitness and athletic performance for adults. J Am Diet Assoc 1993;93:691–6.
12. Horton ES. Metabolic fuels, utilization, and exercise. Am J Clin Nutr 1989;49(suppl):931–2.
13. Satlin B, Astrand P-O. Free fatty acids and exercise. Am J Clin Nutr 1993;57(suppl):752S–8S.
14. Barrett S. Don't buy phoney "ergogenic aids." Nutrition Forum 1997;May/June: 17, 19–21, 24.
15. Nieman DC, et al. Nutrient intake of marathon runners. J Am Diet Assoc 1989;89:1273–8.
16. Kleiner SM. Water: an essential but overlooked nutrient. J Am Diet Assoc 1999;99:200–6.
17. Lakka T, et al. Cardiorespiratory fitness and the progression of carotid atherosclerosis in middle-aged men. Ann Intern Med 2001;134:12–20.
18. Maron Bj. The paradox of exercise. N Engl J Med 2000;343:1409–11.
19. Physical activity trends, United States, 1990–1998. MMWR 2001;50:Mar 9.
20. White AA, Skinner JD. Adolescents' perception about their nutrition education needs. Tenn Farm and Home Sci 1989;Spring:5–19.
21. Health Objectives for the Nation, 2010. Physical Activity. www.cdc.gov/nchs; 8/00.
22. Barbeau P, et al. Correlates of individual differences in body composition resulting from physical training in obese children. Am J Clin Nutr 1999;69:705–11.
23. Kuntzleman CT. Childhood fitness: What's happening? What needs to be done? Prev Med 1993;22:520–32.

Nutrition

UP CLOSE

Exercise: Your Options

Feedback for Unit 26

For those of you who answered "Active" to the first question and "Yes" to the second question, congratulations! You qualify as a mover and shaker. Keep up the good work!

If you answered "Inactive" or "Average" to the first question or "No" to the second, now is a good time to think about increasing your level of physical fitness. Besides promoting a feeling of well-being, regular exercise can help beat stress, weight gain, heart disease, and high blood pressure. Maybe it's time you exercised your options by incorporating more physical activity into your daily routine.

Unit 27

Nutrition and Physical Performance

Photo Disc

Nutrition Scoreboard

	TRUE	FALSE
1. The diet an athlete consumes affects energy substrate availability to muscles during exercise.		
2. Carbohydrate loading is a waste of time.		
3. Commercially available protein powders and bars build muscle if consumed while training.		
4. Scientists can't explain why, but bee pollen actually improves physical performance.		

(Answers on next page)

Answers to

Nutrition Scoreboard

	TRUE	FALSE
1. Carbohydrate and fat availability to muscles during exercise varies based on usual diet.	✓	
2. Carbohydrate loading can work for endurance events if combined with the right training program.		✓
3. Protein powders or bars have not been found to increase muscle mass whether consumed during training or not.		✓
4. In bees, maybe—in humans, absolutely not.		✓

Key Concepts and Facts

- → Endurance is affected by genetics, training, and nutrition.
- → Glycogen stores and endurance can be increased by diet and training.
- → Water is the best fluid replacer for most physical events.
- → Carbohydrate loading, carbohydrate intake during exercise, and fluid replacement improve physical performance in endurance events.
- → Most ergogenic aids that claim to help athletes do *not* work.

Sports Nutrition

glycogen
The storage form of glucose. Glycogen is stored in muscles and the liver.

BAM! Nobody heard it, but Lou felt it. She had "hit the wall." She was ahead of her planned pace, but now her legs felt like lead. She would have to finish the last two miles of the marathon at the slow pace her legs would allow.

From her carefully crafted and scrupulously followed training program to her refined shaping of mental attitude, Lou thought she had done everything right. She had left one thing out, however, and that may have cost her the race. Lou failed to pay attention to her diet while training and ran out of ***glycogen*** too soon (Illustration 27.1).

Illustration 27.1
Hitting the wall. This runner ran out of muscle glycogen before the finish line.

PhotoEdit - Tony Freeman

Three major factors affect physical performance: genetics, training, and nutrition.[1] The first gives some people an innate edge in sprinting or endurance, and nothing can be done about it. The second is acknowledged as a basic truth. Most athletes know a good bit about proper training, and the trick is to follow the right plan. The third is often ignored or, when taken seriously, misunderstood.

Nutrition has important effects on physical performance, but the legitimate role of nutrition is often poorly understood by athletes and coaches alike.[2] Athletes' lack of knowledge may make them vulnerable to phony nutrition claims and bad dietary advice. It leaves some coaches wishing they had taken a college course on nutrition. Only 1 in 10 coaches has had a course in nutrition, but nearly all of them regularly dispense nutritional advice.[3] (To the coaches and future coaches who are taking this nutrition course: Bravo!)

Basic Components of Energy Formation during Exercise

Understanding the role of nutrition and performance, and the potential effects of some ***ergogenic aids,*** can be fostered by basic knowledge of how energy is formed within muscle cells. The processes by which energy for muscle movement is formed are overviewed in Illustration 27.2.

There are two main substrates for energy formation in muscles: glucose from muscle and liver glycogen, and fatty acids released from fat stores. How much of each is used depends on the intensity and duration of the exercise, as well as the body's ability to deliver each along with oxygen to muscle cells. Each substrate is used to form ATP (adenosine triphosphate) from ADP (adenosine diphosphate). ATP serves as the source of energy for muscle contraction.

ergogenic aids (ergo = work, genic = producing) Substances that increase the capacity for muscular work.

Anaerobic Energy Formation Glucose from the liver and muscle glycogen (which is converted to glucose) form ATP without oxygen. This route of energy formation is "anaerobic," or "without oxygen," and fuels intense muscular work. Creatine phosphate (abbreviated CrP in Illustration 27.2), an amino acid containing a high energy phosphate molecule, also converts ADP to ATP to a limited extent. Creatine phosphate stores are limited and decrease rapidly during intensive exercise.

Glucose is converted to "pyruvate" during anaerobic energy formation. In the absence of oxygen, pyruvate is converted to lactate. Lactate can build up in muscles and blood if not reconverted back to pyruvate by the addition of oxygen. Pyruvate yields additional energy when it enters "aerobic" energy formation pathways along with fatty acids from fat stores.

Aerobic Energy Formation The conversion of pyruvate and fatty acids to ATP requires oxygen. Much more ATP is delivered by the breakdown of fatty acids than glucose (fats have nine calories per gram, while glucose only has four). The rate of energy formation from fatty acids is four times slower than that from glucose, however. It's the reason fatty acids are used to fuel low- and moderate-intensity exercise. Unlike glucose, energy formation from fatty acids is not limited by availability. Muscle cells can continue to produce energy from fatty acids as long as delivery of oxygen from the lungs and the circulation is sufficient.[4]

Nutrition and Physical Performance: What's the Match?

Glycogen stores, foods eaten before and between events, water intake, and ***electrolyte*** balance are all related to energy formation and physical performance.

electrolytes Minerals such as sodium and potassium that carry a charge when in solution. Many electrolytes help the body maintain an appropriate amount of fluid.

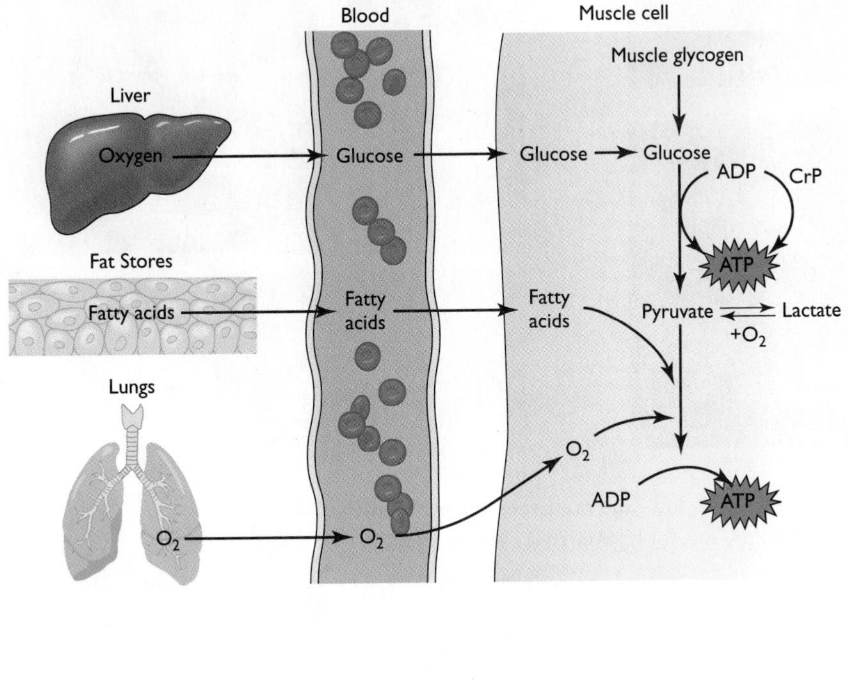

Illustration 27.2 Schematic representation of how ATP is formed for muscular movement.

Glycogen Stores and Performance The vast majority of energy for muscle activity comes from fat and glucose. For most activities, both types of fuel are used, and the more intense the activity, the higher the amount of fat and glucose used for it. Fats, however, are the principal source of fuel for activities of low to moderate intensity, and glucose is the main source of energy for activities of high intensity (Illustration 27.3). Most people have a more than sufficient supply of fat that can be used to fuel activities of low and moderate intensity and enough glucose for high-intensity events of short duration. When participating in endurance events, however, they may need to build up their reserves of glycogen—the storage form of glucose.

Glycogen is stored in muscles and the liver and is much less abundant than body fat stores. Consequently, a person's ability to perform continuous, intense physical activity is limited by the amount of glycogen stored.[5] People who run out of muscle glycogen during an event "hit the wall"—they have to slow down their pace substantially because they can no longer use muscle glycogen as a fuel. The pace they are able to maintain will be dictated by the body's ability to use fat and liver glycogen as fuel for muscular work. If athletes keep pushing themselves after muscle glycogen runs out, they may end up "bonking"—using up the liver's supply of glycogen. That's worse than hitting the wall because hypoglycemia (low blood sugar) develops and the person becomes dizzy and shaky and may pass out.[6] Obviously, endurance athletes do not want to exhaust their glycogen stores before an event is over.

Athletes consuming a typical U.S. diet normally have enough glycogen stores to fuel continuous, intense exercise for about an hour or two.[7] Both glycogen stores and endurance can be increased, however, by loading up muscles with glycogen (Table 27.1).

Carbohydrate Loading The recommended method for increasing glycogen stores for endurance events is much less stringent than in years past. It basically calls for increasing carbohydrate intake to 60 to 70% of calories three days prior

Activities primarily fueled by fat.

Photo Disc

Photo Researchers

Photo Researchers

Activities primarily fueled by glycogen.

PhotoEdit - José Carillo

Photo Disc

Photo Researchers

Illustration 27.3
Fat is the main source of energy for low- and moderate-intensity activities whereas glycogen is the primary fuel for high-intensity events.

Carbohydrate Level of Diet for Previous 3 Days	Muscle Glycogen (Grams per 100 Grams of Muscle)	Time to Exhaustion at 75% VO_2 max
Low	0.6	60 minutes
Average	1.8	126 minutes
High (65–70% of total calories)	3.5	189 minutes

Table 27.1
Effect of diet on muscle glycogen stores during training.[8]

to an event (Illustration 27.4). Exercise time should be decreased during this three-day period, ending with complete rest the day before the event to conserve glycogen stores.[5] Table 27.2 shows how many grams of carbohydrate correspond to 60 or 70% of total calories for people with differing caloric needs. Average diets provide about 50% of calories from carbohydrates.

Breads, pastas, rice, cereals, and starchy vegetables should be the main food sources of carbohydrate during carbohydrate loading (Illustration 27.5). An example of a one-day diet that provides 68% of total calories from carbohydrate is shown in Illustration 27.6.

High-carbohydrate diets may provide low amounts of zinc, iron, vitamin B_{12}, and other nutrients, so care should be taken to consume a wide variety of foods during carbohydrate loading. The Food Guide Pyramid adjusted for the number of servings recommended for active men and women (Illustration 27.7) may provide a useful guide for designing an adequate high-carbohydrate diet.

There is a limit on glycogen storage in muscle. Each gram of glycogen binds with two grams of water, limiting the total amount that can "fit" into muscle tissue. Once the capacity of muscles to store glycogen is reached, any additional carbohdyrate is largely converted to fat and stored.[9]

Not all athletes tolerate the relatively high-carbohydrate, low-fat, carbohydrate-loading diet equally well. Some athletes dislike the stiff, heavy feeling that may

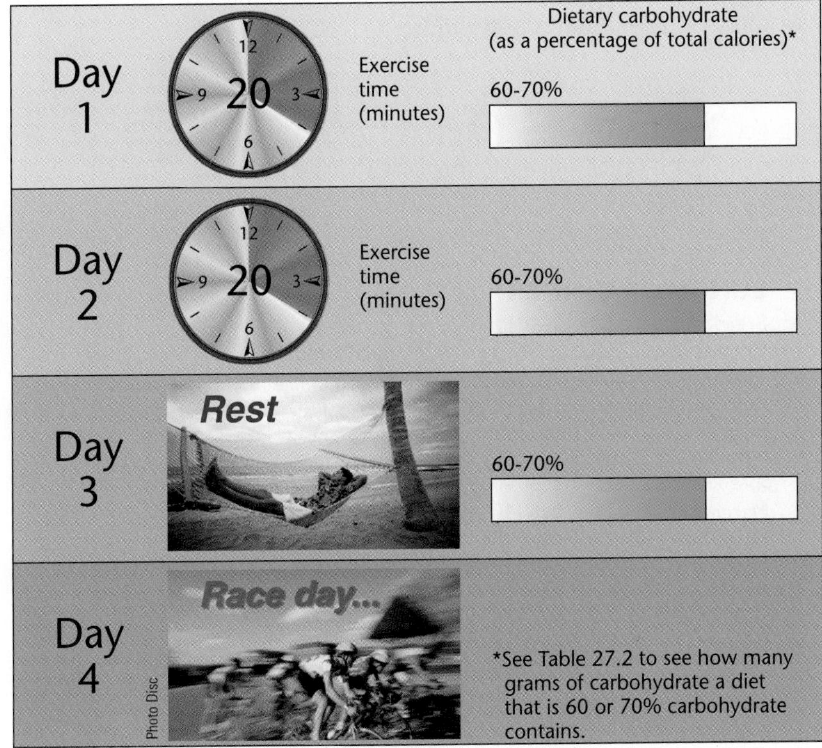

Illustration 27.4
A schedule for carbohydrate loading.

Table 27.2
How many grams of carbohydrate are in a diet that provides 60 or 70% of total calories from carbohydrates? It depends on your caloric intake. (Food sources of carbohydrate are shown in Illustrations 27.5 and 27.6.)

Usual Caloric Intake	Grams of Carbohydrate 60% Carbohydrate Diet	Grams of Carbohydrate 70% Carbohydrate Diet
1600	240	280
1800	270	315
2000	300	350
2200	330	385
2400	360	420
2600	390	455
2800	420	490
3000	450	525
3200*	480	560

*For each additional 200 calories of usual intake, add 30 grams carbohydrate for a 60% carbohydrate diet and 35 grams for a 70% carbohydrate diet.

occur when muscles become "loaded" with glycogen. Carbohydrate loading, therefore, should not be attempted for the first time before an important event. In addition, it should be remembered that carbohydrate loading does not improve performance in athletes participating in events that last less than an hour or two. The body normally has enough glycogen to meet the needs of athletes participating in intense but short-duration activities.

Protein Need Many athletes require no more than their RDA for protein. Individuals undertaking strength training may need up to 15 grams more protein daily, however. Even then, most people (including athletes) consume much more than their RDA for protein so they are unlikely to need extra protein in their diets. Inadequate protein is not the reason athletes tire early or fail to build muscles.[10]

What Are Good Pre-Event Food Choices? Steak and egg pre-event meals may instill confidence, but if you were to ask your body what it needs, the reply would be carbohydrates and fluids.

Illustration 27.5
Foods loaded with carbohydrate. Serving size and grams (g) of carbohydrate are shown in parentheses.

Photo Disc

Illustration 27.6
A high-carbohydrate diet. Carbohydrates provide 68% of the 2420 calories in this day's menus, protein provides 11%, and fats provide 21%. Grams of carbohydrate are shown in parentheses.

BREAKFAST
Orange juice — 1 cup (25 g)
Oatmeal — 1 cup (25 g)
Brown sugar — 1 tablespoon (13 g)
Skim milk — 1 cup (12 g)
Banana — 1 (27 g)

SNACK
Granola — ½ cup (29 g)
Apple juice — 1 cup (30 g)

LUNCH
Lettuce salad — 2 cups (10 g)
Salad dressing — 2 tablespoons (4 g)
Macaroni & cheese — 1 cup (40 g)
Crackers — 4 squares (9 g)
Sliced tomato — ½ (3 g)
Skim milk — 1 cup (12 g)

SNACK
Cornflakes — 1 cup (20 g)
Canned peaches — ½ cup (25 g)
Skim milk — 1 cup (12 g)

DINNER
Spanish rice — 1 cup (41 g)
Baked potato — 1 (35 g)
Corn — ½ cup (17 g)
Apple crisp — ½ cup (43 g)
Iced tea — 2 cups (0 g)

Photo Disc

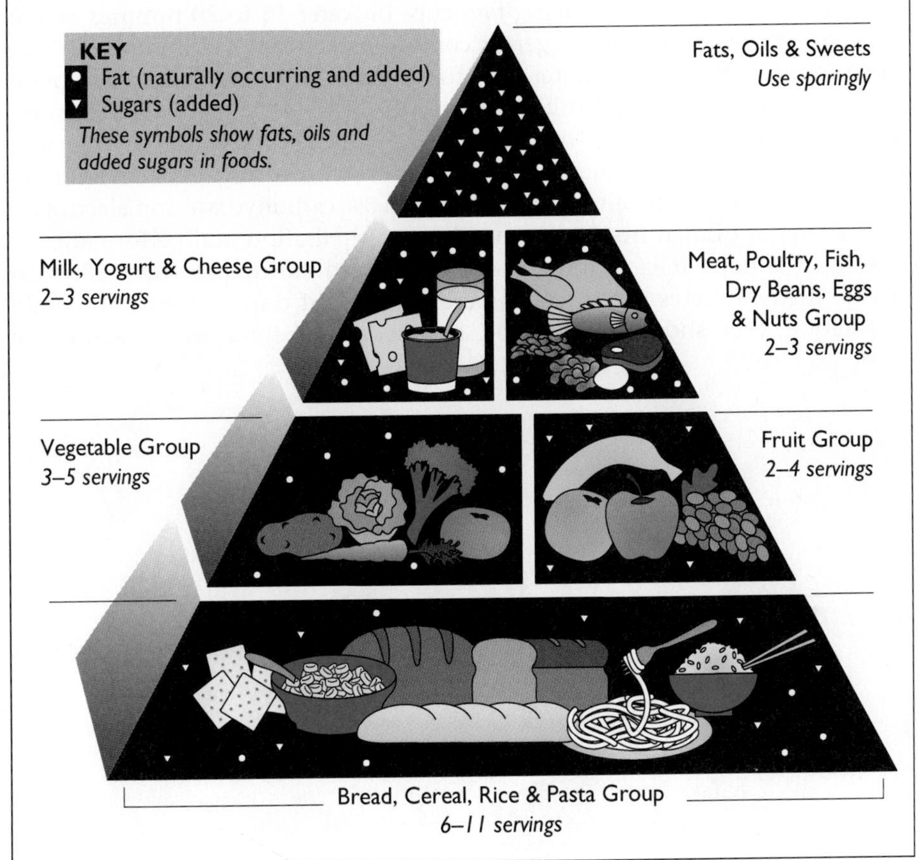

USDA

Illustration 27.7
Recommended number of servings for active women and men from the food groups from the Food Guide Pyramid.

Carbohydrates stay in the stomach a shorter time than protein or fat and are easily digested. Within two hours, carbohydrates have left the stomach, been digested, and are being absorbed into the bloodstream primarily as glucose. Some of the carbohydrate will have been converted to liver glycogen.[11] Illustration 27.8 shows three examples of high-carbohydrate, low-fat pre-event meals.

High-Fat Diets Recent research indicates that athletes who train while consuming relatively high-fat diets may increase their body's ability to use fatty acids for energy formation during prolonged exercise. Fat droplets are stored to some extent in muscle cells, and such stores may be more efficiently utilized by individual athletes accustomed to higher fat intakes.[9,12]

Eating between Events Athletes competing in tournaments and meets should snack between events. Once again, the foods of choice are those high in carbohydrates and low in fat (Illustration 27.9).

Hydration

Muscular activity produces heat that must be eliminated to prevent the body from becoming overheated. To keep the body cool internally, the heat produced by muscles is collected in the blood and then released both through blood that circulates near the surface of the skin and in sweat. Sweating cools the body because heat is released when water evaporates on the skin. People sweat more during physical activity in hot, humid weather because it is harder to add moisture and heat to warm, moist air than to dry, cool air. To stay cool during exercise in hot, humid conditions, the body must release more heat and water than when exercise is undertaken in drier, cooler conditions.

Athletes engaged in short events (less than an hour) can generally stay appropriately hydrated by drinking two cups of water 15 to 20 minutes before the event. When events are longer or conditions are hot and humid, athletes should consume two cups of water prior to the event and enough fluid to replace water losses during the event (Illustration 27.10). This generally means consuming four to eight ounces (one-half to one cup) of water every 15 to 20 minutes during an event. Water is usually sufficient for exercise lasting less than an hour. If it is over an hour, water alternated with a 4 to 8% carbohydrate and electrolyte sports drink, or diluted fruit juice, may increase hydration and performance.[13] Since thirst is not a reliable indicator of water need during prolonged physical activity, many athletes fail to consume enough fluid during events. To avoid dehydration, they should follow a preset schedule for fluid intake before and during exercise.[14]

Illustration 27.8
High-carbohydrate, low-fat pre-event meals.

Dehydration: The Consequences Loss of more than 2% of body weight (2–4 pounds) during an event indicates that the body is becoming dehydrated. Effects of dehydration range from mild to severe depending on how much body water is lost (Illustration 27.11). Any amount of dehydration impairs physical performance. At the extreme, dehydration can lead to heat exhaustion or heat stroke (Table 27.3). People who overexercise in hot weather when they are out of condition are most likely to suffer heat exhaustion or heat stroke, but these conditions occasionally occur among seasoned athletes, too. Heat exhaustion can be remedied by fluids, but heat stroke requires emergency medical care.

Illustration 27.9
Between-event snacks.

What about the Loss of Sodium? A few cases of "hyponatremia"—or sodium deficiency—have occurred in athletes competing in ultramarathons over the past few years. The news hit the sports headlines, causing renewed concern about replacing sodium during exercise.

Hyponatremia is a rare but serious condition that produces some of the same symptoms as heat stroke.[15] It can occur when only water is consumed during very long events, such as marathons and 50-mile runs, 100-mile bike races, and long triathlons.[6] It can be prevented by consuming a sports drink containing sodium, potassium, and other electrolytes along with water during an event.

Sports Drinks Specially formulated sports drinks (Illustration 27.12) that contain a weak solution of carbohydrate (4 to 8% carbohydrate by weight) and sodium (around 50 milligrams per eight ounces) and other electrolytes appear to improve endurance and hydration status during prolonged, intense events that last over an hour.[16,17] Sports drinks should be consumed in addition to water, and not as a replacement for water.[14] One of the main benefits of sports drinks is the water they provide.[18]

Fluids high in sugar such as soft drinks don't quench a thirst and should not be consumed during events. Their high-sugar content may draw fluid from the blood into the intestines, thereby increasing the risk of dehydration, nausea, and bloating.[6] Alcohol-containing beverages such as beer, wine, and gin-and-tonics are not hydrating, either. Coffee and tea are less hydrating than water.[16]

Photo Disc

Illustration 27.10
Hydration is not an abstract concept. It's a must for athletes.

Body Fat and Weight: Heavy Issues for Athletes

As a group, athletes tend to be very concerned about their weight and body shape, and they are leaner than the U.S. population in general (Table 27.4 on p. 27-11).[19] But the advantages of low body weight and body fat disappear when they become too low.[20] Body fat levels of less than 3% for men and less than 12% for women can seriously interfere with health. These percentages represent the obligatory body fat content that people need for the functions of fat unrelated to energy production.[21] Everybody needs some fat to use for hormone production, to maintain normal body temperature in cold weather, and to cushion internal organs.

Exercise, Body Fat, and Health in Women It is not uncommon for female athletes to experience irregular or absent menstrual periods or the late onset of periods during adolescence (Table 27.5 on p. 27-11). These aberrations in menstrual period appear to be related to specific effects of high levels of physical activity or low caloric intakes on hormone production and to low levels of body fat.[24] Female athletes with missing and irregular periods are at risk for

Illustration 27.11
Effects of dehydration.[14,16] Loss of 9 to 12% of body water can be fatal; however, any amount of dehydration impairs physical performance.

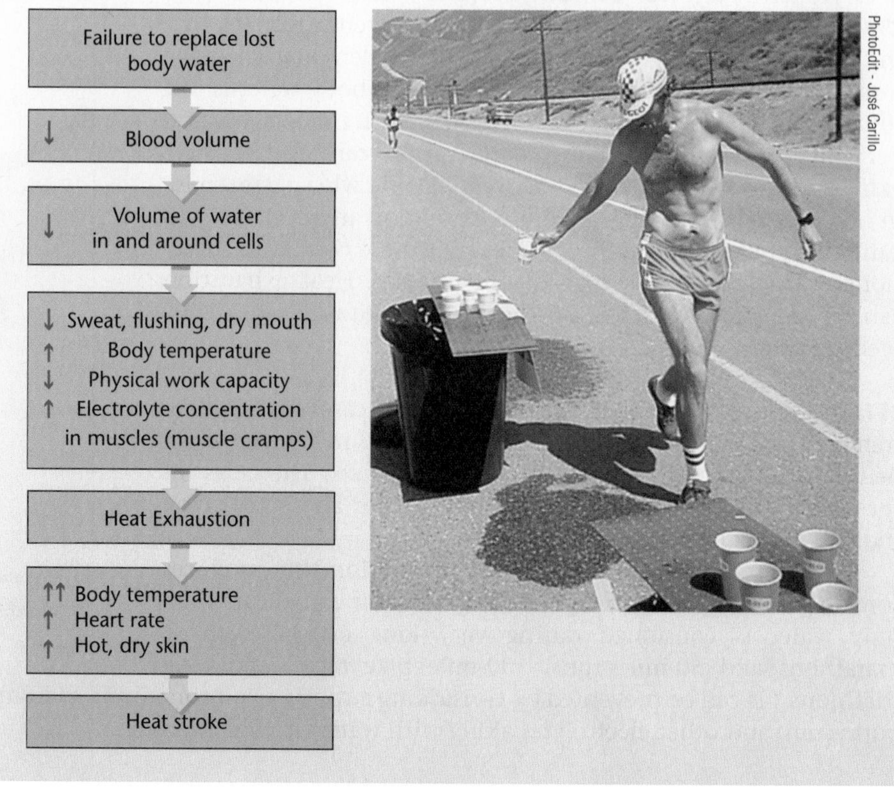

developing low bone density and osteoporosis later in life.[20] Suppressed testosterone levels in male athletes with very low levels of body fat may be the parallel to menstrual irregularities in female athletes.[25]

Wrestling: The Sport of Weight Cycling "Making weight" is a common and recurring practice among wrestlers (Illustration 27.13). Most wrestlers will "cut" 1 to 20 pounds over a period of days between 50 and 100 times during a high school or college career.[26] That makes wrestling more than a test of strength and agility. It makes it a contest of rapid weight loss.

For competitive reasons, wrestlers often want to stay in the lowest weight class possible and may go to great lengths to achieve it. They may fast, "sweat

Table 27.3
A primer on heat exhaustion and heat stroke.

Heat exhaustion: A condition caused by low body water due to excessive loss of water through sweat in hot weather. Symptoms include weakness, paleness, dizziness, nausea, fatigue, and confusion. Fluids, such as water and fruit juices, and a cool place are the remedy. Also called "heat prostration" and "heat collapse."

Heat stroke: A condition requiring emergency medical care. It is characterized by hot, dry skin, labored and rapid breathing, a rapid pulse, nausea, blurred vision, irrational behavior, and, often, collapse. Internal body temperature exceeds 105°F due to a breakdown of the mechanisms for regulating body temperature. Heat stroke is caused by prolonged exposure to high temperatures. Medical help should be sought immediately, and the person affected should be kept cool by any means possible, such as removing clothing and soaking the person in ice-cold water. If conscious, the person should be given water or fruit juice. Also called "sun stroke."

Illustration 27.12
Sports drinks. Many athletes have their favorite. Each type of drink shown provides 4 to 8% carbohydrate and about 50 milligrams of sodium per eight ounces (one cup).

the weight off" in saunas or rubber suits, or vomit after eating to lose weight before the weigh-in. These practices can be dangerous, even life-threatening, if taken too far. After the match is over, the wrestlers may binge and regain the weight they lost.

Wrestlers, like other athletes involved in intense exercise, perform better if they have a good supply of glycogen and a normal amount of body water.

Table 27.4
Average body fat content of various athletes.

	Body Fat Content (as Percentage of Body Weight) Women	Men
Bodybuilders	10%	6%
Long-distance runners	17	6–13
Baseball players	—	12–14
Basketball players	21–27	7–10
Football players	—	9–19
Gymnasts	10	5
Soccer players	—	10
Tennis players	20	16
Wrestlers	—	9

Table 27.5
Incidence of irregular or absent menstrual cycles in female athletes and sedentary women.[22–24]

	Incidence of Irregular or Absent Menstrual Cycles
Joggers (5 to 30 miles per week)	23%
Runners (over 30 miles per week)	34
Long-distance runners (over 70 miles per week)	43
Competitive bodybuilders	86
Noncompetitive bodybuilders	30
Sedentary women	13

Stock, Boston, Charles Gupil

Illustration 27.13
Dropping weight quickly in the days before a match can wreck a wrestler's chances and harm his health.

Fasting before a weigh-in dramatically reduces glycogen stores, and withholding fluids or losing water by sweating puts the wrestler at risk of becoming dehydrated. Trying to stay within a particular weight class too long may also stunt or delay a young wrestler's growth.

The American Medical Association and the Association for Sports Medicine recommend that wrestling weight be determined after six weeks of training and normal eating. In addition, a minimum of 7% body fat should be used as a qualifier for assigning wrestlers to a particular weight class.

Wisconsin high school wrestling programs adopted the medical association's recommendations and found that wrestlers, coaches, and parents supported the changes. Participation in wrestling increased because the athletes didn't have to deal with weight cutting.[27]

Iron Status of Athletes Iron status is an important topic in sports nutrition because iron-deficiency anemia decreases endurance.[28] Iron is a component of hemoglobin, a protein in blood that carries oxygen to cells throughout the body, including exercising muscle cells. When hemoglobin levels are low, as occurs in iron-deficiency anemia, less oxygen is delivered to cells than normal. The reduced supply of oxygen lowers the amount of energy muscles can produce. Low iron stores may also decrease endurance for reasons other than oxygen delivery.[28]

Female athletes are at higher risk of iron-deficiency anemia than other females. Consequently, it is recommended that female athletes especially pay attention to the amount of iron consumed.[20]

Ergogenic Aids: The Athlete's Dilemma

The quest for ownership of a competitive edge has drawn athletes to ergogenic aids throughout much of history. (See Table 27.6 for a historical review of ergogenic aids use.) Relatively few of the hundreds of products available work, and most are sold as "dietary supplements" so they do not have to be tested for safety. Those found to increase muscle mass, strength, or endurance are usually banned for use by competitive athletes. Several of the aids clearly pose serious risks to health, while others simply represent misguided hopes and misspent money. Table 27.7 summarizes claims made for a variety of ergogenic aids, research results, and what is known about adverse effects.

Use of performance-enhancing drugs is expected to decline substantially in the future due to improved blood tests and monitoring of athletes.[29]

Table 27.6
Faster, higher, stronger, longer. A brief history of performance-enhancing substances used by athletes.[29]

B.C.	Large quantities of beef consumed by athletes in Greece to obtain "the strength of 10 men."
1880s	Morphine used to increase performance in (painful) endurance events.
1910s	Strychnine consumed for the same reason as morphine.
1930s	Amphetamines used to increase energy levels and endurance, testosterone taken to increase muscle mass.
1980s	Blood doping, EPO used to increase endurance.

Table 27.7
Ergogenic aids: claims and evidence.[29–35]

Ergogenic Aid—Normal Functions	Claims	Evidence
Amphetamines— · no normal function.	Reduce fatigue, improve concentration, increase aggressiveness, decrease appetite and weight.	Reduce fatigue, appetite, and weight; increase aggressiveness. **Adverse effects** Many, including dizziness, tremors, confusion, paranoia, increased heart rate, chest pain, cardiovascular collapse.
Androgenic anabolic steroids · testosterone and related substances. · promote protein synthesis, decrease protein tissue breakdown, increase blood volume and red blood cell production.	Gradually increasing doses of several forms of steroids increase muscle mass.	Increase strength, muscle mass; decrease recovery time. **Adverse effects** Liver, heart, psychiatric disorders, testicle atrophy, reduced HDL- and elevated LDL-cholesterol.
Bee pollen · pollen collected by bees from flowers and mixed with bee secretions. Provides food for bee larva.	Detoxification of blood, improves performance.	No such effects found. **Adverse effects** Allergic reaction in people who are allergic to bee stings and honey, and who have asthma.
Boron · a mineral that influences calcium, magnesium, and phosphorus utilization.	Increases testosterone and muscle mass.	No such effects found. May improve bone density, arthritis.
Branched chain amino acids · isoleucine, leucine, valine. Are essential amino acids. May lower brain serotonin levels.	Improve performance, delay fatigue, build muscle, increase mental concentration.	May delay muscle breakdown, spare glycogen, increase fat utilization and time to exhaustion. **Adverse effects** High doses (>20 g) may impair performance.
Caffeine · central nervous system stimulant, diuretic, circulatory and respiratory stimulant.	Decreases fatigue in endurance events.	May reduce fatigue and improve endurance. **Adverse effects** Nervousness, sleeplessness, irregular heartbeat; fatal dose 3–10 g. Olympic illegal dose equivalent to 6–8 cups coffee 2–3 hours pre-event.
Carnitine · a compound required for energy formation from fatty acids.	Enhances fatty acids utilization for energy.	Most studies show no effect on fatty acid utilization during exercise in healthy adults.
Cocaine · no normal functions.	Same as for stimulants but much shorter acting.	
Creatine · component of muscle, generates ATP from ADP in muscle.	Increases performance in high-intensity, short-duration exercises.	Increases peak muscle force in short-term, high-intensity exercises, no increase in strength but may increase energy available for training. **Adverse effects** Nausea, vomiting, diarrhea, increased blood pressure, muscle cramps.

continued

Table 27.7
Ergogenic aids: claims and evidence[29–35], continued.

Ergogenic Aid—Normal Functions	Claims	Evidence
DHEA, AD (Dehydroepinandrosterone, Androstendione) · precursors of testosterone.	Promote testosterone production and muscle mass.	Promote testosterone production and increase muscle mass. **Adverse effects** Expected to be the same as for anabolic steroids.
Ephedra, Chinese (Ma huang) · no normal function.	Central nervous system stimulant, enhances endurance, strength, and body fat loss.	No effects on strength, endurance, or aerobic capacity. **Adverse effects** Nervousness, anxiety, rapid heart rate, headache, hypertension, death. Not recommended for use.
EPO (Erythropoietin) · a hormone that increases red blood cell production by bone marrow.	Replaces blood doping, or injection of red blood cells for increasing oxygen availability and endurance.	Increases endurance. **Adverse effects** Increases blood viscosity, blood pressure, blood clot formation; headache.
GBL (Gammabutyrolactone, furanodehydro) · no normal function.	Increases muscle mass and muscle recovery from exercise.	No such effects **Adverse effects** Vomiting, seizures, death. FDA has called for its voluntary removal from the market.
Ginseng, Siberian · no normal function.	Enhances endurance and strength.	Increases muscular strength and aerobic work capacity.
Glycerol · a clear, sweet fluid attached to fatty acids in fat molecules.	Increases hydration.	Conflicting results. **Adverse effects** Bloating, nausea.
HGH (Human growth hormone) · stimulates muscle and bone growth, converted to glucose.	Increases muscle mass and strength; similar effects as anabolic steroids but safer.	Increases muscle mass but probably not strength. **Adverse effects** Growth of facial bones, heart disease, stroke, hypertension, diabetes. Effects generally irreversible.
Insulin · speeds entry of glucose into cells, increases production of glycogen.	Increases muscle mass.	Increases glucose passage into cells and glycogen production. **Adverse effects** Low blood sugar, seizure, brain damage.
Protein powders · can contribute to meeting the body's need for protein.	Increase muscle mass and strength.	No such effects found. **Adverse effects** Dehydration, liver and kidney problems.
Vitamin and mineral supplements · may contribute to meeting need for vitamins and minerals.	Improve energy and stamina.	Such supplements improve energy and stamina in deficient individuals only. **Adverse effects** Toxicity reactions.

The Path to Improved Performance Genetics, training, and nutrition: these are the real keys to physical performance. Although other aids will be sought, those that exceed the boundaries of what is considered fair and safe will not be approved for use by athletes. After all, athletic competition is not a test of drugs or performance aids. It's a test of an individual's ability to excel. Anything less wouldn't be sporting.

WWW links

www.mayohealth.org
The Mayo Clinic's website has a hot button for Fitness and Exercise. Go there for reliable advice on exercise and sports nutrition.

www.acsm.org
The American College of Sports Medicine website offers information on sports nutrition and the opportunity to join the national Coalition for Promoting Physical Activity.

www.nal.usda.gov/fnic/etext/fnic.html
Select the "sport nutrition" link.

www.quackwatch.com
Check out promises made by sports aids.

Nutrition UP CLOSE

Testing Performance Aids

FOCAL POINT: The critical examination of studies on performance aids.

Read the following summary of a study (imaginary) of the effects of a creatine supplement on strength, and then answer the critical thinking questions.

- *Purpose:* To assess the effect of a creatine supplement on strength.
- *Methods:* Twenty volunteers from the crew team were given the creatine supplement for a week. Strength, assessed as the maximum number of push-ups a study participant could do in one session, was assessed before and after supplementation. Participants recorded any supplement side effects.
- *Results:* The number of push-ups increased by an average of 5% after supplementation. Diarrhea was the only side effect consistently noted by participants.

Critical Thinking Questions:

1. Does the study demonstrate that the supplement is safe?

Yes ____ No ____ Give reasons for your answer.

____

2. Does the study demonstrate that the supplement increases strength?

Yes ____ No ____ Give reasons for your answer.

____

FEEDBACK (answers to these questions) can be found at the end of Unit 27.

REFERENCES

1. Bannister R. Special presentation. Health, fitness, and sport. Am J Clin Nutr 1989;49(suppl):927–30.
2. Barrett S. Don't buy phony "ergogenic aids." Nutrition Forum 1997;May/June: 17, 19–21, 24.
3. Corley G, et al. Nutrition knowledge and dietary practices of college coaches. J Am Diet Assoc 1990;90:705–9.
4. Jones NL, Killian KJ. Exercise limitations in health and disease. N Engl J Med 2000;343:632–41.
5. Kleiner SM. The lowdown on carbohydrate loading. Phys Sports Med 1995;23:19–20.
6. Coleman E. Eating for endurance. Palo Alto (CA): Bull Publishing Co.; 1988.
7. Hultman E. Nutritional effects on work performance. Am J Clin Nutr 1989;49(suppl):949–57.
8. Foster C, et al. Effects of preexercise feeding on endurance performance. Med Sci Sports Exer 1979;11:1–5.
9. Brouns F, et al. Utilization of lipids during exercise in human subjects: metabolic and dietary constraints. Brit J Nutr 1998;79:117–28.
10. Kleiner SM. Healthy muscle gains. Phys Sports Med 1995;23:21–22.
11. Hultman E, et al. Work and exercise. In: Shils ME, Olson JA, Shike M, editors. Modern nutrition in health and disease, 8th ed. Philadelphia: Lea & Febiger; 1994: pp 663–80.
12. Brown R, et al. High carbohydrate versus high fat diets: effect on body composition in trained cyclists. Med Sci Sports Exer 2000;32:690–94.
13. Kleiner SM. Water: an essential but overlooked nutrient. J Am Diet Assoc 1999;99:200–6.
14. Clark N. Water. Phys Sports Med 1995; 23:23–24.
15. Barr SI, Costill DL. Water: can the endurance athlete get too much of a good thing? J Am Diet Assoc 1989;89:1629–35.
16. Gisolfi CV. Fluid balance for optimal performance. Nutr Rev 1996;54: S159–S68.
17. Manning EH. Sports drinks proven in exercise labs. Clin Chem News 1991;17(1):10–11.
18. Walsh J. Sports supplements: powerful pills or pricey placebos? Envir Nutr 1997;May:6.
19. Wilmore JH. Body composition in sports and exercise: directions for future research. Med Sci Sports Exer 1983;15:21–31.
20. Manson JE, Lee I-M. Exercise for women: how much pain for optimal gain? N Engl J Med 1996;334:1325–6.
21. Elliot DL, Goldberg L. Nutrition and exercise. Med Clin North Am 1985;69:71–81.
22. Lloyd T, et al. Interrelationships of diet, athletic activity, menstrual status, and bone density in collegiate women. Am J Clin Nutr 1987;46:681–4.
23. Walberg JL, Johnston CS. Menstrual function and eating behavior in female recreational weight lifters and competitive body builders. Med Sci Sports Exer 1991;23:30–36.
24. Loucks AB. Effects of exercise training on the menstrual cycle: existence and mechanisms. Med Sci Sports Exer 1990; 22:275–80.
25. Brownell KP, et al. Weight regulation practices in athletes: analysis of metabolic and health effects. Med Sci Sports Exer 1987;19:546–56.
26. Steen SN, et al. Metabolic effects of repeated weight loss and regain in adolescent wrestlers. JAMA 1988;260:47–50.
27. Oppliger RA, et al. The Wisconsin wrestling minimum weight project: a model for weight control among high school wrestlers. Med Sci Sports Exer 1995;27:1220–24.
28. Zhu YI, Haas JD. Iron depletion without anemia and physical performance in young women. Am J Clin Nutr 1997; 66:334–41.
29. Birchard K. Past, present, and future of drug abuse at the Olympics. Lancet 2000;356:1008.
30. Ergogenic aids: reported facts and claims. SCAN'S Pulse 1999;winter:12–21.
31. Wagner DR. Hyperhydration with glycerol: implications for athletic performance. J Am Diet Assoc 1999;99:207–12.
32. Lund BC, Perry PJ. Androgenic anabolic steroids: an overview for clinicians. Medscape 1999;69:705–11.
33. Lund BC, Perry PJ. Nonsteroid performance-enhancing agents in athletic competition: an overview for clinicians. Medscape Pharmacotherapy:www.medscape.com/Medscape/pharmacology/journal/2000/v02,n05; Sept. 25, 2000.
34. Brass EP. Supplemental carnitine and exercise. Am J Clin Nutr 2000;72(suppl): 618S–23S.
35. Bucci LR. Selected herbals and human exercise performance. Am J Clin Nutr 2000;72(suppl):624S–36S.

Nutrition

UP CLOSE

Testing Performance Aids

Feedback for Unit 27

1. Answer: No. Diarrhea may be a problem. Other side effects, such as blood pressure changes, abnormalities in kidney function, and weight loss, were not assessed. Safety of long-term use was not addressed.
2. Answer: No. A 5% increase in strength is small and may have been due to usual differences in the number of push-ups a person can do from time to time. Any increase in strength may have been due to increased training during the week the supplement was taken. There was no control group. All participants knew they were taking creatine, a supplement believed by many to increase strength. This belief may have changed performance on the push-up test.

Good Nutrition for Life: Pregnancy, Breastfeeding, and Infancy

Photo Disc

Nutrition Scoreboard

	TRUE	FALSE
1. The fetus is a parasite.		
2. The United States has the lowest rate of infant mortality in the world.		
3. Normal health, growth, and development of infants are defined by the experience of breastfed infants.		
4. Drinking alcohol-containing beverages is safe during breastfeeding because the alcohol doesn't pass into the milk.		

(Answers on next page)

Answers to Nutrition Scoreboard

	TRUE	FALSE
1. The fetus does not harm the mother for its own gain. (A true parasite injures the host while benefiting from it.) If there is a shortage of essential nutrients, the mother generally will get access to them before the fetus does.		✓
2. Over 20 economically developed countries have lower infant mortality rates than the United States.		✓
3. That's the American Academy of Pediatrics's standard.	✓	
4. Alcohol does pass into breast milk. Breastfeeding women should limit their intake of alcohol if they drink at all.		✓

Key Concepts and Facts

- There are no "maternal instincts" that draw women to select and consume a good diet during pregnancy.
- The fetus is not a parasite.
- Fetal and infant growth are characterized by "critical periods" during which all essential nutrients needed for growth and development must be available or growth and development will not proceed normally.
- An adult's risk of certain chronic diseases may be partially determined by maternal nutrition during pregnancy and the person's own nutrition early in life.
- Breastfeeding is the optimal method of feeding infants.
- Rates of growth and development are higher during infancy than at any other time in life.

Photo Disc

A Healthy Start

The day has finally arrived, the one Crystal and Tyrone have anticipated for a year. Crystal's pregnancy test is positive!

This is a planned pregnancy, and both Crystal and Tyrone have done all they could to prepare. Crystal's diet has been the picture of perfection, she has kept up her regular exercise schedule, and she has quit drinking alcohol entirely. Now they are ready to have the baby of their dreams. They are convinced that not only will this baby be healthy and strong, but it will be above average in every respect.

All parents want to give their children every advantage in life that they can. Although no one can guarantee that a baby will be born healthy and strong—no matter what the parents do—there are steps parents can take to make the best baby possible.

Unhealthy Starts on Life

Unfortunately, many infants born in the United States do not have the advantage of good health at birth. Among economically developed countries with populations over 2.5 million, the United States ranks twenty-first in terms of infant mortality rate (Table 28.1). The relatively high rate of infant deaths in the United States is due primarily to the proportion of infants born with low birthweights (or a weight of less than 5½ pounds) or preterm (delivered before 37 weeks of pregnancy) (Illustration 28.1). One in 14 U.S. infants is born too small, and over 1 in 10 is born too early.[1] These infants are at particular risk for requir-

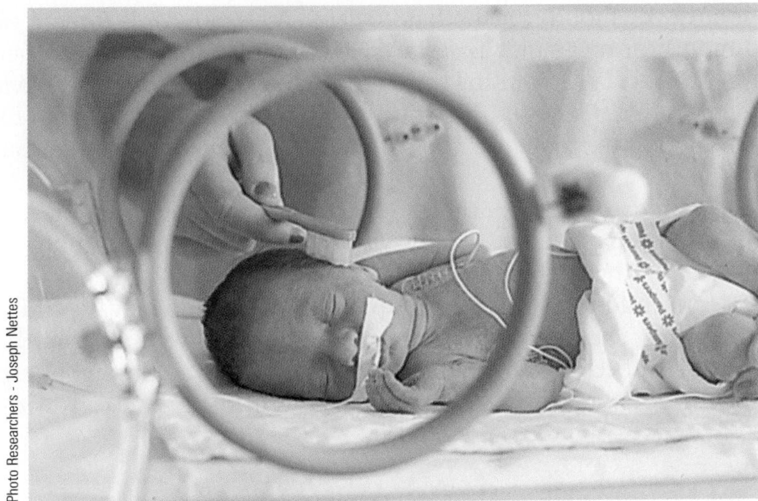
Photo Researchers - Joseph Nettes

Illustration 28.1
In 1998, 7.6% of the infants born in the United States had a low birthweight, and 11.6% were born preterm.[1]

ing intensive and continuing care and of dying within the first year of life.[2] In contrast, infants weighing between 7 pounds, 11 ounces and 8 pounds, 13 ounces (or 3500 to 4000 grams) are least likely to die within the first year.[3] Currently, the average weight of U.S. infants is 7 pounds, 7 ounces (3350 grams).[3]

Improving the Health of U.S. Infants A high proportion of poor infant outcomes in the United States is attributed to a combination of factors including poverty, poor nutrition, limited access to health care, and a maternal lifestyle that includes the use of illicit drugs, cigarettes, and excessive amounts of alcohol.[4] Many infant deaths, preterm infants, and low-birthweight infants can be prevented, however, by optimizing maternal behaviors that influence health. Of the various behaviors that affect maternal health, none offers potentially greater advantage to both pregnant women and their infants than good nutrition.

Nutrition and Pregnancy

Nutrition is important during pregnancy because the ***fetus*** depends on it. Energy and every nutrient the fetus needs for ***growth*** and ***development*** must be supplied by the mother's diet (preferably) or by supplements. If either insufficient or excessive amounts of nutrients are supplied, fetal growth and development may be compromised. The nature and extent of the impairment depend on which organ or tissue is growing most rapidly when the nutritional deficiencies or overdoses occur.[5]

Critical Periods Fetal growth and development proceed in a series of ***critical periods.*** A critical period is an interval of time during which cells of a tissue or organ are genetically programmed to multiply. The period is considered critical because if the cells do not multiply as programmed during this set time interval, they cannot make up the deficiency later. The level of nutrients required for cell multiplication to occur normally must be available during this specific time interval. If the nutrients are not available, the developing tissue or organ will contain fewer cells than normal, will form abnormally, or will function less than optimally.

Table 28.1
Infant mortality per 1000 livebirths in 1995 in economically developed countries with populations over 2.5 million.[1]

Country	Infant Mortality per 1000 Livebirths
Sweden	3.7
Finland	3.9
Singapore	4.0
Japan	4.3
Hong Kong	4.6
France	4.9
Norway	5.2
Denmark	5.3
Germany	5.3
Switzerland	5.3
Netherlands	5.4
Austria	5.5
Spain	5.6
Australia	5.7
Belgium	6.1
Canada	6.1
Italy	6.2
United Kingdom	6.2
Ireland	6.4
New Zealand	6.7
United States	7.6
Greece	7.9

➡ **fetus**
A baby in the womb from the eighth week of pregnancy until birth. (Before then, it is referred to as an *embryo.*)

➡ **growth**
A process characterized by increases in cell number and size.

➡ **development**
Processes involved in enhancing functional capabilities. For example, the brain *grows,* but the ability to reason *develops.*

➡ **critical period**
A specific interval of time during which cells of a tissue or organ are genetically programmed to multiply. If the supply of nutrients needed for cell multiplication is not available during the specific time interval, the growth and development of the tissue or organ are permanently impaired.

The roof of the mouth (or the hard palate), for example, is formed early in the third month of pregnancy when two developing plates fuse together. This process can only occur early in the third month. If excessive amounts of vitamin A are present in fetal tissues during this period, the two plates may fail to combine, resulting in a cleft palate. (The divided hard palate can be surgically corrected after birth.)

Critical periods of cell multiplication are most intensive in the first few months of pregnancy when fetal tissues and organs are forming rapidly (Illustration 28.2)—hence the importance of the nutritional status of women at the time of conception and very early in pregnancy. For some organs and tissues, cell multiplication even continues through the first two years after birth. Most of the growth that occurs in the fetus late in pregnancy and throughout the rest of the growing years, however, is due to increases in the size of cells within tissues and organs.

The Fetal Origins Hypothesis One of the most striking advances in research on pregnancy concerns the potential effects of maternal nutrition on the baby's risk of developing certain chronic diseases later in life. It appears that an important element of increased susceptibility to heart disease, stroke, diabetes, obesity, hypertension, and other disorders may be "programmed" by inadequate or excessive supplies of energy or nutrients during pregnancy.[6]

Given optimal conditions, fetal growth and development proceed according to the genetic blueprint established at conception. Organs and tissues are well developed and ready to function optimally after birth. Under less-than-optimal growing conditions, such as those introduced by maternal weight loss, poor nutrient intakes, or diseases in the mother that alter her ability to supply the fetus with energy or nutrients, fetal growth and development are modified. Fetal tissues undergoing critical phases of development at that time have to make adaptations to cope with the under- or oversupply of nutrients. These adaptations may produce long-term changes in the structure and function of tissues.[7,8]

Illustration 28.2
Baby's first picture. An ultrasound image of a rapidly growing and developing 16-week-old fetus.

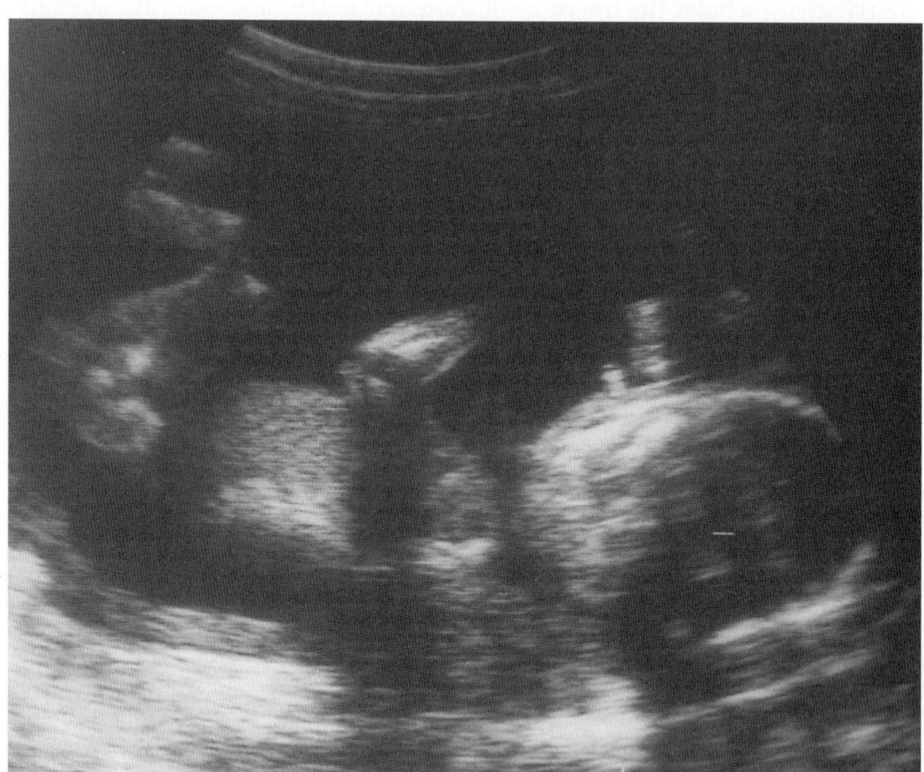

Low maternal energy intake in the last few months of pregnancy, for example, may hinder the development of cells that produce insulin and limit the body's ability to control blood glucose levels later in life. This could increase the risk of developing diabetes later in life.[8]

A large body of evidence supports the fetal origins hypothesis. Nevertheless, much remains to be discovered about the effects of diet during pregnancy on the risk of later disease.

The Fetus Is Not a Parasite From early history to current times, every culture has had myths about diet in pregnancy. According to one of the more common pregnancy myths, the fetus is a parasite, capable of drawing whatever nutrients it needs from the mother at the expense of her health. If this were true, there would be no such thing as a small or poorly nourished newborn. Obviously, it is not true.

The fetus is not a parasite because, with few exceptions, it receives an adequate supply of nutrients only if the mother's intake is sufficient to maintain her own health.[5] The situation makes sense in terms of survival of the species. By meeting the mother's needs first, nature protects the reproducer.

Rather striking evidence that the fetus is not a parasite comes from studies that have identified deficiencies of vitamin B_{12}, thiamin, iodine, folate, zinc, and other nutrients in newborns but not in their mothers.[9,10] Similarly, infants born to women who consumed excessive levels of vitamin or mineral supplements during pregnancy are more likely to display signs of nutrient overdose than are the mothers.[11,12]

On the Side

The ancient Romans believed that if a woman wanted a child with brown eyes, she should eat mice often during pregnancy.

Photo Disc

Prepregnancy Weight Status and Prenatal Weight Gain Are Important

Women who enter pregnancy underweight or who fail to gain a certain minimum of weight during pregnancy are much more likely to deliver low-birthweight and preterm infants than are women who enter pregnancy at normal weight or above and gain an appropriate amount of weight.[13,14] The risk of low birthweight can be reduced substantially by good diets that lead to a desired rate and amount of weight gain during pregnancy. Along with the duration of pregnancy and smoking, prepregnancy weight status and weight gain in pregnancy are the major factors known to influence an infant's birthweight (Table 28.2).

What's the Right Amount of Weight to Gain during Pregnancy?

Weight-gain recommendations for pregnancy vary depending on whether a woman enters pregnancy underweight, normal weight, overweight, or obese. Thus, a woman who is obese when she becomes pregnant will need to gain less weight than a woman who enters pregnancy underweight. It is recommended that underweight women gain 28 to 40 pounds, normal-weight women 25 to 35 pounds, overweight women 15 to 25 pounds, and obese women at least 15 pounds during pregnancy. Women carrying twins should gain 35 to 45 pounds during pregnancy.[14] Table 28.3 can be used to identify prepregnancy weight status and the recommended weight gain during pregnancy. After the goal is identified, a woman may want to plot her prenatal weight gain on a chart like the one in Illustration 28.3. The weight gained should be the result of a high-quality diet that leads to gradual and consistent gains in weight throughout pregnancy.

Table 28.2

Major factors that directly influence birthweight.[14]

- Duration of pregnancy
- Prenatal weight gain
- Prepregnancy weight status
- Smoking

Table 28.3
Identifying prepregnancy weight status and the recommended pregnancy weight-gain goal.

Height		Weight Status (Pounds)			
Feet	Inches	Underweight	Normal Weight	Overweight	Obese
4	9	92 or less	93–113	114–134	135 or more
4	10	94 or less	95–117	118–138	139 or more
4	11	97 or less	98–120	121–142	143 or more
5	0	100 or less	101–123	124–146	147 or more
5	1	103 or less	104–127	128–150	151 or more
5	2	106 or less	107–131	132–155	156 or more
5	3	109 or less	110–134	135–159	160 or more
5	4	113 or less	114–140	141–165	166 or more
5	5	117 or less	118–144	145–170	171 or more
5	6	121 or less	122–149	150–176	177 or more
5	7	124 or less	125–153	154–181	182 or more
5	8	128 or less	129–157	158–186	187 or more
5	9	131 or less	132–162	163–191	192 or more
5	10	135 or less	136–166	167–196	197 or more
5	11	139 or less	140–171	172–202	203 or more
6	0	142 or less	143–175	176–207	208 or more
Prenatal weight-gain goal (pounds):		28–40	25–35	15–25	15+

For twin pregnancies, the weight-gain goal is 35–45 pounds.

Note: Height is measured without shoes; weight includes light indoor clothing. Weight-for-height ranges are calculated from the 1959 Metropolitan Height and Weight Tables for Women over the age of 25 years. A midpoint value was determined from the range of weight for height for women of "medium frame." The cut-point for underweight women is designated as a weight for height that is more than 10% below the midpoint. The normal-weight range is calculated as plus or minus 10% of the midpoint for each height. The overweight range is calculated as greater than 10% through 30% above the midpoint of weight for height. The cut-point in weight for the obese category is calculated as a weight for height that is more than 30% above the midpoint of weight for height.
Source: Courtesy of the Healthy Infant Outcome Program, University of Minnesota, 1989; updated 1998.

Where Does the Weight Gain Go? If healthy infants tend to weigh about eight pounds at birth, why do women need to gain more than that? Where does the rest of the weight go? Many pregnant women ask themselves these questions, especially when they don't relish the thought of gaining weight. Illustration 28.4 shows where the weight gain goes.

Fetal growth is accompanied by marked increases in maternal blood volume, fat stores, and breast and uterus size, all of which contribute to weight gain. In addition, water accumulates in the amniotic fluid, which cushions and protects the fetus, and the volume of fluid that exists outside cells increases. The placenta (the tissue that transfers nutrients in the mother's blood supply to the fetus) also accounts for some of the weight that is gained during pregnancy.

Where Does the Weight Gain Go—after Pregnancy? Delivery is one of the greatest weight-loss plans known to humankind. On average, women lose 15 pounds within the first week after delivery, and weight loss generally continues for up to a year.[15] Women who gain more than the recommended amount of weight, however, and those who gain weight after delivery may have extra pounds to lose to get back to prepregnancy weight.

The Need for Calories and Key Nutrients during Pregnancy

Pregnant women need more calories, protein, and essential nutrients than non-pregnant women need. All of these are important, but calories, folate, vitamin B_6, vitamin A, calcium, iron, and zinc are of paramount concern. Although many pregnant women are concerned about protein, low protein intakes are

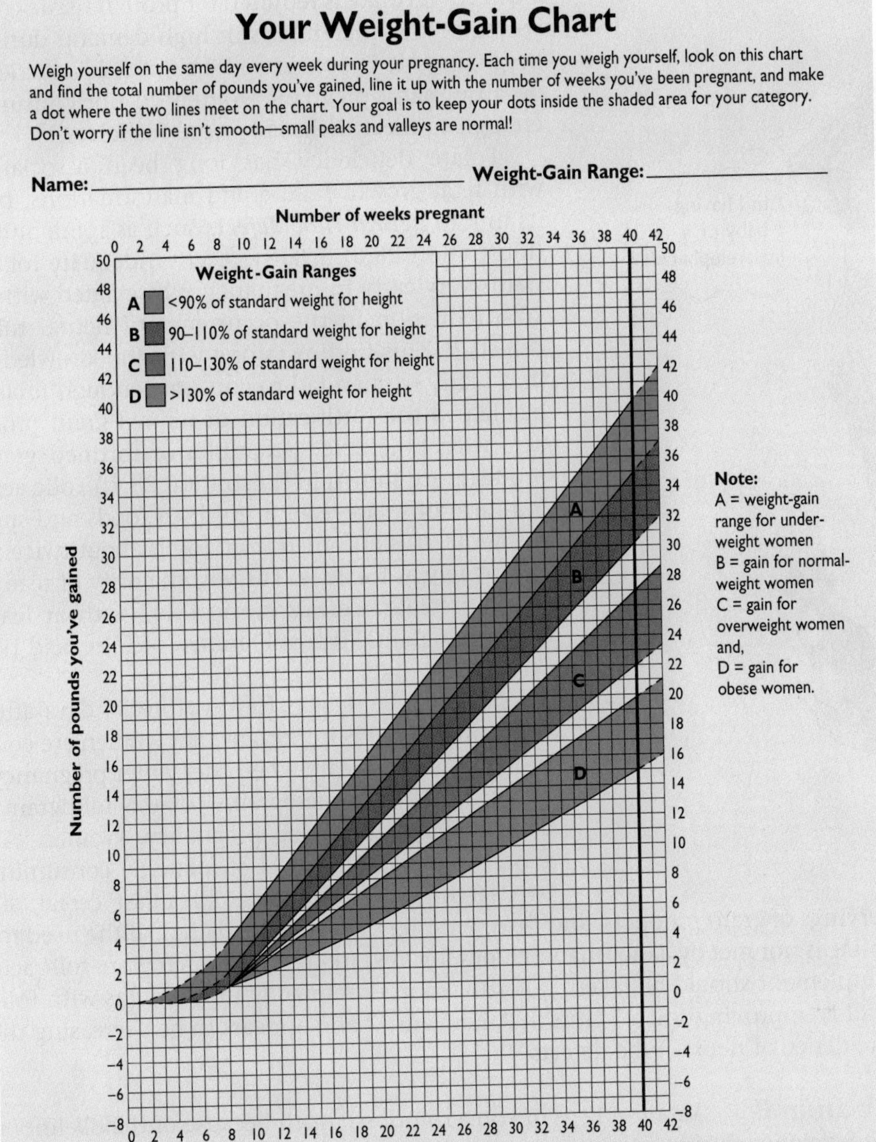

Illustration 28.3
An example of a prenatal weight-gain chart for pregnant women.

rarely a problem in pregnancy. Most women consume 10 to 30 grams more protein daily than they require.[16]

According to dietary intake recommendations, pregnant women need at least 15% more calories and 8 to 100% more of various nutrients than do non-pregnant women (Illustration 28.5). The relatively high requirements for many nutrients mean that pregnant women should increase their intake of nutrient-dense foods more than their consumption of calorie-rich foods.

Calories On average, women need an additional 300 calories a day in the last six months of pregnancy (the second and third ***trimesters***). Women entering pregnancy underweight will need more calories than this, and those entering overweight will need fewer. In addition, physically active pregnant women require higher caloric intakes than average. It is generally easier and more accurate to monitor the adequacy of caloric intake by tracking weight gain than by counting calories.

➡ **trimester**
One-third of the normal duration of pregnancy. The first trimester is 0 to 13 weeks, the second 13 to 26 weeks, and the third trimester is 26 to 40 weeks.

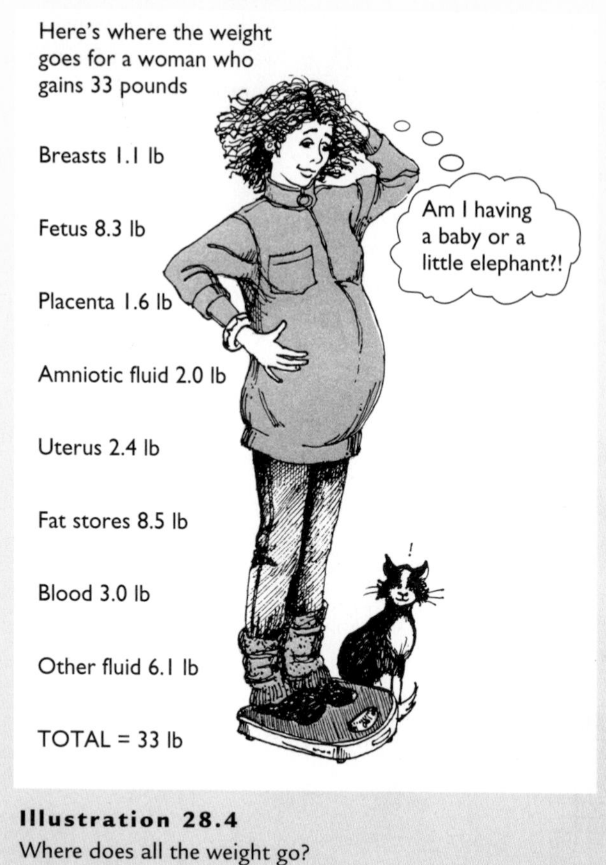

Illustration 28.4
Where does all the weight go?

Folate Folate is required for protein tissue construction and therefore is in high demand during pregnancy. It is considered an "at risk" nutrient because many pregnant women do not consume enough to meet their daily need for folate.[17]

Folate deficiency has long been associated with fetal growth failure and malformations, but its link to ***neural tube defects*** such as spina bifida (Illustration 28.6) is fairly recent. Adequate folate status very early in pregnancy is associated with a 70% reduction in the occurrence of neural tube defects.[18] This finding, along with the knowledge that many women fail to consume enough folate, has led to the fortification of refined grain products with folic acid. (Examples of fortified grain products are shown in Illustration 28.7.) Folic acid is a form of folate used in fortified foods and supplements. It is absorbed from foods about twice as completely as folate is. The vast majority of ready-to-eat breakfast cereals are fortified with at least 100 micrograms (0.1 milligrams) of folic acid per serving.

Neural tube defects form before 30 days after conception, so it is important that women are consuming enough folate when they enter pregnancy. "Enough" is 600 micrograms (or 0.6 milligrams) daily before and during pregnancy. Women can generally obtain this level of folate by consuming two servings of ready-to-eat breakfast cereal, six servings of grain products, and three servings of vegetables daily.[17] If the need for folate is not met by a diet that includes fortified foods, a 400-microgram folic acid supplement should be taken.[19] Fortification of refined grain products with folic acid is improving folate status in Americans,[19] and appears to be decreasing the incidence of neural tube defects.[20]

➡ **neural tube defects** Malformations of the spinal cord and brain. They are among the most common and severe fetal malformations, occurring in approximately 1 in every 1000 pregnancies. Neural tube defects include spina bifida (spinal cord fluid protrudes through a gap in the spinal cord; shown in Illustration 28.6), anencephaly (absence of the brain or spinal cord), and encephalocele (protrusion of the brain through the skull).

Vitamin B_6 Vitamin B_6 is of concern not so much because of the likelihood of deficiency during pregnancy, but because it is often used to treat nausea and vomiting of early pregnancy, or morning sickness. Doses of 75 milligrams of vitamin B_6 a day (a dose well above the RDA of 2.2 milligrams) effectively resolve morning sickness in many women. This dose of vitamin B_6 appears to be safe for pregnant women and their babies.[21]

Vitamin A Both low and high intakes of vitamin A may cause problems during pregnancy. Too little vitamin A is associated with poor fetal growth.[14] Too much vitamin A in the form of retinol from supplements can cause fetal malformations.[7]

The effects of vitamin A overdoses during pregnancy came to public attention in the early 1990s when women taking accutane or retinoic acid for acne were found to deliver far more than the expected number of infants with malformations. Use of these drugs before and very early in pregnancy increases the risk that babies will be born with malformations of facial features and the heart; the intake of more than 10,000 to 15,000 IU of retinol daily during the same period has the same effect. As a precaution, the American College of Obstetrics and Gynecology recommends that women who are or could become pregnant should limit their intake of retinol to less than 5000 IU per day and should not use vitamin A–containing medications. Beta-carotene, a precursor to vitamin A, is not harmful, however.[7]

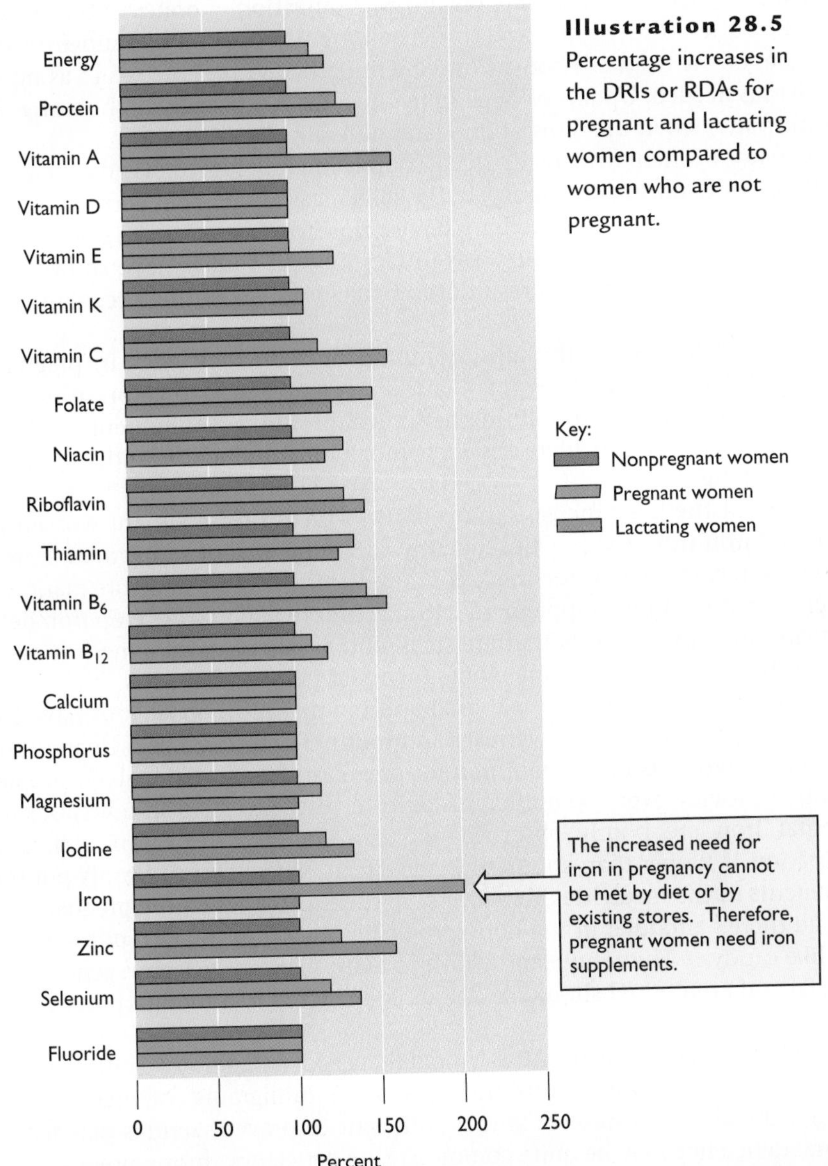

Illustration 28.5
Percentage increases in the DRIs or RDAs for pregnant and lactating women compared to women who are not pregnant.

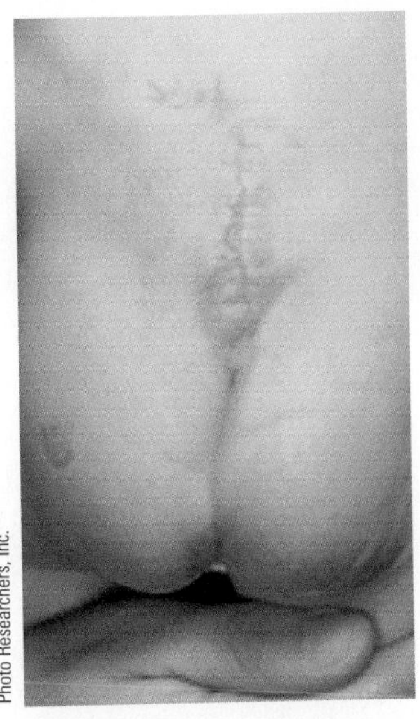

Illustration 28.6
Spina bifida—a neural tube defect. The interruption in the spinal cord results in paralysis below the injury. This photo shows the back after surgery to close the interruption.

Illustration 28.7
Foods fortified with folic acid.

Calcium Calcium used to support the mineralization of bones in the fetus is supplied by the mother's diet and, if needed, by the calcium contained in the long bones of the mother's body. Consequently, the fetus has access to as much calcium as is needed. Uptake of calcium by the fetus is especially high during the third trimester, when the fetus's bones are mineralizing.

Pregnant women who regularly consume low-calcium diets lose calcium from their bones during pregnancy, but usually regain it after delivery.[14] Because any calcium losses are from the long bones, not from the teeth, the old saying "for every baby a tooth" has no basis in fact. Several studies have shown that teeth do not demineralize as a result of low-calcium diets during pregnancy.[5]

Iron Iron deficiency is the most common nutrient deficiency in pregnant women.[22] It develops when a woman enters pregnancy with low iron stores and fails to consume enough iron during her pregnancy. Iron requirements increase during pregnancy due to increases in hemoglobin production and storage of iron by the fetus.

Because of the large increase in need, it is difficult for pregnant women to get enough iron from foods. Consequently, 30 milligrams of supplemental iron per day are recommended for the second and third trimesters of pregnancy.[14] Women who do not take supplemental iron are more likely to develop iron deficiency and to deliver infants that are small and at risk of developing iron deficiency in their first year than are women who do take supplements.[23]

The importance of iron to maternal and infant health appears to have led some health professionals to overreact and routinely prescribe iron in amounts that are too high. It is not uncommon for pregnant women to be given supplements that provide over 100 milligrams of iron daily. Excessive amounts of supplemental iron are poorly absorbed and can produce heartburn, gas, and constipation.[14] Rather than endure the side effects, some women simply put the supplements in the medicine cabinet or on a shelf. Later, when the pregnancy is over and there's a toddler in the house, the child may find the iron pills and eat them like candy.[24] The result can be iron toxicity that requires emergency care. Iron pills that aren't used should be discarded or stored in a toddler-proof place.

Zinc On average, women in the United States consume 8 or 9 milligrams of zinc per day, a level well below the RDA of 15 milligrams for pregnancy.[25] Although the average intake of zinc is sufficient to prevent serious deficiency, marginal deficiency may be quite common. Zinc deficiency during pregnancy is associated with long labors and the delivery of small and malformed infants.[26]

One important way to reduce the risk of zinc deficiency is to limit iron supplements to 30 to 60 milligrams per day. Zinc absorption is reduced substantially when it is consumed with large doses of supplemental iron.[27] Another way is to consume at least two servings of meats and meat alternates each day.

What's a Good Diet for Pregnancy?

Contrary to folklore, women do not instinctively select and consume a healthy diet during pregnancy. A good diet takes planning.

Regardless of a pregnant woman's age and whether or not she is a vegetarian, a healthy diet provides all the nutrients she needs with the possible exception of iron. Foods are recommended over supplements because foods provide fiber, antioxidants, and other beneficial substances that supplements do not. Pregnancy diets should include sufficient fluid and fiber and should consist of regular meals and snacks. A pregnant woman should not consume alcohol and should drink coffee only in moderation. Table 28.4 lists these and other recommendations for the diet during pregnancy. Planning a diet for pregnancy around the basic food groups is the most straightforward approach to meeting nutrient needs (Illustration 28.8).

Table 28.4
Dietary recommendations for pregnancy.

- Consume sufficient calories for adequate weight gain.
- Eat a variety of foods from each food group.
- Eat regular meals and snacks.
- Consume sufficient dietary fiber (about 30 grams per day).
- Consume 10 or more cups of water each day from fluids and foods.
- Use salt to taste (within reason).
- Do not drink alcoholic beverages.
- Limit coffee to four or fewer cups per day.

Why Alcohol and Pregnancy Don't Mix As early as the 1800s, maternal consumption of alcohol during pregnancy was said to cause the birth of "sickly" infants. The ill effects of alcohol on babies were not fully acknowledged, however, until the 1970s, when several research reports described a condition called "Fetal Alcohol Syndrome" (Illustration 28.9). Women who drank heavily or frequently binged on alcohol during pregnancy were found to be at high risk of delivering infants with specific malformations and retarded physical and mental development.[28] The effect of maternal alcohol consumption on the fetus worsens as intake increases. Heavy drinking in the first half of pregnancy is closely associated with the birth of malformed, small, mentally impaired infants. When excessive drinking occurs only in the second half of the pregnancy, infants

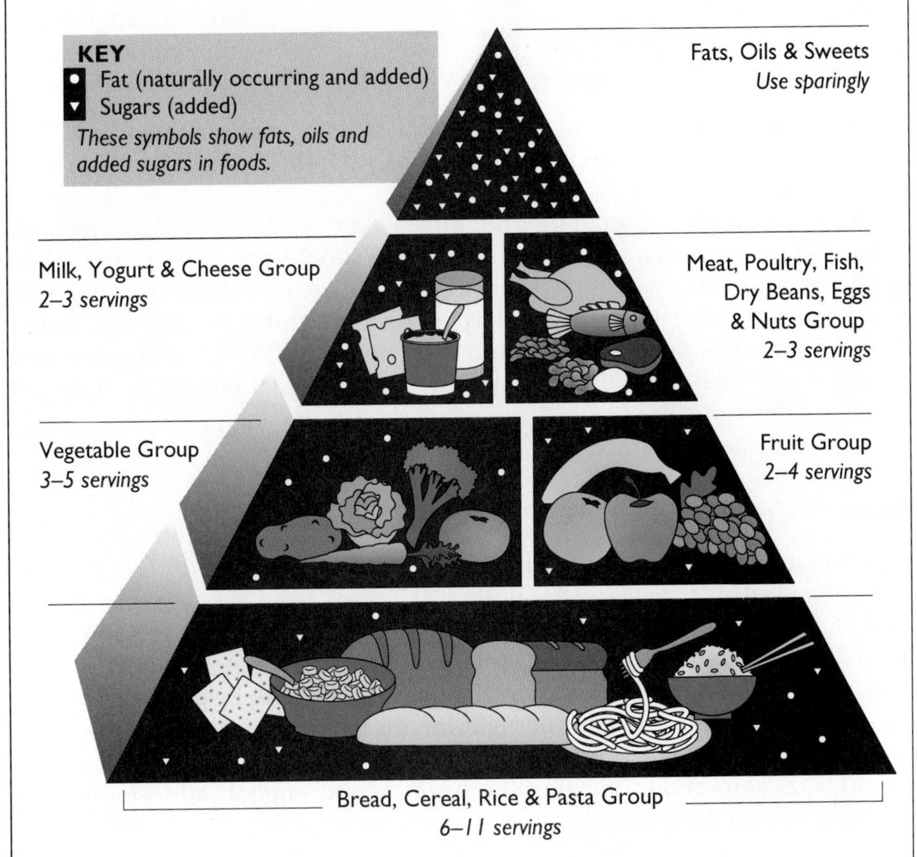

Illustration 28.8
Food group recommendations for pregnant and breastfeeding women based on the Food Guide Pyramid.

Illustration 28.9
Children with Fetal Alcohol Syndrome (FAS) have characteristic facial features, mental retardation, heart defects, and other problems. Less severe cases can include small size and slow development.

are less likely to be malformed, but are still likely to be small and to suffer abnormal mental development.[29] These conditions are permanent; they cannot be fully corrected with special treatment, and the child does not outgrow them.

No amount of alcohol has been found to be absolutely safe during pregnancy. When only an occasional drink is consumed, however, the adverse effects of alcohol on fetal development appear to be small and rare. To exclude the possibility of even small impairments in fetal growth and development, it is recommended that women not drink alcohol at all during pregnancy or when they are attempting to become pregnant.[14]

Vitamin and Mineral Supplements Iron is the only supplement recommended for all pregnant women.[14] About 83% of pregnant women take multiple vitamin and mineral supplements, however. Apparently, supplements are being prescribed as insurance against the possibility of poor diets. Women should be given supplements the same way they are given medications—when they are indicated. Under certain conditions, supplementation with nutrients besides iron is indicated.[14]

A multivitamin-mineral preparation is recommended for pregnant women who do not ordinarily consume an adequate diet and for those in high-risk categories, such as women carrying more than one fetus, heavy cigarette smokers, and alcohol and drug abusers. The preparation should be taken daily beginning in the second trimester and should contain the following nutrients:

Iron	30 milligrams	Vitamin B_6	2	milligrams
Zinc	15 milligrams	Folate	300	micrograms
Copper	2 milligrams	Vitamin C	50	milligrams
Calcium	250 milligrams	Vitamin D	5	micrograms

Vegans (those who consume no animal products at all) should supplement their diet daily with 10 micrograms (400 IU) of vitamin D and 2 micrograms of vitamin B_{12}.[14]

Because a number of vitamins and minerals can produce overdose reactions when taken in excessive amounts, supplements should not contain more than the DRI levels. As little as 15,000 IU of vitamin A (five times the RDA) taken early in pregnancy is associated with the development of central nervous system and bone abnormalities, for example,[30] and daily doses of 2000 IU of supplemental vitamin D (10 times the DRI) are related to mental retardation and heart defects in infants.[7,14]

Teen Pregnancy

Approximately 3 of every 100 females between the ages of 15 and 17 years in the United States deliver babies each year.[31] Although teens who do not smoke, drink alcohol, or use drugs and who enter pregnancy in good physical and nutritional health tend to deliver healthy infants, many teens do not enter pregnancy in such good condition. Poor lifestyle habits are thought to be primarily responsible for the high rate of low-birthweight infants born to teens.[32] Few teens (1%) meet all of the Food Guide Pyramid's food group recommendations.[33] Individualized attention, nutrition intervention, and support all foster healthy outcomes of teen pregnancy.[34]

Breastfeeding

Food is the first enjoyment of life.
—Lin Yutang

A woman's capacity to nourish a growing infant does not end at birth; it continues in the form of breastfeeding (Illustration 28.10). Breast milk from healthy, well-nourished women is ideally suited for infant nutrition and health.[35]

PhotoEdit - Myrleen Ferguson

Illustration 28.10

What's So Special about Breast Milk?

The breastfed infant is the reference model against which all alternative feeding methods must be assessed with regard to growth, health, and development; and other short- and long-term health outcomes.
—HHS Blueprint for Action on Breastfeeding, 2000

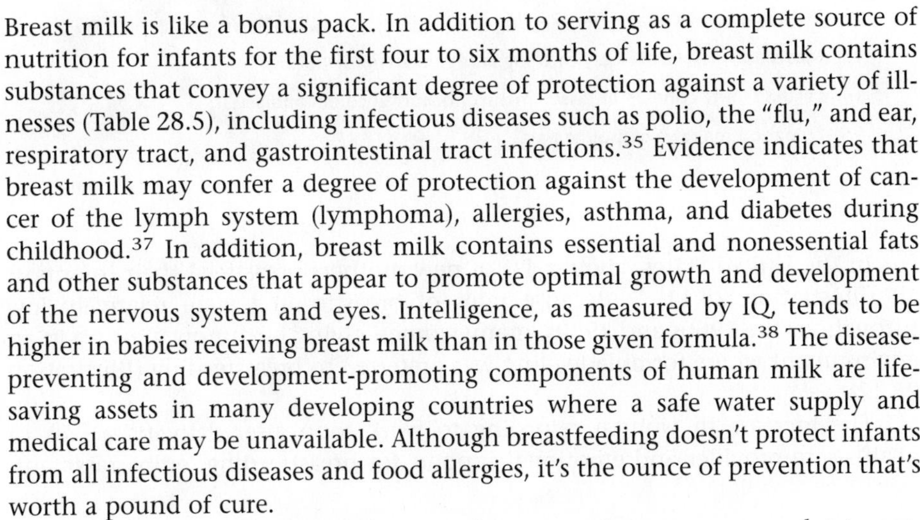

Breast milk is like a bonus pack. In addition to serving as a complete source of nutrition for infants for the first four to six months of life, breast milk contains substances that convey a significant degree of protection against a variety of illnesses (Table 28.5), including infectious diseases such as polio, the "flu," and ear, respiratory tract, and gastrointestinal tract infections.[35] Evidence indicates that breast milk may confer a degree of protection against the development of cancer of the lymph system (lymphoma), allergies, asthma, and diabetes during childhood.[37] In addition, breast milk contains essential and nonessential fats and other substances that appear to promote optimal growth and development of the nervous system and eyes. Intelligence, as measured by IQ, tends to be higher in babies receiving breast milk than in those given formula.[38] The disease-preventing and development-promoting components of human milk are life-saving assets in many developing countries where a safe water supply and medical care may be unavailable. Although breastfeeding doesn't protect infants from all infectious diseases and food allergies, it's the ounce of prevention that's worth a pound of cure.

Breastfeeding also offers other benefits—ones that may mean a lot to parents. Table 28.6 discusses them.

Is Breastfeeding Best for All New Mothers and Infants?

Over 96% of women are biologically capable of breastfeeding, and the vast majority of infants thrive on breast milk.[39] But breastfeeding is not best for every woman and infant. Successful breastfeeding involves more than biology; it is heavily influenced by environmental and psychological conditions.

Table 28.5

A sample of U.S. infants shows fewer significant episodes of illness among those who were breastfed.[36,37]

	Percentage of Infants Fed:	
Illness	**Breast Milk**	**Formula/Other**
Ear infections	3.7%	9.1%
Respiratory illness to age 7	17.0	32.0
Diarrhea, vomiting	3.5	6.9
Hospital admissions	1.0	3.0
Average number of episodes of illness per infant	**8.2**	**21.1**

Table 28.6
Fifteen reasons to breastfeed.

1. The milk container is easy to clean.
2. Breast milk is a renewable resource.
3. There's no packaging to discard.
4. Breast milk comes in an attractive container.
5. The temperature of breast milk is always perfect right out of the container.
6. Breast milk tastes real good.
7. There are *no* leftovers.
8. You don't have to go to the kitchen in the middle of the night to get breast milk ready.
9. It takes just seconds to get a meal ready.
10. There's no bottle to repeatedly pick up off the floor.
11. The price is right.
12. The meal comes in a perfect serving size.
13. Meals and snacks are easy to bring along on a trip or outing.
14. Feeding units come in an assortment of beautiful colors and sizes.
15. One food makes a complete meal.

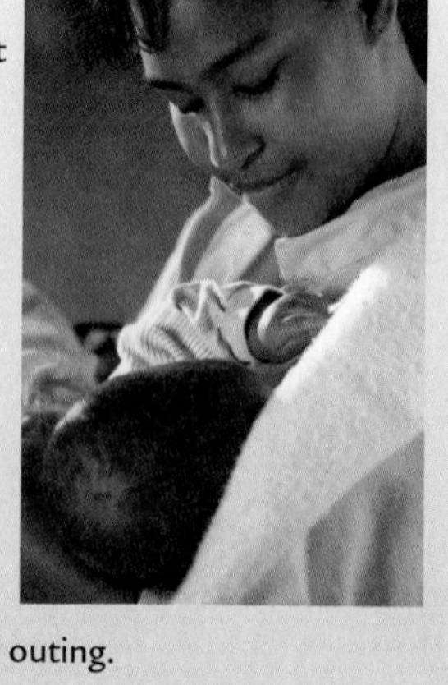

In the United States, about 64% of new mothers breastfeed their infants at all (Illustration 28.11).[35] Because rates of breastfeeding were relatively low throughout the 1960s and 1970s, many women giving birth now grew up in an environment where formula feeding was preferred to breastfeeding (Illustration 28.12).

The increase in women returning to work soon after delivery, a lack of health care provider and emotional support for breastfeeding, embarrassment,

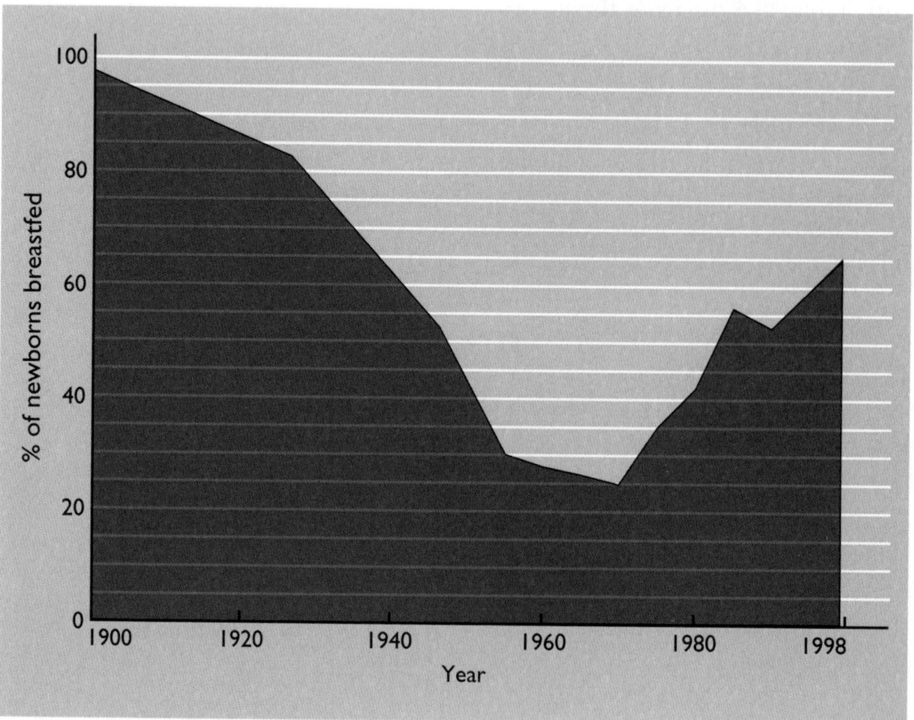

Illustration 28.11
Percentage of breastfed newborns in the United States.[35,40]

early hospital discharge, free formula samples, and inadequate knowledge about how to breastfeed all appear to be deterring U.S. women from breastfeeding.[3] If more women are to have the opportunity to breastfeed, these and other barriers must be broken down. Perhaps a good place to start is with the value our society places on breastfeeding. A number of European countries that actively promote breastfeeding allow women to stay in the hospital after delivery until breastfeeding is going well. The usual practice in some African countries is to relieve a breastfeeding mother's workload so that she can devote nearly full time to feeding and caring for her young infant. A relative may move in with the family and take over household chores, or the mother and baby may live with her parents for a time.

Courtesy of Hans and Libby Schapiro

Illustration 28.12
Said a friend when she saw this photo in 1951: "This is the best advertisement for breastfeeding I've ever seen!"

Public health initiatives are also in place to facilitate breastfeeding. In order to meet national health objectives for improving rates of breastfeeding in the United States by 2010 (see Table 28.7), it is recommended that:

a. Health care workers encourage and facilitate breastfeeding,
b. "How to" advice and problem solving guidance be available for breastfeeding women from qualified health care staff, and
c. Breastfeeding women returning to work have access to on-site child care facilities and private rooms for breastfeeding.

Breastfeeding is at its best when both the mother and infant benefit from the experience. If mutual benefit is not possible, then formula feeding may be necessary. Infant growth and development are well supported by commercially available infant formulas.[41]

colostrum
The milk produced during the first few days after delivery. It contains more antibodies, protein, and certain minerals than the mature milk that is produced later. It is thicker than mature milk and has a yellowish color.

How Breastfeeding Works

The mother's body prepares for breastfeeding during pregnancy. Fat is deposited in breast tissue, and networks of blood vessels and nerves infiltrate the breasts. Ducts that will channel milk from the milk-producing cells forward to the nipple—the milk collection ducts—also mature (Illustration 28.13). Hormonal changes that occur at delivery signal milk production to begin.

Breast milk produced during the first three days or so after delivery is different from the milk produced later. Called ***colostrum***, this early milk contains higher levels of protein, minerals, and antibodies than does "mature" milk.

On the Side

Can adopting moms breastfeed their infants?

Yes, they can. It takes hormones and motivation, however. Breastfeeding success is higher the younger infants are when they begin to breastfeed with their adoptive moms.[42]

Photo Disc

Breast Milk Production While an infant is consuming one meal, she or he is "ordering" the next. The pressure produced inside the breast by the infant's sucking and the emptying of the breasts during a feeding cause a hormone to be released from specific cells in the brain. The hormone stimulates the production of milk so that more

	Newborns	6 Months	12 Months
National goal	75%	50%	25%
Actual (all women)	64	29	25

Table 28.7
Actual versus national goals for breastfeeding in the United States.[3]

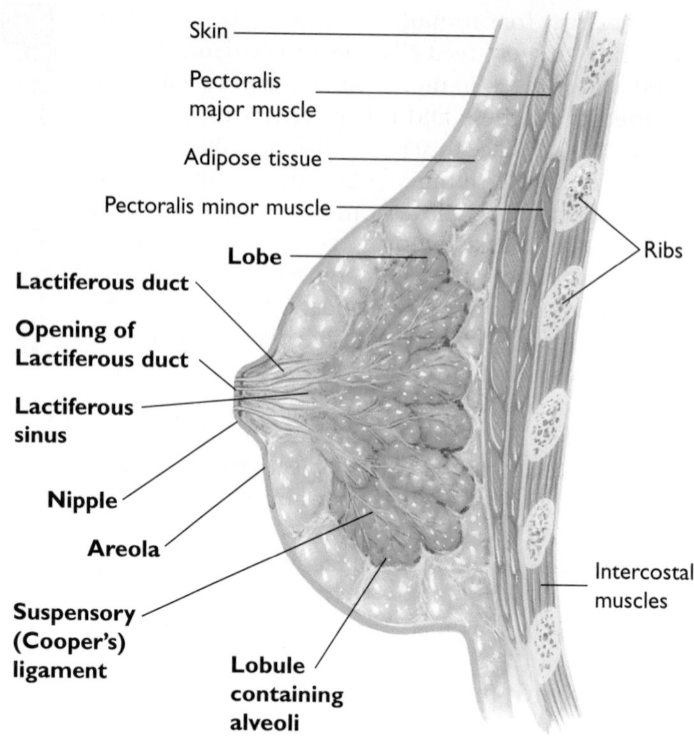

Illustration 28.13
A view of the interior of the breast.

milk is produced for the next feeding. It generally takes about two hours for the milk-producing cells to manufacture enough milk for the next feeding. An important exception to this occurs when an infant enters a growth spurt and consumes more milk than usual. Then milk production takes longer, perhaps a day, to catch up with demand.

Only very rarely is a breastfeeding woman unable to produce enough milk. As long as an infant is allowed to satisfy her or his appetite by breastfeeding as often as desired, milk production will catch up with the baby's need.

Nutrition and Breastfeeding

A breastfeeding woman needs an adequate and balanced diet to replenish her body's nutrient stores, maintain her health, and produce sufficient milk for her baby. Increases in recommended dietary intakes for breastfeeding are generally higher than those for pregnancy. As during pregnancy, proportionately higher amounts of nutrients than calories are required, indicating the need for a nutrient-dense diet (see Illustration 28.5 on p. 28–9).

Calorie and Nutrient Needs The RDA for calories is about 25% higher for breastfeeding women than for other women. Actually, a breastfeeding woman needs about 40% more calories than the RDA for women who are not breastfeeding, but she does not have to consume that level of calories from food. Energy supplied from fat stores that normally accumulate during pregnancy contributes to meeting these needs during breastfeeding, so not all of the calories must come from the mother's diet.[43]

What's a Good Diet for Breastfeeding Women?

The calories and nutrients needed by the breastfeeding woman can be obtained from a varied diet that includes foods from each of the basic food groups. The

Table 28.8
Dietary recommendations for breastfeeding women.[43]

- Diets should supply all of the nutrients breastfeeding women need. (The routine use of vitamin and mineral supplements is not recommended.)
- Fluids should be consumed to thirst (10 cups of fluid each day).
- Weight loss should not exceed 6 to 8 pounds per month after the first month after delivery; caloric intakes should not fall below 1500.
- Alcohol intake should be avoided.

recommended numbers of servings from each food group are shown in Illustration 28.8 (on p. 28–11); they are the same as for pregnant women. Dietary recommendations for breastfeeding women are summarized in Table 28.8.

Increases in hunger and food intake that accompany breastfeeding generally take care of meeting caloric needs. Failure to consume enough calories from food can decrease milk production, however. Low-calorie diets (those providing less than 1500 calories per day) and weight loss that exceeds 1.5 to 2 pounds per week—even in women with a good supply of fat stores—can reduce the amount of milk women produce.[43] Weight loss of about a pound a week starting a month after delivery appears to be safe and helpful in women's attempts to return to prepregnancy weight.[44]

Are Supplements Recommended for Breastfeeding Women? Supplements are not recommended for breastfeeding women. Instead, breastfeeding women should try to get all of the nutrients they need from food.[43] Supplements, if needed, should be prescribed on an individual basis.

Dietary Cautions for Breastfeeding Women

Almost anything a woman consumes may end up in her breast milk. When a breastfeeding woman drinks coffee, her infant receives a small dose of caffeine. Breastfed infants of women who are heavy coffee drinkers (10 or more cups per day) may develop "caffeine jitters."[45] Alcohol also is transferred from a woman's body to breast milk. The development of the brain and nervous system of infants breastfed by chronic, heavy drinkers appears to be retarded.[46] There is also concern that alcohol ingestion by breastfeeding women may decrease milk production.[47]

Environmental contaminants, such as lead, DDT, chlordane, PCBs, and PBBs, for example, are transferred into breast milk. Many environmental contaminants may be stored in a woman's fat or bone tissues. When the fat stores are later broken down for use in breast milk, or the calcium in bone is mobilized, the contaminants stored in the fat or bone may enter the breast milk. The ingestion of fish from contaminated waters in Lake Ontario and Lake Michigan was linked to abnormally high levels of PCBs in breast milk.[48] Infants exposed to PCBs can develop rashes, digestive upsets, and nervous system problems.[49] With the exception of breastfeeding women known to be exposed to excess levels of environmental toxins, however, it is concluded that the benefits of breastfeeding outweigh the risks to infants from harmful substances in the environment.[50]

Infants are much smaller than women, and it takes a smaller dose of caffeine, alcohol, drugs, or environmental contaminants to have an effect on them than on an adult. Whereas breastfeeding mothers may show no adverse effects from these substances, their infants may.

Breastfeeding women are advised to limit their consumption of regular coffee to four or fewer cups a day and to avoid alcohol or limit it to one drink with a meal per day.[43] Drugs or medications should be taken only on the advice of a health care provider.

Illustration 28.14
Infants develop at a truly amazing rate.

Infant Nutrition

He who possesses virtue in abundance may be compared to an infant.
—Lao Tzu, sixteenth century B.C.

At no other time during life outside the womb do growth and development proceed at a faster pace than during infancy. Infants grow out of clothes within weeks, long before they wear them out. Each day infants learn new behaviors, and their minds absorb large chunks of information that will serve them well in the future (Illustration 28.14). The rate at which growth and development proceed in the first year of life is truly amazing. You can get an idea of just how rapidly infants grow and develop by reviewing Illustration 28.15. Just as infants

Illustration 28.15
Growth and developmental characteristics from birth through one year.

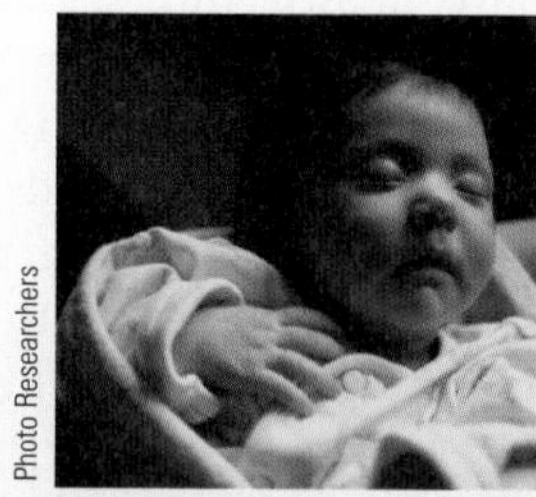

First Month
Generally weighs from 8 to 11 pounds; length is 20 to 23 inches. Head is relatively large and has soft spot on top. Startles and sneezes easily. Jaw may tremble. May hiccup and spit up. Eats every few hours.

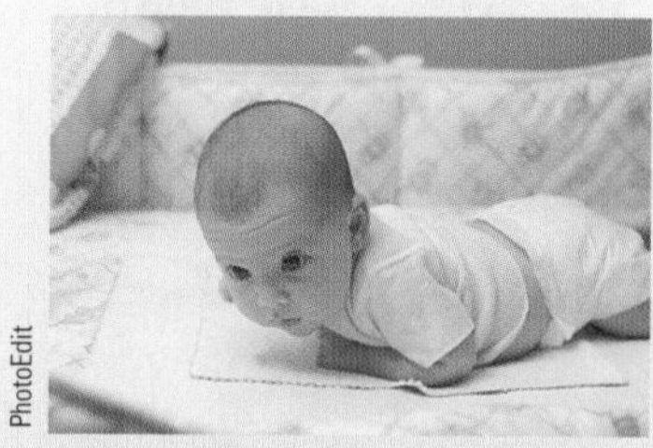

One to Three Months
Lifts head briefly when placed on stomach; smiles, coos, and gurgles. Whole body moves when infant is touched or lifted. Eats every three to four hours.

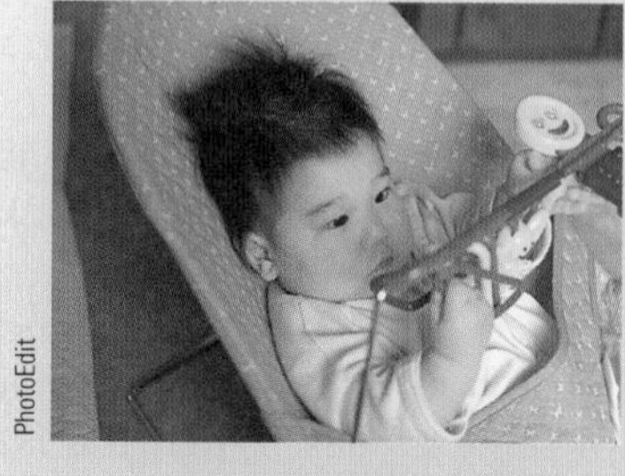

Four to Six Months
Weight nearly doubled. Has grown three to four inches. Follows objects with eyes. Reaches toward objects with both hands; puts fingers and objects into mouth. Turns over, sits unassisted. Awake longer at feeding time. Eats six to seven times per day. Sleeps six to seven hours at night.

Seven to Eleven Months
Gains in weight and height are less rapid, appetite has decreased. Stands up with help, hitches self along the floor. Reaches for, grasps, and examines objects with hands, eyes, and mouth. Has one or two teeth. Takes two naps a day.

Twelve Months
Usually has tripled birthweight and increased length by 50%. Grasps and releases objects with fingers. Holds spoon, but uses it poorly. Begins to walk unassisted.

need security, love, and attention to flourish, so too do they need calories and essential nutrients.

Infant Growth

During the first week or two of life, infants generally lose 5 to 10% of their birthweight while adjusting to the new surroundings. After that, infants grow rapidly. Most infants double their birthweight by four months and triple it by one year. Length usually increases by 50% during the first year. If this rate of growth were to continue, 10-year-old children would be about 10 stories high and weigh over 220 tons! After infancy, the growth rate declines and remains at a fairly low level until the adolescent growth spurt begins.

Growth Charts for Infants In 2000 the Centers for Disease Control released new growth charts for female and male infants, children, and adolescents. Separate charts were not developed by ethnic group or race because growth potential is a shared human trait and varies primarily due to environmental factors such as nutritional status and disease. An example of the growth charts that include infants is shown in Illustration 28.16. The charts should be carefully plotted based on accurate, periodic measures of an infant's size. They are best used for screening growth problems; confirmation of underweight or obesity requires assessment of body fat by skinfold thickness and other measures.[51]

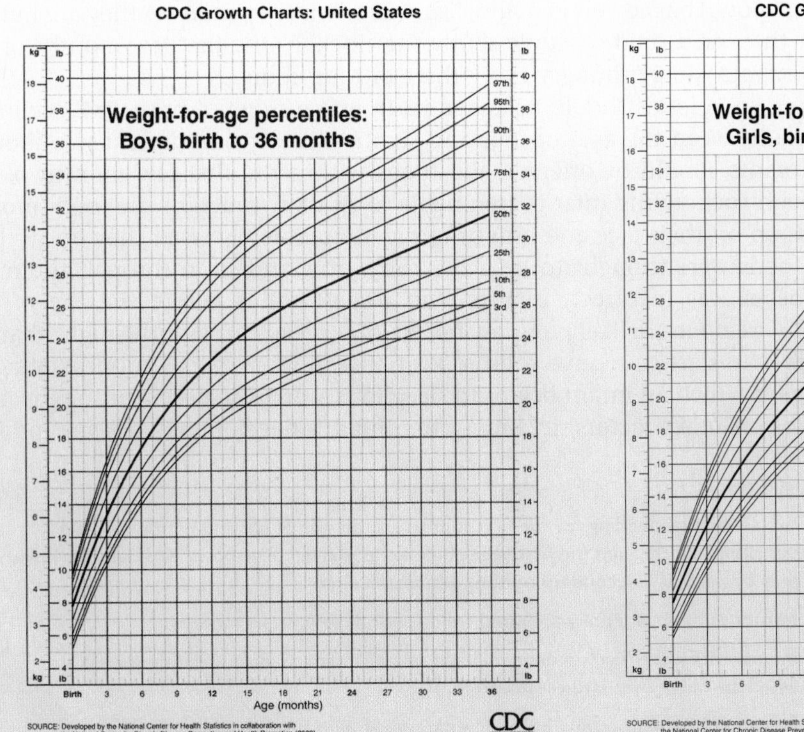

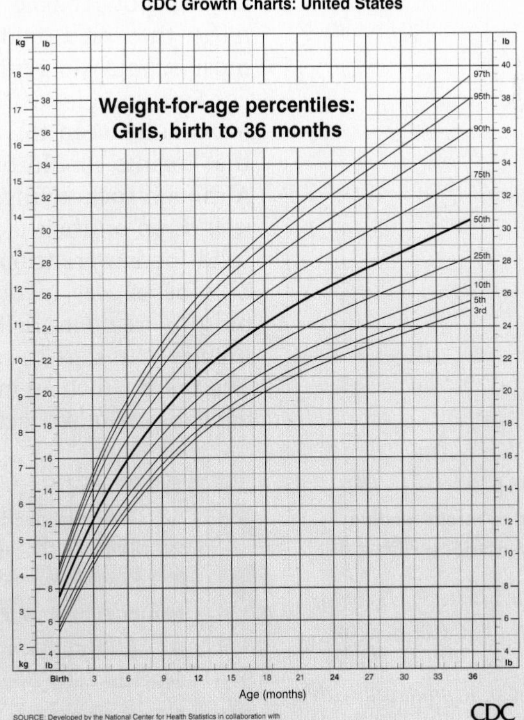

Illustration 28.16
An example of the CDC's growth charts. The charts can be copied from the Internet using the address www.cdc.gov/ growthcharts. *National Center for Chronic Disease Prevention and Health Promotion.*

Body Composition Changes with Growth Humans grow up and out, and their body composition and proportions change as growth progresses. Although generally measured by gains in pounds and inches, growth also reflects changes in bone mass, organ size, body proportions, and composition (the proportionate amounts of water, muscle, and fat). Infants' heads are very large in relation to the rest of their bodies, for example. Brain growth takes precedence over trunk and limb growth early in life, and the body eventually grows to "fit" the head. During the first 10 years of life, a child who initially has the shape of a loaf of bread may normally come to resemble a string bean.

Nutrition and Mental Development

Malnutrition has the greatest impact on mental development when it is severe and occurs during the critical period for brain cell multiplication. For humans, this vulnerable period begins during pregnancy and ends after the first year of life. Impairment in mental development is less severe when malnutrition occurs only during pregnancy or only during infancy than if it occurs throughout both pregnancy and infancy.[52]

Children's mental development is also greatly influenced by the social and psychological environment in which they are raised. Because malnutrition is generally accompanied by both social and psychological deprivation, these factors often contribute jointly to poor mental development. For children in the United States, poverty, neglect, illnesses, and psychological problems appear to be the main cause of undernutrition and poor mental development.[53]

Infant Feeding Recommendations

Current dietary recommendations for infants call, preferably, for breastfeeding or, alternatively, for formula feeding for the first 12 months of life (Table 28.9).[54] Infants should be fed "on demand," that is, when they indicate they are hungry, rather than on a rigid schedule set by the clock. In the first few months of life, most infants will get hungry every three or four hours.

Soft, solid foods should be added to an infant's diet of breast milk or infant formula between the ages of four and six months. At one time, it was thought that infants should be offered solid foods within the first few months of life. Although such young infants are unable to swallow much of the food (most of it ends up on their face and bib) or to digest completely what they do swallow, solid foods were thought to help the baby grow and sleep through the night. This belief was incorrect. Infants who receive solids before the age of four months are no more likely to grow normally or sleep through the night than are infants who start to receive solid foods between four and six months of age.[55] The age at which an infant begins to sleep for six or more hours during the night depends on other factors, including the infant's developmental level and how

Table 28.9
Overview of infant feeding recommendations[35,54]

From The Science of Human Nutrition by Judith E. Brown, copyright © 1990 by Harcourt Brace & Co. Reproduced with permission of the publisher.

- Exclusive breastfeeding for the first four to six months of life and continuation of breastfeeding through the first year of life are preferred. Iron-fortified infant formulas may be used as a secondary option to breastfeeding.
- Introduce basic, specially prepared solid foods at four to six months.
- Infants should be fed "on demand" and not by a set schedule. Feeding should stop when the baby loses interest in eating.
- Breastfed infants receiving little sun exposure should be given a vitamin D supplement (200 IU or 5 micrograms per day).
- Infants not receiving fluoridated water should be given fluoride drops (dose depends on age).

much he or she has slept during the day. Neither the infant nor the infant's parents are likely to get a full night's sleep for at least four months.

Introducing Solid Foods Recommendations for the timing and sequence of offering solid foods to infants are shown in Illustration 28.17. Rice cereal should be offered first. It is easily digested by infants of this age, and allergies to rice are extremely rare. It is recommended that iron-fortified rice cereal be given to all breastfed infants and to bottle-fed infants who are not receiving iron-fortified formula. The iron provided by iron-fortified rice cereal or formula helps to restock the infant's iron stores, which have been drawn upon since birth. Generally, solid foods prepared for infants should consist of single, basic foods such as strained vegetables, fruits, or meats.[56]

Solid foods for infants can be purchased as commercial baby food or prepared at home. When baby foods are prepared at home, care should be taken to avoid contamination and to achieve the right consistency. (Table 28.10 provides instructions.) Infants should be offered new foods one at a time. The new offerings should be separated by several days so that any allergic reactions to a food can be identified. Variety is the key to achieving a healthy diet for infants in their second six months of life, and an assortment of basic foods should be given.

By nine months of age, infants are ready for mashed foods and foods such as yogurt, applesauce, ripe banana pieces, and grits. Most infants have several teeth by this time, and they are able to bite into and chew soft foods.

Infants graduate to adult-type foods after the age of 12 months. Although most foods still need to be mashed or cut up into small pieces for them, one-year-olds are able to eat the same types of food as the rest of the family. They can drink from a cup and nearly feed themselves with a spoon. Infants have come a long way in 12 months!

Foods to Avoid Not all foods can be considered "baby foods." Some foods should be omitted from an infant's diet because they are apt to cause allergic

Illustration 28.17
Introduction of "solid" foods to infants.

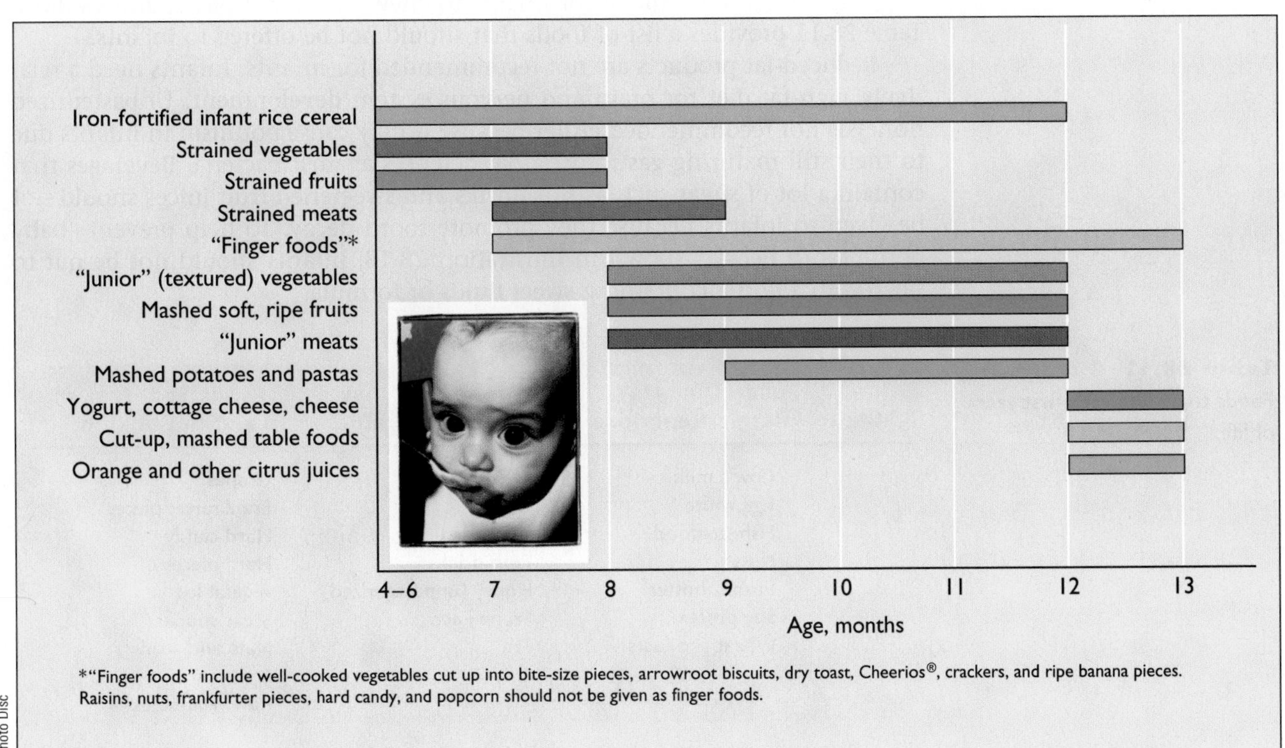

*"Finger foods" include well-cooked vegetables cut up into bite-size pieces, arrowroot biscuits, dry toast, Cheerios®, crackers, and ripe banana pieces. Raisins, nuts, frankfurter pieces, hard candy, and popcorn should not be given as finger foods.

Table 28.10
Directions for making baby foods at home.

Getting Ready

- Equipment needed: A blender or baby-food mill and a strainer.
- Setting up: Clean all utensils, equipment, and counter surfaces thoroughly. Wash hands with soap and water.

Preparation

Use basic foods that do not contain added sugar, salt, spices, margarine, butter, or other additives. After the infant is 7 months old, make the food a bit lumpy, but still soft.

- **Fruits:** Use clean, ripe fruits. Remove all skin, seeds, and cores (or use canned fruits). Puree in blender or food mill thoroughly. (Cook hard fruits like apples first.) Mash if fruit is soft enough.

  Examples: Mashed bananas; applesauce; pureed apricots, pears, and peaches.
- **Vegetables:** Clean thoroughly. Remove stems and any tough skin or seeds. Boil or steam until soft. Puree and serve lukewarm.

  Examples: Pureed green beans, peas, squash, carrots, and potatoes.
- **Meats:** Cook the meat thoroughly. Trim off fat, skin, and gristle. Blend or grind meat with enough water to make a puree (generally ½ cup water to 1 cup cooked meat).

  Examples: Pork, chicken, fish, beef, lamb, and turkey.

Storage

- Freeze extra servings in a tightly covered ice-cube tray or other airtight container, or store tightly covered in refrigerator for no more than three days. Do not thaw and refreeze.
- Discard any leftovers in the baby's dish. When the spoon used to feed the child contacts the food, bacteria are introduced that may cause the food to spoil while stored.

reactions or are too difficult for infants to chew into small pieces and swallow. Table 28.11 provides a list of foods that should not be offered to infants.

Reduced-fat products are not recommended for infants. Infants need a relatively high-fat diet for brain and nervous system development. Unpasteurized honey is not recommended either because it may cause botulism in infants due to their still maturing gastrointestinal defenses against bacteria. Beverages that contain a lot of sugar such as soft drinks and sweetened fruit juices should not be given to infants because they promote tooth decay. To help prevent "baby bottle tooth decay," shown in Illustration 28.18, infants should not be put to sleep with a bottle containing sweet fluids or formula.

Table 28.11
Foods to avoid in the first year of life.

Foods That May Cause Allergic Reactions	Foods That May Cause Other Problems	Foods That May Lead to Choking
Cow's milk	Blueberries	Grapes
Egg white	Coffee	Frankfurter pieces
Fish, seafood	Corn	Hard candy
Nuts	Fruit drinks	Hard pieces of vegetables
Peanut butter	Honey (unpasteurized)	Meat chunks
Soy protein	Prune juice	Nuts and seeds
Wheat products	Tea	Popcorn
		Raw vegetables

Do Infants Need Supplements?

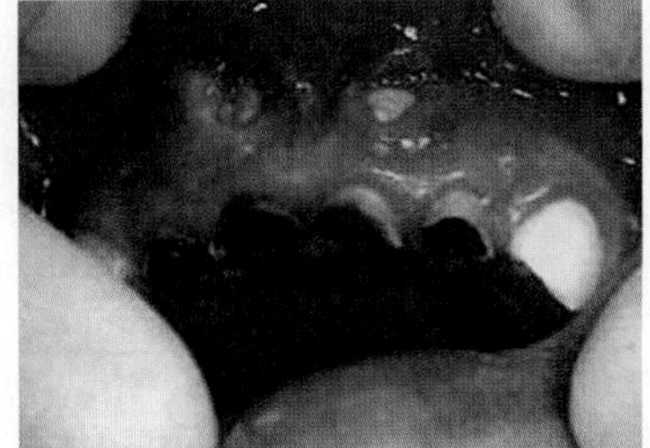
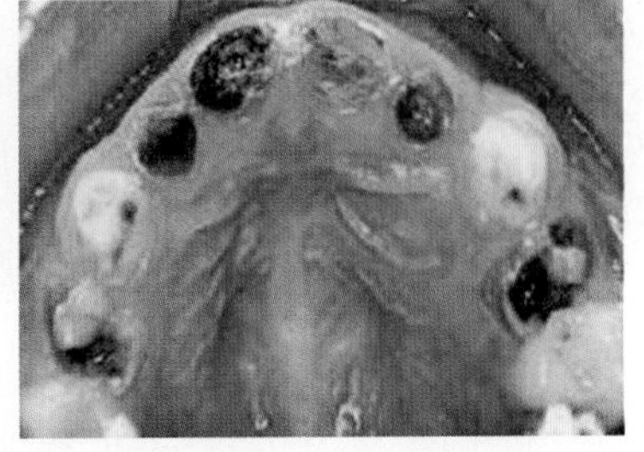

Illustration 28.18
Baby bottle tooth decay.

Two situations call for the use of supplements during infancy (see Table 28.12). Breastfed infants and infants receiving formula from a concentrate that is not diluted with fluoridated water need fluoride supplements after six months of age. Since breast milk contains a low amount of vitamin D, and since cases of rickets are increasing among breastfed infants in the United States,[59] breastfed infants not exposed regularly to sunshine should receive a vitamin D supplement.

The Development of Healthy Eating Habits Begins in Infancy

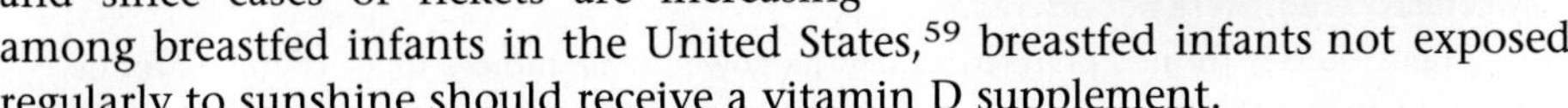

Older infants and young children should be offered a wide variety of nutritious foods in a positive eating environment. They alone should make the decision about how much to eat of any food offered.

Food preferences are primarily learned and are unique to every individual.[60] The learning process begins early in life and is affected by a variety of influences. "Instincts" that draw infants to select certain, nutritious foods are not one of the influences, however. Infants do not instinctively know what to eat.[61] Rather, they are born with mechanisms that help regulate how much they eat. These basic characteristics of infant food intake were beautifully demonstrated by Clara Davis in experiments with 7- to 9-month-old infants in the 1920s and 30s.[62] One of her most famous experiments is described in Illustration 28.19.

Subsequent research reinforced earlier conclusions that infants and young children should be offered a wide variety of nutritious foods, and that decisions about how much to eat should be left up to the infant or child.[63] Beginning at birth, infants who are hungry eat enthusiastically. They stop eating when they are full. Coaxing, cajoling, and pleading by parents or caregivers can override infants' decisions about how much food to eat. That, unfortunately, may upset infants' built-in food intake regulating mechanisms and lead to over- or undereating, and to "fussy" eaters.

Teaching Infants the Right Lessons about Food Recommendations for feeding infants are primarily based on energy and nutrient needs, the developmental readiness of infants for solid foods, and the prevention of food allergies. But infant feeding recommendations also include a large educational component. Many of the lessons infants learn about food and eating make an impression that lasts a lifetime. Later food habits and preferences, appetite, and food intake regulation are all influenced by early learning experiences. Table

Table 28.12
Indications for supplementing infants.

1. Fluoride supplementation is recommended for infants older than 6 months who live in areas with no or low fluoride in the household water supply (0.25 mg of fluoride/day from 6 months to 3 years of age).[57]
2. Vitamin D supplements (200 IU/day) should be given to breastfed infants receiving little exposure to sunshine.
   - Breastfed infants exposed to about 30 minutes of sunshine each week with only diapers on, or those receiving two hours of sun exposure while wearing clothes but no hat are likely meeting their needs for vitamin D by its manufacture in their skin.[58]

Illustration 28.19
Infant food preferences.

Clara Davis conducted her most quoted feeding experiment with older infants living in an orphanage connected to a large hospital in Chicago. Three times a day infants were offered a tray of various foods (such as the one shown below) by a nurse. Infants would pick up or grab for the foods they wanted, and the nurse would bring the food selected to the infant's hands or mouth. Altogether 32 different foods were offered, consisting of fresh, unprocessed, unseasoned, and simply prepared basic foods. Foods included milk, beef, kidney, bone marrow, liver, brain, thymus, fish, whole-wheat cereals, raw eggs, sea salt, 15 fresh fruits, and 10 fresh vegetables. Desserts, sugars, syrups, and other sweetened foods were not offered.

Infants quickly formed preferences and narrowed their choices to 14 of the 32 foods. Vegetables were the least liked, whereas milk, bone marrow, eggs, bananas, apples, oranges, and oatmeal were the best liked. They ate enough to grow normally and to remain in good health.

Dr. Davis concluded that appetite is the best guide to how much food an infant or young child needs, and that self-selection can have but doubtful value if the diet must be chosen from inferior foods.[62]

lactic milk whole milk cooked marrow raw beef cooked beef
chicken sea salt ordinary salt carrots, raw
cooked turnips cauliflower rye krisp

Clara Davis -*American Journal of Diseases of Children*

28.13 provides a lesson plan for teaching your infant the right lessons about food and eating.

Make Feeding Time Pleasurable

> More than any other area of development, feeding is an arena in which parents and baby work out the continuing struggle between dependence (being fed) and independence (feeding oneself). Independence must win. Pushing a child to eat is the surest way to create problems. Feeding has got to be pleasurable.
>
> —T. Berry Brazelton, 1993

Table 28.13
Tips for teaching infants the right lessons about food and eating.

1. Infants learn to eat a variety of healthy foods by being offered an assortment of nutritious food choices. There are *no* inborn mechanisms that direct babies to select a nutritious diet.
2. Infants must be allowed to eat when they are hungry and to stop eating when they are full. Infants, not parents, know when they are hungry or have had enough to eat.
3. Food should be offered in a pleasant environment with positive adult attention.
4. Food should not be used as a reward, punishment, or pacifier.
5. Infants or children should never be coerced into eating anything.
6. Food preferences change throughout infancy. Because an infant rejects a food one time doesn't mean she or he will not accept the food if offered later. Offering a food on a number of occasions often improves acceptance of the food. Babies still may not like strong-flavored vegetables until they are older, however.

As Dr. Brazelton, an expert on child development, points out, there is much more to the feeding of infants than meets the eye. Infants learn about communication, relationships, and independence during pleasurable eating experiences. If undertaken in positive and supportive circumstances, mealtimes for infants can provide lessons that will support their physical and psychological health well into the future.

Nutrition UP CLOSE

You Be the Judge!

FOCAL POINT: Apply your nutrition knowledge to the diet of a pregnant woman.

Gloria's baby is due in five months. Before becoming pregnant, she never gave much thought to what she ate, but now she is trying to eat healthy foods for her baby and herself. Compare her choices with the Food Guide Pyramid recommendations for a pregnant woman to find out how well she is doing. Assume that each food item listed is the equivalent of one serving.

- *Breakfast:* Bran muffin, cereal with strawberries, and a glass of milk.
- *Lunch:* Vegetable soup, carrot sticks, an apple, and a glass of orange juice.
- *Afternoon snack:* Yogurt.
- *Dinner:* Chicken, broccoli, a green salad, and iced tea.

Categorize Gloria's choices into the food groups below to determine if she ate the recommended servings from the Food Guide Pyramid. The breads and cereals group has been done for you.

Breads and Cereals
bran muffin
cereal
Gloria's servings: 2
Recommended servings: 6-11

Vegetables
Gloria's servings:
Recommended servings: 3-5

Fruits
Gloria's servings:
Recommended servings: 2-4

Milk & Milk Products
Gloria's servings:
Recommended servings: 3

Meat and Alternates
Gloria's servings:
Recommended servings: 2

Miscellaneous (fats, sugars)
Gloria's servings:
Recommended servings: in moderation

FEEDBACK (including answers) can be found at the end of Unit 28.

Source: Adapted from Healthy Infant Outcome Program "You Be the Judge," University of Minnesota, 1989.

WWW links

www.ncemch.org
A one-stop shopping site for maternal and child health resources.

www.lalecheleague.org
A comprehensive online resource for breastfeeding information and support resources.

www.aap.org
The American Academy of Pediatrics's website is loaded with information on breastfeeding and infant and child nutrition.

www.4woman.gov
A website from the Public Health Service that focuses on breastfeeding and other health topics of interest to women.

www.cdc.gov/breastfeeding
The Centers for Disease Control website on breastfeeding.

www.cdc.gov/growthcharts
Need more information about the growth charts or copies of them? Get it here.

www.hc-sc.gc.ca.ca
Health Canada's website. Search pregnancy, breastfeeding, infants, or other topics.

www.eatright.org and www.dietitians.ca
The American Dietetic Association and Dietitians of Canada offer sound nutrition advice for pregnant and breastfeeding women, and for infants.

REFERENCES

1. Guyer B, et al. Annual summary of vital statistics. Pediatrics 2000;106:1307–17.
2. McCormick MC. The contribution of low birth weight to infant mortality and childhood morbidity. N Engl J Med 1985;312:82–90.
3. Health, United States 2000. www.cdc.gov/nchs, 11/00.
4. Healthy people 2000: national health promotion and disease prevention objectives. Washington, DC: U.S. Department of Health and Human Services, DHHS publication No. (PHS) 91-50212; 1991.
5. Rosso P. Nutrition and metabolism in pregnancy. New York: Oxford University Press; 1990.
6. Lewis R. New light on fetal origins of adult disease. The Scientist 2000(Oct 30);14:1,16.
7. Brown JE, Kahn ESB. Maternal nutrition and the outcome of pregnancy. Clin Perinatol 1997;24:433–49.
8. Clark PM. Programming of the hypothalamo-pituitary-adrenal axis and the fetal origins of adult disease hypothesis. Eur J Pediatrics 1998;157(suppl):S7–S10.
9. Doyle JJ, et al. Nutritional vitamin B12 deficiency in infancy: three case reports and a review of the literature. Pediatr Hematol Oncol 1989;6:161–72.
10. Winick M. Nutrition and pregnancy. White Plains (NY): March of Dimes Birth Defects Foundation; 1986.
11. Miller DR, Hayes KC. Vitamin excess and toxicity. In: Hathcock JN, editor. Nutritional toxicology. New York: Academic Press; 1982: pp 81–131.
12. Cochrane WA. Overnutrition in prenatal and neonatal life: a problem? Can Med Assoc J 1965;93:893–9.
13. Brown JE, et al. Prenatal weight gains related to the birth of healthy-sized infants to low-income women. J Am Diet Assoc 1986;86:1679–83.
14. National Academy of Science (Institute of Medicine). Nutrition during pregnancy. I. Weight gain. II. Nutrient supplements. Washington, DC: National Academy Press; 1990: p 48.
15. Ohlin A, Rossner S. Maternal body weight development after pregnancy. Int J Obesity 1990;14:159–73.
16. Food and Nutrition Board (National Research Council). Recommended dietary allowances, 10th ed. Washington, DC: National Academy of Sciences; 1989.
17. Brown JE, et al. Predictors of red cell folate level in women attempting pregnancy. JAMA 1997;277:548–52.
18. Daly LE, et al. Folate levels and neural tube defects. JAMA 1995;274:1698–1702.
19. Folate levels among American women. Morbidity and Mortality Weekly Report 2000;49:962–65.
20. Palomaki GE, et al. Prenatal screening for open neural tube defects in Maine (letter). N Engl J Med 1999;340:1049–50.
21. Sahakian V, et al. Vitamin B_6 is effective therapy for nausea and vomiting of pregnancy: a randomized, double-blind placebo-controlled study. Obstet Gynecol 1991;78:33–36.
22. Recommendations to prevent and control iron deficiency in the United States. MMWR 1998;47:1–28.
23. Preziosi P, et al. Effect of iron supplementation on the iron status of pregnant women: consequences for newborns. Am J Clin Nutr 1997;66:1178–82.
24. Litovitz T, Manoguerra A. Comparison of pediatric poisoning hazards: an analysis of 3.8 million exposure incidents. Pediatrics 1992;89:999–1006.
25. Taper LJ, et al. Zinc and copper retention during pregnancy: the adequacy of prenatal diets with and without dietary supplementation. Am J Clin Nutr 1985;41:1184–92.
26. Swanson CA, King JC. Zinc and pregnancy outcome. Am J Clin Nutr 1987;46:763–71.
27. Herbert V. Recommended dietary intake (RDI) of iron in humans. Am J Clin Nutr 1987;45:679–86.
28. Jones KL. Pattern of malformation in offspring of chronic alcoholic mothers. Lancet 1973;i:1267–71; and Spohr H-L, et al. Prenatal alcohol exposure and long-term developmental consequences. Lancet 1993;341:907–10.
29. Mattson SN, et al. Heavy prenatal alcohol exposure with or without physical features of fetal alcohol syndrome leads to IQ deficits. J Pediatr 1997;131:718–21.
30. Hathcock JN, et al. Evaluation of vitamin A toxicity. Am J Clin Nutr 1990; 52:183–202.
31. Births: Final Data, 1998. Washington, DC: National Center for Health Statistics, 2000; www.cdc.gov/nchs.

32. Story M, Alton I. Nutrition and adolescent pregnancy. Nutr Today 1995;30:142–8.
33. Muñoz KA, et al. Food intakes of US children and adolescents compared with recommendations. Pediatrics 1997;100:323–9.
34. Dubois S, et al. Ability of the Higgins Nutrition Intervention Program to improve adolescent pregnancy outcome. J Am Diet Assoc 1997;97:871–8.
35. HHS Blueprint for Action on Breastfeeding. www.4woman.gov. 11/8/00.
36. Cummings AS. Morbidity in breastfed and artificially fed infants. J Pediatr 1979;90:726–9; and Cunningham AS, et al. Breastfeeding and health in the 1980's: A global epidemiologic review. J Pediatr 1991;118:659–65.
37. Wilson AC, et al. Relation of infant diet to childhood health. Brit Med J 1998;316:21–25.
38. U.S. Public Health Service. A review of the medical benefits and contraindications to breastfeeding in the United States. Washington, DC: Maternal and Child Health Bureau; 1997: pp 9–10.
39. Simopoulos AP, Grave GD. Factors associated with the choice and duration of infant-feeding practices. Pediatrics 1984;74(suppl):603–14.
40. Ryan AS, et al. A comparison of breast-feeding data from the National Surveys of Family Growth and the Ross Laboratories Mothers Surveys. Am J Public Health 1991;81:1049–52.
41. Committee on Nutrition, American Academy of Pediatrics. Breastfeeding and the use of human milk. Pediatrics 1997;100:1035–9.
42. Lakhkar BB. Breastfeeding in adoptive babies. Indian Pediatr 2000;37:1114–16.
43. National Academy of Sciences (Institute of Medicine). Nutrition during lactation. Washington, DC: National Academy Press; 1991.
44. Lovelady CA, et al. The effect of weight loss in overweight, lactating women on the growth of their infants. N Engl J Med 2000;342:449–53.
45. Watkinson B, Fried PA. Maternal caffeine use before, during and after pregnancy and effects upon offspring. Neurobehavioral Toxicology and Teratology 1985;7:9–17.
46. Fetal alcohol syndrome and alcohol-related neurodevelopmental disorders. American Academy of Pediatrics. Pediatrics 2000;106:358–61.
47. Menella JA, Beauchamp GK. The transfer of alcohol to human milk. N Engl J Med 1991;325:981–85.
48. Schwartz PM, et al. Lake Michigan fish consumption as a source of polychlorinated biphenyls in human cord serum, maternal serum, and milk. Am J Public Health 1983;73:293–6.
49. Miller SA, Chopra JG. Problems with human milk and infant formulas. Pediatrics 1984;74(suppl):639–47.
50. Rogan WJ. Pollutants in breast milk. Arch Pediatr Adol Med 1996;150:981–90.
51. Thompson D, et al. New CDC growth charts. WIC National Meeting, Salt Lake City, Utah, 9/8/00.
52. Winick M. Malnutrition and brain development. New York: Oxford University Press; 1976.
53. Lloyd-Still JD, et al. Intellectual development after severe malnutrition in infancy. Pediatrics 1974;54:306–11.
54. Rappo PD, et al. Pediatrician's responsibility for infant nutrition. Pediatrics 1997;99:749–50.
55. Beal VA. Termination of night feeding in infancy. J Pediatr 1969;75:690–2.
56. Krebs NK. Dietary zinc and iron sources, physical growth and cognitive development of breastfed infants. J Nutr 2000;130:358S–60S.
57. ADA Council on Access, Prevention, and Interpersonal Relations. J Am Dent Assoc 1995;126 (suppl):19S.
58. Specker BL, et al. Sunshine exposure and serum 25-hydroxyvitamin D concentrations in exclusively breast-fed infants. J Pediatr 1985;107:372–6.
59. Welch TR, et al. Vitamin D–deficient rickets: the reemergence of a once-conquered disease (editorial). J Pediatr 2000;137:152–7.
60. Marianne BM, et al. Genetic and environmental influences on eating patterns of twins ages >50 years. Am J Clin Nutr 1999;70:456–65.
61. Story M, Brown JE. Do young children instinctively know what to eat? The studies of Clara Davis revisited. N Engl J Med 1987;316:103–6.
62. Davis C. Self-selection of diets by newly weaned infants: an experimental study. Am J Dis Child 1928;36:651–79.
63. Birch LL, Deysher M. Caloric compensation and sensory specific satiety: evidence for self-regulation food intake by young children. Appetite 1986;7:323–31.

Nutrition UP CLOSE

You Be the Judge!

Feedback for Unit 28

Gloria's choices:

Breads and cereals	*2 servings* (bran muffin, cereal)
Vegetables	*4 servings* (vegetable soup, carrot sticks, broccoli, green salad)
Fruits	*3 servings* (strawberries, apple, orange juice)
Milk and milk products	*2 servings* (milk, yogurt)
Meat and alternates	*1 serving* (chicken)
Miscellaneous (fats, sugars)	*0 servings!*

All of Gloria's choices are healthy ones, but she is missing some important foods. While concentrating hard on eating enough fruits and vegetables, she has neglected to consume enough high-protein and carbohydrate-rich foods from the remaining food groups. She needs to include in her diet more dairy products, more meat or protein alternates, and more starchy foods like breads, pasta, rice, tortillas, corn, and legumes. With a few additions to her existing choices, the quality of her diet can be greatly improved. Here is a more balanced version of Gloria's menu for a day:

- *Breakfast:* Bran muffin, cereal with strawberries, and a glass of milk.
- *Lunch:* Vegetable soup, carrot sticks, an apple, **tuna sandwich,** and a glass of orange juice.
- *Afternoon snack:* Yogurt, **cheese, and whole-grain crackers.**
- *Dinner:* Chicken, broccoli, green salad, **mashed potatoes, a dinner roll,** and iced tea.

Nutrition for the Growing Years: Childhood through Adolescence

Photo Disc

Nutrition Scoreboard

	TRUE	FALSE
1. Parents should not allow their children to eat "junk" foods.		
2. It is up to parents and caretakers to decide how much children should eat.		
3. Chronic dieting during adolescence can reduce adult height.		
4. Childhood and adolescence represent "grace periods" during which the diet consumed does not influence future health.		

(Answers on next page)

Answers to

Nutrition Scoreboard

	TRUE	FALSE
1. Withholding sweets and other favorite foods makes kids value and want them more. Such foods should not be given special emphasis.		✔
2. Children, not parents or caregivers, should decide how much food is consumed.		✔
3. Failure to gain weight can impair growth in height during adolescence.	✔	
4. Childhood and adolescent diets can affect future health. Heart disease and diabetes, for example, may have "pediatric" origins that are related to dietary intake patterns.		✔

Key Concepts and Facts

- There is no evidence to support the notion that children are born knowing what foods they should eat.
- Children are born with regulatory processes that help them decide how much to eat.
- Parents should decide *what* foods to offer children. Children should decide *how much* to eat.
- Diet and other behaviors of children and adolescents affect health during the adult years.

Photo Researchers - Juneberg Clark

The Span of Growth and Development

Physical and mental development proceed at a high rate from infancy through adolescence (Illustration 29.1). Young children generally enter this phase of life able to take a step or two and to guide a spoon into their mouths sideways. They will likely leave adolescence with the ability to drive, work for pay, and solve complex problems. These are the formative years that lay the foundation for the rest of life.

The Nutritional Foundation

As the twig is bent, so grows the bough.

Good nutrition takes on particular importance during the growing years for many reasons. Growth is an energy- and nutrient-requiring process. It will not proceed normally unless the diet supplies enough of both.

Children learn about food and its importance to health and well-being during these early years of life, and they also establish food preferences and physical activity patterns that may endure into the adult years. Food intake regulatory mechanisms are affected by the lessons children learn early in life. Given control over decisions about how much to eat, children generally become responsive to internal cues that signal when they should eat and when they should stop eating.[1] If the lessons go well, they learn to eat for the right reason.

Eating well and learning the right lessons about food and health have implications that transcend the growing years. Early diets may have long-term effects on the risk of developing a number of diseases later in life.

Characteristics of Growth in Children

Growth slows substantially after the first year of life. Between the ages of 2 and 10 years, children normally gain somewhere around five pounds and two to three

Illustration 29.1
Growth and development characteristics from 1 through 16 years.
(continues on next page)

PhotoEdit

One to Two Years
Gains in height and weight continue at a lower rate; appetite is less. Uses finger and thumb to pick up things. Soft spot grows smaller and then disappears. Baby teeth continue to appear. Usually takes one long nap a day. Drinks from a cup, attempts to feed self with a spoon. Likes to eat with hands. Pulls self up to standing position. Walks alone.

PhotoEdit

Two to Three Years
Slower and more irregular gains in height and weight. Has all 20 teeth. Runs and climbs, pushes, pulls, lugs, walks upstairs one step at a time. Feeds self using fingers, spoon, and cup; spills a lot. At times has one favorite food. Associates the sensation of hunger with a need for food.

PhotoEdit

Three to Four Years
Gains 4–6 pounds and grows about 2–3 inches. Feeds self and drinks from a cup quite neatly, carries things without spilling. May give up sleep at nap time, substituting quiet play.

PhotoEdit

Four to Five Years
Gains in height and weight about same as previous year. Hops and skips, throws ball. Has increasingly good coordination, masters buttons and shoelaces. Can use knife and fork and is a good self-feeder.

PhotoEdit

Five to Six Years
Growth continues at about the same rate. Legs lengthen. Six-year permanent molars usually appear (new teeth, not replacing baby teeth). Begins to lose front baby teeth. Prefers plain, bland, and unmixed foods.

Illustration 29.1
Growth and development characteristics from 1 through 16 years. *(continued)*

PhotoEdit

Six to Nine Years
Slow gains in height and weight (2–3 inches and 4–6 pounds a year). Some additional permanent teeth appear. Likely to have the childhood communicable diseases. Sleeps 11 to 13 hours.

PhotoEdit

Nine to Twelve Years
May be long legged and rangy, but health is generally sturdy. Permanent teeth continue to appear. Appetite good. Needs about 10 hours sleep. Growth spurt in girls usually begins. May get very irritable when hungry.

PhotoEdit

Twelve to Fourteen Years
Wide differences in height and weight in children of either sex of same age. Menstruation usually begins (sometimes earlier). Girls develop breasts, are usually taller and heavier than boys of the same age. Growth spurt of boys begins. Muscle growth rapid. Appetite increases. Likes to spend time with friends.

PhotoEdit

Fourteen to Sixteen Years
Boys are in period of growth spurt. Voice deepens. Girls have usually achieved maximum growth. Menstruation is established though may still be irregular. Enormous appetite. Pimples a common problem. Sleep reaches its adult pattern. All permanent teeth except wisdom teeth.

inches in height per year (Illustration 29.2). Gains in weight and height occur in "spurts," rather than continuously and gradually. Prior to a growth spurt, appetite and food intake increase (given an adequate food supply), and the child puts on a few pounds of fat stores (Illustration 29.3). During the growth spurt, these fat stores are used to supply energy needed for growth in height.

CDC's Growth Charts for Children and Adolescents The Centers for Disease Control (CDC) recently released growth charts for females and males from 0 to 36 months old, and from 2 to 20 years. Growth charts for 2 to 20 year olds consist of graphs of:

- weight for age (see Illustration 29.4),
- height (stature) for age,
- weight for height, and
- body mass index (BMI) for age.

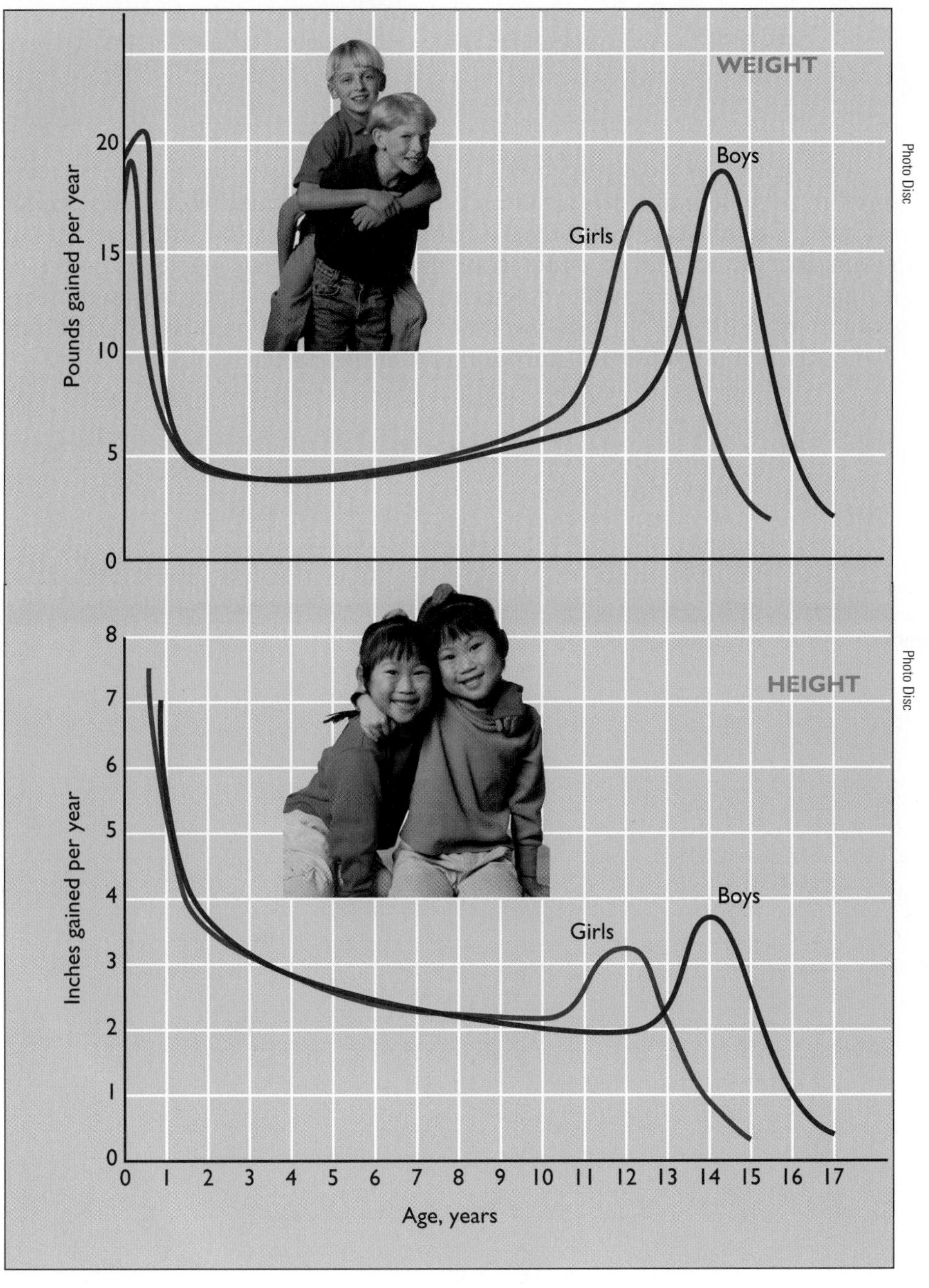

Illustration 29.2
Average yearly growth in weight and height during childhood and adolescence.

Illustration 29.3
Growth occurs in a series of "spurts." Each spurt follows this sequence.

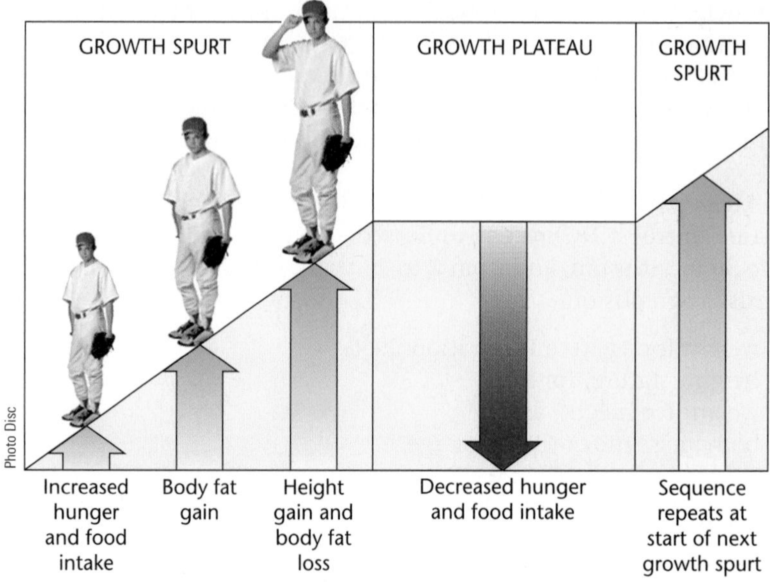

Illustration 29.4
An example of the CDC's growth charts: Weight for age for boys and girls. The charts can be copied from the Internet using the address www.cdc.gov/growthcharts.

Each graph provides percentiles that reflect the distribution of these measures in a representative sample of 2 to 20 year olds in the United States. So, for example, if a child's weight for age is between the 50th and 75th percentiles, it means that 50 to 75% of children of that same age fall between these percentiles, too. Children and adolescents whose weight status measurements place them in the highest and lowest percentile ranges should be evaluated further for potential underlying nutrition and health problems.

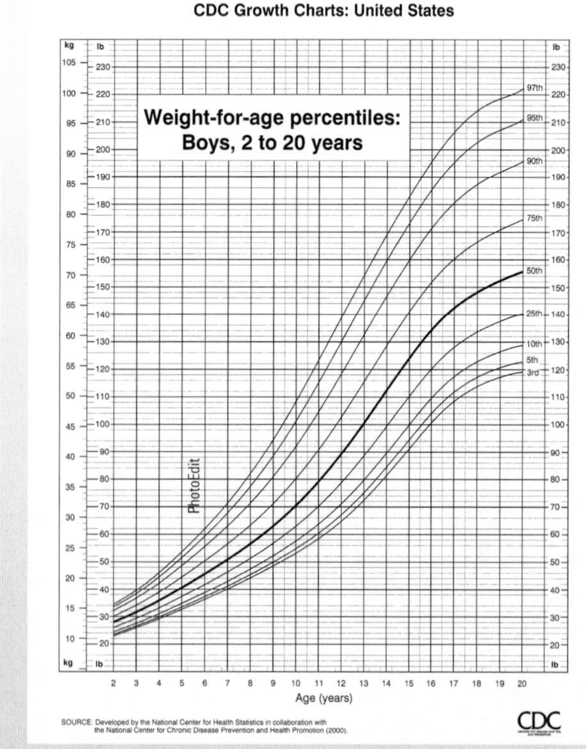

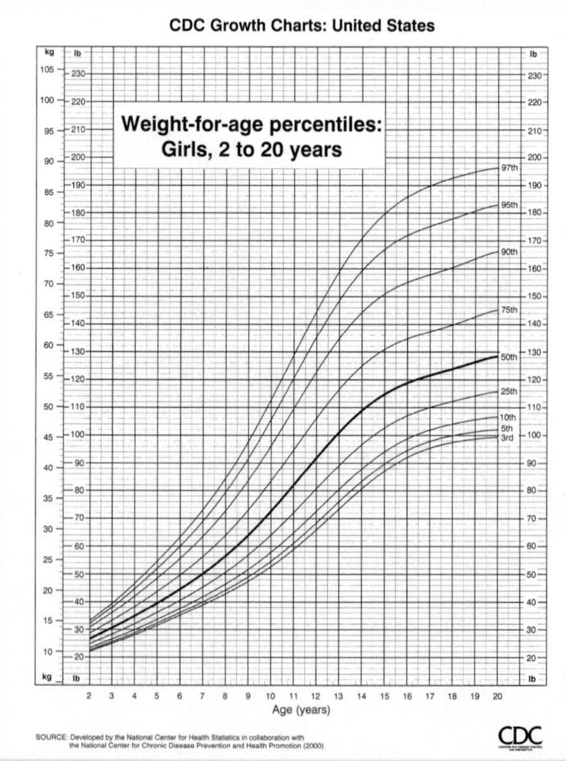

National Center for Chronic Disease Prevention

The BMI charts for age are a new feature of the growth charts. Since BMI increases with age in children and adolescents, ranges of BMI used to classify weight status in adults (for example, BMIs of 25 to 30 are considered "overweight") cannot be used. BMIs for age that fall between the 85th and 95th percentiles are considered "at risk" for overweight, and those above the 95th percentile at risk for obesity.[2] Although now recommended for determining weight status in children and adolescents, use of the CDC's new BMI for age charts requires that body mass index (BMI) be calculated. Illustration 29.5 shows you how to calculate BMI.

How to calculate BMI

Lucy just turned 7 years old and has had her weight and height carefully measured. She weighs 57 pounds (lb), and her height (stature) is 4 feet (ft), 2 inches (in). What's her BMI?

1. Convert her height of 4 ft 2 in to inches:

$$4 \text{ ft} \times 12 \text{ in per foot} = 48 \text{ in}$$
$$+2 \text{ in}$$
$$50 \text{ in}$$

2. Square the inches:

$$50 \text{ in} \times 50 \text{ in} = 2500 \text{ in}^2$$

3. Apply formula:

$$\text{BMI} = \frac{\text{kg}}{\text{m}^2} \quad \text{or} \quad \frac{\text{weight in lbs} \times 700}{\text{height in in}^2}$$

$$\text{BMI} = \frac{57 \text{ lbs} \times 700}{2500 \text{ in}^2} = \frac{39{,}900}{2500 \text{ in}^2} = 15.96 \text{ kg/m}^2$$

In this example, Lucy's BMI for age of 15.96 kg/m² falls between the 50th and 75th percentiles, or well within the normal range.

Photo Disc

Illustration 29.5
How to calculate BMI.

Food Jags and Normal Appetite Changes When growth is not occurring, children are often disinterested in food and eat very little at times. They may go on "food jags," insisting on eating only a few favorite foods like peanut butter and jelly sandwiches or breakfast cereal.

The ups and downs of children's food intake can be unnerving for parents. But as long as growth continues normally and children remain in good health, there's little reason to worry about food jags and fluctuations in appetite and food intake. Only children know when they are hungry or full.[3] (Actually, children appear to regulate caloric intake meal by meal better than adults do. Read about it in Illustration 29.6.)

Illustration 29.6
The pudding study.

Lunch at the day care center the day of the study was a bit different than usual—it started with pudding. Preschoolers were offered a serving of pudding that contained 150 calories or another with 40 calories. Both looked and tasted the same. The amount of pudding consumed was secretly recorded, as was their food intake during the rest of lunch. The experiment was repeated on adults.

Children compensated almost perfectly for the calories in the different puddings. Those given the higher calorie pudding ate less at lunch, and those receiving the low calorie pudding ate more. The adults, however, didn't fare so well. Their caloric intakes for the rest of lunch bore no relationship to calories consumed in the pudding. The study suggests that children are more sensitive to caloric intake on a short term basis than adults.[3]

Hunger Can Make Kids Irritable!

It was a hot summer day, and 10-year old Kim and her friends had spent it in the pool. The fun went on and on . . . until five o'clock. That's when the argument started and Kim left the pool in a huff.

When Kim arrived home, still in a huff, her mom knew what was going on. Kim was cranky because she was hungry. She was so cranky, in fact, that suggesting she'd feel better if she ate something would only make her more irritated! Kim was ravenously hungry, but she didn't want anybody telling her what her problem was!

Kim's mom had the solution to her daughter's grumpiness on the table. "Dinner's ready," her mom called softly. Kim stomped to the table and grudgingly started to eat. Within minutes, she did feel better, but it wasn't because she was no longer hungry!

Only children know when they are hungry, but sometimes when they stay hungry too long, the signal to eat is overpowered by fun or the effects of not eating. Gently offering food to an irritable child who has skipped a meal or played too long can provide the cure for irritability.

The Adolescent Growth Spurt

> It dawned on us that Max was entering his growth spurt when we had to wait for him to finish eating before we could all go out to dinner.
>
> —Max's parents

The adolescent growth spurt usually occurs in girls between the ages of 11 and 15 years. For boys, this period of growth generally begins around the age of 12 and ends near the age of 17. Actually, though, the age when adolescents start their growth spurt normally varies considerably. Pictured in Illustration 29.7 are three friends, all aged 12. It's not hard to tell which one of them has experienced a growth spurt! The difference was also clear at the dinner table. By the time all three were 19 years old, Max (the one in the middle) had caught up with Ben (on the left), and David had grown taller but less than Max or Ben.

During these years of growth, teenagers gain approximately 50% of their adult weight, 20 to 25% of adult height, and 45% of their total bone mass. In the year of peak growth, girls gain 18 pounds on average, and boys 20 pounds.[4]

Can You Predict or Influence Adult Height? Ultimate height is difficult to predict. On average, children tend to achieve adult heights that are between the heights of their biological parents.[5] There are many exceptions to this general finding, however, and that means that heredity is not the only influence on height. The dramatic increases in height of Japanese youth since World War II provide clear evidence that nongenetic factors have the strongest influence on height. Since the late 1940s, Japanese youth have grown an average of 2 inches taller each generation. (The increase in height has meant that everything from shoes to beds must be produced in larger sizes.) The increase in size of Japanese people is largely attributed to higher calorie and protein intakes and more nutrient-dense diets.[6] Indeed, people in most economically developed countries continue to grow taller; a maximal genetically determined height has not yet been reached. If children are less well nourished than their parents, though, they tend to be shorter than their parents as adults.[6]

A healthy diet during the growing years and, likely, exercise and freedom from frequent bouts of illness as well support growth in height. There are no supplements, powders, or special diets that can be used to increase growth rate. Growth hormone injections can increase height somewhat, but the side effects of growth hormone are numerous and its use is limited.

Illustration 29.7
This photo (on the left) was taken when these boys, born two months apart, were 12 years old. Ben (on the left) started his growth spurt at age 11, while Max (in the middle) began his at age 13. David (on the right) began to spring up in height after he turned 14. The photo (on the right) shows the same boys at age 19.

"Overgrowth": Overweight in Children and Adolescents

Approximately one in four children between the ages of 6 and 17 years in the United States is overweight or obese, and the proportion is increasing.[7] A major reason for the increase appears to be the general decline in physical activity among children.[8] Some experts have suggested that the reason physical activity is declining is that children are watching too much television. Recent evidence suggests, however, that watching television by itself doesn't cause obesity. Rather, inadequate exercise seems to be the primary culprit.[9]

Jean Mayer, a noted professor of nutrition, said we shouldn't be surprised that so many children are overweight: "A society where most children sit and watch TV, where nobody walks, and where no domestic chores are required is a society where we may expect to have obese children."[10]

"Undergrowth" in Children and Adolescents Growth failure typically occurs due to undernutrition or prolonged illness. Another reason for impaired growth has been termed "middle-class malnutrition." As the name implies, this undernutrition occurs primarily among children from middle- and upper-income families; it is associated with low-calorie, low-fat diets.[11]

Some parents say that they implement this sort of diet to help their children form good lifelong eating habits and avoid heart disease. Nevertheless, parents' fear that their children will become obese appears to be the major motivation for restricting their children's food intake.[12]

As children become older, they also become concerned about body weight. Approximately 40% of girls aged 9 to 11, most of them Caucasians, are trying to lose weight, even girls who are thin.[13] In addition, though not anorexic or bulimic, some adolescents may moderately and continually restrict their weight gain. Failure to gain weight during periods of growth may halt increments in height. Most adolescents don't want to be short either. Counseling can be effective in getting adolescents to eat more so as to allow growth in height to proceed.[14]

How Do Food Preferences Develop?

> I do not like broccoli, and I haven't liked it since I was a little kid and my mother made me eat it. And I'm President of the United States, and I'm not going to eat any more broccoli.
>
> —George Bush, 1990

Food preferences are a highly individual matter (Illustration 29.8). It's hard to find two people who share the same food likes and dislikes, and many people are very picky about how food is prepared and served. How *do* food preferences form? Are they "shaped" by early learning experiences or are they inborn and beyond anyone's control?

With the exception of genetically determined sensitivity to bitter-tasting foods and an inborn preference for sweet-tasting foods, there is no direct evidence that food preferences are inborn.[15] The belief that young children instinctively know what to eat can be hazardous to a child's health. Parents may take an overly relaxed attitude toward poor food habits and inadvertently contribute to the development of health problems later in life. Informed parents should decide what types of food their child should be offered, but the decision about how much to eat should be left to the child.[1]

Food likes and dislikes appear to be almost totally shaped by the environment in which children learn about food. Which foods are offered, the way they are offered, and how frequently particular foods are offered all influence whether a child will like a given food or not.[16]

Humans are born with a tendency to be cautious about accepting new things, including foods. They may need to get used to a new food before they trust it. Parents may have to offer a new food on five or more occasions before

Illustration 29.8
Why select the apple?
Food likes and dislikes appear to be almost totally shaped by the environment in which children learn about food. Which foods are offered, the way they are offered, and how frequently particular foods are offered all influence whether a child will like a given food or not.

the child makes a decision.[16] Sometimes children decide they like the food, and sometimes they really don't like it. Often infants and young children do not like strong-flavored vegetables, spicy foods, and mixed foods (Illustration 29.9).

Forcing a young child to eat foods she or he does not like, or restricting access to favorite foods, can have lifelong negative effects on food preferences and health. When attempts to get a child to eat a particular food turn the dinner table into a battleground for control, nobody wins. Foods should be offered in an objective, nonthreatening way so that the child has a fair chance to try the food and make a decision about it. Restricting access to, or prohibiting intake of, children's favorite "junk" foods tends to strengthen their interest in the foods and consumption of those foods when they get a chance. Such prohibitions have the opposite effect of that intended because they make kids want the foods even more.[17]

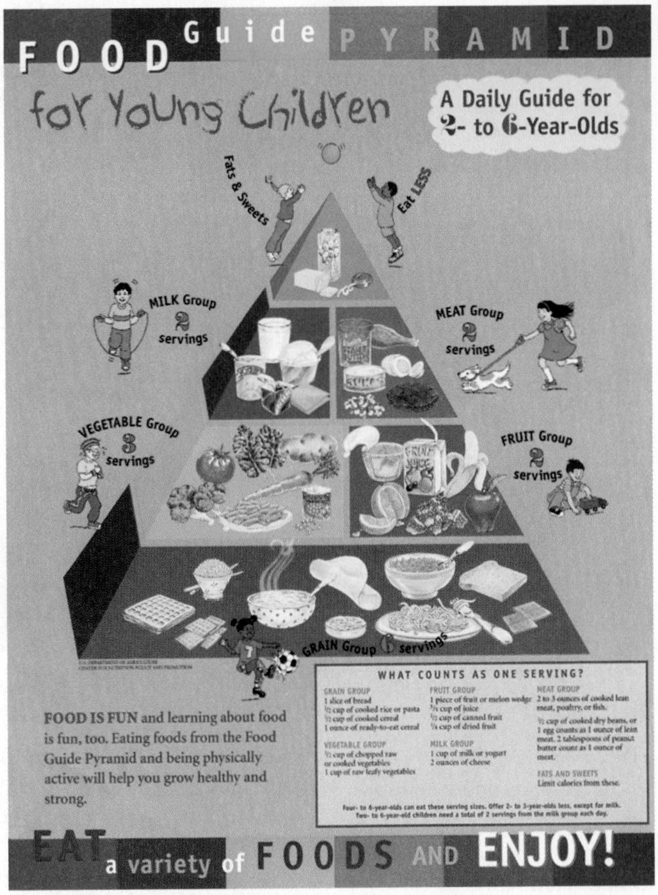

USDA

Illustration 29.9

What's a Good Diet for Children and Adolescents?

> Children today are tyrants. They contradict their parents, gobble their food, and tyrannize their teachers.
>
> —Socrates, 470–399 B.C.

Good diets for the growing years can be achieved by following the food group recommendations shown in Illustration 29.10. The type of foods selected from the food groups is important. Whole-grain breads and cereals should be represented in diets; and boiled, broiled, and baked foods should be consumed more often than fried foods.

Foods from the "miscellaneous" group that contribute mainly calories may be needed by children and teenagers. Although cakes, candy, cookies, potato chips, and so on are generally considered "sometimes" foods rather than "always" foods, they can serve as sources of needed calories. Snacks also are an important source of calories in the diets of children and teenagers (Illustration 29.11). Good choices for snacks include:

Yogurt	Bananas	Carrots
Cheese	Oranges	Cucumbers
Low-fat milk	Apples	Popcorn
Ice milk	Apple juice	Peanuts
Pears	Grapes	Peanut butter
Melons	Raisins	

Recommendation for Fat Intake The Dietary Guidelines for fat intake do not apply to children under the age of five. It is recommended that fat intake gradually be reduced from 40 to 45% during infancy to no more than 30% of total calories by age five. Diets containing 25 to 30% fat do not appear to reduce growth in children if sufficient essential fatty acids from vegetable oils are included in the diet.[18]

Over 70% of children consume more than 30% of total calories from fat. However, intake of fat by children has decreased over recent decades and now represents about 33% of total calories. Saturated fat intake has also decreased: from

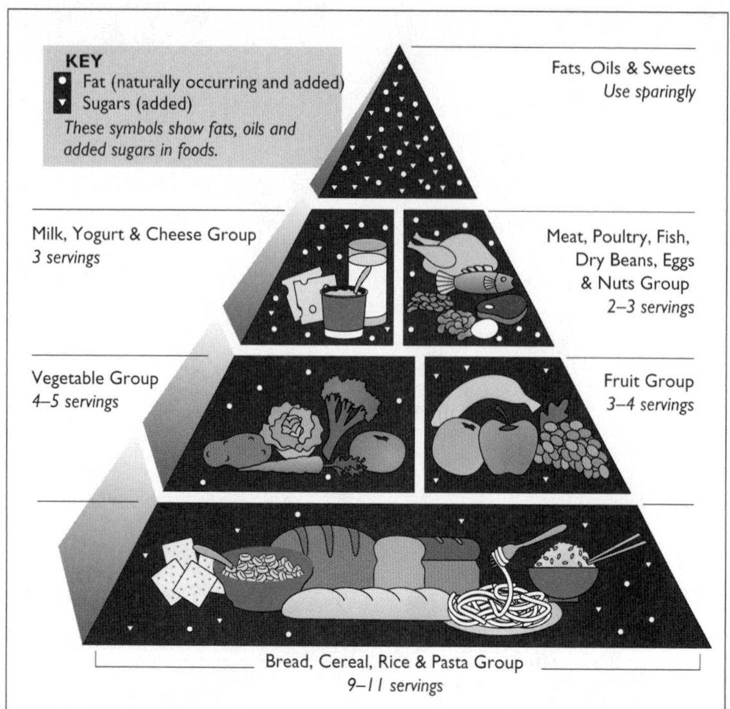

Illustration 29.10
Food group recommendations for children aged 2 to 6 and 7 to 15 years, based on the Food Guide Pyramid.*

*Children between the ages of one and two need the same variety of foods as older children but smaller serving sizes.

16 to 11% of total calories. Some of this benefit is related to increased consumption of low-fat milk and other dairy products.[18,19]

Should Children Drink Milk? Yes! Children certainly should drink milk. If this sounds like a silly question, good. It is, but the notion that children (and adults, for that matter) should not drink milk or consume other animal products erupts in the media from time to time.

Children don't have to consume milk in order to get adequate nutrition. It's just difficult to get *enough* calcium, vitamin D, and some other nutrients from foods other than milk (Illustration 29.12).

Some Children Get Too Much Fruit Juice Children usually love fruit juice, and parents feel good about giving it to them, thinking that it is nutritious. However, consuming too much of a good thing, in this case more than 12 ounces (1½ cups) of fruit juice a day on a regular basis, may be a bad thing. High intakes of fruit juice may decrease the variety of basic foods in the diet and promote weight gain to a greater extent than growth in height.[20]

Status of Children's and Adolescents' Diets

How well are children and teenagers doing at meeting these recommendations? Children over the age of two in the United States tend to eat too few vegetables and fruits; too little calcium, zinc, and vitamins E, D, C, and B_6; and too little dietary fiber.[18,21] Few (9%) children consume 5 servings of vegetables and fruits a day, and most consume 2.5 servings or less.[18] Adolescents are even less likely to consume their "five a day"; only 1% of them meet the vegetable and fruit intake goal.[22] Low vegetable and fruit intakes contribute to less than optimal intakes of vitamins and minerals by children and teens.

Approximately half of all children fail to consume the recommended intake of calcium daily. Low calcium and vitamin D intakes during childhood and adolescence decrease peak bone mass and increase the risk of osteoporosis later in life.[18] A part of the reason children and teens fail to get enough calcium and vitamin D is that they tend to substitute soft drinks for milk during these years.[19]

Illustration 29.11
Snacks are an important source of calories and nutrition in the diets of children and teenagers.

Photo Disc

Diets of children and adolescents generally provide about half of the recommended intake of dietary fiber. This makes constipation a problem for many children and teens. As a general guide, children over the age of five years and teens should consume grams of dietary fiber equal to "5 + age." Thus, for example, a 13 year old following this recommendation would consume 18 grams of dietary fiber.[18,23]

Can Fast Foods Meet Recommendations? Many of the foods sold in fast-food restaurants fit into the basic food groups. A hamburger on a bun contributes to the meat and meat alternates group and to the bread and cereals group. French fries could fit into the vegetable group. Nevertheless, the hamburger and fries are high in fat, and the bun is low in fiber. If these and other high-fat, low-fiber foods are commonly chosen, a diet can become out of balance. Take the example of a 16-year-old boy who eats foods such as hamburgers, cheese pizza, fried chicken, soft drinks, and french fries at fast-food restaurants for two meals a day. That diet will be high in fat and sugar, and low in vegetables and fruits, dietary fiber, vitamins D and C, calcium, zinc, and other minerals.

Fast-food dining doesn't have to be off-limits to the health-conscious, hungry person. Menus frequently include salads, fruit juices, broiled or grilled chicken sandwiches, low-fat milk and yogurt, muffins, and other foods that contribute to healthy diets.

Photo Researchers, Inc.

Illustration 29.12
Children should drink milk in order to meet their needs for calcium and vitamin D. Virtually all school lunch programs include milk as a beverage.

Early Diet and Later Disease

Children and adolescents generally feel vigorous and healthy even when they are accumulating diet-related risk factors that will influence disease development later in life. This seeming health can make the message about good nutrition difficult to convey.

Many studies indicate that in populations with a low incidence of heart disease, blood cholesterol values of both children and adults are lower than in countries with high rates of heart disease.[24] Typical cholesterol levels of children in the United States are higher than those of children in countries where the rate of heart disease is lower. In addition, obesity during adolescence is associated with the presence of risk factors for subsequent heart disease: high LDL-cholesterol, low HDL-cholesterol, and increased blood pressure.[25] High-sodium diets tend to increase blood pressure in children and adolescents who have a hypertensive parent.[26] U.S. children tend to consume far more sodium than is recommended.[27] Children and adolescents who become obese are at higher risk for obesity later in life, and for disorders related to obesity such as diabetes, heart disease, some cancers, and hypertension. Obesity may also lead to lower self-esteem, poor body image, and eating disorders.[18]

Dietary recommendations, and acceptable ways children and adolescents can achieve them, will continue to be refined as more is learned about early diet and health.

Nutrition UP CLOSE

Eating on the Run

FOCAL POINT: Using the Diet Analysis Plus program to check out a teenager's diet.

Tanya is 16 and her life is hectic. She has skating practice at 6:30 A.M., attends school from 7:45 A.M. to 2:30 P.M., and works after school at a supermarket. Here's a list of the foods she eats in a typical day:

Breakfast:	Orange juice, 12 oz
Lunch:	Bagel, 2 oz M and M's candy, 2 oz Diet cola, 12 oz
Snack at work:	Apple, 1 Potato chips, 1 oz
Dinner:	French fries, 1 cup Hamburger, 3 oz Cheddar cheese, 1½ oz Bun, 2 oz Diet cola, 12 oz
Evening snack:	Banana-peanut butter shake (banana, 1; peanut butter, 2 tbs; 2% milk, 1 cup)

Checking It Out Use the Diet Analysis Plus program's Food Guide Pyramid option to determine how Tanya's typical diet compares to that recommended. Fill in the table below with the number of servings of foods from each food group in Tanya's diet. Then list two strengths and two weaknesses of her diet.

	Food Groups					
	Bread, Cereal, Rice, Pasta	**Vegetables**	**Fruits**	**Milk, Yogurt, Cheese**	**Meat, Poultry, etc.**	**Fats, Sweets**
Recommended number of servings	9–11	4–5	3–4	3	2–3	limit
Tanya's number of servings	____	____	____	____	____	____

List two strengths of Tanya's diet:

1. ______________________
2. ______________________

List two weaknesses:

1. ______________________
2. ______________________

FEEDBACK (including answers) can be found at the end of Unit 29.

WWW links

www.cdc.gov/nccdphp/dnpa/kidswalk.htm or 1-800-CDC-4NRG
"Kidswalk" program kit available from the CDC provides guides for encouraging increased physical activity in children.

www.exnet.iastate.edu
Select this National Network for Child Care Nutrition Database for handouts and other information related to child nutrition.

www.aap.org
The American Academy of Pediatrics's website is loaded with information on child nutrition.

www.cdc.gov/growthcharts
Need more information about the growth charts or copies of them? Get it here.

www.hc-sc.gc.ca.ca
Health Canada's website. Search infant and child nutrition for information and resources.

www.eatright.org and www.dietitians.ca
The American Dietetic Association and Dietitians of Canada offer sound nutrition advice for parents and children.

www.usda.gov/cnpp
Provides access to "Tips for Using the Food Guide Pyramid for Young Children 2 to 6 years old," as well as the Food Guide Pyramid booklet and the Dietary Guidelines for Americans.

www.navigator.tufts.edu
Click on lifecycle.

REFERENCES

1. Satter E. The feeding relationship: problems and interventions. J Pediatr 1990;117:S181–S9.
2. Thompson D, et al. New CDC growth charts. WIC National Meeting, Salt Lake City, Utah, 9/8/00.
3. Birch LL. Presentation at the National Conference on Nutrition Education Research, Chicago, September 1986.
4. Tanner JM. Fetus into man: physical growth from conception to maturity. Cambridge (MA): Harvard University Press; 1978.
5. Smith GD. Growth and its disorders. Philadelphia: W. B. Saunders; 1979.
6. Bronner F. Adaptation and nutritional needs. (letter) Am J Clin Nutr 1997;65:1570.
7. Health, United States 2000. www.cdc.gov/nchs, 7/00.
8. Position of the American Dietetic Association on weight management. J Am Diet Assoc 1997;97:71–74.
9. Andersen RE, et al. Relationship of physical activity and television watching with body weight and level of fitness among children. JAMA 1998;279:938–42.
10. Mayer J. Obesity during childhood. In: Winick M, editor. Childhood obesity. New York: John Wiley & Sons; 1975:pp 73–80.
11. Lifshitz F. Nutrition and growth. Clin Nutr 1985;4:40–47.
12. Olson RE. Is it wise to restrict fat in the diets of children? J Am Diet Assoc 2000;100:28–32.
13. Schreiber GB, et al. Weight modification efforts reported by black and white preadolescent girls: National Heart, Lung, and Blood Institute Growth and Health Study. Pediatrics 1996;98:63–70.
14. Pugliese MT, et al. Parental health beliefs as a cause of nonorganic failure to thrive. Pediatrics 1987;80:175–82.
15. Story M, Brown JE. Do young children instinctively know what to eat? The studies of Clara Davis revisited. N Engl J Med 1987;316:103–6.

16. Birch LL. The role of experience in children's food acceptance patterns. J Am Diet Assoc 1987;87(suppl):S36–S40.
17. Fisher JO, Birch LL. Restricting access to palatable foods affects children's behavioral response, food selection, and intake. Am J Clin Nutr 1999;69:1264–72.
18. Dietary guidance for healthy children aged 2 to 11 years: Position of the American Dietetic Association. J Am Diet Assoc 1999;99:93–103.
19. Harnack L, et al. Soft drink consumption among U.S. children and adolescents: nutritional implications. J Am Diet Assoc 1999;99:436–41.
20. Dennison BA, et al. Excess fruit juice consumption by preschool-aged children is associated with short stature and obesity. Pediatrics 1997;99:15–22.
21. Stang J, et al. Relationship between vitamin and mineral supplement, dietary intake, and dietary adequacy among adolescents. J Am Diet Assoc 2000;100:905–10.
22. Muñoz KA, et al. Food intakes of US children and adolescents compared with recommendations. Pediatrics 1997; 100:323–9.
23. Williams CI, et al. A summary of conference recommendations on dietary fiber for children: how much? Pediatrics 1995;96:1019–22.
24. American Academy of Pediatrics (Committee on Nutrition). Prudent lifestyle for children: dietary fat and cholesterol. Pediatrics 1986;78:521–5.
25. Bergstrom E, et al. Insulin resistance syndrome in adolescents. Meta Clin Exp 1996;28:908–14.
26. Roy CC, Galeano N. Childhood antecedents of adult degeneration disease. Pediatr Clin North Am 1985;32:517–33.
27. Frank GC, et al. Sodium, potassium, calcium, magnesium, and phosphorus intakes of infants and children: Bogalusa Heart Study. J Am Diet Assoc 1988;88:801–7.

Nutrition UP CLOSE

Eating on the Run

Feedback for Unit 29

	Bread	Vegetables	Fruits	Milk	Meat	Fats
Tanya's number of servings	4	2	4	2	1⅓	3

Two strengths of Tanya's diet:

1. Six (!) vegetables and fruits
2. Fruit for snacks

Two weaknesses:

1. Includes a number of foods high in saturated fat (hamburger, cheddar cheese)
2. Includes too many fats and sweets (potato chips, candy)

Other: Diet is low in milk products, breads and cereals, and meat and meat alternates.

Unit 30

Nutrition and Health Maintenance for Adults of All Ages

Nutrition Scoreboard

	TRUE	FALSE
1. Most of the gain in life expectancy in the United States since 1900 is due to advances in medical care and technology.		
2. Aging is a kind of disease, and people can do little to prevent their health and quality of life from declining as they age.		
3. The sense of taste declines dramatically with age.		

(Answers on next page)

Answers to

Nutrition Scoreboard

	TRUE	FALSE
1. Advances in medical care and technology have contributed to quality of life and have added a few years to life expectancy. But most of the gain is due to declines in infant mortality, prevention of infectious diseases, and improvements in nutrition, sanitation, and other living conditions.[1]		✓
2. Aging is not a disease process! Many things can be done to extend life and its quality. (This unit discusses some of those things.)		✓
3. Taste remains fairly robust into old age, but can be adversely affected by medications, some diseases, and certain surgeries.[2]		✓

Key Concepts and Facts

- Age does not necessarily predict health status. Healthy adults come in all ages.
- Dietary intake, body weight, and physical activity influence changes in health status with age.
- Aging processes begin at the cellular level.
- Medications, diseases, and biological processes associated with aging influence adults' requirements for certain essential nutrients.

You Never Outgrow Your Need for a Good Diet

One of the most important types of health changes that occurs with age is no change at all.

—Daphne Roe, 1985

Eating right during the adult years is a wise practice. Good nutrition helps adults feel healthy and vigorous as they age and improves their overall sense of well-being.[3] Adults who eat healthfully are at lower risk for developing a number of diseases and disorders that lower the quality of life. Hypertension, heart disease, osteoporosis, diabetes, gallbladder disease, most types of cancer, and other health problems are related to habitual diet and weight status.[3,4]

Aging Is Not a Disease

Aging is a normal process, not a disease. Although the incidence of many diseases increases with age, the causes of the diseases are often unrelated to aging.[5] They may fully or partially result from the cumulative effects of diets high in saturated fat and salt and low in vegetables and fruits, obesity, smoking, physical inactivity, excessive stress, or other habits that insidiously influence health on a day-to-day basis.[5,6] Aging cannot be prevented, but how healthy we are during aging can be influenced by what we do to our bodies.

Maintaining health as we age is becoming an increasingly important concern. There are more older adults in the United States than ever before, and the numbers are growing.

The Age Wave

➡ **life expectancy** The average length of life of people of a given age.

The proportion of people in the United States aged 65 years and older is steadily increasing (Illustration 30.1). Since 1900, ***life expectancy*** at birth has increased

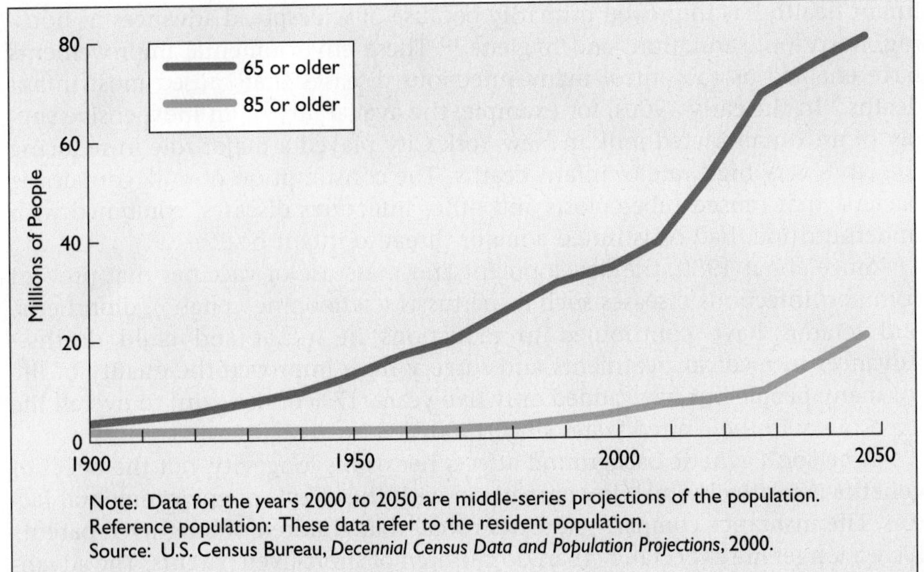

Illustration 30.1 Expected growth in the U.S. population of people aged 65 and 85 years and older.[7]

by 62%, from 47.3 to 76.7 years.[7] By the year 2030, approximately 20% of the U.S. population will be 65 years of age or older. Despite these advances in life expectancy, the United States continues to rank poorly in international comparisons. According to World Health Organization statistics, life expectancy is higher in Japan (79.7 years), Sweden (78.9), Canada (78.2 years), Greece (77.7 years), and nine other industrialized countries than in the United States.[7] Theoretically, the human life span could reach a maximum of 130–135 years.[8]

All groups in the U.S. population are not benefiting equally from the increase in life expectancy. Life expectancy varies a good deal by sex and race. Table 30.1 shows overall life expectancy for Caucasian and African American men and women in 1900, 1940, and 1998. African American men have the shortest life expectancy, followed by Caucasian men. As in other industrialized countries, life expectancy for females exceeds that for males.[7] In the United States, the gender difference in average life span appears to be largely related to behaviors. U.S. males tend to smoke more, consume more alcoholic beverages, pay less attention to what they eat, and seek medical care less often than do females.[9]

Why the Gains in Life Expectancy? Most of the gains in life expectancy in the United States are attributable to improvements in infant health, the prevention of diseases, and improved nutrition.[1] In 1900, 1 out of 10 newborns died in the first year of life; today the proportion is less than 1 in 100.

Table 30.1 Average length of life for Caucasian and African American males and females in the United States in 1900, 1940, and 1998.[7]

	Life Expectancy (Years)		
	1900	1940	1998
Caucasian Americans			
Males	46.6	62.1	74.5
Females	48.7	66.6	80.0
African Americans			
Males	32.5	51.5	67.6
Females	33.5	54.9	74.8

(Data for other ethnic groups were unavailable.)

Infant health has improved primarily because of widespread advances in housing, nutrition, sanitation, and hygiene.[10] These environmental improvements have enabled us to control many infectious diseases that caused most infant deaths.[8] In the early 1900s, for example, the availability of an inexpensive supply of uncontaminated milk in New York City played a major role in reducing the city's very high rate of infant deaths. The consumption of milk containing bacteria that caused tuberculosis and other infectious diseases, combined with undernutrition, had constituted a major threat to infant health.

Since about 1930, the development and mass use of vaccines that prevent common infectious diseases such as pertussis ("whooping cough"), diphtheria, and tetanus have contributed to reductions in infant and child deaths.[8] Advances in medical treatments and surgery have improved the quality of life for many people but have added only five years (17% of the gain) to overall life expectancy in the United States since 1900.[11]

A person's genetic background affects her or his longevity, but the effect of genetics is relatively small in comparison with the effects of environmental factors. Life insurance company statistics reveal that children of long-lived parents have a longer life expectancy than do children of short-lived parents. The advantage amounts to an additional three years.[12] Since none of us can select our parents, it is perhaps fortunate that environmental factors generally are more important than genetic background in determining life expectancy.

Living in the Bonus Round: Diet and Life Expectancy Scientists have yet to identify exactly why some people live longer than others, but diet appears to be related in important ways.[13] Characteristics of diets of adults and older people who experience low disease rates and increased longevity include:

- regular consumption of vegetables and fruits,
- above average intake of whole grain products,
- lower consumption of saturated fats,
- alcohol in moderation, and
- eating breakfast.[4,14,15]

Taking time to socialize and celebrate with friends and regular physical activity also appear to contribute bonus years to life (Illustration 30.2).[16]

Illustration 30.2
Fun with friends may increase longevity.

Nutrition Issues for Adults of All Ages

Researchers and others interested in nutrition and health during middle age and beyond have tended to focus their attention on the cumulative effects of diet on chronic disease. Nutrition exerts its effects on chronic disease development over time, and therefore diseases related to poor diets are most likely to express themselves during the older adult years. The occurrence of diseases related to behavioral traits such as smoking and physical inactivity also increases among adults as they age.

In addition to the health effects of behaviors, people age biologically. The combined effects of poor diets, other risky lifestyle behaviors, and biological aging increase the rates of serious illness during adulthood. How soon a disease develops largely depends on the intensity and duration of exposure to behavioral risks that contribute to disease development. Behavioral factors also affect the progress of biological aging processes.

Breaking the Chains of Chronic Disease Development For the most part, the development of chronic disease in middle-aged and older adults can be viewed as a chain that represents the accumulation over time of problems that impair the functions of cells. Each link that is added to the chain, or each additional insult to cellular function, increases the risk that a chronic disease will develop. The presence of a disease indicates that the chain has gotten too long—that the accumulation of problems is sufficient to interfere noticeably with the normal functions of cells and tissues.

It appears that the chain can be shortened by healthful dietary and other behaviors. For example:

- Correcting obesity and stabilizing weight during the adult years may lengthen life expectancy.[17]
- Dietary intakes that correspond to the Dietary Guidelines for Americans are related to decreased death rates in women around the age of 60.[4]
- Low-fat diets (less than 25% of total calories) and physical activity in overweight women decrease weight, central body fat, and blood triglyceride levels and improve blood glucose control.[18]
- Maintaining adequate calcium and vitamin D intake and engaging in regular physical activity during the adult years may prevent, postpone, or lessen the severity of osteoporosis.[19]
- Adequate intakes of fruit and vegetables (more than five servings a day) appear to delay or prevent the development of a number of types of cancer[20] and the formation of cataracts that "cloud over" the lens of the eye.[21]

The health status of adults is not necessarily "fixed" by age; it can change for the better or the worse, or not much at all.[5]

Nutrient Needs of Middle-Aged and Older Adults The biological processes and lifestyle changes that generally accompany aging affect caloric and nutrient needs. A person's need for calories generally declines with age as physical activity, muscle mass, and basal metabolic rate decrease.[22] However, people who remain physically active into their older years maintain muscle mass and have a higher need for calories than people who are inactive.[23]

While caloric need decreases, the requirements for certain nutrients such as protein, vitamin C, vitamin D, vitamin B_{12}, and calcium may increase with aging.[24] The increased need for protein appears to be a result of decreased efficiency of protein utilization. Dietary requirements for vitamin C and calcium

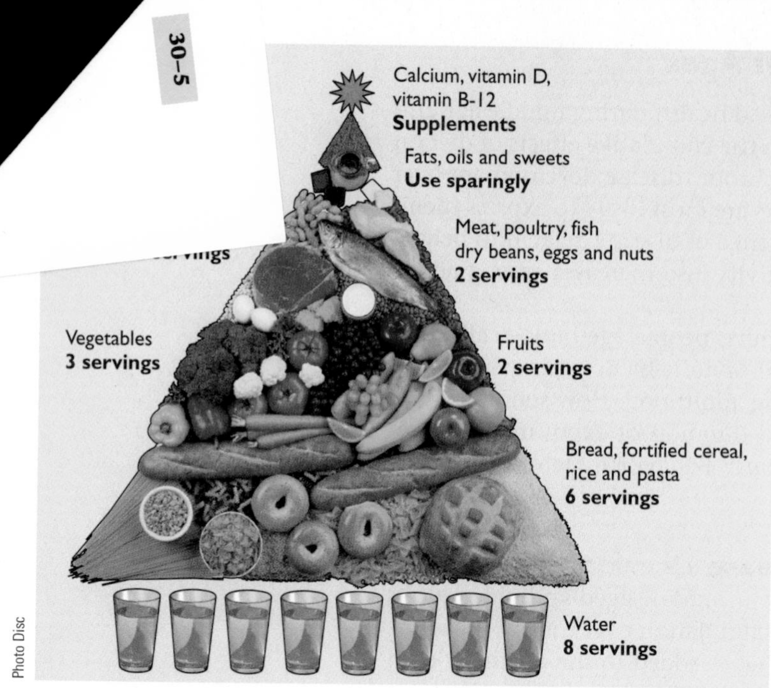

Photo Disc

Illustration 30.3
A Food Guide Pyramid for adults over 70.[26]

Source: Tufts University, © 1999.

may increase due to lower levels of stomach acidity that often occur with advancing age. Decreased stomach acidity reduces the body's ability to absorb vitamins C and B_{12} and calcium. Vitamin D status may become a problem for older adults for two reasons: their intake of milk and milk products is often low, and exposure of the skin to sunlight produces about half of the amount of vitamin D as it does in young adults.[24] An estimated 15% of older adults in the United States may be deficient in vitamin D.[25]

A Food Guide Pyramid constructed to reflect the nutrient needs of people over 70 years is presented in Illustration 30.3. This pyramid takes into account the lower food intakes of most older people by recommending 6 servings of grain products daily rather than the 6–11 recommended in the Food Guide Pyramid. It also highlights nutrient-dense foods that are most likely to provide adequate nutrient intakes in the relatively low-calorie diets of many elderly.[26]

Fluid Needs For reasons that aren't yet entirely clear, many older people don't get thirsty when their bodies are running low on water. The implication of this change is that older people may become dehydrated and need medical assistance.[27] Approximately one million elderly men and women are admitted to hospitals each year due to dehydration.[13] Fluid needs can be met through consumption of water, juices, teas, and other beverages. The food guide for the elderly shown in Illustration 30.3 emphasizes the importance of adequate fluid consumption by placing 8 glasses of water at the base of the pyramid.

Does Taste Change with Age? Poor diets observed in some middle-aged and older adults have been ascribed to "declining taste" with age. Besides being a dreadful thought, it's not true that taste declines with age to the extent that it makes eating less pleasurable than before.[28] Although taste sensitivity does diminish somewhat with age, chances are your favorite foods will taste as good to you in your older years as they did in your youth (Illustration 30.4). Sight, smell, and hearing senses usually decline with age a good deal more than taste does.[29]

Illustration 30.4
Taste is less affected by age than are some other senses.

PhotoEdit - Jeff Greenberg

The senses of taste and smell are affected by medications such as antibiotics, antihistamines, some lipid-lowering drugs, and cancer treatments; diseases including Alzheimer's disease, cancer, and allergies; and surgeries that affect parts of the brain and nasal passages.[2] These treatments and conditions are more prevalent in older adults than in middle-aged populations and are primarily responsible for declines in the senses of taste and smell.[2] Changes in these senses increase the likelihood that the person will experience food poisoning or consume an inadequate diet and a low intake of food.[2]

Social Aspects of Nutrition in Older Adults Preparing meals and eating right may not be as simple as it sounds for many older adults. Consuming an adequate and balanced diet may not be easy when you depend on someone else to take you shopping, when mealtimes involve little social life, or when you "don't feel up" to making a meal. Isolation, loneliness, and poor health can be major contributors to poor diets in older adults. The diets of older people are often lacking in nutrients, and because many older adults do not, or cannot, consume enough nutrients to meet their increased need for them, supplementation may be required.

Eating Right during Middle Age and the Older Years

The best diet for middle-aged adults is one that contains a wide variety of basic foods and provides less than 30% of total calories from fat, about 15% of calories from protein, and 55% of calories from carbohydrates, mainly complex carbohydrates. Furthermore, unsaturated fats from vegetable sources should be the primary type of fat in the diet, and cholesterol intake should be kept below 300 milligrams per day.[4] A balanced and adequate diet can be obtained by judiciously selecting foods from the basic food groups. It is important to select *judiciously* because not all foods within the respective food groups are equally desirable. Food choices should emphasize the members of the food groups that are low in fat, cholesterol, and sodium and high in fiber. Once the recommended numbers of servings from each food group are in the diet, remaining caloric needs can be met by adding servings from the various groups.

Altering food habits in favor of healthy diets appears to be a common practice among adults of all ages. A study of 100 women and men over the age of 60 found that 100% had made some change in their food choices in the recent past. The changes had been for health, taste, convenience, and social reasons.[30] Clearly, a person's food choices and intake change throughout life, and the changes can be for the better.

www.4woman.gov
The National Women's Health Information Center provides information on diet and women's health. Or call 1-800-994-WOMAN.

www.agingwell.state.ny.us
This site provides a "health and wellness village" for mature adults and includes nutrition quizzes and diet and exercise tips.

www.healthfinder.gov
A search engine for middle-aged and older adult nutrition and health topics.

www.nih.gov/nia
Your contact for the National Institute on Aging.

www.healthtouch.com
Links you to specific health organizations.

www.aarp.org
At this site, you can obtain educational materials from the American Association of Retired Persons.

Nutrition UP CLOSE

Does He Who Laughs, Last?

FOCAL POINT: Critically thinking about factors that influence longevity.

The following is an adaptation of a letter printed in the "Dear Abby" newspaper column:

> Dear Abby,
>
> Since I've reached my 80s, my mail is full of ads for health products to help me live longer.
>
> I once had many friends, all of whom were health vigilantes. They shook their heads knowingly as I avoided all health food fads and exercise. They made liquid out of good vegetables and spent fortunes buying all the latest supplements. They argued that "organic" was better and "natural" was best. I would tell them that snake venom, poison ivy and manure were "natural". But they wouldn't listen and they didn't laugh.
>
> Now my friends are all dead and I have no one left to argue with.

Critically think about the contents of this letter, and identify three alternate explanations for why this individual outlived his friends:

____

____

____

FEEDBACK can be found at the end of Unit 30.

REFERENCES

1. Bernarducci MP, Owens NJ. Is there a fountain of youth? A review of current life extension strategies. Pharmacotherapy 1996;16:183–200.
2. Schiffman SS. Taste and smell losses in normal aging and disease. JAMA 1997;278:1357–62.
3. Folsom AR et al. Waist-hip ratio and morbidity and mortality risk in older women. Arch Intern Med 2000;160:2177–2228.
4. Kant AK et al. Consumption of recommended foods and mortality risk in women, JAMA 2000;283:2109–15.
5. Rowe JW, Kahn RL. Human aging: usual and successful. Clin Nutr 1990;9:26–33.
6. Fried LP et al. Risk factors for 5-year mortality in older adults: the Cardiovascular Health Study. JAMA 1998;279:585–92.
7. Health, United States, 2000. www.cdc.gov/nchs, accessed 11/00.
8. Guyer B et al. Annual summary of vital statistics: trends in the health of Americans during the 20th century. Pediatrics 2000;106:307–17.
9. Women's health. Report of the Public Health Service Task Force on Women's Health Issues. Vol. 1. Public Health Rep 1985;100: 73–106.
10. Wegman ME. Annual summary of vital statistics—1991. Pediatrics 1992;90: 835–47.
11. Gold MR, Senior Adviser, Office of Disease Prevention and Health Promotion, Public Health Service, U.S. Department of Health and Human Services, 1997.
12. Cooper R, Sempos C. The price of progress. J National Med Assoc 1984;76:163–6.
13. Position of the American Dietetic Association and Dietitians of Canada: women's health and nutrition. J Am Diet Assoc 1999;99:738–51.
14. Lasheras C et al. Mediterranean diet and age with respect to overall survival in institutionalized, nonsmoking elderly people. Am J Clin Nutr 2000;71:987–92.
15. Johnson MA et al. Nutritional patterns of centenarians. Internatl J Aging Hum Dev 1992;34:57–76.
16. Leaf A. Observations of a peripatetic gerontologist. Nutr Today 1973; (September/October):4–12.
17. Stevens J et al. The effect of age on the association between body-mass index and mortality. New Engl J Med 1998;338:1–7.
18. Kuller LH et al. Lifestyle intervention and heart disease risk factors associated with menopause. Circulation 2001;103:32–7.
19. Dawson-Hughes B et al. Effects of withdrawal of calcium and vitamin D supplements on bone mass in elderly men and women, Am J Clin Nutr 2000;72:745–50.
20. Steinmetz KA, Potter JD. Vegetables, fruit, and cancer prevention: a review. J Am Diet Assoc 1996;96:1027–39.
21. Brunce GE. Antioxidant nutrition and cataract in women: a prospective study. Nutr Rev 1993;51:84–86.
22. Munro HN. Nutrition and the elderly: a general overview. J Am Coll of Nutr 1984;3:341–50.
23. Guo S et al. Aging, body composition, and lifestyle: the Fels Longitudinal Study. Am J Clin Nutr 1999;70:405–11.
24. Blumberg J. Nutritional needs of seniors. J Am Coll Nutr 1997;16:517–23.
25. Drinka PJ, Goodwin JS. Prevalence and consequences of vitamin deficiency in the nursing home: a critical review. J Am Geriatr Soc 1991;39:1008–17.
26. Russell R et al. Elderly food pyramid. J Nutr 1999; March.
27. Vogelzang JL et al. Overview of fluid maintenance/prevention of dehydration. J Am Diet Assoc 1999;99:605–9.
28. Shamburek RD, Farrar JT. Disorders of the digestive system in the elderly. N Engl J Med 1990;322:438.
29. Rolls BJ. Do chemosensory changes influence food intake in the elderly? Physiol Behav 1999;66:193–7.
30. Bilderbeck N et al. Changing food habits among 100 elderly men and women in the United Kingdom. J Hum Nutr 1981;35:448–55.

Nutrition UP CLOSE

Does He Who Laughs, Last?

Feedback for Unit 30

Alternate explanations:

1. The letter writer's friends may have switched to perceived health foods when they found out they were sick.
2. The letter writer may have consumed a life-long healthy diet without choosing organic or "natural" foods.
3. The letter writer may come from a family whose members tend to have long lives.

Other possible explanations:

- A sense of humor may help extend life.
- Many factors in addition to diet, exercise, and genetic background influence longevity.
- One individual's experience may not apply to the masses.

Unit 31

Nutrient-Gene Interactions in Health and Disease

Nutrition Scoreboard

	TRUE	FALSE
1. Most diseases are genetically caused.		
2. Individual differences in genetic traits are responsible for large differences in nutrient needs between individuals.		
3. Heart disease, cancer, obesity, and hypertension primarily result from interactions among environmental and genetic factors.		

(Answers on next page)

Answers to

Nutrition Scoreboard

	TRUE	FALSE
1. That's incorrect. Heart disease, diabetes, most types of cancer, asthma, hypertension, and other common health problems are most strongly related to environmental factors.		✓
2. Individual genetic traits alter nutrient needs to a small extent in many people, and to a large extent in very few. The Dietary Reference Intakes cover the nutrient needs of nearly all healthy people.		✓
3. How true. Common diseases are influenced by interactions of multiple genetic and environmental factors such as high-fat diets, vegetable intake, and activity level.	✓	

Key Concepts and Facts

- Nutrient-gene interactions influence an individual's susceptibility to many diseases.
- Advances in knowledge of nutrient-gene interactions are dramatically changing nutritional approaches to disease prevention and treatment.
- Health problems related to nutrient-gene interactions originate within cells.
- Genetic traits that profoundly increase or decrease nutrient needs affect a very small proportion of the population.

Nutrition, Health, and Your Genes

There have been at least two breakthroughs in the history of the relatively young science of nutrition that have defined the future. The first resulted from experiments in the late 1800s demonstrating that some constituents of foods are essential for life. The second is represented by completion of the draft of the human *genome*. (Find definitions and explanations of many of the genetic terms used in this unit in Table 31.1.) It marked the beginning of a new era in the discovery of contributing factors to health and disease, and the role nutrition and other "nurture" factors play in them. It is being made possible by the identification of genetic codes for enzyme and other protein production in cells. Enzymes and other proteins produced as a result of genetic codes are central to life and health because they determine which chemical changes will take place within the body. These changes influence growth, digestion, nutrient absorption, disease resistance, blood pressure control, and many other functions of the body.

Discoveries will revolutionize nutritional approaches to health promotion by answering such questions as "Who can cut their risk of hypertension by following a low-salt diet? Who would be at lower risk for obesity or certain types of cancer if a low-fat diet is consumed?" and "Who is at high risk for developing diabetes during adulthood if they become overweight?" Knowledge gained will play important roles in disease management by overcoming genetic errors related to disease processes through dietary manipulations.

Table 31.1

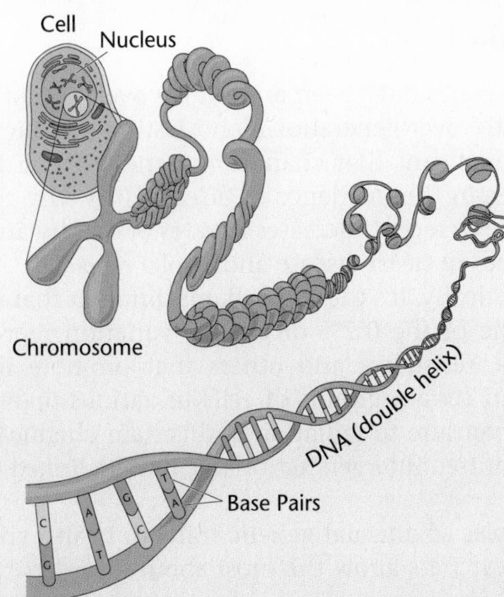

DNA is packed in chromosomes located in cell nuclei

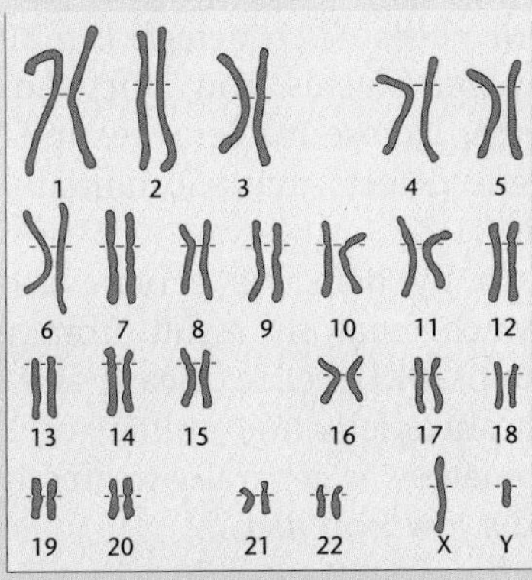

Human chromosomes

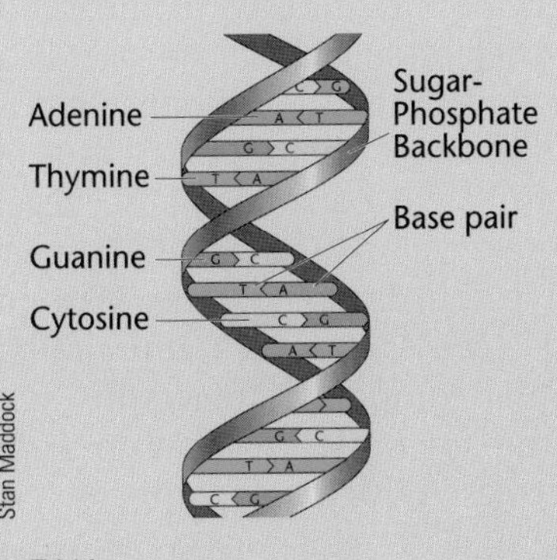

Stan Maddock

DNA

Definitions of Genetic Terms

Genome Combined term for "genes" and "chromosomes." It represents all the genes and DNA contained in an organism, which are principally located in chromosomes. The human genome consists of about 100,000 genes.

Genes The basic units of heredity that occupy specific places (loci) on chromosomes. Genes consist of large DNA molecules, each of which contains the code for the manufacture of a specific enzyme or other protein.

Chromosomes Structures in the nuclei of cells that contain genes. Humans have 23 pairs of chromosomes (shown below); half of each pair comes from the mother and half from the father.

DNA (deoxyribonucleic acid) Segments of genes that provide instructions for the manufacture of enzymes and other proteins by cells. It looks something like an immensely long ladder twisted into a helix, or coil. The sides of the ladder structure of DNA are formed by a backbone of sugar-phosphate molecules, and the "rungs" consist of pairs of bases joined by weak chemical bonds. Bases are the "letters" that spell out the genetic code, and there are over 3 billion of them in human DNA. There are two types of base pairs: adenine-guanine and cytosine-thymine. Each sequence of three base pairs on DNA codes for a specific amino acid. A specific enzyme or other protein is formed when coded amino acids are collected and strung together in the sequence dictated by DNA.

Some sections of DNA (so-called junk DNA) do not transmit genetic information because they don't code for protein production. What functions these segments serve are not yet known, and they are not considered to be part of the human genome. Characteristics of these segments of DNA vary among individuals, making it possible to identify individuals based on "DNA fingerprinting."

DNA fingerprinting The process of identifying specific individuals by their DNA. This is possible because no two individuals have the same genetic makeup. Differences among individuals are due to variations in the sections of DNA molecules that do not transmit genetic information.

Source: www.nhgri.nih.gov/DIR/VIP/Glossary/

Research into the genetic bases of disease risk will also clarify which health characteristics have little to do with genetic traits. It is tempting now to consider that diseases with no known cause are due to yet-to-be-identified "genes." This assumption needs testing because attributing genetic causes to nongenetic health problems may misdirect efforts to reduce disease risk.[1]

Genetic Secrets Unfolded

Our bodies may be new, but our genes have been around for over 40,000 years. Genes replicate themselves exactly over generations, and lasting modifications in them almost never occur. This means that changes in genetic traits do not account for increases or decreases in the incidence of disease. It is why genetic traits cannot be given as the cause of recent increases in rates of obesity and diabetes, nor credit for major declines in heart disease and stroke rates.

Humans are 99.8% alike genetically. It's the 0.2% that is different that makes every individual distinct.[2,3] Some of the 0.2% of genetic variation represents genes that contribute to disease resistance and others that promote disease development. Genetic changes in certain groups of cells in various body parts can be induced by viruses and exposure to radiation and certain chemical toxins. These changes are not transmitted intergenerationally, but are linked to the development of cancer.

Many questions about the roles of unusual genetic traits in health and disease remain to be answered. Researchers know the most about disorders related to ***single gene defects.***

➡ **single gene defects** Disorders resulting from one abnormal gene. Also called "inborn errors of metabolism." Over 800 single gene defects have been cataloged, and most are very rare.

Single Gene Defects

Hundreds of diseases related to one or more defects in a single gene have been identified, and many of these affect nutrient needs. Such defects can alter the absorption or utilization of nutrients such as amino acids, iron, zinc, and the vitamins B_{12}, B_6, or folate.[4] PKU, celiac disease, lactose intolerance, and hemochromatosis are four examples of single gene defects that substantially affect nutrient needs. (These are described in Table 31.2.)

PKU and lactose intolerance are caused by defective genetic codes for enzymes, whereas celiac disease and hemochromatosis result from genetic abnormalities in the formation of other proteins. PKU, celiac disease, and lactose intolerance are treated by diets that limit phenylalanine, gluten, or lactose, respectively. Hemochromatosis is generally treated by periodic blood removal and a low iron diet.

Not all single gene abnormalities produce ill effects. For instance, lack of the gene that codes for an enzyme that helps the body excrete a specific sulfur-containing chemical unique to cruciferous vegetables (such as cabbage, broccoli, and bok choy) may be good. Smokers and nonsmokers who have the gene for the enzyme that helps excrete this beneficial sulfur compound and who consume these vegetables regularly are more likely to develop lung cancer than are people who eat the vegetables but lack the gene. Without the gene for the enzyme that causes its quick elimination, the beneficial sulfur-containing compound lingers in the body, extending the time it can play a role in cancer prevention.[5] Risk of lung cancer is decreased by not smoking.

Although unrelated to disease risk as far as we know, another unusual gene exists that is related to vegetable intake. You have a 1 in 10 chance of being affected. (See On the Side for details.)

On the Side

Are you reminded an hour or two later that you ate asparagus? If yes, you are among the 1 out of 10 people born with a gene for detecting the odor of a sulfur-containing compound excreted in urine by 4 in 10 people after consumption of asparagus.

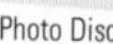

Photo Disc

Table 31.2 Examples of single gene disorders that affect nutrient need.

PKU (phenylketonuria)	A very rare disorder caused by the lack of the enzyme phenylalanine hydroxylase. Lack of this enzyme causes phenylalanine, an essential amino acid, to build up in the blood. High blood levels of phenylalanine during growth lead to mental retardation, poor growth, and other problems. PKU is treated by low phenylalanine diets.
Celiac disease	An intestinal malabsorption disorder caused by an inherited intolerance to gluten in wheat, rye, and to a lesser extent barley and oats. It causes multiple nutrient deficiencies and is treated with gluten-free diets. Celiac disease is also called "nontropical sprue" and "gluten enteropathy."
Lactose intolerance	A common disorder in adults in many countries resulting from lack of the enzyme lactase. Ingestion of lactose in dairy products causes gas, cramps, and nausea due to the presence of undigested lactose in the gut.
Hemochromatosis	A disorder affecting 1 in 200 people that occurs due to a genetic deficiency of a protein that helps regulate iron absorption. Individuals with hemochromatosis absorb more iron than normal and have excessive levels of body iron. High levels of body iron have toxic effects on tissues such as the liver and heart. The disorder can also be produced by excessive levels of iron intake over time and frequent iron injections or blood transfusions.

Most diseases related to genetic traits are not as straightforward as are single gene defects. They are more likely to represent an interwoven mesh of genetic and environmental risk factors.

Chronic Disease: Nurture *and* Nature

A very large part of our health story is more genetically influenced than genetically caused.

—Elbert Branscomb, Director of the Joint Human Genome Institute

Major health problems of our day, including heart disease, cancer, hypertension, obesity, diabetes, and disorders associated with aging, are not due to single gene defects or mainly due to genetic traits. Their development involves complex mixtures of nurture and nature influences.[2] These diseases appear to result from interactions of environmental exposures with multiple genes that confer low or moderate disease risk, or that trigger the expression (or activation) of disadvantageous genetic traits. If this is beginning to sound complicated, that's because it is.

Descriptions of nurture and nature relationships to health and disease will remain incomplete and murky until much more is understood. Enough information exists, however, to demonstrate that many common diseases result from complex environmental and genetic interactions. Several examples of these interactions are given for the diseases discussed next.

Heart Disease

I can't worry about heart disease and my diet. It's genetic.

—Nursing student

It is widely believed that heart disease is inherited from parents or grandparents. Such a belief can have unexpected consequences, such as it did for Elana:

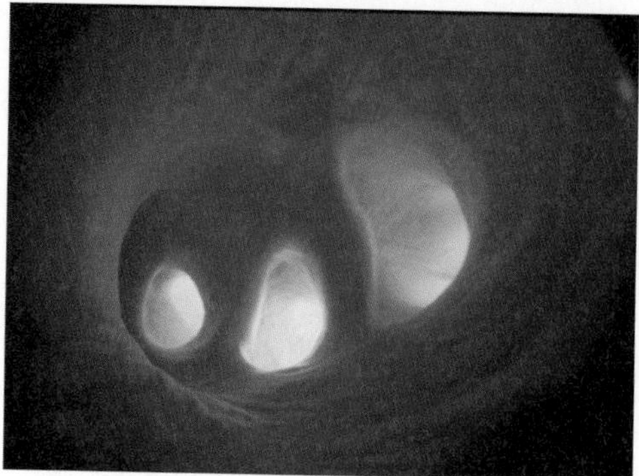
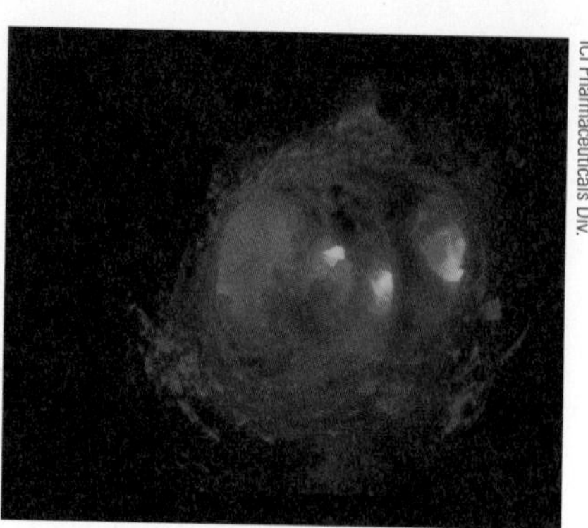

ICI Pharmaceuticals Div.

Illustration 31.1
Healthy (left) versus clogged (right) arteries. How much plaque builds up in arteries is influenced by multiple environmental and genetic factors.

Happily for Elana, both of her parents enjoyed good health, and none of her grandparents had died before the age of 80. She thought she would enjoy a long life regardless of how much she exercised or what she ate—she had good "genes." Hospitalized for a heart attack at the age of 61, Elana began to wonder what had made her so sure in the first place.

Heart disease primarily stems from plaque build-up in arteries near the heart (Illustration 31.1). Plaque build-up "hardens the arteries," narrows artery openings, and increases the risk of artery blockage and subsequent heart attack. For example, some people who consume high cholesterol diets build up plaque in their arteries, whereas others don't. For many others, low folate and vegetable diets increase the risk that plaque will accumulate. Heart disease is also influenced by blood pressure, body weight, and levels of triglycerides and clotting factors in the blood. Each of these is influenced by environmental factors such as high fat or alcohol intake and by genetic traits.[6] Risk for heart disease is the sum of all the pro and con influences of environmental and genetic factors.

As is the case for a number of other common health problems, just because a disease is linked to family history doesn't mean that genes are our destiny or that environmental factors don't matter.[7]

Cancer

Most types of cancer are primarily related to environmental factors such as high fat and alcohol intakes, low vegetable and fruit diets, high levels of body fat, smoking, and toxins in the environment.[8] Cancers often have genetic components, however, and many of these are likely to interact with environmental exposures. Studies of the roles played by genetic and environmental factors, to the extent they can currently be separated, show differences by cancer site (Illustration 31.2). As can be seen by the large contributions of environmental factors to cancer development, the hope for cancer prevention primarily lies in our ability to modify harmful environmental exposures.

Hypertension

The primary causes of most cases of hypertension are unknown, but factors such as high body fat, high alcohol intake, physical inactivity, and genetic traits play a role.[12] High salt (sodium, really) intakes are also related, but not in all studies. Differences in results between studies may be related to the proportion of people in study samples who are genetically predisposed to the effects of salt intake on blood pressure. Such people are ***salt sensitive;*** their blood pressure increases when they consume high amounts of salt. Approximately 51% of people with hypertension, and 26% of people with normal blood pressure, are salt sensitive.[14]

➡ **Salt sensitivity**
A genetically determined condition in which a person's blood pressure rises when high amounts of salt or sodium are consumed. Such individuals are sometimes identified by blood pressure increases of 10% or more when switched from a low salt (1–3 grams) to a high salt (12–15 grams) diet.[13]

Cancer Site	Environmental Factors (%)	Genetic Factors (%)
Endometrial (uterine wall)	100	0
Ovary	78	22
Lung	74	26
Breast	73–95	5–27
Stomach	72	28
Colon	65	35
Pancreas	64	36
Prostate	58	42

Table 31.3
Estimates of the contribution of environmental and genetic factors to development of some types of cancer.[9–11]

Since it is currently difficult to identify who in the population is salt sensitive and who is not, and because most Americans consume much more sodium that needed, official advice is that Americans limit their salt intake.

Obesity

> **All children inherit—along with their parents' genes—their parents, peers, and the community in which they live.**
> —L. Eisenberg, 1999

Obesity is primarily influenced by excessive calorie diets and inadequate physical activity. However, over 200 genetic traits have been associated with its development, and many of these appear to interact strongly with environmental factors (Illustration 31.2). High fat diets may influence gene expression and obesity development in some people, while a genetic predisposition toward inactivity may enhance obesity development in others.[7] When people in developing countries who have subsisted on low calorie diets for generations switch to high calorie intakes, rates of obesity tend to increase dramatically. It appears that genetic traits expressed as a result of high food intakes, or body processes that have adapted to survive on low caloric intakes, promote the development of obesity. Diabetes and heart disease rates in population groups also rise with increasing rates of obesity.[15] Many other nurture and nature interactions on obesity are possible.

Illustration 31.2
Over 200 genetic traits have been associated with the development of obesity.

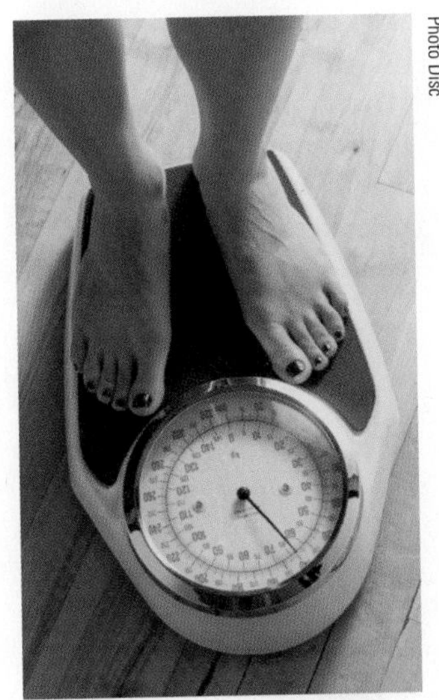

Photo Disc

Diseases Related to Aging

The incidence of disorders such as diabetes, cancer, heart disease, memory impairment, hypertension, and stroke usually increases with age. Such declines in health are neither inevitable nor largely determined by genetic factors. They are heavily influenced by diet, exercise, and other components of lifestyle Illustration 31.3).[16]

Illustration 31.3
Diseases related to aging are neither inevitable nor largely determined by genetic factors. They are influenced more by diet, exercise, and other components of lifestyle.

Photo Disc

Genetics of Food Selection

Food preferences are largely learned, but which vegetables we like or don't like may be influenced by genetic traits. There are over 80 genes that help people taste bitter foods, and some people get the set of genes that make them highly sensitive to bitter-tasting foods (which are mainly vegetables). People born with a high sensitivity to bitter tastes tend to dislike cooked cabbage, collard greens, spinach, brussels sprouts, or other vegetables that taste bitter to them. People who tend to like these vegetables generally don't perceive them to be bitter-tasting. A genetic tendency to reject these vegetables is likely to limit intake and therefore may be linked to diseases associated with low vegetable intake.[17]

Looking Ahead

In our rush to fit medicine with the genetic mantle, we are losing sight of other possibilities for improving public health.

—N. A. Holtzman, Johns Hopkins Medical Institute

The promise of advances in knowledge of the genetic bases of disease is longer, healthier lives. No doubt drugs that counter ill effects of genetic traits will be devised. The search is on for drugs that would lower the amount of a certain protein overproduced by people with rheumatoid arthritis and lupus, for example. High blood levels of this protein lead to joint and connective tissue inflammation and pain. Drugs that lower levels or the activity of the protein could greatly relieve the inflammation and pain associated with these conditions.[18] Just as specific genes are introduced into plants to improve disease resistance, genes will be introduced into human cells to combat disease. Animal and human studies now under way are testing the safety and effectiveness of gene transplants. Genes are given passage into cells by attaching them to harmless viruses or other substances. Once inside the cell, genes produce specific proteins that modify disease processes. (Since the genes introduced do not become part of DNA, the new genes would not be transferred intergenerationally.) This type of research is in its infancy, and many obstacles to success exist.

Illustration 31.4
In the future, advice given by registered dietitians may be tailored to the specific genetic traits of individuals.

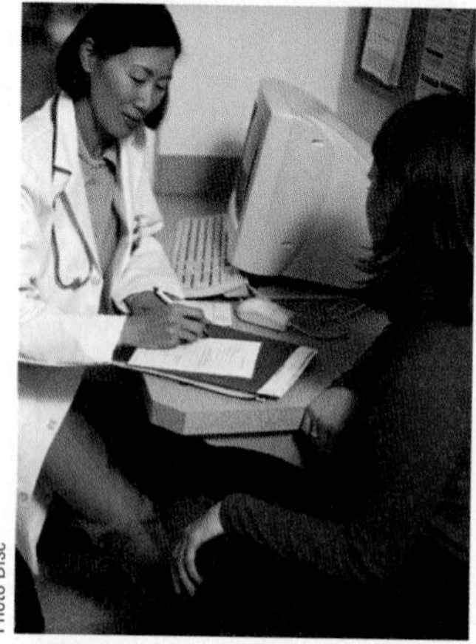

Photo Disc

Individualized dietary interventions will eventually become a routine part of health care (Illustration 31.4). Rather than population-wide advice, recommendations to increase folate intake, to decrease fat or cholesterol consumption, to limit salt intake, or to change certain other dietary practices may be individualized based on genetic characteristics. Such changes would reduce the risk of some diseases, but would not prevent them altogether because multiple environmental factors and genetic traits influence chronic disease development. Even if all genetic traits for common diseases were eventually identified, it would not mean the end of most diseases, because they are largely environmentally determined. Safe and effective methods for modifying genetic traits are under investigation but are not yet available.[19] Larger reductions in disease incidence will come from improvements in diet and exercise and reductions in smoking and other unhealthy habits and exposures. Formidable gains in health will rely on successfully lowering both nurture and nature risks to health.

WWW links

www.nhgri.nih.gov/DIR/VIP/Glossary
A great site from the National Human Genome Research Institute that provides written and spoken definitions of genetic terms and useful illustrations.

www.cdc.gov/genetics/huge.htm
The Centers for Disease Control's genetics Web site. It provides updates on developments in genetic influences on disease risk.

www.healthfinder.gov
Search any disease of interest to find out current information about it.

Nutrition UP CLOSE

Nature and Nurture

FOCAL POINT: Gaining insight into your family health history and your potential risk of common diseases.

Environmental and inherited traits that influence development of chronic diseases are often shared by families. Check out your family's health history by completing the following family tree activity. If you don't know all of your relatives included in the activity, fill out the parts for relatives you do know. If you were raised by adults who are not your biological relatives, behaviors you and your close relatives share will affect your risk of disease more than will shared genetic traits.

Relative	Check each disease that applies:											
	Heart Disease		Cancer		Diabetes		Hyper-tension		Obesity		Alcoholism	
	yes	no	yes	no	yes	no	yes	no	yes	no	yes	no
A. On your mother's side:												
Grandmother	___	___	___	___	___	___	___	___	___	___	___	___
Grandfather	___	___	___	___	___	___	___	___	___	___	___	___
B. On your father's side:												
Grandmother	___	___	___	___	___	___	___	___	___	___	___	___
Grandfather	___	___	___	___	___	___	___	___	___	___	___	___
C. Your mother	___	___	___	___	___	___	___	___	___	___	___	___
D. Your father	___	___	___	___	___	___	___	___	___	___	___	___

FEEDBACK (answers to these questions) can be found at the end of Unit 31.

REFERENCES

1. McDermott R. Ethics, epidemiology and the thrifty gene: biological determinism as a health hazard. Soc Sci Med 1998;47:1189–95.
2. Emmett 2000.
3. Tolstoi LG, Smith CL. Human Genome Project and cystic fibrosis—a symbiotic relationship. J Am Diet Assoc 1999;99:1421–7.
4. Shils ME, et al., eds. Modern nutrition in health and disease. 9th ed. Philadelphia: Lippincott, Williams & Wilkins; 1998.
5. London SJ, Yuan JM, Coetzee GA, et al. CYP1A1 1462V genetic polymorphism and lung cancer risk in a cohort of men in Shanghai, China. Cancer Epidemiol Biomarkers Prev 2000;9:987–91.
6. Krauss RM. Genetic recipes for heart-healthy diets [editorial]. Am J Clin Nutr 2000;71:668–9.
7. Allison DB. Genetic and environmental influences on human body weight. Nutr Today 2000;35:18–21.
8. Willett WC. Nutrition and cancer. Salud Publica Mex 1997;39:298–309.
9. Lichtenstein P, Holm NV, Verkasalo PK, et al. Environmental and heritable factors in the causation of cancer—analyses of cohorts of twins from Sweden, Denmark, and Finland. N Engl J Med 2000;343:78–85.
10. Terry P, Baron JA, Weiderpass E, et al. Lifestyle and endometrial cancer risk: a cohort study from the Swedish Twin Registry. Int'l J Cancer 1999;82:38–42.
11. Hunter DJ, Willett WC. Nutrition and breast cancer. Cancer Causes and Control 1996;7:56–68.
12. Wong ZY, Stebbing M, Ellis JA, et al. Genetic linkage of beta and gamma

subunits of epithelial sodium channel to systolic blood pressure. Lancet 1999;353: 1222–5.

13. Morimoto A, Uzu T, Fujii T, et al. Sodium sensitivity and cardiovascular events in patients with essential hypertension. Lancet 1997;350:1734–7.
14. Kaplan NM. The dietary guideline for sodium: should we shake it up? No. Am J Clin Nutr 2000;71:1020–6.
15. Berdanier CD. Nutrient-gene interactions. Nutr Today 2000;35:8–17.
16. Rowe JW. Geriatrics, prevention, and the remodeling of Medicare [editorial]. N Engl J Med 1999;340:720–1.
17. Drewnowski A, Henderson SA, Hann CS, et al. Genetic taste markers and preferences for vegetables and fruit of female breast care patients. J Am Diet Assoc 2000;100:191–7.
18. Collins FS, McKusick VA. Implications of the Human Genome Project for medical science. JAMA 2001;285:540–4.
19. Holtzman NA, Marteau TM. Will genetics revolutionize medicine? N Engl J Med 2000;343:141–4.

Nutrition

UP CLOSE

Nature and Nurture

Feedback for Unit 31

A history of the same disease in one or more close relatives increases the chances that you, too, are at risk. Taking preventive measures, such as making healthy changes in your diet or physical activity levels, would be advisable. Remember, however, that just because a certain disease tends to run in your family does not mean it's your destiny. It is only part of the array of factors you should consider when making decisions about your health and habits.

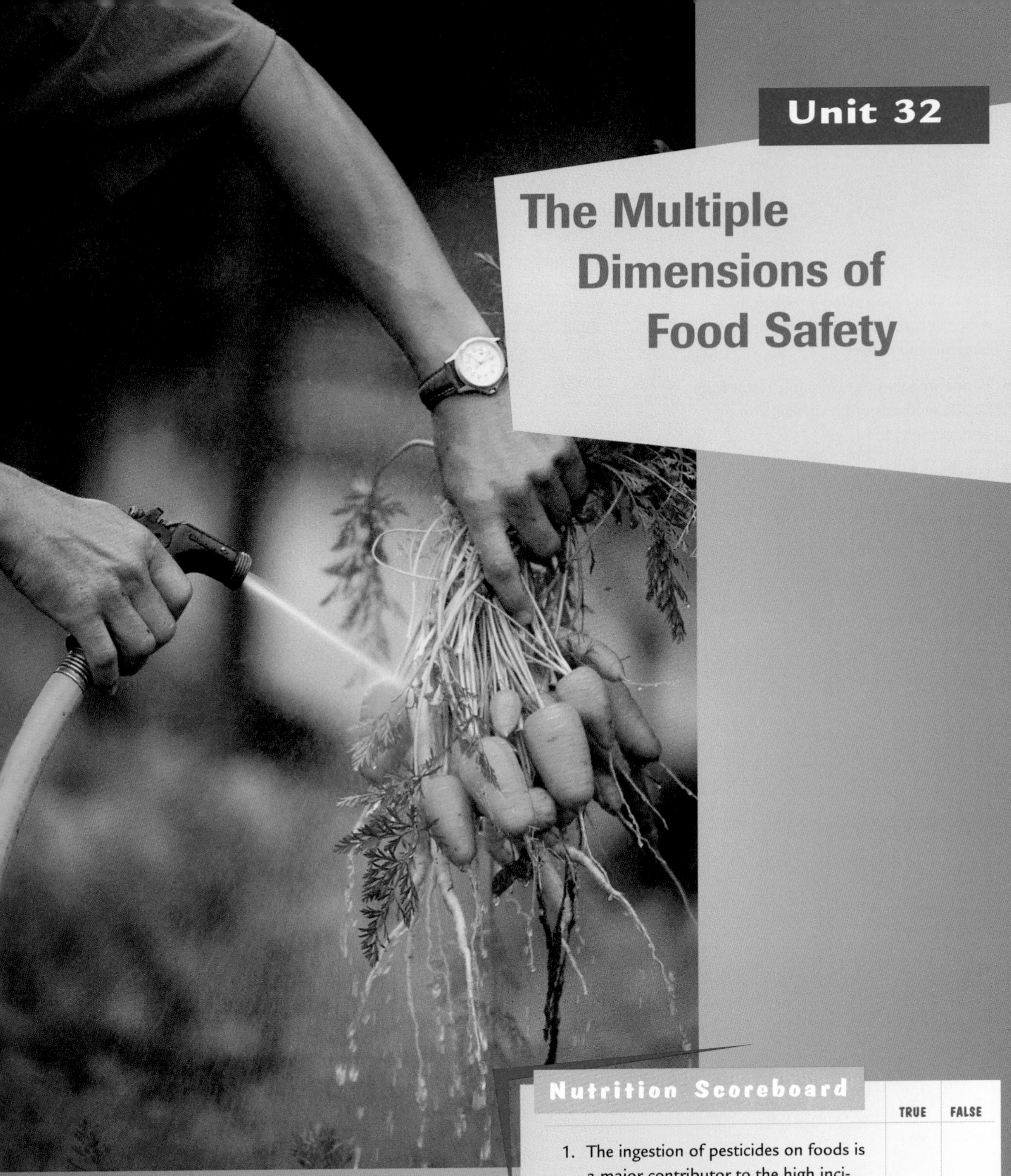
Photo Disc

Unit 32

The Multiple Dimensions of Food Safety

Nutrition Scoreboard

	TRUE	FALSE
1. The ingestion of pesticides on foods is a major contributor to the high incidence of cancer in humans.		
2. Freezing kills bacteria in food.		
3. Of the 50 states, Alaska has the highest incidence of botulism.		

(Answers on next page)

Answers to

Nutrition Scoreboard

	TRUE	FALSE
1. Consumers in general are at low risk of developing cancer due to pesticides on food.[1] People at highest risk for pesticide-related illnesses are farmers and others who work with pesticides and do not follow appropriate safety procedures.		✓
2. Freezing causes most bacteria to cease multiplying, but does not kill them. If a food is contaminated with bacteria before you freeze it, bacteria will be present when you thaw it.		✓
3. Here's a question for your nutrition trivia file. Alaska does have the highest incidence of botulism of all 50 states. The reason why is explained in this unit. Any guesses?	✓	

Key Concepts and Facts

- The leading food safety problem in America is contamination of food by bacteria and viruses.
- Foodborne illnesses primarily result from unsafe methods of producing, storing, and handling food.
- Foodborne illnesses are linked to hundreds of foods but most commonly to raw and undercooked meats and eggs, shellfish, and unpasteurized milk.
- Most cases of foodborne illness are preventable.

Threats to the Safety of the Food Supply

Hamburgers contaminated with *E. coli,* chicken tainted with *Salmonella,* fruit coated with organophosphate, mad cow disease—it's enough to give you food fright. Actually, Americans have good reason to be concerned about the safety of some foods (Illustration 32.1). The Centers for Disease Control (CDC) estimates that each year in the United States foodborne illnesses cause:

- sickness in 76 million people,
- more than 300,000 hospitalizations, and
- over 5000 deaths.[2]

Foodborne illnesses can be caused by bacteria, viruses, marine organisms, fungi, and the toxins they produce, as well as by chemical contaminants in foods or water.[2] They are spread by a wide assortment of foods, including meats, eggs, unpasteurized milk, shellfish, raspberries, sprouts, breakfast cereals, and hummus. Foods most commonly associated with foodborne illness, however, are raw and undercooked meat and eggs, shellfish, and unpasteurized milk.[4]

On the Side

More than 50% of consumers do not realize that foodborne diseases can be transmitted by fruits and vegetables.[3]

Photo Disc

How Good Foods Go Bad

Bacteria and viruses are the most common causes of ***foodborne illnesses*** and largely enter the food supply during food processing, storage, or preparation (Illustration 32.2).[2] They are transferred to humans in foods through many different routes, a major one being the contamination of food with animal feces. The lower intestines of many healthy farm animals are colonized by bacteria that may be harmful

HEALTH

Meat plant recalls 282,000 pounds of ground beef because of E. Coli

said that 50 retail and food-service customers have been contacted and asked to return the meat to IBP. The beef is typically processed further by grocers and restaurants, so there are no product codes for consumers to check.

In fact, Knight said, because the beef was fresh and produced more than two ...

WORLD & NATION

Study links mad cow disease and new strain of fatal brain disease in humans

Associated Press

WASHINGTON, D.C. — A laboratory experiment gives powerful new evidence that an infectious protein that causes mad cow diseases also causes a new type of fatal human brain disease that has killed 51 people in Europe.

that causes the diseases can move more easily between species than once thought.

Although a link betwe... and the new type of CJD ha... experts wh...

Enough to make you sick

A recent Twin Cities outbreak of hepatitis A is a reminder that food-borne bugs are everywhere

Health

Doing battle against E. coli

By Maria Elena Baca
Star Tribune Staff Writer

Just before quitting time on the Friday before Labor Day this year, scientists at the Minnesota Department of Health interviewed ... people

The patients had two things in common: The DNA of the bacteria that had made them sick were identical. And just two weeks earlier, they all had eaten Simek's hamburger patties; one man still had part ... in his freezer. The

Despite, or perhaps because of, the state Health Department's stature in the field, the number of reported O157 cases in Minnesota had been disproportionately higher in recent years than other sta...

Illustration 31.1
Problems with food safety often make the headlines.

to humans. These bacteria contaminate food when unsanitary practices are used to prepare and process meats, and when vegetables and fruits are fertilized with animal manure. They can also be transferred to foods by humans who have colonies of harmful bacteria in their gastrointestinal tracts through the use of human sewage on crops, and food handling by people carrying the bacteria on their hands. Humans also transfer certain harmful microorganisms to food when fluids from infected injuries or body secretions contact food.[2] These types of foodborne illnesses are mainly caused by bacteria that are "on" rather than "in" foods.

➡ **foodborne illness**
An illness related to consumption of foods or beverages containing disease-causing bacteria, viruses, parasites, toxins, or other contaminants.

PhotoEdit - Dennis MacDonald

Illustration 32.2
Can you find three potential opportunities for the spread of foodborne illness in this photograph?

Although less common, bacteria can be present on the inside of foods. They can enter vegetables and fruits, for example, if the protective skin coatings are broken, allowing an entrance for bacteria. *Salmonella* bacteria can infect the ovaries of hens and cause them to lay normal-looking but infected eggs. Shellfish can concentrate microorganisms present in surrounding water, and although the bacteria may be harmless to the shellfish, they can provide large doses of harmful bacteria and viruses to humans.[4]

The source of many cases of foodborne illness is food that has come into contact with contaminated food. This situation, referred to as "cross-contamination," increases the reach of foodborne illnesses.

Cross-Contamination of Foods Microorganisms including bacteria, viruses, toxins, and other harmful substances can contaminate safe foods during processing, shipping, preparation, or storage. Opportunities for cross-contamination of foods during processing, for example, are plentiful. The hamburger you eat may contain meat from hundreds of different cows, an omelet in a restaurant may contain the eggs of hundreds of different chickens, and the chicken you baked may have bathed with hundreds of others when they were all washed in the same vat of water at the meat processing plant.[4] Cross-contamination also occurs on cutting boards across America. The failure to routinely wash cutting boards in between the preparation of different raw foods is a major route to the spread of foodborne illness.[5]

Ed Kashi

Illustration 32.3
Gathering broccoli samples for pesticide testing. Pesticide residues are not a major cause of foodborne illnesses.

Role of Pesticides and Other Chemicals Other potential sources of food contamination include pesticides and other agricultural chemicals, antibiotics and hormones given to animals, and industrial pollutants. Of these, perhaps the greatest health threat is represented by antibiotics routinely used to prevent infectious diseases and promote growth in poultry, cattle, and other food animals.

Antibiotic Resistance Microorganisms are very good at adapting to their environments; they can transform themselves to become resistant to the effects of antibiotics used to kill them. Since many of the same types of antibiotics given to animals are used in humans, these new strains of resistant bacteria produced in animals may be resistant to antibiotics in humans, too. Many common forms of foodborne illnesses today are represented by these new forms of bacteria.[6] Because the routine use of antibiotics in food animals is increasing the proportion of bacterial infections in humans that cannot be treated, the call is going out for tighter regulations on the use of antibiotics in animals.[6]

Hormones Little is known about the effects on human health of hormones given to animals to promote growth. Many consumers are concerned about the effects, however, and would prefer such hormones not be present in food. It is thought that effects of ingested hormones in humans may take many years to develop and will require long-term studies to be identified.

Pesticides and PCBs Pesticides containing organophosphates, mercury-containing fungicides, and DDT remain causes of foodborne illnesses, as do PCBs.[2,7] Use of DDT on insects and PCBs in transformers was phased out over 20 years ago due to links with cancer, but these long-lasting chemicals still contaminate some land, lakes, and streams. Fewer than 1 in 10,000 foods on the market contains an excessive level of pesticide, and two out of three contain no trace of agricultural chemicals (Illustration 32.3).[8] Farmers and food processing employees are most likely to experience illnesses associated with exposure to agricultural chemicals (Illustration 32.4). Health problems due to agricultural chemicals appear to be very rare in other groups of people.[8]

Martha Cooper and Peter Arnold

Illustration 32.4
Protective gear—including a chemical-resistant suit, respirator, and goggles—is required for use with some of the most toxic pesticides, but is not always worn.

Causes and Consequences of Foodborne Illness

Over 250 types of foodborne illnesses caused by infectious agents (bacteria, viruses, and parasites) and noninfectious agents (toxins and chemical contaminants) have been identified. Their impact on health ranges from a day or two of nausea and diarrhea to death within minutes. Effects of foodborne illnesses are generally most severe in people with weakened immune systems or certain chronic illnesses, pregnant women, young children, and older persons (Table 32.1).[9] Symptoms of foodborne illness most commonly consist of nausea, vomiting, abdominal cramps, and diarrhea.[4] Many cases go unreported, making it difficult to identify the most common causes of foodborne illness. Estimates of the causes are usually based on cases reported by health care professionals and health departments.[4]

The most prevalent reported causes of foodborne illness result from *Salmonella, Campylobacter,* and *E. coli* 0157:H7 bacteria and Norwalk-like viruses.[10] Table 32.2 summarizes facts about these top four leading causes of foodborne illness.

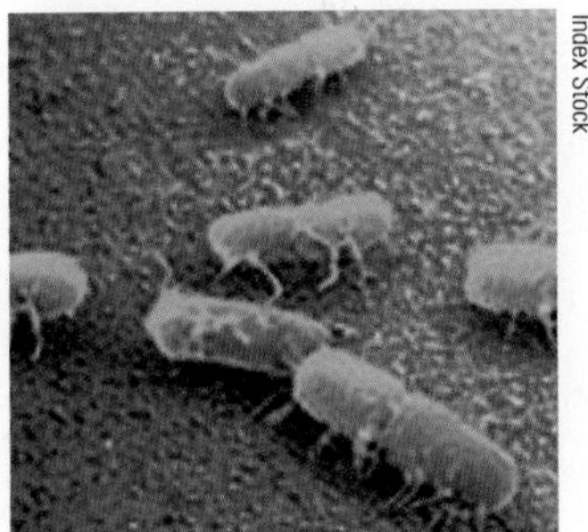

Index Stock

Salmonella.

Salmonella Nearly 35,000 cases of *Salmonella* infection were reported to the CDC in 1998;[4] and it is estimated that 38 times that number, or 1,330,000 cases, actually occurred.[11] An outbreak of *Salmonella* infection that affected approximately 224,000 people was linked to the transport of an ice cream mixture in a tanker that had previously transported *Salmonella*-infected eggs. The infected ice cream was distributed in 41 states.[12] It is estimated that 1% of the U.S. population experiences a *Salmonella* infection every year.[1]

Campylobacter This bacterium is a common contaminant on chicken and is the cause of the greatest number of cases of bacteria-related foodborne illness each year. That figure is a startling 2,453,926![6] The bacteria are found on up to 80% of chickens, and 20% represent a strain of *Campylobacter* resistant to antibiotics.[6] In addition to poultry, *Campylobacter* infection has been traced to unpasteurized milk and contaminated water.

E. Coli *E. coli* 0157:H7 is a new strain of *E. coli* that has evolved into a killer. As few as 10 of these bacteria can lead to death by kidney failure. *E. coli* infections are particularly associated with consumption of undercooked ground beef. A

Table 32.1
High risk groups for severe effects of foodborne illness.[9]

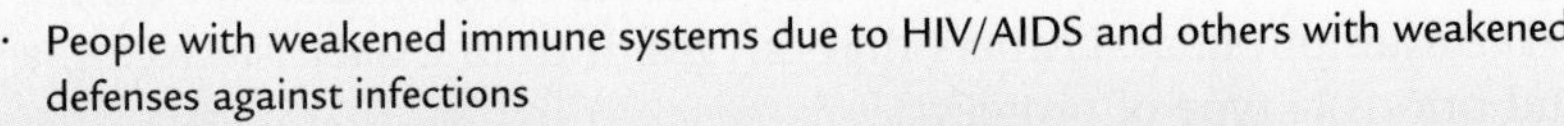

- People with weakened immune systems due to HIV/AIDS and others with weakened defenses against infections
- People with certain chronic illnesses such as diabetes and cancer
- Pregnant women
- Young children
- Older persons

Bacteria	Illness Onset	Illness Duration	Symptoms	Foods Most Commonly Affected	Usual Source of Contamination
Salmonella	1–3 days	4–7 days	Diarrhea, abdominal pain, chills, fever, vomiting, dehydration	Uncooked or undercooked eggs, unpasteurized milk, raw meat and poultry, vegetables and fruits	Infected animals, human feces on food, contaminated water
Campylobacter	2–5 days	2–5 days	Diarrhea (may be bloody), abdominal cramps, fever, vomiting	Undercooked poultry, unpasteurized milk	Infected poultry and other animals
E. coli 0157:H7	1–8 days	5–10 days	Watery, bloody diarrhea, abdominal cramps, little or no fever	Raw or undercooked beef, unpasteurized milk, raw vegetables and fruits, contaminated water	Infected cattle
Norwalk-like viruses	1–2 days	1–3 days	Nausea, vomiting, diarrhea	Undercooked seafood	Human feces contamination of oysters and other shellfish beds

Table 32.2
The top four causes of foodborne illness.[10]

major outbreak of *E. coli* 0157:H7 infection occurred when contaminated ground beef leftover from one day's production was added to the next day's batch of beef. Use of the leftover, contaminated meat kept recontaminating subsequent batches.[13] Some 25 million pounds of ground beef had to be recalled as a result.

On the Side

Due to concerns about *E. coli* infections, some restaurants require patrons who want their meat cooked rare to sign a waiver before it is served to them.

Photo Disc

Norwalk-like Viruses These viruses are an extremely common but underreported cause of foodborne illnesses. The viruses go underreported because laboratory tests required for diagnosis are not widely performed. Norwalk-like viruses usually cause an acute bout of vomiting that resolves within two days. They are thought to be primarily spread by infected kitchen workers and fishermen who have dumped sewage waste into waters above oyster beds. An estimated 23,000,000 cases of Norwalk-like virus–related illnesses occur in the United States each year.[6]

Other Causes of Foodborne Illnesses Of the hundreds of other causes of foodborne illness, six have been selected for brief review here. They represent a sampling of foodborne illnesses stemming from seafood consumption, and others that can result from a bacterial toxin, a parasite, and prions (a type of protein).

Foodborne Illnesses Related to Seafoods Many types of seafood may become contaminated due to industrial or human pollution of fresh and ocean waters. Examples presented relate to mercury contamination, ciguatera fish poisoning, and the neurotoxin (or nervous system toxin) responsible for "red tide."

Mercury Contamination Seafoods have come under fire as a potential source of foodborne illnesses due to mercury contamination of waters and fish by fungicides, fossil fuel exhaust, smelting plants, pulp and paper mills, leather tanning facilities, and chemical manufacturing plants. High levels of mercury are most likely to be present in large, long-lived fish such as shark, swordfish, and tuna. Because mercury can interfere with fetal brain development, consumption of these types of fish by pregnant women should be limited.[14] The Food and Drug Administration (FDA) advises that women who are pregnant or may become pregnant not eat shark, swordfish, king mackerel, or tile fish. Women should limit consumption of fish caught by family or friends to six ounces a week, and may eat an average of twelve ounces of fish from stores and restaurants weekly.[15]

Ciguatera Ciguatera poisoning from fish is caused by a neurotoxin (ciguatoxin) present in microoganisms called dinoflagellates in reefs. The toxin is transferred through herbivorous reef fish to carnivorous fish, and then to humans who eat these fish. Over 200 types of fish may cause ciguatera poisoning, the most common being grouper, red snapper, and barracuda. Primary areas affected by ciguatera poisoning include the Caribbean and South Pacific Islands. Symptoms develop one to 30 hours after ingestion of poisoned fish and cause nausea, vomiting, abdominal cramps, watery diarrhea, and then numbness, shooting pains in the legs, and other symptoms. The toxin causing the poisoning is not destroyed by cooking, freezing, or digestive enzymes, and there is no effective treatment.[16]

Red Tide "Red tide" may occur between June and October on Pacific and New England coasts. It is due to the accumulation of a microorganism that produces a nerve toxin. Oysters and other shellfish that consume the microorganism become contaminated, as do humans who ingest the shellfish. Resistant to cooking, the toxin produced causes a burning or prickling sensation in the mouth five to 30 minutes after contaminated shellfish are consumed. This symptom is followed by nausea, vomiting, muscle weakness, and a loss of feeling in the hands and feet. Recovery is usually complete, but it's easy to understand that the warning not to eat mussels, clams, oysters, scallops, or other shellfish from red tide waters is for real.

Botulism *Botulinum* bacteria produce a toxin that is one of the most deadly known. It can cause nerve damage and respiratory failure. (An antidote is available but must be given soon after the infection begins.) These bacteria are commonly present in soil and ocean and lake sediment and may contaminate crops, honey, animals, and seafood. In humans, botulism usually results from eating underheated, contaminated foods stored in airtight containers. The bacteria thrive in low-oxygen environments and produce gases as they grow. The gases expand the container the food is in. (For a real example of a can that exploded due to gas produced by the growth of bacteria, see Illustration 32.5.) Consequently, foods in cans, plastic bags and wraps, and other airtight containers that have bulged-out areas should *not* be eaten.

Alaska has the highest incidence of botulism in the United States. Why does the 49th state rank first? The reason is that some Alaskans place uncooked or partially cooked salmon eggs, whale blubber, and other seafoods in plastic bags to ferment. The bags are squeezed to expel the air before sealing, and the low-oxygen environment fosters the growth of *Botulinum.*

Illustration 32.5
Pressure caused by bacterial gases in this can of glucose drink made the can explode.

Parasites Various parasitic worms, such as tapeworms, flatworms, and roundworms, may enter food and water through fecal material and soil. They are generally killed by freezing and always killed by high temperature. Once consumed, roundworms may attach to the lining of the intestine and feed on the person's blood. This can lead to anemia. One type of roundworm can bore a hole through the stomach within an hour after the worm's source—raw fish—is eaten (Illustration 32.6). The severe pain that results sends people to their doctors on the double. Hundreds of Japanese people experience this foodborne illness every year.[17]

It's a good thing most parasites are destroyed by freezing. One study in Seattle assessed the parasite content of raw fish used in sushi. About 40% of the raw fish samples contained roundworms, but since the fish had been frozen, all the worms were dead.[17]

Mad Cow Disease Technically called bovine spongiform encephalopathy, this rare disease in cattle is suspected of causing at least 90 human deaths in England and France. The disease appears to have started when cows, who are herbivores by nature, were given sheep intestines and parts of the spinal cord in their feed. Some of the sheep harbored a protein, called a prion, that caused a deadly disease when consumed by cows. A prion is not a bacterium or other microorganism; it's a small protein that can transmit disease when consumed by a similar species. Only one other known foodborne illness is spread by the consumption of otherwise healthy body parts. That disease is kuru, and it is transmitted by cannibalism.[18]

Researchers concluded that mad cow disease was transferred to humans who ate the meat of prion-infected cows. The disease that affects humans is a new form of Creutzfeldt Jakob syndrome. It inevitably leads to death in humans after many years due to brain damage.[19] Needless to say, it is no longer legal to feed cows animal parts that may transmit the disease. Although the risk of consuming beef from cattle with mad cow disease is extremely small, the possibility that animals may develop this or a similar disease exists. Stringent monitoring efforts are in place in the United States and many European countries to prevent the occurrence and spread of prions.[20]

Because the disease was first recognized in 1994 and may take 20 or more years to develop, we may not know the full impact of the exposure to contaminated beef for years to come.

HERMAN

"I need 148 get-well cards."

Illustration 32.6

Source: Jim Unger/dist. by LaughingStock Licensing Inc. Used by permission.

Preventing Foodborne Illnesses

There are two major approaches to the prevention of foodborne illnesses. The first relies on food safety regulations that control food processing and handling practices, and the second involves consumer behaviors that lessen the risk of consuming contaminated foods. (Read what can happen when both of these principles are violated in the nearby Health Action.)

Food Safety Regulations

According to the Federal Food, Drug, and Cosmetic Act, it is illegal to produce or dispense foods that are contaminated with substances that cause illness in humans. Foods are considered "safe" if there is a reasonable certainty that no harm will result from repeated exposure to any substance added to foods. The act governs all substances that are added intentionally or accidentally to foods—except pesticides. According to legislation passed in 1996, pesticides are permitted if their consumption is associated with a "negligible risk" of cancer or other health problems.[21]

HEALTH ACTION

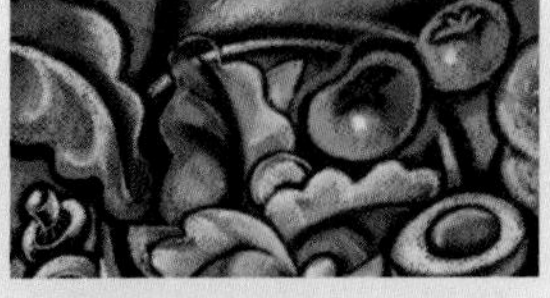

Don't Let This Happen to You!

From the true story department of *Nutrition Now* comes the tale of a man who took a night off with his wife to dine at their favorite restaurant. The man, a physician, ordered his favorite item: the luscious chicken burritos. He was served more than he could eat, so he took the leftovers home and promptly put them in the refrigerator. About 8 hours later his intestines started to explode; he thought he'd rather die than be so sick. Then the light bulb went off—he had probably consumed *C. perfingens* toxin somehow.

By the third day he began to feel better and noticed he was ravenously hungry. Then he remembered there was a leftover burrito in the refrigerator. . . . Happily, the second round of symptoms wasn't as bad as the first.

Irradiation of Foods Increasing rates of foodborne illness have led some health officials to recommend the expanded use of food irradiation. Food irradiation is a safe process when conducted under specified conditions. It destroys bacteria, parasites, and viruses present in or on foods. Availability of irradiated foods is limited due to lingering low consumer acceptance of the process and of the foods.[22]

Although the proposal has merits, food irradiation is not the silver bullet that will prevent all cases of foodborne illness. Prions, toxins, pesticides, mercury, and PCBs are resistant to radiation, so radiation is unlikely to affect the safety of foods with respect to these contaminants.[22] Once irradiated, foods can later become contaminated in such places as a packing plant, grocery store, restaurant, or home. The absence of all microorganisms in irradiated food may enable individual types of bacteria that contact the food after it is sterilized to grow at unusually high rates. Irradiation does not work well on all foods—some fruits and vegetables develop a poor texture and change flavor when irradiated.

The Consumer's Role in Preventing Foodborne Illnesses

Food safety education will never be a substitute for a safe food supply.

—Caroline Smith DeWall, 1997

Due in large measure to the demands of individuals and consumer groups, many mechanisms are in place to help ensure the safety of the food supply (Illustration 32.7). However, existing safeguards in food production and processing, and government regulations and enforcement efforts, are insufficient to guarantee that all foods purchased by consumers will be free from contamination. This situation,

Illustration 32.7

Source: © L. Trepel.

along with the possibility that food can become contaminated in unanticipated ways, means that responsibility for food safety is shared by consumers.

Food Safety Basics The first rule of food safety is to wash your hands thoroughly with soap and water before and after handling food. According to the Centers for Disease Control, this is the single most important means of preventing the spread of foodborne illness caused by bacteria.

There's a Right Way to Wash Your Hands Ever watch a TV dramatic series about doctors and hospitals and wondered why they show surgeons scrubbing their hands and upper arms before surgery for so long? TV got it right. It takes about 20 seconds to sanitize your hands. First you need to lather up with soap and very warm water. The soap makes germs lose their grip on your skin. Next you have to make sure to scrub all the crevices between your fingers and under your fingernails. Rinse your hands thoroughly with really warm water and dry them with a paper towel. (If you use a dish cloth, you may reinfect your hands.)[23]

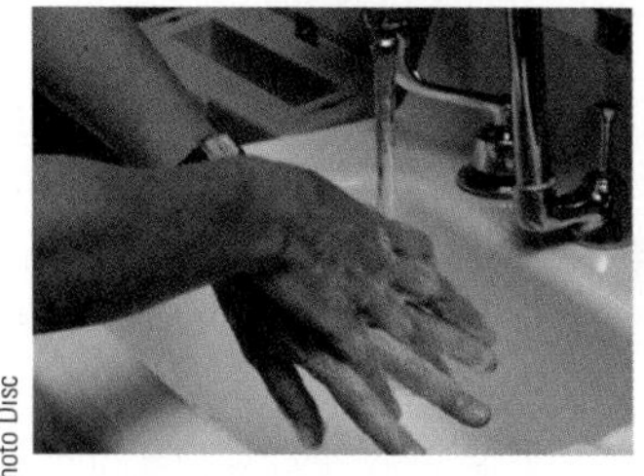

Photo Disc

Keep Hot Foods Hot and Cold Foods Cold Good foods can go bad if stored improperly. One of the most effective ways to prevent foods from spoiling is to store them at, and heat them to, the right temperatures (Illustration 32.8 and Table 32.3). When holding prepared foods before serving, hot foods should be kept hot, and cold foods cold. Freezing foods halts the growth of all of the main types of bacteria that contaminate foods. Once the foods are thawed, however, bacteria growth resumes, and the foods may become contaminated with new bacteria while thawing.

Bacteria grow best at temperatures between 40° and 140°F (Illustration 32.9). The room temperature of homes is within this range, and that's why foods that spoil should not be left outside the refrigerator for more than an hour. If a food has been improperly stored or kept past the expiration date on the label, it should not be eaten even if it tastes all right. This general rule should apply: when in doubt, throw it out!

On the Side

You can't tell if a food is contaminated by staring at it, tasting it, or smelling it. Some, but not all, foods that are contaminated look, taste, or smell bad.[23]

Photo Disc

Food Handling and Storage Particular care needs to be taken when handling and storing foods. Dairy products such as milk and cheese may spoil in about a week or sooner if contaminated with bacteria from the air or hands. If you hold cheese by the wrapper when you take it out of the refrigerator to cut a slice off, it will likely last longer than if you touch it with your hands. Handling foods with clean hands or plastic gloves and using clean utensils on washed surface areas reduce the number of bacteria that come in contact with the food. Raw meats should be separated from other foods, and utensils and surface areas used to prepare meats should be thoroughly cleaned after each use.

The Safety of Canned Foods Commercially canned foods are heated to the point where all bacteria are killed after the can is sealed. Consequently, the contents of canned foods are sterile and will not spoil if left on the shelf for years. Nevertheless, canned foods may spoil if the can develops a pinhole leak or if errors made during the canning process allow bacteria to enter the food. A

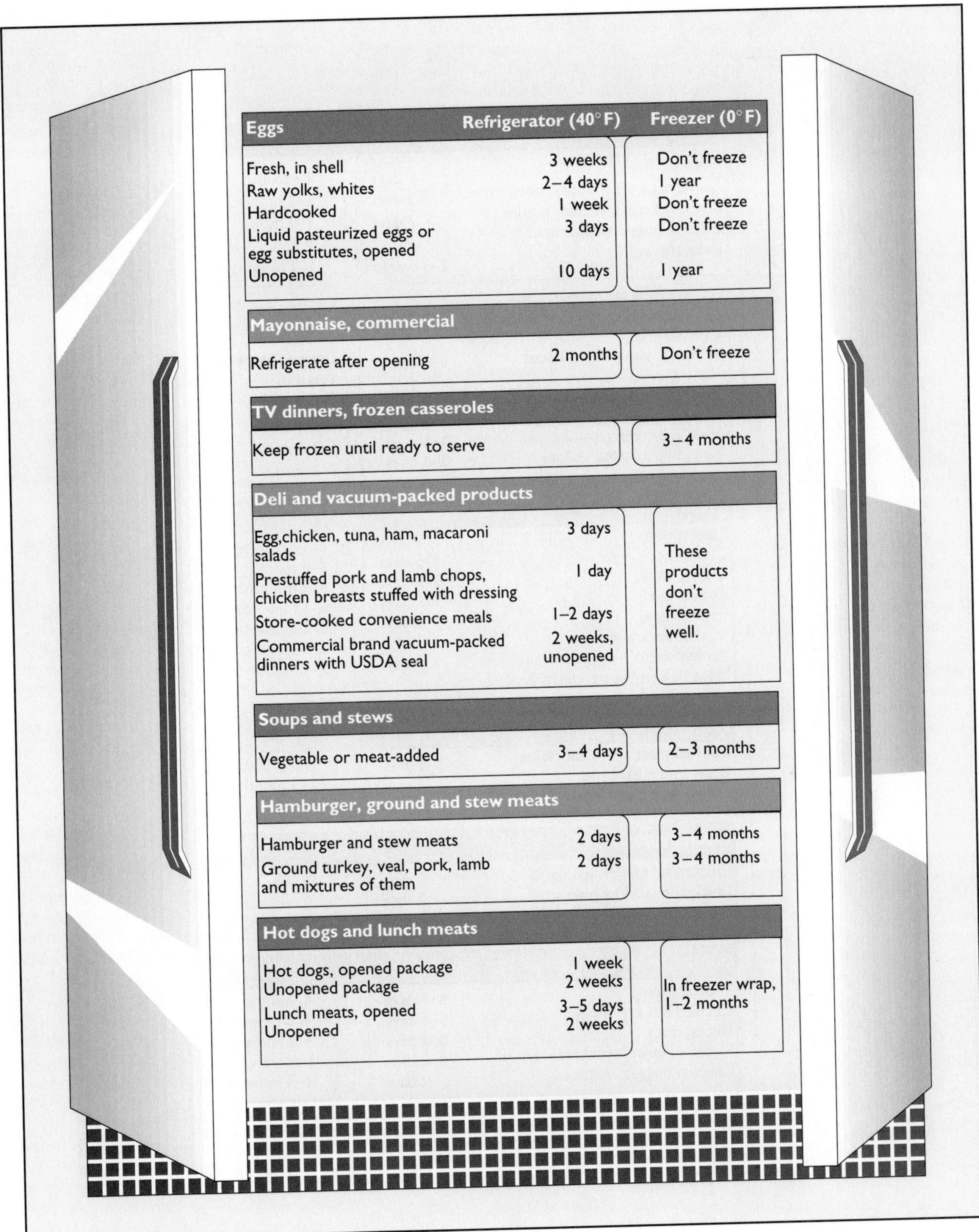

Eggs	Refrigerator (40° F)	Freezer (0° F)
Fresh, in shell	3 weeks	Don't freeze
Raw yolks, whites	2–4 days	1 year
Hardcooked	1 week	Don't freeze
Liquid pasteurized eggs or egg substitutes, opened	3 days	Don't freeze
Unopened	10 days	1 year
Mayonnaise, commercial		
Refrigerate after opening	2 months	Don't freeze
TV dinners, frozen casseroles		
Keep frozen until ready to serve		3–4 months
Deli and vacuum-packed products		
Egg,chicken, tuna, ham, macaroni salads	3 days	These products don't freeze well.
Prestuffed pork and lamb chops, chicken breasts stuffed with dressing	1 day	
Store-cooked convenience meals	1–2 days	
Commercial brand vacuum-packed dinners with USDA seal	2 weeks, unopened	
Soups and stews		
Vegetable or meat-added	3–4 days	2–3 months
Hamburger, ground and stew meats		
Hamburger and stew meats	2 days	3–4 months
Ground turkey, veal, pork, lamb and mixtures of them	2 days	3–4 months
Hot dogs and lunch meats		
Hot dogs, opened package	1 week	In freezer wrap, 1–2 months
Unopened package	2 weeks	
Lunch meats, opened	3–5 days	
Unopened	2 weeks	

Illustration 32.8
Cold storage: these safe time limits will help keep refrigerated food from spoiling or becoming dangerous to eat. These time limits will keep frozen food at top quality.
(Table continues on next page)

Source: U.S. Department of Agriculture, Food Safety and Inspection Service, September 1990 and 1997.

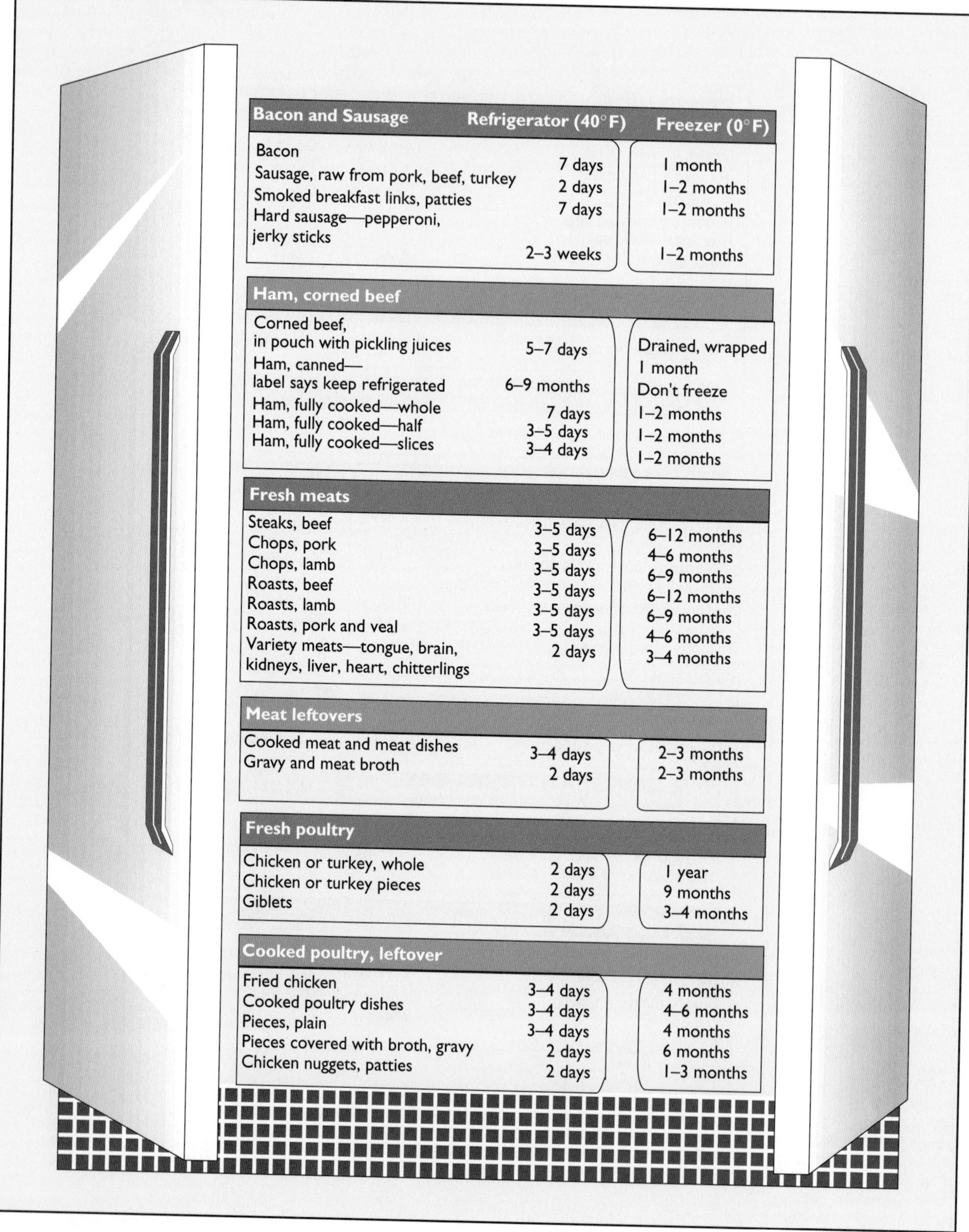

Bacon and Sausage	Refrigerator (40°F)	Freezer (0°F)
Bacon	7 days	1 month
Sausage, raw from pork, beef, turkey	2 days	1–2 months
Smoked breakfast links, patties	7 days	1–2 months
Hard sausage—pepperoni, jerky sticks	2–3 weeks	1–2 months
Ham, corned beef		
Corned beef, in pouch with pickling juices	5–7 days	Drained, wrapped 1 month
Ham, canned—label says keep refrigerated	6–9 months	Don't freeze
Ham, fully cooked—whole	7 days	1–2 months
Ham, fully cooked—half	3–5 days	1–2 months
Ham, fully cooked—slices	3–4 days	1–2 months
Fresh meats		
Steaks, beef	3–5 days	6–12 months
Chops, pork	3–5 days	4–6 months
Chops, lamb	3–5 days	6–9 months
Roasts, beef	3–5 days	6–12 months
Roasts, lamb	3–5 days	6–9 months
Roasts, pork and veal	3–5 days	4–6 months
Variety meats—tongue, brain, kidneys, liver, heart, chitterlings	2 days	3–4 months
Meat leftovers		
Cooked meat and meat dishes	3–4 days	2–3 months
Gravy and meat broth	2 days	2–3 months
Fresh poultry		
Chicken or turkey, whole	2 days	1 year
Chicken or turkey pieces	2 days	9 months
Giblets	2 days	3–4 months
Cooked poultry, leftover		
Fried chicken	3–4 days	4 months
Cooked poultry dishes	3–4 days	4–6 months
Pieces, plain	3–4 days	4 months
Pieces covered with broth, gravy	2 days	6 months
Chicken nuggets, patties	2 days	1–3 months

Illustration 32.8
Cold storage (Continued)

Raw Food	Internal Temperature
Ground Products	
Hamburger	160°F
Beef, veal, lamb, pork	160°F
Chicken, turkey	165°F
Beef, Veal, Lamb	
Roasts and steaks	
Medium-rare	145°F
Medium	160°F
Well-done	170°F
Pork	
Chops, roasts, ribs	
Medium	160°F
Well-done	170°F
Ham, fresh	160°F
Sausage, fresh	160°F
Poultry	
Chicken, whole and pieces	180°F
Duck	180°F
Turkey (unstuffed)	180°F
Whole	180°F
Breast	170°F
Dark meat	180°F
Stuffing (cooked separately)	165°F
Eggs	
Fried, poached	Yolk and white are firm
Casseroles	160°F
Sauces, custards	160°F

*Fish should be cooked until the flesh is not clear-looking and flakes easily with a fork; shrimp, lobster, and crab until shells are red and the flesh is opaque and not clear- or raw-looking inside; and clams, oysters, and mussels until the shells open.

Table 32.3
A guide for cooking foods to safe temperatures.

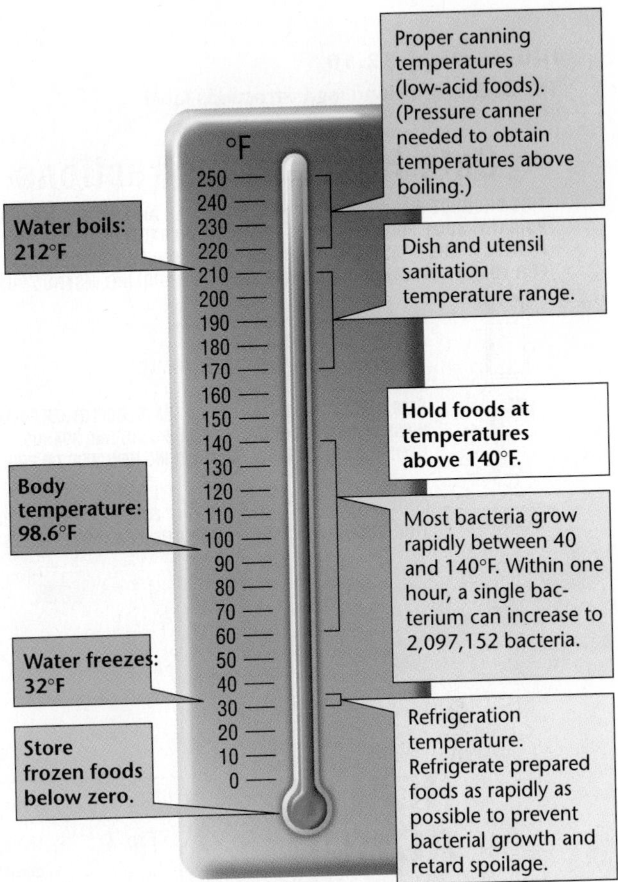

Illustration 32.9
Temperature guide for safe handling of food in the home.

Source: Adapted from Minnesota Extension Service, University of Minnesota.

cardinal sign that canned foods have spoiled is a buildup of pressure inside the can. The pressure will cause the top of the can to bulge out instead of curving in. Because a buildup of pressure inside a can may be due to bacteria that cause botulism, the food should be thrown away. These and other tips for safe food handling are listed on the USDA's Safe Handling Instructions label (Illustration 32.10) and in the Health Action.

Because so many consumers have questions about food safety, the U.S. Department of Agriculture (USDA) has established a hotline for meat and poultry. When in doubt, call the hotline during weekdays at 1-800-535-4555 from 10 A.M. to 4 P.M. (EST). If you are planning to visit a foreign country and want to avoid "traveler's diarrhea," contact the Centers for Disease Control for safety tips (http://www.cdc.gov).

Illustration 32.10
The USDA Safe Handling Instructions label.

Safe Handling Instructions

THIS PRODUCT WAS PREPARED FROM INSPECTED AND PASSED MEAT AND/OR POULTRY. SOME FOOD PRODUCTS MAY CONTAIN BACTERIA THAT CAN CAUSE ILLNESS IF THE PRODUCT IS MISHANDLED OR COOKED IMPROPERLY. FOR YOUR PROTECTION, FOLLOW THESE SAFE HANDLING INSTRUCTIONS.

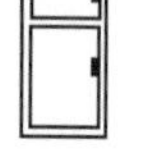

KEEP REFRIGERATED OR FROZEN. THAW IN REFRIGERATOR OR MICROWAVE.

KEEP RAW MEAT AND POULTRY SEPARATE FROM OTHER FOODS. WASH WORKING SURFACES (INCLUDING CUTTING BOARDS), UTENSILS, AND HANDS AFTER TOUCHING RAW MEAT OR POULTRY.

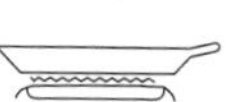

COOK THOROUGHLY.

KEEP HOT FOODS HOT. REFRIGERATE LEFTOVERS IMMEDIATELY OR DISCARD.

There Is a Limit to What the Consumer Can Do

In the end, it is up to consumers to demand a safe food supply, up to industry to produce it, up to researchers to develop better ways of doing so, and up to government to see that it happens, to make sure it works, and to identify problems still in need of resolution.

—CDC's Report on Foodborne Illnesses[4]

Keeping food safe from the farm, lake, or ocean to the table represents a major challenge to industry, government, and consumers. Until contamination of food is prevented, consumers will continue to play an important role in ensuring food's safety.

HEALTH ACTION

Tips for Safe Food Handling

Tip 1. Wash hands with soap and water before handling foods.

Tip 2. Wash hands, surfaces, and utensils with soap and water after contact with raw meat and meat juices. Avoid contact between raw and cooked foods. Sanitize cutting boards and utensils in contact with raw meats by washing them in water that contains 1 tablespoon of chlorine (i.e., Clorox) per quart.

Tip 3. Cook meats to and store foods at safe temperatures (see Table 32.3 and Illustration 32.8). Eat thoroughly cooked hamburger only.

Tip 4. Never leave perishable foods such as potato and egg salad, meats, and dairy products at room temperature for more than an hour. Keep hot foods hot (over 140°F) and cold foods cold (under 40°F).

Tip 5. Freeze or refrigerate leftovers promptly in sealed containers. Use flat containers that will allow food to cool quickly if storing large amounts of food.

Tip 6. Do not buy or eat food in cans or jars with bulging tops or loose lids.

Tip 7. Do not consume foods that are past their expiration date.

Tip 8. Follow label instructions for food storage (for example: "Refrigerate after opening").

Tip 9. Thaw foods such as turkey and roasts in the refrigerator to reduce bacterial growth.

Tip 10. When in doubt, throw it out. If a food has a strange color or odor, throw it away.

Tip 11. Wash fresh vegetables and fruits thoroughly with water before eating. Peel vegetables and fruits covered with a waxy film.

Tip 12. Avoid "traveler's diarrhea" by getting safety tips from the Centers for Disease Control before you travel (http://www.cdc.gov).

Nutrition UP CLOSE

Food Safety Detective

FOCAL POINT: Investigating paths to foodborne illness.

Assume an outbreak of foodborne illness occurred among people who ate foods from the onion and lettuce platter shown being prepared in Illustration 32.2. Review the food safety tips in the Health Action on page 32–14, and identify three potential causes of the outbreak related to the preparation of the foods as shown in the illustration:

1. ____________________

2. ____________________

3. ____________________

FEEDBACK (answers to these questions) can be found at the end of Unit 32.

WWW links

www.fightbac.org
Educational materials, workbooks, posters about food safety, and "mug shots" of the 10 least wanted foodborne pathogens are available from this site.

www.foodsafety.gov
The national gateway to food safety information, resources, outbreak alerts, and consumer advice. Or call 1-888-SAFEFOOD.

www.cdc.gov
Search foodborne illnesses, food safety, or international travel, and find a plethora of information.

vm.cfsan.fda.gov
The FDA's Center for Food Safety and Applied Nutrition offers press releases, fact sheets, food safety warnings, and consumer advice.

www.epa.gov/ost/fish
The place to go for local fish advisories.

www.cdc.gov/epo/mmwr
Sends you to the Centers for Disease Control's Morbidity and Mortality Monthly Report. Includes updates on foodborne illnesses.

bluehen.ags.udel.edu/deces/fnf/fnf-list.htm
The Delaware Cooperative Extension Service provides consumer advice on safe food preparation, preservation, and storage techniques.

REFERENCES

1. CDC issues new 5-year report of foodborne illness outbreaks. Food Safety Educator 1997;2:1–6.
2. Anderson ER et al. Diagnosis and management of foodborne illness: a primer for physicians. MMWR, 2001;Jan 26.
3. www.fightbac.org, accessed 1/27/01.
4. Foodborne illnesses. www.cdc.gov, accessed 1/26/01.
5. Zhao P et al. Development of a model for evaluation of microbial cross-contamination in the kitchen. J Food Protection 1998;61:960–3.
6. Wegener HC. The consequences for food safety of the use of fluoroquinolones in food animals. N Engl J Med 1999;340:1581–2.
7. Blaser MJ. How safe is our food? N Engl J Med 1996;334:1324–5.
8. Position of the American Dietetic Association: food and water safety. J Am Diet Assoc 1997;97:184–9.
9. Dietary Guideline for Americans. U.S. Dept. of HHS and USDA, 2000.
10. Food safety: a challenge for everyone in public health. Webcast from the University of North Carolina, 1/26/01.
11. Mead PS et al. Food-related illness in the United States. Emerging Infect Dis 1999;5:607–25.
12. Hennessy TW et al. A national outbreak of *salmonella enteritidis* infection from ice cream. N Engl J Med 1996;334:1281–6.
13. Welland D. *E. coli*. . . . *salmonella* . . . *Listeria* . . . unwelcome house guests. Envir Nutr 1997;20(Oct):1, 6.
14. Mahaffey KR. Methylmercury: a new look at the risks. Public Health Rept 1999;114:396–415.
15. EPA national advice on mercury in freshwater fish for women who are or may become pregnant, nursing mothers, and young children. www.epd.gov, 1/2001.
16. Treatment of ciguatera poisoning with gabapentin. N Engl J Med 2001;344: 692–3.
17. Tighter reins sought for imported produce. Envir Nutr 1997;20(Nov):1, 7.
18. Lewis R. Prion's changeability. The New Scientist 1999;13:1,8.
19. Bovine spongiform encephalopathy: "mad cow disease." Nutr Rev 1996;54:208–10.
20. Tragedy of variant Creutzfeldt-Jakob disease. The Lancet 1999;353:939.
21. Madigan D. After Delaney: how will the new pesticide law work? CNI Weekly Report 1996;Oct 11:4–5.
22. Shea KM. Technical report: irradiation of food. Pediatrics 2000;106:1505–9.
23. Henneman A. Don't mess with food safety myths. Nutr Today 1999;34:23–8.

Nutrition

UP CLOSE

Food Safety Detective

Feedback for Unit 32

Potential causes of the foodborne illness outbreak:

1. Disease spread from hands to foods.
2. Use of contaminated cutting board, knife, or food platter.
3. Cross-contamination between onions and lettuce or between other foods stored together.

Unit 33

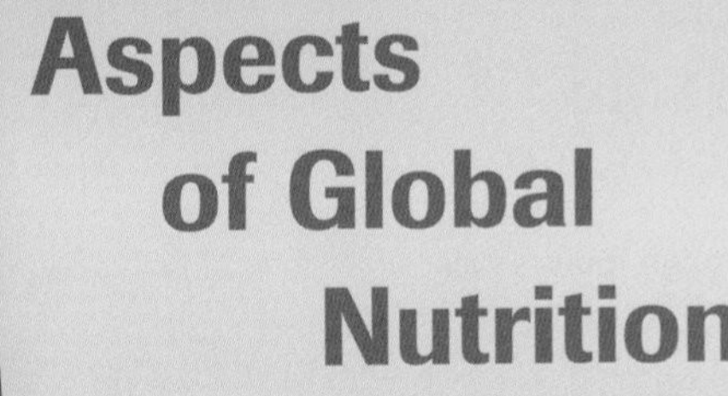

Aspects of Global Nutrition

Photo Disc

Nutrition Scoreboard

	TRUE	FALSE
1. Approximately 30% of the world's population does not have access to a safe supply of water.		
2. At most, 25% of deaths of young children in developing countries are related to malnutrition and infection.		
3. The main cause of famine in developing countries is that too little food is available.		

(Answers on next page)

Answers to

Nutrition Scoreboard

	TRUE	FALSE
1. Contaminated water supplies are a major source of infection in many developing countries.	✓	
2. Nearly 50% of deaths of children in developing countries are associated with malnutrition and infection.[1]		✓
3. Poverty is the primary cause of famine. There are many other important causes as well.		✓

Key Concepts and Facts

- ➔ The world produces enough food for all its people.
- ➔ Poverty, civil war, natural disasters, overpopulation, low rates of breastfeeding, unsafe water supplies, and discrimination against females all contribute to malnutrition.
- ➔ Malnutrition early in life has long-term effects on mental and physical development.
- ➔ Rates of diseases such as heart disease, diabetes, and hypertension increase in developing countries that adopt Western lifestyles and eating habits.

State of the World's Health

The global community consists of over 100 countries grouped into the categories of "industrialized nations," "developing nations," and "least developed nations." (Countries within each category are listed in Table 33.1.) Only 31 countries are considered industrialized, most are developing (113), and, of the developing countries, 49 are least developed.[1] All countries in the global community have interests and issues in common, such as the adequacy of the food supply, availability of health care and education, and safe water. But the countries of the world are also dissimilar in many ways. Differences in financial resources, population growth, and political and other systems translate into large disparities in health status and leading causes of death (Table 33.2). People in the least developed and developing countries are more likely to have substantially shorter life expectancies, to die from infectious rather than chronic diseases, and to experience malnutrition than individuals living in industrialized countries.[2]

On the Side

If you represented the world's population with a group of 100 people, 70 of the people would be nonwhite, 57 would be Asian, 21 European, 14 North and South American, and 8 African. One person would have a college education, 80 would live in poor housing, and 50 would be malnourished.

Photo Disc

The general state of health of populations in various countries is monitored by tracking key environmental, health, and behavioral characteristics. Pregnancy outcome and child growth, rates of breastfeeding, and access to safe drinking water, for example, are key indicators of the health status of a population.[1] As you can see in Illustration 33.1 and Table 33.3 (p. 33–5), rates of these key indicators vary substantially worldwide. When percentages of low birthweight in newborns and underweight in young children are high, and rates of breastfeeding and access to a safe water supply are low, one can rightfully assume that malnutrition, infection, and other health problems are common. When these key indicators show improvement, the health status of the population and longevity improve.

Table 33.1
Countries classified by the United Nations International Children's Emergency Fund (UNICEF) as industrialized, developing, and least developed, 1998.*

Industrialized Countries	Developing Countries	Least Developed Countries
Andorra; Australia; Austria; Belgium; Canada; Denmark; Finland; France; Germany; Greece; Holy See; Iceland; Ireland; Israel; Italy; Japan; Liechtenstein; Luxembourg; Malta; Monaco; Netherlands; New Zealand; Norway; Portugal; San Marino; Slovenia; Spain; Sweden; Switzerland; United Kingdom; United States.	Afghanistan; Algeria; Angola; Antigua and Barbuda; Argentina; Armenia; Azerbaijan; Bahamas; Bahrain; Bangladesh; Barbados; Belize; Benin; Bhutan; Bolivia; Botswana; Brazil; Brunei Darussalam; Burkina Faso; Burundi; Cambodia; Cameroon; Cape Verde; Central African Rep.; Chad; Chile; China; Colombia; Comoros; Congo; Congo, Dem. Rep.; Cook Islands; Costa Rica; Côte d'Ivoire; Cuba; Cyprus; Djibouti; Dominica; Dominican Rep.; Ecuador; Egypt; El Salvador; Equatorial Guinea; Eritrea; Ethiopia; Fiji; Gabon; Gambia; Georgia; Ghana; Grenada; Guatemala; Guinea; Guinea-Bissau; Guyana; Haiti; Honduras; India; Indonesia; Iran; Iraq; Jamaica; Jordan; Kazakstan; Kenya; Kiribati; Korea, Dem. People's Rep.; Korea, Rep. of; Kuwait; Kyrgyzstan; Lao People's Dem. Rep.; Lebanon; Lesotho; Liberia; Libya; Madagascar; Malawi; Malaysia; Maldives; Mali; Marshall Islands; Mauritania; Mauritius; Mexico; Micronesia, Fed. States of; Mongolia; Morocco; Mozambique; Myanmar; Namibia; Nauru; Nepal; Nicaragua; Niger; Nigeria; Niue; Oman; Pakistan; Palau; Panama; Papua New Guinea; Paraguay; Peru; Philippines; Qatar; Rwanda; Saint Kitts and Nevis; Saint Lucia; Saint Vincent/Grenadines; Samoa; Sao Tome and Principe; Saudi Arabia; Senegal; Seychelles; Sierra Leone; Singapore; Solomon Islands; Somalia; South Africa; Sri Lanka; Sudan; Suriname; Swaziland; Syria; Tajikistan; Tanzania; Thailand; Togo; Tonga; Trinidad and Tobago; Tunisia; Turkey; Turkmenistan; Tuvalu; Uganda; United Arab Emirates; Uruguay; Uzbekistan; Vanuatu; Venezuela; Viet Nam; Yemen; Zambia; Zimbabwe.	Afghanistan; Angola; Bangladesh; Benin; Bhutan; Burkina Faso; Burundi; Cambodia; Cape Verde; Central African Rep.; Chad; Comoros; Congo, Dem. Rep.; Djibouti; Equatorial Guinea; Eritrea; Ethiopia; Gambia; Guinea; Guinea-Bissau; Haiti; Kiribati; Lao People's Dem. Rep.; Lesotho; Liberia; Madagascar; Malawi; Maldives; Mali; Mauritania; Mozambique; Myanmar; Nepal; Niger; Rwanda; Samoa; Sao Tome and Principe; Sierra Leone; Solomon Islands; Somalia; Sudan; Tanzania; Togo; Tuvalu; Uganda; Vanuatu; Yemen; Zambia.

Photo Disc

Photo Disc

*Data for some eastern European countries, former Soviet republics, and the Baltic states are not available. Least developed countries represent a subgroup of the developing countries.

The health status of populations in developing countries is monitored by the World Health Organization (WHO) and organizations such as the United Nations International Children's Emergency Fund (UNICEF). Recent reports, such as the 1998 WHO World Health Fact Sheet highlighted in Table 33.4 (p. 33–6), describe improvements in the health of the world's most vulnerable

Table 33.2
Leading causes of death worldwide compared to U.S. causes.[2]

Cause of Death	Worldwide Deaths (%)	U.S. Deaths (%)
1. Infectious and parasitic diseases	33	5
Lower respiratory infection	21	
Tuberculosis	17	
Diarrhea	14	
HIV/AIDS	13	
Malaria	9–16	
2. Cardiovascular diseases	29	32
Heart disease	47	
Stroke	30	
Other	17	
3. Cancer	12	23
Lung	18	
Stomach	12	
Colon/rectum	8	
Liver	8	
Breast	6	
Other	48	
4. Health problems originating in pregnancy and the first month of life	7	1
5. Respiratory disease	6	5
6. Other	13	34

people and identify lingering problems. Important progress is being made, and rates of malnutrition, poor growth, and infectious disease are generally decreasing worldwide. Much progress remains to be made, however.

Food and Nutrition: The Global Challenge

He who is healthy has hope; and he who has hope has everything.
—Arab proverb

We have all heard about starvation in Somalia, Ethiopia, Sudan, and Bangladesh. The pictures of starving children with desperation in their big eyes and heads too large for their frail bodies burn an image in our minds. What we don't hear about in the news, however, is the extent of starvation and malnutrition in the world. These are not just problems of isolated areas experiencing civil war or crop failures. In the developing regions, malnutrition is an ongoing problem for 10 to 51% of children under the age of five, depending on the region (Illustration 33.1). Approximately 50% of women worldwide are iron deficient, and millions suffer from iodine deficiency. Vitamin A deficiency exists among millions of children. These children are at risk of frequent and severe infectious disease, decreased growth, and blindness because of the deficiency.[1,4]

The children we usually see in pictures from famine areas are victims of ***marasmus,*** a disease caused primarily by a lack of calories and protein. These children look starved (Illustration 33.2) and consume far fewer calories and far less protein and other essential nutrients than they need. In victims of marasmus, the body uses its own muscle and other tissues as an energy source.

marasmus
Malnutrition caused by a lack of calories and protein. Also called *protein-calorie malnutrition.*

kwashiorkor
Malnutrition resulting from inadequate protein intake in children.

Sometimes malnutrition is due to an inadequate protein intake rather than a deficiency of calories. Children who lack protein develop ***kwashiorkor.*** These children may appear fat due to massive swelling that occurs with the disease

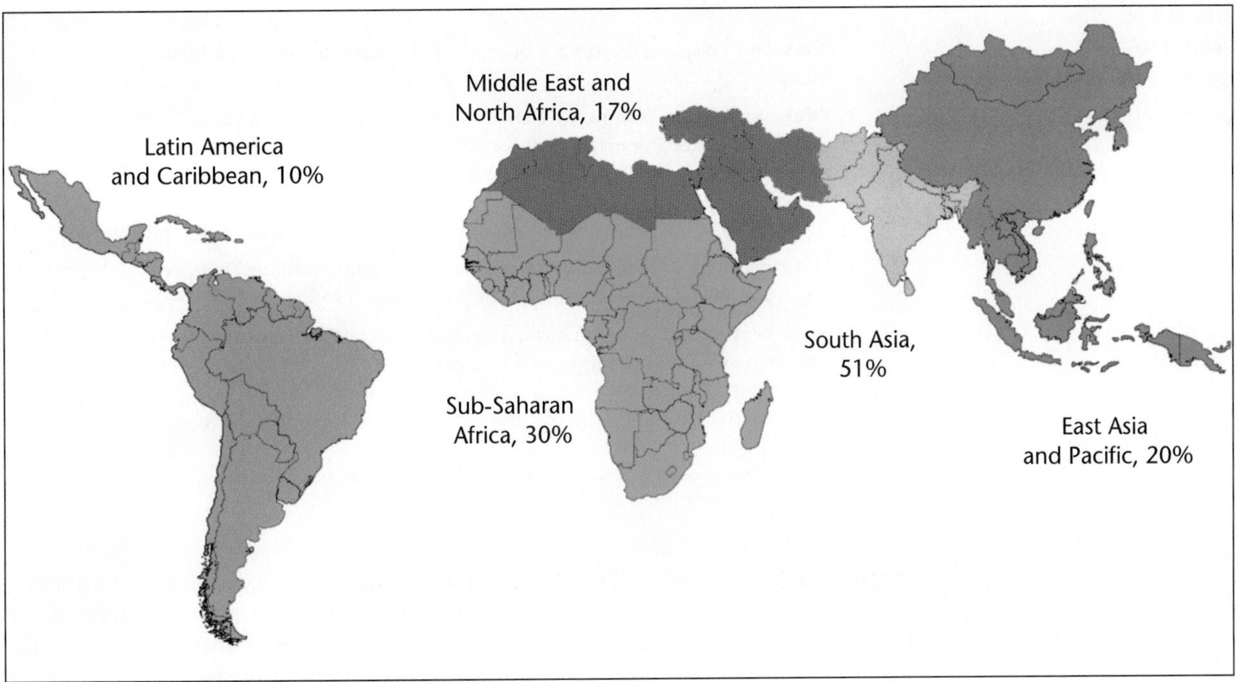

Illustration 33.1
Underweight children are an indicator of malnutrition. This map shows the percentage of children who were underweight in developing regions of the world in 1997.[3]

(Illustration 33.3). Most victims of kwashiorkor are somewhat marasmic as well. Children with either type of severe malnutrition are at high risk of dying from infectious diseases because their bodies cannot mount an adequate defense against them.

Survivors of Malnutrition

Malnutrition that occurs before a child's brain has completely grown has lasting effects on development.[4] Brain growth occurs during pregnancy through the age of two years. If children suffer severe malnutrition during these years, they will

Table 33.3
Percentage of low-birthweight infants, exclusive breastfeeding, and access to safe water in different areas of the world.[1]

Region[a]	Low-Birthweight Infants (%)	Exclusive Breastfeeding, 0–3 Months (%)	Access to Safe Water (%)
Sub-Saharan Africa	16	32	49
Middle East and North Africa	11	43	81
South Asia	33	46	80
East Asia and Pacific	11	56	67
Latin America and Caribbean	10	38	77
Developing countries	18	44	70
Least developed countries	23	46	54
Industrialized countries[b]	6	—	100
World	17	44	71

[a]Data for the former Soviet republics and the Baltic states are not available.
[b]Data for exclusive breastfeeding are not available; the figure is approximately 50%.

Table 33.4
Key global health facts and trends from the World Health Organization, 1998.*

1. The world's population was 5.8 billion in 1997. It will increase to 8 billion by the year 2025.
2. Of the world's population, 11% are children under the age of 5 years.
   - ➡Children under 5 years account for 21% of all deaths.
   - ➡Deaths among children under 5 years have decreased by more than half since 1955.
   - ➡Over 50% of the deaths of children under 5 years are associated with malnutrition.
3. Life expectancy at birth in 1995 was 65 years; in 2025 it will be 73 years.
   - ➡In 16 countries, life expectancy declined between 1955 and 1995.
4. Rates of chronic disease are increasing in many developing countries, largely due to adoption of Western lifestyles and dietary habits.
   - ➡Diabetes will more than double between 1997 and 2025.
   - ➡Cancer and cardiovascular diseases will increase during the same period.

*Access these and other health facts from WHO through www.who.org/whr/1998/ factse.htm.

experience permanent delays in mental development even if refeeding improves their physical growth and health. The severity of the mental delays depends on the timing and duration of the malnutrition. The longer malnutrition exists, the harder it becomes for young children to achieve a brighter future. In addition, the state of hunger and starvation leads people to become self-centered and to lose any sense of well-being. The need for food for survival may prompt unethical behaviors such as stealing and injuring others to obtain food. The devastating psychological effects of starvation may follow children into adulthood and have a lasting effect on behavior throughout life. Persistent malnutrition saps the physical and mental energy of people and, ultimately, compromises the economic and social progress of nations.[2]

Malnutrition and Infection There is a very close relationship between malnutrition and infection: malnutrition weakens the immune system and increases the likelihood and severity of infection. Simultaneously, repeated infection and the bouts of diarrhea that often accompany it can produce undernutrition.[5] Poor sanitary conditions, contaminated water supplies, and lack of refrigeration contribute to the spread of infectious diseases and thus to the malnutrition the diseases promote. The majority of deaths of children under the age of five years in developing countries are related to the presence of both malnutrition and infection.[5]

Low rates of breastfeeding promote the spread of infection, especially in countries where contaminated water is used to prepare formula. In addition to providing optimal nutrition, breast milk contains a number of substances that protect babies from infection.

Vitamin A deficiency in children also plays an important role in the spread and severity of infections. Children who are deficient in vitamin A are much more likely to die from a measles infection, for example, than are children whose vitamin A status is sufficient. HIV infection progresses more rapidly into AIDS in vitamin A–deficient and malnourished individuals.[6] Over 36 million people in the world have HIV infection and the incidence is increasing.[3]

Illustration 33.2
The look of marasmus. Children with marasmus look like "skin and bones." They don't get enough calories and protein in their diet. This child is suffering from the extreme emaciation of marasmus.

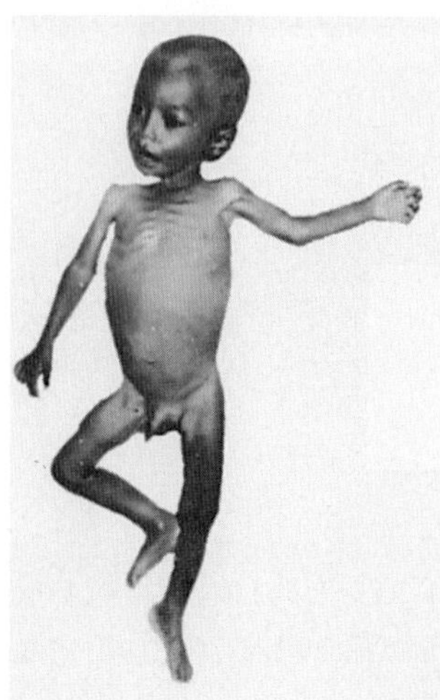

Why Do Starvation and Malnutrition Happen?

> *It is very much an oversimplification to assert, without qualification, that people starve because they do not have enough to eat.*
> —M. Pyke, 1972

Malnutrition and starvation don't have to happen anywhere. The world's agricultural systems have the capacity to produce enough food to feed everyone on earth. People become malnourished or starve primarily because of poverty. Human-made disasters, including civil war, the use of food as a weapon against the enemy, and corrupt governments that buy favors with food aid and ignore the social and economic needs of their people, also contribute to malnutrition. In some instances, starvation and malnutrition are due to natural disasters that lead to crop failures or the inability to distribute food to those in need.[3]

The famine in Somalia that led to the death of over 1.5 million people was due to the collapse of the government, fighting among rival clans, and general lawlessness. Farmers and their families were driven off their land by bandits who stole their crops and possessions. Many had nowhere to go for help. Mass starvation in Ethiopia and Sudan was similarly initiated by violent conflict within the country. Seizing the opposing side's food and preventing them from obtaining food aid are among the primary weapons used to fight these civil wars. In Bangladesh where poor people have no choice but to live on floodplains, cyclones wipe out crops and regularly result in the death of thousands due to floods, starvation, and infectious disease. None of these cases of mass famine was due primarily to the ravages of nature. Starvation and malnutrition are largely initiated by humans, and it is within the power of humans to end them.

Women and female children are at particular risk for malnutrition in some societies because cultural practices call for food to be allocated to men and boys first (Illustration 33.4). If too little food is available, what remains after meals for women and daughters may be too little to support health and growth.[7] In many developing countries, discrimination against women in education and employment, sanctions against the use of birth control, and violence toward women place women at high risk of developing malnutrition and having a low quality of life.[7]

The leading causes of malnutrition in developing and least developed countries are summarized in Table 33.5.

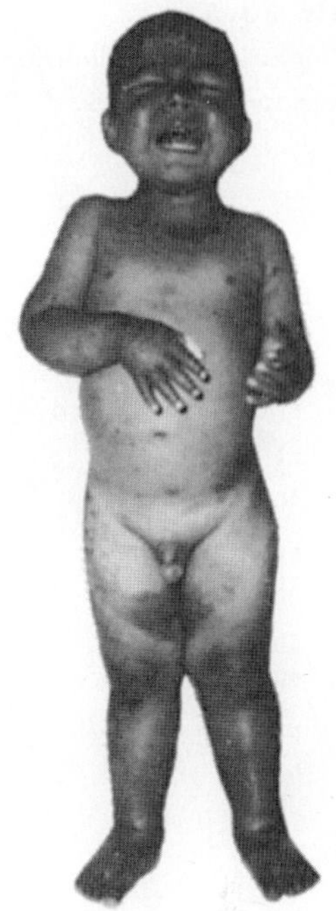

Illustration 33.3
The look of kwashiorkor. This child has the characteristic "moon face" (edema), swollen belly, and patchy dermatitis often seen with kwashiorkor. Children with kwashiorkor may not look starved because of the massive swelling. Kwashiorkor results from protein deficiency.

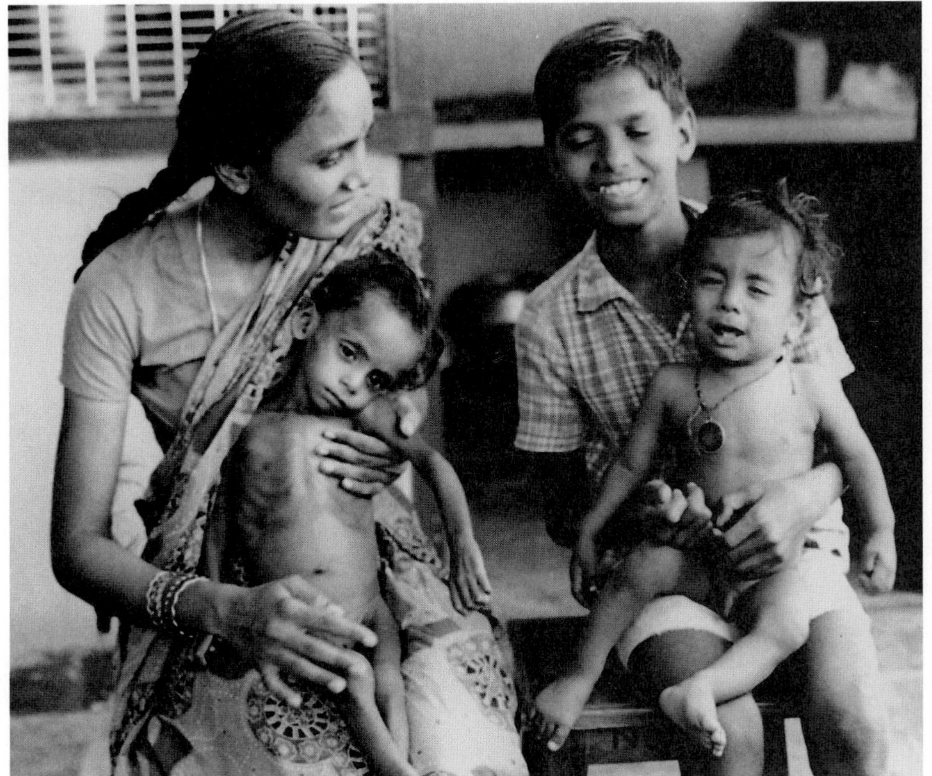

Illustration 33.4
Women and children in some developing countries are particularly vulnerable to malnutrition because they, and their nutritional needs, may be considered less important than men and their needs. Similarly, boys' needs may take precedence over girls' needs. These children are two-year-old twins: the child on the left is a girl, and the child on the right is a boy.

Table 33.5
Leading causes of malnutrition in developing countries.[3,8]

- Poverty
- Civil war, corrupt governments
- Poor sanitation and infection
- Natural disasters
- Discrimination against women
- Overpopulation, high birthrates
- Poor education levels

Ending Malnutrition

Adequate food is the cradle of normal resistance, the playground of normal immunity, the workshop of good health, and the laboratory of long life.

—Charles Mayo

The long-term solution to malnutrition will depend on the ability of humans to work together to achieve educational and economic development, peace, population growth control, improved sanitation, social equity for women and children, and environmentally sound and productive agricultural policies and practices in developing countries (Illustration 33.5). "Bottom-up" approaches to problem solving work better than "top-down" approaches. The people most affected by malnutrition must be active participants in the planning and implementation of improvement programs. Efforts to improve the nutritional status of populations can and have paid off.

Success Stories Examples of strikingly successful efforts to reduce malnutrition in various countries around the world are available for telling. A repeated theme is the reduction in malnutrition as economic, educational, nutritional, and sanitary conditions improve. Improvements in these same areas were responsible for the dramatic declines in malnutrition and infectious diseases and the increases in life expectancy experienced in the United States, Canada, and many European countries around 100 years ago.[8] In other instances, specific nutrient deficiencies have been greatly reduced or eliminated by food fortification and nutrition education programs. Here are several specific examples:

- Vitamin A supplements and education on vitamin A–rich foods were related to a dramatic decrease in cases of severe and moderate vitamin A deficiency and infection in children in Indonesia.
- Food supplements given to groups of infants in Russia, Brazil, South Africa, and China were associated with higher IQ scores at age eight.

Photo Researchers - Porterfield/Chickering

Illustration 33.5
Safe water supplies represent a major advance in quality of life for people in developing countries.

- Iodization of salt has eliminated iodine deficiency in Bolivia and Ecuador. Worldwide, iodization of salt is associated with a drop in the number of iodine-deficient children from 48 million in 1990 to 28 million in 1997.
- Fortification of flour with iron has led to a decrease in iron-deficiency anemia in the Philippines.[9]

UNICEF

Illustration 33.6
After a rooming-in policy that enabled mothers to breastfeed their infants on demand was introduced in one hospital in the Philippines, the incidence of early infant deaths due to infection dropped by 95%.

Source: First Call for Children, A UNICEF Quarterly 1992;(January/March):1.

In Thailand, a country with a serious iodine-deficiency problem, a highly creative approach to increasing iodine intake was implemented by the king. To celebrate his birthday, salt producers gave the king 10,000 tons of iodized salt. The king had the salt packaged in small plastic bags and, with the help of the Red Cross and the army, delivered a bag to every household in Thailand. A message from the king about the importance of using iodized salt was attached to each bag.[10]

Deaths of young children from malnutrition and its related diseases have dropped substantially in countries experiencing a resurgence of breastfeeding (Illustration 33.6). Breast milk protects infants and young children from a variety of infectious diseases, supports the growth and health of infants, and protects them from the hazards of formulas reconstituted with contaminated water. Worldwide health initiatives have led to the development of the International Code of Marketing Breast Milk Substitutes and to "Baby Friendly Hospitals." The international marketing code calls for the prohibition of both free formula samples and the promotion of infant formulas by health care professionals. More than 12,700 hospitals have adopted Baby Friendly Hospital policies that promote and facilitate breastfeeding.[3]

Health status of people in developing countries has also been improved by the widespread use of oral rehydration fluids that protect children with diarrhea from dehydration. Broadly based vaccination programs have protected many from various diseases.[3]

The World Food Summit The 1996 World Food Summit ended with a pledge that countries would reduce by half the number of people in the world who are hungry and undernourished by 2015. Progress is being made toward the goal (Illustration 33.7), but meeting it will take sustained efforts on the part of both developed and developing countries.[2]

Countries in "Nutrition Transition" As countries develop economically and food supplies increase, the incidence of undernutrition usually decreases, and chronic health problems such as heart disease, diabetes, and hypertension tend to increase. Elevated rates of chronic disease appear to be closely related to high-fat and high-meat diets, low levels of physical activity, and other characteristics of Western lifestyles and dietary habits.[12]

Countries undergoing the nutrition transition often experience rising rates of obesity in children.[13] Children born to poorly nourished women may be biologically programmed while in the womb to conserve energy. When exposed to a generous food supply later in life, they may gain fat more readily than height.[14] Excess accumulation of fat later in life may place the children at increased risk for hypertension, diabetes, heart disease, and a number of other "diseases of civilization."[15] Although infections such as malaria and tuberculosis will predominate

On the Side

Approximately 14% of the U.S. population lives in poverty, but there are sizable differences in poverty rates by age and racial group. Twenty percent of children, 30% of Hispanics, 29% of African Americans, and 11% of whites live in poverty.[11]

Photo Disc

Illustration 33.7
Estimates of the number of undernourished in developing countries and progress toward the World Food Summit target of 50% reduction by 2015.

Source: Food and Agriculture Organization, United Nations, 2000.

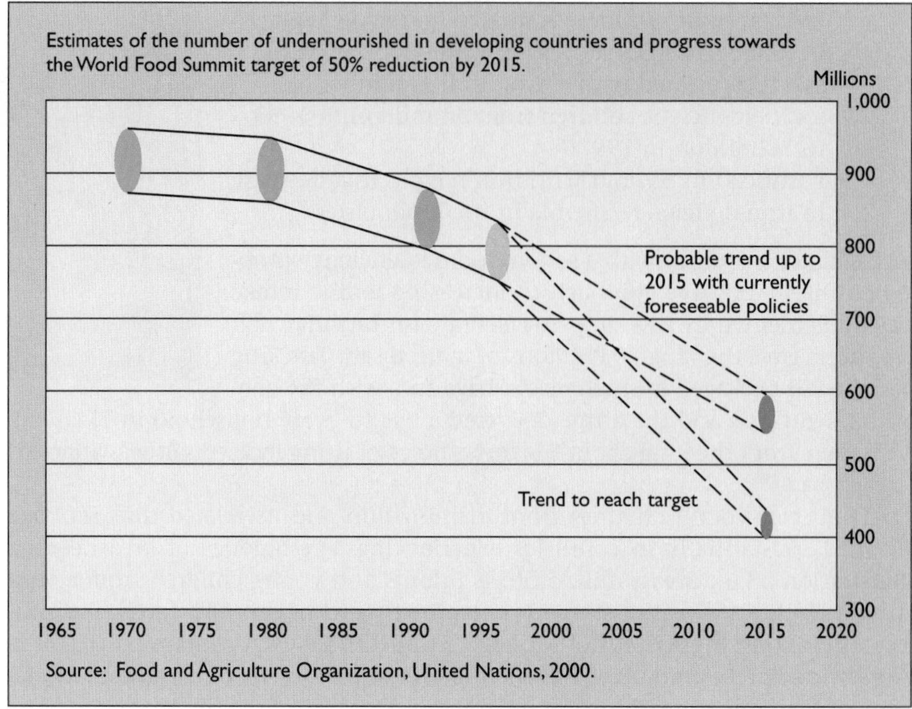

through the year 2025, the incidence of diseases of "Western civilization" will continue to expand.[3]

Problems and Programs in the United States

Some people in every country experience hunger and malnutrition for a variety of reasons, including poverty. In the United States, approximately 12 million households are "food insecure."[16,17] Food and nutrition assistance programs, such as Food Stamps, School Breakfast and Lunch, and the WIC Program (Special Supplemental Nutrition Program for Women, Infants, and Children), are available in the United States to persons with low incomes. (Income eligibility standards for these programs are shown in Table 33.6.) These and other programs, for example, the Nutrition Program for the Elderly, the Disaster Feeding Program, soup kitchens (Illustration 33.8), and emergency food shelves, help prevent problems associated with food insecurity and malnutrition. Some developing countries have infant feeding and school lunch programs and distribute vitamin

Table 33.6
Federal income eligibility standards for food and nutrition programs such as WIC and Food Stamps in 2000–01. In some states, families whose income is 100% of the poverty level or lower are eligible for assistance; in other states, families whose income is 185% of the poverty level or lower are eligible.

Family Size	Family Income Level*	
	100% of Poverty Level	185% of Poverty Level
1	$10,062	$18,615
2	13,468	24,916
3	20,280	31,217
4	23,686	37,518
Each additional member	+3,406	+6,301

*The poverty guidelines for Alaska and Hawaii are higher. Income eligibility guidelines are announced each summer by the U.S. Department of Agriculture.
Source: U.S. Department of Agriculture, 2000–01.

PhotoEdit - Michael Newman

Illustration 33.8
A visit to a soup kitchen.

and mineral supplements to pregnant women and children. Most, however, cannot afford such programs. The U.S. government has recently begun to support international school lunches through donations of corn, wheat, rice, powdered milk, and other surplus commodities to nations with hungry children.

If you would like to help lessen the burden of food insecurity, hunger, or malnutrition, consider volunteering at a food or meals program, in community courses for language or self-sufficiency skills, or in other programs. Many people learn a lot about what it is like to be poor and without food. The experience will likely enrich your life as well as those you serve.

WWW links

www.who.org
Leads to causes of death in industrialized and developing countries, information about WHO programs, and much more.

www.nutrition.gov
Provides updates on government-sponsored nutrition programs.

www.fao.org
The United Nations' Food and Agriculture Organization address. Provides updates on worldwide malnutrition and meeting the World Food Summit's goal of reducing hunger by half by 2015.

www.travel.state.gov
Traveling abroad? This site offers travel alerts and public announcements.

www.cdc.gov
Check out international foodborne illness and other infectious disease warnings here.

www.wfp.org
The address for the international World Food Programme.

Nutrition UP CLOSE

Ethnic Foods Treasure Hunt

FOCAL POINT: Values placed on foods are culturally specific.

The following questions deal with ethnic food preferences. Answer any *three* of the questions. Resources that will help you find the answers include knowledgeable individuals, the Internet, travel and anthropology books, and encyclopedias.

1. Name three foods commonly served at celebrations by the Igbo in Nigeria (Igbos were formerly referred to as Biafrans):
a. ______________
b. ______________
c. ______________

2. List two special foods given to Hmong women during pregnancy:
a. ______________
b. ______________

3. List the five foods that make up the traditional Costa Rican breakfast:
a. ______________
b. ______________
c. ______________
d. ______________
e. ______________

4. List three foods commonly sold at soccer matches in Italy:
a. ______________
b. ______________
c. ______________

5. Name two traditional foods of the Zulu in South Africa:
a. ______________
b. ______________

6. Name one of the staple foods of people in Nentsi (it's near Siberia):
____

7. List two foods considered to be "yin" and two considered to be "yang" in Chinese culture:

Yin	Yang
a. ______________	a. ______________
b. ______________	b. ______________

FEEDBACK (answers to these questions) can be found at the end of Unit 33.

REFERENCES

1. Darton-Hill I, Coyne ET. Feast and famine: socioeconomic disparities in global nutrition and health. Public Health Nutr 1998;1:23–31.
2. The state of food insecurity in the world in 2000. FAO, United Nations Report, Nov. 2000, www.fao.org, accessed 2/01.
3. United Nations International Children's Emergency Fund. The state of the world's children, 1998. New York: United Nations; 1998.
4. Parvanta I. The hidden hunger: micronutrient malnutrition. CDC Report, 2000, www.cdc.gov.
5. Schroeder DG, Martorell R. Enhancing child survival by preventing malnutrition. Am J Clin Nutr 1997;65:1080–1.
6. Global HIV infections exceed predictions by 50%. WHO estimates reported by Reuters Medical news, 2000 Dec 5.
7. Patten-Hitt E. UN calls gender discrimination a global health issue. www.womenshealth.medscape.com, Sept. 26, 2000, accessed 11/00.
8. Gold MR. Presentation at the Tenth Chronic Disease Conference, Washington, DC, 1996.
9. Decimosexta Conferencia Internacional de Nutrición, Montreal. Nutr View 1997; fall:1–36.
10. Food fortification to end micronutrient malnutrition: state of the art. Nutr View 1997; special issue:1–8.
11. Health, United States, 2000. www.cdc.gov/nchs, accessed 8/00.
12. Popkin BM. The nutrition transition and its health implications in lower-income countries. Public Health Nutr 1998;1:5–21.
13. De Onis M et al. Prevalence and trends of overweight among preschool children in developing countries. Am J Clin Nutr 2000;72:1032–9.
14. Popkin BM. Stunting is associated with overweight in children of four nations that are undergoing the nutrition transition. J Nutr 1996;126:3009–16.
15. Mi J et al. Effects of infant birthweight and maternal body mass index in pregnancy on components of the insulin resistance syndrome in China. Ann Int Med 2000;132:253–60.
16. Household food security in the United States. Alexandria (VA): USDA Food and Consumer Service; 1997.
17. Kendall A, et al. Position of the American Dietetic Association: domestic food and nutrition security. J Am Diet Assoc 1998;98:337–42.

Nutrition

UP CLOSE

Ethnic Foods Treasure Hunt

Feedback for Unit 33

1. a. yams; b. kola nuts; c. breadfruit with corn and yams; d. cassava.
2. a. chicken; b. rice; c. ripe mango; d. grapes; e. ginger; f. milk; g. many more.
3. "Gallo pinto" includes the following: a. rice and black beans; b. eggs; c. tortillas; d. coffee; e. fruit; f. sometimes steak.
4. a. Italian ices; b. calzones; c. Italian sodas; d. pizza; e. pasta; f. hot dogs.
5. a. beef; b. cassava; c. milk; d. yams; e. pumpkin; f. millet; g. corn; h. dried beans; i. beer; j. honey; k. yogurt; l. cow's blood.
6. a. reindeer; b. polar bear; c. sour milk; d. caribou; e. tea.
7. Yin: a. raw kelp (seaweed); b. boiled rice; c. raw vegetables; d. fruit; e. dairy products.
   Yang: a. cooked kelp; b. cooked vegetables; c. fried rice; d. fish; e. eggs; f. chicken.

A

APPENDIX A

Table of Food Composition

Index to Appendix A

A

This edition of the table of food composition contains more complete values for several nutrients than any comparable table.[1] These include dietary fiber; saturated, monounsaturated, and polyunsaturated fat; vitamin B_6; vitamin E; folate; magnesium; and zinc. The table includes a wide variety of foods from all food groups and is updated yearly to reflect current food patterns. For example, this edition includes many new nonfat items; several new ethnic items such as adzuki beans, Tahitian taro, and gai choy Chinese mustard; and a new selection of vegetarian foods.

• ***Sources of Data*** • To achieve a complete and reliable listing of nutrients for all the foods, over 1200 sources of information are researched. Government sources are the primary base for all data for most foods. In addition to USDA data (from Release 12 and surveys), provisional USDA information—both published and unpublished—is included.

Even with all the government sources available, however, some nutrient values are still missing; and as the USDA updates various data, it sometimes reports conflicting values for the same items. To fill in the missing values and resolve discrepancies, other reliable sources of information are used. These sources include journal articles, food composition tables from Canada and England, information from other nutrient data banks and publications, unpublished scientific data, and manufacturers' data.

The data for brand foods are listed as provided by the food manufacturers and the food chain restaurants. This information changes often because recipes and formulations are modified to meet consumer preferences, and the data are usually limited to those nutrients required for food labels. To provide more complete information, values for several nutrients are often estimated based on known values for major ingredients.

• ***Accuracy*** • The energy and nutrients in recipes and combination foods vary widely, depending on the ingredients. The amounts of various fatty acids and cholesterol are influenced by the type of fat used (the specific type of oil, vegetable shortening, butter, margarine, etc.).

Estimates of nutrient amounts for foods and nutrients include all possible adjustments in the interest of accuracy. When multiple values are reported for a nutrient, the numbers are averaged and weighted with consideration of the original number of analyses in the separate sources. Whenever water percentages are available, estimates of nutrient amounts are adjusted for water content. When no water is given, water percentage is assumed to be that shown in the table. Whenever a reported weight appeared inconsistent, many kitchen tests were made, and the average weight of the typical product was given as tested.

When estimates of nutrient amounts in cooked foods are derived from reported amounts in raw foods, published retention factors are applied. Data for combination foods are modified to include newer data for major ingredients.

Considerable effort has been made to report the most accurate data available. The table is revised annually, and the author welcomes any suggestions or comments for future editions.

• ***Average Values*** • It is important to know that many different nutrient values can be reported for foods, even by reliable sources. Many factors influence the amounts of nutrients in foods, including the mineral content of the soil, the method of processing, genetics, the diet of the animal or the fertilizer of the plant, the season of the year, methods of analysis, the difference in moisture content of the samples analyzed, the length and method of storage, and methods of cooking the food. The mineral content of water also varies according to the source.

Although each nutrient is presented as a single number, each number is actually an average of a range of data. More detailed reports from the USDA, for example, indicate the number of samples and the standard deviation of the data. One can also find different reported values for foods as older data are replaced with newer data from more recent analyses using newer analytical techniques. Therefore, nutrient data should be viewed and used only as a guide, a close approximation of nutrient content.

• ***Dietary Fiber*** • There can be many different reported values for dietary fiber in foods because information depends on the type of analytical technique used. The fiber data in this table are primarily from the USDA/ARS Human Nutrition Information Service in Hyattsville, Maryland; Composition of Foods by Southgate and Paul (England); and many journal articles.

• ***Vitamin A*** • Vitamin A is reported in retinol equivalents (RE). The amount of vitamin A can vary by the season of the year and the maturity of the plant. Reported values in both dairy products and plants are higher in summer and early fall than in winter. The values reported here represent year-round averages. The organ meats of all animal products (liver especially) contain large amounts of vitamin A, which vary widely, depending on the background of the animal. The vitamin is also present in very small amounts in regular meat and is often reported as a trace.

• ***Vitamin E*** • Vitamin E is actually a combination of various forms of this nutrient, and the measure of alpha tocopherol equivalents (a-TE) summarizes the activity of the various types of tocopherols and tocotrienols into one measure.

A

• ***Fats*** • Total fats, as well as the breakdown of total fats to saturated, monounsaturated, and polyunsaturated fats, are listed in the table. The fatty acids seldom add up to the total due to rounding and to other fatty acid components that are not included in these basic categories, such as trans fatty acids and glycerol. Trans fatty acids can comprise a large share of the total fat in margarine and shortening (hydrogenated oils) and in of foods that include them as ingredients.

• ***Enrichment-Fortification*** • The mandatory enrichment values for foods are presented as appropriate, including the new values for folate enrichment of grain products.

• ***Niacin*** • Niacin values are for preformed niacin and do not include additional niacin that may form in the body from the conversion of tryptophan.

• ***Using the Table*** • The items in this table have been organized into several categories, which are listed at the head of each right-hand page. As the key shows, each group has been color-coded to make it easier to find individual items.

In an effort to conserve space, the following abbreviations have been used in the food descriptions and nutrient breakdowns:

- diam = diameter
- ea = each
- enr = enriched
- f/ = from
- frzn = frozen
- g = grams
- liq = liquid
- pce = piece
- pkg = package
- w/ = with
- w/o = without
- t = trace
- 0 = zero (no nutrient value)
- blank space = information not available

1. This food composition table has been prepared for Wadsworth Publishing Company and is copyrighted by ESHA Research in Salem, Oregon—the developer and publisher of the Food Processor®, Genesis® R&D, and the Computer Chef® nutrition software systems. The major sources for the data are from the USDA, supplemented by more than 1200 additional sources of information. Because the list of references is so extensive, it is not provided here, but is available from the publisher.

Table A-1

Food Composition (Computer code number is for Wadsworth Diet Analysis Plus program) (For purposes of calculations, use "0" for t, <1, <.1, <.01, etc.)

Computer Code Number	Food Description	Measure	Wt (g)	H_2O (%)	Ener (kcal)	Prot (g)	Carb (g)	Dietary Fiber (g)	Fat (g)	Fat Breakdown (g) Sat	Mono	Poly
	BEVERAGES											
	Alcoholic:											
	Beer:											
1	Regular (12 fl oz)	1½ c	356	92	146	1	13	1	0	0	0	0
2	Light (12 fl oz)	1½ c	354	95	99	1	5	0	0	0	0	0
1506	Nonalcoholic (12 fl oz)	1½ c	360	98	32	1	5	0	0	0	0	0
	Gin, rum, vodka, whiskey:											
3	80 proof	1½ fl oz	42	67	97	0	0	0	0	0	0	0
4	86 proof	1½ fl oz	42	64	105	0	<1	0	0	0	0	0
5	90 proof	1½ fl oz	42	62	110	0	0	0	0	0	0	0
	Liqueur:											
1359	Coffee liqueur, 53 proof	1½ fl oz	52	31	175	<1	24	0	<1	.1	t	.1
1360	Coffee & cream liqueur, 34 proof	1½ fl oz	47	46	154	1	10	0	7	4.5	2.1	.3
1361	Crème de menthe, 72 proof	1½ fl oz	50	28	186	0	21	0	<1	t	t	.1
	Wine, 4 fl oz:											
6	Dessert, sweet	½ c	118	72	181	<1	14	0	0	0	0	0
7	Red	½ c	118	88	85	<1	2	0	0	0	0	0
8	Rosé	½ c	118	89	84	<1	2	0	0	0	0	0
9	White medium	½ c	118	90	80	<1	1	0	0	0	0	0
1592	Nonalcoholic	1 c	232	98	14	1	3	0	0	0	0	0
1593	Nonalcoholic light	1 c	232	98	14	1	3	0	0	0	0	0
1409	Wine cooler, bottle (12 fl oz)	1½ c	340	90	169	<1	20	<1	<1	t	t	t
1595	Wine cooler, cup	1 c	227	90	113	<1	13	<1	<1	t	t	t
	Carbonated:											
10	Club soda (12 fl oz)	1½ c	355	100	0	0	0	0	0	0	0	0
11	Cola beverage (12 fl oz)	1½ c	372	89	153	0	39	0	0	0	0	0
12	Diet cola w/aspartame (12 fl oz)	1½ c	355	100	4	<1	<1	0	0	0	0	0
13	Diet soda pop w/saccharin (12 fl oz)	1½ c	355	100	0	0	<1	0	0	0	0	0
14	Ginger ale (12 fl oz)	1½ c	366	91	124	0	32	0	0	0	0	0
15	Grape soda (12 fl oz)	1½ c	372	89	160	0	42	0	0	0	0	0
16	Lemon-lime (12 fl oz)	1½ c	368	89	147	0	38	0	0	0	0	0
17	Orange (12 fl oz)	1½ c	372	88	179	0	46	0	0	0	0	0
18	Pepper-type soda (12 fl oz)	1½ c	368	89	151	0	38	0	<1	.1	0	0
19	Root beer (12 fl oz)	1½ c	370	89	152	0	39	0	0	0	0	0
20	Coffee, brewed	1 c	237	99	5	<1	1	0	<1	t	0	t
21	Coffee, prepared from instant	1 c	238	99	5	<1	1	0	<1	t	0	t
	Fruit drinks, noncarbonated:											
22	Fruit punch drink, canned	1 c	248	88	117	0	29	<1	<1	t	t	t
1358	Gatorade	1 c	241	93	60	0	15	0	0	0	0	0
23	Grape drink, canned	1 c	250	87	125	<1	32	<1	0	0	0	0
1304	Koolade sweetened with sugar	1 c	262	90	97	0	25	0	<1	t	t	t
1356	Koolade sweetened with nutrasweet	1 c	240	95	43	0	11	0	0	0	0	0
26	Lemonade,frzn concentrate (6-oz can)	¾ c	219	52	396	1	103	1	<1	.1	t	.1
27	Lemonade, from concentrate	1 c	248	89	99	<1	26	<1	<1	t	t	t
28	Limeade, frzn concentrate (6-oz can)	¾ c	218	50	408	<1	108	1	<1	t	t	t
29	Limeade, from concentrate	1 c	247	89	101	0	27	<1	<1	t	t	.1
24	Pineapple grapefruit, canned	1 c	250	88	118	<1	29	<1	<1	t	t	t
25	Pineapple orange, canned	1 c	250	87	125	3	29	<1	0	0	0	0
	Fruit and vegetable juices: see Fruit and Vegetable sections											
	Ultra Slim Fast, ready to drink, can:											
30411	Chocolate Royale	1 ea	350	84	220	10	38	5	3	1	1	.5
30415	French Vanilla	1 ea	350	84	220	10	38	5	3	1	1.5	.5
30413	Strawberries n' cream	1 ea	350	83	220	10	42	5	3	1	1.5	.5
1357	Water, bottled: Perrier (6½ fl oz)	1 ea	192	100	0	0	0	0	0	0	0	0
1594	Water, bottled: Tonic water	1½ c	366	91	124	0	32	0	0	0	0	0
	Tea:											
30	Brewed, regular	1 c	237	100	2	0	1	0	<1	t	0	t
1662	Brewed, herbal	1 c	237	100	2	0	<1	0	<1	t	t	t
32	From instant, sweetened	1 c	259	91	88	<1	22	0	<1	t	t	t
31	From instant, unsweetened	1 c	237	100	2	<1	<1	0	0	0	0	0

Chol (mg)	Calc (mg)	Iron (mg)	Magn (mg)	Pota (mg)	Sodi (mg)	Zinc (mg)	VT-A (RE)	Thia (mg)	VT-E (α-TE)	Ribo (mg)	Niac (mg)	V-B6 (mg)	Fola (μg)	VT-C (mg)
0	18	.11	21	89	18	.07	0	.02	0	.09	1.61	.18	21	0
0	18	.14	18	64	11	.11	0	.03	0	.11	1.39	.12	14	0
0	25	.04	32	90	18	.04	0	.02	0	.09	1.63	.18	22	0
0	0	.02	0	1	<1	.02	0	<.01	0	<.01	<.01	0	0	0
0	0	.02	0	1	<1	.02	0	<.01	0	<.01	<.01	0	0	0
0	0	.02	0	1	<1	.02	0	<.01	0	<.01	<.01	0	0	0
0	1	.03	2	16	4	.02	0	<.01	0	.01	.07	0	0	0
7	8	.06	1	15	43	.07	20	0	.12	.03	.04	.01	0	0
0	0	.03	0	0	2	.02	0	0	0	0	<.01	0	0	0
0	9	.28	11	109	11	.08	0	.02	0	.02	.25	0	<1	0
0	9	.51	15	132	6	.11	0	.01	0	.03	.1	.04	2	0
0	9	.45	12	117	6	.07	0	<.01	0	.02	.09	.03	1	0
0	11	.38	12	94	6	.08	0	<.01	0	.01	.08	.02	<1	0
0	21	.93	23	204	16	.19	0	0	0	.02	.23	.05	2	0
0	21	.93	23	204	16	.19	0	0	0	.02	.23	.05	2	0
0	19	.92	18	152	29	.2	<1	.02	.02	.02	.15	.04	4	6
0	13	.61	12	102	19	.13	<1	.01	.02	.02	.1	.03	3	4
0	18	.04	4	7	75	.35	0	0	0	0	0	0	0	0
0	11	.11	4	4	15	.04	0	0	0	0	0	0	0	0
0	14	.11	4	0	21	.28	0	.02	0	.08	0	0	0	0
0	14	.14	4	7	57	.18	0	0	0	0	0	0	0	0
0	11	.66	4	4	26	.18	0	0	0	0	0	0	0	0
0	11	.3	4	4	56	.26	0	0	0	0	0	0	0	0
0	7	.26	4	4	40	.18	0	0	0	0	.05	0	0	0
0	19	.22	4	7	45	.37	0	0	0	0	0	0	0	0
0	11	.15	0	4	37	.15	0	0	0	0	0	0	0	0
0	18	.18	4	4	48	.26	0	0	0	0	0	0	0	0
0	5	.12	12	128	5	.05	0	0	0	0	.53	0	<1	0
0	7	.12	10	86	7	.07	0	0	0	<.01	.67	0	0	0
0	20	.52	5	62	55	.3	3	0	0	.06	.05	0	3	73
0	0	.12	2	26	96	.05	0	.01	0	0	0	0	0	0
0	7	.25	10	87	2	.07	<1	.02	0	.02	.25	.05	2	40
0	42	.13	3	3	37	.08	0	0	0	<.01	<.01	0	<1	31
0	17	.65	5	50	50	.26	2	.02	0	.05	.05	0	5	77
0	15	1.58	11	147	9	.17	22	.06	0	.21	.16	.05	22	39
0	7	.4	5	37	7	.1	5	.01	0	.05	.04	.01	5	10
0	11	.22	9	129	0	.09	0	.02	0	.02	.22	0	9	26
0	7	.07	2	32	5	.05	0	<.01	0	<.01	.05	0	2	7
0	17	.77	15	153	35	.15	9	.07	0	.04	.67	.1	26	115
0	12	.67	15	115	7	.15	13	.07	0	.05	.52	.12	27	56
5	400	2.7	140	530	220	2.24	525	.52	7	.59	7	.7	120	21
5	400	2.8	140	450	460	2.1	525	.52	7	.59	7	.7	120	21
5	400	2.7	140	450	460	2.24	525	.52	7	.59	7	.7	120	21
0	27	0	0	0	2	0	0	0	0	0	0	0	0	0
0	4	.04	0	0	15	.37	0	0	0	0	0	0	0	0
0	0	.05	7	88	7	.05	0	0	0	.03	0	0	12	0
0	5	.19	2	21	2	.09	0	.02	0	.01	0	0	1	0
0	5	.05	5	49	8	.08	0	0	0	.05	.09	<.01	10	0
0	5	.05	5	47	7	.07	0	0	0	<.01	.09	<.01	1	0

Table A-1

Food Composition (Computer code number is for Wadsworth Diet Analysis Plus program) (For purposes of calculations, use "0" for t, <1, <.1, <.01, etc.)

Computer Code Number	Food Description	Measure	Wt (g)	H_2O (%)	Ener (kcal)	Prot (g)	Carb (g)	Dietary Fiber (g)	Fat (g)	Fat Breakdown (g) Sat	Mono	Poly
	DAIRY											
	Butter: see Fats and Oils, #158,159,160											
	Cheese, natural:											
33	Blue	1 oz	28	42	99	6	1	0	8	5.2	2.2	.2
34	Brick	1 oz	28	41	104	6	1	0	8	5.3	2.4	.2
35	Brie	1 oz	28	48	93	6	<1	0	8	4.9	2.2	.2
36	Camembert	1 oz	28	52	84	6	<1	0	7	4.3	2	.2
37	Cheddar:	1 oz	28	37	113	7	<1	0	9	5.9	2.6	.3
38	1" cube	1 ea	17	37	68	4	<1	0	6	3.6	1.6	.2
39	Shredded	1 c	113	37	455	28	1	0	37	24	10.6	1.1
1406	Low fat, low sodium	1 oz	28	65	48	7	1	0	2	1.2	.6	.1
	Cottage:											
984	Low sodium, low fat	1 c	225	83	162	28	6	0	2	1.4	.7	.1
40	Creamed, large curd	1 c	225	79	232	28	6	0	10	6.4	2.9	.3
41	Creamed, small curd	1 c	210	79	216	26	6	0	9	6	2.7	.3
42	With fruit	1 c	226	72	280	22	30	0	8	4.9	2.2	.2
43	Low fat 2%	1 c	226	79	203	31	8	0	4	2.8	1.2	.1
44	Low fat 1%	1 c	226	82	164	28	6	0	2	1.5	.7	.1
46	Cream	1 tbs	15	54	52	1	<1	0	5	3.3	1.5	.2
983	low fat	1 tbs	15	64	35	2	1	0	3	1.7	.9	.1
47	Edam	1 oz	28	42	100	7	<1	0	8	4.9	2.3	.2
48	Feta	1 oz	28	55	74	4	1	0	6	4.2	1.3	.2
49	Gouda	1 oz	28	41	100	7	1	0	8	4.9	2.2	.2
50	Gruyère	1 oz	28	33	116	8	<1	0	9	5.3	2.8	.5
51	Gorgonzola	1 oz	28	43	97	6	1	0	8	5		
1676	Limburger	1 oz	28	48	92	6	<1	0	8	4.7	2.4	.1
53	Monterey Jack	1 oz	28	41	104	7	<1	0	8	5.3	2.4	.3
54	Mozzarella, whole milk	1 oz	28	54	79	5	1	0	6	3.7	1.8	.2
55	Mozzarella, part-skim milk, low moisture	1 oz	28	49	78	8	1	0	5	3	1.4	.1
56	Muenster	1 oz	28	42	103	7	<1	0	8	5.3	2.4	.2
2422	Neufchatel	1 oz	28	62	73	3	1	0	7	4.1	1.9	.2
1399	Nonfat cheese (Kraft Singles)	1 oz	28	61	44	6	4	0	0	0	0	0
59	Parmesan, grated:	1 oz	28	18	128	12	1	0	8	5.5	2.4	.2
57	Cup, not pressed down	1 c	100	18	456	42	4	0	30	19.7	8.7	.7
58	Tablespoon	1 tbs	6	18	27	2	<1	0	2	1.2	.5	t
60	Provolone	1 oz	28	41	98	7	1	0	7	4.9	2.1	.2
61	Ricotta, whole milk	1 c	246	72	428	28	7	0	32	20.4	8.9	.9
62	Ricotta, part-skim milk	1 c	246	74	339	28	13	0	19	12.1	5.7	.6
63	Romano	1 oz	28	31	108	9	1	0	8	4.8	2.2	.2
64	Swiss	1 oz	28	37	105	8	1	0	8	5	2	.3
976	low fat	1 oz	28	60	50	8	1	0	1	.9	.4	t
	Pasteurized processed cheese products:											
65	American	1 oz	28	39	105	6	<1	0	9	5.5	2.5	.3
66	Swiss	1 oz	28	42	93	7	1	0	7	4.5	2	.2
67	American cheese food, jar	½ c	57	43	187	11	4	0	14	9	4.1	.4
68	American cheese spread	1 tbs	15	48	43	2	1	0	3	2	.9	.1
982	Velveeta cheese spread, low fat, low sodium, slice	1 pce	34	62	61	9	1	0	2	1.5	.7	.1
	Cream, sweet:											
69	Half & half (cream & milk)	1 c	242	81	315	7	10	0	28	17.3	8	1
70	Tablespoon	1 tbs	15	81	19	<1	1	0	2	1.1	.5	.1
71	Light, coffee or table:	1 c	240	74	468	6	9	0	46	28.8	13.4	1.7
72	Tablespoon	1 tbs	15	74	29	<1	1	0	3	1.8	.8	.1
73	Light whipping cream, liquid:	1 c	239	63	698	5	7	0	74	46.1	21.7	2.1
74	Tablespoon	1 tbs	15	63	44	<1	<1	0	5	2.9	1.4	.1
75	Heavy whipping cream, liquid:	1 c	238	58	821	5	7	0	88	54.7	25.5	3.3
76	Tablespoon	1 tbs	15	58	52	<1	<1	0	6	3.4	1.6	.2
77	Whipped cream, pressurized:	1 c	60	61	154	2	7	0	13	8.3	3.8	.5
78	Tablespoon	1 tbs	4	61	10	<1	<1	0	1	.6	.3	t
79	Cream, sour, cultured:	1 c	230	71	492	7	10	0	48	29.9	13.9	1.8
80	Tablespoon	1 tbs	14	71	30	<1	1	0	3	1.8	.8	.1

PAGE KEY: A–4 = Beverages A–6 = Dairy A–10 = Eggs A–10 = Fat/Oil A–14 = Fruit A–20 = Bakery A–28 = Grain A–32 = Fish A–34 = Meats A–38= Poultry A–40 = Sausage A–42 = Mixed/Fast A–46 = Nuts/Seeds A–50 = Sweets A–52 = Vegetables/Legumes A–62 = Vegetarian A–62 = Misc A–66 = Soups/Sauces A–68 = Fast A–84 = Convenience A–88 = Baby foods

Chol (mg)	Calc (mg)	Iron (mg)	Magn (mg)	Pota (mg)	Sodi (mg)	Zinc (mg)	VT-A (RE)	Thia (mg)	VT-E (α-TE)	Ribo (mg)	Niac (mg)	V-B6 (mg)	Fola (µg)	VT-C (mg)
21	148	.09	6	72	391	.74	64	.01	.18	.11	.29	.05	10	0
26	189	.12	7	38	157	.73	85	<.01	.14	.1	.03	.02	6	0
28	51	.14	6	43	176	.67	51	.02	.18	.15	.11	.07	18	0
20	109	.09	6	52	236	.67	71	.01	.18	.14	.18	.06	17	0
29	202	.19	8	28	174	.87	85	.01	.1	.1	.02	.02	5	0
18	123	.12	5	17	106	.53	51	<.01	.06	.06	.01	.01	3	0
119	815	.77	31	111	702	3.51	342	.03	.41	.42	.09	.08	21	0
6	197	.2	8	31	6	.86	17	.01	.05	.01	.02	.02	5	0
9	137	.31	11	194	29	.85	25	.04	.25	.36	.29	.16	27	0
33	135	.31	12	190	911	.83	108	.05	.27	.37	.28	.15	27	0
31	126	.29	11	177	851	.78	101	.04	.26	.34	.26	.14	26	0
25	108	.25	9	151	915	.65	81	.04	.21	.29	.23	.12	22	0
19	155	.36	14	217	918	.95	45	.05	.13	.42	.32	.17	30	0
10	138	.32	12	193	918	.86	25	.05	.25	.37	.29	.15	28	0
16	12	.18	1	18	44	.08	66	<.01	.14	.03	.01	.01	2	0
8	17	.25	1	25	44	.11	33	<.01	.07	.04	.02	.01	3	0
25	205	.12	8	53	270	1.05	71	.01	.21	.11	.02	.02	5	0
25	138	.18	5	17	312	.81	36	.04	.01	.24	.28	.12	9	0
32	196	.07	8	34	229	1.09	49	.01	.1	.09	.02	.02	6	0
31	283	.05	10	23	94	1.09	84	.02	.1	.08	.03	.02	3	0
30	170	.18			280		43							0
25	139	.04	6	36	224	.59	88	.02	.18	.14	.04	.02	16	0
25	209	.2	8	23	150	.84	71	<.01	.09	.11	.03	.02	5	0
22	145	.05	5	19	104	.62	67	<.01	.1	.07	.02	.02	2	0
15	205	.07	7	26	148	.88	53	.01	.13	.1	.03	.02	3	0
27	201	.11	8	37	176	.79	88	<.01	.13	.09	.03	.02	3	0
21	21	.08	2	32	112	.15	74	<.01	.26	.05	.03	.01	3	0
4	221	0		81	427		126		0	.1				0
22	385	.27	14	30	521	.89	48	.01	.22	.11	.09	.03	2	0
79	1375	.95	51	107	1861	3.19	173	.04	.8	.39	.31	.1	8	0
5	82	.06	3	6	112	.19	10	<.01	.05	.02	.02	.01	<1	0
19	212	.15	8	39	245	.9	74	<.01	.1	.09	.04	.02	3	0
124	509	.93	28	258	207	2.85	330	.03	.86	.48	.26	.11	30	0
76	669	1.08	36	308	308	3.3	278	.05	.53	.45	.19	.05	32	0
29	298	.22	11	24	336	.72	39	.01	.2	.1	.02	.02	2	0
26	269	.05	10	31	73	1.09	71	.01	.14	.1	.03	.02	2	0
10	269	.05	10	31	73	1.09	18	.01	.05	.1	.02	.02	2	0
26	172	.11	6	45	400	.84	81	.01	.13	.1	.02	.02	2	0
24	216	.17	8	60	384	1.01	64	<.01	.19	.08	.01	.01	2	0
36	327	.48	17	159	678	1.7	125	.02	.4	.25	.08	.08	4	0
8	84	.05	4	36	202	.39	28	.01	.11	.06	.02	.02	1	0
12	233	.15	8	61	2	1.13	22	.01	.17	.13	.03	.03	3	0
89	254	.17	25	315	98	1.23	259	.08	.27	.36	.19	.09	6	2
6	16	.01	2	19	6	.08	16	<.01	.02	.02	.01	.01	<1	<1
159	231	.1	21	293	95	.65	437	.08	.36	.35	.14	.08	6	2
10	14	.01	1	18	6	.04	27	<.01	.02	.02	.01	<.01	<1	<1
265	166	.07	17	231	82	.6	705	.06	1.43	.3	.1	.07	9	1
17	10	<.01	1	14	5	.04	44	<.01	.09	.02	.01	<.01	1	<1
326	154	.07	17	179	89	.55	1001	.05	1.5	.26	.09	.06	9	1
21	10	<.01	1	11	6	.03	63	<.01	.09	.02	.01	<.01	1	<1
46	61	.03	6	88	78	.22	124	.02	.36	.04	.04	.02	2	0
3	4	<.01		6	5	.01	8	<.01	.02	<.01	<.01	<.01	<1	0
102	267	.14	26	331	123	.62	449	.08	1.3	.34	.15	.04	25	2
6	16	.01	2	20	7	.04	27	<.01	.08	.02	.01	<.01	2	<1

Table A–1

Food Composition (Computer code number is for Wadsworth Diet Analysis Plus program) (For purposes of calculations, use "0" for t, <1, <.1, <.01, etc.)

Computer Code Number	Food Description	Measure	Wt (g)	H_2O (%)	Ener (kcal)	Prot (g)	Carb (g)	Dietary Fiber (g)	Fat (g)	Fat Breakdown (g) Sat	Mono	Poly
	DAIRY—Continued											
	Cream products—imitation and part dairy:											
81	Coffee whitener, frozen or liquid	1 tbs	15	77	20	<1	2	0	1	1.4	t	0
82	Coffee whitener, powdered	1 tsp	2	2	11	<1	1	0	1	.6	t	0
83	Dessert topping, frozen, nondairy:	1 c	75	50	239	1	17	0	19	16.4	1.2	.4
84	Tablespoon	1 tbs	5	50	16	<1	1	0	1	1.1	.1	t
85	Dessert topping, mix with whole milk:	1 c	80	67	151	3	13	0	10	8.6	.7	.2
86	Tablespoon	1 tbs	5	67	9	<1	1	0	1	.5	t	t
88	Dessert topping, pressurized	1 c	70	60	185	1	11	0	16	13.3	1.3	.2
87	Tablespoon	1 tbs	4	60	11	<1	1	0	1	.8	.1	t
91	Sour cream, imitation:	1 c	230	71	478	6	15	0	45	40.9	1.3	.1
92	Tablespoon	1 tbs	14	71	29	<1	1	0	3	2.5	.1	t
89	Sour dressing, part dairy:	1 c	235	75	418	8	11	0	39	31.3	4.6	1.1
90	Tablespoon	1 tbs	15	75	27	1	1	0	2	2	.3	.1
	Milk, fluid:											
93	Whole milk	1 c	244	88	150	8	11	0	8	5.1	2.7	.3
94	2% reduced-fat milk	1 c	244	89	121	8	12	0	5	2.9	1.3	.2
95	2% milk solids added	1 c	245	89	125	9	12	0	5	2.9	1.4	.2
96	1% lowfat milk	1 c	244	90	102	8	12	0	3	1.6	.7	.1
97	1% milk solids added	1 c	245	90	104	9	12	0	2	1.5	.7	.1
98	Nonfat milk, vitamin A added	1 c	245	91	85	8	12	0	<1	.3	.1	t
99	Nonfat milk solids added	1 c	245	90	90	9	12	0	1	.4	.2	t
100	Buttermilk, skim	1 c	245	90	99	8	12	0	2	1.3	.6	.1
	Milk, canned:											
101	Sweetened condensed	1 c	306	27	982	24	166	0	27	16.8	7.4	1
103	Evaporated, nonfat	1 c	256	79	199	19	29	0	1	.3	.2	t
	Milk, dried:											
104	Buttermilk, sweet	1 c	120	3	464	41	59	0	7	4.3	2	.3
105	Instant, nonfat, vit A added (makes 1 qt)	1 ea	91	4	326	32	47	0	1	.4	.2	t
106	Instant nonfat, vit A added	1 c	68	4	243	24	35	0	<1	.3	.1	t
107	Goat milk	1 c	244	87	168	9	11	0	10	6.5	2.7	.4
108	Kefir	1 c	233	88	149	8	11	0	8			
	Milk beverages and powdered mixes:											
	Chocolate:											
109	Whole	1 c	250	82	209	8	26	2	8	5.3	2.5	.3
110	2% fat	1 c	250	84	179	8	26	1	5	3.1	1.5	.2
111	1% fat	1 c	250	84	158	9	26	1	2	1.5	.7	.1
	Chocolate-flavored beverages:											
112	Powder containing nonfat dry milk:	1 oz	28	1	101	3	22	<1	1	.7	.4	t
113	Prepared with water	1 c	275	86	138	4	30	3	2	.9	.5	t
114	Powder without nonfat dry milk:	1 oz	28	1	98	1	25	2	1	.5	.3	t
115	Prepared with whole milk	1 c	266	81	226	9	31	1	9	5.5	2.6	.3
116	Eggnog, commercial	1 c	254	74	343	10	34	0	19	11.3	5.7	.9
974	Eggnog, 2% reduced-fat	1 c	254	85	189	12	17	0	8	3.7	2.7	.7
1027	Instant Breakfast, envelope,powder only:	1 ea	37	7	131	7	24	<1	1	.3	.1	t
1028	Prepared with whole milk	1 c	281	77	280	15	36	<1	9	5.4	2.5	.3
1029	Prepared with 2% milk	1 c	281	78	252	15	36	<1	5	3.3	1.5	.2
1283	Prepared with 1% milk	1 c	281	79	233	15	36	<1	3	1.9	.9	.1
1284	Prepared with nonfat milk	1 c	282	80	216	16	36	<1	1	.7	.3	t
117	Malted milk, chocolate, powder:	3 tsp	21	1	79	1	18	<1	1	.5	.2	.1
118	Prepared with whole milk	1 c	265	81	228	9	30	<1	9	5.5	2.6	.4
1661	Ovaltine with whole milk	1 c	265	81	225	9	29	<1	9	5.5	2.5	.4
119	Malted mix powder, natural:	3 tsp	21	2	87	2	16	<1	2	.9	.4	.3
120	Prepared with whole milk	1 c	265	81	236	10	27	<1	10	6	2.8	.6
121	Milk shakes, chocolate	1 c	166	71	211	6	34	1	6	3.8	1.8	.2
122	Milk shakes, vanilla	1 c	166	75	184	6	30	1	5	3.1	1.4	.2
	Milk desserts:											
134	Custard, baked	1 c	282	79	296	14	30	0	13	6.6	4.3	1
1548	Low-fat frozen dessert bars	1 ea	81	72	88	2	19	0	1	.2	.1	.4
	Ice cream, vanilla (about 10% fat):											
123	Hardened: ½ gallon	1 ea	1064	61	2138	37	251	0	117	72.4	33.8	4.4
124	Cup	1 c	132	61	265	5	31	0	14	9	4.2	.5
126	Soft serve	1 c	172	60	370	7	38	0	22	12.9	6	.8

PAGE KEY: A–4 = Beverages A–6 = Dairy A–10 = Eggs A–10 = Fat/Oil A–14 = Fruit A–20 = Bakery A–28 = Grain
A–32 = Fish A–34 = Meats A–38= Poultry A–40 = Sausage A–42 = Mixed/Fast A–46 = Nuts/Seeds A–50 = Sweets
A–52 = Vegetables/Legumes A–62 = Vegetarian A–62 = Misc A–66 = Soups/Sauces A–68 = Fast A–84 = Convenience A–88 = Baby foods

Chol (mg)	Calc (mg)	Iron (mg)	Magn (mg)	Pota (mg)	Sodi (mg)	Zinc (mg)	VT-A (RE)	Thia (mg)	VT-E (α-TE)	Ribo (mg)	Niac (mg)	V-B6 (mg)	Fola (μg)	VT-C (mg)
0	1	<.01		29	12	<.01	1	0	.24	0	0	0	0	0
0		.02		16	4	.01	<1	0	<.01	<.01	0	0	0	0
0	5	.09	1	14	19	.02	64	0	.14	0	0	0	0	0
0		.01		1	1	<.01	4	0	.01	0	0	0	0	0
8	72	.03	8	121	53	.22	39	.02	.11	.09	.05	.02	3	1
<1	5	<.01		8	3	.01	2	<.01	.01	.01	<.01	<.01	<1	<1
0	4	.01	1	13	43	.01	33	0	.12	0	0	0	0	0
0		<.01		1	2	0	2	0	.01	0	0	0	0	0
0	6	.9	15	370	235	2.71	0	0	.34	0	0	0	0	0
0		.05	1	22	14	.16	0	0	.02	0	0	0	0	0
13	266	.07	23	381	113	.87	5	.09	.29	.38	.17	.04	28	2
1	17	<.01	1	24	7	.06	<1	.01	.02	.02	.01	<.01	2	<1
33	290	.12	33	371	120	.93	76	.09	.24	.39	.2	.1	12	2
18	298	.12	33	376	122	.95	139	.09	.17	.4	.21	.1	12	2
18	314	.12	35	397	128	.98	140	.1	.17	.42	.22	.11	13	2
10	300	.12	34	381	123	.95	144	.09	.1	.41	.21	.1	12	2
10	314	.12	35	397	128	.98	145	.1	.1	.42	.22	.11	13	2
4	301	.1	28	407	126	.98	149	.09	.1	.34	.22	.1	13	2
5	316	.12	35	419	130	1	149	.1	.1	.43	.22	.11	13	2
9	284	.12	27	370	257	1.03	20	.08	.15	.38	.14	.08	12	2
104	869	.58	79	1135	389	2.88	248	.27	.65	1.27	.64	.16	34	8
9	742	.74	69	850	294	2.3	300	.11	.01	.79	.44	.14	22	3
83	1420	.36	132	1910	620	4.82	65	.47	.48	1.9	1.05	.41	57	7
17	1119	.28	106	1551	500	4.01	646	.38	.02	1.58	.81	.31	45	5
12	836	.21	80	1159	373	3	483	.28	.01	1.18	.61	.23	34	4
28	327	.12	34	498	122	.73	137	.12	.22	.34	.68	.11	1	3
		.3	33	373	107									
30	280	.6	32	418	149	1.03	72	.09	.23	.4	.31	.1	12	2
17	285	.6	33	423	151	1.03	143	.09	.13	.41	.31	.1	12	2
7	288	.6	33	425	152	1.03	148	.09	.06	.41	.32	.1	12	2
1	91	.33	23	199	141	.41	1	.03	.04	.16	.16	.03	0	<1
3	129	.47	33	270	198	.6	1	.04	.06	.21	.22	.04	0	1
0	10	.88	27	165	59	.43	1	.01	.11	.04	.14	<.01	2	<1
32	301	.8	53	497	165	1.28	77	.1	.21	.43	.32	.1	12	2
149	330	.51	47	419	138	1.17	203	.09	.58	.48	.27	.13	2	4
194	269	.71	32	367	155	1.26	197	.11	1.01	.55	.21	.15	30	2
4	105	4.74	84	350	142	3.16	554	.31	5.31	.07	5.27	.42	105	28
38	396	4.86	117	719	262	4.09	630	.41	5.51	.47	5.46	.52	118	31
23	401	4.86	118	726	264	4.12	693	.41	5.41	.48	5.46	.53	118	31
14	406	4.86	118	731	266	4.12	698	.41	5.36	.48	5.47	.53	118	31
9	407	4.83	112	755	268	4.14	703	.4	5.3	.42	5.47	.52	118	31
1	13	.48	15	130	53	.17	4	.04	.08	.04	.42	.03	4	<1
34	305	.61	48	498	172	1.09	79	.13	.26	.44	.62	.13	16	3
34	384	3.76	53	620	244	1.17	901	.73	.32	1.26	10.9	1.02	32	34
4	63	.15	19	159	104	.21	18	.11	.08	.19	1.1	.09	10	1
37	355	.26	53	530	223	1.14	95	.2	.32	.59	1.31	.19	22	3
22	188	.51	28	332	161	.68	38	.1	.11	.41	.27	.08	6	1
18	203	.15	20	289	136	.6	53	.07	.1	.3	.31	.09	5	1
245	316	.85	39	431	217	1.49	169	.09	.68	.64	.24	.14	28	1
1	82	.07	10	111	47	.26	38	.03	.07	.11	.06	.03	3	1
468	1361	.96	149	2117	851	7.34	1244	.44	0	2.55	1.23	.51	53	6
58	169	.12	18	263	106	.91	154	.05	0	.32	.15	.06	7	1
157	225	.36	21	304	105	.89	265	.08	.64	.31	.16	.08	15	1

Table A-1

Food Composition (Computer code number is for Wadsworth Diet Analysis Plus program) (For purposes of calculations, use "0" for t, <1, <.1, <.01, etc.)

Computer Code Number	Food Description	Measure	Wt (g)	H_2O (%)	Ener (kcal)	Prot (g)	Carb (g)	Dietary Fiber (g)	Fat (g)	Fat Breakdown (g) Sat	Mono	Poly
	DAIRY—Continued											
	Ice cream, rich vanilla (16% fat):											
127	Hardened: ½ gallon	1 ea	1188	60	2554	49	264	0	154	88.9	41.5	5.5
128	Cup	1 c	148	57	357	5	33	0	24	14.8	6.9	.9
1724	Ben & Jerry's	½ c	108		230	4	21	0	17	10		
	Ice milk, vanilla (about 4% fat):											
129	Hardened: ½ gallon	1 ea	1048	68	1456	40	238	0	45	27.7	12.9	1.7
130	Cup	1 c	132	68	183	5	30	0	6	3.5	1.6	.2
131	Soft serve (about 3% fat)	1 c	176	70	222	9	38	0	5	2.9	1.3	.2
	Pudding, canned (5 oz can = .55 cup):											
135	Chocolate	1 ea	142	69	189	4	32	1	6	1	2.4	2
136	Tapioca	1 ea	142	74	169	3	27	<1	5	.9	2.3	1.9
137	Vanilla	1 ea	142	71	185	3	31	<1	5	.8	2.2	1.9
	Puddings, dry mix with whole milk:											
138	Chocolate, instant	1 c	294	74	326	9	55	3	9	5.4	2.7	.5
139	Chocolate, regular, cooked	1 c	284	74	315	9	51	3	10	5.9	2.8	.4
140	Rice, cooked	1 c	288	72	351	9	60	<1	8	5.1	2.4	.3
141	Tapioca, cooked	1 c	282	74	321	8	55	0	8	5.1	2.3	.3
142	Vanilla, instant	1 c	284	73	324	8	56	0	8	4.9	2.4	.4
143	Vanilla, regular, cooked	1 c	280	75	311	8	52	0	8	5.1	2.4	.4
	Sherbet (2% fat):											
132	½ gallon	1 ea	1542	66	2127	17	469	8	31	17.9	8.3	1.2
133	Cup	1 c	198	66	273	2	60	1	4	2.3	1.1	.2
144	Soy milk	1 c	245	93	81	7	4	3	5	.7	1	2.6
2301	Soy milk, fortified, fat free	1 c	240	88	110	6	22	1	0	0	0	0
	Yogurt, frozen, low-fat											
1584	Cup	1 c	144	65	229	6	35	0	8	4.9	2.3	.3
1512	Scoop	1 ea	79	74	78	4	15	0	<1	.1	t	t
	Yogurt, lowfat:											
1172	Fruit added with low-calorie sweetener	1 c	241	86	122	12	19	1	<1	.2	.1	t
145	Fruit added	1 c	245	74	250	11	47	<1	3	1.7	.7	.1
146	Plain	1 c	245	85	155	13	17	0	4	2.4	1	.1
147	Vanilla or coffee flavor	1 c	245	79	209	13	34	0	3	2	.8	.1
148	Yogurt, made with nonfat milk	1 c	245	85	137	15	19	0	<1	.3	.1	t
149	Yogurt, made with whole milk	1 c	245	88	150	9	11	0	8	5.1	2.2	.2
	EGGS											
	Raw, large:											
150	Whole, without shell	1 ea	50	75	74	6	1	0	5	1.5	1.9	.7
151	White	1 ea	33	88	16	3	<1	0	0	0	0	0
152	Yolk	1 ea	17	49	61	3	<1	0	5	1.6	2	.7
	Cooked:											
153	Fried in margarine	1 ea	46	69	91	6	1	0	7	1.9	2.8	1.3
154	Hard-cooked, shell removed	1 ea	50	75	77	6	1	0	5	1.6	2	.7
155	Hard-cooked, chopped	1 c	136	75	211	17	2	0	14	4.4	5.5	1.9
156	Poached, no added salt	1 ea	50	75	74	6	1	0	5	1.5	1.9	.7
157	Scrambled with milk & margarine	1 ea	61	73	101	7	1	0	7	2.2	2.9	1.3
1681	Egg substitute, liquid:	½ c	126	83	106	16	1	0	4	.8	1.1	2
1254	Egg Beaters, Fleischmann's	½ c	122		60	12	2	0	0	0	0	0
1262	Egg substitute, liquid, prepared	½ c	105	80	100	14	1	0	4	.8	1.1	1.9
	FATS and OILS											
158	Butter: Stick	½ c	114	16	817	1	<1	0	92	57.7	27.4	3.4
159	Tablespoon:	1 tbs	14	16	100	<1	<1	0	11	7.1	3.4	.4
8025	Unsalted	1 tbs	14	18	100	<1	<1	0	11	7.1	3.4	.4
160	Pat (about 1 tsp)	1 ea	5	16	36	<1	<1	0	4	2.5	1.2	.2
1682	Whipped	1 tsp	3	16	21	<1	<1	0	2	1.5	.7	.1
	Fats, cooking:											
1363	Bacon fat	1 tbs	14		125	0	0	0	14	6.3	5.9	1.1
1362	Beef fat/tallow	1 c	205	0	1849	0	0	0	205	103	87.3	8.2
1364	Chicken fat	1 c	205		1845	0	0	0	205	61.1	91.6	42.8
161	Vegetable shortening:	1 c	205	0	1812	0	0	0	205	52.1	91.2	53.5
162	Tablespoon	1 tbs	13	0	115	0	0	0	13	3.3	5.8	3.4

PAGE KEY: A–4 = Beverages A–6 = Dairy A–10 = Eggs A–10 = Fat/Oil A–14 = Fruit A–20 = Bakery A–28 = Grain
A–32 = Fish A–34 = Meats A–38= Poultry A–40 = Sausage A–42 = Mixed/Fast A–46 = Nuts/Seeds A–50 = Sweets
A–52 = Vegetables/Legumes A–62 = Vegetarian A–62 = Misc A–66 = Soups/Sauces A–68 = Fast A–84 = Convenience A–88 = Baby foods

Chol (mg)	Calc (mg)	Iron (mg)	Magn (mg)	Pota (mg)	Sodi (mg)	Zinc (mg)	VT-A (RE)	Thia (mg)	VT-E (α-TE)	Ribo (mg)	Niac (mg)	V-B6 (mg)	Fola (µg)	VT-C (mg)
1081	1556	2.49	143	2102	725	6.18	1829	.58	4.4	2.16	1.13	.57	107	9
90	173	.07	16	235	83	.59	272	.06	0	.24	.12	.06	7	1
95	150	.36			55		214		0					0
147	1456	1.05	157	2211	891	4.61	493	.61	0	2.78	.94	.68	63	8
18	183	.13	20	279	112	.58	62	.08	0	.35	.12	.09	8	1
21	276	.11	25	389	123	.93	51	.09	0	.35	.21	.08	11	2
4	128	.72	30	256	183	.6	16	.04	.18	.22	.49	.04	4	3
1	119	.33	11	148	168	.38	0	.03	.13	.14	.44	.14	4	1
10	125	.18	11	160	192	.35	9	.03	.18	.2	.36	.02	0	0
32	300	.85	53	488	835	1.23	62	.1	.18	.41	.28	.11	12	3
34	315	1.02	43	463	293	1.28	74	.09	.17	.49	.29	.1	11	2
32	297	1.09	37	372	314	1.09	58	.22	.17	.4	1.28	.1	11	2
34	293	.17	34	372	341	.96	76	.08	.23	.4	.21	.11	11	2
31	287	.2	34	364	812	.94	71	.09	.18	.39	.21	.1	11	2
34	300	.14	36	381	448	.98	76	.08	.17	.4	.21	.09	11	2
77	833	2.16	123	1480	709	7.4	216	.39	.88	1.05	1.48	.52	77	66
10	107	.28	16	190	91	.95	28	.05	.11	.13	.19	.07	10	9
0	10	1.42	47	345	29	.56	7	.39	.02	.17	.36	.1	4	0
0	400	1.44		20	60		0	.07		.1	3			0
3	206	.43	20	304	125	.6	82	.05	.07	.32	.41	.11	9	1
1	137	.07	13	175	53	.67	1	.03	<.01	.16	.08	.04	8	1
3	369	.61	41	550	139	1.83	6	.1	.17	.45	.5	.11	32	26
10	372	.17	36	478	143	1.81	27	.09	.07	.44	.23	.1	23	2
15	448	.2	43	573	172	2.18	39	.11	.1	.52	.28	.12	27	2
12	419	.17	40	537	161	2.03	32	.1	.08	.49	.26	.11	26	2
4	488	.22	47	625	187	2.38	5	.12	.01	.57	.3	.13	30	2
31	296	.12	28	380	114	1.45	73	.07	.22	.35	.18	.08	18	1
213	24	.72	5	60	63	.55	95	.03	.52	.25	.04	.07	23	0
0	2	.01	4	47	54	<.01	0	<.01	0	.15	.03	<.01	1	0
218	23	.6	2	16	7	.53	99	.03	.54	.11	<.01	.07	25	0
211	25	.72	5	61	162	.55	114	.03	.75	.24	.03	.07	17	0
212	25	.59	5	63	62	.52	84	.03	.52	.26	.03	.06	22	0
577	68	1.62	14	171	169	1.43	228	.09	1.43	.7	.09	.16	60	0
212	24	.72	5	60	140	.55	95	.02	.52	.21	.03	.06	17	0
215	43	.73	7	84	171	.61	119	.03	.8	.27	.05	.07	18	<1
1	67	2.65	11	416	223	1.64	272	.14	.61	.38	.14	<.01	19	0
0	80	2.16		170	200		80		.59					
1	63	2.51	10	394	211	1.55	258	.11	.58	.34	.12	<.01	13	0
250	27	.18	2	30	942	.06	860	.01	1.8	.04	.05	<.01	3	0
31	3	.02		4	116	.01	106	<.01	.22	<.01	.01	0	<1	0
31	3	.02		4	2	.01	106	<.01	.22	<.01	.01	0	<1	0
11	1	.01		1	41	<.01	38	0	.08	<.01	<.01	0	<1	0
7	1	<.01		1	25	<.01	23	0	.05	<.01	<.01	0	<1	0
14		0			76	<.01	0	0	.31	0	0	0	0	0
223	0	0	0		<1	0	0	0	3.08	0	0	0	0	0
174	0	0	0	0	0	0	0	0	5.54	0	0	0	0	0
0	0	0	0	0	0	0	0	0	17	0	0	0	0	0
0	0	0	0	0	0	0	0	0	1.08	0	0	0	0	0

Table A–1

Food Composition (Computer code number is for Wadsworth Diet Analysis Plus program) (For purposes of calculations, use "0" for t, <1, <.1, <.01, etc.)

Computer Code Number	Food Description	Measure	Wt (g)	H_2O (%)	Ener (kcal)	Prot (g)	Carb (g)	Dietary Fiber (g)	Fat (g)	Fat Breakdown (g) Sat	Mono	Poly
	FATS and OILS—Continued											
163	Lard:	1 c	205	0	1849	0	0	0	205	81.1	87	28.3
164	Tablespoon	1 tbs	13	0	117	0	0	0	13	5.1	5.5	1.8
	Margarine:											
165	Imitation (about 40% fat), soft:	1 c	232	58	800	1	1	0	90	17.9	36.4	32
166	Tablespoon	1 tbs	14	58	48	<1	<1	0	5	1.1	2.2	1.9
167	Regular, hard (about 80% fat):	½ c	114	16	820	1	1	0	92	18	40.8	29
168	Tablespoon	1 tbs	14	16	101	<1	<1	0	11	2.2	5	3.6
169	Pat	1 ea	5	16	36	<1	<1	0	4	.8	1.8	1.3
170	Regular, soft (about 80% fat):	1 c	227	16	1625	2	1	0	183	31.3	64.7	78.5
171	Tablespoon	1 tbs	14	16	100	<1	<1	0	11	1.9	4	4.8
2056	Saffola, unsalted	1 tbs	14	20	100	0	0	0	11	2	3	4.5
2057	Saffola, reduced fat	1 tbs	14	37	60	0	0	0	8	1.3	2.7	4.4
172	Spread (about 60% fat), hard:	1 c	227	37	1225	1	0	0	138	32	59	41.1
173	Tablespoon	1 tbs	14	37	76	<1	0	0	9	2	3.6	2.5
174	Pat	1 ea	5	37	27	<1	0	0	3	.7	1.2	1
175	Spread (about 60% fat), soft:	1 c	227	37	1225	1	0	0	138	29.3	71.5	31.3
176	Tablespoon	1 tbs	14	37	76	<1	0	0	9	1.8	4.4	1.9
2160	Touch of Butter (47% fat)	1 tbs	14		60	0	0	0	7	1.5	3.1	1.5
	Oils:											
1585	Canola:	1 c	218	0	1927	0	0	0	218	15.5	128	64.5
1586	Tablespoon	1 tbs	14	0	124	0	0	0	14	1	8.2	4.1
177	Corn:	1 c	218	0	1927	0	0	0	218	29.4	54.1	131
178	Tablespoon	1 tbs	14	0	124	0	0	0	14	1.9	3.5	8.4
179	Olive:	1 c	216	0	1909	0	0	0	216	29.4	159	21.3
180	Tablespoon	1 tbs	14	0	124	0	0	0	14	1.9	10.3	1.4
1683	Olive, extra virgin	1 tbs	14		126	0	0	0	14	2	10.8	1.3
181	Peanut:	1 c	216	0	1909	0	0	0	216	40	99.8	71.3
182	Tablespoon	1 tbs	14	0	124	0	0	0	14	2.6	6.5	4.6
183	Safflower:	1 c	218	0	1927	0	0	0	218	19.8	26.4	162
184	Tablespoon	1 tbs	14	0	124	0	0	0	14	1.3	1.7	10.4
185	Soybean:	1 c	218	0	1927	0	0	0	218	32	50.8	126
186	Tablespoon	1 tbs	14	0	124	0	0	0	14	2.1	3.3	8.1
187	Soybean/cottonseed:	1 c	218	0	1927	0	0	0	218	39.5	64.3	105
188	Tablespoon	1 tbs	14	0	124	0	0	0	14	2.5	4.1	6.7
189	Sunflower:	1 c	218	0	1927	0	0	0	218	25.3	42.5	143
190	Tablespoon	1 tbs	14	0	124	0	0	0	14	1.6	2.7	9.2
	Salad dressings/sandwich spreads:											
191	Blue cheese, regular	1 tbs	15	32	76	1	1	0	8	1.5	1.8	4.2
1040	Low calorie	1 tbs	15	79	15	1	<1	<1	1	.2	.5	.4
1684	Caesar's	1 tbs	12	36	56	1	<1	<1	5	1	3.7	.5
192	French, regular	1 tbs	16	38	69	<1	3	0	7	1.5	1.3	3.5
193	Low calorie	1 tbs	16	71	21	<1	3	0	1	.1	.2	.5
194	Italian, regular	1 tbs	15	40	70	<1	2	0	7	1	1.7	4.2
195	Low calorie	1 tbs	15	84	16	<1	1	<1	1	.2	.3	.9
	Kraft, Deliciously Right:											
2150	1000 Island	1 tbs	16		35	0	4	0	2	.5		
2153	Bacon & tomato	1 tbs	16		31	1	2	0	3	.5		
2154	Cucumber ranch	1 tbs	16		31	0	1	0	3	.5		
2151	French	1 tbs	16		25	0	3	0	1	.2		
2152	Ranch	1 tbs	16		52	0	3	0	5	.8		
199	Mayo type, regular	1 tbs	15	40	58	<1	4	0	5	.7	1.3	2.7
1030	Low calorie	1 tbs	14	54	36	<1	3	0	3	.4	.7	1.4
	Mayonnaise:											
197	Imitation, low calorie	1 tbs	15	63	35	<1	2	0	3	.5	.7	1.6
196	Regular (soybean)	1 tbs	14	17	100	<1	<1	0	11	1.7	3.2	5.8
1488	Regular, low calorie, low sodium	1 tbs	14	63	32	<1	2	0	3	.5	.6	1.4
1493	Regular, low calorie	1 tbs	15	63	35	<1	2	0	3	.5	.7	1.6
198	Ranch, regular	1 tbs	15	39	80	0	<1	0	8	1.2		
2251	Low calorie	1 tbs	14	69	30	<1	1	0	2	.5		
1685	Russian	1 tbs	15	34	74	<1	2	0	8	1.1	1.8	4.4
1502	Salad dressing, low calorie, oil free	1 tbs	15	88	4	<1	1	<1	<1	0	0	0

PAGE KEY: A–4 = Beverages A–6 = Dairy A–10 = Eggs A–10 = Fat/Oil A–14 = Fruit A–20 = Bakery A–28 = Grain A–32 = Fish A–34 = Meats A–38= Poultry A–40 = Sausage A–42 = Mixed/Fast A–46 = Nuts/Seeds A–50 = Sweets A–52 = Vegetables/Legumes A–62 = Vegetarian A–62 = Misc A–66 = Soups/Sauces A–68 = Fast A–84 = Convenience A–88 = Baby foods

Chol (mg)	Calc (mg)	Iron (mg)	Magn (mg)	Pota (mg)	Sodi (mg)	Zinc (mg)	VT-A (RE)	Thia (mg)	VT-E (α-TE)	Ribo (mg)	Niac (mg)	V-B6 (mg)	Fola (μg)	VT-C (mg)
195		0			<1	.23	0	0	2.46	0	0	0	0	0
12		0			<1	.01	0	0	.16	0	0	0	0	0
0	41	0	4	59	2227	0	1853	.01	5.41	.05	.03	.01	2	<1
0	2	0		4	134	0	112	<.01	.33	<.01	<.01	<.01	<1	<1
0	34	.07	3	48	1075	0	911	.01	14.6	.04	.03	.01	1	<1
0	4	.01		6	132	0	112	<.01	1.79	<.01	<.01	<.01	<1	<1
0	1	<.01		2	47	0	40	<.01	.64	<.01	<.01	0	<1	<1
0	60	0	5	86	2447	0	1813	.02	27.2	.07	.04	.02	2	<1
0	4	0		5	151	0	112	<.01	1.68	<.01	<.01	<.01	<1	<1
	0	0			0		51							0
	0	0			115		51							0
0	47	0	4	68	2256	0	1813	.02	11.4	.06	.04	.01	2	<1
0	3	0		4	139	0	112	<.01	.7	<.01	<.01	<.01	<1	<1
0	1	0		1	50	0	40	0	.25	<.01	<.01	0	<1	<1
0	47	0	4	68	2256	0	1813	.02	20.5	.06	.04	.01	2	<1
0	3	0		4	139	0	112	<.01	1.26	<.01	<.01	<.01	<1	<1
0	0	0		0	110		100		1.27					0
0	0	0	0	0	0	0	0	0	45.8	0	0	0	0	0
0	0	0	0	0	0	0	0	0	2.94	0	0	0	0	0
0	0	0	0	0	0	0	0	0	46	0	0	0	0	0
0	0	0	0	0	0	0	0	0	2.95	0	0	0	0	0
0		.82		0	<1	.13	0	0	26.8	0	0	0	0	0
0		.05		0	<1	.01	0	0	1.74	0	0	0	0	0
									1.74					
0		.06			<1	.02	0	0	27.9	0	0	0	0	0
0		<.01			<1	<.01	0	0	1.81	0	0	0	0	0
0	0	0	0	0	0	0	0	0	94	0	0	0	0	0
0	0	0	0	0	0	0	0	0	6.03	0	0	0	0	0
0		.04		0	0	0	0	0	39.7	0	0	0	0	0
0		<.01		0	0	0	0	0	2.55	0	0	0	0	0
0	0	0	0	0	0	0	0	0	61.5	0	0	0	0	0
0	0	0	0	0	0	0	0	0	3.95	0	0	0	0	0
0	0	0	0	0	0	0	0	0	110	0	0	0	0	0
0	0	0	0	0	0	0	0	0	7.08	0	0	0	0	0
3	12	.03	0	6	164	.04	10	<.01	1.4	.01	.01	.01	1	<1
<1	13	.07	1	1	180	.04	<1	<.01	.13	.01	.01	<.01	<1	<1
12	23	.2	3	21	207	.13	7	<.01	.72	.02	.5	.01	2	1
0	2	.06	0	13	219	.01	21	<.01	1.35	<.01	0	<.01	1	0
0	2	.06	0	13	126	.03	21	0	.19	0	0	0	0	0
0	1	.03		2	118	.02	4	<.01	1.56	<.01	0	<.01	1	0
1		.03	0	2	118	.02	0	0	.22	0	0	0	0	0
2	0	0		27	160		0		.19					0
1	0	0		21	155		0		.75					0
0	0	0		10	232		0		.73					0
0	0	0		7	130		50		.42					0
0	0	0		5	165		0		1.31					0
4	2	.03		1	107	.03	13	<.01	.6	<.01	<.01	<.01	1	0
4	2	.03		1	99	.02	9	<.01	.6	<.01	0	<.01	1	0
4		0		1	75	.02	0	0	.96	0	0	0	0	0
8	3	.07		5	79	.02	12	0	1.65	0	<.01	.08	1	0
3	0	0	0	1	15	.01	1	0	.53	<.01	0	0	<1	0
4		0		1	75	.02	0	0	.96	0	0	0	0	0
5	0	0			105		0							0
5	5	0			120		0		.7					0
3	3	.09		24	130	.06	31	.01	1.53	.01	.09	<.01	2	1
0	1	.04	2	7	256	<.01	<1	0	0	0	<.01	<.01	<1	<1

Table A–1

Food Composition (Computer code number is for Wadsworth Diet Analysis Plus program) (For purposes of calculations, use "0" for t, <1, <.1, <.01, etc.)

Computer Code Number	Food Description	Measure	Wt (g)	H_2O (%)	Ener (kcal)	Prot (g)	Carb (g)	Dietary Fiber (g)	Fat (g)	Fat Breakdown (g) Sat	Mono	Poly
	FATS and OILS—Continued											
	Salad dressing, no cholesterol											
1605	Miracle Whip	1 tbs	15	57	48	0	2	0	4	1.1	1.1	2.1
203	Salad dressing, from recipe, cooked	1 tbs	16	69	25	1	2	0	2	.5	.6	.3
200	Tartar sauce, regular	1 tbs	14	34	74	<1	1	<1	8	1.5	2.6	4.1
1503	Low calorie	1 tbs	14	63	31	<1	2	<1	2	.4	.6	1.3
201	Thousand island, regular	1 tbs	16	46	60	<1	2	0	6	1	1.3	3.2
202	Low calorie	1 tbs	15	69	24	<1	2	<1	2	.2	.4	.9
204	Vinegar and oil	1 tbs	16	47	72	0	<1	0	8	1.5	2.4	3.9
	Wishbone:											
2180	Creamy Italian, lite	1 tbs	15		26	<1	2		2	.4	.9	.7
2166	Italian, lite	1 tbs	16	90	6	0	1		<1	0	.2	.1
8427	Ranch, lite	1 tbs	15	56	50	0	2	0	4	.7		
	FRUITS and FRUIT JUICES											
	Apples:											
	Fresh, raw, with peel:											
205	2¾" diam (about 3 per lb w/cores)	1 ea	138	84	81	<1	21	4	<1	.1	t	.1
206	3¼" diam (about 2 per lb w/cores)	1 ea	212	84	125	<1	32	6	1	.1	t	.2
207	Raw, peeled slices	1 c	110	84	63	<1	16	2	<1	.1	t	.1
208	Dried, sulfured	10 ea	64	32	156	1	42	6	<1	t	t	.1
209	Apple juice, bottled or canned	1 c	248	88	117	<1	29	<1	<1	t	t	.1
210	Applesauce, sweetened	1 c	255	80	194	<1	51	3	<1	.1	t	.1
211	Applesauce, unsweetened	1 c	244	88	105	<1	28	3	<1	t	t	t
	Apricots:											
212	Raw, w/o pits (about 12 per lb w/pits)	3 ea	105	86	50	1	12	3	<1	t	.2	.1
	Canned (fruit and liquid):											
213	Heavy syrup	1 c	240	78	199	1	52	4	<1	t	.1	t
214	Halves	3 ea	120	78	100	1	26	2	<1	t	t	t
215	Juice pack	1 c	244	87	117	2	30	4	<1	t	t	t
216	Halves	3 ea	108	87	52	1	13	2	<1	t	t	t
217	Dried, halves	10 ea	35	31	83	1	22	3	<1	t	.1	t
218	Dried, cooked, unsweetened, w/liquid	1 c	250	76	213	3	55	8	<1	t	.2	.1
219	Apricot nectar, canned	1 c	251	85	141	1	36	2	<1	t	.1	t
	Avocados, raw, edible part only:											
220	California	1 ea	173	73	306	4	12	8	30	4.5	19.5	3.5
221	Florida	1 ea	304	80	340	5	27	16	27	5.3	14.8	4.5
222	Mashed, fresh, average	1 c	230	74	370	5	17	11	35	5.6	22.1	4.5
	Bananas, raw, without peel:											
223	Whole, 8¾" long (175g w/peel)	1 ea	118	74	109	1	28	3	1	.2	t	.1
224	Slices	1 c	150	74	138	2	35	4	1	.3	.1	.1
1285	Bananas, dehydrated slices	½ c	50	3	173	2	44	4	1	.3	.1	.2
225	Blackberries, raw	1 c	144	86	75	1	18	8	1	t	.1	.3
	Blueberries:											
226	Fresh	1 c	145	85	81	1	20	4	1	t	.1	.2
227	Frozen, sweetened	10 oz	284	77	230	1	62	6	<1	t	.1	.2
228	Frozen, thawed	1 c	230	77	186	1	51	5	<1	t	t	.1
	Cherries:											
229	Sour, red pitted, canned water pack	1 c	244	90	88	2	22	3	<1	.1	.1	.1
230	Sweet, red pitted, raw	10 ea	68	81	49	1	11	2	1	.1	.2	.2
231	Cranberry juice cocktail,vitamin C added	1 c	253	85	144	0	36	<1	<1	t	t	.1
1411	Cranberry juice, low calorie	1 c	237	95	45	0	11	<1	0	0	0	0
232	Cranberry-apple juice,vitamin C added	1 c	245	83	164	<1	42	<1	0	0	0	0
233	Cranberry sauce, canned, strained	1 c	277	61	418	1	108	3	<1	t	.1	.2
234	Dates, whole, without pits	10 ea	83	22	228	2	61	6	<1	.2	.1	t
235	Dates, chopped	1 c	178	22	490	4	131	13	1	.3	.3	.1
236	Figs, dried	10 ea	190	28	485	6	124	18	2	.4	.5	1.1
	Fruit cocktail, canned, fruit and liq:											
237	Heavy syrup pack	1 c	248	80	181	1	47	2	<1	t	t	.1
238	Juice pack	1 c	237	87	109	1	28	2	<1	t	t	t
	Grapefruit:											
	Raw 3¾" diam (half w/rind = 241g)											
239	Pink/red, half fruit, edible part	1 ea	123	91	37	1	9	2	<1	t	t	t
240	White, half fruit, edible part	1 ea	118	90	39	1	10	1	<1	t	t	t
241	Canned sections with light syrup	1 c	254	84	152	1	39	1	<1	t	t	.1

PAGE KEY: A–4 = Beverages A–6 = Dairy A–10 = Eggs A–10 = Fat/Oil A–14 = Fruit A–20 = Bakery A–28 = Grain A–32 = Fish A–34 = Meats A–38= Poultry A–40 = Sausage A–42 = Mixed/Fast A–46 = Nuts/Seeds A–50 = Sweets A–52 = Vegetables/Legumes A–62 = Vegetarian A–62 = Misc A–66 = Soups/Sauces A–68 = Fast A–84 = Convenience A–88 = Baby foods

Chol (mg)	Calc (mg)	Iron (mg)	Magn (mg)	Pota (mg)	Sodi (mg)	Zinc (mg)	VT-A (RE)	Thia (mg)	VT-E (α-TE)	Ribo (mg)	Niac (mg)	V-B6 (mg)	Fola (μg)	VT-C (mg)
0	0	<.01	0	0	102	0	2	0	.64	0	0	0	0	0
9	13	.08	1	19	117	.06	20	.01	.3	.02	.04	<.01	1	<1
7	3	.13		11	99	.02	9	<.01	2.24	<.01	0	<.01	1	<1
3	2	.09		5	83	.02	2	0	.83	<.01	.01	<.01	<1	<1
4	2	.1		18	112	.02	15	<.01	.18	<.01	<.01	<.01	1	0
2	2	.09		17	150	.02	14	<.01	.18	<.01	0	<.01	1	0
0	0	0	0	1	<1	0	0	0	1.41	0	0	0	0	0
<1	0	0			148		0	0	.56	0	0			0
0	1	0			255		2	0	.24	0	0			<1
2	0	0			120		0							0
0	10	.25	7	159	0	.05	7	.02	.44	.02	.11	.07	4	8
0	15	.38	11	244	0	.08	11	.04	.68	.03	.16	.1	6	12
0	4	.08	3	124	0	.04	4	.02	.09	.01	.1	.05	<1	4
0	9	.9	10	288	56	.13	4	0	.35	.1	.59	.08	0	2
0	17	.92	7	295	7	.07	<1	.05	.02	.04	.25	.07	<1	2
0	10	.89	8	156	8	.1	3	.03	.03	.07	.48	.07	2	4
0	7	.29	7	183	5	.07	7	.03	.02	.06	.46	.06	1	3
0	15	.57	8	311	1	.27	274	.03	.93	.04	.63	.06	9	10
0	22	.72	17	336	10	.26	295	.05	2.14	.05	.9	.13	4	7
0	11	.36	8	168	5	.13	148	.02	1.07	.03	.45	.06	2	4
0	29	.73	24	403	10	.27	412	.04	2.17	.05	.84	.13	4	12
0	13	.32	11	178	4	.12	183	.02	.96	.02	.37	.06	2	5
0	16	1.65	16	482	3	.26	253	<.01	.52	.05	1.05	.05	4	1
0	40	4.18	42	1222	7	.65	590	.01	1.25	.07	2.36	.28	0	4
0	18	.95	13	286	8	.23	331	.02	.2	.03	.65	.05	3	2
0	19	2.04	71	1096	21	.73	106	.19	2.32	.21	3.32	.48	113	14
0	33	1.61	103	1483	15	1.28	185	.33	2.37	.37	5.84	.85	162	24
0	25	2.35	90	1377	23	.97	140	.25	3.08	.28	4.42	.64	142	18
0	7	.37	34	467	1	.19	9	.05	.32	.12	.64	.68	22	11
0	9	.46	43	594	1	.24	12	.07	.4	.15	.81	.87	29	14
0	11	.57	54	746	1	.3	15	.09	0	.12	1.4	.22	7	3
0	46	.82	29	282	0	.39	23	.04	1.02	.06	.58	.08	49	30
0	9	.25	7	129	9	.16	14	.07	1.45	.07	.52	.05	9	19
0	17	1.11	6	170	3	.17	11	.06	2.02	.15	.72	.17	19	3
0	14	.9	5	138	2	.14	9	.05	1.63	.12	.58	.14	15	2
0	27	3.34	15	239	17	.17	183	.04	.32	.1	.43	.11	19	5
0	10	.26	7	152	0	.04	14	.03	.09	.04	.27	.02	3	5
0	8	.38	5	45	5	.18	1	.02	0	.02	.09	.05	<1	90
0	21	.09	5	52	7	.05	1	.02	0	.02	.08	.04	<1	76
0	17	.15	5	66	5	.1	1	.01	0	.05	.15	.05	<1	78
0	11	.61	8	72	80	.14	6	.04	.28	.06	.28	.04	3	6
0	27	.95	29	541	2	.24	4	.07	.08	.08	1.83	.16	10	0
0	57	2.05	62	1160	5	.52	9	.16	.18	.18	3.92	.34	22	0
0	274	4.24	112	1352	21	.97	25	.13	9.5	.17	1.32	.43	14	2
0	15	.72	12	218	15	.2	50	.04	.72	.05	.93	.12	6	5
0	19	.5	17	225	9	.21	73	.03	.47	.04	.95	.12	6	6
0	13	.15	10	159	0	.09	32	.04	.31	.02	.23	.05	15	47
0	14	.07	11	175	0	.08	1	.04	.29	.02	.32	.05	12	39
0	36	1.02	25	328	5	.2	0	.1	.63	.05	.62	.05	22	54

A

Table A-1

Food Composition

(Computer code number is for Wadsworth Diet Analysis Plus program) (For purposes of calculations, use "0" for t, <1, <.1, <.01, etc.)

Computer Code Number	Food Description	Measure	Wt (g)	H_2O (%)	Ener (kcal)	Prot (g)	Carb (g)	Dietary Fiber (g)	Fat (g)	Fat Breakdown (g) Sat	Mono	Poly
	FRUITS and FRUIT JUICES											
	Grapefruit juice:											
242	Fresh, white, raw	1 c	247	90	96	1	23	<1	<1	t	t	.1
243	Canned, unsweetened	1 c	247	90	94	1	22	<1	<1	t	t	.1
244	Sweetened	1 c	250	87	115	1	28	<1	<1	t	t	.1
	Frozen concentrate, unsweetened:											
245	Undiluted, 6-fl-oz can	¾ c	207	62	302	4	72	1	1	.1	.1	.2
246	Diluted with 3 cans water	1 c	247	89	101	1	24	<1	<1	t	t	.1
	Grapes, raw European (adherent skin):											
247	Thompson seedless	10 ea	50	81	35	<1	9	<1	<1	.1	t	.1
248	Tokay/Emperor, seeded types	10 ea	50	81	35	<1	9	<1	<1	.1	t	.1
	Grape juice:											
249	Bottled or canned	1 c	253	84	154	1	38	<1	<1	.1	t	.1
	Frozen concentrate, sweetened:											
250	Undiluted, 6-fl-oz can, vit C added	¾ c	216	54	387	1	96	1	1	.2	t	.2
251	Diluted with 3 cans water, vit C added	1 c	250	87	128	<1	32	<1	<1	.1	t	.1
1410	Low calorie	1 c	253	84	154	1	38	<1	<1	.1	t	.1
252	Kiwi fruit, raw, peeled (88 g with peel)	1 ea	76	83	46	1	11	3	<1	t	t	.2
253	Lemons, raw, without peel and seeds (about 4 per lb whole)	1 ea	58	89	17	1	5	2	<1	t	t	.1
	Lemon juice:											
254	Fresh:	1 c	244	91	61	1	21	1	0	0	0	0
255	Tablespoon	1 tbs	15	91	4	<1	1	<1	0	0	0	0
256	Canned or bottled, unsweetened:	1 c	244	92	51	1	16	1	1	.1	t	.2
257	Tablespoon	1 tbs	15	92	3	<1	1	<1	<1	t	t	t
258	Frozen, single strength, unsweetened:	1 c	244	92	54	1	16	1	1	.1	t	.2
2298	Tablespoon	1 tbs	15	92	3	<1	1	<1	<1	t	t	t
	Lime juice:											
260	Fresh:	1 c	246	90	66	1	22	1	<1	t	t	.1
261	Tablespoon	1 tbs	15	90	4	<1	1	<1	<1	t	t	t
262	Canned or bottled, unsweetened	1 c	246	92	52	1	16	1	1	.1	.1	.2
263	Mangos, raw, edible part (300 g w/skin & seeds)	1 ea	207	82	135	1	35	4	1	.1	.2	.1
	Melons, raw, without rind and contents:											
264	Cantaloupe, 5" diam (2⅓ lb whole with refuse), orange flesh	½ ea	276	90	97	2	23	2	1	.2	t	.3
265	Honeydew, 6½" diam (5¼ lb whole with refuse), slice = ⅒ melon	1 pce	160	90	56	1	15	1	<1	t	t	.1
266	Nectarines, raw, w/o pits, 2¼" diam	1 ea	136	86	67	1	16	2	1	.1	.2	.3
	Oranges, raw:											
267	Whole w/o peel and seeds, 2⅝" diam (180 g with peel and seeds)	1 ea	131	87	62	1	15	3	<1	t	t	t
268	Sections, without membranes	1 c	180	87	85	2	21	4	<1	t	t	t
	Orange juice:											
269	Fresh, all varieties	1 c	248	88	112	2	26	<1	<1	.1	.1	.1
270	Canned, unsweetened	1 c	249	89	105	1	24	<1	<1	t	.1	.1
271	Chilled	1 c	249	88	110	2	25	<1	1	.1	.1	.2
	Frozen concentrate:											
272	Undiluted (6-oz can)	¾ c	213	58	339	5	81	2	<1	.1	.1	.1
273	Diluted w/3 parts water by volume	1 c	249	88	112	2	27	<1	<1	t	t	t
1345	Orange juice, from dry crystals	1 c	248	88	114	0	29	0	0	0	0	0
274	Orange and grapefruit juice, canned	1 c	247	89	106	1	25	<1	<1	t	t	t
	Papayas, raw:											
275	½" slices	1 c	140	89	55	1	14	3	<1	.1	.1	t
276	Whole, 3½" diam by 5⅛" w/o seeds and skin (1 lb w/refuse)	1 ea	304	89	119	2	30	5	<1	.1	.1	.1
1031	Papaya nectar, canned	1 c	250	85	143	<1	36	1	<1	.1	.1	.1
	Peaches:											
277	Raw, whole, 2½" diam, peeled, pitted (about 4 per lb whole)	1 ea	98	88	42	1	11	2	<1	t	t	t
278	Raw, sliced	1 c	170	88	73	1	19	3	<1	t	.1	.1
	Canned, fruit and liquid:											
279	Heavy syrup pack:	1 c	262	79	194	1	52	3	<1	t	.1	.1
280	Half	1 ea	98	79	72	<1	19	1	<1	t	t	t

Chol (mg)	Calc (mg)	Iron (mg)	Magn (mg)	Pota (mg)	Sodi (mg)	Zinc (mg)	VT-A (RE)	Thia (mg)	VT-E (α-TE)	Ribo (mg)	Niac (mg)	V-B6 (mg)	Fola (μg)	VT-C (mg)
0	22	.49	30	400	2	.12	2	.1	.12	.05	.49	.11	25	94
0	17	.49	25	378	2	.22	2	.1	.12	.05	.57	.05	26	72
0	20	.9	25	405	5	.15	0	.1	.12	.06	.8	.05	26	67
0	56	1.01	79	1001	6	.37	6	.3	.37	.16	1.6	.32	26	248
0	20	.35	27	336	2	.12	2	.1	.12	.05	.54	.11	9	83
0	5	.13	3	92	1	.02	3	.05	.35	.03	.15	.05	2	5
0	5	.13	3	92	1	.02	3	.05	.35	.03	.15	.05	2	5
0	23	.61	25	334	8	.13	3	.07	0	.09	.66	.16	7	<1
0	28	.78	32	160	15	.28	6	.11	.38	.2	.93	.32	9	179
0	10	.25	10	52	5	.1	2	.04	.12	.06	.31	.1	3	60
0	23	.61	25	334	8	.13	3	.07	0	.09	.66	.16	7	<1
0	20	.31	23	252	4	.13	14	.01	.85	.04	.38	.07	29	74
0	15	.35	5	80	1	.03	2	.02	.14	.01	.06	.05	6	31
0	17	.07	15	303	2	.12	5	.07	.22	.02	.24	.12	31	112
0	1	<.01	1	19	<1	.01	<1	<.01	.01	<.01	.01	.01	2	7
0	27	.32	19	249	51	.15	5	.1	.22	.02	.48	.1	25	60
0	2	.02	1	15	3	.01	<1	.01	.01	<.01	.03	.01	2	4
0	19	.29	19	217	2	.12	2	.14	.22	.03	.33	.15	23	77
0	1	.02	1	13	<1	.01	<1	.01	.01	<.01	.02	.01	1	5
0	22	.07	15	268	2	.15	2	.05	.22	.02	.25	.11	20	72
0	1	<.01	1	16	<1	.01	<1	<.01	.01	<.01	.01	.01	1	4
0	29	.57	17	185	39	.15	5	.08	.17	.01	.4	.07	19	16
0	21	.27	19	323	4	.08	805	.12	2.32	.12	1.21	.28	29	57
0	30	.58	30	853	25	.44	889	.1	.41	.06	1.58	.32	47	116
0	10	.11	11	434	16	.11	6	.12	.24	.03	.96	.09	10	40
0	7	.2	11	288	0	.12	101	.02	1.21	.06	1.35	.03	5	7
0	52	.13	13	237	0	.09	27	.11	.31	.05	.37	.08	40	70
0	72	.18	18	326	0	.13	38	.16	.43	.07	.51	.11	54	96
0	27	.5	27	496	2	.12	50	.22	.22	.07	.99	.1	75	124
0	20	1.1	27	436	5	.17	45	.15	.22	.07	.78	.22	45	86
0	25	.42	27	473	2	.1	20	.28	.47	.05	.7	.13	45	82
0	68	.75	72	1435	6	.38	60	.6	.68	.14	1.53	.33	330	294
0	22	.25	25	473	2	.12	20	.2	.47	.04	.5	.11	109	97
0	62	.2	2	50	12	.1	551	<.01	0	.04	0	0	143	121
0	20	1.14	25	390	7	.17	30	.14	.17	.07	.83	.06	35	72
0	34	.14	14	360	4	.1	39	.04	1.57	.04	.47	.03	53	86
0	73	.3	30	781	9	.21	85	.08	3.4	.1	1.03	.06	116	188
0	25	.85	7	77	12	.37	27	.01	.05	.01	.37	.02	5	7
0	5	.11	7	193	0	.14	53	.02	.69	.04	.97	.02	3	6
0	8	.19	12	335	0	.24	92	.03	1.19	.07	1.68	.03	6	11
0	8	.71	13	241	16	.24	86	.03	2.33	.06	1.61	.05	8	7
0	3	.26	5	90	6	.09	32	.01	.87	.02	.6	.02	3	3

Table A-1

Food Composition (Computer code number is for Wadsworth Diet Analysis Plus program) (For purposes of calculations, use "0" for t, <1, <.1, <.01, etc.)

Computer Code Number	Food Description	Measure	Wt (g)	H_2O (%)	Ener (kcal)	Prot (g)	Carb (g)	Dietary Fiber (g)	Fat (g)	Fat Breakdown (g) Sat	Mono	Poly
	FRUITS and FRUIT JUICES—Continued											
281	Juice pack:	1 c	248	87	109	2	29	3	<1	t	t	t
282	Half	1 ea	98	87	43	1	11	1	<1	t	t	t
283	Dried, uncooked	10 ea	130	32	311	5	80	11	1	.1	.4	.5
284	Dried, cooked, fruit and liquid	1 c	258	78	199	3	51	7	1	.1	.2	.3
	Frozen, slice, sweetened:											
285	10-oz package,vitamin C added	1 ea	284	75	267	2	68	5	<1	t	.1	.2
286	Cup, thawed measure, vitamin C added	1 c	250	75	235	2	60	4	<1	t	.1	.2
1032	Peach nectar, canned	1 c	249	86	134	1	35	1	<1	t	t	t
	Pears:											
	Fresh, with skin, cored:											
287	Bartlett, 2½" diam (about 2½ per lb)	1 ea	166	84	98	1	25	4	1	t	.1	.2
288	Bosc, 2⅕" diam (about 3 per lb)	1 ea	139	84	82	1	21	3	1	t	.1	.1
289	D'Anjou, 3" diam (about 2 per lb)	1 ea	209	84	123	1	32	5	1	t	.2	.2
	Canned, fruit and liquid:											
290	Heavy syrup pack:	1 c	266	80	197	1	51	4	<1	t	.1	.1
291	Half	1 ea	76	80	56	<1	15	1	<1	t	t	t
292	Juice pack:	1 c	248	86	124	1	32	4	<1	t	t	t
293	Half	1 ea	76	86	38	<1	10	1	<1	t	t	t
294	Dried halves	10 ea	175	27	459	3	122	13	1	.1	.2	.3
1033	Pear nectar, canned	1 c	250	84	150	<1	39	1	<1	t	t	t
	Pineapple:											
295	Fresh chunks, diced	1 c	155	86	76	1	19	2	1	t	.1	.2
	Canned, fruit and liquid:											
	Heavy syrup pack:											
296	Crushed, chunks, tidbits	½ c	127	79	99	<1	26	1	<1	t	t	.1
297	Slices	1 ea	49	79	38	<1	10	<1	<1	t	t	t
298	Juice pack, crushed, chunks, tidbits	1 c	250	83	150	1	39	2	<1	t	t	.1
299	Juice pack, slices	1 ea	47	83	28	<1	7	<1	<1	t	t	t
300	Pineapple juice, canned, unsweetened	1 c	250	85	140	1	34	<1	<1	t	t	.1
	Plantains, yellow flesh, without peel:											
301	Raw slices (whole=179 g w/o peel)	1 c	148	65	181	2	47	3	1	.2	t	.1
302	Cooked, boiled, sliced	1 c	154	67	179	1	48	4	<1	.1	t	.1
	Plums:											
303	Fresh, medium, 2⅛" diam	1 ea	66	85	36	1	9	1	<1	t	.3	.1
304	Fresh, small, 1½" diam	1 ea	28	85	15	<1	4	<1	<1	t	.1	t
	Canned, purple, with liquid:											
305	Heavy syrup pack:	1 c	258	76	230	1	60	3	<1	t	.2	.1
306	Plums	3 ea	138	76	123	<1	32	1	<1	t	.1	t
307	Juice pack:	1 c	252	84	146	1	38	3	<1	t	t	t
308	Plums	3 ea	138	84	80	1	21	1	<1	t	t	t
1698	Pomegranate, fresh	1 ea	154	81	105	1	26	1	<1	.1	.1	.1
	Prunes, dried, pitted:											
309	Uncooked (10 = 97 g w/pits, 84 g w/o pits)	10 ea	84	32	201	2	53	6	<1	t	.3	.1
310	Cooked, unsweetened, fruit & liq (250 g w/pits)	1 c	248	70	265	3	70	16	1	t	.4	.1
311	Prune juice, bottled or canned	1 c	256	81	182	2	45	3	<1	t	.1	t
	Raisins, seedless:											
312	Cup, not pressed down	1 c	145	15	435	5	115	6	1	.2	t	.2
313	One packet, ½ oz	½ oz	14	15	42	<1	11	1	<1	t	t	t
	Raspberries:											
314	Fresh	1 c	123	87	60	1	14	8	1	t	.1	.4
315	Frozen, sweetened	10 oz	284	73	293	2	74	12	<1	t	t	.3
316	Cup, thawed measure	1 c	250	73	258	2	65	11	<1	t	t	.2
317	Rhubarb, cooked, added sugar	1 c	240	68	278	1	75	5	<1	t	t	.1
	Strawberries:											
318	Fresh, whole, capped	1 c	144	92	43	1	10	3	1	t	.1	.3
	Frozen, sliced, sweetened:											
319	10-oz container	10 oz	284	73	273	2	74	5	<1	t	.1	.2
320	Cup, thawed measure	1 c	255	73	245	1	66	5	<1	t	t	.2
	Tangerines, without peel and seeds:											
321	Fresh (2⅜" whole) 116 g w/refuse	1 ea	84	88	37	1	9	2	<1	t	t	t
322	Canned, light syrup, fruit and liquid	1 c	252	83	154	1	41	2	<1	t	t	t

Chol (mg)	Calc (mg)	Iron (mg)	Magn (mg)	Pota (mg)	Sodi (mg)	Zinc (mg)	VT-A (RE)	Thia (mg)	VT-E (α-TE)	Ribo (mg)	Niac (mg)	V-B6 (mg)	Fola (µg)	VT-C (mg)
0	15	.67	17	317	10	.27	94	.02	3.72	.04	1.44	.05	8	9
0	6	.26	7	125	4	.11	37	.01	1.47	.02	.57	.02	3	4
0	36	5.28	55	1294	9	.74	281	<.01	0	.28	5.69	.09	<1	6
0	23	3.38	33	826	5	.46	52	.01	0	.05	3.92	.1	<1	10
0	9	1.05	14	369	17	.14	79	.04	2.53	.1	1.85	.05	9	268
0	7	.92	12	325	15	.12	70	.03	2.23	.09	1.63	.04	8	236
0	12	.47	10	100	17	.2	65	.01	.2	.03	.72	.02	3	13
0	18	.41	10	208	0	.2	3	.03	.83	.07	.17	.03	12	7
0	15	.35	8	174	0	.17	3	.03	.69	.06	.14	.02	10	6
0	23	.52	12	261	0	.25	4	.04	1.05	.08	.21	.04	15	8
0	13	.58	11	173	13	.21	0	.03	1.33	.06	.64	.04	3	3
0	4	.17	3	49	4	.06	0	.01	.38	.02	.18	.01	1	1
0	22	.72	17	238	10	.22	2	.03	1.24	.03	.5	.03	3	4
0	7	.22	5	73	3	.07	1	.01	.38	.01	.15	.01	1	1
0	59	3.68	58	933	10	.68	1	.01	0	.25	2.4	.13	0	12
0	12	.65	7	32	10	.17	<1	<.01	.25	.03	.32	.03	3	3
0	11	.57	22	175	2	.12	3	.14	.15	.06	.65	.13	16	24
0	18	.48	20	132	1	.15	1	.11	.13	.03	.36	.09	6	9
0	7	.19	8	51	<1	.06	<1	.04	.05	.01	.14	.04	2	4
0	35	.7	35	305	2	.25	10	.24	.25	.05	.71	.18	12	24
0	7	.13	7	57	<1	.05	2	.04	.05	.01	.13	.03	2	4
0	42	.65	32	335	2	.27	1	.14	.05	.05	.64	.24	58	27
0	4	.89	55	739	6	.21	167	.08	.4	.08	1.02	.44	33	27
0	3	.89	49	716	8	.2	140	.07	.22	.08	1.16	.37	40	17
0	3	.07	5	114	0	.07	21	.03	.4	.06	.33	.05	1	6
0	1	.03	2	48	0	.03	9	.01	.17	.03	.14	.02	1	3
0	23	2.17	13	235	49	.18	67	.04	1.81	.1	.75	.07	6	1
0	12	1.16	7	126	26	.1	36	.02	.97	.05	.4	.04	3	1
0	25	.86	20	388	3	.28	255	.06	1.76	.15	1.19	.07	7	7
0	14	.47	11	213	1	.15	139	.03	.97	.08	.65	.04	4	4
0	5	.46	5	399	5	.18	0	.05	.85	.05	.46	.16	9	9
0	43	2.08	38	626	3	.44	167	.07	1.22	.14	1.65	.22	3	3
0	57	2.75	50	828	5	.59	77	.06	<.01	.25	1.79	.54	<1	7
0	31	3.02	36	707	10	.54	1	.04	.03	.18	2.01	.56	1	10
0	71	3.02	48	1088	17	.39	1	.23	1.02	.13	1.19	.36	5	5
0	7	.29	5	105	2	.04	<1	.02	.1	.01	.11	.03	<1	<1
0	27	.7	22	187	0	.57	16	.04	.55	.11	1.11	.07	32	31
0	43	1.85	37	324	3	.51	17	.05	1.28	.13	.65	.1	74	47
0	37	1.63	32	285	2	.45	15	.05	1.13	.11	.57	.08	65	41
0	348	.5	29	230	2	.19	17	.04	.48	.05	.48	.05	13	8
0	20	.55	14	239	1	.19	4	.03	.2	.09	.33	.08	25	82
0	31	1.68	20	278	9	.17	6	.04	.4	.14	1.14	.08	42	118
0	28	1.5	18	250	8	.15	5	.04	.36	.13	1.02	.08	38	106
0	12	.08	10	132	1	.2	77	.09	.2	.02	.13	.06	17	26
0	18	.93	20	197	15	.6	212	.13	.86	.11	1.12	.11	12	50

Table A–1

Food Composition (Computer code number is for Wadsworth Diet Analysis Plus program) (For purposes of calculations, use "0" for t, <1, <.1, <.01, etc.)

Computer Code Number	Food Description	Measure	Wt (g)	H_2O (%)	Ener (kcal)	Prot (g)	Carb (g)	Dietary Fiber (g)	Fat (g)	Fat Breakdown (g) Sat	Mono	Poly
	FRUITS and FRUIT JUICES—Continued											
323	Tangerine juice, canned, sweetened	1 c	249	87	125	1	30	<1	<1	t	t	.1
	Watermelon, raw, without rind and seeds:											
324	Piece, 1/16 wedge	1 pce	286	91	91	2	20	1	1	.1	.3	.4
325	Diced	1 c	152	91	49	1	11	1	1	.1	.2	.2
	BAKED GOODS: BREADS, CAKES, COOKIES, CRACKERS, PIES											
326	Bagels, plain, enriched, 3½" diam.	1 ea	71	33	195	7	38	2	1	.2	.1	.5
1663	Bagel, oat bran	1 ea	71	33	181	8	38	3	1	.1	.2	.3
	Biscuits:											
327	From home recipe	1 ea	60	29	212	4	27	1	10	2.6	4.2	2.5
328	From mix	1 ea	57	29	191	4	28	1	7	1.6	2.4	2.5
329	From refrigerated dough	1 ea	74	27	276	4	34	1	13	8.7	3.4	.5
330	Bread crumbs, dry, grated (see # 364, 365 for soft crumbs)	1 c	108	6	427	13	78	3	6	1.4	2.3	1.7
2087	Bread sticks, brown & serve	1 ea	57	34	150	7	28	1	1	.5	.5	.5
	Breads:											
331	Boston brown, canned, 3¼" slice	1 pce	45	47	88	2	19	2	1	.1	.1	.3
332	Cracked wheat (¼ cracked-wheat & ¾ enr wheat flour): 1-lb loaf	1 ea	454	36	1180	39	225	25	18	4.2	8.6	3.1
333	Slice (18 per loaf)	1 pce	25	36	65	2	12	1	1	.2	.5	.2
334	Slice, toasted	1 pce	23	30	65	2	12	1	1	.2	.5	.2
335	French/Vienna, enriched: 1-lb loaf	1 ea	454	34	1243	40	236	14	14	2.9	5.5	3.1
337	Slice, 4¾ x 4 x ½"	1 pce	25	34	68	2	13	1	1	.2	.3	.2
336	French, slice, 5 x 2½"	1 pce	25	34	68	2	13	1	1	.2	.3	.2
	French toast: see Mixed Dishes, and Fast Foods, #691											
2083	Honey wheatberry	1 pce	38	38	100	3	18	2	1	0	.5	
338	Italian, enriched: 1-lb loaf	1 ea	454	36	1230	40	227	12	16	3.9	3.7	6.3
339	Slice, 4½ x 3¼ x ¾"	1 pce	30	36	81	3	15	1	1	.3	.2	.4
340	Mixed grain, enriched: 1-lb loaf	1 ea	454	38	1135	45	211	29	17	3.7	6.9	4.2
341	Slice (18 per loaf)	1 pce	26	38	65	3	12	2	1	.2	.4	.2
342	Slice, toasted	1 pce	24	32	65	3	12	2	1	.2	.4	.2
343	Oatmeal, enriched: 1-lb loaf	1 ea	454	37	1221	38	220	18	20	3.2	7.2	7.7
344	Slice (18 per loaf)	1 pce	27	37	73	2	13	1	1	.2	.4	.5
345	Slice, toasted	1 pce	25	31	73	2	13	1	1	.2	.4	.5
346	Pita pocket bread, enr, 6½" round	1 ea	60	32	165	5	33	1	1	.1	.1	.3
	Pumpernickel (⅔ rye & ⅓ enr wheat flr):											
347	1-lb loaf	1 ea	454	38	1135	40	216	29	14	2	4.2	5.6
348	Slice, 5 x 4 x ⅜"	1 pce	26	38	65	2	12	2	1	.1	.2	.3
349	Slice, toasted	1 pce	29	32	80	3	15	2	1	.1	.3	.4
350	Raisin, enriched: 1-lb loaf	1 ea	454	34	1243	36	237	19	20	4.9	10.5	3.1
351	Slice (18 per loaf)	1 pce	26	34	71	2	14	1	1	.3	.6	.2
352	Slice, toasted	1 pce	24	28	71	2	14	1	1	.3	.6	.2
353	Rye, light (⅓ rye & ⅔ enr wheat flr): 1-lb loaf	1 ea	454	37	1175	39	219	26	15	2.9	6	3.6
354	Slice, 4¾ x 3¾ x 7/16"	1 pce	32	37	83	3	15	2	1	.2	.4	.3
355	Slice, toasted	1 pce	24	31	68	2	13	2	1	.2	.3	.2
356	Wheat (enr wheat & whole-wheat flour): 1-lb loaf	1 ea	454	37	1160	43	213	25	19	3.9	7.3	4.5
357	Slice (18 per loaf)	1 pce	25	37	65	2	12	1	1	.2	.4	.2
358	Slice, toasted	1 pce	23	32	65	2	12	1	1	.2	.4	.2
359	White, enriched: 1-lb loaf	1 ea	454	35	1293	36	225	9	26	5.4	5.9	12.6
360	Slice	1 pce	42	35	120	3	21	1	2	.5	.5	1.2
361	Slice, toasted	1 pce	38	29	119	3	21	1	2	.5	.5	1.2
366	Whole-wheat: 1-lb loaf	1 ea	454	38	1116	44	209	31	19	4.2	7.6	4.5
367	Slice (16 per loaf)	1 pce	28	38	69	3	13	2	1	.3	.5	.3
368	Slice, toasted	1 pce	25	30	69	3	13	2	1	.3	.5	.3
	Bread stuffing, prepared from mix:											
369	Dry type	1 c	200	65	356	6	43	6	17	3.5	7.6	5.2
370	Moist type, with egg and margarine	1 c	232	65	390	9	51	5	17	3.4	7.4	4.9

Chol (mg)	Calc (mg)	Iron (mg)	Magn (mg)	Pota (mg)	Sodi (mg)	Zinc (mg)	VT-A (RE)	Thia (mg)	VT-E (α-TE)	Ribo (mg)	Niac (mg)	V-B6 (mg)	Fola (μg)	VT-C (mg)
0	45	.5	20	443	2	.07	105	.15	.22	.05	.25	.08	11	55
0	23	.49	31	332	6	.2	106	.23	.43	.06	.57	.41	6	27
0	12	.26	17	176	3	.11	56	.12	.23	.03	.3	.22	3	15
0	52	2.53	21	72	379	.62	0	.38	.02	.22	3.24	.04	62	0
0	9	2.19	40	145	360	1.48	<1	.23	.17	.24	2.1	.14	57	<1
2	141	1.74	11	73	348	.32	14	.21	1.45	.19	1.77	.02	37	<1
2	105	1.17	14	107	544	.35	15	.2	.23	.2	1.72	.04	3	<1
5	89	1.64	9	87	584	.29	24	.27	.44	.18	1.63	.03	6	0
0	245	6.61	50	239	931	1.32	<1	.83	.95	.47	7.4	.11	118	0
0	60	2.7			290		0	.22		.1	1.6			0
<1	31	.94	28	143	284	.22	5	.01	.13	.05	.5	.04	5	0
														0
0	195	12.8	236	804	2442	5.63	0	1.63	2.56	1.09	16.7	1.38	277	
0	11	.7	13	44	135	.31	0	.09	.14	.06	.92	.08	15	0
0	11	.7	13	44	135	.31	0	.07	.14	.05	.83	.07	7	0
0	341	11.5	123	513	2764	3.95	0	2.36	1.07	1.49	21.6	.19	431	0
0	19	.63	7	28	152	.22	0	.13	.06	.08	1.19	.01	24	0
0	19	.63	7	28	152	.22	0	.13	.06	.08	1.19	.01	24	0
0	20	.72			200		0	.12	.24	.07	.8			0
0	354	13.3	123	499	2651	3.9	0	2.15	1.26	1.33	19.9	.22	431	0
0	23	.88	8	33	175	.26	0	.14	.08	.09	1.31	.01	28	0
0	413	15.8	241	926	2210	5.77	0	1.85	2.79	1.55	19.8	1.51	363	1
0	24	.9	14	53	127	.33	0	.11	.16	.09	1.14	.09	21	<1
0	24	.9	14	53	127	.33	0	.08	.16	.08	1.02	.08	16	<1
0	300	12.3	168	645	2719	4.63	9	1.81	1.56	1.09	14.3	.31	281	2
0	18	.73	10	38	162	.27	1	.11	.09	.06	.85	.02	17	<1
0	18	.73	10	38	163	.28	<1	.09	.09	.06	.77	.02	13	<1
0	52	1.57	16	72	322	.5	0	.36	.02	.2	2.78	.02	57	0
0	309	13	245	944	3046	6.72	0	1.48	2.3	1.38	14	.57	363	0
0	18	.75	14	54	174	.38	0	.08	.13	.08	.8	.03	21	0
0	21	.91	17	66	214	.47	0	.08	.17	.09	.89	.04	20	0
0	300	13.2	118	1030	1770	3.27	1	1.54	3.44	1.81	15.8	.31	395	2
0	17	.75	7	59	101	.19	<1	.09	.2	.1	.9	.02	23	<1
0	17	.76	7	59	102	.19	<1	.07	.2	.09	.81	.02	18	<1
0	331	12.8	182	754	2996	5.18	2	1.97	2.51	1.52	17.3	.34	390	1
0	23	.91	13	53	211	.36	<1	.14	.18	.11	1.22	.02	27	<1
0	19	.74	10	44	174	.3	<1	.09	.15	.08	.9	.02	17	<1
0	572	15.8	209	627	2447	4.77	0	2.09	3	1.45	20.5	.49	204	0
0	26	.83	11	50	133	.26	0	.1	.14	.07	1.03	.02	19	0
0	26	.83	11	50	132	.26	0	.08	.14	.06	.93	.02	15	0
14	259	13.5	86	663	1629	2.91	100	1.84	4.95	1.74	16.3	.23	413	1
1	24	1.25	8	61	151	.27	9	.17	.46	.16	1.51	.02	38	<1
1	24	1.24	8	61	150	.27	8	.13	.46	.14	1.35	.02	12	<1
0	327	15	390	1144	2392	8.81	0	1.59	4.72	.93	17.4	.81	227	0
0	20	.92	24	71	148	.54	0	.1	.29	.06	1.08	.05	14	0
0	20	.93	24	71	148	.54	0	.08	.23	.05	.97	.04	9	0
0	64	2.18	24	148	1086	.56	162	.27	2.8	.21	2.96	.08	202	0
0	148	3.8	35	304	1069	.74	160	.39	2.78	.33	3.69	.12	39	4

Table A-1

Food Composition (Computer code number is for Wadsworth Diet Analysis Plus program) (For purposes of calculations, use "0" for t, <1, <.1, <.01, etc.)

Computer Code Number	Food Description	Measure	Wt (g)	H_2O (%)	Ener (kcal)	Prot (g)	Carb (g)	Dietary Fiber (g)	Fat (g)	Fat Breakdown (g) Sat	Mono	Poly
	BAKED GOODS: BREADS, CAKES, COOKIES, CRACKERS, PIES—Continued											
	Cakes, prepared from mixes using enriched flour and veg shortening, w/frostings made from margarine:											
	Angel food:											
371	Whole cake, 9 ¾" diam tube	1 ea	340	33	877	20	197	5	3	.4	.2	1.2
372	Piece, 1/12 of cake	1 pce	28	33	72	2	16	<1	<1	t	t	.1
373	Boston cream pie, ⅛ of cake	1 pce	123	45	310	3	53	2	10	3.1	5.4	1.2
	Coffee cake:											
374	Whole cake, 7¾ x 5⅝ x 1¼"	1 ea	336	30	1068	18	177	4	32	6.3	13	10.7
375	Piece, ⅙ of cake	1 pce	56	30	178	3	30	1	5	1	2.2	1.8
	Devil's food, chocolate frosting:											
376	Whole cake, 2 layer, 8 or 9" diam	1 ea	1021	23	3747	42	557	29	167	47.9	91.9	19.5
377	Piece, 1/16 of cake	1 pce	64	23	235	3	35	2	10	3	5.8	1.2
378	Cupcake, 2½" diam	1 ea	42	23	154	2	23	1	7	2	3.8	.8
	Gingerbread:											
379	Whole cake, 8" square	1 ea	603	33	1863	24	306	7	61	15.8	34	8.1
380	Piece, 1/9 of cake	1 pce	67	33	207	3	34	1	7	1.8	3.8	.9
	Yellow, chocolate frosting, 2 layer:											
381	Whole cake, 8 or 9" in diam	1 ea	1024	22	3880	39	567	18	178	49	99	21.4
382	Piece, 1/16 of cake	1 pce	64	22	243	2	35	1	11	3.1	6.2	1.3
	Cakes from home recipes w/enr flour:											
	Carrot cake, made with veg oil, cream cheese frosting:											
383	Whole, 9 x 13" cake	1 ea	1776	21	7743	82	838	21	469	86.8	116	242
384	Piece, 1/16 of cake, 2¼ x 3¼" slice	1 pce	111	21	484	5	52	1	29	5.4	7.2	15.1
	Fruitcake, dark:											
385	Whole cake, 7½"diam tube, 2¼"high	1 ea	1376	25	4458	40	848	51	125	15.4	57.4	44.6
386	Piece, 1/32 of cake, ⅔" arc	1 pce	43	25	139	1	26	2	4	.5	1.8	1.4
	Sheet, plain, made w/veg shortening, no frosting:											
387	Whole cake, 9" square	1 ea	774	24	2817	35	433	3	108	29.9	51.5	25.5
388	Piece, 1/9 of cake	1 pce	86	24	313	4	48	<1	12	3.3	5.7	2.8
	Sheet, plain, made w/margarine, uncooked white frosting:											
389	Whole cake, 9" square	1 ea	576	22	2148	20	339	2	83	13.8	35.5	29.5
390	Piece, 1/9 of cake	1 pce	64	22	239	2	38	<1	9	1.5	3.9	3.3
	Cakes, commerical:											
	Cheesecake:											
401	Whole cake, 9" diam	1 ea	960	46	3081	53	245	4	216	111	74.4	13.2
402	Piece, 1/12 of cake	1 pce	80	46	257	4	20	<1	18	9.2	6.2	1.1
	Pound cake:											
393	Loaf, 8½ x 3½ x 3"	1 ea	340	25	1319	19	166	2	68	38.1	19	3.7
394	Slice, 1/17 of loaf, 2" slice	1 pce	28	25	109	2	14	<1	6	3.1	1.6	.3
	Snack cakes:											
395	Chocolate w/creme filling,Ding Dong	1 ea	50	20	188	2	30	<1	7	1.6	2.7	2.1
396	Sponge cake w/creme filling,Twinkie	1 ea	43	20	157	1	27	<1	5	1.2	1.9	1.5
1677	Sponge cake, 1/12 of 12" cake	1 pce	38	30	110	2	23	<1	1	.3	.4	.2
	White, white frosting, 2 layer:											
397	Whole cake, 8 or 9" diam	1 ea	1136	20	4260	37	716	11	153	68.2	60.1	15.4
398	Piece, 1/16 of cake	1 pce	71	20	266	2	45	1	10	4.3	3.8	1
	Yellow, chocolate frosting, 2 layer:											
399	Whole cake, 8 or 9" in diam	1 ea	1024	22	3880	39	567	18	178	49	99	21.4
400	Piece, 1/16 of cake	1 pce	64	22	243	2	35	1	11	3.1	6.2	1.3
1332	Bagel chips	5 pce	70	3	298	6	52	6	7	1.2	2	3.4
2225	Bagel chips, onion garlic, toasted	1 oz	28		193	5	31	3	8	1.7	5.2	0
1035	Cheese puffs/Cheetos	1 c	20	1	111	2	11	<1	7	1.3	4.1	1
	Cookies made with enriched flour:											
	Brownies with nuts:											
403	Commercial w/frosting, 1½ x 1¾ x ⅞"	1 ea	61	14	247	3	39	1	10	2.6	5.1	1.6
1902	Fat free fudge, Entenmann's	1 pce	40	24	110	2	27	1	0	0	0	0

PAGE KEY: A–4 = Beverages A–6 = Dairy A–10 = Eggs A–10 = Fat/Oil A–14 = Fruit A–20 = Bakery A–28 = Grain A–32 = Fish A–34 = Meats A–38= Poultry A–40 = Sausage A–42 = Mixed/Fast A–46 = Nuts/Seeds A–50 = Sweets A–52 = Vegetables/Legumes A–62 = Vegetarian A–62 = Misc A–66 = Soups/Sauces A–68 = Fast A–84 = Convenience A–88 = Baby foods

Chol (mg)	Calc (mg)	Iron (mg)	Magn (mg)	Pota (mg)	Sodi (mg)	Zinc (mg)	VT-A (RE)	Thia (mg)	VT-E (α-TE)	Ribo (mg)	Niac (mg)	V-B6 (mg)	Fola (μg)	VT-C (mg)
0	476	1.77	41	316	2546	.24	0	.35	.34	1.67	3	.1	119	0
0	39	.15	3	26	210	.02	0	.03	.03	.14	.25	.01	10	0
45	28	.47	7	48	177	.2	28	.5	1.3	.33	.23	.03	18	<1
165	457	4.8	60	376	1414	1.51	134	.56	5.58	.59	5.11	.17	228	1
27	76	.8	10	63	236	.25	22	.09	.93	.1	.85	.03	38	<1
470	439	22.5	347	2042	3410	7.04	286	.28	17.3	1.36	5.89	.32	174	1
29	27	1.41	22	128	214	.44	18	.02	1.08	.08	.37	.02	11	<1
19	18	.92	14	84	140	.29	12	.01	.71	.06	.24	.01	7	<1
211	416	20	96	1453	2761	2.47	96	1.14	8.26	1.12	9.41	.23	60	1
23	46	2.22	11	161	307	.27	11	.13	.92	.12	1.05	.02	7	<1
563	379	21.3	307	1822	3450	6.35	276	1.23	27.6	1.61	12.8	.3	225	1
35	24	1.33	19	114	216	.4	17	.08	1.73	.1	.8	.02	14	<1
959	444	22.2	320	1989	4368	8.7	6819	2.42	74.9	2.77	17.9	1.35	213	19
60	28	1.39	20	124	273	.54	426	.15	4.68	.17	1.12	.08	13	1
69	454	28.5	220	2105	3715	3.72	261	.69	42.9	1.36	10.9	.63	261	5
2	14	.89	7	66	116	.12	8	.02	1.34	.04	.34	.02	8	<1
503	495	11.7	108	611	2322	2.74	372	1.24	11	1.39	10.1	.26	54	2
56	55	1.3	12	68	258	.3	41	.14	1.22	.15	1.12	.03	6	<1
323	357	6.16	35	305	1981	1.44	109	.58	10.9	.4	2.88	.2	156	1
36	40	.68	4	34	220	.16	12	.06	1.22	.04	.32	.02	17	<1
528	490	6.05	106	864	1987	4.9	1545	.27	10.1	1.85	1.87	.5	173	6
44	41	.5	9	72	166	.41	129	.02	.84	.15	.16	.04	14	<1
751	119	4.69	37	405	1353	1.56	530	.47	2.24	.78	4.45	.12	139	<1
62	10	.39	3	33	111	.13	44	.04	.18	.06	.37	.01	11	<1
8	36	1.68	20	61	213	.28	2	.11	1.01	.15	1.22	.01	14	<1
7	19	.55	3	39	157	.13	2	.07	.83	.06	.52	.01	12	<1
39	27	1.03	4	38	93	.19	17	.09	.17	.1	.73	.02	15	0
91	545	9.09	60	659	2658	1.76	368	1.14	20.4	1.48	10.2	.16	64	1
6	34	.57	4	41	166	.11	23	.07	1.28	.09	.64	.01	4	<1
563	379	21.3	307	1822	3450	6.35	276	1.23	27.6	1.61	12.8	.3	225	1
35	24	1.33	19	114	216	.4	17	.08	1.73	.1	.8	.02	14	<1
0	9	1.38	41	167	419	.9	0	.13	.46	.12	1.57	.19	58	0
0	0	2.52			490		0	.39	<.01	.24	3.5			0
1	12	.47	4	33	210	.08	7	.05	1.02	.07	.65	.03	24	<1
10	18	1.37	19	91	190	.44	12	.16	1.3	.13	1.05	.02	13	<1
0	0	1.08		90	140		0		.01					0

Table A-1

Food Composition (Computer code number is for Wadsworth Diet Analysis Plus program) (For purposes of calculations, use "0" for t, <1, <.1, <.01, etc.)

Computer Code Number	Food Description	Measure	Wt (g)	H_2O (%)	Ener (kcal)	Prot (g)	Carb (g)	Dietary Fiber (g)	Fat (g)	Fat Breakdown (g) Sat	Mono	Poly
	BAKED GOODS: BREADS, CAKES, COOKIES, CRACKERS, PIES—Continued											
	Chocolate chip cookies:											
405	Commercial, 2¼" diam	4 ea	60	12	275	2	35	2	15	4.5	7.8	1.6
406	Home recipe, 2¼" diam	4 ea	64	6	312	4	37	2	18	5.2	6.7	5.4
407	From refrigerated dough, 2¼" diam	4 ea	64	13	284	3	39	1	13	4.5	6.5	1.3
408	Fig bars	4 ea	64	16	223	2	45	3	5	.9	2.6	.8
2052	Fruit bar, no fat	1 ea	28		90	2	21	0	0	0	0	0
2162	Fudge, fat free, Snackwell	1 ea	16	14	53	1	12	<1	<1	.1	.1	t
409	Oatmeal raisin, 2⅝" diam	4 ea	60	6	261	4	41	2	10	1.9	4.1	3
410	Peanut butter, home recipe, 2⅝"diam	4 ea	80	6	380	7	47	2	19	3.5	8.7	5.8
411	Sandwich-type, all	4 ea	40	2	189	2	28	1	8	1.7	4.7	1.1
412	Shortbread, commercial, small	4 ea	32	4	161	2	21	1	8	2	4.3	1
413	Shortbread, home recipe, large	2 ea	22	3	120	1	12	<1	7	4.5	2.1	.3
414	Sugar from refrigerated dough, 2" diam	4 ea	48	5	232	2	31	<1	11	2.8	6.2	1.4
1874	Vanilla sandwich, Snackwell's	2 ea	26	4	109	1	21	1	2	.5	.8	.2
415	Vanilla wafers	10 ea	40	5	176	2	29	1	6	1.4	2.4	1.5
416	Corn chips	1 c	26	1	140	2	15	1	9	1.2	2.5	4.3
	Crackers (enriched):											
417	Cheese	10 ea	10	3	50	1	6	<1	3	.9	.9	.5
418	Cheese with peanut butter	4 ea	28	4	135	4	16	1	6	1.4	3.4	1.2
	Fat free, enriched:											
2161	Cracked pepper, Snackwell	1 ea	15	2	60	2	12	<1	<1	.1	t	.1
2159	Wheat, Snackwell	7 ea	15	1	60	2	12	1	<1	.1	.1	.1
2075	Whole wheat, herb seasoned	5 ea	14	5	50	2	11	2	0	0	0	0
2077	Whole wheat, onion	5 ea	14	4	50	2	11	2	0	0	0	0
419	Graham, enriched	2 ea	14	4	59	1	11	<1	1	.4	.7	.2
420	Melba toast, plain, enriched	1 pce	5	5	19	1	4	<1	<1	t	t	.1
1514	Rice cakes, unsalted, enriched	2 ea	18	6	70	1	15	1	<1	.1	.2	.2
421	Rye wafer, whole grain	2 ea	22	5	73	2	18	5	<1	t	t	.1
422	Saltine-enriched	4 ea	12	4	52	1	9	<1	1	.3	.8	.2
1971	Saltine, unsalted tops, enriched	2 ea	6		25	1	4	0	<1	0	0	0
423	Snack-type, round like Ritz, enriched	3 ea	9	3	45	1	5	<1	2	.4	1	.7
424	Wheat, thin, enriched	4 ea	8	3	38	1	5	<1	2	.7	.8	.2
425	Whole-wheat wafers	2 ea	8	3	35	1	5	1	1	.2	.8	.2
426	Croissants, 4½ x 4 x 1¾"	1 ea	57	23	231	5	26	1	12	6.7	3.2	.7
1699	Croutons, seasoned	½ c	20	4	93	2	13	1	4	1	1.9	.5
	Danish pastry:											
427	Packaged ring, plain, 12 oz	1 ea	340	21	1349	19	181	1	65	13.5	40.9	6.4
428	Round piece, plain, 4¼" diam, 1" high	1 ea	88	21	349	5	47	<1	17	3.5	10.6	1.6
429	Ounce, plain	1 oz	28	21	111	2	15	<1	5	1.1	3.4	.5
430	Round piece with fruit	1 ea	94	29	335	5	45		16	3.3	10.1	1.6
	Desserts, 3 x 3" piece:											
1348	Apple crisp	1 pce	78	61	127	1	25	1	3	.6	1.2	.9
1353	Apple cobbler	1 pce	104	57	199	2	35	2	6	1.2	2.8	2
1349	Cherry crisp	1 pce	138	77	146	2	24	1	5	.9	2.5	1.8
1352	Cherry cobbler	1 pce	129	66	198	2	34	1	6	1.2	2.8	1.9
1350	Peach crisp	1 pce	139	75	155	2	27	2	5	.9	2.5	1.7
1351	Peach cobbler	1 pce	130	64	204	2	36	2	6	1.2	2.8	1.9
	Doughnuts:											
431	Cake type, plain, 3¼" diam	1 ea	47	21	198	2	23	1	11	1.8	4.5	3.8
432	Yeast-leavened, glazed, 3¾" diam	1 ea	60	25	242	4	27	1	14	3.5	7.8	1.7
	English muffins:											
433	Plain, enriched	1 ea	57	42	134	4	26	2	1	.2	.2	.5
434	Toasted	1 ea	52	37	133	4	26	2	1	.1	.2	.5
1504	Whole wheat	1 ea	66	46	134	6	27	4	1	.2	.3	.6
1414	Granola bar, soft	1 ea	28	6	124	2	19	1	5	2	1.1	1.5
1415	Granola bar, hard	1 ea	25	4	118	3	16	1	5	.6	1.1	3
1985	Granola bar, fat free, all flavors	1 ea	42	10	140	2	35	3	0	0	0	0
	Muffins, 2½" diam, 1½" high:											
	From home recipe:											
435	Blueberry	1 ea	57	39	165	4	23	1	6	1.4	1.6	3.1
436	Bran, wheat	1 ea	57	35	164	4	24	4	7	1.5	1.8	3.6
437	Cornmeal	1 ea	57	32	183	4	25	2	7	1.6	1.8	3.5

PAGE KEY: A–4 = Beverages A–6 = Dairy A–10 = Eggs A–10 = Fat/Oil A–14 = Fruit A–20 = Bakery A–28 = Grain A–32 = Fish A–34 = Meats A–38= Poultry A–40 = Sausage A–42 = Mixed/Fast A–46 = Nuts/Seeds A–50 = Sweets A–52 = Vegetables/Legumes A–62 = Vegetarian A–62 = Misc A–66 = Soups/Sauces A–68 = Fast A–84 = Convenience A–88 = Baby foods

Chol (mg)	Calc (mg)	Iron (mg)	Magn (mg)	Pota (mg)	Sodi (mg)	Zinc (mg)	VT-A (RE)	Thia (mg)	VT-E (α-TE)	Ribo (mg)	Niac (mg)	V-B6 (mg)	Fola (μg)	VT-C (mg)
0	9	1.45	21	56	196	.28	1	.07	1.74	.12	.97	.1	23	0
20	25	1.57	35	143	231	.59	105	.12	1.86	.11	.87	.05	21	<1
15	16	1.44	15	115	134	.32	11	.12	1.31	.12	1.27	.03	36	0
0	41	1.86	17	132	224	.25	3	.1	.45	.14	1.2	.05	17	<1
0	0	.36			95		0		.01					0
0	3	.29	5	26	71	.08	<1	.02	<.01	.02	.26	<.01		0
20	60	1.59	25	143	323	.52	98	.15	1.5	.1	.76	.04	18	<1
25	31	1.78	31	185	414	.66	125	.18	3.04	.17	2.81	.07	44	<1
0	10	1.55	18	70	242	.32	<1	.03	1.21	.07	.83	.01	17	0
6	11	.88	5	32	146	.17	4	.11	.98	.1	1.07	.01	19	0
20	4	.58	3	15	102	.09	67	.08	.18	.06	.64	<.01	2	0
15	43	.88	4	78	225	.13	5	.09	1.54	.06	1.16	.01	25	0
<1	17	.61	5	28	95	.16	<1	.05		.07	.69	.01		0
23	19	.95	6	39	125	.14	7	.11	.54	.13	1.24	.03	20	0
0	33	.34	20	37	164	.33	2	.01	.35	.04	.31	.06	5	0
1	15	.48	4	14	99	.11	3	.06	.1	.04	.47	.05	8	0
1	22	.82	16	69	278	.3	10	.11	1.24	.1	1.83	.42	25	0
<1	26	.73	4	19	148	.14	<1	.05		.06	.78	.01		<1
<1	28	.58	7	43	169	.21	<1	.04		.07	.73	.02		0
0	0				80		100							2
0	0	0			80		100							2
0	3	.52	4	19	85	.11	0	.03	.27	.04	.58	.01	8	0
0	5	.18	3	10	41	.1	0	.02	.01	.01	.21	<.01	6	0
0	2	.27	24	52	5	.54	1	.01	.02	.03	1.41	.03	4	0
0	9	1.31	27	109	175	.62	<1	.09	.44	.06	.35	.06	3	<1
0	14	.65	3	15	156	.09	0	.07	.2	.05	.63	<.01	15	0
0		.36		5	50				.1					
0	11	.32	2	12	76	.06	0	.04	.4	.03	.36	<.01	7	0
2	3	.28	5	16	70	.13	<1	.04	.02	.03	.34	.01	1	0
0	4	.25	8	24	53	.17	0	.02	.31	.01	.36	.01	3	0
43	21	1.16	9	67	424	.43	78	.22	.24	.14	1.25	.03	35	<1
1	19	.56	8	36	248	.19	1	.1	.32	.08	.93	.02	18	0
105	143	6.94	54	371	1261	1.87	20	.99	3.06	.75	8.5	.2	211	10
27	37	1.8	14	96	326	.48	5	.25	.79	.19	2.2	.05	55	3
9	12	.57	4	30	104	.15	2	.08	.25	.06	.7	.02	17	1
19	22	1.4	14	110	333	.48	24	.29	.85	.21	1.8	.06	31	2
0	22	.58	5	76	142	.12	24	.07		.06	.6	.03	4	2
1	21	.79	6	106	288	.16	76	.1	1.11	.09	.74	.04	3	<1
0	26	2.14	11	154	74	.15	150	.06	.93	.08	.6	.06	11	3
1	28	1.81	9	133	294	.2	135	.1	1.01	.11	.85	.05	9	2
0	20	.89	12	189	70	.19	108	.06	1.13	.05	1.06	.03	6	5
1	24	.91	10	159	291	.23	105	.09	1.16	.09	1.19	.03	6	3
17	21	.92	9	60	257	.26	8	.1	1.63	.11	.87	.03	22	<1
4	26	1.22	13	65	205	.46	6	.22	1.75	.13	1.71	.03	26	0
0	99	1.43	12	75	264	.4	0	.25	.07	.16	2.21	.02	46	<1
0	98	1.41	11	74	262	.39	0	.2	.07	.14	1.98	.02	38	<1
0	175	1.62	47	139	420	1.06	0	.2	.46	.09	2.25	.11	28	0
<1	29	.72	21	91	78	.42	0	.08	.34	.05	.14	.03	7	0
0	15	.74	24	84	73	.51	4	.07	.33	.03	.39	.02	6	<1
0	0	3.6			5		100							0
22	107	1.29	9	69	251	.31	16	.15	1.03	.16	1.26	.02	7	1
20	106	2.39	44	181	335	1.57	136	.19	1.31	.25	2.29	.18	30	4
26	147	1.49	13	82	333	.35	23	.17	1.08	.18	1.36	.05	10	<1

Table A-1

Food Composition (Computer code number is for Wadsworth Diet Analysis Plus program) (For purposes of calculations, use "0" for t, <1, <.1, <.01, etc.)

Computer Code Number	Food Description	Measure	Wt (g)	H_2O (%)	Ener (kcal)	Prot (g)	Carb (g)	Dietary Fiber (g)	Fat (g)	Fat Breakdown (g) Sat	Mono	Poly
	BAKED GOODS: BREADS, CAKES, COOKIES, CRACKERS, PIES—Continued											
	From commercial mix:											
438	Blueberry	1 ea	50	36	150	3	24	1	4	.7	1.8	1.5
439	Bran, wheat	1 ea	50	35	138	3	23	2	5	1.2	2.3	.7
440	Cornmeal	1 ea	50	30	161	4	25	1	5	1.4	2.6	.6
1864	Nabisco Newtons, fat free, all flavors	1 ea	23		69	1	16		0	0	0	0
	Pancakes, 4" diam:											
441	Buckwheat, from mix w/ egg and milk	1 ea	30	54	62	2	8	1	2	.6	.6	.8
442	Plain, from home recipe	1 ea	38	53	86	2	11	1	4	.8	.9	1.7
443	Plain, from mix; egg, milk, oil added	1 ea	38	53	74	2	14	<1	1	.2	.3	.3
1468	Pan dulce, sweet roll w/topping	1 ea	79	21	291	5	48	1	9	2	3.9	2.7
	Piecrust,with enriched flour, vegetable shortening, baked:											
444	Home recipe, 9" shell	1 ea	180	10	949	12	85	3	62	15.5	27.4	16.4
	From mix:											
445	Piecrust for 2-crust pie	1 ea	320	10	1686	21	152	5	111	27.6	48.6	29.2
446	1 pie shell	1 ea	160	11	802	11	81	3	49	12.3	27.7	6.2
	Pies, 9" diam; pie crust made with vegetable shortening, enriched flour:											
447	Apple: Whole pie	1 ea	1000	52	2370	19	340	16	110	21.1	59.4	20.9
448	Piece, ⅙ of pie	1 pce	167	52	396	3	57	3	18	3.5	9.9	3.5
449	Banana cream: Whole pie	1 ea	1152	48	3098	51	379	8	157	43.3	65.9	38
450	Piece, ⅙ of pie	1 pce	192	48	516	8	63	1	26	7.2	11	6.3
451	Blueberry: Whole pie	1 ea	1176	51	2881	32	394	16	140	34.3	60.2	36.2
452	Piece, ⅙ of pie	1 pce	196	51	480	5	66	3	23	5.7	10	6
453	Cherry: Whole pie	1 ea	1140	46	3078	32	439	17	139	34.1	60.5	37.1
454	Piece, ⅙ of pie	1 pce	240	46	648	7	92	4	29	7.2	12.7	7.8
455	Chocolate cream: Whole pie	1 ea	1194	63	2150	49	281	12	97	35.5	38.2	18.6
456	Piece, ⅙ of pie	1 pce	199	63	358	8	47	2	16	5.9	6.4	3.1
457	Custard: Whole pie	1 ea	630	61	1323	35	131	10	73	17.5	36.3	12.1
458	Piece, ⅙ of pie	1 pce	105	61	221	6	22	2	12	2.9	6	2
459	Lemon meringue: Whole pie	1 ea	678	42	1817	10	320	8	59	10.6	24.6	19.6
460	Piece, ⅙ of pie	1 pce	113	42	303	2	53	1	10	1.8	4.1	3.3
461	Peach: Whole pie	1 ea	1111	45	2994	26	443	16	130	31.1	55.7	37.4
462	Piece, ⅙ of pie	1 pce	139	45	375	3	55	2	16	3.9	7	4.7
463	Pecan: Whole pie	1 ea	678	19	2712	27	388	24	125	25.5	73.2	20.1
464	Piece, ⅙ of pie	1 pce	113	19	452	5	65	4	21	4.2	12.2	3.4
465	Pumpkin: Whole pie	1 ea	654	58	1373	25	179	18	62	13.2	32.8	10.5
466	Piece, ⅙ of pie	1 pce	109	58	229	4	30	3	10	2.2	5.5	1.7
467	Pies, fried, commercial: Apple	1 ea	85	40	266	2	33	1	14	6.5	5.8	1.2
468	Pies, fried, commercial: Cherry	1 ea	128	38	404	4	54	3	21	3.1	9.5	6.9
	Pretzels, made with enriched flour:											
469	Thin sticks, 2¼" long	1 oz	28	3	107	3	22	1	1	.2	.4	.3
470	Dutch twists	10 pce	60	3	229	5	47	2	2	.4	.8	.7
471	Thin twists, 3¼ x 2¼ x ¼"	10 pce	60	3	229	5	47	2	2	.4	.8	.7
	Rolls & buns, enriched, commercial:											
472	Cloverleaf rolls, 2½" diam, 2" high	1 ea	28	32	84	2	14	1	2	.5	1	.3
473	Hot dog buns	1 ea	43	34	123	4	22	1	2	.5	1.1	.4
474	Hamburger buns	1 ea	43	34	123	4	22	1	2	.5	1.1	.4
475	Hard roll, white, 3¾" diam, 2" high	1 ea	57	31	167	6	30	1	2	.3	.6	1
476	Submarine rolls/hoagies, 11¼ x 3 x 2½"	1 ea	135	31	392	12	75	4	4	.9	1.3	1.4
	Rolls & buns, enriched, home recipe:											
477	Dinner rolls 2½" diam, 2" high	1 ea	35	29	112	3	19	1	3	.7	1.1	.7
	Sports/fitness bar:											
2043	Forza energy bar	1 ea	70	18	231	10	45	4	1			
2042	Power bar	1 ea	65		230	10	45	3	2			
2041	Tiger sports bar	1 ea	65	17	229	11	40	4	2			
478	Toaster pastries, fortified (Poptarts)	1 ea	52	12	204	2	37	1	5	.8	2.1	2
2132	Toaster strudel pastry—cream cheese	1 ea	54	32	188	3	24	<1	9	2.7		
2134	Toaster strudel pastry—french toast	1 ea	54	32	188	3	24	<1	9	2.9		

PAGE KEY: A–4 = Beverages A–6 = Dairy A–10 = Eggs A–10 = Fat/Oil A–14 = Fruit A–20 = Bakery A–28 = Grain A–32 = Fish A–34 = Meats A–38= Poultry A–40 = Sausage A–42 = Mixed/Fast A–46 = Nuts/Seeds A–50 = Sweets A–52 = Vegetables/Legumes A–62 = Vegetarian A–62 = Misc A–66 = Soups/Sauces A–68 = Fast A–84 = Convenience A–88 = Baby foods

Chol (mg)	Calc (mg)	Iron (mg)	Magn (mg)	Pota (mg)	Sodi (mg)	Zinc (mg)	VT-A (RE)	Thia (mg)	VT-E (α-TE)	Ribo (mg)	Niac (mg)	V-B6 (mg)	Fola (μg)	VT-C (mg)
23	12	.56	5	39	219	.19	11	.07	.7	.16	1.12	.04	5	<1
34	16	1.27	28	73	234	.57	15	.1	.75	.12	1.44	.09	8	0
31	37	.97	10	65	398	.32	22	.12	.75	.14	1.05	.05	5	<1
					77									
20	77	.56	17	70	160	.35	20	.05	.62	.08	.4	.04	5	<1
22	83	.68	6	50	167	.21	20	.08	.36	.11	.6	.02	14	<1
5	48	.59	8	66	239	.15	3	.08	.32	.08	.65	.03	3	<1
26	13	1.82	10	57	140	.35	87	.23	1.35	.21	1.98	.04	22	<1
0	18	5.2	25	121	976	.79	0	.7	9.94	.5	5.96	.04	121	0
0	32	9.25	45	214	1734	1.41	0	1.25	17.7	.89	10.6	.08	214	0
0	96	3.44	24	99	1166	.62	0	.48	8.83	.3	3.79	.09	19	0
0	110	4.5	70	650	2660	1.6	300	.28	16.5	.27	2.63	.38	220	32
0	18	.75	12	109	444	.27	50	.05	2.76	.04	.44	.06	37	5
588	864	12	184	1900	2764	5.53	806	1.6	16.9	2.38	12.1	1.53	311	18
98	144	2	31	317	461	.92	134	.27	2.82	.4	2.02	.25	52	3
0	82	14.5	94	588	2175	2.35	47	1.8	24.7	1.55	14	.4	270	8
0	14	2.41	16	98	363	.39	8	.3	4.12	.26	2.33	.07	45	1
0	114	21.1	103	878	2177	2.28	547	1.69	21.7	1.43	14.6	.39	308	11
0	24	4.44	22	185	458	.48	115	.35	4.56	.3	3.07	.08	65	2
109	1028	8.84	170	1705	2085	4.93	235	1.03	11.4	2.43	7.34	.37	59	6
18	171	1.47	28	284	348	.82	39	.17	1.9	.41	1.22	.06	10	1
208	504	3.65	69	668	1512	3.28	315	.25	7.5	1.31	1.84	.3	126	2
35	84	.61	12	111	252	.55	52	.04	1.25	.22	.31	.05	21	<1
305	380	4.14	102	603	990	3.32	353	.42	9.7	1.42	4.4	.2	88	22
51	63	.69	17	101	165	.55	59	.07	1.62	.24	.73	.03	15	4
0	59	12.1	79	1047	2025	1.88	386	1.41	26.2	1.19	15.3	.2	58	589
0	7	1.52	10	131	253	.23	48	.18	3.28	.15	1.91	.02	7	74
217	115	7.05	122	502	2874	3.86	319	.62	17.2	.83	1.69	.14	183	7
36	19	1.18	20	84	479	.64	53	.1	2.86	.14	.28	.02	30	1
131	392	5.17	98	1007	1844	2.94	3139	.36	10.5	1	1.22	.37	131	10
22	65	.86	16	168	307	.49	523	.06	1.75	.17	.2	.06	22	2
13	13	.88	8	51	325	.17	33	.1	.37	.08	.98	.03	4	1
0	28	1.56	13	83	479	.29	22	.18	.55	.14	1.83	.04	23	2
0	10	1.21	10	41	480	.24	0	.13	.06	.17	1.47	.03	48	0
0	22	2.59	21	88	1029	.51	0	.28	.13	.37	3.15	.07	103	0
0	22	2.59	21	88	1029	.51	0	.28	.13	.37	3.15	.07	103	0
<1	33	.88	6	37	146	.22	0	.14	.22	.09	1.13	.01	27	<1
0	60	1.36	9	61	241	.27	0	.21	.2	.13	1.69	.02	41	0
0	60	1.36	9	61	241	.27	0	.21	.2	.13	1.69	.02	41	0
0	54	1.87	15	62	310	.54	0	.27	.1	.19	2.42	.03	54	0
0	122	3.78	27	122	783	.85	0	.54	.1	.33	4.47	.05	40	0
13	21	1.04	7	53	145	.24	28	.14	.35	.14	1.21	.02	15	<1
0	300	6.3	160	220	65	5.25		1.5	20	1.7	20	2	400	60
0	300	5.4	140	150	110	5.25		1.5		1.7	20	2	400	60
	349	4.49	140	279	100		50	1.5	19.9	1.69	19.9	1.99	399	60
0	13	1.81	9	58	218	.34	149	.15	.97	.19	2.05	.2	34	<1
12	12	.97			217		17		1					0
12	12	.97			217		17		1					0

Table A–1

Food Composition (Computer code number is for Wadsworth Diet Analysis Plus program) (For purposes of calculations, use "0" for t, <1, <.1, <.01, etc.)

Computer Code Number	Food Description	Measure	Wt (g)	H_2O (%)	Ener (kcal)	Prot (g)	Carb (g)	Dietary Fiber (g)	Fat (g)	Fat Breakdown (g) Sat	Mono	Poly
	BAKED GOODS: BREADS, CAKES, COOKIES, CRACKERS, PIES—Continued											
	Tortilla chips:											
1271	Plain	10 pce	18	2	90	1	11	1	5	.9	2.8	.7
1036	Nacho flavor	1 c	26	2	129	2	16	1	7	1.3	3.9	.9
1037	Taco flavor	1 pce	18	2	86	1	11	1	4	.8	2.6	.6
	Tortillas:											
479	Corn, enriched, 6" diam	1 ea	26	44	58	2	12	1	1	.1	.2	.3
480	Flour, 8" diam	1 ea	49	27	159	4	27	2	3	.6	1.4	1.4
1301	Flour, 10" diam	1 ea	72	27	234	6	40	2	5	.8	2.1	2
481	Taco shells	1 ea	14	4	63	1	9	1	3	.4	1.5	.6
	Waffles, 7" diam:											
482	From home recipe	1 ea	75	42	218	6	25	1	11	2.2	2.6	5.1
483	From mix, egg/milk added	1 ea	75	42	218	5	26	1	10	1.7	2.7	5.2
1510	Whole grain, prepared from frozen	1 ea	39	43	107	4	13	1	5	1.6	1.9	.9
	GRAIN PRODUCTS: CEREAL, FLOUR, GRAIN, PASTA and NOODLES, POPCORN											
484	Barley, pearled, dry, uncooked	1 c	200	10	704	20	155	31	2	.5	.3	1.1
485	Barley, pearled, cooked	1 c	157	69	193	4	44	6	1	.1	.1	.3
2009	Breakfast bars, fat free, all flavors	1 ea	38	25	110	2	26	3	0	0	0	0
	Breakfast bar, Snackwell:											
2165	Apple-cinnamon	1 ea	37	16	119	1	29	1	<1	.1	t	.1
2164	Blueberry	1 ea	37	16	121	1	29	1	<1	t	t	.1
2163	Strawberry	1 ea	37	16	120	1	29	1	<1	t	t	.1
	Breakfast cereals, hot, cooked w/o salt added:											
	Corn grits (hominy) enriched:											
486	Regular/quick prep w/o salt, yellow:	1 c	242	85	145	3	31	<1	<1	.1	.1	.2
487	Instant, prepared from packet, white	1 ea	137	82	89	2	21	1	<1	t	t	.1
	Cream of wheat:											
488	Regular, quick, instant	1 c	239	87	129	4	27	1	<1	.1	.1	.3
489	Mix and eat, plain, packet	1 ea	142	82	102	3	21	<1	<1	t	t	.2
1664	Farina cereal, cooked w/o salt	1 c	233	88	117	3	25	3	<1	t	t	.1
490	Malt-O-Meal, cooked w/o salt	1 c	240	88	122	4	26	1	<1	.1	.1	t
494	Maypo	1 c	216	83	153	5	29	5	2	.4	.7	.8
	Oatmeal or rolled oats:											
491	Regular, quick, instant,nonfortified cooked w/o salt	1 c	234	85	145	6	25	4	2	.4	.7	.9
	Instant, fortified:											
492	Plain, from packet	½ c	118	85	70	4	12	2	1	.2	.4	.4
493	Flavored, from packet	½ c	109	76	106	3	21	2	1	.2	.5	.5
	Breakfast cereals, ready to eat:											
495	All-Bran	1 c	62	3	160	8	46	20	2	.4	.4	1.3
1306	Alpha Bits	1 c	28	1	110	2	24	1	1	.1	.2	.2
1307	Apple Jacks	1 c	33	3	120	2	30	1	<1	.1	.1	.2
1308	Bran Buds	1 c	90	3	240	8	72	36	2	.4	.4	1.4
1305	Bran Chex	1 c	49	2	156	5	39	8	1	.2	.3	.7
1309	Honey BucWheat Crisp	1 c	38	5	147	4	31	3	1	.2	.3	.6
1310	C.W. Post, plain	1 c	97	2	421	9	73	7	13	1.7	6	4.7
1311	C.W. Post, with raisins	1 c	103	4	446	9	74	14	15	11	1.7	1.4
496	Cap'n Crunch	1 c	37	2	147	2	32	1	2	.5	.4	.3
1312	Cap'n Crunchberries	1 c	35	2	140	2	30	1	2	.5	.3	.3
1313	Cap'n Crunch, peanut butter	1 c	35	2	146	3	28	1	3	.7	1.1	.7
497	Cheerios	1 c	23	3	84	2	17	2	1	.3	.5	.2
1314	Cocoa Krispies	1 c	41	2	159	3	36	1	1	.7	.2	.2
1316	Cocoa Pebbles	1 c	32	2	131	1	27	1	2	1.1	.4	.1
1315	Corn Bran	1 c	36	3	120	2	30	6	1	.3	.3	.4
1317	Corn Chex	1 c	28	2	110	2	25	<1	<1	t	t	t
498	Corn Flakes, Kellogg's	1 c	28	3	100	2	24	1	<1	.1	t	.1
499	Corn Flakes, Post Toasties	1 c	24	3	93	2	21	1	<1	t	t	t
1340	Corn Pops	1 c	31	3	120	1	28	<1	<1	.1	.1	t
1318	Cracklin' Oat Bran	1 c	65	4	252	5	48	8	8	3.4	3.8	.9
1038	Crispy Wheat 'N Raisins	1 c	43	7	150	3	35	3	1	.1	.1	.2

PAGE KEY: A–4 = Beverages A–6 = Dairy A–10 = Eggs A–10 = Fat/Oil A–14 = Fruit A–20 = Bakery A–28 = Grain A–32 = Fish A–34 = Meats A–38= Poultry A–40 = Sausage A–42 = Mixed/Fast A–46 = Nuts/Seeds A–50 = Sweets A–52 = Vegetables/Legumes A–62 = Vegetarian A–62 = Misc A–66 = Soups/Sauces A–68 = Fast A–84 = Convenience A–88 = Baby foods

Chol (mg)	Calc (mg)	Iron (mg)	Magn (mg)	Pota (mg)	Sodi (mg)	Zinc (mg)	VT-A (RE)	Thia (mg)	VT-E (α-TE)	Ribo (mg)	Niac (mg)	V-B6 (mg)	Fola (μg)	VT-C (mg)
0	28	.27	16	35	95	.27	4	.01	.24	.03	.23	.05	2	0
1	38	.37	21	56	184	.31	11	.03	.35	.05	.37	.07	4	<1
1	28	.36	16	39	142	.23	16	.04	.24	.04	.36	.05	4	<1
0	45	.36	17	40	42	.24	6	.03	.04	.02	.39	.06	30	0
0	61	1.62	13	64	234	.35	0	.26	.62	.14	1.75	.02	60	0
0	90	2.38	19	94	344	.51	0	.38	.91	.21	2.57	.04	89	0
0	35	.36	15	34	25	.19	6	.04	.59	.02	.24	.04	4	0
52	191	1.73	14	119	383	.51	49	.2	1.73	.26	1.55	.04	34	<1
38	93	1.22	15	134	458	.35	19	.15	1.5	.19	1.23	.07	9	<1
39	84	.69	15	91	150	.45	25	.08	.53	.13	.75	.04	7	<1
0	58	5	158	560	18	4.26	4	.38	.26	.23	9.2	.52	46	0
0	17	2.09	34	146	5	1.29	2	.13	.08	.1	3.23	.18	25	0
0	20	.72			25		20							1
<1	17	5	6	68	103	3.88	260	.39		.44	5.2	.52		<1
<1	14	4.83	5	43	107	3.85	260	.39		.44	5.2	.52		<1
<1	14	4.82	6	47	102	3.83	260	.39		.44	5.2	.52		2
0	0	1.55	10	53	0	.17	14	.24	.12	.14	1.96	.06	75	0
0	8	8.19	11	38	289	.21	0	.15	.03	.08	1.38	.05	47	0
0	50	10.3	12	45	139	.33	0	.24	.03	0	1.43	.03	108	0
0	20	8.09	7	38	241	.24	125	.43	.02	.28	4.97	.57	101	0
0	5	1.17	5	30	0	.16	0	.19	.03	.12	1.28	.02	54	0
0	5	9.6	5	31	2	.17	0	.48	.03	.24	5.76	.02	5	0
0	112	7.56	45	190	233	1.34	633	.65	1.51	.65	8.42	.86	9	26
0	19	1.59	56	131	2	1.15	5	.26	.23	.05	.3	.05	9	0
0	109	4.2	28	66	190	.58	302	.35	.14	.19	3.65	.49	100	0
0	112	4.45	34	91	169	.66	306	.35	.14	.25	3.92	.51	100	<1
0	200	9	280	620	560	7.5	450	.75	1.14	.85	10	1	186	30
0	8	2.66	16	54	178	1.48	371	.36	.02	.42	4.93	.5	99	0
0	0	4.5	8	35	150	3.75	225	.38	.05	.43	5	.5	116	15
0	60	13.5	240	809	599	11.3	676	1.17	1.42	1.26	15	1.53	270	45
0	29	14	69	216	345	6.47	11	.64	.56	.26	8.62	.88	173	26
0	54	10.9	43	142	361	.68	913	.9	8.99	1.03	12.1	1.88	11	36
<1	47	15.4	67	198	167	1.64	1284	1.26	.68	1.46	17.1	1.75	342	0
<1	50	16.4	74	261	161	1.64	1363	1.34	.72	1.55	18.1	1.85	364	0
0	7	6.18	13	47	286	5.14	5	.51	.18	.58	6.85	.68	137	0
0	9	6.06	13	49	256	5.39	6	.5	.25	.57	6.72	.67	135	<1
0	3	5.85	24	80	264	4.87	5	.49	.19	.55	6.48	.65	130	0
0	42	6.21	25	68	218	2.88	288	.29	.16	.33	3.84	.38	77	11
0	0	2.38	11	79	278	1.97	298	.49	.19	.57	6.6	.66	123	20
0	5	2.02	13	53	180	1.7	424	.42	.04	.48	5.63	.58	113	0
0	27	10.1	19	75	338	5	5	.1	.19	.56	6.66	.67	134	0
0	3	8.01	4	23	306	.1	14	.36	.07	.07	4.93	.5	99	15
0	0	8.68	3	25	300	.17	225	.36	.03	.43	5	.5	99	15
0	1	.63	4	28	252	.07	318	.31	.06	.36	4.22	.43	85	0
0	0	1.86	2	25	120	1.55	225	.4	.03	.43	5.18	.5	109	15
0	26	2.41	79	305	226	1.95	299	.5	.43	.56	6.63	.66	181	20
0	54	3.52	33	180	223	.85	293	.29	.45	.33	3.91	.39	78	0

A

Table A–1

Food Composition (Computer code number is for Wadsworth Diet Analysis Plus program) (For purposes of calculations, use "0" for t, <1, <.1, <.01, etc.)

Computer Code Number	Food Description	Measure	Wt (g)	H_2O (%)	Ener (kcal)	Prot (g)	Carb (g)	Dietary Fiber (g)	Fat (g)	Fat Breakdown (g) Sat	Mono	Poly
	GRAIN PRODUCTS: CEREAL, FLOUR, GRAIN, PASTA and NOODLES, POPCORN—Continued											
1319	Fortified Oat Flakes	1 c	48	3	180	8	36	1	1	.2	.3	.4
500	40% Bran Flakes, Kellogg's	1 c	39	4	121	4	32	7	1	.2	.2	.5
501	40% Bran Flakes, Post	1 c	47	3	152	5	37	9	1	.1	.1	.4
502	Froot Loops	1 c	32	2	120	2	28	1	1	.4	.2	.3
518	Frosted Flakes	1 c	41	3	159	2	37	1	<1	.1	t	.1
1320	Frosted Mini-Wheats	1 c	51	5	170	5	41	5	1	.2	.1	.6
1321	Frosted Rice Krispies	1 c	35	2	135	2	32	<1	<1	.1	.1	.1
1324	Fruit & Fibre w/dates	1 c	57	9	193	5	43	8	3	.4	1.3	1
1322	Fruity Pebbles	1 c	32	3	130	1	28	<1	2	1.4	.1	.1
503	Golden Grahams	1 c	39	3	150	2	33	1	1	.2	.4	.2
504	Granola, homemade	½ c	61	5	285	9	32	6	15	2.9	4.8	6.5
505	Granola, low fat	½ c	47	3	181	5	38	3	3	0		
1670	Granola, low fat, commercial	½ c	45	5	165	4	35	2	2	.6	.7	.9
505	Grape Nuts	½ c	55	3	196	7	45	5	<1	t	t	.1
1326	Grape Nuts Flakes	1 c	39	3	144	4	32	4	1	.6	.1	.2
1665	Heartland Natural with raisins	1 c	110	5	468	11	76	6	16	4	4.2	6.2
1327	Honey & Nut Corn Flakes	1 c	37	2	148	3	31	1	2	.3	.7	.6
506	Honey Nut Cheerios	1 c	33	2	126	3	27	2	1	.3	.5	.2
1328	HoneyBran	1 c	35	2	119	3	29	4	1	.3	.1	.3
1329	HoneyComb	1 c	22	1	86	1	20	1	<1	.2	.1	.1
1330	King Vitaman	1 c	21	2	81	2	18	1	1	.2	.3	.2
1039	Kix	1 c	19	2	72	1	16	1	<1	.1	.1	t
1331	Life	1 c	44	4	167	4	35	3	2	.3	.6	.8
507	Lucky Charms	1 c	32	2	124	2	27	1	1	.2	.4	.2
1323	Mueslix Five Grain	1 c	82	5	279	7	63	7	3	.5	1	1.2
508	Nature Valley Granola	1 c	113	4	510	12	74	7	20	2.6	13.3	3.8
1666	Nutri Grain Almond Raisin	1 c	40	6	147	3	31	3	2	.1	1	1.2
1336	100% Bran	1 c	66	3	178	8	48	19	3	.6	.6	1.9
509	100% Natural cereal, plain	1 c	104	2	462	11	71	8	17	7.5	7.5	2.3
1337	100% Natural with apples & cinnamon	1 c	104	2	477	11	70	7	20	15.5	1.8	1.3
1338	100% Natural with raisins & dates	1 c	110	3	496	12	72	7	20	13.6	3.7	1.7
510	Product 19	1 c	33	4	110	2	28	1	<1	t	.2	.2
1339	Quisp	1 c	30	3	121	2	25	1	2	.5	.4	.2
511	Raisin Bran, Kellogg's	1 c	61	9	200	6	47	8	1	.1	.1	.4
512	Raisin Bran, Post	1 c	56	9	172	5	42	8	1	.2	.1	.5
1667	Raisin Squares	1 c	71	9	241	6	55	7	2	.2	.2	.6
1041	Rice Chex	1 c	33	3	130	2	29	1	<1	t	t	t
513	Rice Krispies, Kellogg's	1 c	28	2	111	2	25	<1	<1	t	t	t
514	Rice, puffed	1 c	14	4	54	1	12	<1	<1	t	t	t
515	Shredded Wheat	1 c	43	5	154	5	35	4	1	.1	.1	.4
516	Special K	1 c	31	3	110	6	22	1	<1	t	t	.2
517	Super Golden Crisp	1 c	33	1	123	2	30	<1	<1	.1	.1	.1
519	Honey Smacks	1 c	36	3	133	3	32	1	1	.4	.1	.3
1341	Tasteeos	1 c	24	2	94	3	19	3	1	.2	.2	.2
1342	Team	1 c	42	4	164	3	36	1	1	.1	.2	.3
520	Total, wheat, with added calcium	1 c	40	3	140	4	32	4	1	.2	.2	.1
521	Trix	1 c	28	2	114	1	24	1	2	.4	.9	.3
1344	Wheat Chex	1 c	46	2	169	5	38	4	1	.2	.1	.5
1043	Wheat cereal, puffed, fortified	1 c	12	4	44	2	9	1	<1	t	t	.1
522	Wheaties	1 c	29	3	106	3	23	2	1	.2	.2	.1
523	Buckwheat flour, dark	1 c	120	11	402	15	85	12	4	.8	1.1	1.1
525	Buckwheat, whole grain, dry	1 c	170	10	583	23	122	17	6	1.3	1.8	1.8
526	Bulgar, dry, uncooked	1 c	140	9	479	17	106	26	2	.3	.2	.8
527	Bulgar, cooked	1 c	182	78	151	6	34	8	<1	.1	.1	.2
	Cornmeal:											
528	Whole-ground, unbolted, dry	1 c	122	10	442	10	94	9	4	.6	1.2	2
530	Degermed, enriched, dry	1 c	138	12	505	12	107	10	2	.3	.6	1
38041	Degermed, enriched, baked	1 c	138	12	505	12	107	10	2	.3	.6	1
	Macaroni, cooked:											
532	Enriched	1 c	140	66	197	7	40	2	1	.1	.1	.4
533	Whole wheat	1 c	140	67	174	7	37	4	1	.1	.1	.3

PAGE KEY: A–4 = Beverages A–6 = Dairy A–10 = Eggs A–10 = Fat/Oil A–14 = Fruit A–20 = Bakery A–28 = Grain A–32 = Fish A–34 = Meats A–38= Poultry A–40 = Sausage A–42 = Mixed/Fast A–46 = Nuts/Seeds A–50 = Sweets A–52 = Vegetables/Legumes A–62 = Vegetarian A–62 = Misc A–66 = Soups/Sauces A–68 = Fast A–84 = Convenience A–88 = Baby foods

Chol (mg)	Calc (mg)	Iron (mg)	Magn (mg)	Pota (mg)	Sodi (mg)	Zinc (mg)	VT-A (RE)	Thia (mg)	VT-E (α-TE)	Ribo (mg)	Niac (mg)	V-B6 (mg)	Fola (μg)	VT-C (mg)
0	68	13.7	58	228	220	2.54	636	.62	.34	.72	8.45	.86	169	0
0	0	10.9	81	229	309	5.03	505	.51	7.22	.58	6.71	.66	138	20
0	21	13.4	102	251	431	2.49	622	.61	.54	.7	8.27	.85	166	0
0	0	4.51	8	35	150	3.75	225	.37	.12	.43	5	.5	96	15
0	0	6.15	4	26	264	.2	298	.5	.05	.56	6.61	.66	123	20
0	0	15	60	170	0	1.5	0	.37	.46	.42	4.64	.5	102	0
0	0	2.42	8	27	256	.42	303	.49	.03	.56	6.72	.66	140	20
0	30	10.1	81	335	270	3.02	725	.75	1.32	.85	10.1	1	201	0
0	4	2.02	9	24	178	1.7	424	.42	.03	.48	5.63	.58	113	0
0	19	5.85	12	69	357	4.88	293	.49	.29	.55	6.51	.65	130	19
0	49	2.56	109	328	15	2.48	2	.45	7.87	.17	1.25	.19	52	1
0		2.71	36	143	90	5.64	226	.56	7.57	.64	7.52	.75	151	
0	15	1.35	30	127	101	2.84	169	.27	4.03	.31	3.74	.36	90	0
0	5	15.7	37	184	382	1.21	728	.71	.14	.82	9.68	.99	194	0
0	16	11.2	43	136	220	.78	516	.51	.1	.58	6.86	.7	138	0
0	66	4.02	141	415	226	2.83	7	.32	.77	.14	1.54	.2	44	1
0	0	3.03	3	40	249	.2	152	.26	.09	.3	3.37	.33	74	10
0	22	4.95	32	94	285	4.13	248	.41	.34	.47	5.51	.55	110	16
0	16	5.57	46	151	202	.9	463	.45	.81	.52	6.16	.63	23	19
0	4	2.09	7	25	124	1.17	291	.29	.09	.33	3.87	.4	78	0
0	3	5.92	18	58	176	2.65	212	.26	1.42	.3	3.53	.35	71	8
0	28	5.13	6	26	167	2.38	238	.24	.05	.27	3.17	.32	63	9
0	134	12.3	43	109	240	5.5	2	.55	.22	.62	7.35	.73	147	0
0	35	4.8	21	58	217	4	240	.4	.14	.45	5.34	.53	107	16
0	38	8.94	82	369	107	7.46	747	.75	8.94	.84	9.84	.99	197	1
0	85	3.53	107	375	183	2.27	0	.35	7.97	.12	1.25	.16	17	0
0	122	1.14	13	147	139	3.06	0	.32	4.38	.35	4.08	.41	80	0
0	46	8.12	312	652	457	5.74	0	1.58	1.53	1.78	20.9	2.11	47	63
1	100	3.11	109	457	28	2.5	1	.36	1.19	.17	1.84	.19	26	<1
0	157	2.89	72	514	52	2	6	.33	.73	.57	1.87	.11	17	1
0	160	3.12	124	538	47	2.11	7	.31	.77	.65	2.09	.16	45	0
0	0	19.8	18	55	308	16.5	248	1.65	24.4	1.88	22	2.21	429	66
0	6	5.1	15	40	216	4.26	4	.42	.15	.48	5.67	.56	113	0
0	40	4.5	80	350	390	3.75	225	.37	.56	.43	5	.5	122	0
0	26	8.9	95	345	365	2.97	741	.73	1.3	.84	9.86	1.01	198	0
0	0	21.7	54	335	4	1.99	0	.5	.38	.57	6.67	.64	142	0
0	5	9.44	8	38	276	.45	2	.43	.04	.01	5.81	.59	116	17
0	5	.7	12	27	206	.46	371	.52	.03	.59	6.92	.69	138	15
0	1	.41	4	16	1	.15	0	.06	.01	.01	.87	0	1	0
0	16	1.81	57	155	4	1.42	0	.11	.23	.12	2.26	.11	21	0
0	0	8.4	16	55	250	3.75	225	.53	.08	.59	7.01	.71	93	15
0	7	2.08	20	48	51	1.75	437	.43	.12	.49	5.81	.59	116	0
0	0	2.4	21	53	67	.4	300	.5	.18	.58	6.66	.68	133	20
0	11	6.86	26	71	183	.69	318	.31	.17	.36	4.22	.43	85	13
0	6	12	12	71	260	.58	556	.55	.1	.63	7.39	.76	7	22
0	344	24	43	129	265	20	500	2	31.3	2.27	26.8	2.67	533	80
0	30	4.2	3	16	184	3.5	210	.35	.56	.4	4.68	.47	93	14
0	18	13.2	58	173	308	1.23	0	.6	.17	.17	8.1	.83	162	24
0	3	.56	16	44	1	.37	<1	.05	.08	.03	1.43	.02	4	0
0	53	7.83	31	101	215	.68	218	.36	.36	.41	4.84	.48	97	14
0	49	4.87	301	692	13	3.74	0	.5	1.24	.23	7.38	.7	65	0
0	31	3.74	393	782	2	4.08	0	.17	1.75	.72	11.9	.36	51	0
0	49	3.44	230	574	24	2.7	0	.32	.22	.16	7.15	.48	38	0
0	18	1.75	58	124	9	1.04	0	.1	.05	.05	1.82	.15	33	0
0	7	4.21	155	350	43	2.22	57	.47	.82	.24	4.43	.37	31	0
0	7	5.7	55	224	4	.99	57	.99	.45	.56	6.94	.35	258	0
0	7	5.7	55	224	4	.99	57	.79	.5	.5	6.25	.32	181	0
0	10	1.96	25	43	1	.74	0	.29	.04	.14	2.34	.05	98	0
0	21	1.48	42	62	4	1.13	0	.15	.14	.06	.99	.11	7	0

Table A-1

Food Composition (Computer code number is for Wadsworth Diet Analysis Plus program) (For purposes of calculations, use "0" for t, <1, <.1, <.01, etc.)

Computer Code Number	Food Description	Measure	Wt (g)	H_2O (%)	Ener (kcal)	Prot (g)	Carb (g)	Dietary Fiber (g)	Fat (g)	Fat Breakdown (g) Sat	Mono	Poly
	GRAIN PRODUCTS: CEREAL, FLOUR, GRAIN, PASTA and NOODLES, POPCORN—Continued											
534	Vegetable, enriched	1 c	134	68	172	6	36	2	<1	t	t	.1
535	Millet, cooked	1 c	240	71	286	8	57	3	2	.4	.4	1.2
	Noodles (see also Pasta and Spaghetti):											
1507	Cellophane noodles, cooked	1 c	190	79	160	<1	39	<1	<1	t	t	t
1995	Cellophane noodles, dry	1 c	140	13	491	<1	121	1	<1	t	t	t
537	Chow mein, dry	1 c	45	1	237	4	26	2	14	2	3.5	7.8
536	Egg noodles, cooked, enriched	1 c	160	69	213	8	40	2	2	.5	.7	.7
538	Spinach noodles, dry	3½ oz	100	8	372	13	75	11	2	.2	.2	.6
1343	Oat bran, dry	¼ c	24	7	59	4	16	4	2	.3	.6	.7
	Pasta, cooked:											
1418	Fresh	2 oz	57	69	75	3	14	1	1	.1	.1	.2
1417	Linguini/Rotini	1 c	140	66	197	7	40	4	1	.1	.1	.4
	Popcorn:											
539	Air popped, plain	1 c	8	4	31	1	6	1	<1	t	.1	.2
1042	Microwaved, low fat, low sodium	1 c	6	3	25	1	4	1	1	.1	.2	.3
540	Popped in vegetable oil/salted	1 c	11	3	55	1	6	1	3	.5	.9	1.5
541	Sugar-syrup coated	1 c	35	3	151	1	28	2	4	1.3	1	1.6
	Rice:											
542	Brown rice, cooked	1 c	195	73	216	5	45	4	2	.4	.6	.6
2215	Mexican rice, cooked	1 c	226		820	16	180	6	30	4	1	1
2216	Spanish rice, cooked	1 c	246	85	130	3	28	2	1			
	White, enriched, all types:											
543	Regular/long grain, dry	1 c	185	12	675	13	148	2	1	.3	.4	.3
544	Regular/long grain, cooked	1 c	158	68	205	4	45	1	<1	.1	.1	.1
545	Instant, prepared without salt	1 c	165	76	162	3	35	1	<1	.1	.1	.1
	Parboiled/converted rice:											
546	Raw, dry	1 c	185	10	686	13	151	3	1	.3	.3	.3
547	Cooked	1 c	175	72	200	4	43	1	<1	.1	.1	.1
1486	Sticky rice (glutinous), cooked	1 c	174	77	169	4	37	2	<1	.1	.1	.1
548	Wild rice, cooked	1 c	164	74	166	7	35	3	1	.1	.1	.4
1700	Rice and pasta (Rice-a-Roni), cooked	1 c	202	72	246	5	43	1	6	1.1	2.3	1.9
549	Rye flour, medium	1 c	102	10	361	10	79	15	2	.2	.2	.8
1044	Soy flour, low-fat	1 c	88	3	325	45	30	9	6	.9	1.3	3.3
	Spaghetti pasta:											
550	Without salt, enriched	1 c	140	66	197	7	40	4	1	.1	.1	.4
551	With salt, enriched	1 c	140	66	197	7	40	2	1	.1	.1	.4
552	Whole-wheat spaghetti, cooked	1 c	140	67	174	7	37	6	1	.1	.1	.3
1302	Tapioca-pearl, dry	1 c	152	11	544	<1	135	1	<1	t	t	t
553	Wheat bran, crude	1 c	58	10	125	9	37	25	2	.4	.4	1.3
554	Wheat germ, raw	1 c	115	11	414	27	60	15	11	1.9	1.6	6.9
555	Wheat germ, toasted	1 c	113	6	432	33	56	15	12	2.1	1.7	7.5
1669	Wheat germ, with brown sugar & honey	1 c	113	3	420	30	66	11	9	1.5	1.2	5.5
556	Rolled wheat, cooked	1 c	240	84	149	5	33	4	1	.1	.1	.5
557	Whole-grain wheat, cooked	1 c	150	86	84	4	20	3	<1	.1	.1	.2
	Wheat flour (unbleached):											
	All-purpose white flour, enriched:											
558	Sifted	1 c	115	12	419	12	88	3	1	.2	.1	.5
559	Unsifted	1 c	125	12	455	13	95	3	1	.2	.1	.5
560	Cake or pastry, enriched, sifted	1 c	96	12	348	8	75	2	1	.1	.1	.4
561	Self-rising, enriched, unsifted	1 c	125	11	443	12	93	3	1	.2	.1	.5
562	Whole wheat, from hard wheats	1 c	120	10	407	16	87	15	2	.4	.3	.9
	MEATS: FISH and SHELLFISH											
1045	Bass, baked or broiled	4 oz	113	69	165	27	0	0	5	1.4	1.5	2.3
1046	Bluefish, baked or broiled	4 oz	113	63	180	29	0	0	6	1.4	2	2.7
1686	Catfish, breaded/flour fried	4 oz	113	49	325	21	14	1	20	5	9	4.7
	Clams:											
563	Raw meat only	1 ea	145	82	107	19	4	0	1	.3	.4	.7
564	Canned, drained	1 c	160	64	237	41	8	0	3	.7	.9	1.5
1290	Steamed, meat only	10 ea	95	64	141	24	5	0	2	.4	.5	.9

PAGE KEY: A–4 = Beverages A–6 = Dairy A–10 = Eggs A–10 = Fat/Oil A–14 = Fruit A–20 = Bakery A–28 = Grain A–32 = Fish A–34 = Meats A–38= Poultry A–40 = Sausage A–42 = Mixed/Fast A–46 = Nuts/Seeds A–50 = Sweets A–52 = Vegetables/Legumes A–62 = Vegetarian A–62 = Misc A–66 = Soups/Sauces A–68 = Fast A–84 = Convenience A–88 = Baby foods

Chol (mg)	Calc (mg)	Iron (mg)	Magn (mg)	Pota (mg)	Sodi (mg)	Zinc (mg)	VT-A (RE)	Thia (mg)	VT-E (α-TE)	Ribo (mg)	Niac (mg)	V-B6 (mg)	Fola (μg)	VT-C (mg)
0	15	.66	25	41	8	.59	7	.15	.05	.08	1.43	.03	87	0
0	7	1.51	106	149	5	2.18	0	.25	.43	.2	3.19	.26	46	0
0	14	1	3	5	9	.23	0	.07	.06	0	.09	.02	1	0
0	35	3.04	4	14	14	.57	0	.21	.18	0	.28	.07	3	0
0	9	2.13	23	54	198	.63	4	.26	.07	.19	2.68	.05	40	0
53	19	2.54	30	45	11	.99	10	.3	.08	.13	2.38	.06	102	0
0	58	2.13	174	376	36	2.76	46	.37	.04	.2	4.55	.32	48	0
0	14	1.3	56	136	1	.75	0	.28	.41	.05	.22	.04	12	0
19	3	.65	10	14	3	.32	3	.12	.09	.09	.56	.02	36	0
0	10	1.96	25	43	1	.74	0	.29	.08	.14	2.34	.05	98	0
0	1	.21	10	24	<1	.27	2	.02	.01	.02	.15	.02	2	0
0	1	.14	9	14	29	.23	1	.02	.06	.01	.12	.01	1	0
0	1	.31	12	25	97	.29	2	.01	.03	.01	.17	.02	2	<1
2	15	.61	12	38	72	.2	3	.02	.42	.02	.77	.01	1	0
0	19	.82	84	84	10	1.23	0	.19	.53	.05	2.98	.28	8	0
0	300	9			2700		120							96
0		.72			1340									
0	52	7.97	46	213	9	2.02	0	1.07	.24	.09	7.75	.3	427	0
0	16	1.9	19	55	2	.77	0	.26	.08	.02	2.34	.15	92	0
0	13	1.04	8	7	5	.4	0	.12	.08	.08	1.45	.02	68	0
0	111	6.59	57	222	9	1.78	0	1.1	.24	.13	6.72	.65	427	0
0	33	1.98	21	65	5	.54	0	.44	.09	.03	2.45	.03	87	0
0	3	.24	9	17	9	.71	0	.03	.07	.02	.5	.04	2	0
0	5	.98	52	166	5	2.2	0	.08	.38	.14	2.12	.22	43	0
2	16	1.9	24	85	1147	.57	0	.25	.27	.16	3.6	.2	89	<1
0	24	2.16	76	347	3	2.03	0	.29	1.36	.12	1.76	.27	19	0
0	165	5.27	202	2261	16	1.04	4	.33	.17	.25	1.9	.46	361	0
0	10	1.96	25	43	1	.74	0	.29	.08	.14	2.34	.05	98	0
0	10	1.96	25	43	140	.74	0	.29	.38	.14	2.34	.05	98	0
0	21	1.48	42	62	4	1.13	0	.15	.07	.06	.99	.11	7	0
0	30	2.4	2	17	2	.18	0	.01	0	0	0	.01	6	0
0	42	6.15	354	686	1	4.22	0	.3	1.35	.33	7.89	.75	46	0
0	45	7.2	275	1025	14	14.1	0	2.16	20.7	.57	7.83	1.5	323	0
0	51	10.3	362	1070	5	18.9	0	1.89	20.5	.93	6.32	1.11	398	7
0	56	9.1	307	1089	12	15.7	11	1.51	24.9	.78	5.34	.56	376	0
0	17	1.49	53	170	0	1.15	0	.17	.48	.12	2.14	.17	26	0
0	9	.88	35	99	1	.73	0	.12	.3	.03	1.5	.08	12	0
0	17	5.34	25	123	2	.8	0	.9	.07	.57	6.79	.05	177	0
0	19	5.8	27	134	2	.87	0	.98	.07	.62	7.38	.05	193	0
0	13	7.03	15	101	2	.59	0	.86	.06	.41	6.52	.03	148	0
0	423	5.84	24	155	1587	.77	0	.84	.07	.52	7.29	.06	193	0
0	41	4.66	166	486	6	3.52	0	.54	1.48	.26	7.64	.41	53	0
98	116	2.16	43	515	102	.94	40	.1	.84	.1	1.72	.16	19	2
86	10	.7	47	539	87	1.18	156	.08	.71	.11	8.19	.52	2	0
92	41	1.44	34	376	598	1.05	33	.4	2.48	.18	3.37	.21	19	1
49	67	20.3	13	455	81	1.99	131	.12	1.45	.31	2.57	.09	23	19
107	147	44.8	29	1004	179	4.37	274	.24	3.04	.68	5.36	.18	46	35
64	87	26.6	17	597	106	2.59	162	.14	1.86	.4	3.18	.1	27	21

Table A–1

Food Composition (Computer code number is for Wadsworth Diet Analysis Plus program) (For purposes of calculations, use "0" for t, <1, <.1, <.01, etc.)

Computer Code Number	Food Description	Measure	Wt (g)	H_2O (%)	Ener (kcal)	Prot (g)	Carb (g)	Dietary Fiber (g)	Fat (g)	Fat Breakdown (g) Sat	Mono	Poly
	MEATS: FISH and SHELLFISH—Continued											
	Cod:											
565	Baked	4 oz	113	76	119	26	0	0	1	.2	.1	.5
566	Batter fried	4 oz	113	67	196	20	8	<1	9	2.2	3.6	2.6
567	Poached, no added fat	4 oz	113	77	116	25	0	0	1	.2	.1	.3
	Crab, meat only:											
1048	Blue crab, cooked	1 c	118	77	120	24	0	0	2	.3	.3	.8
1049	Dungeness crab, cooked	1 c	118	73	130	26	1	0	1	.2	.3	.5
568	Blue crab, canned	1 c	135	76	134	28	0	0	2	.4	.3	.6
1587	Crab, imitation, from surimi	4 oz	113	74	115	14	11	0	1	.3	.2	.8
569	Fish sticks, breaded pollock	2 ea	56	46	152	9	13	<1	7	1.8	2.8	1.8
572	Flounder/sole, baked	4 oz	113	73	132	27	0	0	2	.5	.4	.9
1599	Grouper, baked or broiled	4 oz	113	73	133	28	0	0	1	.4	.4	.6
573	Haddock, breaded, fried	4 oz	113	55	264	22	14	1	13	3.2	5.4	3.3
1050	Haddock, smoked	4 oz	113	71	131	28	0	0	1	.3	.3	.5
	Halibut:											
17291	Baked	4 oz	113	72	158	30	0	0	3	.7	1	1.4
1051	Smoked	4 oz	113	64	203	34			4	.6	1.2	1.5
1054	Raw	4 oz	113	78	124	23	0	0	3	.7	.8	1.1
575	Herring, pickled	4 oz	113	55	296	16	11	0	20	4.4	11	4.8
1052	Lobster meat, cooked w/moist heat	1 c	145	76	142	30	2	0	1	.2	.2	.5
1687	Ocean perch, baked/broiled	4 oz	113	73	137	27	0	0	2	.4	1	.7
576	Ocean perch, breaded/fried	4 oz	113	59	249	22	9	<1	13	3.2	5.7	3.4
1056	Octopus, raw	4 oz	113	80	93	17	2	0	1	.3	.2	.3
	Oysters:											
577	Raw, Eastern	1 c	248	85	169	17	10	0	6	2	.9	2.8
578	Raw, Pacific	1 c	248	82	201	23	12	0	6	1.3	.9	2.2
	Cooked:											
579	Eastern, breaded, fried, medium	5 ea	73	65	144	6	8	<1	9	2.7	1.7	4.6
580	Western, simmered	5 ea	125	64	204	24	12	0	6	1.9	.9	2.7
581	Pollock, baked, broiled, or poached	4 oz	113	74	128	27	0	0	1	.3	.2	.6
	Salmon:											
582	Canned pink, solids and liquid	4 oz	113	69	157	22	0	0	7	1.7	2.1	2.3
583	Broiled or baked	4 oz	113	62	244	31	0	0	12	2.2	6	2.7
584	Smoked	4 oz	113	72	132	21	0	0	5	1.2	2.3	1.1
585	Atlantic sardines, canned, drained, 2 = 24 g	4 oz	113	60	235	28	0	0	13	1.9	4.4	6.4
586	Scallops, breaded, cooked from frozen	6 ea	93	58	200	17	9	<1	10	2	2.5	5.3
1588	Scallops, imitation, from surimi	4 oz	113	74	112	14	12	0	1	.1	.1	.3
1688	Scallops, steamed/boiled	½ c	60	76	64	10	1	0	2	.3	.7	.6
	Shrimp:											
587	Cooked, boiled, 2 large = 11g	16 ea	88	77	87	18	0	0	1	.2	.2	.5
588	Canned, drained	½ c	64	73	77	15	1	0	1	.3	.3	.7
589	Fried, 2 large = 15 g, breaded	12 ea	90	53	218	19	10	<1	11	2	3	5.9
1057	Raw, large, about 7g each	14 ea	98	76	104	20	1	0	2	.3	.3	1
1589	Shrimp, imitation, from surimi	4 oz	113	75	114	14	10	0	2	.3	.2	1
1053	Snapper, baked or broiled	4 oz	113	70	145	30	0	0	2	.4	.4	.7
1060	Squid, fried in flour	4 oz	113	64	198	20	9	0	8	2.1	3.1	2.4
1590	Surimi	4 oz	113	76	112	17	8	0	1	.2	.2	.6
1058	Swordfish, raw	4 oz	113	76	137	22	0	0	5	1.3	1.7	1
1059	Swordfish, baked or broiled	4 oz	113	69	175	29	0	0	6	1.7	2.2	1.3
590	Trout, baked or broiled	4 oz	113	70	170	26	0	0	7	1.8	2	2.2
	Tuna, light, canned, drained solids:											
591	Oil pack	1 c	145	60	287	42	0	0	12	2.2	4.3	4.2
592	Water pack	1 c	154	74	179	39	0	0	1	.4	.2	.5
1061	Bluefin tuna, fresh	4 oz	113	68	163	26	0	0	6	1.4	1.8	1.6
	MEATS: BEEF, LAMB, PORK and others											
	BEEF, cooked, trimmed to ½" outer fat:											
	Braised, simmered, pot roasted:											
	Relatively fat, choice chuck blade:											
593	Lean and fat, piece 2½ x 2½ x ¾"	4 oz	113	47	393	30	0	0	29	13	14.8	1.2
594	Lean only	4 oz	113	55	297	35	0	0	16	7.3	8.2	.7

PAGE KEY: A–4 = Beverages A–6 = Dairy A–10 = Eggs A–10 = Fat/Oil A–14 = Fruit A–20 = Bakery A–28 = Grain A–32 = Fish A–34 = Meats A–38= Poultry A–40 = Sausage A–42 = Mixed/Fast A–46 = Nuts/Seeds A–50 = Sweets A–52 = Vegetables/Legumes A–62 = Vegetarian A–62 = Misc A–66 = Soups/Sauces A–68 = Fast A–84 = Convenience A–88 = Baby foods

Chol (mg)	Calc (mg)	Iron (mg)	Magn (mg)	Pota (mg)	Sodi (mg)	Zinc (mg)	VT-A (RE)	Thia (mg)	VT-E (α-TE)	Ribo (mg)	Niac (mg)	V-B6 (mg)	Fola (μg)	VT-C (mg)
62	16	.55	47	276	88	.65	16	.1	.39	.09	2.84	.32	9	1
64	43	.92	36	443	124	.62	17	.13	.92	.13	2.58	.23	10	1
61	23	.54	41	496	69	.63	14	.09	.32	.08	2.48	.28	8	1
118	123	1.07	39	382	329	4.98	2	.12	1.18	.06	3.89	.21	60	4
90	70	.51	68	481	446	6.45	37	.07	1.33	.24	4.27	.2	50	4
120	136	1.13	53	505	450	5.43	2	.11	1.35	.11	1.85	.2	57	4
23	15	.44	49	102	950	.37	23	.04	.12	.03	.2	.03	2	0
63	11	.41	14	146	326	.37	17	.07	.77	.1	1.19	.03	10	0
77	20	.38	65	389	119	.71	12	.09	2.6	.13	2.46	.27	10	0
53	24	1.29	42	537	60	.58	56	.09	.71	.01	.43	.4	11	0
96	63	1.92	46	345	523	.59	33	.08	1.56	.14	4.49	.28	19	<1
87	55	1.58	61	469	862	.56	25	.05	.56	.05	5.73	.45	17	0
46	68	1.21	121	651	78	.6	61	.08	1.23	.1	8.05	.45	16	0
59	87	1.56	154	833	2260	.78	86	.11	1.11	.14	10.8	.64	22	0
36	53	.95	94	509	61	.47	53	.07	.96	.08	6.61	.39	14	0
15	87	1.38	9	78	983	.6	292	.04	1.81	.16	3.73	.19	3	0
104	88	.57	51	510	551	4.23	38	.01	2.1	.1	1.55	.11	16	0
61	155	1.33	44	396	108	.69	16	.15	1.84	.15	2.76	.3	12	1
71	136	1.57	38	323	431	.67	23	.14	2.41	.18	2.68	.24	15	1
54	60	5.99	34	396	260	1.9	51	.03	1.36	.04	2.37	.41	18	6
131	112	16.5	117	387	523	225	74	.25	1.98	.24	3.42	.15	25	9
124	20	12.7	55	417	263	41.2	201	.17	2.11	.58	4.98	.12	25	20
59	45	5.07	42	178	304	63.6	66	.11	1.66	.15	1.2	.05	23	3
125	20	11.5	55	378	265	41.5	183	.16	2.21	.55	4.53	.11	19	16
108	7	.32	82	437	131	.68	26	.08	.32	.09	1.86	.08	4	0
62	241	.95	38	368	626	1.04	19	.03	1.53	.21	7.39	.34	17	0
98	8	.62	35	424	75	.58	71	.24	1.42	.19	7.54	.25	6	0
26	12	.96	20	198	886	.35	29	.03	1.53	.11	5.33	.31	2	0
160	432	3.3	44	449	571	1.48	76	.09	.34	.26	5.93	.19	13	0
57	39	.76	55	310	432	.99	20	.04	1.77	.1	1.4	.13	34	2
25	9	.35	49	116	898	.37	23	.01	.12	.02	.35	.03	2	0
19	15	.15	32	168	246	.55	27	.01	.81	.04	.6	.08	7	1
172	34	2.72	30	160	197	1.37	58	.03	.66	.03	2.28	.11	3	2
111	38	1.75	26	134	108	.81	11	.02	.59	.02	1.77	.07	1	1
159	60	1.13	36	203	310	1.24	50	.12	1.35	.12	2.76	.09	7	1
149	51	2.36	36	181	145	1.09	53	.03	.8	.03	2.5	.1	3	2
41	21	.68	49	101	797	.37	23	.03	.12	.04	.19	.03	2	0
53	45	.27	42	590	64	.5	40	.06	.71	<.01	.39	.52	7	2
294	44	1.14	43	315	346	1.97	12	.06	2.09	.52	2.94	.07	16	5
34	10	.29	49	127	162	.37	23	.02	.28	.02	.25	.03	2	0
44	5	.91	30	325	102	1.3	41	.04	.56	.11	10.9	.37	2	1
56	7	1.18	38	417	130	1.66	46	.05	.71	.13	13.3	.43	3	1
78	97	.43	35	506	63	.58	17	.17	.57	.11	6.52	.39	21	2
26	19	2.02	45	300	513	1.31	33	.05	1.74	.17	18	.16	8	0
46	17	2.36	42	365	521	1.19	26	.05	.82	.11	20.5	.54	6	0
43	9	1.15	56	285	44	.68	740	.27	1.13	.28	9.77	.51	2	0
112	11	3.45	21	275	67	7.57	0	.08	.26	.27	3.54	.32	10	0
120	15	4.16	26	297	80	11.6	0	.09	.16	.32	3.02	.33	7	0

Table A-1

Food Composition (Computer code number is for Wadsworth Diet Analysis Plus program) (For purposes of calculations, use "0" for t, <1, <.1, <.01, etc.)

Computer Code Number	Food Description	Measure	Wt (g)	H_2O (%)	Ener (kcal)	Prot (g)	Carb (g)	Dietary Fiber (g)	Fat (g)	Fat Breakdown (g) Sat	Mono	Poly
	MEATS: BEEF, LAMB, PORK and others—Continued											
	Relatively lean, like choice round:											
595	Lean and fat, pce 4⅛ x 2½ x ¾"	4 oz	113	52	311	32	0	0	19	8.5	9.7	.8
596	Lean only	4 oz	113	57	249	36	0	0	11	4.8	5.4	.5
	Ground beef, broiled, patty 3 x ⅝":											
597	Extra lean, about 16% fat	4 oz	113	54	299	32	0	0	18	8	9	.8
598	Lean, 21% fat	4 oz	113	53	316	32	0	0	20	8.9	10.1	.8
	Roasts, oven cooked, no added liquid:											
	Relatively fat, prime rib:											
601	Lean and fat, piece 4⅛ x 2¼ x ½"	4 oz	113	46	425	25	0	0	35	15.8	17.9	1.5
602	Lean only	4 oz	113	58	271	31	0	0	16	7	7.9	.7
	Relatively lean, choice round:											
603	Lean and fat, piece 2½ x 2½ x ¾"	4 oz	113	59	272	30	0	0	16	7.1	8.1	.7
604	Lean only	4 oz	113	65	198	33	0	0	6	2.9	3.3	.3
1701	Steak, rib, broiled, lean	4 oz	113	58	250	32	0	0	13	5.7	6.4	.5
	Steak, broiled, relatively lean,											
606	choice sirloin, lean only	4 oz	113	62	228	34	0	0	9	4	4.6	.4
	Steak, broiled, relatively fat,											
	choice T-bone:											
1063	Lean and fat	4 oz	113	52	349	26	0	0	26	11.8	13.3	1.1
1064	Lean only	4 oz	113	61	232	30	0	0	11	5.1	5.8	.5
	Variety meats:											
1086	Brains, panfried	4 oz	113	71	221	14	0	0	18	6.7	7	3.9
599	Heart, simmered	4 oz	113	64	198	32	<1	0	6	2.9	1.5	1.5
600	Liver, fried	4 oz	113	56	245	30	9	0	9	3.1	1.8	1.9
1062	Tongue, cooked	4 oz	113	56	320	25	<1	0	23	10.1	10.7	.9
607	Beef, canned, corned	4 oz	113	58	283	31	0	0	17	7.5	8.5	.7
608	Beef, dried, cured	1 oz	28	56	46	8	<1	0	1	.5	.5	.1
	LAMB, domestic, cooked:											
	Chop, arm, braised (5.6 oz raw w/bone):											
609	Lean and fat	1 ea	70	44	242	21	0	0	17	7.8	7.3	1.4
610	Lean only	1 ea	55	49	153	19	0	0	8	3.6	3.4	.6
	Chop, loin, broiled (4.2 oz raw w/bone):											
611	Lean and fat	1 ea	64	52	202	16	0	0	15	6.8	6.4	1.2
612	Lean only	1 ea	46	61	99	14	0	0	4	2.1	1.9	.4
1067	Cutlet, avg of lean cuts, cooked	4 oz	113	54	330	28	0	0	23	10.9	10.2	1.9
	Leg, roasted, 3 oz = 4⅛ x 2¼ x ½":											
613	Lean and fat	4 oz	113	57	292	29	0	0	19	8.7	8.1	1.6
614	Lean only	4 oz	113	64	216	32	0	0	9	4.1	3.8	.7
615	Rib, roasted, lean and fat	4 oz	113	48	406	24	0	0	34	15.7	14.7	2.8
616	Rib, roasted, lean only	4 oz	113	60	262	30	0	0	15	7	6.6	1.2
1065	Shoulder, roasted, lean and fat	4 oz	113	56	312	25	0	0	23	10.5	9.8	1.9
1066	Shoulder, roasted, lean only	4 oz	113	63	231	28	0	0	12	5.7	5.3	1.1
	Variety meats:											
1069	Brains, panfried	4 oz	113	76	164	14	0	0	11	4.4	3.7	1.9
1068	Heart, braised	4 oz	113	64	209	28	2	0	9	3.9	2.7	1.1
1070	Sweetbreads, cooked	4 oz	113	60	264	26	0	0	17	8.1	6.5	1.4
1071	Tongue, cooked	4 oz	113	58	311	24	0	0	23	8.8	11.6	1.4
	PORK, cured, cooked (see also Sausages and Lunch Meats)											
617	Bacon, medium slices	3 pce	19	13	109	6	<1	0	9	3.3	4.5	1.1
1087	Breakfast strips, cooked	2 pce	23	27	106	7	<1	0	8	2.9	3.8	1.3
618	Canadian-style bacon	2 pce	47	62	87	11	1	0	4	1.3	1.9	.4
	Ham, roasted:											
619	Lean and fat, 2 pces 4⅛ x 2¼ x ¼"	4 oz	113	64	201	25	0	0	10	3.4	5	1.7
620	Lean only	4 oz	113	68	164	24	2	0	6	2.1	3	.6
621	Ham, canned, roasted, 8% fat	4 oz	113	69	154	24	1	0	6	1.8	2.8	.5
	PORK, fresh, cooked:											
	Chops, loin (cut 3 per lb with bone):											
1291	Braised, lean and fat	1 ea	89	58	213	24	0	0	12	4.5	5.4	1
1292	Lean only	1 ea	80	61	163	23	0	0	7	2.7	3.3	.6
622	Broiled, lean and fat	1 ea	82	58	197	23	0	0	11	3.9	4.8	.8
623	Broiled, lean only	1 ea	74	61	149	22	0	0	6	2.2	2.7	.4

PAGE KEY: A–4 = Beverages A–6 = Dairy A–10 = Eggs A–10 = Fat/Oil A–14 = Fruit A–20 = Bakery A–28 = Grain A–32 = Fish A–34 = Meats A–38= Poultry A–40 = Sausage A–42 = Mixed/Fast A–46 = Nuts/Seeds A–50 = Sweets A–52 = Vegetables/Legumes A–62 = Vegetarian A–62 = Misc A–66 = Soups/Sauces A–68 = Fast A–84 = Convenience A–88 = Baby foods

Chol (mg)	Calc (mg)	Iron (mg)	Magn (mg)	Pota (mg)	Sodi (mg)	Zinc (mg)	VT-A (RE)	Thia (mg)	VT-E (α-TE)	Ribo (mg)	Niac (mg)	V-B6 (mg)	Fola (μg)	VT-C (mg)
108	7	3.53	25	319	56	5.55	0	.08	.21	.27	4.21	.37	11	0
108	6	3.91	28	348	58	6.19	0	.08	.2	.29	4.61	.41	12	0
112	10	3.13	28	417	93	7.27	0	.08	.2	.36	6.61	.36	12	0
114	14	2.77	27	394	101	7.01	0	.07	.23	.27	6.75	.34	12	0
96	12	2.61	21	334	71	5.92	0	.08	.27	.19	3.8	.26	8	0
91	11	2.95	28	425	84	7.84	0	.09	.14	.24	4.64	.34	9	0
81	7	2.07	27	406	67	4.87	0	.09	.23	.18	3.92	.4	7	0
78	6	2.2	30	446	70	5.36	0	.1	.12	.19	4.24	.43	8	0
90	15	2.9	30	445	78	7.9	0	.11	.16	.25	5.42	.45	9	0
101	12	3.8	36	455	75	7.37	0	.15	.16	.33	4.84	.51	11	0
76	9	3.06	26	363	72	5.03	0	.1	.24	.24	4.46	.37	8	0
67	7	3.58	32	427	80	6	0	.12	.16	.28	5.23	.44	9	0
2254	10	2.51	17	400	179	1.53	0	.15	2.37	.29	4.27	.44	7	4
218	7	8.49	28	263	71	3.54	0	.16	.81	1.74	4.6	.24	2	2
545	12	7.1	26	411	120	6.16	12123	.24	.72	4.68	16.3	1.62	249	26
121	8	3.83	19	203	68	5.42	0	.03	.4	.4	2.43	.18	6	1
97	14	2.35	16	154	1136	4.03	0	.02	.17	.17	2.75	.15	10	0
12	2	1.26	9	124	972	1.47	0	.02	.04	.06	1.53	.1	3	0
84	17	1.67	18	214	50	4.26	0	.05	.1	.17	4.66	.08	13	0
67	14	1.49	16	186	42	4.02	0	.04	.1	.15	3.48	.07	12	0
64	13	1.16	15	209	49	2.23	0	.06	.08	.16	4.54	.08	11	0
44	9	.92	13	173	39	1.9	0	.05	.07	.13	3.15	.07	11	0
110	12	2.26	25	340	77	4.67	0	.12	.15	.32	7.48	.16	19	0
105	12	2.24	27	354	75	4.97	0	.11	.17	.3	7.45	.17	23	0
101	9	2.4	29	382	77	5.58	0	.12	.2	.33	7.16	.19	26	0
110	25	1.81	23	306	82	3.94	0	.1	.11	.24	7.63	.12	17	0
99	24	2	26	356	91	5.05	0	.1	.17	.26	6.96	.17	25	0
104	23	2.23	26	284	75	5.91	0	.1	.16	.27	6.95	.15	24	0
98	21	2.41	28	299	77	6.83	0	.1	.2	.29	6.51	.17	28	0
2308	14	1.9	16	232	151	1.54	0	.12	1.73	.27	2.79	.12	6	14
281	16	6.24	27	212	71	4.16	0	.19	.79	1.34	4.93	.34	2	8
452	14	2.4	21	329	59	3.03	0	.02	.78	.24	2.89	.06	15	23
214	11	2.97	18	179	76	3.38	0	.09	.36	.47	4.17	.19	3	8
16	2	.31	5	92	303	.62	0	.13	.1	.05	1.39	.05	1	0
24	3	.45	6	107	483	.85	0	.17	.08	.08	1.75	.08	1	0
27	5	.38	10	183	727	.8	0	.39	.15	.09	3.25	.21	2	0
67	9	1.51	25	462	1695	2.79	0	.82	.45	.37	6.95	.35	3	0
60	9	1.67	16	324	1359	3.25	0	.85	.29	.23	4.54	.45	3	0
34	7	1.04	24	393	1282	2.52	0	1.18	.29	.28	5.53	.51	6	0
71	19	.95	17	333	43	2.12	2	.56	.3	.23	3.93	.33	3	1
63	14	.9	16	310	40	1.98	2	.53	.3	.21	3.67	.31	3	<1
67	27	.66	20	294	48	1.85	2	.88	.27	.24	4.3	.35	5	<1
61	23	.63	20	278	44	1.76	2	.85	.31	.23	4.1	.35	4	<1

Table A-1

Food Composition (Computer code number is for Wadsworth Diet Analysis Plus program) (For purposes of calculations, use "0" for t, <1, <.1, <.01, etc.)

Computer Code Number	Food Description	Measure	Wt (g)	H_2O (%)	Ener (kcal)	Prot (g)	Carb (g)	Dietary Fiber (g)	Fat (g)	Fat Breakdown (g) Sat	Mono	Poly
	MEATS: BEEF, LAMB, PORK and others—Continued											
624	Panfried, lean and fat	1 ea	78	53	216	23	0	0	13	4.7	5.5	1.5
625	Panfried, lean only	1 ea	63	59	152	16	0	0	10	3.2	3.9	1.2
626	Leg, roasted, lean and fat	4 oz	113	55	308	30	0	0	20	7.3	8.9	1.9
627	Leg, roasted, lean only	4 oz	113	61	233	35	0	0	9	3.2	4.3	.9
628	Rib, roasted, lean and fat	4 oz	113	56	288	31	0	0	17	6.7	7.9	1.4
629	Rib, roasted, lean only	4 oz	113	59	252	32	0	0	13	4.9	5.9	1
630	Shoulder, braised, lean and fat	4 oz	113	48	372	32	0	0	26	9.6	11.8	2.6
631	Shoulder, braised, lean only	4 oz	113	54	280	36	0	0	14	4.7	6.5	1.3
1088	Spareribs, cooked, yield from 1 lb raw with bone	4 oz	113	40	449	33	0	0	34	12.5	15.3	3.1
1095	Rabbit, roasted (1 cup meat = 140 g)	4 oz	113	61	223	33	0	0	9	4	2	3
	VEAL, cooked:											
632	Cutlet, braised or broiled, 4⅛ x 2¼ x ½"	4 oz	113	52	321	34	0	0	19	7.6	7.6	1.3
633	Rib roasted, lean, 2 pieces 4⅛ x 2¼ x ¼"	4 oz	113	60	258	27	0	0	16	6.1	6.1	1.1
634	Liver, panfried	4 oz	113	67	186	24	3	0	8	3.3	1.9	2.2
1096	Venison (deer meat), roasted	4 oz	113	65	179	34	0	0	4	1.4	1	.7
	MEATS: POULTRY and POULTRY PRODUCTS											
	CHICKEN, cooked:											
	Fried, batter dipped:											
635	Breast	1 ea	280	52	728	69	25	1	37	9.9	15.3	8.6
636	Drumstick	1 ea	72	53	193	16	6	<1	11	3	4.7	2.7
637	Thigh	1 ea	86	51	238	19	8	<1	14	3.8	5.9	3.3
638	Wing	1 ea	49	46	159	10	5	<1	11	2.9	4.5	2.5
	Fried, flour coated:											
639	Breast	1 ea	196	57	435	62	3	<1	17	4.9	7.1	3.8
1212	Breast, without skin	1 ea	86	60	161	29	<1	<1	4	1.1	1.5	.9
640	Drumstick	1 ea	49	57	120	13	1	<1	7	1.8	2.7	1.6
641	Thigh	1 ea	62	54	162	17	2	<1	9	2.5	3.7	2.1
1099	Thigh, without skin	1 ea	52	59	113	15	1	<1	5	1.4	2	1.3
642	Wing	1 ea	32	49	103	8	1	<1	7	1.9	2.9	1.6
	Roasted:											
643	All types of meat	1 c	140	64	266	40	0	0	10	2.9	3.8	2.4
644	Dark meat	1 c	140	63	287	38	0	0	14	3.7	5.2	3.2
645	Light meat	1 c	140	65	242	43	0	0	6	1.8	2.2	1.4
646	Breast, without skin	1 ea	172	65	284	53	0	0	6	1.8	2.2	1.3
647	Drumstick, without skin	1 ea	44	67	76	12	0	0	2	.7	.8	.6
1703	Leg, without skin	1 ea	95	65	181	26	0	0	8	2.2	2.9	1.9
648	Thigh	1 ea	62	59	153	16	0	0	10	2.7	3.9	2.1
1100	Thigh, without skin	1 ea	52	63	109	13	0	0	6	1.6	2.2	1.3
649	Stewed, all types	1 c	140	67	248	38	0	0	9	2.6	3.5	2.2
656	Canned, boneless chicken	4 oz	113	69	186	25	0	0	9	2.5	3.6	2
1102	Gizzards, simmered	1 c	145	67	222	39	2	0	5	1.5	1.3	1.5
1101	Hearts, simmered	1 c	145	65	268	38	<1	0	11	3.3	2.9	3.3
2300	Liver, simmered: Ounce	3 oz	85	68	133	21	1	0	5	1.6	1.1	.8
1098	Liver, simmered: Piece = 20 g	6 ea	120	68	188	29	1	0	7	2.2	1.6	1.1
	DUCK, roasted:											
1293	Meat with skin, about 2.7 cups	½ ea	382	52	1287	73	0	0	108	36.9	49.3	13.9
651	Meat only, about 1.5 cups	½ ea	221	64	444	52	0	0	25	9.2	8.2	3.2
	GOOSE, domesticated, roasted:											
1294	Meat only, about 4.2 cups	½ ea	591	57	1406	173	0	0	75	23.6	40.2	10.9
1295	Meat with skin, about 5.5 cups	½ ea	774	52	2360	195	0	0	170	53.2	80.5	24.8
	TURKEY:											
	Roasted, meat only:											
652	Dark meat	4 oz	113	63	211	33	0	0	8	2.7	1.8	2.5
653	Light meat	4 oz	113	66	177	34	0	0	4	1.2	.6	1
654	All types, chopped or diced	1 c	140	65	238	42	0	0	7	2.3	1.5	2
1103	Ground, cooked	4 oz	113	59	266	31	0	0	15	4.1	5.5	3.6
1106	Gizzard, cooked	2 ea	134	65	218	39	1	0	5	1.5	1	1.5
1107	Heart, cooked	4 ea	64	64	113	17	1	0	4	1.1	.8	1.1
1108	Liver, cooked	1 ea	75	66	127	18	3	0	4	1.4	1.1	.8

PAGE KEY: A–4 = Beverages A–6 = Dairy A–10 = Eggs A–10 = Fat/Oil A–14 = Fruit A–20 = Bakery A–28 = Grain A–32 = Fish A–34 = Meats A–38= Poultry A–40 = Sausage A–42 = Mixed/Fast A–46 = Nuts/Seeds A–50 = Sweets A–52 = Vegetables/Legumes A–62 = Vegetarian A–62 = Misc A–66 = Soups/Sauces A–68 = Fast A–84 = Convenience A–88 = Baby foods

Chol (mg)	Calc (mg)	Iron (mg)	Magn (mg)	Pota (mg)	Sodi (mg)	Zinc (mg)	VT-A (RE)	Thia (mg)	VT-E (α-TE)	Ribo (mg)	Niac (mg)	V-B6 (mg)	Fola (μg)	VT-C (mg)
72	21	.71	23	332	62	1.8	2	.89	.32	.24	4.37	.37	5	1
52	14	.67	16	230	49	2.44	1	.46	.3	.23	2.8	.26	3	<1
106	16	1.14	25	398	68	3.34	3	.72	.34	.35	5.16	.45	11	<1
108	8	1.29	33	442	73	3.4	3	.91	.46	.4	5.56	.38	3	<1
82	32	1.06	24	476	52	2.33	2	.82	.41	.34	6.92	.37	3	<1
80	29	1.11	25	494	53	2.41	2	.86	.55	.36	7.25	.38	3	<1
123	20	1.82	21	417	99	4.72	3	.61	.5	.35	5.89	.4	5	<1
129	9	2.2	25	458	115	5.62	3	.68	.58	.41	6.71	.46	6	<1
137	53	2.09	27	362	105	5.2	3	.46	.52	.43	6.19	.4	5	0
93	21	2.57	24	433	53	2.57	0	.1	.96	.24	9.53	.53	12	0
133	32	1.23	27	316	90	4.1	0	.04	.45	.34	10.2	.29	16	0
124	12	1.1	25	333	104	4.62	0	.06	.4	.3	7.89	.28	15	0
634	8	2.96	21	232	60	10.8	9095	.15	.42	2.19	9.58	.55	858	35
127	8	5.05	27	379	61	3.11	0	.2	.28	.68	7.58	.42	5	0
238	56	3.5	67	563	770	2.66	56	.32	2.97	.41	29.4	1.2	42	0
62	12	.97	14	134	194	1.68	19	.08	.88	.15	3.67	.19	13	0
80	15	1.25	18	165	248	1.75	25	.1	1.05	.19	4.92	.22	16	0
39	10	.63	8	68	157	.68	17	.05	.52	.07	2.58	.15	9	0
174	31	2.33	59	508	149	2.16	29	.16	1.12	.26	26.9	1.14	12	0
78	14	.98	27	237	68	.93	6	.07	.36	.11	12.7	.55	3	0
44	6	.66	11	112	44	1.42	12	.04	.41	.11	2.96	.17	5	0
60	9	.92	15	147	55	1.56	18	.06	.52	.15	4.31	.2	7	0
53	7	.76	13	135	49	1.45	11	.05	.3	.13	3.7	.2	5	0
26	5	.4	6	57	25	.56	12	.02	.18	.04	2.14	.13	2	0
125	21	1.69	35	340	120	2.94	22	.1	.58	.25	12.8	.66	8	0
130	21	1.86	32	336	130	3.92	31	.1	.81	.32	9.17	.5	11	0
119	21	1.48	38	346	108	1.72	13	.09	.37	.16	17.4	.84	6	0
146	26	1.79	50	440	127	1.72	10	.12	.66	.2	23.6	1.03	7	0
41	5	.57	11	108	42	1.4	8	.03	.25	.1	2.68	.17	4	0
89	11	1.24	23	230	86	2.72	18	.07	.55	.22	6	.35	8	0
58	7	.83	14	138	52	1.46	30	.04	.35	.13	3.95	.19	4	0
49	6	.68	12	124	46	1.34	10	.04	.3	.12	3.4	.18	4	0
116	20	1.64	29	252	98	2.79	21	.07	.42	.23	8.57	.36	8	0
70	16	1.79	14	156	568	1.59	38	.02	.24	.15	7.15	.4	5	2
281	14	6.02	29	260	97	6.35	81	.04	2.29	.35	5.77	.17	77	2
351	28	13.1	29	191	70	10.6	13	.1	2.32	1.07	4.06	.46	116	3
536	12	7.2	18	119	43	3.69	4176	.13	1.45	1.49	3.78	.49	655	13
757	17	10.2	25	168	61	5.21	5895	.18	2.04	2.1	5.34	.7	924	19
321	42	10.3	61	779	225	7.11	241	.66	2.5	1.03	18.5	.69	23	0
197	26	5.97	44	557	144	5.75	51	.57	1.55	1.04	11.3	.55	22	0
567	83	17	148	2293	449	18.7	71	.54	9.16	2.3	24.1	2.78	71	0
704	101	21.9	170	2546	542	20.3	163	.6	13.5	2.5	32.3	2.86	15	0
96	36	2.63	27	328	89	5.04	0	.07	.94	.28	4.12	.41	10	0
78	21	1.53	32	345	72	2.31	0	.07	.12	.15	7.73	.61	7	0
106	35	2.49	36	417	98	4.34	0	.09	.59	.25	7.62	.64	10	0
115	28	2.18	27	305	121	3.23	0	.06	.45	.19	5.45	.44	8	0
311	20	7.29	25	283	72	5.57	74	.04	.27	.44	4.11	.16	70	2
145	8	4.41	14	117	35	3.37	5	.04	.13	.56	2.08	.2	51	1
470	8	5.85	11	146	48	2.32	2805	.04	2.41	1.07	4.46	.39	500	1

Table A–1

Food Composition (Computer code number is for Wadsworth Diet Analysis Plus program) (For purposes of calculations, use "0" for t, <1, <.1, <.01, etc.)

Computer Code Number	Food Description	Measure	Wt (g)	H_2O (%)	Ener (kcal)	Prot (g)	Carb (g)	Dietary Fiber (g)	Fat (g)	Fat Breakdown (g) Sat	Mono	Poly
	MEATS: POULTRY and POULTRY PRODUCTS—Continued											
	POULTRY FOOD PRODUCTS (see also items in Sausages & Lunchmeats section):											
1567	Chicken patty, breaded, cooked	1 ea	75	49	213	12	11	<1	13	4.1	6.4	1.6
659	Turkey and gravy, frozen package	3 oz	85	85	57	5	4	<1	2	.8	.8	.4
	Turkey breast, Louis Rich:											
1104	Barbecued	2 oz	56	72	58	12	2	0	<1	.2	.2	.1
1943	Hickory smoked	1 pce	80		80	16	2	0	1	0		
1947	Honey roasted	1 pce	80		80	16	3	0	1	.5		
1945	Oven roasted	1 pce	80		70	16		0	1	0		
661	Turkey patty, breaded, fried	2 oz	57	50	161	8	9	<1	10	2.7	4.3	2.7
662	Turkey, frozen, roasted, seasoned	4 oz	113	68	175	24	3	0	7	2.1	1.4	1.9
1704	Turkey roll, light meat	1 pce	28	72	41	5	<1	0	2	.6	.7	.5
	MEATS: SAUSAGES and LUNCHMEATS (see also Poultry Food Products)											
1072	Beerwurst/beer salami, beef	1 oz	28	53	92	3	<1	0	8	3.6	3.9	.3
1074	Beerwurst/beer salami, pork	1 oz	28	61	67	4	1	0	5	1.8	2.5	.7
1075	Berliner sausage	1 oz	28	61	64	4	1	0	5	1.7	2.2	.4
	Bologna:											
1297	Beef	1 pce	23	55	72	3	<1	0	7	2.8	3.2	.3
2115	Beef, light, Oscar Mayer	1 pce	28	65	56	3	2	0	4	1.6	2	.1
663	Beef & pork	1 pce	28	54	88	3	1	0	8	3	3.7	.7
2155	Healthy Favorites	1 pce	23		22	3	1	0	<1	0		
1298	Pork	1 pce	23	61	57	4	<1	0	5	1.6	2.2	.5
2114	Regular, light, Oscar Mayer	1 pce	28	65	56	3	2	0	4	1.6	2	.4
664	Turkey	1 pce	28	65	56	4	<1	0	4	1.4	1.3	1.2
1970	Turkey, Louis Rich	1 pce	28	67	57	3	<1	0	5	1.5	1.8	1.3
665	Braunschweiger sausage	2 pce	57	48	205	8	2	0	18	6.2	8.5	2.1
1073	Bratwurst, link	1 ea	70	51	226	10	2	0	19	6.9	9.3	2
666	Brown & serve sausage links, cooked	2 ea	26	45	102	4	1	0	10	3.4	4.4	1
1089	Cheesefurter/cheese smokie	2 ea	86	52	281	12	1	0	25	9	11.8	2.6
2157	Chicken breast, Healthy Favorites	4 pce	52		40	9	1	0	0	0	0	0
1556	Chorizo, pork & beef	1 ea	60	32	273	15	1	0	23	8.6	11	2.1
1090	Corned beef loaf, jellied	1 pce	28	69	43	6	0	0	2	.7	.7	.1
	Frankfurters:											
1077	Beef, large link, 8/package	1 ea	57	55	180	7	1	0	16	6.9	7.9	.8
1078	Beef and pork, large link, 8/package	1 ea	57	54	182	6	1	0	17	6.2	8	1.6
667	Beef and pork, small link, 10/pkg	1 ea	45	54	144	5	1	0	13	4.9	6.3	1.2
668	Turkey frankfurter, 10/package	1 ea	45	63	102	6	1	0	8	2.7	2.5	2.2
1968	Turkey/chicken frank 8/pkg	1 ea	43		80	6	1	0	6	2		
	Ham:											
669	Ham lunchmeat, canned, 3 x 2 x ½"	1 pce	21	52	70	3	<1	0	6	2.3	3	.7
670	Chopped ham, packaged	2 pce	42	64	96	7	0	0	7	2.4	3.4	.9
2156	Honey ham, Healthy Favorites	4 pce	52	73	55	9	2	0	1	.4	.8	.1
2113	Oscar Mayer lower sodium ham	1 pce	21	73	23	3	1	0	1	.3	.4	.1
673	Turkey ham lunchmeat	2 pce	57	71	73	11	<1	0	3	1	.7	.9
1091	Kielbasa sausage	1 pce	26	54	81	3	1	0	7	2.6	3.4	.8
1092	Knockwurst sausage, link	1 ea	68	55	209	8	1	0	19	6.9	8.7	2
1093	Mortadella lunchmeat	2 pce	30	52	93	5	1	0	8	2.8	3.4	.9
1097	Olive loaf lunchmeat	2 pce	57	58	134	7	5	<1	9	3.3	4.5	1.1
1952	Turkey breast, fat free	1 pce	28	77	22	4	1	0	<1	.1	.1	t
1080	Turkey pastrami	2 pce	57	71	80	10	1	0	4	1	1.2	.9
1969	Turkey salami	1 pce	28	72	41	4	<1	0	3	.9	1	.8
1081	Pepperoni sausage	2 pce	11	27	55	2	<1	0	5	1.8	2.3	.5
1094	Pickle & pimento loaf	2 pce	57	57	149	7	3	<1	12	4.5	5.5	1.5
1082	Polish sausage	1 oz	28	53	91	4	<1	0	8	2.9	3.8	.9
674	Pork sausage, cooked, link, small	2 ea	26	45	96	5	<1	0	8	2.8	4.1	.8
1079	Pork sausage, cooked, patty	4 oz	113	45	417	22	1	0	35	12.1	17.7	3.3
675	Salami, pork and beef	2 pce	57	60	143	8	1	0	11	4.6	5.2	1.1
677	Salami, pork and beef, dry	3 pce	30	35	125	7	1	0	10	3.7	5.1	1
676	Salami, turkey	2 pce	57	66	112	9	<1	0	8	2.3	2.6	2
	Sandwich spreads:											
1300	Ham salad spread	2 tbs	30	63	65	3	3	0	5	1.5	2.2	.8
678	Pork and beef	2 tbs	30	60	70	2	4	<1	5	1.8	2.3	.8
1296	Chicken/turkey	2 tbs	26	66	52	3	2	0	4	.9	.8	1.6

Chol (mg)	Calc (mg)	Iron (mg)	Magn (mg)	Pota (mg)	Sodi (mg)	Zinc (mg)	VT-A (RE)	Thia (mg)	VT-E (α-TE)	Ribo (mg)	Niac (mg)	V-B6 (mg)	Fola (µg)	VT-C (mg)
45	12	.94	15	185	399	.78	22	.07	1.46	.1	5.04	.23	8	<1
15	12	.79	7	52	471	.59	11	.02	.3	.11	1.53	.08	3	0
25	14	.62	16	175	599	.59	0	.02		.06	5.35	.22	2	0
35	0	.72			1060		0							0
35	0	.72			940		0							0
35	0				910		0							0
35	8	1.25	9	157	456	.82	6	.06	1.36	.11	1.31	.11	16	0
60	6	1.84	25	337	768	2.87	0	.05	.43	.18	7.09	.3	6	0
12	11	.36	4	70	137	.44	0	.02	.04	.06	1.96	.09	1	0
17	3	.42	3	49	288	.68	0	.02	.05	.03	.95	.05	1	0
16	2	.21	4	71	347	.48	0	.15	.06	.05	.91	.1	1	0
13	3	.32	4	79	363	.69	0	.11	.06	.06	.87	.06	1	0
13	3	.38	3	36	226	.5	0	.01	.04	.02	.55	.03	1	0
13	4	.34	4	44	314	.53	0							0
15	3	.42	3	50	285	.54	0	.05	.06	.04	.72	.05	1	0
7		.18			255									
14	3	.18	3	65	272	.47	0	.12	.06	.04	.9	.06	1	0
15	14	.39	5	46	312	.45	0							0
28	23	.43	4	56	246	.49	0	.01	.15	.05	.99	.06	2	0
22	34	.45	5	51	242	.57	0	.01		.05	1.08	.05		0
89	5	5.34	6	113	652	1.6	2405	.14	.2	.87	4.77	.19	25	0
44	34	.72	11	197	778	1.47	0	.17	.19	.16	2.31	.09	3	0
16	2	.62	4	70	248	.3	0	.21	.06	.09	.96	.06	1	0
58	50	.93	11	177	931	1.94	33	.21	.27	.14	2.49	.11	3	0
25		.72			620									
53	5	.95	11	239	741	2.05	0	.38	.13	.18	3.08	.32	1	0
13	3	.57	3	28	267	1.15	0	0	.05	.03	.49	.03	2	0
35	11	.81	2	95	585	1.24	0	.03	.11	.06	1.38	.07	2	0
28	6	.66	6	95	638	1.05	0	.11	.14	.07	1.5	.07	2	0
22	5	.52	4	75	504	.83	0	.09	.11	.05	1.18	.06	2	0
48	48	.83	6	81	642	1.4	0	.02	.28	.08	1.86	.1	4	0
40	60	1.08			480		0							0
13	1	.15	2	45	271	.31	0	.08	.05	.04	.66	.04	1	<1
21	3	.35	7	134	576	.81	0	.26	.11	.09	1.63	.15	<1	0
24	6	.7	18	144	635	1.02	0							0
9	1	.3	5	197	174	.42	0							0
32	6	1.57	9	185	568	1.68	0	.03	.36	.14	2.01	.14	3	0
17	11	.38	4	70	280	.52	0	.06	.06	.06	.75	.05	1	0
39	7	.62	7	135	687	1.13	0	.23	.39	.09	1.86	.12	1	0
17	5	.42	3	49	374	.63	0	.04	.07	.05	.8	.04	1	0
22	62	.31	11	169	846	.79	34	.17	.14	.15	1.05	.13	1	0
9	3	.34	8	59	387	.24	0							0
31	5	.95	8	148	596	1.23	0	.03	.12	.14	2.01	.15	3	0
21	11	.35	6	61	281	.65	0							0
9	1	.15	2	38	224	.27	0	.03	.02	.03	.55	.03	<1	0
21	54	.58	10	194	792	.8	4	.17	.14	.14	1.17	.11	3	0
20	3	.4	4	66	245	.54	0	.14	.06	.04	.96	.05	1	<1
22	8	.33	4	94	336	.65	0	.19	.07	.07	1.18	.09	1	<1
94	36	1.42	19	408	1462	2.84	0	.84	.29	.29	5.11	.37	2	2
37	7	1.52	9	113	607	1.22	0	.14	.12	.21	2.02	.12	1	0
24	2	.45	5	113	558	.97	0	.18	.08	.09	1.46	.15	1	0
47	11	.92	9	139	572	1.03	0	.04	.32	.1	2.01	.14	2	0
11	2	.18	3	45	274	.33	0	.13	.52	.04	.63	.04	<1	0
11	4	.24	2	33	304	.31	3	.05	.52	.04	.52	.04	1	0
8	3	.16	3	48	98	.27	11	.01	.57	.02	.43	.03	1	<1

A

Table A-1

Food Composition (Computer code number is for Wadsworth Diet Analysis Plus program) (For purposes of calculations, use "0" for t, <1, <.1, <.01, etc.)

Computer Code Number	Food Description	Measure	Wt (g)	H_2O (%)	Ener (kcal)	Prot (g)	Carb (g)	Dietary Fiber (g)	Fat (g)	Fat Breakdown (g) Sat	Mono	Poly
	MEATS: SAUSAGES and LUNCHMEATS—Continued											
1084	Smoked link sausage, beef and pork	1 ea	68	52	228	9	1	0	21	7.2	9.7	2.2
1083	Smoked link sausage, pork	1 ea	68	39	265	15	1	0	22	7.7	9.9	2.6
1085	Summer sausage	2 pce	46	51	154	7	<1	0	14	5.5	6	.6
1076	Turkey breakfast sausage	1 pce	28	60	64	6	0	0	5	1.6	1.8	1.2
679	Vienna sausage, canned	2 ea	32	60	89	3	1	0	8	3	4	.5
	MIXED DISHES and FAST FOODS											
	MIXED DISHES:											
1445	Almond Chicken	1 c	242	77	275	20	18	4	14	2	5.3	5.8
1981	Baked beans, fat free, honey	½ c	120	73	110	7	24	7	0	0	0	0
1454	Bean cake	1 ea	32	23	130	2	16	1	7	1	2.9	2.6
680	Beef stew w/ vegetables, homemade	1 c	245	82	218	16	15	2	10	4.9	4.5	.5
1109	Beef stew w/ vegetables, canned	1 c	245	82	194	14	17	2	8	2.4	3.1	.3
1116	Beef, macaroni, tomato sauce casserole	1 c	226	76	255	16	26	2	10	3.8	4.1	.5
2295	Beef fajita	1 ea	223	63	409	17	46	4	17	5.1	7.6	3.9
1265	Beef flauta	1 ea	113	49	360	16	13	2	27	4.9	11.6	9.1
681	Beef pot pie, homemade	1 pce	210	55	517	21	39	3	30	8.4	14.7	7.3
1898	Broccoli, batter fried	1 c	85	74	123	3	9	2	9	1.3	2.2	4.9
1462	Buffalo wings/spicy chicken wings	2 pce	32	53	98	8	<1	<1	7	1.8	2.8	1.6
1675	Carrot raisin salad	½ c	88	58	204	1	21	2	14	2	3.9	7.3
2248	Cheeseburger deluxe	1 ea	219	52	563	28	38		33	15	12.6	2
682	Chicken à la king, homemade	1 c	245	68	468	27	12	1	34	12.7	14.3	6.2
683	Chicken & noodles, homemade	1 c	240	71	367	22	26	2	18	5.9	7.1	3.5
684	Chicken chow mein, canned	1 c	250	89	95	6	18	2	1	0	.1	.8
685	Chicken chow mein, homemade	1 c	250	78	255	31	10	1	10	2.4	4.3	3.1
1266	Chicken fajita	1 ea	223	61	405	22	50	4	13	2.4	6	3.5
1264	Chicken flauta	1 ea	113	52	343	14	13	2	27	4.3	11.1	9.6
686	Chicken pot pie, homemade (⅓)	1 pce	232	57	545	23	42	3	31	10.9	14.5	5.8
1672	Chili con carne	½ c	127	77	128	12	11	2	4	1.7	1.7	.3
1112	Chicken salad with celery	½ c	78	53	268	11	1	<1	25	3.1	4.5	15.8
1382	Chicken teriyaki, breast	1 ea	128	67	176	26	7	<1	4	.9	1	.9
687	Chili with beans, canned	1 c	256	75	287	15	30	11	14	6	6	.9
1479	Chinese pastry	1 oz	28	46	67	1	13	<1	1	.2	.4	.8
688	Chop suey with beef & pork	1 c	220	63	425	22	31	3	24	5	8.6	9.3
690	Coleslaw	1 c	132	74	195	2	17	2	15	2.1	3.2	8.5
689	Corn pudding	1 c	250	76	273	11	32	4	13	6.4	4.3	1.8
1110	Corned beef hash, canned	1 c	220	67	398	19	23	1	25	11.9	10.9	.9
1255	Deviled egg (½ egg + filling)	1 ea	31	69	62	4	<1	0	5	1.2	1.7	1.5
	Egg foo yung patty:											
1467	Meatless	1 ea	86	78	113	6	3	1	8	1.9	3.3	2.1
1458	With beef	1 ea	86	74	129	9	3	<1	9	2.2	3.2	2.4
1465	With chicken	1 ea	86	74	130	9	4	<1	9	2.1	3.1	2.5
1602	Egg roll, meatless	1 ea	64	70	102	3	10	1	6	1.2	2.5	1.6
1550	Egg roll, with meat	1 ea	64	66	114	5	9	1	6	1.5	2.7	1.5
1113	Egg salad	1 c	183	57	586	17	3	0	56	10.6	17.4	24.2
691	French toast w/wheat bread, homemade	1 pce	65	54	151	5	16	<1	7	2	3	1.7
1355	Green pepper, stuffed	1 ea	172	75	229	11	20	2	11	5	4.9	.5
1487	Hot & sour soup (Chinese)	1 c	244	88	133	12	5	<1	6	2	2.5	1
2242	Hamburger deluxe	1 ea	110	49	279	13	27		13	4.1	5.3	2.6
1997	Hummous/hummus	¼ c	62	65	106	3	12	3	5	.8	2.2	2
	Lasagna:											
1346	With meat, homemade	1 pce	245	67	382	22	39	3	15	7.7	5	.9
1111	Without meat, homemade	1 pce	218	69	298	15	39	3	9	5.4	2.4	.6
1117	Frozen entree	1 ea	340	75	390	24	42	4	14	6.7	5.5	.8
1606	Lo mein, meatless	1 c	200	82	134	6	27	3	1	.1	.1	.3
1607	Lo mein, with meat	1 c	200	70	285	17	31	2	10	1.9	2.9	4.5
692	Macaroni & cheese, canned	1 c	240	80	228	9	26	1	10	4.2	3.1	1.4
693	Macaroni & cheese, homemade	1 c	200	58	430	17	40	1	22	8.9	8.8	3.6
1115	Macaroni salad, no cheese	1 c	177	60	461	5	28	2	37	4	6	25.5
1120	Meat loaf, beef	1 pce	87	63	182	16	4	<1	11	4.4	4.7	.5
1119	Meat loaf, beef and pork (⅓)	1 pce	87	60	205	15	4	<1	14	5.2	6.3	.9
1303	Moussaka (lamb & eggplant)	1 c	250	82	237	16	13	4	13	4.6	5.4	1.9

PAGE KEY: A–4 = Beverages A–6 = Dairy A–10 = Eggs A–10 = Fat/Oil A–14 = Fruit A–20 = Bakery A–28 = Grain
A–32 = Fish A–34 = Meats A–38= Poultry A–40 = Sausage A–42 = Mixed/Fast A–46 = Nuts/Seeds A–50 = Sweets
A–52 = Vegetables/Legumes A–62 = Vegetarian A–62 = Misc A–66 = Soups/Sauces A–68 = Fast A–84 = Convenience A–88 = Baby foods

Chol (mg)	Calc (mg)	Iron (mg)	Magn (mg)	Pota (mg)	Sodi (mg)	Zinc (mg)	VT-A (RE)	Thia (mg)	VT-E (α-TE)	Ribo (mg)	Niac (mg)	V-B6 (mg)	Fola (μg)	VT-C (mg)
48	7	.99	8	129	643	1.43	0	.18	.15	.12	2.2	.12	1	0
46	20	.79	13	228	1020	1.92	0	.48	.17	.17	3.08	.24	3	1
34	6	1.17	6	125	571	1.18	0	.07	.1	.15	1.98	.12	1	0
23	5	.51	6	75	188	.96	0	.03	.14	.08	1.4	.08	1	0
17	3	.28	2	32	305	.51	0	.03	.07	.03	.51	.04	1	0
35	81	2	59	551	615	1.54	75	.08	2.64	.19	8.59	.42	31	10
0	40	2.7			135		450							12
0	3	.67	6	57	55	.16	0	.07	1.14	.05	.55	.02	9	0
64	29	2.94	40	613	292	5.29	568	.15	.49	.17	4.66	.28	37	17
34	29	2.21	39	426	1006	4.24	262	.07	.34	.12	2.45	.2	31	7
39	26	2.7	40	522	862	3.14	97	.22	.57	.22	4.31	.29	20	14
26	76	3.69	38	427	850	2.38	52	.46	2.08	.3	4.73	.32	25	29
45	50	2.15	29	292	187	4.18	15	.07	4	.15	2.13	.25	10	14
44	29	3.78	6	334	596	3.17	519	.29	3.78	.29	4.83	.24	29	6
16	67	.94	20	242	62	.38	102	.08	2.1	.13	.75	.11	43	53
26	5	.4	6	59	61	.56	17	.01	.23	.04	2.06	.13	1	<1
10	26	.75	14	317	118	.19	1452	.08	5.03	.05	.64	.22	9	5
88	206	4.66	44	445	1108	4.6	129	.39	1.18	.46	7.38	.28	81	8
186	127	2.45	20	404	760	1.8	272	.1	.98	.42	5.39	.23	11	12
96	26	2.16	26	149	600	1.53	10	.05		.17	4.32	.19	10	0
7	45	1.25	14	418	725	1.3	28	.05	.05	.1	1	.09	12	12
77	57	2.5	28	473	718	2.12	50	.07	.75	.22	4.25	.41	19	10
41	83	3.7	51	532	439	1.77	55	.48	2.04	.37	6.64	.35	41	22
37	52	.97	27	243	189	1.18	21	.05	4.06	.1	3.21	.22	8	14
72	70	3.02	25	343	594	2	735	.32	3.25	.32	4.87	.46	29	5
67	34	2.6	23	347	505	1.79	84	.06	.81	.57	1.24	.16	23	1
48	16	.62	11	138	201	.79	31	.03	6.27	.07	3.28	.34	8	1
80	27	1.75	36	309	1866	1.94	16	.08	.35	.2	8.69	.46	13	3
43	120	8.78	115	934	1336	5.12	87	.12	1.88	.27	.92	.34	58	4
0	8	.51	6	28	3	.18	<1	.05	.25	<.01	.41	.02	1	0
46	39	4.16	54	515	818	3.52	134	.36	1.82	.37	5.63	.44	44	20
7	45	.96	12	236	356	.26	66	.05	5.28	.04	.11	.14	51	11
250	100	1.4	37	403	138	1.25	90	1.03	.52	.32	2.47	.29	63	7
73	29	4.4	36	440	1188	3.3	0	.02	.48	.2	4.62	.43	20	0
121	15	.35	3	36	94	.3	49	.02	.86	.14	.02	.05	13	0
184	31	1.04	12	118	310	.7	86	.04	1.57	.25	.44	.09	30	5
180	26	1.11	11	145	184	1.16	92	.05	1.79	.24	.74	.15	22	3
182	27	.86	12	144	187	.81	95	.05	1.87	.25	.96	.13	22	3
30	12	.76	9	98	306	.25	15	.08	.81	.1	.81	.05	12	3
37	13	.78	10	124	304	.46	14	.16	.78	.13	1.31	.1	9	2
574	74	1.8	13	180	665	1.42	260	.08	8.87	.66	.09	.46	62	0
76	64	1.09	11	86	311	.44	81	.13	.31	.21	1.06	.05	15	<1
34	16	1.77	20	233	201	2.28	44	.15	.75	.1	2.74	.3	17	55
23	29	1.83	27	351	1562	1.17	2	.19	.12	.22	4.58	.15	12	1
26	63	2.63	22	227	504	2.06	9	.23	.82	.2	3.69	.12	52	2
0	31	.97	18	108	151	.68	1	.06	.62	.03	.25	.25	37	5
56	258	3.22	50	461	745	3.25	158	.23	1.15	.33	3.97	.21	19	15
31	252	2.5	44	375	714	1.77	156	.22	1.07	.27	2.49	.17	17	15
55	263	3.44	64	752	823	3.7	248	.29	3.45	.39	5.07	.32	28	41
0	47	2.06	33	389	623	.92	130	.23	.35	.24	2.83	.19	49	12
30	25	2.11	39	246	276	1.63	6	.37	1.51	.24	4.25	.28	41	8
24	199	.96	31	139	730	1.2	73	.12	.14	.24	.96	.02	8	<1
42	362	1.8	37	240	1086	1.2	234	.2	.12	.4	1.8	.05	10	1
27	31	1.56	20	170	352	.53	44	.18	10.3	.1	1.43	.33	20	4
84	29	1.61	14	187	145	3.23	23	.05	.31	.22	2.61	.15	11	1
84	33	1.42	14	213	381	2.68	23	.19	.32	.22	2.68	.17	10	1
97	68	1.79	40	557	432	2.56	105	.15	.81	.31	4.14	.23	45	6

Table A-1

Food Composition (Computer code number is for Wadsworth Diet Analysis Plus program) (For purposes of calculations, use "0" for t, <1, <.1, <.01, etc.)

Computer Code Number	Food Description	Measure	Wt (g)	H_2O (%)	Ener (kcal)	Prot (g)	Carb (g)	Dietary Fiber (g)	Fat (g)	Fat Breakdown (g) Sat	Mono	Poly
	MIXED DISHES and FAST FOODS—Continued											
1899	Mushrooms, batter fried	5 ea	70	66	148	2	8	1	12	2.1	3	6.4
715	Potato salad with mayonnaise and eggs	½ c	125	76	179	3	14	2	10	1.8	3.1	4.7
1674	Pizza, combination, 1⁄12 of 12″ round	1 pce	79	48	184	13	21		5	1.5	2.5	.9
1673	Pizza, pepperoni, 1⁄12 of 12″ round	1 pce	71	46	181	10	20		7	2.2	3.1	1.2
694	Quiche Lorraine ⅛ of 8″ quiche	1 pce	176	54	508	20	20	1	39	17.6	13.8	4.9
1449	Ramen noodles, cooked	1 c	227	82	156	6	29	3	2	.4	.5	.5
1671	Ravioli, meat	½ c	125	68	194	10	18	1	9	2.9	3.6	1
1597	Fried rice (meatless)	1 c	166	68	264	5	34	1	12	1.7	3	6.3
2142	Roast beef hash	½ c	117	66	230	9	11	1	16	7	5.8	3.2
	Spaghetti (enriched) in tomato sauce											
	With cheese:											
695	Canned	1 c	250	80	190	5	38	2	1	0	.4	.5
696	Home recipe	1 c	250	77	260	9	37	2	9	2	5.4	1.2
	With meatballs:											
697	Canned	1 c	250	78	258	12	28	6	10	2.1	3.9	3.9
698	Home recipe	1 c	248	70	332	19	39	8	12	3.3	6.3	2.2
716	Spinach soufflé	1 c	136	74	219	11	3	3	19	9.5	5.7	2.2
1553	Sweet & sour pork	1 c	226	77	231	15	25	1	8	2.2	2.9	2.4
1263	Sweet & sour chicken breast	1 ea	131	79	117	8	15	1	3	.6	.8	1.5
1515	Three bean salad	1 c	150	82	139	4	13	3	8	1.2	1.9	4.9
717	Tuna salad	1 c	205	63	383	33	19	0	19	3.2	5.9	8.4
1121	Tuna noodle casserole, homemade	1 c	202	75	237	17	25	1	7	1.9	1.5	3.2
1270	Waldorf salad	1 c	137	58	408	4	13	2	40	4.1	7.3	27
	FAST FOODS and SANDWICHES (see end of this appendix for additional Fast Foods)											
699	Burrito, beef & bean	1 ea	116	52	255	11	33	3	9	4.2	3.5	.6
700	Burrito, bean	1 ea	109	52	225	7	36	4	7	3.5	2.4	.6
2106	Burrito, chicken con queso	1 ea	306	76	280	12	53	5	6	1.5		
701	Cheeseburger with bun, regular	1 ea	154	55	359	18	28		20	9.2	7.2	1.5
702	Cheeseburger with bun, 4-oz patty	1 ea	166	51	417	21	35		21	8.7	7.8	2.7
703	Chicken patty sandwich	1 ea	182	47	515	24	39	1	29	8.5	10.4	8.4
704	Corndog	1 ea	175	47	460	17	56		19	5.2	9.1	3.5
1922	Corndog, chicken	1 ea	113	52	271	13	26		13			
705	Enchilada	1 ea	163	63	319	10	28		19	10.6	6.3	.8
706	English muffin with egg, cheese, bacon	1 ea	146	49	383	20	31	1	20	9	6.8	2.1
	Fish sandwich:											
707	Regular, with cheese	1 ea	183	45	523	21	48	<1	28	8.1	8.9	9.4
708	Large, no cheese	1 ea	158	47	431	17	41	<1	23	5.2	7.7	8.2
709	Hamburger with bun, regular	1 ea	107	45	275	14	33	1	10	3.5	3.7	1.8
710	Hamburger with bun, 4-oz patty	1 ea	215	50	576	32	39		32	12	14.1	2.8
711	Hotdog/frankfurter with bun	1 ea	98	54	242	10	18		14	5.1	6.8	1.7
	Lunchables:											
2129	Bologna & American cheese	1 ea	128		450	18	19	0	34	15		
2130	Ham & cheese	1 ea	128		320	22	19	0	17	8		
2117	Honey ham & Amer. w/choc pudding	1 ea	176		390	18	34	<1	20	9		
2118	Honey turkey & cheddar w/Jello	1 ea	163		320	17	27	<1	16	9		
2131	Pepperoni & American cheese	1 ea	128		480	20	19	0	36	17		
2125	Salami & American cheese	1 ea	128		430	18	18	0	32	15		
2127	Turkey & cheddar cheese	1 ea	128		360	20	20	1	22	11		
712	Pizza, cheese, ⅛ of 15″ round	1 pce	63	48	140	8	20	1	3	1.5	1	.5
	SANDWICHES:											
	Avocado, chesse, tomato & lettuce:											
1276	On white bread, firm	1 ea	210	58	478	15	41	5	29	8.8	11.3	7.2
1278	On part whole wheat	1 ea	201	59	444	14	34	7	29	8.6	11.4	7.3
1277	On whole wheat	1 ea	214	58	468	16	40	8	30	8.7	11.6	7.5
	Bacon, lettuce & tomato sandwich:											
1137	On white bread, soft	1 ea	124	53	308	10	28	2	18	4.5	6.1	6.1
1139	On part whole wheat	1 ea	124	54	303	10	26	3	17	4.3	6.2	6.1
1138	On whole wheat	1 ea	137	53	328	12	32	4	18	4.4	6.4	6.3

PAGE KEY: A–4 = Beverages A–6 = Dairy A–10 = Eggs A–10 = Fat/Oil A–14 = Fruit A–20 = Bakery A–28 = Grain A–32 = Fish A–34 = Meats A–38= Poultry A–40 = Sausage A–42 = Mixed/Fast A–46 = Nuts/Seeds A–50 = Sweets A–52 = Vegetables/Legumes A–62 = Vegetarian A–62 = Misc A–66 = Soups/Sauces A–68 = Fast A–84 = Convenience A–88 = Baby foods

Chol (mg)	Calc (mg)	Iron (mg)	Magn (mg)	Pota (mg)	Sodi (mg)	Zinc (mg)	VT-A (RE)	Thia (mg)	VT-E (α-TE)	Ribo (mg)	Niac (mg)	V-B6 (mg)	Fola (μg)	VT-C (mg)
14	54	.76	8	180	121	.42	10	.07	.92	.22	1.65	.05	8	1
85	24	.81	19	318	661	.39	41	.1	2.33	.07	1.11	.18	8	12
20	101	1.53	18	179	382	1.11	101	.21		.17	1.96	.09	32	2
14	65	.94	9	153	267	.52	55	.13		.23	3.05	.06	37	2
205	201	1.9	27	271	549	1.66	243	.23	1.91	.44	4.71	.19	17	3
38	20	1.89	24	51	1349	.76	204	.22	.09	.1	1.75	.06	9	<1
84	32	2.03	20	259	619	1.67	94	.15	1.52	.22	2.95	.14	14	11
42	30	1.84	24	134	286	.89	21	.21	2.46	.11	2.25	.15	22	4
40	10	.9	22	362	695	2.99	0	.09		.12	2.33	.3	12	0
7	40	2.75	21	303	955	1.12	120	.35	2.13	.27	4.5	.13	6	10
7	80	2.25	26	408	955	1.3	140	.25	2.75	.17	2.25	.2	8	12
22	52	3.25	20	245	1220	2.39	100	.15	1.5	.17	2.25	.12	5	5
74	124	3.72	40	665	1009	2.45	159	.25	1.64	.3	3.97	.2	10	22
184	230	1.35	38	201	763	1.29	675	.09	1.22	.3	.48	.12	80	3
38	28	1.36	34	390	1219	1.46	28	.55	.62	.21	3.6	.41	10	20
23	16	.79	21	187	732	.66	20	.06	.39	.08	3.06	.18	6	12
0	35	1.42	25	224	514	.54	23	.07	1.96	.09	.4	.04	53	4
27	35	2.05	39	365	824	1.15	55	.06	1.95	.14	13.7	.17	16	5
41	34	2.3	30	182	772	1.2	13	.18	1.18	.15	7.78	.2	10	1
21	43	.88	39	270	236	.63	39	.1	8.67	.05	.36	.36	27	6
24	53	2.46	42	329	670	1.93	32	.27	.7	.42	2.71	.19	58	1
2	57	2.27	44	328	495	.76	16	.32	.87	.3	2.04	.15	44	1
10	40	.72			600		40							15
52	182	2.65	26	229	976	2.62	71	.32	1.34	.23	6.38	.15	65	2
60	171	3.42	30	335	1050	3.49	65	.35		.28	8.05	.18	61	2
60	60	4.68	35	353	957	1.87	31	.33	.55	.24	6.81	.2	100	9
79	102	6.18	17	263	973	1.31	37	.28	.7	.7	4.17	.09	103	0
64					668									
44	324	1.32	50	240	784	2.51	186	.08	1.47	.42	1.91	.39	65	1
234	207	3.29	34	213	784	1.81	158	.48	.6	.53	3.93	.16	47	1
68	185	3.5	37	353	939	1.17	97	.46	1.83	.42	4.23	.11	91	3
55	84	2.61	33	340	615	.99	30	.33	.87	.22	3.4	.11	85	3
43	51	2.46	22	215	564	2.05	13	.26	.43	.32	4.7	.13	52	3
103	92	5.55	45	527	742	5.81	4	.34	1.61	.41	6.73	.37	84	1
44	23	2.31	13	143	670	1.98	0	.23	.27	.27	3.65	.05	48	<1
85	300	2.7			1620		60							0
60	300	1.8			1770		80							
55	250	2.7			1540		40							
50	20	6			1360		80							
95	250	2.7			1840		60							
80	250	2.7			1740		60							
70	300	1.8			1650		60							
9	117	.58	16	110	336	.81	74	.18		.16	2.48	.04	35	1
34	294	3.06	54	581	550	1.71	140	.37	4.55	.39	3.77	.32	80	11
31	291	3.1	67	617	525	1.91	140	.35	4.55	.39	3.94	.35	80	11
31	281	3.53	102	679	593	2.68	140	.36	4.17	.37	4.29	.42	92	11
21	52	1.99	20	233	590	.96	31	.35	2.34	.19	3.2	.14	34	12
20	63	2.27	33	283	604	1.19	31	.36	2.7	.21	3.62	.17	37	12
20	55	2.68	64	342	670	1.9	31	.37	2.36	.2	3.97	.24	48	12

Table A-1

Food Composition (Computer code number is for Wadsworth Diet Analysis Plus program) (For purposes of calculations, use "0" for t, <1, <.1, <.01, etc.)

Computer Code Number	Food Description	Measure	Wt (g)	H_2O (%)	Ener (kcal)	Prot (g)	Carb (g)	Dietary Fiber (g)	Fat (g)	Fat Breakdown (g) Sat	Mono	Poly
	MIXED DISHES and FAST FOODS—Continued											
	Cheese, grilled:											
1140	On white bread, soft	1 ea	119	37	400	18	30	1	24	13.2	7.5	2
1142	On part whole wheat	1 ea	119	37	396	18	28	3	24	13	7.6	2.1
1141	On whole wheat	1 ea	132	38	420	20	33	4	24	13.1	7.8	2.2
1596	Chicken fillet	1 ea	182	47	515	24	39	1	29	8.5	10.4	8.4
	Chicken salad:											
1143	On white bread, soft	1 ea	110	40	369	11	31	1	23	3.7	6.1	12.1
1145	On part whole wheat	1 ea	110	41	364	11	29	4	23	3.5	6.1	12.1
1144	On whole wheat	1 ea	123	41	387	13	34	5	23	3.6	6.3	12.2
1146	Corned beef & swiss on rye	1 ea	156	49	420	28	22	<1	26	9.4	7.4	6.3
	Egg salad:											
1147	On white bread, soft	1 ea	117	43	380	10	31	1	25	4.4	6.8	12
1149	On part whole wheat	1 ea	116	43	374	10	29	3	25	4.1	6.8	12
1148	On whole wheat	1 ea	130	43	400	12	35	5	25	4.3	7.1	12.2
	Ham:											
1279	On rye bread	1 ea	150	60	283	22	21	<1	13	2.4	3.8	6
1151	On white bread, soft	1 ea	157	55	334	22	30	1	14	3	4.4	5.9
1153	On part whole wheat	1 ea	156	55	328	22	28	3	14	2.7	4.4	6
1152	On whole wheat	1 ea	169	54	352	24	34	4	15	2.9	4.7	6.1
	Ham & cheese:											
1280	On white bread, soft	1 ea	157	49	403	23	30	1	22	8.4	5.9	6.1
1282	On part whole wheat	1 ea	156	50	397	23	28	3	21	8.1	5.9	6.2
1281	On whole wheat	1 ea	170	49	424	24	34	4	22	8.3	6.2	6.4
1150	Ham & swiss on rye	1 ea	150	54	339	22	22	<1	19	6.5	5.1	6
	Ham salad:											
1154	On white bread, soft	1 ea	131	47	362	11	37	1	20	4.8	6.7	7.4
1156	On part whole wheat	1 ea	131	48	357	11	35	3	20	4.5	6.8	7.5
1155	On whole wheat	1 ea	144	47	380	12	40	4	20	4.7	7	7.6
1157	Patty melt: Ground beef & cheese on rye	1 ea	182	46	561	37	22	3	37	13.2	11.7	8.4
	Peanut butter & jelly:											
1158	On white bread, soft	1 ea	101	26	351	12	47	3	15	3.1	6.7	3.9
1160	On part whole wheat	1 ea	101	27	346	12	45	5	15	2.9	6.7	4
1159	On whole wheat	1 ea	114	28	370	13	50	6	15	3	7	4.1
1161	Reuben, grilled: Corned beef, swiss cheese, sauerkraut on rye	1 ea	239	64	462	28	25	2	29	9.9	9.5	7.1
	Roast beef:											
713	On a bun	1 ea	139	49	346	21	33		14	3.6	6.8	1.7
1162	On white bread, soft	1 ea	157	46	404	29	34	1	17	3.4	4.2	8.2
1164	On part whole wheat	1 ea	156	47	398	29	32	3	17	3.2	4.3	8.3
1163	On whole wheat	1 ea	169	46	422	31	38	4	17	3.3	4.5	8.4
	Tuna salad:											
1165	On white bread, soft	1 ea	122	46	327	14	35	2	15	2.5	3.8	7.9
1167	On part whole wheat	1 ea	122	47	322	14	33	4	15	2.2	3.8	8
1166	On whole wheat	1 ea	135	46	346	16	39	5	15	2.3	4.1	8.1
	Turkey:											
1168	On white bread, soft	1 ea	156	54	346	24	29	1	15	2.4	3.2	8.3
1170	On part whole wheat	1 ea	155	54	338	24	27	3	14	2.1	3.2	8.3
1169	On whole wheat	1 ea	169	53	365	26	33	4	15	2.3	3.5	8.5
	Turkey ham:											
1272	On rye bread	1 ea	150	60	280	21	20	<1	14	2.5	2.8	6.9
1273	On white bread, soft	1 ea	156	55	331	21	29	1	14	3	3.4	6.8
1275	On part whole wheat	1 ea	156	56	326	21	28	3	14	3	4.2	5.6
1274	On whole wheat	1 ea	169	55	350	23	33	4	15	2.9	3.7	7
714	Taco	1 ea	171	58	369	21	27		20	11.4	6.6	1
	Tostada:											
1114	With refried beans	1 ea	144	66	223	10	26	7	10	5.4	3	.7
1118	With beans & beef	1 ea	225	70	333	16	30	4	17	11.5	3.5	.6
1354	With beans & chicken	1 ea	156	68	248	19	18	3	11	5.3	3.9	1.6
	NUTS, SEEDS, and PRODUCTS											
	Almonds:											
1365	Dry roasted, salted	1 c	138	3	810	22	33	19	71	6.7	46.2	14.9

PAGE KEY: A–4 = Beverages A–6 = Dairy A–10 = Eggs A–10 = Fat/Oil A–14 = Fruit A–20 = Bakery A–28 = Grain A–32 = Fish A–34 = Meats A–38= Poultry A–40 = Sausage A–42 = Mixed/Fast A–46 = Nuts/Seeds A–50 = Sweets A–52 = Vegetables/Legumes A–62 = Vegetarian A–62 = Misc A–66 = Soups/Sauces A–68 = Fast A–84 = Convenience A–88 = Baby foods

Chol (mg)	Calc (mg)	Iron (mg)	Magn (mg)	Pota (mg)	Sodi (mg)	Zinc (mg)	VT-A (RE)	Thia (mg)	VT-E (α-TE)	Ribo (mg)	Niac (mg)	V-B6 (mg)	Fola (μg)	VT-C (mg)
55	399	1.81	25	154	1143	2.05	212	.24	1.13	.34	1.91	.06	24	<1
54	412	2.12	39	209	1160	2.31	212	.25	1.53	.36	2.39	.1	28	<1
54	402	2.54	73	271	1226	3.08	212	.26	1.14	.35	2.74	.17	40	<1
60	60	4.68	35	353	957	1.87	31	.33	.55	.24	6.81	.2	100	9
32	60	2.04	18	139	460	.8	24	.25	6.16	.18	3.68	.25	26	1
31	73	2.35	33	195	475	1.06	24	.26	6.57	.2	4.16	.29	30	1
30	63	2.79	68	259	543	1.85	24	.27	6.14	.19	4.5	.36	42	1
82	268	3.12	28	225	1392	3.65	81	.19	2.59	.33	2.72	.17	19	1
157	71	2.18	16	113	526	.74	76	.26	4.52	.31	1.98	.2	37	0
155	83	2.47	31	169	539	1	75	.27	4.91	.33	2.45	.23	41	0
155	73	2.92	66	234	611	1.8	76	.28	4.53	.32	2.82	.3	53	0
47	48	2.3	26	364	1566	2.11	8	.99	2.36	.31	5.45	.47	15	23
47	60	2.39	29	368	1619	2.04	8	1.02	2.34	.33	5.99	.47	24	22
45	71	2.68	43	421	1630	2.29	8	1.03	2.73	.35	6.45	.5	27	22
45	62	3.1	77	483	1696	3.05	8	1.04	2.34	.33	6.78	.57	39	22
61	232	2.28	30	315	1620	2.34	90	.76	2.53	.36	4.64	.36	25	15
59	244	2.58	45	368	1630	2.59	90	.77	2.93	.39	5.09	.39	28	15
59	236	3.02	79	432	1707	3.37	90	.78	2.56	.37	5.47	.46	40	15
57	258	2.25	29	344	1602	2.59	79	.72	2.52	.36	4.06	.35	16	15
30	56	2.07	19	159	921	1.06	8	.51	3.29	.22	3.25	.17	22	4
29	69	2.38	33	216	936	1.32	8	.52	3.69	.24	3.74	.21	26	4
29	59	2.81	69	279	1001	2.11	8	.53	3.29	.23	4.09	.28	38	4
113	222	4.19	36	391	701	7.11	123	.25	3.5	.46	6.14	.35	25	<1
2	60	2.25	56	245	293	1.07	<1	.27	.12	.17	5.33	.13	40	<1
0	72	2.55	70	299	308	1.32	<1	.28	.51	.19	5.8	.17	44	<1
0	63	2.97	104	361	375	2.09	<1	.29	.14	.17	6.14	.24	56	<1
80	288	4.24	38	361	1949	3.73	130	.21	4.46	.34	2.79	.27	38	13
51	54	4.23	31	316	792	3.39	21	.37	.19	.31	5.87	.26	57	2
45	60	3.98	28	432	1595	3.78	12	.29	3.3	.3	6.39	.39	30	12
43	72	4.27	42	485	1607	4.02	12	.31	3.69	.32	6.84	.42	34	12
43	62	4.7	77	547	1672	4.79	12	.31	3.31	.31	7.18	.49	45	12
14	60	2.24	22	161	567	.67	22	.25	2.72	.18	5.53	.11	25	1
13	73	2.55	37	217	582	.93	22	.26	3.12	.2	6	.16	29	1
12	63	2.98	72	280	649	1.72	22	.27	2.71	.19	6.34	.23	41	1
45	56	2	29	302	1585	1.33	12	.26	3.46	.23	8.97	.4	24	0
43	68	2.28	43	354	1589	1.57	11	.27	3.82	.25	9.38	.44	28	0
43	59	2.73	77	418	1665	2.35	12	.28	3.47	.23	9.77	.51	39	0
55	51	4.06	25	342	1185	3	8	.22	2.8	.33	4.3	.29	17	<1
55	62	4.09	28	346	1248	2.9	8	.27	2.75	.35	4.87	.28	25	0
53	74	4.37	42	400	1262	3.15	8	.28	3.57	.37	5.34	.32	29	0
53	65	4.81	76	462	1329	3.91	8	.29	2.76	.35	5.68	.39	41	0
56	221	2.41	70	474	802	3.93	147	.15	1.88	.44	3.21	.24	68	2
30	210	1.89	59	403	543	1.9	85	.1	1.15	.33	1.32	.16	43	1
74	189	2.45	67	491	871	3.17	173	.09	1.8	.49	2.86	.25	85	4
53	168	1.79	47	365	433	2.28	86	.11	1.87	.2	4.52	.32	53	3
0	389	5.24	420	1062	1076	6.76	0	.18	7.66	.83	3.89	.1	88	1

A

Table A-1

Food Composition (Computer code number is for Wadsworth Diet Analysis Plus program) (For purposes of calculations, use "0" for t, <1, <.1, <.01, etc.)

Computer Code Number	Food Description	Measure	Wt (g)	H_2O (%)	Ener (kcal)	Prot (g)	Carb (g)	Dietary Fiber (g)	Fat (g)	Fat Breakdown (g) Sat	Mono	Poly
	NUTS, SEEDS, and PRODUCTS—Continued											
	Almonds:											
718	Slivered, packed, unsalted	1 c	108	4	636	22	22	12	56	5.3	36.6	11.9
719	Whole, dried, unsalted	1 c	142	4	836	28	29	15	74	7	48.1	15.6
720	Ounce	1 oz	28	4	165	6	6	3	15	1.4	9.5	3.1
721	Almond butter:	1 tbs	16	1	101	2	3	1	9	.9	6.1	2
4572	Salted	1 tbs	16	1	101	2	3	1	9	.9	6.1	2
722	Brazil nuts, dry (about 7)	1 c	140	3	918	20	18	8	93	22.7	32.2	33.7
	Cashew nuts, dry roasted:											
723	Salted:	1 c	137	2	786	21	45	4	64	12.8	37.4	10.7
724	Ounce	1 oz	28	2	161	4	9	1	13	2.6	7.6	2.2
4621	Unsalted:	1 c	137	2	786	21	45	4	64	12.8	37.4	10.7
4621	Ounce	1 oz	28	2	161	4	9	1	13	2.6	7.6	2.2
725	Oil roasted:	1 c	130	4	749	23	37	5	63	12.6	36.9	10.6
726	Ounce	1 oz	28	4	161	5	8	1	13	2.7	7.9	2.3
4622	Unsalted:	1 c	130	4	749	21	37	5	63	12.6	36.9	10.6
4622	Ounce	1 oz	28	4	161	5	8	1	13	2.7	7.9	2.3
727	Cashew butter, unsalted	1 tbs	16	3	94	3	4	<1	8	1.6	4.7	1.3
4662	Cashew butter, salted	1 tbs	16	3	94	3	4	<1	8	1.6	4.7	1.3
728	Chestnuts, European, roasted (1 cup = approx 17 kernels)	1 c	143	40	350	5	76	7	3	.6	1.1	1.2
	Coconut, raw:											
729	Piece 2 x 2 x ½"	1 pce	45	47	159	2	7	4	15	13.5	.6	.2
730	Shredded/grated, unpacked	½ c	40	47	142	1	6	4	13	12	.6	.1
	Coconut, dried, shredded/grated:											
731	Unsweetened	1 c	78	3	515	6	19	13	50	45.1	2.1	.6
732	Sweetened	1 c	93	13	466	3	44	4	33	29.6	1.4	.4
733	Filberts/hazelnuts, chopped:	1 c	135	5	853	18	21	8	84	6.2	66.3	8.1
734	Ounce	1 oz	28	5	177	4	4	2	17	1.3	13.7	1.7
735	Macadamias, oil roasted, salted:	1 c	134	2	962	10	17	12	103	15.4	80.9	1.8
736	Ounce	1 oz	28	2	201	2	4	3	21	3.2	16.9	.4
1368	Macadamias, oil roasted, unsalted	1 c	134	2	962	10	17	12	103	15.4	80.9	1.8
	Mixed nuts:											
737	Dry roasted, salted	1 c	137	2	814	24	35	12	71	9.4	43	14.8
738	Oil roasted, salted	1 c	142	2	876	24	30	13	80	12.4	45	18.9
1369	Oil roasted, unsalted	1 c	142	2	876	27	30	14	80	12.4	45	18.9
	Peanuts:											
739	Oil roasted, salted	1 c	144	2	837	38	27	13	71	9.8	35.3	22.5
740	Ounce	1 oz	28	2	163	7	5	3	14	1.9	6.9	4.4
1370	Oil roasted, unsalted	1 c	144	2	837	38	27	10	71	9.8	35.3	22.5
741	Dried, salted	1 c	146	2	854	35	31	12	73	10.1	36.1	22.9
742	Ounce	1 oz	28	2	164	7	6	2	14	1.9	6.9	4.4
743	Peanut butter:	½ c	128	1	759	33	25	8	65	14.3	31.1	17.7
1371	Tablespoon	2 tbs	32	1	190	8	6	2	16	3.6	7.8	4.4
744	Pecan halves, dried, unsalted:	1 c	108	5	720	9	20	8	73	5.8	45.6	18
745	Ounce	1 oz	28	5	187	2	5	2	19	1.5	11.8	4.7
1372	Pecan halves, dry roasted, salted	¼ c	28	1	185	2	6	3	18	1.4	11.3	4.5
746	Pine nuts/piñons, dried	1 oz	28	6	176	3	5	3	17	2.6	6.4	7.2
747	Pistachios, dried, shelled	1 oz	28	4	162	6	7	3	14	1.8	9.2	2
1373	Pistachios,dry roasted,salted,shelled	1 c	128	2	776	19	35	14	68	8.8	45.7	10.2
748	Pumpkin kernels, dried, unsalted	1 oz	28	7	151	7	5	1	13	2.4	4	5.8
1374	Pumpkin kernels, roasted, salted	1 c	227	7	1184	75	30	9	96	18.1	29.7	43.6
749	Sesame seeds, hulled, dried	¼ c	38	5	223	10	4	3	21	2.9	7.9	9.1
	Sunflower seed kernels:											
750	Dry	¼c	36	5	205	8	7	4	18	1.9	3.4	11.8
751	Oil roasted	¼c	34	3	209	7	5	2	20	2	3.7	12.9
752	Tahini (sesame butter)	1 tbs	15	3	91	3	3	1	8	1.2	3.2	3.7
1334	Trail mix w/chocolate chips	1 c	146	7	707	21	66	8	47	9.3	19.8	16.5
753	Black walnuts, chopped:	1 c	125	4	759	31	15	6	71	4.8	15.9	46.9
754	Ounce	1 oz	28	4	170	7	3	1	16	1.1	3.6	10.5
755	English walnuts, chopped:	1 c	120	4	770	17	22	6	74	7.2	17	46.9
756	Ounce	1 oz	28	4	180	4	5	1	17	1.7	4	10.9

Chol (mg)	Calc (mg)	Iron (mg)	Magn (mg)	Pota (mg)	Sodi (mg)	Zinc (mg)	VT-A (RE)	Thia (mg)	VT-E (α-TE)	Ribo (mg)	Niac (mg)	V-B6 (mg)	Fola (μg)	VT-C (mg)
0	287	3.95	320	791	12	3.15	0	.23	25.9	.84	3.63	.12	63	1
0	378	5.2	420	1039	16	4.15	0	.3	34.1	1.11	4.77	.16	83	1
0	74	1.02	83	205	3	.82	0	.06	6.72	.22	.94	.03	16	<1
0	43	.59	48	121	2	.49	0	.02	3.25	.1	.46	.01	10	<1
0	43	.59	48	121	72	.49	0	.02	3.25	.1	.46	.01	10	<1
0	246	4.76	315	840	3	6.43	0	1.4	10.6	.17	2.27	.35	6	1
0	62	8.22	356	774	877	7.67	0	.27	.78	.27	1.92	.35	95	0
0	13	1.68	73	158	179	1.57	0	.06	.16	.06	.39	.07	19	0
0	62	8.22	356	774	22	7.67	0	.27	.78	.27	1.92	.35	95	0
0	13	1.68	73	158	4	1.57	0	.06	.16	.06	.39	.07	19	0
0	53	5.33	332	689	814	6.18	0	.55	2.03	.23	2.34	.32	88	0
0	11	1.15	71	148	175	1.33	0	.12	.44	.05	.5	.07	19	0
0	53	5.33	332	689	22	6.18	0	.55	2.03	.23	2.34	.32	88	0
0	11	1.15	71	148	5	1.33	0	.12	.44	.05	.5	.07	19	0
0	7	.8	41	87	2	.83	0	.05	.25	.03	.26	.04	11	0
0	7	.8	41	87	98	.83	0	.05	.25	.03	.26	.04	11	0
0	41	1.3	47	847	3	.81	3	.35	1.72	.25	1.92	.71	100	37
0	6	1.09	14	160	9	.49	0	.03	.33	.01	.24	.02	12	1
0	6	.97	13	142	8	.44	0	.03	.29	.01	.22	.02	11	1
0	20	2.59	70	424	29	1.57	0	.05	1.05	.08	.47	.23	7	1
0	14	1.79	46	313	244	1.69	0	.03	1.26	.02	.44	.25	8	1
0	254	4.41	385	601	4	3.24	9	.67	32.3	.15	1.54	.83	97	1
0	53	.92	80	125	1	.67	2	.14	6.69	.03	.32	.17	20	<1
0	60	2.41	157	441	348	1.47	1	.28	.55	.15	2.71	.26	21	0
0	13	.5	33	92	73	.31	<1	.06	.11	.03	.57	.05	4	0
0	60	2.41	157	441	9	1.47	1	.28	.55	.15	2.71	.26	21	0
0	96	5.07	308	818	917	5.21	1	.27	8.22	.27	6.44	.41	69	1
0	153	4.56	334	825	926	7.21	3	.71	8.52	.31	7.19	.34	118	1
0	153	4.56	334	825	16	7.21	3	.71	8.52	.31	7.19	.34	118	1
0	127	2.64	266	982	624	9.55	0	.36	10.7	.16	20.6	.37	181	0
0	25	.51	52	191	121	1.86	0	.07	2.07	.03	4	.07	35	0
0	127	2.64	266	982	9	9.55	0	.36	10.7	.16	20.6	.37	181	0
0	79	3.3	257	961	1186	4.83	0	.64	10.8	.14	19.7	.37	212	0
0	15	.63	49	184	228	.93	0	.12	2.07	.03	3.78	.07	41	0
0	49	2.36	204	856	598	3.74	0	.11	12.8	.13	17.2	.58	95	0
0	12	.59	51	214	149	.93	0	.03	3.2	.03	4.29	.14	24	0
0	39	2.3	138	423	1	5.91	14	.92	3.35	.14	.96	.2	42	2
0	10	.6	36	110	<1	1.53	4	.24	.87	.04	.25	.05	11	1
0	10	.61	37	104	218	1.59	4	.09	.84	.03	.26	.05	11	1
0	2	.86	65	176	20	1.2	1	.35	.98	.06	1.22	.03	16	1
0	38	1.9	44	306	2	.37	6	.23	1.46	.05	.3	.07	16	2
0	90	4.06	166	1241	998	1.74	31	.54	8.26	.31	1.8	.33	76	9
0	12	4.2	150	226	5	2.09	11	.06	.28	.09	.49	.06	16	1
0	98	33.8	1212	1829	1305	16.9	86	.48	2.27	.72	3.95	.2	130	4
0	50	2.96	132	155	15	3.91	3	.27	.86	.03	1.78	.05	36	0
0	42	2.44	127	248	1	1.82	2	.82	18.1	.09	1.62	.28	82	<1
0	19	2.28	43	164	1	1.77	2	.11	17.1	.09	1.4	.27	80	<1
0	21	.95	53	69	<1	1.58	1	.24	.34	.02	.85	.02	15	0
6	159	4.95	235	946	177	4.58	7	.6	15.6	.33	6.44	.38	95	2
0	72	3.84	253	655	1	4.28	37	.27	3.28	.14	.86	.69	82	4
0	16	.86	57	147	<1	.96	8	.06	.73	.03	.19	.15	18	1
0	113	2.93	203	602	12	3.28	14	.46	3.14	.18	1.25	.67	79	4
0	26	.68	47	141	3	.76	3	.11	.73	.04	.29	.16	18	1

Table A–1

Food Composition (Computer code number is for Wadsworth Diet Analysis Plus program) (For purposes of calculations, use "0" for t, <1, <.1, <.01, etc.)

Computer Code Number	Food Description	Measure	Wt (g)	H_2O (%)	Ener (kcal)	Prot (g)	Carb (g)	Dietary Fiber (g)	Fat (g)	Fat Breakdown (g) Sat	Mono	Poly
	SWEETENERS and SWEETS (see also Dairy [milk desserts] and Baked Goods)											
757	Apple butter	2 tbs	36	52	66	<1	17	<1	<1	t	t	t
1124	Butterscotch topping	2 tbs	41	32	103	1	27	<1	<1	t	t	0
1125	Caramel topping	2 tbs	41	32	103	1	27	<1	<1	t	t	0
	Cake frosting, creamy vanilla:											
1127	Canned	2 tbs	39	13	163	<1	27	<1	7	1.9	3.4	.9
1123	From mix	2 tbs	39	12	165	<1	28	<1	6	1.3	2.6	2.2
	Cake frosting, lite:											
2061	Milk chocolate	1 tbs	16	18	58	<1	11	<1	1	.4		
2062	Vanilla	1 tbs	16	15	60	0	12	<1	1	.4		
	Candy:											
1128	Almond Joy candy bar	1 oz	28	10	131	1	16	1	7	4.8	1.8	.4
2069	Butterscotch morsels	¼ c	43	1	246	0	31	0	12	12.5	0	0
758	Caramel, plain or chocolate	1 pce	10	8	38	<1	8	<1	1	.7	.1	t
1961	Chewing gum, sugarless	1 pce	3		6	0	2		0	0	0	0
	Chocolate (see also #784, 785, 971):											
	Milk chocolate:											
759	Plain	1 oz	28	1	144	2	17	1	9	5.2	2.8	.3
760	With almonds	1 oz	28	1	147	3	15	2	10	4.8	3.8	.6
761	With peanuts	1 oz	28	1	155	5	11	2	11	3.4	5.1	2.5
762	With rice cereal	1 oz	28	2	139	2	18	1	7	4.4	2.4	.2
763	Semisweet chocolate chips	1 c	168	1	805	7	106	10	50	29.9	16.8	1.6
764	Sweet dark chocolate (candy bar)	1 ea	41	1	226	2	25	2	13	8.3	4.6	.4
765	Fondant candy, uncoated (mints, candy corn, other)	1 pce	16	7	57	0	15	0	0	0	0	0
1697	Fruit Roll-Up (small)	1 ea	14	11	49	<1	12	<1	<1	.1	.2	.1
766	Fudge, chocolate	1 pce	17	10	65	<1	13	<1	1	.9	.4	.1
767	Gumdrops	1 c	182	1	703	0	180	0	0	0	0	0
768	Hard candy, all flavors	1 pce	6	1	22	0	6	0	<1	0	0	0
769	Jellybeans	10 pce	11	6	40	0	10	0	<1	t	t	t
1134	M&M's plain chocolate candy	10 pce	7	2	34	<1	5	<1	1	.9	.5	t
1135	M&M's peanut chocolate candy	10 pce	20	2	103	2	12	1	5	2.1	2.2	.8
1130	Mars almond bar	1 ea	50	4	234	4	31	1	11	2.7	5.5	2.8
1129	Milky Way candy bar	1 ea	60	6	254	3	43	1	10	4.7	3.6	.4
1708	Milk chocolate-coated peanuts	1 c	149	2	773	19	74	7	50	21.8	19.4	6.4
1709	Peanut brittle, recipe	1 c	147	2	666	11	102	3	28	7.4	12.5	6.9
1132	Reese's peanut butter cup	2 ea	50	3	271	5	27	2	16	5.5	6.5	2.8
1133	Skor English toffee candy bar	1 ea	39	3	217	2	22	1	13	8.5	4.3	.5
1131	Snickers candy bar (2.2oz)	1 ea	62	5	297	5	37	2	15	5.6	6.5	.3
1482	Fruit juice bar (2.5 fl oz)	1 ea	77	78	63	1	16	0	<1	t	0	t
771	Gelatin dessert/Jello, prepared	½ c	135	85	80	2	19	0	0	0	0	0
1702	SugarFree	½ c	117	98	8	1	1	0	0	0	0	0
772	Honey:	1 c	339	17	1030	1	279	1	0	0	0	0
773	Tablespoon	1 tbs	21	17	64	<1	17	<1	0	0	0	0
774	Jams or preserves:	1 tbs	20	29	54	<1	14	<1	<1	0	t	t
775	Packet	1 ea	14	34	34	<1	9	<1	<1	t	t	0
776	Jellies:	1 tbs	19	28	51	<1	13	<1	<1	t	t	t
777	Packet	1 ea	14	28	38	<1	10	<1	<1	t	t	t
1136	Marmalade	1 tbs	20	33	49	<1	13	<1	0	0	0	0
770	Marshmallows	1 ea	7	16	22	<1	6	<1	<1	t	t	t
1126	Marshmallow creme topping	2 tbs	38	18	118	1	30	<1	<1	t	t	t
778	Popsicle/ice pops	1 ea	128	80	92	0	24	0	0	0	0	0
	Sugars:											
779	Brown sugar	1 c	220	2	827	0	214	0	0	0	0	0
780	White sugar, granulated:	1 c	200		774	0	200	0	0	0	0	0
781	Tablespoon	1 tbs	12		46	0	12	0	0	0	0	0
782	Packet	1 ea	6		23	0	6	0	0	0	0	0
783	White sugar, powdered, sifted	1 c	100		389	0	99	0	<1	t	t	t
	Sweeteners:											
1711	Equal, packet	1 ea	1	12	4	<1	1	0	<1	0	t	t
1712	Sweet 'N Low, packet	1 ea	1		0	0	1	0	0	0	0	0

PAGE KEY: A–4 = Beverages A–6 = Dairy A–10 = Eggs A–10 = Fat/Oil A–14 = Fruit A–20 = Bakery A–28 = Grain A–32 = Fish A–34 = Meats A–38= Poultry A–40 = Sausage A–42 = Mixed/Fast A–46 = Nuts/Seeds A–50 = Sweets A–52 = Vegetables/Legumes A–62 = Vegetarian A–62 = Misc A–66 = Soups/Sauces A–68 = Fast A–84 = Convenience A–88 = Baby foods

Chol (mg)	Calc (mg)	Iron (mg)	Magn (mg)	Pota (mg)	Sodi (mg)	Zinc (mg)	VT-A (RE)	Thia (mg)	VT-E (α-TE)	Ribo (mg)	Niac (mg)	V-B6 (mg)	Fola (μg)	VT-C (mg)
0	2	.05	1	33	0	.02	0	<.01	.01	<.01	.03	.01	<1	1
<1	22	.08	3	34	143	.08	11	<.01	0	.04	.02	.01	1	<1
<1	22	.08	3	34	143	.08	11	<.01	0	.04	.02	.01	1	<1
0	1	.04		14	35	0	88	0	.79	<.01	<.01	0	0	0
0	4	.09	1	9	87	.04	42	.01	.79	.01	.13	<.01	0	0
0	1	.24			40		<1							0
0		.02			29		0							0
1	17	.39	18	69	41	.22	1	.01	.63	.04	.13	.02		<1
0	0	0		80	46		0	.03		.04	.03			0
1	14	.01	2	21	24	.04	1	<.01	.05	.02	.02	<.01	<1	<1
				0	0									
6	53	.39	17	108	23	.39	15	.02	.35	.08	.09	.01	2	<1
5	63	.46	25	124	21	.37	4	.02	.53	.12	.21	.01	3	<1
3	32	.52	34	150	11	.68	6	.08	1.3	.05	2.12	.04	23	0
5	48	.21	14	96	41	.31	3	.02	.35	.08	.13	.02	3	<1
0	54	5.26	193	613	18	2.72	3	.09	2	.15	.72	.06	5	0
<1	11	.98	47	139	3	.61	1	.01	.41	.1	.27	.02	1	0
0		.01		3	6	.01	<1	0	0	<.01	0	0	0	0
0	4	.14	3	41	9	.03	2	.01	.04	<.01	.01	.04	1	1
2	7	.08	4	17	10	.07	8	<.01	.02	.01	.02	<.01	<1	<1
0	5	.73	2	9	80	0	0	0	0	<.01	<.01	0	0	0
0		.02			2	<.01	0	0	0	0	0	0	0	0
0		.12		4	3	.01	0	0	0	0	0	0	0	0
1	7	.08	3	19	4	.07	4	<.01	.06	.01	.02	<.01	<1	<1
2	20	.23	12	69	10	.27	5	.03	.43	.04	.41	.02	7	<1
4	84	.55	36	163	85	.55	22	.02	.3	.16	.47	.03	9	<1
8	78	.46	20	145	144	.43	34	.02	.39	.13	.21	.03	6	1
13	155	1.95	134	748	61	2.8	0	.17	3.8	.26	6.33	.31	12	0
19	44	2.03	73	306	664	1.43	69	.28	2.41	.08	5.15	.15	103	0
2	39	.6	42	176	159	.7	9	.02	.66	.1	1.99	.04	27	<1
20	51	.02	13	93	108	.3	27	.01	.53	.13	.03	.01		<1
8	58	.47	42	209	165	.88	24	.13	.95	.1	2.26	.07	25	<1
0	4	.15	3	41	3	.04	2	.01	0	.01	.12	.02	5	7
0	3	.04	1	1	57	.04	0	0	0	<.01	<.01	<.01	0	0
0	2	.01	1	0	56	.03	0	0	0	<.01	<.01	<.01	0	0
0	20	1.42	7	176	14	.75	0	0	0	.13	.41	.08	7	2
0	1	.09		11	1	.05	0	0	0	.01	.02	<.01	<1	<1
0	4	.2	1	18	2	.01	<1	<.01	.02	.01	.04	<.01	2	<1
0	3	.07	1	11	6	.01	<1	0	0	<.01	<.01	<.01	5	1
0	2	.04	1	12	7	.01	<1	0	0	<.01	.01	<.01	<1	<1
0	1	.03	1	9	5	.01	<1	0	0	<.01	<.01	<.01	<1	<1
0	8	.03		7	11	.01	1	<.01	0	<.01	.01	<.01	7	1
0		.02			3	<.01	<1	0	0	0	<.01	0	<1	0
0	1	.08	1	2	17	.01	<1	0	0	0	.03	<.01	<1	0
0	0	0	1	5	15	.03	0	0	0	0	0	0	0	0
0	187	4.2	64	761	86	.4	0	.02	0	.01	.18	.06	2	0
0	2	.12	0	4	2	.06	0	0	0	.04	0	0	0	0
0		.01	0		<1	<.01	0	0	0	<.01	0	0	0	0
0		<.01	0		<1	<.01	0	0	0	<.01	0	0	0	0
0	1	.06	0	2	1	.03	0	0	0	0	0	0	0	0
0		<.01			<1	0	0	0	0	0	0	0	0	0
0	0	0			0		0							0

Table A–1

Food Composition (Computer code number is for Wadsworth Diet Analysis Plus program) (For purposes of calculations, use "0" for t, <1, <.1, <.01, etc.)

Computer Code Number	Food Description	Measure	Wt (g)	H_2O (%)	Ener (kcal)	Prot (g)	Carb (g)	Dietary Fiber (g)	Fat (g)	Fat Breakdown (g) Sat	Mono	Poly
	SWEETENERS and SWEETS—Continued											
	Syrups, chocolate:											
785	Hot fudge type	2 tbs	43	22	149	2	27	1	4	2.4	1.6	1.4
784	Thin type	2 tbs	38	29	93	1	25	1	<1	.3	.2	t
786	Molasses, blackstrap	2 tbs	41	29	96	0	25	0	0	0	0	0
1710	Light cane syrup	2 tbs	41	24	103	0	27	0	0	0	0	0
787	Pancake table syrup (corn and maple)	2 tbs	40	24	115	0	30	0	0	0	0	0
	VEGETABLES and LEGUMES											
788	Alfalfa sprouts	1 c	33	91	10	1	1	1	<1	t	.1	t
1815	Amaranth leaves, raw, chopped	1 c	28	92	7	1	1	<1	<1	t	t	t
1816	Amaranth leaves, raw, each	1 ea	14	92	4	<1	1	<1	<1	t	t	t
1817	Amaranth leaves, cooked	1 c	132	91	28	3	5	2	<1	.1	.1	.1
1987	Arugula, raw, chopped	½ c	10	92	2	<1	<1	<1	<1	t	t	t
789	Artichokes, cooked globe (300 g with refuse)	1 ea	120	84	60	4	13	6	<1	t	t	.1
1177	Artichoke hearts, cooked from frozen	1 c	168	86	76	5	15	8	1	.2	t	.4
1176	Artichoke hearts, marinated	1 c	130	81	128	3	10	6	10	1.5	2.3	5.9
2021	Artichoke hearts, in water	½ c	100	91	37	2	6	0	0	0	0	0
	Asparagus, green, cooked:											
	From fresh:											
790	Cuts and tips	½ c	90	92	22	2	4	1	<1	.1	t	.1
791	Spears, ½" diam at base	4 ea	60	92	14	2	3	1	<1	t	t	.1
	From frozen:											
792	Cuts and tips	½ c	90	91	25	3	4	1	<1	.1	t	.2
793	Spears, ½" diam at base	4 ea	60	91	17	2	3	1	<1	.1	t	.1
794	Canned, spears, ½" diam at base	4 ea	72	94	14	2	2	1	<1	.1	t	.2
795	Bamboo shoots, canned, drained slices	1 c	131	94	25	2	4	2	1	.1	t	.2
1795	Bamboo shoots, raw slices	1 c	151	91	41	4	8	3	<1	.1	t	.2
1798	Bamboo shoots, cooked slices	1 c	120	96	14	2	2	1	<1	.1	t	.1
	Beans (see also alphabetical listing this section):											
1990	Adzuki beans, cooked	½ c	115	66	147	9	28	1	<1	t	t	t
796	Black beans, cooked	½ c	86	66	114	8	20	7	<1	.1	t	.2
	Canned beans (white/navy):											
803	With pork and tomato sauce	½ c	127	73	124	7	25	6	1	.5	.6	.2
804	With sweet sauce	½ c	130	71	144	7	27	7	2	.7	.8	.2
805	With frankfurters	½ c	130	69	185	9	20	9	9	3.1	3.7	1.1
	Lima beans:											
797	Thick seeded (Fordhooks), cooked from frozen	½ c	85	73	85	5	16	5	<1	.1	t	.1
798	Thin seeded (Baby), cooked from frozen	½ c	90	72	94	6	18	5	<1	.1	t	.1
799	Cooked from dry, drained	½ c	94	70	108	7	20	7	<1	.1	t	.2
1998	Red Mexican, cooked f/dry	½ c	112	70	126	8	23	9	<1	.1	.1	.2
	Snap bean/green string beans cuts and french style:											
800	Cooked from fresh	½ c	63	89	22	1	5	2	<1	t	t	.1
801	Cooked from frozen	½ c	68	91	19	1	4	2	<1	t	t	.1
802	Canned, drained	½ c	68	93	14	1	3	1	<1	t	t	t
1713	Snap bean, yellow, cooked f/fresh	½ c	63	89	22	1	5	2	<1	t	t	.1
	Bean sprouts (mung):											
806	Raw	½ c	52	90	16	2	3	1	<1	t	t	t
807	Cooked, stir fried	½ c	62	84	31	3	7	1	<1	t	t	t
808	Cooked, boiled, drained	½ c	62	93	13	1	3	<1	<1	t	t	t
1788	Canned, drained	½ c	63	96	8	1	1	<1	<1	t	t	t
	Beets, cooked from fresh:											
809	Sliced or diced	½ c	85	87	37	1	8	2	<1	t	t	.1
810	Whole beets, 2" diam	2 ea	100	87	44	2	10	2	<1	t	t	.1
	Beets, canned:											
811	Sliced or diced	½ c	79	91	24	1	6	1	<1	t	t	t
812	Pickled slices	½ c	114	82	74	1	19	2	<1	t	t	t
813	Beet greens, cooked, drained	½ c	72	89	19	2	4	2	<1	t	t	.1

PAGE KEY: A–4 = Beverages A–6 = Dairy A–10 = Eggs A–10 = Fat/Oil A–14 = Fruit A–20 = Bakery A–28 = Grain A–32 = Fish A–34 = Meats A–38= Poultry A–40 = Sausage A–42 = Mixed/Fast A–46 = Nuts/Seeds A–50 = Sweets A–52 = Vegetables/Legumes A–62 = Vegetarian A–62 = Misc A–66 = Soups/Sauces A–68 = Fast A–84 = Convenience A–88 = Baby foods

Chol (mg)	Calc (mg)	Iron (mg)	Magn (mg)	Pota (mg)	Sodi (mg)	Zinc (mg)	VT-A (RE)	Thia (mg)	VT-E (α-TE)	Ribo (mg)	Niac (mg)	V-B6 (mg)	Fola (μg)	VT-C (mg)
5	43	.52	21	92	56	.34	9	.01	0	.09	.09	.01	2	<1
0	5	5.17	25	183	58	.28	494	<.01	.01	.31	12.8	.01	2	<1
0	353	7.18	88	1021	23	.41	0	.01	0	.02	.44	.29	<1	0
0	68	1.76	100	376	6	.12	0	.03	0	.02	.08	.27	0	0
0		.04	1	1	33	.02	0	<.01	0	<.01	.01	0	0	0
0	11	.32	9	26	2	.3	5	.02	—	.04	.16	.01	12	3
0	60	.65	15	171	6	.25	82	.01	.22	.04	.18	.05	24	12
0	30	.32	8	85	3	.13	41	<.01	.11	.02	.09	.03	12	6
0	276	2.98	73	846	28	1.16	366	.03	.66	.18	.74	.23	75	54
0	16	.15	5	37	3	.05	24	<.01	.04	.01	.03	.01	10	1
0	54	1.55	72	425	114	.59	22	.08	.23	.08	1.2	.13	61	12
0	35	.94	52	444	89	.6	27	.1	.32	.26	1.54	.15	200	8
0	30	1.24	37	335	688	.41	21	.05	1.43	.13	1.06	.11	114	40
0	0	1.35		0	250		12							4
0	18	.66	9	144	10	.38	49	.11	.34	.11	.97	.11	131	10
0	12	.44	6	96	7	.25	32	.07	.23	.08	.65	.07	88	6
0	21	.58	12	196	4	.5	74	.06	1.13	.09	.94	.02	122	22
0	14	.38	8	131	2	.34	49	.04	.75	.06	.62	.01	81	15
0	11	1.32	7	124	207	.29	38	.04	.31	.07	.69	.08	69	13
0	10	.42	5	105	9	.85	1	.03	.5	.03	.18	.18	4	1
0	20	.75	5	805	6	1.66	3	.23	1.51	.11	.91	.36	11	6
0	14	.29	4	640	5	.56	0	.02	.8	.06	.36	.12	3	1
0	32	2.3	60	612	9	2.04	1	.13	.11	.07	.82	.11	139	0
0	23	1.81	60	305	1	.96	1	.21	.07	.05	.43	.06	128	0
9	71	4.17	44	381	559	7.44	15	.07	.69	.06	.63	.09	29	4
9	79	2.16	44	346	437	1.95	14	.06	.7	.08	.46	.11	49	4
8	62	2.25	36	306	559	2.43	19	.07	.61	.07	1.17	.06	39	3
0	19	1.16	29	347	45	.37	16	.06	.25	.05	.91	.1	18	11
0	25	1.76	50	370	26	.49	15	.06	.58	.05	.69	.1	14	5
0	16	2.25	40	478	2	.89	0	.15	.17	.05	.4	.15	78	0
0	42	1.86	48	369	240	.87	<1	.13	.08	.07	.37	.11	94	2
0	29	.81	16	188	2	.23	42	.05	.09	.06	.39	.03	21	6
0	33	.6	16	86	6	.33	27	.02	.09	.06	.26	.04	16	3
0	18	.61	9	74	178	.2	24	.01	.09	.04	.14	.02	22	3
0	29	.81	16	188	2	.23	5	.05	.18	.06	.39	.03	21	6
0	7	.47	11	77	3	.21	1	.04	.02	.06	.39	.05	32	7
0	8	1.18	20	136	6	.56	2	.09	.01	.11	.74	.08	43	10
0	7	.4	9	63	6	.29	1	.03	.01	.06	.51	.03	18	7
0	9	.27	6	17	88	.18	1	.02	.01	.04	.14	.02	6	<1
0	14	.67	20	259	65	.3	3	.02	.25	.03	.28	.06	68	3
0	16	.79	23	305	77	.35	4	.03	.3	.04	.33	.07	80	4
0	12	1.44	13	117	153	.17	1	.01	.24	.03	.12	.04	24	3
0	12	.47	17	169	301	.3	1	.01	.15	.05	.29	.06	30	3
0	82	1.37	49	654	174	.36	367	.08	.22	.21	.36	.09	10	18

A

Table A–1

Food Composition (Computer code number is for Wadsworth Diet Analysis Plus program) (For purposes of calculations, use "0" for t, <1, <.1, <.01, etc.)

Computer Code Number	Food Description	Measure	Wt (g)	H_2O (%)	Ener (kcal)	Prot (g)	Carb (g)	Dietary Fiber (g)	Fat (g)	Fat Breakdown (g) Sat	Mono	Poly
	VEGETABLES and LEGUMES—Continued											
	Broccoli, raw:											
817	Chopped	½ c	44	91	12	1	2	1	<1	t	t	.1
818	Spears	1 ea	31	91	9	1	2	1	<1	t	t	.1
	Broccoli, cooked from fresh:											
819	Spears	1 ea	180	91	50	5	9	5	1	.1	t	.3
820	Chopped	½ c	78	91	22	2	4	2	<1	t	t	.1
	Broccoli, cooked from frozen:											
821	Spear, small piece	½ c	92	91	26	3	5	3	<1	t	t	.1
822	Chopped	½ c	92	91	26	3	5	3	<1	t	t	.1
1603	Broccoflower, steamed	½ c	78	90	25	2	5	2	<1	t	t	.1
823	Brussels sprouts, cooked from fresh	½ c	78	87	30	2	7	2	<1	.1	t	.2
824	Brussels sprouts, cooked from frozen	½ c	78	87	33	3	6	3	<1	.1	t	.2
	Cabbage, common varieties:											
825	Raw, shredded or chopped	1 c	70	92	17	1	4	2	<1	t	t	.1
826	Cooked, drained	1 c	150	94	33	2	7	3	1	.1	t	.3
	Cabbage, Chinese:											
1178	Bok choy, raw, shredded	1 c	70	95	9	1	2	1	<1	t	t	.1
827	Bok choy, cooked, drained	1 c	170	96	20	3	3	3	<1	t	t	.1
1937	Kim chee style	1 c	150	92	31	2	6	2	<1	t	t	.2
828	Pe Tsai, raw, chopped	1 c	76	94	12	1	2	2	<1	t	t	.1
1796	Pe Tsai, cooked	1 c	119	95	17	2	3	3	<1	t	t	.1
	Cabbage, red, coarsely chopped:											
829	Raw	1 c	89	92	24	1	5	2	<1	t	t	.1
830	Cooked, drained	1 c	150	94	31	2	7	3	<1	t	t	.1
831	Cabbage, savoy, coarsely chopped, raw	1 c	70	91	19	1	4	2	<1	t	t	t
1785	Cabbage, savoy, cooked	1 c	145	92	35	3	8	4	<1	t	t	.1
1896	Capers	1 ea	5	86		<1	<1	<1	<1			
	Carrots, raw:											
832	Whole, 7½ x 1⅛"	1 ea	72	88	31	1	7	2	<1	t	t	.1
833	Grated	½ c	55	88	24	1	6	2	<1	t	t	t
	Carrots, cooked, sliced, drained:											
834	From raw	½ c	78	87	35	1	8	3	<1	t	t	.1
835	From frozen	½ c	73	90	26	1	6	3	<1	t	t	t
836	Carrots, canned, sliced, drained	½ c	73	93	17	<1	4	1	<1	t	t	.1
837	Carrot juice, canned	1 c	236	89	94	2	22	2	<1	.1	t	.2
	Cauliflower, flowerets:											
838	Raw	½ c	50	92	12	1	3	1	<1	t	t	t
839	Cooked from fresh, drained	½ c	62	93	14	1	3	2	<1	t	t	.1
840	Cooked, from frozen, drained	½ c	90	94	17	1	3	2	<1	t	t	.1
	Celery, pascal type, raw:											
841	Large outer stalk, 8 x 1½"(root end)	1 ea	40	95	6	<1	1	1	<1	t	t	t
842	Diced	1 c	120	95	19	1	4	2	<1	t	t	.1
1789	Celeriac/celery root, cooked	1 c	155	92	39	1	9	2	<1	.1	.1	.2
1179	Chard, swiss, raw, chopped	1 c	36	93	7	1	1	1	<1	t	t	t
1180	Chard, swiss, cooked	1 c	175	93	35	3	7	4	<1	t	t	t
1855	Chayote fruit, raw	1 ea	203	94	39	2	9	3	<1	.1	t	.1
1856	Chayote fruit, cooked	1 c	160	93	38	1	8	4	1	.2	.1	.3
	Chickpeas (see Garbanzo Beans #854)											
	Collards, cooked, drained:											
843	From raw	½ c	95	92	25	2	5	3	<1	t	t	.1
844	From frozen	½ c	85	88	31	3	6	3	<1	.1	t	.2
	Corn, yellow, cooked, drained:											
845	From raw, on cob, 5" long	1 ea	77	73	72	2	17	2	1	.1	.2	.3
846	From frozen, on cob, 3½" long	1 ea	63	73	59	2	14	2	<1	.1	.1	.2
847	Kernels, cooked from frozen	½ c	82	77	66	2	16	2	<1	.1	.1	.2
	Corn, canned:											
848	Cream style	½ c	128	79	92	2	23	2	1	.1	.2	.3
849	Whole kernel, vacuum pack	½ c	105	77	83	3	20	2	1	.1	.2	.2
	Cowpeas (see Black-eyed peas #814-816)											
850	Cucumber slices with peel	7 pce	28	96	4	<1	1	<1	<1	t	t	t
1948	Cucumber, kim chee style	1 c	150	91	31	2	7	2	<1	t	0	.1

Chol (mg)	Calc (mg)	Iron (mg)	Magn (mg)	Pota (mg)	Sodi (mg)	Zinc (mg)	VT-A (RE)	Thia (mg)	VT-E (α-TE)	Ribo (mg)	Niac (mg)	V-B6 (mg)	Fola (μg)	VT-C (mg)
0	21	.39	11	143	12	.18	68	.03	.73	.05	.28	.07	31	41
0	15	.27	8	101	8	.12	48	.02	.51	.04	.2	.05	22	29
0	83	1.51	43	526	47	.68	250	.1	3.04	.2	1.03	.26	90	134
0	36	.65	19	228	20	.3	108	.04	1.32	.09	.45	.11	39	58
0	47	.56	18	166	22	.28	174	.05	.95	.07	.42	.12	28	37
0	47	.56	18	166	22	.28	174	.05	1.52	.07	.42	.12	52	37
0	25	.55	16	251	18	.39	5	.06	.23	.07	.59	.14	38	49
0	28	.94	16	247	16	.26	56	.08	.66	.06	.47	.14	47	48
0	19	.58	19	254	18	.28	46	.08	.45	.09	.42	.22	79	36
0	33	.41	10	172	13	.13	9	.03	.07	.03	.21	.07	30	22
0	46	.25	12	146	12	.13	19	.09	.16	.08	.42	.17	30	30
0	73	.56	13	176	45	.13	210	.03	.08	.05	.35	.14	46	31
0	158	1.77	19	631	58	.29	437	.05	.2	.11	.73	.28	69	44
0	145	1.28	27	375	995	.35	426	.07	.24	.1	.75	.34	88	80
0	58	.24	10	181	7	.17	91	.03	.09	.04	.3	.18	60	20
0	38	.36	12	268	11	.21	115	.05	.14	.05	.59	.21	63	19
0	45	.44	13	183	10	.19	4	.04	.09	.03	.27	.19	18	51
0	55	.52	16	210	12	.22	4	.05	.18	.03	.3	.21	19	52
0	24	.28	20	161	20	.19	70	.05	.07	.02	.21	.13	56	22
0	43	.55	35	267	35	.33	129	.07	.15	.03	.03	.22	67	25
0	2	.05			105		1							0
0	19	.36	11	233	25	.14	2025	.07	.33	.04	.67	.11	10	7
0	15	.27	8	178	19	.11	1547	.05	.25	.03	.51	.08	8	5
0	24	.48	10	177	51	.23	1914	.03	.33	.04	.39	.19	11	2
0	20	.34	7	115	43	.17	1292	.02	.31	.03	.32	.09	8	2
0	18	.47	6	131	177	.19	1005	.01	.31	.02	.4	.08	7	2
0	57	1.09	33	689	68	.42	2584	.22	.02	.13	.91	.51	9	20
0	11	.22	7	152	15	.14	1	.03	.02	.03	.26	.11	28	23
0	10	.2	6	88	9	.11	1	.03	.02	.03	.25	.11	27	27
0	15	.37	8	125	16	.12	2	.03	.04	.05	.28	.08	37	28
0	16	.16	4	115	35	.05	5	.02	.14	.02	.13	.03	11	3
0	48	.48	13	344	104	.16	16	.05	.43	.05	.39	.1	34	8
0	40	.67	19	268	95	.31	0	.04	.31	.06	.66	.16	5	6
0	18	.65	29	136	77	.13	119	.01	.68	.03	.14	.04	5	11
0	102	3.96	151	961	313	.58	550	.06	3.31	.15	.63	.15	15	31
0	34	.69	24	254	4	1.5	12	.05	.24	.06	.95	.15	189	16
0	21	.35	19	277	2	.5	8	.04	.19	.06	.67	.19	29	13
0	113	.44	16	247	9	.4	297	.04	.84	.1	.55	.12	88	17
0	179	.95	25	213	42	.23	508	.04	.42	.1	.54	.1	65	22
0	2	.47	22	193	3	.48	16	.13	.07	.05	1.17	.17	23	4
0	2	.38	18	158	3	.4	13	.11	.06	.04	.96	.14	19	3
0	3	.29	16	121	4	.33	18	.07	.07	.06	1.07	.11	25	3
0	4	.49	22	172	365	.68	13	.03	.11	.07	1.23	.08	57	6
0	5	.44	24	195	286	.48	25	.04	.09	.08	1.23	.06	52	9
0	4	.07	3	40	1	.06	6	.01	.02	.01	.06	.01	4	1
0	13	7.23	12	176	1531	.76	49	.04	.24	.04	.69	.16	34	5

Table A–1

Food Composition (Computer code number is for Wadsworth Diet Analysis Plus program) (For purposes of calculations, use "0" for t, <1, <.1, <.01, etc.)

Computer Code Number	Food Description	Measure	Wt (g)	H_2O (%)	Ener (kcal)	Prot (g)	Carb (g)	Dietary Fiber (g)	Fat (g)	Fat Breakdown (g) Sat	Mono	Poly
	VEGETABLES and LEGUMES—Continued											
	Dandelion Greens:											
851	Raw	1 c	55	86	25	1	5	2	<1	.1	t	.2
852	Chopped, cooked, drained	1 c	105	90	35	2	7	3	1	.2	t	.3
853	Eggplant, cooked	1 c	99	92	28	1	7	2	<1	t	t	.1
1714	Endive, fresh, chopped	1 c	50	94	8	1	2	2	<1	t	t	t
856	Escarole/curly endive, chopped	1 c	50	94	8	1	2	2	<1	t	t	t
854	Garbanzo beans (chickpeas), cooked	1 c	164	60	269	14	45	12	4	.4	1	1.9
1939	Grape leaf, raw:	1 ea	3	73	3	<1	1	<1	<1	t	t	t
7914	Cup	1 c	14	73	13	1	2	2	<1	t	t	.1
855	Great northern beans, cooked	1 c	177	69	209	15	37	12	1	.2	t	.3
857	Jerusalem artichoke, raw slices	1 c	150	78	114	3	26	2	<1	0	t	t
1794	Jicama	1 c	120	90	46	1	11	6	<1	t	t	t
	Kale, cooked, drained:											
858	From raw	1 c	130	91	36	2	7	3	1	.1	t	.3
859	From frozen	1 c	130	90	39	4	7	3	1	.1	t	.3
860	Kidney beans, canned	1 c	256	77	218	13	40	16	1	.1	.1	.5
1181	Kohlrabi, raw slices	1 c	135	91	36	2	8	5	<1	t	t	.1
861	Kohlrabi, cooked	1 c	165	90	48	3	11	2	<1	t	t	.1
1183	Leeks, raw, chopped	1 c	89	83	54	1	13	2	<1	t	t	.1
1182	Leeks, cooked, chopped	1 c	104	91	32	1	8	1	<1	t	t	.1
862	Lentils, cooked from dry	1 c	198	70	230	18	40	16	1	.1	.1	.3
1288	Lentils, sprouted, stir-fried	1 c	124	69	125	11	26	5	1	.1	.1	.2
1289	Lentils, sprouted, raw	1 c	77	67	82	7	17	3	<1	t	.1	.2
	Lettuce:											
	Butterhead/Boston types:											
863	Head, 5″ diameter	¼ ea	41	96	5	1	1	<1	<1	t	t	t
864	Leaves, inner or outer	4 ea	30	96	4	<1	1	<1	<1	t	t	t
	Iceberg/crisphead:											
867	Chopped or shredded	1 c	55	96	7	1	1	1	<1	t	t	.1
865	Head, 6″ diameter	1 ea	539	96	65	5	11	8	1	.1	t	.5
866	Wedge, ¼ head	1 ea	135	96	16	1	3	2	<1	t	t	.1
868	Looseleaf, chopped	½ c	28	94	5	<1	1	1	<1	t	t	t
869	Romaine, chopped	½ c	28	95	4	<1	1	<1	<1	t	t	t
870	Romaine, inner leaf	3 pce	30	95	4	<1	1	1	<1	t	t	t
1930	Luffa, cooked (Chinese okra)	1 c	178	89	57	3	13	6	<1	.1	t	.1
	Mushrooms:											
871	Raw, sliced	½ c	35	92	9	1	2	<1	<1	t	t	.1
872	Cooked from fresh, pieces	½ c	78	91	21	2	4	2	<1	t	t	.1
1962	Stir fried, shitake slices	½ c	73	83	40	1	10	2	<1	t	t	t
873	Canned, drained	½ c	78	91	19	1	4	2	<1	t	t	.1
1951	Mushroom caps, pickled	8 ea	47	92	11	1	2	1	<1	t	t	.1
	Mustard greens:											
874	Cooked from raw	½ c	70	94	10	2	1	1	<1	t	.1	t
875	Cooked from frozen	½ c	75	94	14	2	2	2	<1	t	.1	t
876	Navy beans, cooked from dry	1 c	182	63	258	16	48	12	1	.3	.1	.4
	Okra, cooked:											
877	From fresh pods	8 ea	85	90	27	2	6	2	<1	t	t	t
878	From frozen slices	1 c	184	91	51	4	11	5	1	.1	.1	.1
1236	Batter fried from fresh	1 c	92	69	175	3	11	2	13	2.1	3.4	7.1
1930	Chinese, (Luffa), cooked	1 c	178	89	57	3	13	6	<1	.1	t	.1
	Onions:											
879	Raw, chopped	½ c	80	90	30	1	7	1	<1	t	t	t
880	Raw, sliced	½ c	58	90	22	1	5	1	<1	t	t	t
881	Cooked, drained, chopped	½ c	105	88	46	1	11	1	<1	t	t	.1
882	Dehydrated flakes	¼ c	14	4	49	1	12	1	<1	t	t	t
1934	Onions, pearl, cooked	½ c	93	87	41	1	9	1	<1	t	t	.1
883	Spring/green onions, bulb and top, chopped	½ c	50	90	16	1	4	1	<1	t	t	t
884	Onion rings, breaded, heated f/frozen	2 ea	20	28	81	1	8	<1	5	1.7	2.2	1
1917	Palm hearts, cooked slices	1 c	146	69	150	4	39	2	<1	.1	.1	t
885	Parsley, raw, chopped	½ c	30	88	11	1	2	1	<1	t	.1	t
888	Parsnips, sliced, cooked	½ c	78	78	63	1	15	3	<1	t	.1	t

PAGE KEY: A–4 = Beverages A–6 = Dairy A–10 = Eggs A–10 = Fat/Oil A–14 = Fruit A–20 = Bakery A–28 = Grain A–32 = Fish A–34 = Meats A–38= Poultry A–40 = Sausage A–42 = Mixed/Fast A–46 = Nuts/Seeds A–50 = Sweets A–52 = Vegetables/Legumes A–62 = Vegetarian A–62 = Misc A–66 = Soups/Sauces A–68 = Fast A–84 = Convenience A–88 = Baby foods

Chol (mg)	Calc (mg)	Iron (mg)	Magn (mg)	Pota (mg)	Sodi (mg)	Zinc (mg)	VT-A (RE)	Thia (mg)	VT-E (α-TE)	Ribo (mg)	Niac (mg)	V-B6 (mg)	Fola (μg)	VT-C (mg)
0	103	1.71	20	218	42	.23	770	.1	1.38	.14	.44	.14	15	19
0	147	1.89	25	244	46	.29	1228	.14	2.63	.18	.54	.17	13	19
0	6	.35	13	246	3	.15	6	.07	.03	.02	.59	.08	14	1
0	26	.41	7	157	11	.39	103	.04	.22	.04	.2	.01	71	3
0	26	.41	7	157	11	.39	103	.04	.22	.04	.2	.01	71	3
0	80	4.74	79	477	11	2.51	5	.19	.57	.1	.86	.23	282	2
0	11	.08	3	8	<1	.02	81	<.01	.06	.01	.07	.01	2	<1
0	51	.37	13	38	1	.09	378	.01	.28	.05	.33	.06	12	2
0	120	3.77	88	692	4	1.56	<1	.28	.53	.1	1.21	.21	181	2
0	21	5.1	25	644	6	.18	3	.3	.28	.09	1.95	.12	20	6
0	14	.72	14	180	5	.19	2	.02	5.48	.03	.24	.05	14	24
0	94	1.17	23	296	30	.31	962	.07	1.11	.09	.65	.18	17	53
0	179	1.22	23	417	19	.23	826	.06	.23	.15	.87	.11	19	33
0	61	3.23	72	658	873	1.41	0	.27	.13	.22	1.17	.06	130	3
0	32	.54	26	473	27	.04	5	.07	.65	.03	.54	.2	22	84
0	41	.66	31	561	35	.51	7	.07	2.76	.03	.64	.25	20	89
0	52	1.87	25	160	18	.11	9	.05	.82	.03	.36	.21	57	11
0	31	1.14	15	90	10	.06	5	.03	.63	.02	.21	.12	25	4
0	38	6.59	71	731	4	2.51	2	.33	.22	.14	2.1	.35	358	3
0	17	3.84	43	352	12	1.98	5	.27	.11	.11	1.49	.2	83	16
0	19	2.47	28	248	8	1.16	4	.18	.07	.1	.87	.15	77	13
0	13	.12	5	105	2	.07	40	.02	.18	.02	.12	.02	30	3
0	10	.09	4	77	1	.05	29	.02	.13	.02	.09	.01	22	2
0	10	.27	5	87	5	.12	18	.02	.15	.02	.1	.02	31	2
0	102	2.7	48	852	48	1.19	178	.25	1.51	.16	1.01	.22	302	21
0	26	.67	12	213	12	.3	45	.06	.38	.04	.25	.05	76	5
0	19	.39	3	74	3	.08	53	.01	.12	.02	.11	.01	14	5
0	10	.31	2	81	2	.07	73	.03	.12	.03	.14	.01	38	7
0	11	.33	2	87	2	.07	78	.03	.13	.03	.15	.01	41	7
0	112	.8	101	570	420	.97	103	.23	1.22	.1	1.54	.33	81	29
0	2	.43	3	130	1	.26	0	.04	.04	.16	1.44	.03	7	1
0	5	1.36	9	278	2	.68	0	.06	.09	.23	3.48	.07	14	3
0	2	.32	10	85	3	.97	0	.03	.09	.12	1.1	.12	15	<1
0	9	.62	12	101	332	.56	0	.07	.09	.02	1.24	.05	10	0
0	2	.5	5	139	95	.28	0	.03	.05	.16	1.42	.03	6	1
0	52	.49	10	141	11	.08	212	.03	1.41	.04	.3	.07	51	18
0	76	.84	10	104	19	.15	335	.03	1.31	.04	.19	.08	52	10
0	127	4.51	107	670	2	1.93	<1	.37	.73	.11	.97	.3	255	2
0	54	.38	48	274	4	.47	49	.11	.59	.05	.74	.16	39	14
0	177	1.23	94	431	6	1.14	94	.18	1.27	.23	1.44	.09	269	22
15	104	.77	37	214	137	.5	43	.13	3.08	.1	.75	.13	37	10
0	112	.8	101	570	420	.97	103	.23	1.22	.1	1.54	.33	81	29
0	16	.18	8	126	2	.15	0	.03	.1	.02	.12	.09	15	5
0	12	.13	6	91	2	.11	0	.02	.07	.01	.09	.07	11	4
0	23	.25	12	174	3	.22	0	.04	.14	.02	.17	.13	16	5
0	36	.22	13	227	3	.26	0	.07	.19	.01	.14	.22	23	10
0	21	.22	10	154	218	.19	0	.04	.12	.02	.15	.12	14	5
0	36	.74	10	138	8	.19	19	.03	.06	.04	.26	.03	32	9
0	6	.34	4	26	75	.08	5	.06	.14	.03	.72	.01	13	<1
0	26	2.47	15	2636	20	5.45	10	.07	.73	.25	1.25	1.06	30	10
0	41	1.86	15	166	17	.32	156	.03	.54	.03	.39	.03	46	40
0	29	.45	23	286	8	.2	0	.06	.78	.04	.56	.07	45	10

Table A-1

Food Composition (Computer code number is for Wadsworth Diet Analysis Plus program) (For purposes of calculations, use "0" for t, <1, <.1, <.01, etc.)

Computer Code Number	Food Description	Measure	Wt (g)	H_2O (%)	Ener (kcal)	Prot (g)	Carb (g)	Dietary Fiber (g)	Fat (g)	Fat Breakdown (g) Sat	Mono	Poly
	VEGETABLES and LEGUMES—Continued											
	Peas:											
	Black-eyed, cooked:											
814	From dry, drained	½ c	86	70	100	7	18	6	<1	.1	t	.2
815	From fresh, drained	½ c	82	75	79	3	17	4	<1	.1	t	.1
816	From frozen, drained	½ c	85	66	112	7	20	5	1	.1	.1	.2
889	Edible pod peas, cooked	½ c	80	89	34	3	6	2	<1	t	t	.1
890	Green, canned, drained:	½ c	85	82	59	4	11	3	<1	.1	t	.1
5267	Unsalted	½ c	124	86	66	4	12	4	<1	.1	t	.2
891	Green, cooked from frozen	½ c	80	79	62	4	11	4	<1	t	t	.1
1786	Snow peas, raw	½ c	49	89	21	1	4	1	<1	t	t	t
1787	Snow peas, raw	10 ea	34	89	14	1	3	1	<1	t	t	t
892	Split, green, cooked from dry	½ c	98	69	116	8	21	8	<1	.1	.1	.2
1187	Peas & carrots, cooked from frozen	½ c	80	86	38	2	8	2	<1	.1	t	.2
1186	Peas & carrots, canned w/liquid	½ c	128	88	49	3	11	3	<1	.1	t	.2
	Peppers, hot:											
893	Hot green chili, canned	½ c	68	92	14	1	3	1	<1	t	t	t
894	Hot green chili, raw	1 ea	45	88	18	1	4	1	<1	t	t	t
1715	Hot red chili, raw, diced	1 tbs	9	88	4	<1	1	<1	<1	t	t	t
1988	Jalapeno, raw	1 ea	45	90	11	<1	2		<1			
895	Jalapeno, chopped, canned	½ c	68	89	18	1	3	2	1	.1	t	.3
1918	Jalapeno wheels, in brine (Ortega)	2 tbs	29		10	0	2		0	0	0	0
	Peppers, sweet, green:											
896	Whole pod (90 g with refuse), raw	1 ea	74	92	20	1	5	1	<1	t	t	.1
897	Cooked, chopped (1 pod cooked = 73g)	½ c	68	92	19	1	5	1	<1	t	t	.1
	Peppers, sweet, red:											
1286	Raw, chopped	½ c	75	92	20	1	5	1	<1	t	t	.1
1807	Raw, each	1 ea	74	92	20	1	5	1	<1	t	t	.1
1287	Cooked, chopped	½ c	68	92	19	1	5	1	<1	t	t	.1
	Peppers, sweet, yellow:											
1872	Raw, large	1 ea	186	92	50	2	12	2	<1	.1	t	.2
1873	Strips	10 pce	52	92	14	1	3	<1	<1	t	t	.1
898	Pinto beans, cooked from dry	½ c	85	64	116	7	22	7	<1	.1	.1	.2
1191	Poi, two finger	½ c	120	72	134	<1	33	<1	<1	t	t	.1
	Potatoes:											
	Baked in oven, 4¾"x2⅓" diam											
899	With skin	1 ea	202	71	220	5	51	5	<1	.1	t	.1
900	Flesh only	1 ea	156	75	145	3	34	2	<1	t	t	.1
901	Skin only	1 ea	58	47	115	2	27	5	<1	t	t	t
	Baked in microwave, 4¾"x 2⅓"dm:											
902	With skin	1 ea	202	72	212	5	49	5	<1	.1	t	.1
903	Flesh only	1 ea	156	74	156	3	36	2	<1	t	t	.1
904	Skin only	1 ea	58	63	77	3	17	3	<1	t	t	t
	Boiled, about 2½" diam:											
905	Peeled after boiling	1 ea	136	77	118	3	27	2	<1	t	t	.1
906	Peeled before boiling	1 ea	135	77	116	2	27	2	<1	t	t	.1
	French fried, strips 2–3½" long:											
907	Oven heated	10 ea	50	35	167	2	20	2	9	3	5.7	.7
908	Fried in vegetable oil	10 ea	50	38	158	2	20	2	8	1.9	4.7	.7
1188	Fried in veg and animal oil	10 ea	50	38	158	2	20	2	8	1.9	4.7	.7
909	Hashed browns from frozen	1 c	156	56	340	5	44	3	18	7	8	2.1
	Mashed:											
910	Home recipe with whole milk	½ c	105	78	81	2	18	2	1	.4	.2	.1
911	Home recipe with milk and marg	½ c	105	76	111	2	17	2	4	1.1	1.9	1.3
912	Prepared from flakes; water, milk, margarine, salt added	½ c	110	76	124	2	16	3	6	1.6	2.5	1.7
	Potato products, prepared:											
	Au gratin:											
913	From dry mix	½ c	123	79	114	3	16	1	5	3.5	1.5	.2
914	From home recipe, using butter	½ c	122	74	161	7	14	2	9	4.8	3.2	1.3
	Scalloped:											
915	From dry mix	½ c	122	79	113	3	16	1	5	3.2	1.5	.2
916	From home recipe, using butter	½ c	123	81	106	4	13	2	5	1.7	1.7	.9

PAGE KEY: A–4 = Beverages A–6 = Dairy A–10 = Eggs A–10 = Fat/Oil A–14 = Fruit A–20 = Bakery A–28 = Grain A–32 = Fish A–34 = Meats A–38= Poultry A–40 = Sausage A–42 = Mixed/Fast A–46 = Nuts/Seeds A–50 = Sweets A–52 = Vegetables/Legumes A–62 = Vegetarian A–62 = Misc A–66 = Soups/Sauces A–68 = Fast A–84 = Convenience A–88 = Baby foods

Chol (mg)	Calc (mg)	Iron (mg)	Magn (mg)	Pota (mg)	Sodi (mg)	Zinc (mg)	VT-A (RE)	Thia (mg)	VT-E (α-TE)	Ribo (mg)	Niac (mg)	V-B6 (mg)	Fola (μg)	VT-C (mg)
0	21	2.16	46	239	3	1.11	2	.17	.24	.05	.43	.09	179	<1
0	105	.92	43	343	3	.84	65	.08	.18	.12	1.15	.05	104	2
0	20	1.8	42	319	4	1.21	7	.22	.33	.05	.62	.08	120	2
0	34	1.58	21	192	3	.3	10	.1	.31	.06	.43	.11	23	38
0	17	.81	14	147	214	.6	65	.1	.32	.07	.62	.05	38	8
0	22	1.26	21	124	11	.87	47	.14	.47	.09	1.04	.08	35	12
0	19	1.26	23	134	70	.75	54	.23	.14	.08	1.18	.09	47	8
0	21	1.02	12	98	2	.13	7	.07	.19	.04	.29	.08	20	29
0	15	.71	8	68	1	.09	5	.05	.13	.03	.2	.05	14	20
0	14	1.26	35	355	2	.98	1	.19	.38	.05	.87	.05	64	<1
0	18	.75	13	126	54	.36	621	.18	.26	.05	.92	.07	21	6
0	29	.96	18	128	333	.74	739	.09	.24	.07	.74	.11	23	8
0	5	.34	10	127	798	.12	41	.01	.47	.03	.54	.1	7	46
0	8	.54	11	153	3	.13	35	.04	.31	.04	.43	.12	10	109
0	2	.11	2	31	1	.03	97	.01	.06	.01	.09	.02	2	22
				2	2		30		.37					53
0	16	1.28	10	131	1136	.23	116	.03	.47	.03	.27	.13	10	7
0				55	390		10		.2					21
0	7	.34	7	131	1	.09	47	.05	.51	.02	.38	.18	16	66
0	6	.31	7	113	1	.08	40	.04	.47	.02	.32	.16	11	51
0	7	.34	7	133	1	.09	428	.05	.52	.02	.38	.19	16	143
0	7	.34	7	131	1	.09	422	.05	.51	.02	.38	.18	16	141
0	6	.31	7	113	1	.08	256	.04	.47	.02	.32	.16	11	116
0	20	.86	22	394	4	.32	45	.05	1.28	.05	1.66	.31	48	342
0	6	.24	6	110	1	.09	12	.01	.36	.01	.46	.09	13	96
0	41	2.22	47	398	2	.92	<1	.16	.8	.08	.34	.13	146	2
0	19	1.06	29	220	14	.26	2	.16	.22	.05	1.32	.33	26	5
0	20	2.75	54	844	16	.65	0	.22	.1	.07	3.33	.7	22	26
0	8	.55	39	610	8	.45	0	.16	.06	.03	2.18	.47	14	20
0	20	4.08	25	332	12	.28	0	.07	.02	.06	1.78	.36	12	8
0	22	2.5	54	903	16	.73	0	.24	.1	.06	3.45	.69	24	30
0	8	.64	39	641	11	.51	0	.2	.06	.04	2.54	.5	19	24
0	27	3.45	21	377	9	.3	0	.04	.02	.04	1.29	.28	10	9
0	7	.42	30	515	5	.41	0	.14	.07	.03	1.96	.41	14	18
0	11	.42	27	443	7	.36	0	.13	.07	.03	1.77	.36	12	10
0	6	.83	11	270	307	.2	0	.04	.25	.02	1.34	.11	11	3
0	9	.38	17	366	108	.19	0	.09	.25	.01	1.63	.12	14	5
6	9	.38	17	366	108	.19	0	.09	.25	.01	1.63	.12	14	5
0	23	2.36	26	680	53	.5	0	.17	.3	.03	3.78	.2	10	10
2	27	.28	19	314	318	.3	6	.09	.05	.04	1.18	.24	9	7
2	27	.27	19	303	310	.28	21	.09	.31	.04	1.13	.23	8	6
4	54	.24	20	256	365	.2	23	.12	.77	.05	.74	.01	8	11
18	102	.39	18	269	540	.29	38	.02	1.48	.1	1.15	.05	8	4
18	145	.78	24	483	528	.84	46	.08	.64	.14	1.21	.21	13	12
13	44	.46	17	248	416	.3	26	.02	.18	.07	1.26	.05	12	4
7	70	.7	23	465	412	.49	23	.08	.4	.11	1.29	.22	13	13

Table A-1

Food Composition (Computer code number is for Wadsworth Diet Analysis Plus program) (For purposes of calculations, use "0" for t, <1, <.1, <.01, etc.)

Computer Code Number	Food Description	Measure	Wt (g)	H_2O (%)	Ener (kcal)	Prot (g)	Carb (g)	Dietary Fiber (g)	Fat (g)	Fat Breakdown (g) Sat	Mono	Poly
	VEGETABLES and LEGUMES—Continued											
	Potato Salad (see Mixed Dishes #715)											
1192	Potato puffs, cooked from frozen	½ c	64	53	142	2	19	2	7	3.3	2.8	.5
918	Pumpkin, cooked from fresh, mashed	½ c	123	94	25	1	6	1	<1	t	t	t
919	Pumpkin, canned	½ c	123	90	42	1	10	4	<1	.2	t	t
1891	Radicchio, raw, shredded	½ c	20	93	5	<1	1	<1	<1	t	t	t
1894	Radicchio, raw, leaf	10 ea	80	93	18	1	4	1	<1	t	t	.1
920	Red radishes	10 ea	45	95	8	<1	2	1	<1	t	t	t
1793	Daikon radishes (Chinese) raw	½ c	44	95	8	<1	2	1	<1	t	t	t
921	Refried beans, canned	½ c	126	76	118	7	19	7	2	.6	.7	.2
1375	Rutabaga, cooked cubes	½ c	85	89	33	1	7	2	<1	t	t	.1
922	Sauerkraut, canned with liquid	½ c	118	92	22	1	5	3	<1	t	t	.1
923	Seaweed, kelp, raw	½ c	40	82	17	1	4	1	<1	.1	t	t
924	Seaweed, spirulina, dried	½ c	8	5	23	5	2	<1	1	.2	.1	.2
1866	Shallots, raw, chopped	1 tbs	10	80	7	<1	2	<1	<1	t	t	t
1557	Snow peas, stir-fried	½ c	83	89	35	2	6	2	<1	t	t	.1
925	Soybeans, cooked from dry	½ c	86	63	149	15	9	5	8	1.1	1.7	4.4
1996	Soybeans, dry roasted	½ c	86	1	387	34	28	7	19	2.7	4.1	10.6
	Soybean products:											
926	Miso	½ c	138	41	284	17	39	7	8	1.2	1.9	4.7
	Soy milk (see #144 and #2301 under Dairy)											
	Tofu (soybean curd):											
7540	Extra firm, silken	½ c	126	88	69	9	3	<1	2	.4	.4	1.3
7542	Firm, silken	½ c	126	87	77	9	3	<1	3	.5	.7	1.9
927	Regular	½ c	124	87	76	8	2	<1	5	.7	1	2.6
7541	Soft, silken	½ c	124	89	68	6	4	<1	3	.4	.6	1.9
	Spinach:											
928	Raw, chopped	½ c	28	92	6	1	1	1	<1	t	t	t
929	Cooked, from fresh, drained	½ c	90	91	21	3	3	2	<1	t	t	.1
930	Cooked from frozen (leaf)	½ c	95	90	27	3	5	3	<1	t	t	.1
931	Canned, drained solids:	½ c	107	92	25	3	4	3	1	.1	t	.2
5149	Unsalted	½ c	107	92	25	3	4	3	1	.1	t	.2
	Spinach soufflé (see Mixed Dishes)											
	Squash, summer varieties,cooked w/skin:											
932	Varieties averaged	½ c	90	94	18	1	4	1	<1	.1	t	.1
933	Crookneck	½ c	90	94	18	1	4	1	<1	.1	t	.1
934	Zucchini	½ c	90	95	14	1	4	1	<1	t	t	t
	Squash, winter varieties, cooked:											
	Average of all varieties, baked:											
935	Mashed	1 c	245	89	96	2	21	7	2	.3	.1	.6
936	Cubes	1 c	205	89	80	2	18	6	1	.3	.1	.5
937	Acorn, baked, mashed	½ c	123	83	69	1	18	5	<1	t	t	.1
1218	Acorn, boiled, mashed	½ c	122	90	41	1	11	3	<1	t	t	t
	Butternut squash:											
938	Baked cubes	½ c	103	88	41	1	11	3	<1	t	t	t
1219	Baked, mashed	½ c	103	88	41	1	11	3	<1	t	t	t
1193	Cooked from frozen	½ c	120	88	47	1	12	3	<1	t	t	t
1194	Hubbard, baked, mashed	½ c	120	85	60	3	13	3	1	.2	.1	.3
1195	Hubbard, boiled, mashed	½ c	118	91	35	2	8	3	<1	.1	t	.2
1196	Spaghetti, baked or boiled	½ c	77	92	22	<1	5	1	<1	t	t	.1
1189	Succotash, cooked from frozen	½ c	85	74	79	4	17	3	1	.1	.1	.4
	Sweet potatoes:											
939	Baked in skin, peeled, 5 x 2" diam	1 ea	114	73	117	2	28	3	<1	t	t	.1
940	Boiled without skin, 5 x 2" diam	1 ea	151	73	159	2	37	3	<1	.1	t	.2
941	Candied, 2½ x 2"	1 pce	105	67	144	1	29	3	3	1.4	.7	.2
	Canned:											
942	Solid pack	½ c	128	74	129	3	30	2	<1	.1	t	.1
943	Vacuum pack, mashed	½ c	127	76	116	2	27	2	<1	.1	t	.1
944	Vacuum pack, 3¾ x 1"	2 pce	80	76	73	1	17	1	<1	t	t	.1
1940	Taro shoots, cooked slices	1 c	140	95	20	1	4	1	<1	t	t	t
1941	Taro, tahitian, cooked slices	1 c	137	86	60	6	9	1	1	.2	.1	.4
	Tomatillos:											
1877	Raw, each	1 ea	34	92	11	<1	2	1	<1	t	.1	.1
1875	Raw, chopped	1 c	132	92	42	1	8	3	1	.2	.2	.6

Chol (mg)	Calc (mg)	Iron (mg)	Magn (mg)	Pota (mg)	Sodi (mg)	Zinc (mg)	VT-A (RE)	Thia (mg)	VT-E (α-TE)	Ribo (mg)	Niac (mg)	V-B6 (mg)	Fola (μg)	VT-C (mg)
0	19	1	12	243	477	.19	1	.12	.03	.05	1.38	.15	11	4
0	18	.7	11	283	1	.28	1330	.04	1.3	.1	.51	.05	10	6
0	32	1.71	28	253	6	.21	2713	.03	1.3	.07	.45	.07	15	5
0	4	.11	3	60	4	.12	1	<.01	.45	.01	.05	.01	12	2
0	15	.45	10	242	18	.5	2	.01	1.81	.02	.2	.05	48	6
0	9	.13	4	104	11	.13	<1	<.01	0	.02	.13	.03	12	10
0	12	.18	7	100	9	.07	0	.01	0	.01	.09	.02	12	10
10	44	2.09	42	336	377	1.47	0	.03	.39	.02	.4	.18	14	8
0	41	.45	20	277	17	.3	48	.07	.13	.03	.61	.09	13	16
0	35	1.73	15	201	780	.22	2	.02	.12	.03	.17	.15	28	17
0	67	1.14	48	36	93	.49	5	.02	.35	.06	.19	<.01	72	1
0	10	2.28	16	109	84	.16	5	.19	.4	.29	1.02	.03	8	1
0	4	.12	2	33	1	.04	125	.01	.01	<.01	.02	.03	3	1
0	36	1.73	20	166	3	.22	11	.11	.32	.06	.47	.13	28	42
0	88	4.42	74	443	1	.99	1	.13	1.68	.24	.34	.2	46	1
0	120	3.4	196	1173	2	4.1	2	.37	3.96	.65	.91	.19	176	4
0	91	3.78	58	226	5032	4.58	12	.13	.01	.34	1.19	.3	45	0
0	39	1.5	34	195	80	.76	0	.1	.18	.04	.3	.01		0
0	41	1.3	34	244	45	.77	0	.13	.24	.05	.31	.01		0
0	138	1.38	33	149	10	.79	1	.06	.01	.05	.66	.06	55	<1
0	38	1.02	36	223	6	.64	0	.12	.25	.05	.37	.01		0
0	28	.76	22	156	22	.15	188	.02	.53	.05	.2	.05	54	8
0	122	3.21	78	419	63	.68	737	.09	.86	.21	.44	.22	131	9
0	139	1.44	66	283	82	.66	739	.06	.91	.16	.4	.14	103	12
0	136	2.46	81	370	29	.49	939	.02	1.39	.15	.41	.11	105	15
0	136	2.46	81	370	29	.49	939	.02	1.39	.15	.41	.11	105	15
0	24	.32	22	173	1	.35	26	.04	.11	.04	.46	.06	18	5
0	24	.32	22	173	1	.35	26	.04	.11	.04	.46	.08	18	5
0	12	.31	20	228	3	.16	22	.04	.11	.04	.38	.07	15	4
0	34	.81	20	1070	2	.64	872	.21	.29	.06	1.72	.18	69	23
0	29	.68	16	896	2	.53	730	.17	.25	.05	1.44	.15	57	20
0	54	1.14	53	538	5	.21	53	.2	.15	.02	1.08	.24	23	13
0	32	.68	32	321	4	.13	32	.12	.15	.01	.65	.14	14	8
0	42	.62	30	293	4	.13	721	.07	.17	.02	1	.13	20	16
0	42	.62	30	293	4	.13	721	.07	.17	.02	1	.13	20	16
0	23	.7	11	160	2	.14	401	.06	.16	.05	.56	.08	20	4
0	20	.56	26	430	10	.18	725	.09	.14	.06	.67	.21	19	11
0	12	.33	15	253	6	.12	473	.05	.14	.03	.39	.12	11	8
0	16	.26	8	90	14	.15	8	.03	.09	.02	.62	.08	6	3
0	13	.76	20	225	38	.38	20	.06	.31	.06	1.11	.08	28	5
0	32	.51	23	397	11	.33	2487	.08	.32	.14	.69	.27	26	28
0	32	.85	15	278	20	.41	2574	.08	.42	.21	.97	.37	17	26
8	27	1.19	12	198	73	.16	440	.02	3.99	.04	.41	.04	12	7
0	38	1.7	31	269	96	.27	1936	.03	.35	.11	1.22	.3	14	7
0	28	1.13	28	396	67	.23	1013	.05	.32	.07	.94	.24	21	33
0	18	.71	18	250	42	.14	638	.03	.2	.05	.59	.15	13	21
0	20	.57	11	482	3	.76	7	.05	1.4	.07	1.13	.16	4	26
0	204	2.14	70	854	74	.14	241	.06	3.7	.27	.66	.16	10	52
0	2	.21	7	91	<1	.07	4	.01	.13	.01	.63	.02	2	4
0	9	.82	26	354	1	.29	14	.06	.5	.05	2.44	.07	9	15

Table A–1

Food Composition (Computer code number is for Wadsworth Diet Analysis Plus program) (For purposes of calculations, use "0" for t, <1, <.1, <.01, etc.)

Computer Code Number	Food Description	Measure	Wt (g)	H_2O (%)	Ener (kcal)	Prot (g)	Carb (g)	Dietary Fiber (g)	Fat (g)	Fat Breakdown (g) Sat	Mono	Poly
	VEGETABLES and LEGUMES—Continued											
	Tomatoes:											
945	Raw, whole, 2 ⅗" diam	1 ea	123	94	26	1	6	1	<1	.1	.1	.2
946	Raw, chopped	1 c	180	94	38	2	8	2	1	.1	.1	.2
947	Cooked from raw	1 c	240	92	65	3	14	2	1	.1	.2	.4
948	Canned, solids and liquid:	1 c	240	94	46	2	10	2	<1	t	t	.1
5741	Unsalted	1 c	240	94	46	2	10	2	<1	t	t	.1
1879	Tomatoes, sundried:	1 c	54	15	139	8	30	7	2	.2	.3	.6
1881	Pieces	10 pce	20	15	52	3	11	2	1	.1	.1	.2
1885	Oil pack, drained	10 pce	30	54	64	2	7	2	4	.6	2.6	.6
2020	Tomato, raw	1 ea	123	94	26	1	6	1	<1	.1	.1	.2
949	Tomato juice, canned:	1 c	243	94	41	2	10	1	<1	t	t	.1
5397	Unsalted	1 c	243	94	41	2	10	2	<1	t	t	.1
	Tomato products, canned:											
950	Paste, no added salt	1 c	262	74	215	10	51	11	1	.2	.2	.6
951	Puree, no added salt	1 c	250	87	100	4	24	5	<1	.1	.1	.2
952	Sauce, with salt	1 c	245	89	73	3	18	3	<1	.1	.1	.2
953	Turnips, cubes, cooked from fresh	1 c	156	94	33	1	8	3	<1	t	t	.1
	Turnip greens, cooked:											
954	From fresh, leaves and stems	1 c	144	93	29	2	6	5	<1	.1	t	.1
955	From frozen, chopped	1 c	164	90	49	6	8	6	1	.2	t	.3
956	Vegetable juice cocktail, canned	1 c	242	93	46	2	11	2	<1	t	t	.1
	Vegetables, mixed:											
957	Canned, drained	½ c	81	87	38	2	7	2	<1	t	t	.1
958	Frozen, cooked, drained	½ c	91	83	54	3	12	4	<1	t	t	.1
1818	Water chestnuts, Chinese, raw	½ c	62	73	60	1	15	2	<1	t	t	t
	Water chestnuts, canned:											
959	Slices	½ c	70	86	35	1	9	2	<1	t	t	t
960	Whole	4 ea	28	86	14	<1	3	1	<1	t	t	t
1190	Watercress, fresh, chopped	½ c	17	95	2	<1	<1	<1	<1	t	t	t
	VEGETARIAN FOODS:											
7509	Bacon strips, meatless	3 ea	15	49	46	2	1	<1	4	.7	1.1	2.3
1511	Baked beans, canned	½ c	127	73	118	6	26	6	1	.1	t	.2
7526	Bakon crumbles	¼ c	7	16	28	2	1	<1	2			
7548	Chicken, breaded, fried, meatless	1 pce	57	70	97	6	3	3	7	1	2.9	2.5
7547	Chicken slices, meatless	2 ea	60	59	132	10	4	3	8	1.3	2	4.4
7557	Chili w/meat substitute	½ c	107	64	141	19	15	4	2	.3	.6	.9
7549	Fish stick, meatless	2 ea	57	45	165	13	5	3	10	1.6	2.5	5.4
7550	Frankfurter, meatless	1 ea	51	58	102	10	4	2	5	.8	1.2	2.7
7504	GardenBurger, patty	1 ea	45	53	87	5	13	3	1	.4	.3	.7
7505	GardenSausage, patty	1 ea	35	15	117	4	22	5	1	.7	.3	t
7551	Luncheon slice, meatless	1 sl	67	46	188	17	6	3	11	1.7	2.6	5.6
7560	Meatloaf, meatless	1 ea	71	58	142	15	6	3	6	1	1.5	3.3
1171	Nuteena	1 ea	55	58	162	6	6	2	13	5.1	5.8	1.7
7556	Pot pie, meatless	1 ea	227	59	524	15	41	5	34	9.5	12.6	9.8
7554	Soyburger, patty	1 ea	71	58	142	15	6	3	6	1	1.5	3.3
7562	Soyburger w/cheese, patty	1 ea	135	50	316	21	29	4	13	4.2	3.9	3.7
7564	Tempeh	1 c	166	55	330	31	28	9	13	1.9	2.9	7.2
7670	Vegan burger, patty	1 ea	78	71	75	11	6	4	<1	.1	.3	.2
	Vegetarian foods, Green Giant:											
7677	Breakfast links	3 ea	68	65	114	12	5	4	5	.7	1.2	3.1
7676	Breakfast patties	2 ea	57	65	95	10	5	3	4	.6	1	2.6
	Burger, harvest, patty:											
7673	Italian	1 ea	90	65	139	17	8	5	4	1.4	.3	.4
7674	Original	1 ea	90	65	137	18	8	5	4	1.3	.1	.4
7675	Southwestern	1 ea	90	65	135	16	9	5	4	1.4	.2	.4
	Vegetarian foods, Loma Linda											
7727	Chik nuggets, frozen	5 pce	85	47	245	12	13	5	16	2.5	4	8.8
7753	Chik-fried, frozen	1 pce	57	51	178	11	1	1	15	1.9	3.7	8.7
7744	Franks, big, canned	1 ea	51	59	110	10	2	2	7	1.1	1.7	3.8
7747	Linketts, canned	1 ea	35	60	72	7	1		4	.7	1.2	2.5
1173	Redi-burger, patty	1 ea	85	59	172	16	5		10	1.5	2.4	5.8
7755	Swiss steak w/gravy, canned	1 pce	92	71	120	9	8	4	6	.8	1.5	3.3

PAGE KEY: A–4 = Beverages A–6 = Dairy A–10 = Eggs A–10 = Fat/Oil A–14 = Fruit A–20 = Bakery A–28 = Grain A–32 = Fish A–34 = Meats A–38= Poultry A–40 = Sausage A–42 = Mixed/Fast A–46 = Nuts/Seeds A–50 = Sweets A–52 = Vegetables/Legumes A–62 = Vegetarian A–62 = Misc A–66 = Soups/Sauces A–68 = Fast A–84 = Convenience A–88 = Baby foods

Chol (mg)	Calc (mg)	Iron (mg)	Magn (mg)	Pota (mg)	Sodi (mg)	Zinc (mg)	VT-A (RE)	Thia (mg)	VT-E (α-TE)	Ribo (mg)	Niac (mg)	V-B6 (mg)	Fola (μg)	VT-C (mg)
0	6	.55	13	273	11	.11	76	.07	.47	.06	.77	.1	18	23
0	9	.81	20	400	16	.16	112	.11	.68	.09	1.13	.14	27	34
0	14	1.34	34	670	26	.26	178	.17	.91	.14	1.8	.23	31	55
0	72	1.32	29	530	355	.38	144	.11	.77	.07	1.76	.22	19	34
0	72	1.32	29	545	24	.38	144	.11	.91	.07	1.76	.22	19	34
0	59	4.91	105	1850	1131	1.07	47	.28	<.01	.26	4.89	.18	37	21
0	22	1.82	39	685	419	.4	17	.11	<.01	.1	1.81	.07	14	8
0	14	.8	24	470	80	.23	39	.06	.16	.11	1.09	.1	7	31
0	6	.55	13	273	11	.11	76	.07	.47	.06	.77	.1	18	23
0	22	1.41	27	535	877	.34	136	.11	2.21	.07	1.64	.27	48	44
0	22	1.41	27	535	24	.34	136	.11	2.21	.07	1.64	.27	48	44
0	92	5.08	134	2454	231	2.1	639	.41	11.3	.5	8.44	1	59	111
0	42	3.1	60	1065	85	.55	320	.18	6.3	.13	4.3	.38	27	26
0	34	1.89	47	909	1482	.61	240	.16	3.43	.14	2.82	.38	23	32
0	34	.34	12	211	78	.31	0	.04	.05	.04	.47	.1	14	18
0	197	1.15	32	292	42	.2	792	.06	2.48	.1	.59	.26	170	39
0	249	3.18	43	367	25	.67	1308	.09	4.79	.12	.77	.11	65	36
0	27	1.02	27	467	653	.48	283	.1	.77	.07	1.76	.34	51	67
0	22	.85	13	236	121	.33	944	.04	.49	.04	.47	.06	19	4
0	23	.75	20	154	32	.45	389	.06	.33	.11	.77	.07	17	3
0	7	.04	14	362	9	.31	0	.09	.74	.12	.62	.2	10	2
0	3	.61	3	83	6	.27	0	.01	.35	.02	.25	.11	4	1
0	1	.24	1	33	2	.11	0	<.01	.14	.01	.1	.04	2	<1
0	20	.03	4	56	7	.02	80	.01	.17	.02	.03	.02	2	7
0	3	.36	3	25	220	.06	1	.66	1.04	.07	1.13	.07	6	0
0	63	.37	41	376	504	1.78	22	.19	.67	.08	.54	.17	30	4
	8	.44	11	120	172	.25	0	.06		.02	.12	.02	7	0
0	13	.97	7	171	228	.37	0	.4	1.11	.27	2.68	.28	32	0
0	21	.78	10	198	474	.42	0	.66	1.61	.24	3.18	.42	46	0
0	53	4.24	36	362	527	1.26	78	.12	1.25	.07	1.21	.15	82	16
0	54	1.14	13	342	279	.8	0	.63	2.25	.51	6.84	.85	58	0
0	17	.92	9	76	219	.61	0	.56	.98	.61	8.16	.5	40	0
0	36	1.35		129	112		18	.05		.09		.06		<1
0	181	.33		307	78		3	.08		.2		.13		<1
0	27	1.54	15	188	576	1.07	0	.64	2.01	.37	7.37	.74	67	0
0	21	1.49	13	128	391	1.28	0	.64	1.23	.43	7.1	.85	55	0
0	9	.27	33	166	119	.46	0	.1		.35	1.04	.45	49	0
20	66	2.9	31	331	538	1.05	729	.65	4	.4	4.47	.31	40	10
0	21	1.49	13	128	391	1.28	0	.64	1.23	.43	7.1	.85	55	0
13	146	2.71	26	211	931	1.97	45	.77	1.43	.55	8.14	.86	69	1
0	154	3.75	116	609	10	3	115	.22	.03	.18	7.69	.5	86	0
0	79	2.66	15	398	351	.69	0	.23	.01	.51	3.78	.18	225	0
0	65	1.84			340	4.56	0	.18		.09	.27	.18		0
0	54	2			285	3.82	0	.15		.07	2.28	.15		0
0	74	2.61			374	6.93	3	.28		.14	4.05	.28		0
0	76	2.7			378	7.2	0	.29		.14	4.32	.29		0
0	71	2.52			371	6.66	3	.27		.13	4.05	.27		0
2	40	1.4		153	709	.43	0	.67		.3	2.89	.45		0
4	2	.63		76	503	.2	0	.98		.46	2.1	.35		0
2	8	.77		51	243	.89	0	.26		.46	1.98	.14		0
1	4	.39		29	160	.46	0	.13		.22	.64	.29		0
1	12	1.06	16	121	455	1.11	0	.14		.3	1.9	.51	21	0
2	24	.31		225	433	.41	0	1.25		.65	5.41	1		0

Table A-1

Food Composition (Computer code number is for Wadsworth Diet Analysis Plus program) (For purposes of calculations, use "0" for t, <1, <.1, <.01, etc.)

Computer Code Number	Food Description	Measure	Wt (g)	H_2O (%)	Ener (kcal)	Prot (g)	Carb (g)	Dietary Fiber (g)	Fat (g)	Fat Breakdown (g) Sat	Mono	Poly
	VEGETARIAN FOODS:—Continued											
1174	Vege-Burger, patty	1 ea	55	71	66	10	2	2	2	.4	.6	.5
	Vegetarian foods, Morningstar Farms:											
7672	Better-n-burgers, svg	1 ea	78	71	75	11	6	4	<1	.1	.3	.2
7766	Better-n-eggs	¼ c	57	88	23	5	<1	0	<1	.1	.1	.1
57436	Breakfast links	2 pce	45	60	63	8	2	2	2	.5	.7	1.3
7752	Breakfast strips	2 pce	16	43	56	2	2	<1	4	.7	1.1	2.6
7725	Burger crumbles, svg	1 ea	55	60	116	11	3	3	6	1.6	2.3	2.5
7726	Burger, spicy black bean	1 ea	78	60	113	11	15	5	1	.2	.3	.4
7665	Chik pattie	1 ea	71	51	177	7	15	2	10	1.3	2.6	5.9
7724	Frank, deli	1 ea	45	52	109	10	3	3	7	1	2.1	3.5
7722	Garden vege pattie	1 ea	67	60	104	11	9	4	4	.5	1.1	2.2
7746	Grillers	1 ea	64	55	139	14	5	3	7	1.7	2.2	3
7664	Prime pattie	1 ea	64	64	94	16	4	3	2	.2	.4	.6
	Vegetarian foods, Worthington:											
7634	Beef style, meatless, frzn	3 pce	55	58	113	9	4	3	7	1.2	2.7	2.6
7732	Burger, meatless, patty	¼ c	55	71	60	9	2	1	2	.3	.5	1.1
1846	Chik slices, canned	2 pce	60	78	62	6	1	1	4	.6	.9	2.3
1833	Chili, canned	½ c	106	73	136	9	10	4	7	1.1	1.7	4.1
1835	Choplets, slices, canned	2 pce	92	72	93	17	3	2	2	.9	.3	.3
7608	Corned beef style, meatless, frzn	4 pce	57	55	138	10	5	2	9	1.9	4.1	3.1
1831	Country stew, canned	1 c	240	81	208	13	20	5	9	1.6	2.3	4.8
7632	Egg roll, meatless, frzn	1 ea	85	53	181	6	20	2	8	1.7	4.5	2.3
1838	Numete, slices, canned	1 pce	55	58	132	6	5	3	10	2.4	4.4	2.7
1839	Prime steaks, slices, canned	1 pce	92	71	136	9	4	4	9	1.4	2.9	4.9
1840	Protose, slices, canned	1 pce	55	53	131	13	5	3	7	1	3	2.4
7606	Roast, dinner, meatless, frzn	1 ea	85	63	180	12	5	3	12	2.2	5	5.2
1842	Saucette links, canned	1 pce	38	62	86	6	1		6	1.1	1.6	3.8
1844	Savory slices, canned	1 pce	28	66	48	3	2	1	3	1.2	1.3	.6
7735	Steaklets, frzn	1 pce	71	58	145	12	6	2	8	1.4	2.7	3.9
1847	Turkee slices, canned	1 pce	33	64	68	5	1	1	5	.8	1.9	2.1
	MISCELLANEOUS											
	Baking powders for home use:											
	Sodium aluminum sulfate:											
962	With monocalcium phosphate monohydrate	1 tsp	5	2	6	<1	2	0	0	0	0	0
963	With monocalcium phosphate monohydrate, calcium sulfate	1 tsp	5	5	3	0	1	<1	0	0	0	0
964	Straight phosphate	1 tsp	5	4	3	<1	1	<1	0	0	0	0
965	Low sodium	1 tsp	5	6	5	<1	2	<1	<1	t	0	t
1204	Baking soda	1 tsp	5		0	0	0	0	0	0	0	0
966	Basil, dried	1 tbs	5	6	13	1	3	2	<1	t	t	.1
2068	Cajun seasoning	1 tsp	3	5	6	<1	1	<1	<1			
961	Carob flour	1 c	103	4	185	5	92	41	1	.1	.2	.2
967	Catsup:	¼ c	61	67	63	1	17	1	<1	t	t	.1
968	Tablespoon	1 tbs	15	67	16	<1	4	<1	<1	t	t	t
1200	Cayenne/red pepper	1 tbs	5	8	16	1	3	1	1	.2	.1	.4
969	Celery seed	1 tsp	2	6	8	<1	1	<1	<1	t	.3	.1
1203	Chili powder:	1 tbs	8	8	25	1	4	3	1	.2	.3	.6
970	Teaspoon	1 tsp	3	8	9	<1	2	1	<1	.1	.1	.2
	Chocolate:											
971	Baking, unsweetened, square	1 oz	28	1	146	3	8	4	15	9.1	5.2	.5
	For other chocolate items, see Sweeteners & Sweets											
972	Cilantro/coriander, fresh	1 tbs	1	93		<1	<1	<1	<1	0	t	0
2287	Cinnamon	1 tsp	2	10	5	<1	2	1	<1	t	t	t
1197	Cornstarch	1 tbs	8	8	30	<1	7	<1	<1	t	t	t
2239	Curry powder	1 tsp	2	10	6	<1	1	1	<1	t	.1	.1
1202	Dill weed, dried	1 tbs	3	7	8	1	2	<1	<1	t	.1	t
975	Garlic cloves	1 ea	3	59	4	<1	1	<1	<1	t	0	t
2238	Garlic powder	1 tsp	3	6	10	<1	2	<1	<1	t	t	t
977	Gelatin, dry, unsweetened: Envelope	1 ea	7	13	23	6	0	0	<1	t	t	t
978	Ginger root, slices, raw	2 pce	5	82	3	<1	1	<1	<1	t	t	t

Chol (mg)	Calc (mg)	Iron (mg)	Magn (mg)	Pota (mg)	Sodi (mg)	Zinc (mg)	VT-A (RE)	Thia (mg)	VT-E (α-TE)	Ribo (mg)	Niac (mg)	V-B6 (mg)	Fola (μg)	VT-C (mg)
0	8	.5	12	30	114	.58	0	.2		.25	.78	.31	15	0
0	79	2.66	15	398	351	.69	0	.23	.01	.51	3.78	.18	225	0
2	7	.63		68	90	.51	64	.01		.26	0	.11		0
1	15	2.14	16	59	338	.36	0	6.95		.22	5.19	.33	12	0
<1	7	.27		15	220	.05	0	.75		.04	.6	.07		0
0	40	3.2	1	89	238	.82	0	4.96	.34	.18	1.49	.27		0
1	56	1.84	44	269	499	.93	14	8.03	.36	.14	0	.21		0
1	11	1.02		163	536	.31	0	2.15		.16	1.51	.14		0
2	16	.26	4	50	524	.38	0	.14	1.26	.02	0	.01		0
1	34	.72	29	180	382	.59	20	6.47	.98	.1	0	0	29	0
2	43	1.16		127	256	.49	0	11.7		.24	2.99	.37		0
1	46	2.14		142	247	.74	0	.51		.25	.92	.41		2
0	4	2.63		44	624	.22	0	.89		.34	6.46	.56		0
0	4	1.73		25	269	.38	0	.13		.1	1.96	.24		0
1	9	.73		111	257	.26	0	.06		.05	.37	.08		0
0	20	1.49		195	523	.57	0	.02		.03	1.04	.31		0
0	6	.37		40	500	.65	0	.05		.05	0	.05		0
1	6	1.17		58	524	.26	0	10.6		.07	1.36	.3		0
2	51	5.09		270	826	1.03	216	1.85		.29	4.22	.86		0
1	15	.57		96	384	.31	0	1.22		.19	0	.03		0
0	10	1.12		155	272	.56	0	.08		.06	.54	.2		0
2	12	.38		82	445	.38	0	.12		.13	1.98	.38		0
<1	1	1.84		50	283	.7	0	.18		.13	1.34	.24		0
2	36	2.87		38	566	.64	0	2.13		.25	6.02	.6		0
1	9	1.15		25	205	.26	0	.59		.08	.09	.13		0
<1		.47		14	179	.08	0	.08		.06	.48	.1		0
2	49	.99		95	484	.5	0	1.51		.12	3.1	.26		0
1	3	.47		16	203	.11	0	1.13		.05	.39	.09		0
0	97	0		7	547	0	0	0	0	0	0	0	0	0
0	294	.55	1	1	530	<.01	0	0	0	0	0	0	0	0
0	368	.56	2		395	<.01	0	0	0	0	0	0	0	0
0	217	.41	1	505	4	.04	0	0	<.01	0	0	0	0	0
0	0	0	0	0	1368	0	0	0	0	0	0	0	0	0
0	106	2.1	21	172	2	.29	47	.01	.08	.02	.35	.06	14	3
				29	474									
0	358	3.03	56	852	36	.95	1	.05	.65	.47	1.96	.38	30	<1
0	12	.43	13	293	723	.14	62	.05	.9	.04	.84	.11	9	9
0	3	.1	3	72	178	.03	15	.01	.22	.01	.21	.03	2	2
0	7	.39	8	101	1	.12	208	.02	.24	.05	.43	.1	5	4
0	35	.9	9	28	3	.14	<1	.01	.02	.01	.06	.01	<1	<1
0	22	1.14	14	153	81	.22	279	.03	.08	.06	.63	.15	8	5
0	8	.43	5	57	30	.08	105	.01	.03	.02	.24	.06	3	2
0	21	1.77	87	233	4	1.12	3	.02	.34	.05	.31	.03	2	0
0	1	.02		5	<1	<.01	3	<.01	.02	<.01	.01	<.01	<1	<1
0	25	.76	1	10	1	.04	1	<.01	0	<.01	.03	<.01	1	1
0		.04			1	<.01	0	0	0	0	0	0	0	0
0	10	.59	5	31	1	.08	2	<.01	.01	.01	.07	.01	3	<1
0	53	1.46	13	99	6	.1	18	.01		.01	.08	.04		1
0	5	.05	1	12	1	.03	0	.01	0	<.01	.02	.04	<1	1
0	2	.08	2	33	1	.08	0	.01	0	<.01	.02	.08	<1	1
0	4	.08	2	1	14	.01	0	<.01	0	.02	.01	0	2	0
0	1	.02	2	21	1	.02	0	<.01	.01	<.01	.03	.01	1	<1

Table A-1

Food Composition (Computer code number is for Wadsworth Diet Analysis Plus program) (For purposes of calculations, use "0" for t, <1, <.1, <.01, etc.)

Computer Code Number	Food Description	Measure	Wt (g)	H_2O (%)	Ener (kcal)	Prot (g)	Carb (g)	Dietary Fiber (g)	Fat (g)	Fat Breakdown (g) Sat	Mono	Poly
	MISCELLANEOUS—Continued											
1198	Horseradish, prepared	1 tbs	15	85	7	<1	2	<1	<1	t	t	.1
1997	Hummous/hummus	1 c	246	65	421	12	50	12	21	3.1	8.8	7.8
1909	Mustard, country dijon	1 tsp	5		5	<1	<1	0	0	0	0	0
2019	Mustard, gai choy Chinese	1 tbs	16	94	3	<1	1		<1			
979	Mustard, prepared (1 packet = 1 tsp)	1 tsp	5	80	4	<1	<1	<1	<1	t	.2	t
	Miso (see #926 under Vegetables and Legumes, Soybean products)											
980	Olives, green	5 ea	20	78	23	<1	<1	<1	3	.3	1.9	.2
981	Olives, ripe, pitted	5 ea	22	80	25	<1	1	1	2	.3	1.7	.2
26008	Onion powder	1 tsp	2	5	7	<1	2	<1	<1	t	t	t
2237	Oregano, ground	1 tsp	2	7	6	<1	1	1	<1	.1	t	.1
2236	Paprika	1 tsp	2	10	6	<1	1	<1	<1	t	t	.2
887	Parsley, freeze dried	¼ c	1	2	3	<1	<1	<1	<1	t	t	t
	Parsley, fresh (see #885 and #886)											
985	Pepper, black	1 tsp	2	10	5	<1	1	1	<1	t	t	t
	Pickles:											
986	Dill, medium, 3¾ x 1¼" diam	1 ea	65	92	12	<1	3	1	<1	t	t	t
987	Fresh pack, slices, 1½" diam x ¼"	2 pce	15	79	11	<1	3	<1	<1	0	0	t
988	Sweet, medium	1 ea	35	65	41	<1	11	<1	<1	t	t	t
989	Pickle relish, sweet	1 tbs	15	63	21	<1	5	<1	<1	t	t	t
	Popcorn (see Grain Products #539-541)											
917	Potato chips:	10 pce	20	2	107	1	11	1	7	2.2	2	2.4
44076	Unsalted	1 oz	28	2	150	2	15	1	10	3.1	2.8	3.4
1201	Sage, ground	1 tsp	1	8	3	<1	1	<1	<1	.1	t	t
1347	Salsa, from recipe	1 tbs	15	93	3	<1	1	<1	<1	t	t	t
2218	Salsa, pico de gallo, medium	1 tbs	15	92	2	0	1	<1	0	0	0	0
990	Salt	1 tsp	6		0	0	0	0	0	0	0	0
	Salt Substitutes:											
1205	Morton, salt substitute	1 tsp	6			0	<1		0	0	0	0
1207	Morton, light salt	1 tsp	6			0	<1		0	0	0	0
2067	Seasoned salt, no MSG	1 tsp	5	5	4	<1	1	<1	<1			
991	Vinegar, cider	½ c	120	94	17	0	7	0	0	0	0	0
2172	Balsamic	1 tbs	15	64	21	0	4	0	0	0	0	0
2176	Malt	1 tbs	15	90	5	0	<1	0	0	0	0	0
2182	Tarragon	1 tbs	15	95	3	0	<1	0	0	0	0	0
2181	White wine	1 tbs	15	89	5	0	<1	0	0	0	0	0
	Yeast:											
992	Baker's, dry, active, package	1 ea	7	8	21	3	3	1	<1	t	.2	t
993	Brewer's, dry	1 tbs	8	5	23	3	3	3	<1	t	t	0
	SOUPS, SAUCES, and GRAVIES											
	SOUPS, canned, condensed:											
	Unprepared, condensed:											
1210	Cream of celery	1 c	251	85	181	3	18	2	11	2.8	2.6	5
1215	Cream of chicken	1 c	251	82	233	7	18	<1	15	4.2	6.5	3
1216	Cream of mushroom	1 c	251	81	259	4	19	1	19	5.1	3.6	8.9
1220	Onion	1 c	246	86	113	8	16	2	3	.5	1.5	1.3
	Prepared w/equal volume of whole milk:											
994	Clam chowder, New England	1 c	248	85	164	9	17	1	7	2.9	2.3	1.1
1209	Cream of celery	1 c	248	86	164	6	14	1	10	3.9	2.5	2.6
995	Cream of chicken	1 c	248	85	191	8	15	<1	11	4.6	4.5	1.6
996	Cream of mushroom	1 c	248	85	203	6	15	<1	14	5.1	3	4.6
1214	Cream of potato	1 c	248	87	149	6	17	<1	6	3.8	1.7	.6
1213	Oyster stew	1 c	245	89	135	6	10	0	8	5	2.1	.3
997	Tomato	1 c	248	85	161	6	22	3	6	2.9	1.6	1.1
	Prepared with equal volume of water:											
998	Bean with bacon	1 c	253	84	172	8	23	9	6	1.5	2.2	1.8
999	Beef broth/bouillon/consommé	1 c	240	98	17	3	<1	0	1	.3	.2	t
1000	Beef noodle	1 c	244	92	83	5	9	1	3	1.1	1.2	.5
1001	Chicken noodle	1 c	241	92	75	4	9	1	2	.7	1.1	.6
1002	Chicken rice	1 c	241	94	60	4	7	1	2	.5	.9	.4
1208	Chili beef	1 c	250	85	170	7	21	9	7	3.3	2.8	.3
1003	Clam chowder, Manhattan	1 c	244	92	78	2	12	1	2	.4	.4	1.3

PAGE KEY: A–4 = Beverages A–6 = Dairy A–10 = Eggs A–10 = Fat/Oil A–14 = Fruit A–20 = Bakery A–28 = Grain A–32 = Fish A–34 = Meats A–38= Poultry A–40 = Sausage A–42 = Mixed/Fast A–46 = Nuts/Seeds A–50 = Sweets A–52 = Vegetables/Legumes A–62 = Vegetarian A–62 = Misc A–66 = Soups/Sauces A–68 = Fast A–84 = Convenience A–88 = Baby foods

Chol (mg)	Calc (mg)	Iron (mg)	Magn (mg)	Pota (mg)	Sodi (mg)	Zinc (mg)	VT-A (RE)	Thia (mg)	VT-E (α-TE)	Ribo (mg)	Niac (mg)	V-B6 (mg)	Fola (μg)	VT-C (mg)
0	8	.06	4	37	47	.12	<1	<.01	<.01	<.01	.06	.01	9	4
0	123	3.86	71	428	600	2.71	5	.23	2.46	.13	1.01	.98	146	19
0				10	120									
0	4	.1	2	6	63	.03	0	0	.09	0	0	<.01	0	0
0	12	.32	4	11	480	.01	6	0	.6	0	0	<.01	<1	0
0	19	.73	1	2	192	.05	9	<.01	.66	0	.01	<.01	0	<1
0	7	.05	2	19	1	.05	0	.01	<.01	<.01	.01	.03	3	<1
0	31	.88	5	33	<1	.09	14	.01	.03	.01	.12	.02	5	1
0	4	.47	4	47	1	.08	121	.01	.01	.03	.31	.04	2	1
0	2	.54	4	63	4	.06	63	.01	.06	.02	.1	.01	15	1
0	9	.58	4	25	1	.03	<1	<.01	.02	<.01	.02	.01	<1	<1
0	6	.34	7	75	833	.09	21	.01	.1	.02	.04	.01	1	1
0	5	.27	1	30	101	0	2	0	.02	<.01	0	<.01	0	1
0	1	.21	1	11	329	.03	5	<.01	.06	.01	.06	<.01	<1	<1
0	3	.12	1	30	107	.01	1	0	.02	<.01	0	0	0	1
0	5	.33	13	255	119	.22	0	.03	.98	.04	.77	.13	9	6
0	7	.46	19	357	2	.3	0	.05	1.37	.05	1.07	.18	13	9
0	16	.28	4	11	<1	.05	6	.01	.02	<.01	.06	.01	3	<1
0	1	.06	1	24	58	.02	22	.01	.04	<.01	.06	.01	2	5
0					130									
0	1	.02			2325	.01	0	0	0	0	0	0	0	0
	33			3018	<1									
	2		4	1560	1170									
				15	1542									
0	7	.72	26	120	1	0	0	0	0	0	0	0	0	0
	2	.07		10	3		<1	.07		.07	.07			<1
	2	.07		13	4		<1	.07		.07	.07			1
		.07		2	1		<1	.07		.07	.07			<1
	1	.07		12	1		<1	.07		.07	.07			<1
0	4	1.16	7	140	3	.45	<1	.16	.01	.38	2.79	.11	164	<1
0	17	1.38	18	151	10	.63	0	1.25		.34	3.03	.4	313	0
28	80	1.26	13	246	1900	.3	60	.06	.38	.1	.66	.02	5	<1
20	68	1.2	5	176	1972	1.26	113	.06	.33	.12	1.64	.03	3	<1
3	65	1.05	10	168	1736	1.19	0	.06	2.61	.17	1.62	.02	8	2
0	54	1.35	5	138	2115	1.23	0	.07	.57	.05	1.21	.1	30	2
22	186	1.49	22	300	992	.8	40	.07	.15	.24	1.03	.13	10	3
32	186	.69	22	310	1009	.2	67	.07	.97	.25	.44	.06	8	1
27	181	.67	17	273	1046	.67	94	.07	.24	.26	.92	.07	8	1
20	179	.59	20	270	918	.64	37	.08	1.34	.28	.91	.06	10	2
22	166	.55	17	322	1061	.67	67	.08	.1	.24	.64	.09	9	1
32	167	1.05	20	235	1041	10.3	44	.07	.49	.23	.34	.06	10	4
17	159	1.81	22	449	744	.29	109	.13	2.6	.25	1.52	.16	21	68
3	81	2.05	45	402	951	1.03	89	.09	.08	.03	.57	.04	32	2
0	14	.41	5	130	782	0	0	<.01	0	.05	1.87	.02	5	0
5	15	1.1	5	100	952	1.54	63	.07	<.01	.06	1.07	.04	19	<1
7	17	.77	5	55	1106	.39	72	.05	.07	.06	1.39	.03	22	<1
7	17	.75	0	101	815	.26	65	.02	.05	.02	1.13	.02	1	<1
12	42	2.13	30	525	1035	1.4	150	.06	.17	.07	1.07	.16	17	4
2	27	1.63	12	188	578	.98	98	.03	.73	.04	.82	.1	10	4

Table A–1

Food Composition (Computer code number is for Wadsworth Diet Analysis Plus program) (For purposes of calculations, use "0" for t, <1, <.1, <.01, etc.)

Computer Code Number	Food Description	Measure	Wt (g)	H_2O (%)	Ener (kcal)	Prot (g)	Carb (g)	Dietary Fiber (g)	Fat (g)	Fat Breakdown (g) Sat	Mono	Poly
	SOUPS, SAUCES, and GRAVIES—Continued											
1004	Cream of chicken	1 c	244	91	117	3	9	<1	7	2.1	3.3	1.5
1005	Cream of mushroom	1 c	244	90	129	2	9	<1	9	2.4	1.7	4.2
1006	Minestrone	1 c	241	91	82	4	11	1	3	.6	.7	1.1
1211	Onion	1 c	241	93	58	4	8	1	2	.3	.7	.7
1007	Split pea & ham	1 c	253	82	190	10	28	2	4	1.8	1.8	.6
1008	Tomato	1 c	244	90	85	2	17	<1	2	.4	.4	1
1009	Vegetable beef	1 c	244	92	78	6	10	<1	2	.9	.8	.1
1010	Vegetarian vegetable	1 c	241	92	72	2	12	<1	2	.3	.8	.7
	Ready to serve:											
1707	Chunky chicken soup	1 c	251	84	178	13	17	2	7	2	3	1.4
	SOUPS, dehydrated:											
	Prepared with water:											
1299	Beef broth/bouillon	1 c	244	97	19	1	2	0	1	.3	.3	t
1376	Chicken broth	1 c	244	97	22	1	1	0	1	.3	.4	.4
1013	Chicken noodle	1 c	252	94	53	3	7	1	1	.3	.5	.4
1122	Cream of chicken	1 c	261	91	107	2	13	<1	5	3.4	1.2	.4
1014	Onion	1 c	246	96	27	1	5	1	1	.1	.3	.1
1217	Split pea	1 c	255	87	125	7	21	3	1	.4	.7	.3
1015	Tomato vegetable	1 c	253	93	56	2	10	<1	1	.4	.3	.1
	Unprepared, dry products:											
1011	Beef bouillon, packet	1 ea	6	3	14	1	1	0	1	.3	.2	t
1012	Onion soup, packet	1 ea	39	4	115	5	21	4	2	.5	1.4	.3
	SAUCES											
	From dry mixes, prepared with milk:											
1016	Cheese sauce	1 c	279	77	307	17	23	1	17	9.3	5.3	1.6
1017	Hollandaise	1 c	259	84	240	5	14	<1	20	11.6	5.9	.9
1018	White sauce	1 c	264	81	240	10	21	<1	13	6.4	4.7	1.7
	From home recipe:											
1206	Lowfat cheese sauce	¼ c	61	73	85	6	4	<1	5	2.1	1.9	.9
1019	White sauce, medium	¼ c	72	77	102	2	6	<1	8	2.3	3.2	2
	Ready to serve:											
2202	Alfredo sauce, reduced fat	¼ c	69		170	5	16	0	10	6		
1020	Barbeque sauce	1 tbs	16	81	12	<1	2	<1	<1	t	.1	.1
1706	Chili sauce, tomato base	1 tbs	17	68	18	<1	4	<1	<1	t	t	t
2126	Creole sauce	¼ c	62	89	25	1	4	1	1	.1	.2	.3
2124	Hoisin sauce	1 tbs	17	47	35	<1	7	0	1	0		
2199	Pesto sauce	2 tbs	16		83	2	1	0	8	1.8	5.4	.7
1021	Soy sauce	1 tbs	16	71	8	1	1	<1	<1	t	t	t
2123	Szechuan sauce	1 tbs	16	71	21	<1	3	<1	1	.1	.3	.4
1380	Teriyaki sauce	1 tbs	18	68	15	1	3	<1	0	0	0	0
	Spaghetti sauce, canned:											
1377	Plain	1 c	249	75	271	5	40	8	12	1.7	6.1	3.3
1378	With meat	1 c	250	74	300	8	37	8	14	2.7	7	3.2
1379	With mushrooms	½ c	123	84	108	2	13	1	3	.4	1.5	.8
	GRAVIES											
	Canned:											
1022	Beef	1 c	233	87	123	9	11	1	5	2.7	2.2	.2
1023	Chicken	1 c	238	85	188	5	13	1	14	3.4	6.1	3.6
1024	Mushroom	1 c	238	89	119	3	13	1	6	1	2.8	2.4
1025	From dry mix, brown	1 c	258	92	75	2	13	<1	2	.8	.7	.1
1026	From dry mix, chicken	1 c	260	91	83	3	14	<1	2	.5	.9	.4
	FAST FOOD RESTAURANTS											
	ARBY'S											
1402	Bac'n cheddar deluxe	1 ea	231	59	512	21	39	<1	31	8.7	12.7	10.1
	Roast beef sandwiches:											
1403	Regular	1 ea	155	47	383	22	35	1	18	7	8	3.5
1404	Junior	1 ea	89	48	233	11	23	<1	11	4.1	5.2	2.5
1405	Super	1 ea	254	58	552	24	54	1	28	7.6	12.2	8.4
1407	Beef 'n cheddar	1 ea	194	50	508	25	43		26	7.7	12	6.8
1408	Chicken breast sandwich	1 ea	204	52	445	22	52	1	22	3	9.7	10.1
1412	Ham'n cheese sandwich	1 ea	169	54	355	25	34	<1	14	5.1	5.8	3.8
1726	Italian sub sandwich	1 ea	297	58	671	34	47		39	12.8	15.7	8.5

PAGE KEY: A–4 = Beverages A–6 = Dairy A–10 = Eggs A–10 = Fat/Oil A–14 = Fruit A–20 = Bakery A–28 = Grain A–32 = Fish A–34 = Meats A–38= Poultry A–40 = Sausage A–42 = Mixed/Fast A–46 = Nuts/Seeds A–50 = Sweets A–52 = Vegetables/Legumes A–62 = Vegetarian A–62 = Misc A–66 = Soups/Sauces A–68 = Fast A–84 = Convenience A–88 = Baby foods

Chol (mg)	Calc (mg)	Iron (mg)	Magn (mg)	Pota (mg)	Sodi (mg)	Zinc (mg)	VT-A (RE)	Thia (mg)	VT-E (α-TE)	Ribo (mg)	Niac (mg)	V-B6 (mg)	Fola (μg)	VT-C (mg)
10	34	.61	2	88	986	.63	56	.03	.2	.06	.82	.02	2	<1
2	46	.51	5	100	881	.59	0	.05	1.24	.09	.72	.01	5	1
2	34	.92	7	313	911	.73	234	.05	.07	.04	.94	.1	36	1
0	26	.67	2	67	1053	.61	0	.03	.29	.02	.6	.05	15	1
8	23	2.28	48	400	1006	1.32	45	.15	.15	.08	1.47	.07	3	2
0	12	1.76	7	264	695	.24	68	.09	2.49	.05	1.42	.11	15	66
5	17	1.12	5	173	791	1.54	190	.04	.32	.05	1.03	.08	10	2
0	22	1.08	7	210	822	.46	301	.05	.79	.05	.92	.05	11	1
30	25	1.73	8	176	889	1	131	.08	.18	.17	4.42	.05	5	1
0	10	.02	7	37	1361	.07	<1	<.01	.02	.02	.36	0	0	0
0	15	.07	5	24	1483	.01	12	.01	.02	.03	.19	0	2	0
3	33	.5	8	30	1282	.2	5	.07	.02	.06	.88	.01	18	<1
3	76	.26	5	214	1184	1.57	123	.1	.15	.2	2.61	.05	5	1
0	12	.15	5	64	849	.06	<1	.03	.1	.06	.48	0	1	<1
3	20	.94	43	224	1147	.56	5	.21	.13	.14	1.26	.05	39	0
0	8	.63	20	104	1146	.17	20	.06	.81	.05	.79	.05	10	6
1	4	.06	3	27	1018	0	<1	<.01	.01	.01	.27	.01	2	0
2	55	.58	25	260	3493	.23	1	.11	.42	.24	1.99	.04	6	1
53	569	.28	47	552	1565	.97	117	.15	.33	.56	.32	.14	13	2
52	124	.9	8	124	1564	.7	220	.04	.26	.18	.06	.5	22	<1
34	425	.26	264	444	797	.55	92	.08	1.58	.45	.53	.07	16	3
11	166	.25	10	100	389	.73	58	.03	.55	.14	.16	.03	4	<1
8	75	.21	9	100	82	.26	89	.05	.98	.12	.28	.03	4	1
30	150	0	8	80	600		80	0		.1	0			0
0	3	.14	3	28	130	.03	14	<.01	.18	<.01	.14	.01	1	1
0	3	.14	2	63	227	.05	24	.01	.05	.01	.27	.02	1	3
0	35	.31	9	187	339	.1	24	.03	.61	.02	.53	.07	9	0
0	0	0			250		0							0
4	64	.09	6	15	129	.29	39	.01		.03	0	.02	<1	0
0	3	.32	5	29	914	.06	0	.01	0	.02	.54	.03	2	0
0	2	.12	2	13	218	.02	10	<.01	.07	<.01	.1	.01	1	<1
0	4	.31	11	40	690	.02	0	<.01	0	.01	.23	.02	4	0
0	70	1.62	60	956	1235	.52	306	.14	4.98	.15	3.76	.88	54	28
15	68	1.94	60	952	1179	1.37	577	.13	5.91	.17	4.51	.87	52	26
0	15	1	15	332	494	.34	241	.08	1.35	.08	.93	.16	12	9
7	14	1.63	5	189	1304	2.33	0	.07	.15	.08	1.54	.02	5	0
5	48	1.12	5	259	1373	1.9	264	.04	.37	.1	1.05	.02	5	0
0	17	1.57	5	252	1356	1.67	0	.08	.19	.15	1.6	.05	29	0
3	67	.23	10	57	1075	.31	0	.04	.05	.08	.81	0	0	0
3	39	.26	10	62	1133	.32	0	.05	.05	.15	.78	.03	3	3
38	110	4.32		491	1094	3	40	.34		.46	9.6			11
43	60	4.86	16	422	936	3.75	0	.28		.48	11	.2	14	1
22	40	2.7	8	201	519	1.5		.18		.25	6.6	.1	7	
43	90	6.48	25	533	1174	3.75	30	.39		.58	12.4	.3	21	9
52	150	6.12		321	1166	3		.42		.63	9.8			1
45	60	2.88	30	330	1019	.15		.22		.54	9	.38	18	5
55	170	2.7	31	382	1400	.9	40	.82		.37	7.8	.31	26	24
69	410	4.32		565	2062		100	.91		.49	8.2			11

A

Table A–1

Food Composition (Computer code number is for Wadsworth Diet Analysis Plus program) (For purposes of calculations, use "0" for t, <1, <.1, <.01, etc.)

Computer Code Number	Food Description	Measure	Wt (g)	H_2O (%)	Ener (kcal)	Prot (g)	Carb (g)	Dietary Fiber (g)	Fat (g)	Fat Breakdown (g) Sat	Mono	Poly
	FAST FOOD RESTAURANTS											
	ARBY'S—Continued											
1413	Turkey sandwich, deluxe	1 ea	195	69	260	20	33	<1	6	1.6	2.3	2.4
1680	Turkey sub sandwich	1 ea	277	62	486	33	46		19	5.3	6	7
	Milkshakes:											
1419	Chocolate	1 ea	340	74	451	10	76	<1	12	2.8	7	1.7
1420	Jamocha	1 ea	326	75	368	9	59	0	10	2.5	6.4	1.6
1421	Vanilla	1 ea	312	77	330	10	46	0	11	3.9	5.3	2.3
1728	Salad, roast chicken	1 ea	400	88	204	24	12		7	3.3	.9	.9
1729	Sports drink, Upper Ten	1 ea	358	88	169	0	42		0	0	0	0
	Source: Arby's											
	BURGER KING											
1423	Croissant sandwich, egg, sausage & cheese	1 ea	176	46	600	22	25	1	46	16		
	Whopper sandwiches:											
1425	Whopper	1 ea	270	58	640	27	45	3	39	11		
1426	Whopper with cheese	1 ea	294	57	730	33	46	3	46	16		
	Sandwiches:											
1629	BK broiler chicken sandwich	1 ea	248	59	550	30	41	2	29	6		
1432	Cheeseburger	1 ea	138	48	380	23	28	1	19	9		
1434	Chicken sandwich	1 ea	229	45	710	26	54	2	43	9		
1427	Double beef	1 ea	351	57	870	46	45	3	56	19		
1428	Double beef & cheese	1 ea	375	56	960	52	46	3	63	24		
1433	Double cheeseburger with bacon	1 ea	218	48	640	44	28	1	39	18		
1431	Hamburger	1 ea	126	48	330	20	28	1	15	6		
1437	Ocean catch fish fillet	1 ea	255	51	700	26	56	3	41	6		
1435	Chicken tenders	1 ea	88	50	230	16	14	2	12	3		
1439	French fries (salted)	1 svg	116	40	370	5	43	3	20	5		
1630	French toast sticks	1 svg	141	33	500	4	60	1	27	7		
1440	Onion rings	1 svg	124	51	310	4	41	6	14	2	8	4
1441	Milk shakes, chocolate	1 ea	284	75	320	9	54	3	7	4		
1442	Milk shakes, vanilla	1 ea	284	75	300	9	53	1	6	4		
1443	Fried apple pie	1 ea	113	47	300	3	39	2	15	3		
	Source: Burger King Corporation											
	CHICK-FIL-A											
	Sandwiches:											
69153	Chargrilled chicken	1 ea	150	54	280	27	36	1	3	1		
69152	Chicken	1 ea	167	61	290	24	29	1	9	2		
69155	Chicken salad	1 ea	167	55	320	25	42	1	5	2		
69154	Chicken salad club	1 ea	232	62	390	33	38	2	12	5		
	Salads:											
52139	Carrot and raisin	1 ea	76	53	150	5	28	2	2	0		
52136	Chicken plate	1 ea	468	85	290	21	40	6	5	0		
52134	Chicken garden, charbroiled	1 ea	397	89	170	26	10	5	3	1		
52135	Chick-n-strips	1 ea	451	86	290	32	21	5	9	2		
52138	Cole slaw	1 ea	79	70	130	6	11	1	6	1		
52137	Tossed salad	1 ea	130	85	70	5	13	1	0	0	0	0
15263	Chicken nuggets, svg	1 ea	110	51	290	28	12	60	12	3		
15262	Chicken-n-strips, svg	1 ea	119	59	230	29	10	0	8	2		
50885	Hearty breast of chicken soup, svg	1 ea	215	86	110	16	10	1	1	0		
7973	Waffle potato fries, svg	1 ea	85	28	290	1	49	0	10	4		
46489	Cheesecake, svg	1 ea	88	52	270	13	7	0	21	9		
49134	Fudge nut brownie, svg	1 ea	88	8	416	12	49	0	19	3.6		
20601	Icedream, svg	1 ea	127	74	140	11	16	0	4	1		
48214	Lemon pie, svg	1 ea	99	56	280	1	19	0	22	6		
	Source: Chick-Fil-A											

PAGE KEY: A–4 = Beverages A–6 = Dairy A–10 = Eggs A–10 = Fat/Oil A–14 = Fruit A–20 = Bakery A–28 = Grain A–32 = Fish A–34 = Meats A–38= Poultry A–40 = Sausage A–42 = Mixed/Fast A–46 = Nuts/Seeds A–50 = Sweets A–52 = Vegetables/Legumes A–62 = Vegetarian A–62 = Misc A–66 = Soups/Sauces A–68 = Fast A–84 = Convenience A–88 = Baby foods

Chol (mg)	Calc (mg)	Iron (mg)	Magn (mg)	Pota (mg)	Sodi (mg)	Zinc (mg)	VT-A (RE)	Thia (mg)	VT-E (α-TE)	Ribo (mg)	Niac (mg)	V-B6 (mg)	Fola (μg)	VT-C (mg)
33	130	3.42	30	353	1262	1.5	40	.08		.41	15.4	.52	20	12
51	400	4.68		500	2033		20	13.2		.54	18.8			
36	250	.72	48	410	341	1.5	60	.12		.68	.8	.14	14	5
35	250	2.7	36	525	262	1.5	60	.12		.68	.8	.14	14	2
32	300	2.7	36	686	281	1.5	60	.12		.68	4	.14	37	2
43	170	1.98		877	508		485	.33		.54	5.6			51
0				0	40									
260	150	3.6			1140		80							0
90	80	4.5			870		100	.33		.41	7	.35		9
115	250	4.5			1350		150	.34		.48	7	.33		9
80	60	5.4			480		60							6
65	100	2.7			770		60							0
60	100	3.6			1400		0							0
170	80	7.2			940		100	.34		.56	10			9
195	250	7.2			1420		150	.35		.63	10			9
145	200	4.5			1240		80	.31		.42	6			0
55	40	1.8			530		20	.28		.31	4.89			0
90	60	2.7			980		20							1
35	0	.72			530		0							0
0	0	1.08			240		0							4
0	60	2.7			490		0							0
0	100	1.44			810		0							0
20	200	1.8			230		60	.13		.55	.13			0
20	300	0			230		60	.11		.57	.13			4
0	0	1.44			230		0							6
40					640									
50					870									
10					810									
70					980									
6					650									
35					570									
25					650									
20					430									
15					430									
0					0									
14					770									
20					380									
45					760									
5					960									
10					510									
36					773									
40					240									
5					550									

Table A-1

Food Composition (Computer code number is for Wadsworth Diet Analysis Plus program) (For purposes of calculations, use "0" for t, <1, <.1, <.01, etc.)

Computer Code Number	Food Description	Measure	Wt (g)	H_2O (%)	Ener (kcal)	Prot (g)	Carb (g)	Dietary Fiber (g)	Fat (g)	Fat Breakdown (g) Sat	Mono	Poly
	FAST FOOD RESTAURANTS—Continued											
	DAIRY QUEEN											
	Ice cream cones:											
1446	Small vanilla	1 ea	142	63	230	6	38	0	7	4.5		
1447	Regular vanilla	1 ea	213	64	350	8	57	0	10	7		
1448	Large vanilla	1 ea	253	65	410	10	65	0	12	8		
1450	Chocolate dipped	1 ea	234	59	510	9	63	1	25	13		
1453	Chocolate sundae	1 ea	241	62	410	8	73	0	10	6		
1455	Banana split	1 ea	369	67	510	8	96	3	12	8		
1456	Peanut buster parfait	1 ea	305	51	730	16	99	2	31	17		
1457	Hot fudge brownie delight	1 ea	305	52	710	11	102	1	29	14	12	2
1459	Buster bar	1 ea	149	45	450	10	41	2	28	12		
1645	Breeze, strawberry, regular	1 ea	383	70	460	13	99	1	1	1	0	0
1460	Dilly bar	1 ea	85	55	210	3	21	0	13	7	3	3
1461	DQ ice cream sandwich	1 ea	61	46	150	3	24	1	5	2		
1463	Milk shakes, regular	1 ea	397	71	520	12	88	<1	14	8	2	2
1464	Milk shakes, large	1 ea	461	71	600	13	101	<1	16	10	2	2
1466	Milk shakes, malted	1 ea	418	68	610	13	106	<1	14	8	2	2
1470	Misty slush, small	1 ea	454	88	220	0	56	0	0	0	0	0
2250	Starkiss	1 ea	85	75	80	0	21	0	0	0	0	0
	Yogurt:											
1641	Yogurt cone, regular	1 ea	213	66	280	9	59	0	1	.5		
1643	Yogurt sundae, strawberry	1 ea	255	69	300	9	66	1	<1	.5	0	0
	Sandwiches:											
1481	Cheeseburger, double	1 ea	219	55	540	35	30	2	31	16		
1480	Cheeseburger, single	1 ea	152	55	340	20	29	2	17	8		
1474	Chicken	1 ea	191	56	430	24	37	2	20	4		
1647	Chicken fillet, grilled	1 ea	184	64	310	24	30	3	10	2.5		
1475	Fish fillet sandwich	1 ea	170	57	370	16	39	2	16	3.5		
1476	Fish fillet with cheese	1 ea	184	56	420	19	40	2	21	6	7	8
1477	Hamburger, single	1 ea	128	56	269	16	27	2	11	4.6	5.6	.9
1478	Hamburger, double	1 ea	212	62	440	30	29	2	22	10		
	Hotdog:											
1483	Regular	1 ea	99	57	240	9	19	1	14	5		
1484	With cheese	1 ea	113	55	290	12	20	1	18	8	8	2
1485	With chili	1 ea	128	61	280	12	21	2	16	6		
1489	French fries, small	1 ea	71	39	210	3	29	3	10	2	5	3
1490	French fries, large	1 ea	128	40	390	5	52	6	18	4	8	6
1491	Onion rings	1 ea	85	46	240	4	29	2	12	2.5		
	Source: International Dairy Queen											
	HARDEES											
1734	Frisco burger hamburger	1 ea	242		760	36	43		50	18		
1736	Frisco grilled chicken salad	1 ea	278		120	18	2		4	1		
1737	Peach shake	1 ea	345		390	10	77		4	3		
	Source: Hardees											
	JACK IN THE BOX											
	Breakfast items:											
1492	Breakfast Jack sandwich	1 ea	121	49	300	18	30	0	12	5	5	2.5
1494	Sausage crescent	1 ea	156	39	580	22	28	0	43	16		
1495	Supreme crescent	1 ea	153	40	530	23	34	0	33	10	18.9	7.8
1496	Pancake platter	1 ea	231	45	610	15	87	0	22	9	7.6	3.5
1497	Scrambled egg platter	1 ea	213	52	560	18	50	0	32	9	16.6	4.4
	Sandwiches:											
1654	Bacon cheeseburger	1 ea	242	49	710	35	41	0	45	15	15.7	8.7
1499	Cheeseburger	1 ea	110	41	330	16	32	0	15	6	5.9	2.3
1739	Chicken caesar pita sandwich	1 ea	237	59	520	27	44	4	26	6		
1655	Chicken sandwich	1 ea	160	52	400	20	38	0	18	4		
1656	Chicken sandwich, sourdough ranch	1 ea	225		490	29	45	1	21	6		
1505	Chicken supreme	1 ea	245	55	620	25	48	0	36	11	14.8	11.4
1583	Double cheeseburger	1 ea	152	44	450	24	35	0	24	12	11.6	3.1

PAGE KEY: A–4 = Beverages A–6 = Dairy A–10 = Eggs A–10 = Fat/Oil A–14 = Fruit A–20 = Bakery A–28 = Grain A–32 = Fish A–34 = Meats A–38= Poultry A–40 = Sausage A–42 = Mixed/Fast A–46 = Nuts/Seeds A–50 = Sweets A–52 = Vegetables/Legumes A–62 = Vegetarian A–62 = Misc A–66 = Soups/Sauces A–68 = Fast A–84 = Convenience A–88 = Baby foods

Chol (mg)	Calc (mg)	Iron (mg)	Magn (mg)	Pota (mg)	Sodi (mg)	Zinc (mg)	VT-A (RE)	Thia (mg)	VT-E (α-TE)	Ribo (mg)	Niac (mg)	V-B6 (mg)	Fola (μg)	VT-C (mg)
20	200	1.08		250	115		122	.05		.28				
30	300	1.8		390	170		150	.09		.38	.16	.13		2
40	350	1.8		451	200		200	.11		.4	.2			2
30	300	1.8		435	200		150	.09		.38	.16	.13		2
30	250	1.44		394	210		150	.08		.35	.4	.19		0
30	250	1.8		860	180		200	.15		.25	.4	.2		15
35	300	1.8		660	400		150	.15		.51	3	.22		1
35	300	5.4		510	340		80	.15		.68	.3	.18		1
15	150	1.08		400	280		80	.09		.17	3	.08		0
10	450	2.7		530	270		0	.13		.73				9
10	100	.36		170	75		60	.03		.14		.06		0
5	60	.72		105	115		40	.03		.25	.4	.05		0
45	400	1.44		570	230		80	.12		.59	.8	.19		<1
50	450	1.44		660	260		200	.15		.68	.8			<1
45	400	1.44		570	230		80	.12		.59	.8	.19		<1
0	0	0			20		0							0
0	0	0			10		0							0
5	300	1.8		285	170		0	.09		.38				2
5	300	1.8		352	180		0	.09		.49				6
115	250	4.5		426	1130		150	.29		.49	6.78			4
55	150	3.6		263	850		60	.29		.33	3.89			4
55	40	1.8		350	760		0	.37		.34	11			0
50	200	2.7		330	1040		0	.3		1.02	12			0
45	40	1.8		280	630		0	.3		.22	3			0
60	100	1.8		290	850		80	.3		.25	5			0
42	56	2.5		234	584		37	.27		.23	3.6			3
90	60	4.5		444	680		60	.32		.45	7.49			6
25	60	1.8		170	730		20	.22		.14	2			4
40	150	1.8		180	950		60	.22		.17	2			4
35	60	1.8		262	870		80	.23		.14	3			4
0	0	.72		430	115		0	.09		.03	2			5
0	0	1.44		780	200		0	.15		.07	3			9
0	0	1.08		90	135		0	.09		.05	.4			0
70					1280									
60					520									
25					290									
185	200	2.7		220	890		80	.47		.41	3			9
185	150	2.7		260	1010		100	.6		.51	4.6			0
210	150	3.6		270	930		150	.65		.54	4.2			12
100	100	1.8		310	890		80	.03		.85	7			6
380	150	4.5		450	1060		150			.66	5			9
110	250	5.4		540	1240		80	.24		.48	8.8	.39		9
35	200	2.7		200	510		60	.23		.23	3			1
55	250	2.7		490	1050		80							2
45	150	1.8		180	1290		40							0
65	150	1.8		340	1060									0
75	200	2.7		190	1520		100	.39		.32	11			2
75	250	3.6		320	900		100	.15		.34	6			0

Table A–1

Food Composition (Computer code number is for Wadsworth Diet Analysis Plus program) (For purposes of calculations, use "0" for t, <1, <.1, <.01, etc.)

Computer Code Number	Food Description	Measure	Wt (g)	H_2O (%)	Ener (kcal)	Prot (g)	Carb (g)	Dietary Fiber (g)	Fat (g)	Fat Breakdown (g) Sat	Mono	Poly
	FAST FOOD RESTAURANTS—Continued											
	JACK IN THE BOX—Continued											
1651	Grilled sourdough burger	1 ea	223	48	670	32	39	0	43	16	17.8	7.9
1498	Hamburger	1 ea	97	42	280	13	31	0	11	4	4.9	2
1500	Jumbo Jack burger	1 ea	229	55	560	26	41	0	32	10	13	8
1501	Jumbo Jack burger with cheese	1 ea	242	55	610	29	41	0	36	12	15	9
1740	Monterey roast beef sandwich	1 ea	238	57	540	30	40	3	30	9		
1508	Tacos, regular	1 ea	78	57	190	7	15	2	11	4		
1509	Tacos, super	1 ea	126	59	280	12	22	3	17	6		
	Teriyaki bowl:											
1679	Beef	1 ea	440	62	640	28	124	7	3	1		
1668	Chicken	1 ea	440	62	580	28	115	6	1			
1516	French fries	1 ea	109	38	350	4	45	4	17	4		
1517	Hash browns	1 ea	57	53	160	1	14	1	11	2.5	6.8	.3
1518	Onion rings	1 ea	103	34	380	5	38	0	23	6	15.2	.9
	Milkshakes:											
1519	Chocolate	1 ea	322	72	390	9	74	0	6	3.5	2.1	
1520	Strawberry	1 ea	298	74	330	9	60	0	7	4	2	
1521	Vanilla	1 ea	304	73	350	9	62	0	7	4	1.8	
1522	Apple turnover	1 ea	110	34	350	3	48	0	19	4	10.6	1.5
	Source: Jack in the Box Restaurant, Inc											
	KENTUCKY FRIED CHICKEN											
	Rotisserie gold:											
1472	Dark qtr, no skin	1 ea	117	66	217	27	0	0	12	3.5		
1473	Dark qtr, w/skin	1 ea	146	62	333	30	1		24	6.6		
1513	White qtr with wing, w/skin	1 ea	176	65	335	40	1		19	5.4		
1525	White qtr with wing, no skin	1 ea	117	63	199	37	0	0	6	1.7		
	Original Recipe:											
1253	Center breast	1 ea	103	52	260	25	9	<1	14	3.8	7.8	2
1251	Side breast	1 ea	83	47	245	18	9	<1	15	4.2	8.8	2.2
1250	Drumstick	1 ea	57	54	152	14	3	<1	8	2.2	4.1	1.3
1252	Thigh	1 ea	95	49	287	18	8	<1	21	5.3	9.4	3.1
1249	Wing	1 ea	53	45	172	12	5	<1	12	3	6	1.8
	Hot & spicy:											
1451	Center breast	1 ea	125	48	360	28	13		22	5		
1452	Side breast	1 ea	120	43	400	22	16		28	6		
1430	Thigh	1 ea	119	47	370	24	10		27	6		
1471	Wing	1 ea	61	38	220	14	5		16	4		
	Extra crispy recipe:											
1261	Center breast	1 ea	118	48	330	26	14	<1	20	4.8	10.8	2.1
1259	Side breast	1 ea	116	40	400	21	19	<1	27	5.5	12.9	2.3
1258	Drumstick	1 ea	65	49	190	14	6	<1	12	3.4	7.7	1.7
1260	Thigh	1 ea	109	43	380	23	7	<1	30	7.7	16	4.2
1257	Wing	1 ea	59	34	240	13	8	<1	17	4.4	10.7	2.5
1390	Baked beans	½ c	167	70	200	8	36	6	3	1.5	.7	.4
1526	Breadstick	1 ea	33	30	110	3	17	0	3	0		
1388	Buttermilk biscuit	1 ea	65	28	234	5	28	<1	13	3.4	6.2	2.3
1391	Chicken little sandwich	1 ea	47	35	169	6	14	<1	10	2	4.7	3.4
1269	Coleslaw	1 svg	90	75	114	1	13	<1	6	1	1.7	3.4
1527	Cornbread	1 ea	56	26	228	3	25	1	13	2		
1268	Corn-on-the-cob	1 ea	151	70	222	4	27	8	12	2	1	1.5
1429	Chicken, hot wings	1 svg	135	38	471	27	18		33			
1386	Kentucky fries	1 svg	77	42	228	3	26	3	12	3.2		
1381	Kentucky nuggets	6 ea	95	48	284	16	15	<1	18	4		
1534	Macaroni & cheese	1 svg	114	71	162	7	15	0	8	3		
1387	Mashed potatoes & gravy	1 svg	120	80	103	1	16	<1	5	.4	.5	.2
1530	Pasta salad	1 svg	108	78	135	2	14	1	8	1		
1389	Potato salad	½ c	188	74	271	5	27	3	16	3	4.2	7.3
1383	Potato wedges	1 svg	92	59	192	3	25	3	9	3		
1535	Red beans & rice	1 svg	112	76	114	4	18	3	3	1		
1529	Vegetable medley salad	1 ea	114	77	126	1	21	3	4	1		
	Source: Kentucky Fried Chicken Corp											

Chol (mg)	Calc (mg)	Iron (mg)	Magn (mg)	Pota (mg)	Sodi (mg)	Zinc (mg)	VT-A (RE)	Thia (mg)	VT-E (α-TE)	Ribo (mg)	Niac (mg)	V-B6 (mg)	Fola (μg)	VT-C (mg)
110	200	4.5		510	1140		150	.65		.48	8	.33		6
25	100	2.7		190	430		20	.15		.26	2			1
65	100	4.5		450	700		40	.36		.29	1.8			6
80	200	5.4		460	780		60	.36		.44	1.6			6
75	300	3.6		500	1270		80							5
20	100	1.08	35	240	410	1.2	0	.07		.17	1	.13		0
30	150	1.8	45	370	720	1.8	0	.12		.08	1.4	.18		2
25	150	4.5		430	930		1000							6
30	100	1.8		380	1220		1100							9
0	0	1.08		690	190		0	.18		.03	3.8			24
0	0	.36		190	310		0	.05			1			6
0	20	1.8		130	450		0	.29		.17	2.6			2
25	300	.72		680	210		0	.15		.6	.4			0
30	300	0		550	180		0	.15		.43	.4			0
30	300	0		570	180		0	.15		.34	.4			0
0	0	1.8		80	460		0	.2		.12	1.8			9
128	10	.18			772		15							1
163	10	.18			980		15							1
157	10	.18			1104		15							1
97	10	.18			667		15							1
92	30	.72			609		15	.09		.17	11.5			
78	68	1.2			604		15	.06		.13	6.9			
75	21	1.1			269		15	.05		.12	3.2			
112	40	1.08			591		31	.08		.3	5.5			
59	30	.54			383		15	.03		.08	3.7			
80	20	.72			750		15							6
80	40	1.08			850		15							6
100	20	1.08			670		15							6
65	20	.72			440		30							
75	33	.8			740		15	.11		.13	13.1			
75	20	.72			710		15	.09		.1	8.5			
65	20	.36			310		30	.06		.12	3.7			
90	49	1.2			520		30	.1		.21	6.5			
65	20	.36			320		30			.04	.06	3.3		
5	61	2.17	44	348	812	1.96	38	.09		.06	.76	.11	49	3
0	30	.18			15		0							0
3	43	1.92			565		28	.26		.2	2.77			
18	23	1.7			331		5	.16		.12	2.2			
5	30	.36			177		32	.03		.03	.2			27
42	60	.72			194		10							
0	0	.36			76		20	.14		.11	1.8			2
150	40	3.24			1230		15							6
4	11	.98			535		0							0
66	2	.1			865		15	.02		.02	1	.05		<1
16	120	.72			531		190							0
<1	20	.4			388		15			.04	1.2			
1	20	1.08			663		110							7
16	16	3.25	23	385	636	.44	120	.1		.03	.9	.29	11	
3					428		0							
4	10	.72			315									
0	20	.36			240		375							5

Table A-1

Food Composition (Computer code number is for Wadsworth Diet Analysis Plus program) (For purposes of calculations, use "0" for t, <1, <.1, <.01, etc.)

Computer Code Number	Food Description	Measure	Wt (g)	H_2O (%)	Ener (kcal)	Prot (g)	Carb (g)	Dietary Fiber (g)	Fat (g)	Fat Breakdown (g) Sat	Mono	Poly
	FAST FOOD RESTAURANTS—Continued											
	LONG JOHN SILVER'S											
1528	Chicken plank dinner, 3 piece	1 ea	399	56	890	32	101		44	9.5	24.8	9.4
1531	Clam chowder	1 ea	198	86	140	11	10	1	6	1.8	2.5	1.7
1532	Clam dinner	1 ea	361	46	990	24	114		52	10.9	31.4	9.9
	Fish, batter fried:											
1523	Fish & fryes (fries), 3 piece	1 ea	384	54	980	31	92		50	11.3	28.4	9.7
1524	Fish & fryes, 2 piece	1 ea	261	54	610	27	52		37	7.9	23.5	5.3
2240	Fish and lemon crumb dinner, 3 piece	1 ea	493	71	610	39	86		13	2.2	3.9	5.3
2241	Fish and lemon crumb dinner, 2 piece	1 ea	334	77	330	24	46		5	.9	1.6	1.2
1533	Fish & chicken dinner	1 ea	431	55	950	36	102		49	10.6	28.8	9.5
1537	Shrimp dinner, batter fried	1 ea	331	54	840	18	88		47	9.7	27.2	9.1
	Salads:											
1541	Cole slaw	1 ea	98	70	140	1	20	1	6	1	1.5	3.5
1539	Ocean chef salad	1 ea	234	89	110	12	13	2	1	.4	.4	.2
1540	Seafood salad	1 ea	278	79	380	15	12	2	31	5.1	8.2	17.5
1542	Fryes (fries) serving	1 ea	85	43	250	3	28	1	15	2.5	7.4	5.1
1543	Hush puppies	1 ea	24	40	70	2	10	<1	2	.4	1.3	.2
	Source: Long John Silver's, Lexington KY											
	McDONALD'S											
	Sandwiches:											
1221	Big mac	1 ea	216	53	510	25	46	3	26	9.3	7.5	4.1
1226	Cheeseburger	1 ea	122	46	318	15	36	2	13	5.6	3.8	1.1
1224	Filet-o-fish	1 ea	145	49	364	14	41	1	16	3.7	3.8	5.6
1225	Hamburger	1 ea	108	49	266	12	35	2	9	3.2	2.8	.9
1444	McChicken	1 ea	189	52	491	17	42	2	29	5.4	8.5	10.2
1591	McLean deluxe	1 ea	214	64	345	23	37	2	12	4.4	3.6	1.2
1438	McLean deluxe with cheese	1 ea	228	63	398	26	38	2	16	6.8	4.6	1.3
1222	Quarter-pounder	1 ea	171	52	415	23	36	2	20	7.8	6.7	1.3
1223	Quarter-pounder with cheese	1 ea	199	50	520	28	37	2	29	12.6	8.7	1.6
1227	French fries, small serving	1 ea	68	40	207	3	26	2	10	1.7	3.1	2.5
1228	Chicken McNuggets	4 pce	73	51	198	12	10	0	12	2.5	3.7	2.4
	Sauces (packet):											
1229	Hot mustard	1 ea	30	60	63	1	7	1	4	.5	1.1	2
1230	Barbecue	1 ea	32	58	53	<1	12	<1	<1	.1	.1	.2
1231	Sweet & sour	1 ea	32	57	55	<1	12	<1	<1	.1	.1	.3
	Low-fat (frozen yogurt) milk shakes:											
1232	Chocolate	1 ea	295		348	13	62	1	6	3.5	.1	.7
1233	Strawberry	1 ea	294		343	12	63	<1	5	3.4	.1	.6
1234	Vanilla	1 ea	293		308	12	54	<1	5	3.3	.1	.6
	Low-fat (frozen yogurt) sundaes:											
1237	Hot caramel	1 ea	182	56	307	7	62	1	3	2	.3	1
1235	Hot fudge	1 ea	179	60	293	8	53	2	5	4.7	.1	.4
1267	Strawberry	1 ea	178	65	239	6	51	1	1	.7	.1	.2
1238	Vanilla	1 ea	90	68	118	4	24	<1	1	.5	.2	t
1241	Cookies, McDonaldland	1 ea	56	3	258	4	41	1	9	1.7	6.4	.8
1242	Cookies, chocolaty chip	1 ea	56	3	282	3	36	1	14	3.9	4.4	1
1240	Muffin, apple bran, fat-free	1 ea	75	39	182	4	40	1	1	.2	.1	.3
1239	Pie, apple	1 ea	84	35	289	3	37	1	14	3.7	4.5	.3
	Breakfast items:											
1243	English muffin with spread	1 ea	63	33	189	5	30	2	6	2.4	1.5	1.3
1244	Egg McMuffin	1 ea	137	57	289	17	27	1	13	.7	4.5	1.6
1245	Hotcakes with marg & syrup	1 ea	222	44	557	8	100	2	14	2.4	4.6	5.8
1246	Scrambled eggs	1 ea	102	73	170	13	1	0	12	3.6	5.3	1.7
1247	Pork sausage	1 ea	43	45	173	6	<1	0	16	5.5	6.4	2.1
1248	Hashbrown potatoes	1 ea	53	55	130	1	13	1	8	1.3	2.3	1.9
1392	Sausage McMuffin	1 ea	112	42	361	13	26	1	23	8.3	8.2	2.8
1393	Sausage McMuffin with egg	1 ea	163	52	443	19	27	1	29	10	10.7	3.6
1394	Biscuit with biscuit spread	1 ea	76	32	260	4	32	1	13	3.8	3.7	.8

PAGE KEY: A–4 = Beverages A–6 = Dairy A–10 = Eggs A–10 = Fat/Oil A–14 = Fruit A–20 = Bakery A–28 = Grain A–32 = Fish A–34 = Meats A–38= Poultry A–40 = Sausage A–42 = Mixed/Fast A–46 = Nuts/Seeds A–50 = Sweets A–52 = Vegetables/Legumes A–62 = Vegetarian A–62 = Misc A–66 = Soups/Sauces A–68 = Fast A–84 = Convenience A–88 = Baby foods

Chol (mg)	Calc (mg)	Iron (mg)	Magn (mg)	Pota (mg)	Sodi (mg)	Zinc (mg)	VT-A (RE)	Thia (mg)	VT-E (α-TE)	Ribo (mg)	Niac (mg)	V-B6 (mg)	Fola (μg)	VT-C (mg)
55	200	4.5		1170	2000	3	40	.52		.51	16			9
20	200	1.8		380	590	.6	150	.09		.25	2			
75	200	4.5		910	1830	3	40	.75		.42	12			12
70	200	4.5		1120	1530	3	40	.45		.42	8			15
60	40	1.8		900	1480	1.2		.37		.34	8			9
125	200	5.4		990	1420	2.25	700	.75		.59	24			6
75	80	1.8		440	640	.9	1000	.3		.25	14			18
75	200	4.5		1280	2090	3	40	.6		.59	14			9
100	200	3.6		840	1630	3	40	.45		.42	9			9
15	60	.72		190	260	.6	40	.06		.07	2			
40	100	3.6		95	730	.3	500	.12		.14	3			21
55	150	4.5		130	980	.9	200	.15		.25	3			21
0	200	.72		370	500	.3	0	.09			1.6			6
	40	.72		65	25	.3		.06		.03	.8			
76	202	4.32	46	456	932	4.81	66	.49	1.01	.44	6.08	.25	49	3
42	134	2.73	27	281	766	2.62	64	.33	.46	.31	3.81	.15	24	2
37	123	1.85	32	266	708	.7	21	.32	1.52	.23	2.58	.07	30	0
28	126	2.73	24	260	531	2.25	22	.33	.23	.26	3.81	.14	21	2
52	128	2.5	32	319	797	1.06	29	.91	6.16	.24	7.74	.38	37	1
59	131	4.29	40	537	811	4.9	74	.42	.63	.34	7.16	.29	44	8
73	139	4.29	43	559	1046	5.26	115	.42	.85	.39	7.16	.3	47	8
70	127	4.33	33	405	692	4.66	33	.39	.36	.32	6.78	.24	27	3
97	143	4.5			1160		115	.39	.81	.43	6.78	.26	33	3
0	9	.53	26	469	135	.32	0	.05	.83	0	1.94	.24	26	8
42	9	.65	17	210	353	.69	0	.08	.96	.11	5.15	.21		0
3	7	.78		29	85		4	.01		.01	.15			0
0	4	0		51	277		0	.01		.01	.17			4
0	2	.16		7	158		74	0		.01	.08			0
24	372	1.04		543	241		46	.12		.51	.4	.1		3
24	366	.29		542	170		46	.12		.51	.4	.11		3
24	360	.29		533	193		45	.12		.51	.31			3
7	246	.15		344	197		18	.09		.34	.27			2
5	258	.59		441	190		7	.09		.34	.29			2
5	221	.25		325	115		6	.06		.34	.25			2
3	132	.23		175	84		4	<.01		.01	.23			1
0	10	1.73	11	62	267	.38	0	.24	.99	.16	2.01	.03		0
3	28	1.78	24	142	229	.4	0	.14	.92	.16	1.48			0
0	34	1.29	13	77	215	.33	0	.14	0	.14	1.32	.03	5	1
0	17	1.23	7	69	221	.23		.19	1.5	.12	1.55	.04	9	27
13	103	1.59	13	69	386	.42	33	.25	.13	.31	2.61	.04	57	1
234	151	2.44	24	199	730	1.56	100	.49	.85	.45	3.33	.15	33	2
11	108	1.98	27	285	746	.53	119	.24	1.2	.26	1.86	.09	<1	<1
424	50	1.19	10	126	143	1.06	168	.07	.92	.51	.06	.12	44	0
33	7	.5	7	102	292	.78	0	.18	.26	.06	1.7	.09		0
0	7	.3	11	213	332	.15	0	.08	.58	.02	.9	.08	8	3
46	132	2.07	22	191	751	1.51	48	.56	.66	.27	3.76	.14	16	0
257	156	2.8	26	251	821	2.07	117	.59	1.11	.49	3.79	.19	33	0
0	68	1.85	9	105	836	.3	2	.29	.81	.23	2.23	.03	5	0

Table A–1

Food Composition (Computer code number is for Wadsworth Diet Analysis Plus program) (For purposes of calculations, use "0" for t, <1, <.1, <.01, etc.)

Computer Code Number	Food Description	Measure	Wt (g)	H_2O (%)	Ener (kcal)	Prot (g)	Carb (g)	Dietary Fiber (g)	Fat (g)	Fat Breakdown (g) Sat	Mono	Poly
	FAST FOOD RESTAURANTS—Continued											
	McDONALD'S—Continued											
1395	Biscuit with sausage	1 ea	119	37	433	10	32	1	29	8.6	10.1	2.8
1396	Biscuit with sausage & egg	1 ea	170	48	518	16	33	1	35	10.5	12.7	3.7
1397	Biscuit with bacon, egg, cheese	1 ea	152	46	450	17	33	1	27	8.7	8.9	2.3
	Salads:											
1398	Chef salad	1 ea	313	86	206	19	9	3	11	4.2	3	1.2
1400	Garden salad	1 ea	234	92	84	6	7	3	4	1.1	1.4	.7
1401	Chunky chicken salad	1 ea	296	87	164	23	8	3	5	1.3	1.6	1
	Source: McDonald's Corporation											
	PIZZA HUT											
	Pan pizza:											
1657	Cheese	2 pce	216	51	522	24	56	4	22	10	6.8	3.4
1658	Pepperoni	2 pce	208	49	531	22	56	4	24	8	9.9	3.7
1659	Supreme	2 pce	273	56	622	30	56	6	30	12	12	4.2
1660	Super supreme	2 pce	286	56	645	30	56	6	34	12		
	Thin 'n crispy pizza:											
1649	Cheese	2 pce	174	52	411	22	42	4	16	8	4.4	2.3
1623	Pepperoni	2 pce	168	48	431	22	42	2	20	8		
1622	Supreme	2 pce	232	57	514	28	42	4	26	10		
1620	Super supreme	2 pce	247	57	541	28	44	4	28	12		
	Hand tossed pizza:											
1619	Cheese	2 pce	216	53	470	26	58	4	14	7.9		
1618	Pepperoni	2 pce	208	51	477	24	58	4	16	8		
1648	Supreme	2 pce	273	56	568	32	60	6	24	10		
1617	Super supreme	2 pce	286	57	591	32	60	6	26	10		
	Personal pan pizza:											
1610	Pepperoni	1 ea	255	50	637	27	69	5	28	10	11.8	4.5
1609	Supreme	1 ea	327	57	721	33	70	6	34	12	14.7	5.6
	Source: Pizza Hut											
	SUBWAY											
	Deli style sandwich:											
69104	Bologna	1 ea	171	64	292	10	38	2	12	4		
69102	Ham	1 ea	171	69	234	11	37	2	4	1		
69103	Roast beef	1 ea	180	69	245	13	38	2	4	1		
69105	Seafood and crab:	1 ea	178	66	298	12	37	2	11	2		
69106	With light mayo	1 ea	178	68	256	12	37	2	7	2		
69108	Tuna:	1 ea	178		354	11	37	2	18	3		
69107	With light mayo	1 ea	178	67	279	11	38	2	9	2		
69101	Turkey	1 ea	180	69	235	12	38	2	4	1		
	Sandwiches, 6 inch:											
	B.L.T.:											
69135	On white bread	1 ea	191	67	311	14	38	3	10	3		
69136	On wheat bread	1 ea	198	65	327	14	44	3	10	3		
	Chicken taco sub:											
69131	On white bread	1 ea	286	70	421	24	43	3	16	5		
69132	On wheat bread	1 ea	293	69	436	25	49	4	16	5		
	Club :											
69117	On white bread	1 ea	246	73	297	21	40	3	5	1		
69118	On wheat bread	1 ea	253	71	312	21	46	3	5	1		
	Cold cut trio:											
69113	On white bread	1 ea	246	71	362	19	39	3	13	4		
69114	On wheat bread	1 ea	253	68	378	20	46	3	13	4		
	Ham:											
69115	On white bread	1 ea	232	73	287	18	39	3	5	1		
69115	On wheat bread	1 ea	239	71	302	19	45	3	5	1		

PAGE KEY: A–4 = Beverages A–6 = Dairy A–10 = Eggs A–10 = Fat/Oil A–14 = Fruit A–20 = Bakery A–28 = Grain A–32 = Fish A–34 = Meats A–38= Poultry A–40 = Sausage A–42 = Mixed/Fast A–46 = Nuts/Seeds A–50 = Sweets A–52 = Vegetables/Legumes A–62 = Vegetarian A–62 = Misc A–66 = Soups/Sauces A–68 = Fast A–84 = Convenience A–88 = Baby foods

Chol (mg)	Calc (mg)	Iron (mg)	Magn (mg)	Pota (mg)	Sodi (mg)	Zinc (mg)	VT-A (RE)	Thia (mg)	VT-E (α-TE)	Ribo (mg)	Niac (mg)	V-B6 (mg)	Fola (μg)	VT-C (mg)
33	75	2.35	15	207	1128	1.08	2	.48	1.07	.29	3.93	.12	5	0
245	100	2.95	20	271	1199	1.61	59	.51	1.53	.55	3.96	.18	27	0
238	103	2.6	20	245	1315	1.64	99	.39	1.49	.57	3.32	.13	30	0
179	157	1.81	40	605	727	2.16	1179	.33	1.45	.37	4.32	.36	100	22
139	52	1.34	24	407	61	.73	1114	.12	.95	.24	.65	.16	96	22
76	54	1.62	44	673	318	1.52	1973	.51	1.28	.21	8.46	.52	83	30
50	288	3	63	337	1002	4.32	211	.6		.64	5.48	.18		7
48	206	3.21	55	399	1140	4.14	190	.62		.48	5.31	.16	0	8
60	234	4.6	81	620	1529	6	195	.86		.84	6.4	.33		11
68	236	4.39	80	592	1649	5.99	201	.83		.73	7.13			12
50	291	2.06	56	307	1070	4.23	217	.46		.46	5.65	.18		6
50	208	2.2	51	330	1255	4.02	199	.48		.49	5.97			7
62	238	3.6	79	631	1591	5.4	197	.7		.57	6.27			12
70	238	3.41	73	563	1762	5.47	208	.72		.53	6.57			9
50	284	3	71	388	1242	4.6	198	.48		.48	5.3			10
48	202	3.21	84	610	1380	6.01	187	.72		.56	7.59			13
60	232	4.6	87	589	1769	5.48	192	.82		.66	8.45			14
68	232	4.39	89	607	1889	5.65	198	.84		.68	8.71			14
55	250	4	60	406	1338	3.8	233	.56		.66	8.16	.2		10
66	276	5.19	74	603	1757	4.69	240	.73		.82	9.91	.4		14
20	39	3			744		113							14
14	24	3			773		113							14
13	23	3			638		113							14
17	24	3			544		113							14
16	24	3			556		118							14
18	26	3			557		116							14
16	26	3			583		126							14
12	26	3			944		113							14
16	27	3			945		120							15
16	33	3			957		120							15
52	118	4			1264		209							18
52	124	4			1275		209							18
26	29	4			1341		120							15
26	35	4			1352		120							15
64	49	4			1401		130							16
64	55	4			1412		130							16
28	28	3			1308		120							15
28	35	3			1319		120							15

Table A-1

Food Composition (Computer code number is for Wadsworth Diet Analysis Plus program) (For purposes of calculations, use "0" for t, <1, <.1, <.01, etc.)

Computer Code Number	Food Description	Measure	Wt (g)	H_2O (%)	Ener (kcal)	Prot (g)	Carb (g)	Dietary Fiber (g)	Fat (g)	Fat Breakdown (g) Sat	Mono	Poly
	FAST FOOD RESTAURANTS—Continued											
	SUBWAY—Continued											
	Italian B.M.T.											
69139	On white bread	1 ea	246	66	445	21	39	3	21	8		
69140	On wheat bread	1 ea	253	64	460	21	45	3	22	7		
	Meatball:											
69129	On white bread	1 ea	260	70	404	18	44	3	16	6		
69130	On wheat bread	1 ea	267	67	419	19	51	3	16	6		
	Melt with turkey, ham, bacon, cheese:											
69127	On white bread	1 ea	251	70	366	22	40	3	12	5		
69128	On wheat bread	1 ea	258	68	382	23	46	3	12	5		
	Pizza sub:											
69133	On white bread	1 ea	250	66	448	19	41	3	22	9		
69134	On wheat bread	1 ea	257	65	464	19	48	3	22	9		
	Roast beef:											
69121	On white bread	1 ea	232	72	288	19	39	3	5	1		
69122	On wheat bread	1 ea	239	70	303	20	45	3	5	1		
	Roasted chicken breast:											
69125	On white bread	1 ea	246	70	332	26	41	3	6	1		
69126	On wheat bread	1 ea	253	68	348	27	47	3	6	1		
	Seafood and crab:											
69145	On white bread:	1 ea	246	69	415	19	38	3	19	3		
69147	With light mayo	1 ea	246	72	332	19	39	3	10	2		
69146	On wheat bread:	1 ea	253	67	430	20	44	3	19	3		
69148	With light mayo	1 ea	253	70	347	20	45	3	10	2		
	Spicy italian:											
69123	On white bread	1 ea	232	64	467	20	38	3	24	9		
69124	On wheat bread	1 ea	239	62	482	21	44	3	25	9		
	Steak and cheese:											
69119	On white bread	1 ea	257	68	383	29	41	3	10	6		
69120	On wheat bread	1 ea	264	67	398	30	47	3	10	6		
	Tuna:											
69141	On white bread:	1 ea	246	62	527	18	38	3	32	5		
69143	With light mayo	1 ea	246	70	376	18	39	3	15	2		
69142	On wheat bread:	1 ea	253	62	542	19	44	3	32	5		
69144	With light mayo	1 ea	253	68	391	19	46	3	15	2		
	Turkey:											
69111	On white bread	1 ea	232	73	273	17	40	3	4	1		
69112	On wheat bread	1 ea	239	71	289	18	46	3	4	1		
	Turkey breast and ham:											
69137	On white bread	1 ea	232	73	280	18	39	3	5	1		
69138	On wheat bread	1 ea	239	71	295	18	46	3	5	1		
	Veggie delite:											
69109	On white bread	1 ea	175	71	222	9	38	3	3	0		
69110	On wheat bread	1 ea	182	69	237	9	44	3	3	0		
	Salads:											
52128	B.L.T.	1 ea	276	91	140	7	10	2	8	3		
52124	B.M.T., classic Italian	1 ea	331	86	274	14	11	1	20	7		
52127	Chicken taco	1 ea	370	87	250	18	15	2	14	5		
52115	Club	1 ea	331	91	126	14	12	1	3	1		
52120	Cold cut trio	1 ea	330	89	191	13	11	1	11	3		
52123	Ham	1 ea	316	91	116	12	11	1	3	1		
52129	Meatball	1 ea	345	88	233	12	16	2	14	5		
52131	Melt	1 ea	336	88	195	16	12	1	10	4		
52121	Pizza	1 ea	335	86	277	12	13	2	20	8		
52126	Roast beef	1 ea	316	92	117	12	11	1	3	1		
52119	Roasted chicken breast	1 ea	331	89	162	20	13	1	4	1		
52117	Seafood and crab:	1 5	331	88	244	13	10	2	17	3		
52116	With light mayo	1 5	331	90	161	13	11	2	8	1		
52130	Steak and cheese	1 ea	342	87	212	22	13	1	8	5		
52122	Tuna:	1 ea	331	84	356	12	10	1	30	5		
52118	With light mayo	1 ea	331	89	205	12	11	1	13	2		
52114	Turkey breast	1 ea	316	92	102	11	12	1	2	1		
52125	With ham	1 ea	316	92	109	11	11	1	3	1		

PAGE KEY: A–4 = Beverages A–6 = Dairy A–10 = Eggs A–10 = Fat/Oil A–14 = Fruit A–20 = Bakery A–28 = Grain A–32 = Fish A–34 = Meats A–38= Poultry A–40 = Sausage A–42 = Mixed/Fast A–46 = Nuts/Seeds A–50 = Sweets A–52 = Vegetables/Legumes A–62 = Vegetarian A–62 = Misc A–66 = Soups/Sauces A–68 = Fast A–84 = Convenience A–88 = Baby foods

Chol (mg)	Calc (mg)	Iron (mg)	Magn (mg)	Pota (mg)	Sodi (mg)	Zinc (mg)	VT-A (RE)	Thia (mg)	VT-E (α-TE)	Ribo (mg)	Niac (mg)	V-B6 (mg)	Fola (μg)	VT-C (mg)
56	44	4			1652		151							15
56	50	4			1664		151							15
33	32	4			1035		142							16
33	39	4			1046		142							16
42	93	4			1735		155							15
42	100	3			1746		156							15
50	103	4			1609		238							16
50	110	3			1621		238							16
20	25	4			928		120							15
20	32	3			939		120							15
48	35	3			967		123							15
48	42	3			978		123							15
34	28	3			849		121							15
32	28	3			873		131							15
34	34	3			860		121							15
32	34	3			884		131							15
57	40	4			1592		169							15
57	47	4			1604		169							15
70	88	5			1106		175							18
70	95	5			1117		176							18
36	32	3			875		125							15
32	32	3			928		146							15
36	38	3			886		126							15
32	38	3			940		146							15
19	30	4			1391		120							15
19	37	3			1403		120							15
24	29	3			1350		120							15
24	36	3			1361		120							15
0	25	3			582		120							15
0	32	3			593		120							15
16	24	1			672		273							32
56	41	2			1379		303							32
52	115	3			990		361							35
26	26	2			1067		273							32
64	46	2			1127		282							33
28	25	2			1034		273							32
33	30	2			761		295							33
42	90	2			1461		308							32
50	100	2			1336		390							33
20	23	2			654		273							32
48	32	2			693		276							32
34	25	2			575		273							32
32	25	2			599		284							32
70	86	3			832		328							35
36	29	2			601		278							32
32	29	2			654		298							32
19	28	2			1117		273							32
24	27	2			1076		273							32

Table A-1
Food Composition

(Computer code number is for Wadsworth Diet Analysis Plus program) (For purposes of calculations, use "0" for t, <1, <.1, <.01, etc.)

Computer Code Number	Food Description	Measure	Wt (g)	H_2O (%)	Ener (kcal)	Prot (g)	Carb (g)	Dietary Fiber (g)	Fat (g)	Fat Breakdown (g) Sat	Mono	Poly
	FAST FOOD RESTAURANTS—Continued											
	SUBWAY—Continued											
52113	Veggie delite	1 ea	260	94	51	2	10	1	1	0		
	Cookies:											
47662	Brazil nut and chocolate chip	1 ea	48	12	229	3	27	1	12	3.5		
47655	Chocolate chip:	1 ea	48	14	209	2	29	1	10	3.5		
47658	With M&M's	1 ea	48	14	209	2	29	1	10	3		
47659	Chocolate chunk	1 ea	48	14	209	2	29	1	10	3.5		
47656	Oatmeal raisin	1 ea	48	15	199	3	29	1	8	2		
47657	Peanut butter	1 ea	48	13	219	3	26	1	12	2.5		
47660	Sugar	1 ea	48	11	229	2	28	0	12	3		
47661	White chip macademia	1 ea	48	12	229	2	28	1	12	2.5		
	Source: Subway International											
	TACO BELL											
	Breakfast burrito:											
1601	Bacon breakfast burrito	1 ea	99	48	291	11	23		17	4		
1627	Country breakfast burrito	1 ea	113	55	220	8	26	2	14	5		
1626	Fiesta breakfast burrito	1 ea	92	44	280	9	25	2	16	6		
1625	Grande breakfast burrito	1 ea	177	56	420	13	43	3	22	7		
1604	Sausage breakfast burrito	1 ea	106	49	303	11	23		19	6		
	Burritos:											
1544	Bean with red sauce	1 ea	198	58	380	13	55	13	12	4		
1545	Beef with red sauce	1 ea	198	57	432	22	42	4	19	8	6.7	.7
1546	Beef & bean with red sauce	1 ea	198	57	412	17	50	5	16	6	6.1	2.1
1569	Big beef supreme	1 ea	298	64	520	24	54	11	23	10		
1552	Chicken burrito	1 ea	171	58	345	17	41		13	5		
1547	Supreme with red sauce	1 ea	248	64	428	16	50		18	7.8		
1571	7 layer burrito	1 ea	234	61	438	13	55	10	19	5.8		
1538	Chilito	1 ea	156	49	391	17	41	11	18	9		
1549	Chilito, steak	1 ea	257	62	496	26	47		23	10		
	Tacos:											
1551	Taco	1 ea	78	58	180	9	12	3	10	4		
1554	Soft taco	1 ea	99	63	242	10	13	3	11	4.4		
1536	Soft taco supreme	1 ea	128	64	234	11	21	3	13	6.3		
1568	Soft taco, chicken	1 ea	128	63	212	15	22	2	7	2.6		
1572	Soft taco, steak	1 ea	100	63	180	12	16	2	8	1.9		
1555	Tostada with red sauce	1 ea	156	67	264	9	27	11	13	4.4		
1558	Mexican pizza	1 ea	223	53	578	21	43	8	35	10.1		
1559	Taco salad with salsa	1 ea	585	71	923	33	70	17	56	16.3		
1560	Nachos, regular	1 ea	106	40	343	5	36	3	19	4.3		
1561	Nachos, bellgrande	1 ea	287	51	708	19	77	16	36	10.1		
1562	Pintos & cheese with red sauce	1 ea	128	68	203	10	19	11	10	4.3		
1563	Taco sauce, packet	1 ea	9	94	2	<1	<1	<1	<1	0	0	0
1564	Salsa	1 ea	10	28	27	1	6		<1	0	0	0
1565	Cinnamon twists	1 ea	35	6	175	1	24	0	7	0		
1628	Caramel roll	1 ea	85	19	353	6	46		16	4		
	Source: Taco Bell Corporation											
	WENDY'S											
	Hamburgers:											
1566	Single on white bun, no toppings	1 ea	133	44	360	24	31	2	16	6		
1570	Cheeseburger, bacon	1 ea	166	55	380	20	34	2	19	7	10.3	1.4
1730	Chicken sandwich, grilled	1 ea	189	62	310	27	35	2	8	1.5		
	Baked potatoes:											
1573	Plain	1 ea	284	71	310	7	71	7	0	0	0	0
1574	With bacon & cheese	1 ea	380	69	530	17	78	7	18	4	10.7	3.3
1575	With broccoli & cheese	1 ea	411	74	470	9	80	9	14	2.5	8	2.5
1576	With cheese	1 ea	383	68	570	14	78	7	23	8	9.2	4.8

Chol (mg)	Calc (mg)	Iron (mg)	Magn (mg)	Pota (mg)	Sodi (mg)	Zinc (mg)	VT-A (RE)	Thia (mg)	VT-E (α-TE)	Ribo (mg)	Niac (mg)	V-B6 (mg)	Fola (μg)	VT-C (mg)
0	23	1			308		136							32
10	32	1.99			115		0							0
10	16	1.99			139		0							0
15	16	1			139		0							0
10	16	1			139		0							0
15	32	1			159		0							0
0	16	1			179		0							0
20	0	.72			179		0							0
10	16	1			139		0							0
181	80	1.8			652		310							
195	80	1.08			690		250							0
25	80	.72			580		150							0
205	100	1.8			1050		500							0
183	80	1.8			661		320							
10	150	2.7		495	1100		450	.04		2.02	1.98	.31		0
57	160	3.96		380	1303		530	.4		2.14	3.44	.32		1
32	170	3.78	50	442	1221	2.67	450	.49		.41	3.09	.59	38	1
55	150	2.7			1520		600							5
57	140	2.52			854		440							1
34	146	8.75	48	410	1196		486	.39		2.04	2.81	.34		5
21	165	2.98			1058		248							5
47	300	3.06			980		950							
78	200	2.7			1313		970							2
25	80	1.08		159	330		100	.05		.14	1.2	.12		0
27	88	1.19		211	363		110	.42		.24	2.95	1.08		0
31	90	1.62			532		135							3
37	85	1.52			571		63							1
19	62	1.13			797		31							0
13	132	1.59		401	573		441	.05		.17	.63	.26		1
46	253	3.65	80	408	1054	5.37	405	.32		.33	2.96	1.12	60	5
65	326	6.84		1048	1931	1.67	1736	.51		.76	4.8	.56	10	26
5	107	.77		160	610	1.68	64	.17		.16	.68	.19	10	0
32	184	3.31		674	1205		138	.1		.34	2.17			3
16	160	1.92	110	384	693	2.17	267	.05		.15	.43	.21	68	0
0	0	.07		9	75		30	0			.02			<1
0	50	.6		376	709		168	.02		.14	0			10
0	0	.45		27	238		50	.1		.04	.71	.04		0
15	60	1.44			312		330							4
65	110	4.14		296	580		0	.43		.38	6.71			0
60	170	3.42	38	375	850	5.9	80	.3		.31	6.43	.26	28	6
65	100	2.7			790		40							6
0	30	3.78	75	1187	25	.74	0	.31	.14	.12	4.3	.8	31	36
20	180	4.32	87	1498	1390	2.75	100	.24		.19	5.04	.94	36	36
5	210	4.5	93	1745	470	.97	350	.34		.29	4.5	.97	74	72
30	380	4.14	85	1510	640	.67	200	.25		.28	3.6	.88	36	36

Table A-1

Food Composition (Computer code number is for Wadsworth Diet Analysis Plus program) (For purposes of calculations, use "0" for t, <1, <.1, <.01, etc.)

Computer Code Number	Food Description	Measure	Wt (g)	H_2O (%)	Ener (kcal)	Prot (g)	Carb (g)	Dietary Fiber (g)	Fat (g)	Fat Breakdown (g) Sat	Mono	Poly
	FAST FOOD RESTAURANTS—Continued											
	WENDY'S—Continued											
1577	With chili & cheese	1 ea	439	69	630	20	83	9	24	9		
1578	With sour cream & chives	1 ea	314	71	380	8	74	8	6	4		
1579	Chili	1 ea	227	81	210	15	21	5	7	2.5		
1582	Chocolate chip cookies	1 ea	57	6	270	3	36	1	13	6		
1580	French fries	1 ea	130	41	390	5	50	5	19	3	11.9	2.4
1581	Frosty dairy dessert	1 ea	227	68	330	8	56	0	8	5		
	Source: Wendy's International											
	CONVENIENCE FOODS and MEALS											
	BUDGET GOURMET											
1695	Chicken cacciatore	1 ea	312	80	300	20	27		13			
1692	Linguini & shrimp	1 ea	284	77	330	15	33		15			
1691	Scallops & shrimp	1 ea	326	79	320	16	43		9			
2245	Seafood newburg	1 ea	284	74	350	17	43		12			
1693	Sirloin tips with country gravy	1 ea	284	80	310	16	21		18			
1694	Sweet & sour chicken with rice	1 ea	284	72	350	18	53		7			
1689	Teriyaki chicken	1 ea	340	77	360	20	44		12			
1690	Veal parmigiana	1 ea	340	75	440	26	39		20			
1696	Yankee pot roast	1 ea	312	77	380	27	22		21			
	Source: The All American Gourmet Co.											
	HAAGEN DAZS											
1755	Ice cream bar, vanilla almond	1 ea	107		371	6	26		27	14.1	10	3
	Sorbet:											
1758	Lemon	½ c	113		140	0	35		0	0	0	0
1760	Orange	½ c	113		140	0	36		0	0	0	0
1759	Raspberry	½ c	113		110	0	27		0	0	0	0
	Yogurt, frozen:											
1753	Chocolate	½ c	98		171	8	26		4	2	2	0
1754	Strawberry	½ c	98		171	6	27		4	2	2	0
	Yogurt extra, frozen:											
1752	Brownie nut	½ c	101		220	8	29		9	4	4	1
1751	Raspberry rendezvous	½ c	101		132	4	26		2	1	1	0
	Source: Pillsbury											
	HEALTHY CHOICE											
	Entrees:											
2112	Fish, lemon pepper	1 ea	303	78	290	14	47	7	5	1		
1624	Lasagna	1 ea	383	76	390	26	60	9	5	2		
2111	Meatloaf, traditional	1 ea	340	79	320	16	46	7	8	4		
2104	Zucchini lasagna	1 ea	396	80	329	20	58	11	1	1		
2110	Dinner, pasta shells marinara	1 ea	340	74	360	25	59	5	3	1.5		
	Low-fat ice cream:											
973	Brownie	½ c	71	60	120	3	22	2	2	1	.3	.7
259	Butter pecan	½ c	71	60	120	3	22	1	2	1	.3	.7
650	Chocolate chip	½ c	71	62	120	3	21	<1	2	1	1	0
1608	Cookie & cream	½ c	71	62	120	3	21	<1	2	1.5	.5	0
650	Chocolate chip	½ c	71	62	120	3	21	<1	2	1	1	0
45	Rocky road	½ c	71	53	140	3	28	2	1	1	.5	0
1621	Vanilla	½ c	71	66	100	3	18	1	2	.5	1.5	0
391	Vanilla fudge	½ c	71	62	120	3	21	1	2	1.5		
	Source: ConAgra Frozen Foods, Omaha, NE											

PAGE KEY: A–4 = Beverages A–6 = Dairy A–10 = Eggs A–10 = Fat/Oil A–14 = Fruit A–20 = Bakery A–28 = Grain A–32 = Fish A–34 = Meats A–38= Poultry A–40 = Sausage A–42 = Mixed/Fast A–46 = Nuts/Seeds A–50 = Sweets A–52 = Vegetables/Legumes A–62 = Vegetarian A–62 = Misc A–66 = Soups/Sauces A–68 = Fast A–84 = Convenience A–88 = Baby foods

Chol (mg)	Calc (mg)	Iron (mg)	Magn (mg)	Pota (mg)	Sodi (mg)	Zinc (mg)	VT-A (RE)	Thia (mg)	VT-E (α-TE)	Ribo (mg)	Niac (mg)	V-B6 (mg)	Fola (µg)	VT-C (mg)
40	330	5.04	122	1745	770	4.15	200	.33		.29	4.5	.99	55	36
15	80	4.32	71	1438	40	.91	300	.23		.14	3.04	.8	32	48
30	80	2.9		501	800		80	.11		.15	2.66			4
30	10	1.8	13	89	120	.41	0	.05		.06	.36	.03	5	0
0	20	1.08	55	845	120	.62	0	.18		.04	3.6	.33	40	6
35	310	1.08	46	544	200	.97	150	.11		.47	.32	.13	17	0
60	150	1.8			810		40	.23		.51	5			21
75	10	3.6			1250		1000	.3		.17	3			2
70	150	.72			690		150			.26	3			12
70	100	.72			660		40	.23		.26	2			
40	60	.36			570		150	.15		.17	4	.28		2
40	60	.72			640		80	.12		.34	3			2
55	80	1.4			610		300	.15		.34	6			12
165	30	4.5			1160		1000	.45		.6	6			6
70	150	1.8			690		600	.15		.43	7			6
90	161	.38		221	85		161			.18				
0				30	20									7
0				80	20									20
0				60	15									7
40	147	.71		241	45		20			.17				
50	147			141	45		20	.03		.17				5
55	152	.73		250	60		20			.14				
20	81			97	25		0			.1				5
25	20	1.08			360		100							30
15	150	3.6		500	550		100	.3		.26	2			6
35	40	1.8			460		150							54
10	199	2.69			309		249							0
25	400	1.8			390		100							4
2	80	0		268	55		40							0
2	100	0		211	60		40							0
2	100	0		240	50		40							0
2	100			254	90		60	.03		.15				2
2	100	0		240	50		40							0
2	100	0		168	60		40	.03		.15				0
5	100			254	50		60	.05		.22				2
2	100	0		296	50		40							0

Table A–1
Food Composition (Computer code number is for Wadsworth Diet Analysis Plus program) (For purposes of calculations, use "0" for t, <1, <.1, <.01, etc.)

Computer Code Number	Food Description	Measure	Wt (g)	H_2O (%)	Ener (kcal)	Prot (g)	Carb (g)	Dietary Fiber (g)	Fat (g)	Fat Breakdown (g) Sat	Mono	Poly
	CONVENIENCE FOODS and MEALS—Continued											
	HEALTH VALLEY											
	Soups, fat-free:											
2001	Beef broth, no salt added	1 c	240	98	18	5	0	0	0	0	0	0
2073	Beef broth, w/salt	1 c	240	98	30	5	2	0	0	0	0	0
2016	Black bean & vegetable	1 c	240	85	110	11	24	12	0	0	0	0
2017	Chicken broth	1 c	240	97	30	6	0	0	0	0	0	0
2018	14 garden vegetable	1 c	240	90	80	6	17	4	0	0	0	0
2015	Lentil & carrot	1 c	240	85	90	10	25	14	0	0	0	0
2014	Split pea & carrot	1 c	240	89	110	8	17	4	0	0	0	0
2013	Tomato vegetable	1 c	240	90	80	6	17	5	0	0	0	0
	Source: Health Valley											
	LA CHOY											
2100	Egg rolls, mini, chicken	1 svg	106	53	220	8	35	3	6	1.5		
2099	Egg rolls, mini, shrimp	1 svg	106	56	210	7	35	3	4	1		
	Source: Beatrice/Hunt Wesson											
	LEAN CUISINE											
	Dinners:											
1639	Baked cheese ravioli	1 ea	241	77	250	12	32	4	8	3	2	1
1632	Chicken chow mein	1 ea	255	81	210	13	28	2	5	1	2	1
1633	Lasagna	1 ea	291	79	270	19	34	5	6	2.5	1.5	.5
1634	Macaroni & cheese	1 ea	255	78	270	13	39	2	7	3.5	1.5	.5
1631	Spaghetti w/meatballs	1 ea	269	74	290	17	40	4	7	2	3	1.5
	Pizza:											
1636	French bread sausage pizza	1 ea	170	53	420	19	41	4	20	5	13.9	1.1
	Source: Stouffer's Foods Corp, Solon, OH											
	TASTE ADVENTURE SOUPS											
1905	Black bean	1 c	242		139	6	28	6	1			
1904	Curry lentil	1 c	241		138	6	30	5	1			
1906	Lentil chili	1 c	242		181	11	33	6	1			
1903	Split pea	1 c	244		140	5	27	5	1			
	Source: Taste Adventure Soups											
	WEIGHT WATCHERS											
	Cheese, fat-free slices:											
1978	Cheddar, sharp	2 pce	21	65	30	5	2	0	0	0	0	0
1980	Swiss	2 pce	21	65	30	5	2	0	0	0	0	0
1977	White	2 pce	21	65	30	5	2	0	0	0	0	0
1979	Yellow	2 pce	21	65	30	5	2	0	0	0	0	0
	Dinners:											
2029	Chicken chow mein	1 ea	255	81	200	12	34	3	2	.5		
1646	Oven fried fish	1 ea	218	78	230	15	25	2	8	2.5	5	2
1972	Margarine, reduced fat	1 tbs	14	49	59	0	0	0	7	1.5		
	Pizza:											
1653	Cheese	1 ea	163	48	390	23	49	6	12	4	3	1
1650	Deluxe combination pizza	1 ea	186	56	380	23	47	6	11	3.5	5	2
1652	Pepperoni pizza	1 ea	158	48	390	23	46	4	12	4	5	2
	Desserts:											
1644	Chocolate brownie	1 ea	182	75	190	6	35	4	4	1	2	1
2024	Chocolate eclair	1 ea	60	45	151	3	24	2	5	1.5		
2247	Chocolate mousse	1 ea	78	44	190	6	33	3	4	1.5		
1642	Strawberry cheesecake	1 ea	111	62	180	7	28	2	5	2	1	2

PAGE KEY: A–4 = Beverages A–6 = Dairy A–10 = Eggs A–10 = Fat/Oil A–14 = Fruit A–20 = Bakery A–28 = Grain A–32 = Fish A–34 = Meats A–38= Poultry A–40 = Sausage A–42 = Mixed/Fast A–46 = Nuts/Seeds A–50 = Sweets A–52 = Vegetables/Legumes A–62 = Vegetarian A–62 = Misc A–66 = Soups/Sauces A–68 = Fast A–84 = Convenience A–88 = Baby foods

Chol (mg)	Calc (mg)	Iron (mg)	Magn (mg)	Pota (mg)	Sodi (mg)	Zinc (mg)	VT-A (RE)	Thia (mg)	VT-E (α-TE)	Ribo (mg)	Niac (mg)	V-B6 (mg)	Fola (μg)	VT-C (mg)
0				196	74						.98			
0	0	0		196	160		0				.98			5
0	40	3.6		676	280		2000	.34		.11	1.35	.22	135	9
0	20	1.8		147	170		0			.03	2.45			1
0	40	1.8		406	250		2000	.26		.08	2.25	.18	27	15
0	60	5.4		439	220		2000	.1		.16	5.63	.45	27	2
0	40	5.4		439	230		2000	.1		.16	5.63	.45		9
0	40	5.4		609	240		2000	.1		.08	2.25	.13	<1	9
5	20	1.44			460		20							0
5	20	1.44			510		20							0
55	200	1.08	42	400	500	1.5	150	.06		.25	1.2	.2	48	6
35	20	.36	30	300	510	1.1	20	.15		.17	5			6
25	150	1.8	44	620	560	2.9	100	.15		.25	3	.32		12
20	250	.72		170	550		20	.12		.25	1.2			0
30	100	2.7	47	480	520	2.5	80	.15		.25	3	.2		4
35	250	2.7	39	340	900	2.2	80	.45		.51	5	.07		6
				650	565									
				467	584									
				650	448									
				484	591									
0	99	0		64	306		56							0
0	99	0		74	276		56							0
0	99	0		64	306		56							0
0	99	0		64	306		56							0
25	40	.72		360	430		300							36
25	20	1.44		370	450		40	.09		.14	1.6			0
0	0	0		5	128		49							0
35	700	1.8		290	590		80	.3		.51	3	.06		6
40	500	3.6		370	550		150	.3		.51	3	.2		5
45	450	1.8		320	650		80	.23		.51	3			5
5	80	1.08		230	160		0	.06		.03	.2	.03		0
0	40	0		65	151		0							0
5	60	1.8		320	150		0							0
15	80	.36		115	230		40	.06		.07	1.6			2

Table A–1

Food Composition (Computer code number is for Wadsworth Diet Analysis Plus program) (For purposes of calculations, use "0" for t, <1, <.1, <.01, etc.)

Computer Code Number	Food Description	Measure	Wt (g)	H_2O (%)	Ener (kcal)	Prot (g)	Carb (g)	Dietary Fiber (g)	Fat (g)	Fat Breakdown (g) Sat	Mono	Poly
	CONVENIENCE FOODS and MEALS—Continued											
	WEIGHT WATCHERS—CONTINUED											
2027	Triple chocolate cheesecake	1 ea	89	52	199	7	32	1	5	2.5		
	Source: Weight Watchers											
	SWEET SUCCESS:											
	Drinks, prepared:											
1776	Chocolate chip	1 c	265	81	180	15	30	6	3	1.6		
1777	Chocolate fudge	1 c	265	81	180	15	30	6	2			
1774	Chocolate mocha	1 c	265	81	180	15	30	6	1	1		
1778	Milk chocolate	1 c	265	81	180	15	30	6	2	1		
1775	Vanilla	1 c	265	81	180	15	33	6	1	.6		
	Drinks, ready to drink:											
2147	Chocolate mint	1 c	297	82	187	11	36	6	3	0		
2148	Strawberry	1 c	265	82	167	10	32	5	3	0		
	Shakes:											
1771	Chocolate almond	1 c	250	82	158	9	30	5	2	0	2.1	.2
1773	Chocolate fudge	1 c	250	82	158	9	30	5	2	0	2.1	.2
1768	Chocolate mocha	1 c	250	82	158	9	30	5	2	0	.6	1.8
1769	Chocolate raspberry truffle	1 c	250	82	158	9	30	5	2	0	2.2	.2
1770	Vanilla creme	1 c	250	82	158	9	30	5	2	0	2.1	.3
	Snack bars:											
1767	Chocolate brownie	1 ea	33	9	120	2	23	3	4	2	.5	.6
1766	Chocolate chip	1 ea	33	9	120	2	23	3	4	2	.4	.5
1921	Oatmeal raisin	1 ea	33	9	120	2	23	3	4	2		
1765	Peanut butter	1 ea	33	9	120	2	23	3	4	2	.6	.6
	Source: Foodway National Inc, Boise, ID											
	BABY FOODS											
1720	Apple juice	½ c	125	88	59	0	15	<1	<1	t	t	t
1721	Applesauce, strained	1 tbs	16	89	7	<1	2	<1	<1	t	t	t
1716	Carrots, strained	1 tbs	14	92	4	<1	1	<1	<1	t	t	t
1718	Cereal, mixed, milk added	1 tbs	15	75	17	1	2	<1	1	.3		
1719	Cereal, rice, milk added	1 tbs	15	75	17	<1	3	<1	1	.3		
1723	Chicken and noodles, strained	1 tbs	16	88	8	<1	1	<1	<1	.1	.1	t
1722	Peas, strained	1 tbs	15	87	6	1	1	<1	<1	t	t	t
1717	Teething biscuits	1 ea	11	6	43	1	8	<1	<1	.2	.2	.1

PAGE KEY: A–4 = Beverages A–6 = Dairy A–10 = Eggs A–10 = Fat/Oil A–14 = Fruit A–20 = Bakery A–28 = Grain A–32 = Fish A–34 = Meats A–38= Poultry A–40 = Sausage A–42 = Mixed/Fast A–46 = Nuts/Seeds A–50 = Sweets A–52 = Vegetables/Legumes A–62 = Vegetarian A–62 = Misc A–66 = Soups/Sauces A–68 = Fast A–84 = Convenience A–88 = Baby foods

Chol (mg)	Calc (mg)	Iron (mg)	Magn (mg)	Pota (mg)	Sodi (mg)	Zinc (mg)	VT-A (RE)	Thia (mg)	VT-E (α-TE)	Ribo (mg)	Niac (mg)	V-B6 (mg)	Fola (μg)	VT-C (mg)
10	80	1.08		169	199		0							0
6	500	6.3	140	600	288	5.25	350	.52	7.05	.59	7	.7	140	21
6	500	6.3	140	750	336	5.25	350	.52	7.05	.59	7	.7	140	21
6	500	6.3	140	800	336	5.25	350	.52	7.05	.59	7	.7	140	21
6	500	6.3	140	750	336	5.25	350	.52	7.05	.59	7	.7	140	21
6	500	6.3	140	830	312	5.25	250	.52	7.05	.59	7	.7	140	21
6	470	5.94	131	526	226	5.05	329	.5	6.56	.56	6.53	.65	131	20
5	419	5.3	117	310	175	4.51	294	.45	5.86	.5	5.83	.58	117	17
5	396	5	110	443	190	4.25	277	.42	5.53	.47	5.5	.55	110	16
5	396	5	110	443	175	4.25	277	.42	5.53	.47	5.5	.55	110	16
5	396	5	110	403	175	4.25	277	.42	5.53	.47	5.5	.55	110	16
5	383	5	110	443	175	4.25	277	.42	5.53	.47	5.5	.55	110	16
5	396	5	110	293	175	4.25	277	.42	5.53	.47	5.5	.55	110	16
3	150	2.71	60	140	45	.59	150	.22	3.01	.25	3	.3	60	9
3	150	2.71	60	110	40	.59	150	.22	3.01	.25	3	.3	60	9
3	150	2.71	60		30	.59	150	.22	3.01	.25	3	.3	60	9
3	150	2.71	60	125	35	.59	150	.22	3.01	.25	3	.3	60	9
0	5	.71	4	114	4	.04	2	.01	.75	.02	.1	.04	<1	72
0	1	.03		11	<1	<.01	<1	<.01	.1	<.01	.01	<.01	<1	6
0	3	.05	1	27	5	.02	160	<.01	.07	.01	.06	.01	2	1
2	33	1.56	4	30	7	.11	4	.06		.09	.87	.01	2	<1
2	36	1.83	7	28	7	.1	4	.07		.07	.78	.02	1	<1
3	4	.07	1	6	3	.05	18	<.01	.04	.01	.07	<.01	2	<1
0	3	.14	2	17	1	.05	8	.01	.08	.01	.15	.01	4	1
0	29	.39	4	35	40	.1	1	.03	.05	.06	.48	.01	5	1

APPENDIX B

Reliable Sources of Nutrition Information

Many sources of nutrition information are available to consumers, but the quality of the information they provide varies widely. All of the sources listed here provide scientifically based information.

Expert Advice

Registered dietitians (hospitals and the yellow pages) Public health nutritionists (public health departments) College nutrition instructors/professors (colleges and universities)

Extension Service home economists (state and county U.S. Department of Agriculture Extension Service offices) Consumer affairs staff of the Food and Drug Administration (national, regional, and state FDA offices)

You can find hundreds of toll-free telephone numbers for health information through the following Website: **http://www.healthfinder.gov/**. After you are connected, search the term toll-free numbers.

U.S. Government

- Federal Trade Commission (FTC)
  Public Reference Branch
  (202) 326-2222
  www.ftc.gov
- Food and Drug Administration (FDA)
  Office of Consumer Affairs, HFE 1
  Room 16-85
  5600 Fishers Lane
  Rockville, MD 20857
  (301) 443-1544
  www.fda.gov
- FDA Consumer Information Line
  (301) 827-4420
- FDA Office of Food Labeling, HFS 150
  200 C Street SW
  Washington, DC 20204
  (202) 205-4561; fax (202) 205-4564
  www.cfsan.fda.gov
- FDA Office of Plant and Dairy Foods and Beverages
  HFS 300
  200 C Street SW
  Washington, DC 20204
  (202) 205-4064; fax (202) 205-4422
- FDA Office of Special Nutritionals,
  HFS 450
  200 C Street SW
  Washington, DC 20204
  (202) 205-4168; fax (202) 205-5295
- Food and Nutrition Information Center
  National Agricultural Library, Room 304
  10301 Baltimore Avenue
  Beltsville, MD 20705-2351
  (301) 504-5719; fax (301) 504-6409
  www.nal.usda.gov/fnic
- Food Research Action Center (FRAC)
  1875 Connecticut Avenue NW, Suite 540
  Washington, DC 20009
  (202) 986-2200; fax (202) 986-2525
- Superintendent of Documents
  U.S. Government Printing Office
  Washington, DC 20402
  (202) 512-1071
  www.access.gpo.gov/su_docs
- U.S. Department of Agriculture (USDA)
  14th Street SW and Independence Avenue
  Washington, DC 20250
  (202) 720-2791
  www.usda.gov/fcs
- USDA Center for Nutrition Policy and Promotion
  1120 20th Street NW, Suite 200
  North Lobby
  Washington, DC 20036
  (202) 208-2417
  www.usda.gov/fcs/cnpp.htm
- USDA Food Safety and Inspection Service
  Food Safety Education Office,
  Room 1180-S
  Washington, DC 20250
  (202) 690-0351
  www.usda.gov/fsis
- U.S. Department of Education (DOE)
  Accreditation Agency Evaluation Branch
  7th and D Street SW
  ROB 3, Room 3915
  Washington, DC 20202-5244
  (202) 708-7417
- U.S. Department of Health and Human Services
  200 Independence Avenue SW
  Washington, DC 20201
  (202) 619-0257
  www.os.dhhs.gov

B

- U.S. Environmental Protection Agency (EPA)
  401 Main Street SW
  Washington, DC 20460
  (202) 260-2090
  www.epa.gov
- U.S. Public Health Service
  Assistant Secretary of Health
  Humphrey Building, Room 725-H
  200 Independence Avenue SW
  Washington, DC 20201
  (202) 690-7694

Canadian Government

Federal

- Bureau of Nutritional Sciences
  Food Directorate
  Health Protection Branch
  3-West
  Sir Frederick Banting Research Centre, 2203A
  Tunney's Pasture
  Ottawa, Ontario K1A 0L2
  www.hc-sc.gc.ca
- Canadian Food Inspection Agency
  Agriculture and Agri-Food Canada
  59 Camelot Drive
  Nepean, Ontario K1A 0Y9
  (613) 225-CFIA or (613) 225-2342
  www.agr.ca
- Nutrition & Healthy Eating Unit
  Strategies and Systems for Health Directorate, 1917C
  17th Floor—Jeanne Mance Bldg.
  Tunney's Pasture
  Ottawa, Ontario K1A 1B4
  www.hc-sc.gc.ca
- Nutrition Specialist
  Health Support Services
  Indian and Northern Health Services Directorate
  Medical Services Branch
  20th Floor—Jeanne Mance Bldg., 1920B
  Tunney's Pasture
  Ottawa, Ontario K1A 0L3
  www.hc-sc.gc.ca

Provincial and Territorial

- Population Health Strategies Branch
  Alberta Health
  23rd Floor, TELUS Plaza, North Tower
  10025 Jasper Avenue
  Edmonton, AB T5J 2N3
- Nutritionist
  Preventive Services Branch
  Ministry of Health
  1520 Blanshard Street
  Victoria BC V8W 3C8
- Executive Director
  Health Programs
  2nd Floor 800 Portage Avenue
  Winnipeg, MB R3G 0P4
- Project Manager
  Public Health Management Services
  Health and Community Services
  P.O. Box 5100
  520 King Street
  Fredericton, NB E3B 5G8
- Director, Health Promotion
  Department of Health
  Government of Newfoundland and Labrador
  P.O. Box 8700
  Confederation Building, West Block
  St. John's, NF A1B 4J6
- Consultant, Nutrition
  Health & Wellness Promotion
  Population Health
  Department of Health and Social Services
  Government of the Northwest Territories
  Centre Square Tower, 6th Floor
  P.O. Box 1320
  Yellowknife, NT X1A 2L9
- Public Health Nutritionist
  Central Health Region
  201 Brownlow Avenue, Unit 4
  Dartmouth, NS B3B 1W2
- Senior Consultant, Nutrition
  Public Health Branch
  Ministry of Health, 8th Floor
  5700 Yonge St.
  North York, ON M2M 4K5
- Coordinator, Health Information Resource Centre
  Department of Health and Social Services
  1 Rochford Street, Box 2000
  Charlottetown, PEI C1A 7N8
- Responsables de la santé cardio-vasculaire et de la nutrition
  Ministère de la Santé et des Services sociaux, Service de la Prévention en Santé
  3e étage
  1075, chemin Sainte-Foy
  Quèbec (Quèbec) G1S 2M1
- Health Promotion Unit
  Population Health Branch
  Saskatchewan Health
  3475 Albert Street
  Regina, SK S4S 6X6
- Director, Nutrition Services
  Yukon Hospital Corporation
  #5 Hospital Road
  Whitehorse, YT Y1A 3H7

International Agencies

- Food and Agriculture Organization of the United Nations (FAO)
  Liaison Office for North America
  2175 K Street, Suite 300
  Washington, DC 20437
  (202) 653-2400
  www.fao.org
- International Food Information Council Foundation
  1100 Connecticut Avenue NW, Suite 430
  Washington, DC 20036
  (202) 296-6540
  ificinfo.health.org
- UNICEF
  3 United Nations Plaza
  New York, NY 10017
  (212) 326-7000
  www.unicef.com
- World Health Organization (WHO)
  Regional Office
  525 23rd Street NW
  Washington, DC 20037
  (202) 974-3000
  www.who.org

Professional Nutrition Organizations

- American Dietetic Association (ADA)
  216 West Jackson Boulevard, Suite 800
  Chicago, IL 60606-6995
  (800) 877-1600; (312) 899-0040
  www.eatright.org
- ADA, The Nutrition Hotline
  (800) 366-1655
- American Society for Clinical Nutrition
  9650 Rockville Pike
  Bethesda, MD 20814-3998
  (301) 530-7110; fax (301) 571-1863
  www.faseb.org/ascn
- Dietitians of Canada
  480 University Avenue, Suite 604
  Toronto, Ontario M5G 1V2, Canada
  (416) 596-0857; fax (416) 596-0603
  www.dietitians.ca

- Human Nutrition Institute (INACG)
1126 Sixteenth Street NW
Washington, DC 20036
(202) 659-0789
www.ilsi.org

- National Academy of Sciences/
National Research Council (NAS/NRC)
2101 Constitution Avenue, NW
Washington, DC 20418
(202) 334-2000
www.nas.edu

- National Institute of Nutrition
265 Carling Avenue, Suite 302
Ottawa, Ontario K1S 2E1
(613) 235-3355; fax (613) 235-7032
www.nin.ca

- Society for Nutrition Education
7101 Wisconsin Avenue, Suite 901
Bethesda, MD 20814-4805
(301) 656-4938

Aging

- Administration on Aging
330 Independence Avenue SW
Washington, DC 20201
(202) 619-0724
www.aoa.dhhs.gov

- American Association of Retired Persons (AARP)
601 E Street NW
Washington, DC 20049
(202) 434-2277
www.aarp.org

- National Aging Information Center
330 Independence Avenue SW
Washington, DC 20201
(202) 619-7501
www.aoa.dhhs.gov/naic

- National Institute on Aging
Public Information Office
31 Center Drive, MSC 2292
Bethesda, MD 20892
(301) 496-1752
www.nih.gov/nia

Alcohol and Drug Abuse

- Al-Anon Family Group Headquarters, Inc.
1600 Corporate Landing Parkway
Virginia Beach, VA 23454-5617
(800) 356-9996
www.al-anon.alateen.org

- Alateen
1600 Corporate Landing Parkway
Virginia Beach, VA 23454-5617
(800) 356-9996
www.al-anon.alateen.org

- Alcoholics Anonymous (AA)
General Service Office
475 Riverside Drive
New York, NY 10115
(212) 870-3400
www.aa.org

- Narcotics Anonymous (NA)
P.O. Box 9999
Van Nuys, CA 91409
(818) 773-9999; fax (818) 700-0700
www.wsoinc.com

- National Clearinghouse for Alcohol and Drug Information (NCADI)
P.O. Box 2345
Rockville, MD 20847-2345
(800) 729-6686
www.health.org

- National Council on Alcoholism and Drug Dependence (NCADD)
12 West 21st Street
New York, NY 10010
(800) NCA-CALL or (800) 622-2255
(212) 206-6770; fax (212) 645-1690
www.ncadd.org

- U.S. Center for Substance Abuse Prevention
1010 Wayne Avenue, Suite 850
Silver Spring, MD 20910
(301) 459-1591 ext. 244; fax (301) 495-2919
www.covesoft.com/csap.html

Consumer Organizations

- Center for Science in the Public Interest (CSPI)
1875 Connecticut Avenue NW,
Suite 300
Washington, DC 20009-5728
(202) 332-9110; fax (202) 265-4954
www.cspinet.org

- Choice in Dying, Inc.
1035 30th Street NW
Washington, DC 20007
(202) 338-9790; fax (202) 338-0242
www.choices.org

- Consumer Information Center
Pueblo, CO 81009
(888) 8 PUEBLO or (888) 878-3256
www.pueblo.gsa.gov

- Consumers Union of US Inc.
101 Truman Avenue
Yonkers, NY 10703-1057
(914) 378-2000
www.consunion.org

- National Council Against Health Fraud, Inc. (NCAHF)
P.O. Box 1276
Loma Linda, CA 92354
(909) 824-4690
www.ncahf.org

Fitness

- American College of Sports Medicine
P.O. Box 1440
Indianapolis, IN 46206-1440
(317) 637-9200
www.acsm.org/sportsmed

- American Council on Exercise (ACE)
5820 Oberlin Drive, Suite 102
San Diego, CA 92121
(800) 529-8227
www.acefitness.org

- President's Council on Physical Fitness and Sports
Humphrey Building, Room 738
200 Independence Avenue SW
Washington, DC 20201
(202) 690-9000; fax (202) 690-5211
www.indiana.edu/~preschal

- Shape Up America!
6707 Democracy Boulevard, Suite 306
Bethesda, MD 20817
(301) 493-5368
www.shapeup.org

- Sport Medicine and Science Council of Canada
1600 James Naismith Drive, Suite 314
Gloucester, Ontario K1B 5N4, Canada
(613) 748-5671; fax (613) 748-5729
www.smscc.ca

Food Safety

- Alliance for Food & Fiber
Food Safety Hotline
(800) 266-0200

- FDA Center for Food Safety and Applied Nutrition
200 C Street SW
Washington, DC 20204
(800) FDA-4010 or (800) 332-4010
vm.cfsan.fda.gov

B

- National Lead Information Center
(800) LEAD-FYI or (800) 532-3394
(800) 424-LEAD or (800) 424-5323
- National Pesticide Telecommunications Network (NPTN)
Oregon State University
333 Weniger Hall
Corvallis, OR 97331-6502
(541) 737-6091
www.ace.orst.edu/info/nptn
- USDA Meat and Poultry Hotline
(800) 535-4555
- U.S. EPA Safe Drinking Water Hotline
(800) 426-4791

Health and Disease

- Alzheimer's Disease Education and Referral Center
P. O. Box 8250
Silver Spring, MD 20907-8250
(800) 438-4380
www.alzheimers.org
- Alzheimer's Disease Information and Referral Service
919 North Michigan Avenue, Suite 1000
Chicago, IL 60611
(800) 272-3900
www.alz.org
- American Academy of Allergy, Asthma, and Immunology
611 East Wells Street
Milwaukee, WI 53202
(414) 272-6071; fax (414) 276-3349
www.aaaai.org
- American Cancer Society
National Home Office
1599 Clifton Road NE
Atlanta, GA 30329-4251
(800) ACS-2345 or (800) 227-2345
www.cancer.org
- American Council on Science and Health
1995 Broadway, 2nd Floor
New York, NY 10023-5860
(212) 362-7044; fax (212) 362-4919
www.acsh.org
- American Dental Association
211 East Chicago Avenue
Chicago, IL 60611
(312) 440-2800
www.ada.org
- American Diabetes Association
1660 Duke Street
Alexandria, VA 22314
(800) 232-3472 or (703) 549-1500
www.diabetes.org
- American Heart Association
Box BHG, National Center
7320 Greenville Avenue
Dallas, TX 75231
(800) 275-0448 or (214) 373-6300
www.amhrt.org
- American Institute for Cancer Research
1759 R Street NW
Washington, DC 20009
(800) 843-8114 or (202) 328-7744; fax (202) 328-7226
www.aicr.org
- American Medical Association
515 North State Street
Chicago, IL 60610
(312) 464-5000
www.ama-assn.org
- American Public Health Association (APHA)
1015 Fifteenth Street NW, Suite 300
Washington, DC 20005
(202) 789-5600
www.apha.org
- American Red Cross
National Headquarters
8111 Gatehouse Road
Falls Church, VA 22042
(703) 206-7180
www.redcross.org
- Canadian Diabetes Association
15 Toronto Street, Suite 800
Toronto, ON M5C 2E3
(800) BANTING or (800) 226-8464
(416) 363-3373
www.diabetes.ca
- Canadian Public Health Association
400-1565 Carling Avenue
Ottawa, Ontario K1Z 8R1
(613) 725-3769; fax (613) 725-9826
www.cpha.ca
- Centers for Disease Control and Prevention (CDC)
1600 Clifton Road NE
Atlanta, GA 30333
(404) 639-3311
www.cdc.gov
- The Food Allergy Network
10400 Eaton Place, Suite 107
Fairfax, VA 22030-2208
(800) 929-4040 or (703) 691-3179
www.foodallergy.org
- Internet Health Resources
www.ihr.com
- National AIDS Hotline (CDC)
(800) 342-AIDS (English)
(800) 344-SIDA (Spanish)
(800) 2437-TTY (Deaf)
(900) 820-2437
- National Cancer Institute
Office of Cancer Communications
Building 31, Room 10824
Bethesda, MD 20892
(800) 4-CANCER or (800) 422-6237
www.nci.nih.gov
- National Diabetes Information Clearinghouse
1 Information Way
Bethesda, MD 20892-3560
(301) 654-3327
www.niddk.nih.gov
- National Digestive Disease Information Clearinghouse (NDDIC)
2 Information Way
Bethesda, MD 20892-3570
(301) 654-3810
www.niddk.nih.gov
- National Health Information Center (NHIC)
Office of Disease Prevention and Health Promotion
(800) 336-4797
nhic-nt.health.org
- National Heart, Lung, and Blood Institute
Information Center
P.O. Box 30105
Bethesda, MD 20824-0105
(301) 251-1222
www.nhlbi.nih.gov/nhlbi/nhlbi.htm
- National Institute of Allergy and Infectious Diseases
Office of Communications
Building 31, Room 7A50
31 Center Drive, MSC2520
Bethesda, MD 20892-2520
(301) 496-5717
www.niaid.nih.gov
- National Institute of Dental Research (NIDR)
National Institute of Health
Bethesda, MD 20892-2190
(301) 496-4261
www.nidr.nih.gov
- National Institutes of Health (NIH)
9000 Rockville Pike
Bethesda, MD 20892
(301) 496-2433
www.nih.gov

- National Osteoporosis Foundation
  1150 17th Street NW, Suite 500
  Washington, DC 20036
  (202) 223-2226
  www.nof.org
- Office of Disease Prevention and Health Promotion
  odphp.osophs.dhhs.gov
- Office on Smoking and Health (OSH)
  www.americanheart.org/heart.org/Heart_and_stroke_A_Z_Guide/osh.html

Infancy and Childhood

- American Academy of Pediatrics
  141 Northwest Point Boulevard
  Elk Grove Village, IL 60007-1098
  (847) 228-5005
  www.aap.org
- Association of Birth Defect Children, Inc.
  930 Woodcock Road, Suite 225
  Orlando, FL 32803
  (407) 245-7035
  www.birthdefects.org
- Canadian Paediatric Society
  100-2204 Walkley Road
  Ottawa, ON K1G 4G8
  (613) 526-9397; fax (613) 526-3332
  www.cps.ca
- National Center for Education in Maternal & Child Health
  2000 15th Street North, Suite 701
  Arlington, VA 22201-2617
  (703) 524-7802
  www.ncemch.org

Pregnancy and Lactation

- American College of Obstetricians and Gynecologists Resource Center
  409 12th Street SW
  Washington, DC 20024-2188
  (202) 638-5577
  www.acog.org
- La Leche International, Inc.
  1400 N. Meacham Road
  Schaumburg, IL 60173
  (847) 519-7730
  www.lalecheleague.org
- March of Dimes Birth Defects Foundation
  1275 Mamaroneck Avenue
  White Plains, NY 10605
  (914) 428-7100
  www.modimes.org

Trade and Industry Organizations

- Beech-Nut Nutrition Corporation
  P.O. 618
  St. Louis, MO 63188-0618
  (800) 523-6633
  www.beechnut.com
- Borden Inc.
  180 East Broad Street
  Columbus, OH 43215
  (800) 426-7336
- Campbell Soup Company
  Consumer Response Center
  Campbell Place, Box 26B
  Camden, NJ 08103-1701
  (800) 257-8443
  www.campbellssoup.com
- Elan Pharma/Hi Chem Diagnostics
  2 Thurber Boulevard
  Smithfield, RI 02917
  (401) 233-3526
  www.hi-chem.com
- General Mills, Inc.
  Number One General Mills Boulevard
  Minneapolis, MN 55426
  (800) 328-6787
  www.generalmills.com
- Hoffmann-LaRoche, Inc.
  340 Kingsland Street
  Nutley, NJ 07110
  (973) 235-5000
- Kellogg Company
  P.O. Box 3599
  Battle Creek, MI 49016-3599
  (616) 961-2000
  www.kelloggs.com
- Kraft Foods
  Consumer Response and Information Center
  One Kraft Court
  Glenview, IL 60025
  (800) 323-0768
  www.kraftfoods.com
- Mead Johnson Nutritionals
  2400 West Lloyd Expressway
  Evansville, IN 47721
  (800) 247-7893
  www.meadjohnson.com
- Nabisco Consumer Affairs
  100 DeForest Avenue
  East Hanover, NJ 07936
  (800) NABISCO or (800) 932-7800
  www.nabisco.com
- National Dairy Council
  10255 West Higgins Road, Suite 900
  Rosemond, IL 60018-5616
  (847) 803-2000
  www.dairyinfo.com
- NutraSweet/KELCO
  P.O. Box 2986
  Chicago, IL 60654-0986
  www.equal.com
- Pillsbury Company
  Consumer Relations
  P.O. Box 550
  Minneapolis, MN 55440
  (800) 767-4466
  www.pillsbury.com
- Procter and Gamble Company
  One Procter and Gamble Plaza
  Cincinnati, OH 45202
  (513) 983-1100
  www.pg.com/info
- Ross Laboratories, Abbot Laboratory
  625 Cleveland Avenue
  Columbus, OH 43215
  (800) 227-5767
  www.abbot.com
- Sherwood Medical
  1915 Olive Street
  St. Louis, MO 63103
  (800) 428-4400
- Sunkist Growers
  Consumer Affairs
  Fresh Fruit Division
  14130 Riverside Drive
  Sherman Oaks, CA 91423
  (800) CITRUS-5 or (800) 248-7875
  www.sunkist.com
- United Fresh Fruit and Vegetable Association
  727 North Washington Street
  Alexandria, VA 22314
  (703) 836-3410
- USA Rice Federation
  4301 North Fairfax Drive, Suite 305
  Arlington, VA 22203
  Phone: (703) 351-8161
  www.usarice.com
- Weight Watchers International, Inc.
  Consumer Affairs Department/IN
  175 Crossways Park West
  Woodbury, NY 11797
  (516) 390-1400; fax (516) 390-1632.
  www.weightwatchers.com

Weight Control and Eating Disorders

- American Anorexia & Bulimia Association, Inc.
  165 West 46th Street #1108
  New York, NY 10036
  (212) 575-6200
  members.aol.com/amanbu

B

- Anorexia Nervosa and Related Eating Disorders (ANRED)
  P.O. Box 5102
  Eugene, OR 97405
  (541) 344-1144
  www.anred.com
- National Association of Anorexia Nervosa and Associated Disorders, Inc. (ANAD)
  P.O. Box 7
  Highland Park, IL 60035
  (847) 831-3438
  members.aol.com/anad20/index.htm l
- National Eating Disorder Information Centre
  200 Elizabeth Street, College Wing 1-304
  Toronto, Ontario M5G 2C4
  (519) 253-7421; fax (519) 253-7545
- Overeaters Anonymous (OA)
  World Service Office
  6075 Zenith Court NE
  Rio Rancho, NM 87124
  (505) 891-2664; fax (505) 891-4320
  www.overeatersanonymous.org
- TOPS (Take Off Pounds Sensibly)
  4575 South Fifth Street
  P.O. Box 07360
  Milwaukee, WI 53207-0360
  (800) 932-8677 or (414) 482-4620
  www.tops.org
- Weight-Control Information Network
  1 Win Way
  Bethesda, MD 20892-3665
  (800) WIN-8098

World Hunger

- Bread for the World
  1100 Wayne Avenue, Suite 1000
  Silver Spring, MD 20910
  (301) 608-2400
  www.bread.org
- Center on Hunger, Poverty and Nutrition Policy
  Tufts University School of Nutrition
  11 Curtis Avenue
  Medford, MA 02155
  (617) 627-3956
- Freedom from Hunger
  P.O. Box 2000
  1644 DaVinci Court
  Davis, CA 95617
  (530) 758-6200
  www.freefromhunger.org
- Oxfam America
  26 West Street
  Boston, MA 02111
  (617) 482-1211
  www.oxfamamerica.org
- SEEDS Magazine
  P.O. Box 6170
  Waco, TX 76706
  (254) 755-7745
  www.helwys.com/seedhome.htm
- Worldwatch Institute
  1776 Massachusetts Avenue NW, Suite 800
  Washington, DC 20036
  (202) 452-1999
  www.worldwatch.org

Scientific Literature

Nutrition Journals

American Journal of Clinical Nutrition
British Journal of Nutrition
Human Nutrition, Applied Nutrition
Journal of the American College of Nutrition
Journal of the American Dietetic Association
Journal of the Canadian Dietetic Association
Journal of Food Composition and Analysis
Journal of Nutrition
Journal of Nutrition Education
Nutrition Abstracts and Reviews
Nutrition and Metabolism
Nutrition Reports International
Nutrition Research
Nutrition Reviews
Nutrition Today

Other Journals

American Journal of Epidemiology
American Journal of Nursing
American Journal of Public Health
Annals of Internal Medicine
Annals of Surgery
Canadian Journal of Public Health
Caries Research
Food Technology
Gastroenterology
International Journal of Obesity
Journal of the American Dental Association
Journal of the American Medical Association
Journal of Clinical Investigation
Journal of Food Science
Journal of Home Economics
Journal of Pediatrics
Lancet
New England Journal of Medicine
Pediatrics
Science

Appendix C

The U.S. Food Exchange System

The U.S. exchange system divides the foods suitable for use in planning a healthy diet into six lists—the starch/bread, meat/meat alternate, vegetable, fruit, milk, and fat lists.[a] These lists are shown in Tables C.1 through C.6. Following these lists are three other sets of foods: free foods, combination foods, and foods for occasional use (Tables C.7, C.8, and C.9).

[a]The Exchange Lists are the basis of a meal planning system designed by a committee of the American Diabetes Association and The American Dietetic Association. While designed primarily for people with diabetes and others who must follow special diets, the Exchange Lists are based on principles of good nutrition that apply to everyone.

Table C.1

The U.S.. Exchange System: Starch/Bread List

15 g carbohydrate, 3 g protein, trace fat, 80 cal

Amount	Food	Amount	Food
Cereals/Grains/Pasta		**Bread**	
½ c	Bran cereals, concentrated 🌾	½ (1 oz)	Bagels
½ c	Bran cereals, flaked 🌾	2 (⅔ oz)	Bread sticks, crisp, 4" × ½"
½ c	Bulgur, cooked	1 c	Croutons, low-fat
½ c	Cooked cereals	½	English muffins
2½ tbs	Cornmeal, dry	½ (1 oz)	Frankfurter or hamburger buns
3 tbs	Grapenuts	½ loaf	Pita, 6" across
½ c	Grits, cooked	1 (1 oz)	Plain rolls, small
¾ c	Other ready-to-eat unsweetened cereals	1 slice (1 oz)	Raisin, unfrosted
½ c	Pasta, cooked	1 slice (1 oz)	Rye, pumpernickel 🌾
1½ c	Puffed cereals	1	Tortillas, 6" across
⅓ c	Rice, white or brown, cooked	1 slice (1 oz)	White (including French, Italian)
½ c	Shredded wheat	1 slice (1 oz)	Whole-wheat
3 tbs	Wheat germ 🌾	**Crackers/Snacks**	
Dried Beans/Peas/Lentils		8	Animal crackers
¼ c	Baked beans 🌾	3	Graham crackers, 2½" square
⅓ c	Beans and peas, cooked, such as kidney, white, split, black-eyed 🌾	¾ oz	Matzoth
		5 slices	Melba toast
⅓ c	Lentils, cooked 🌾	24	Oyster crackers
Starchy Vegetables		3 c	Popcorn, popped, no fat added
½ c	Corn 🌾	¾ oz	Pretzels
1 cob	Corn on the cob, 6" long 🌾	4	Rye crisp, 2" × 3½"
½ c	Lima beans 🌾	6	Saltine-type crackers
½ c	Peas, green, canned or frozen 🌾	2 to 4 (¾ oz)	Whole-wheat crackers, no fat added (crisp breads)
½ c	Plantains 🌾		
1 small (3 oz)	Potatoes, baked	**Starch Foods Prepared with Fat**	
½ c	Potatoes, mashed	(Count as 1 starch/bread serving, plus 1 fat serving.)	
¾ c	Squash, winter (acorn, butternut)	1	Biscuits, 2½" across
⅓ c	Yams, sweet potatoes, plain	½ c	Chow mein noodles
		1 (2 oz)	Cornbread, 2" cube
		6	Crackers, round butter type
		10 (1½ oz)	French fries, 2" to 3½" long
		1	Muffins, plain, small
		2	Pancakes, 4" across
		¼ c	Stuffing, bread, prepared
		2	Taco shells, 6" across
		1	Waffles, 4½" square
		4 to 6 (1 oz)	Whole-wheat crackers, fat added

🌾 3 grams or more dietary fiber per serving. Average fiber contents of whole-grain products is 2 grams per serving. For starch foods not on this list, the general rule is that ½ cup cereal, grain, or pasta is 1 serving: 1 ounce of a bread product is 1 serving.

Table C.2
U.S. Exchange System: Meat/Meat Alternate Lists

Lean meat = 7 g protein, 3 g fat, 55 cal; medium-fat meat = 7 g protein, 5 g fat, 75 cal; high-fat meat = 7 g protein, 8 g fat, 100 cal.

Category	Amount	Food
Lean Meat and Alternates		
Beef	1 oz	USDA Good or Choice grades of lean beef, such as round, sirloin, and flank steak; tenderloin; chipped beef 🧂
Pork	1 oz	Leak pork, such as fresh ham; canned, cured, or boiled ham 🧂, Canadian bacon 🧂, tenderloin
Veal	1 oz	All cuts are lean except veal cutlets (ground or cubed); examples of lean veal: chops and roasts
Poultry	1 oz	Chicken, turkey, Cornish hen (without skin)
Fish	1 oz	All fresh and frozen fish
	2 oz	Crab, lobster, scallops, shrimp, clams (fresh or canned in water 🧂)
	6 medium	Oysters
	¼ c	Tuna 🧂, canned in water
	1 oz	Herring, uncreamed or smoked
	2 medium	Sardines, canned
Wild game	1 oz	Venison, rabbit, squirrel
	1 oz	Pheasant, duck, goose (without skin)
Cheese	¼ c	Any cottage cheese
	2 tbs	Grated parmesan
	1 oz	Diet cheeses 🧂 (with less than 55 cal/oz)
Other	1 oz	95% fat-free lunch meats
	3 whites	Egg whites
	¼ c	Egg substitutes with less than 55 cal per ¼ c
Medium-Fat Meat and Alternates		
Beef	1 oz	Most beef products fall into this category; examples: all ground beef, roasts (rib, chuck, rump), steak (cubed, porterhouse, T-bone), meatloaf
Pork	1 oz	Most pork products fall into this category; examples: chops, loin roast, Boston butt, cutlets
Lamb	1 oz	Most lamb products fall into this category; examples: chops, leg, roast
Medium-Fat Meat and Alternates *(continued)*		
Veal	1 oz	Cutlet, ground or cubed, unbreaded
Poultry	1 oz	Chicken (with skin), domestic duck or goose (well-drained of fat), ground turkey
Fish	¼ c	Tuna 🧂, canned in oil and drained
	¼ c	Salmon 🧂, canned
Cheese		Skim or part-skim milk cheeses, such as:
	¼ c	Ricotta
	1 oz	Mozzarella
	1 oz	Diet cheeses 🧂 (with 56 to 80 cal/oz)
Other	1 oz	86% fat-free lunch meat 🧂
	1	Eggs (high in cholesterol, limit to 3 per week)
	¼ c	Egg substitutes with 56 to 80 cal per ¼ c
	4 oz	Tofu, 2½" × 2¾" × 1"
	1 oz	Liver, hearts, kidneys, sweetbreads (high in cholesterol)
High-Fat Meat and Alternates[a]		
Beef	1 oz	Most USDA Prime cuts of beef, such as ribs, corned beef 🧂
Pork	1 oz	Spareribs, ground pork, pork sausages 🧂 (patties or links)
Lamb	1 oz	Patties, ground lamb
Fish	1 oz	Any fried fish product
Cheese	1 oz	All regular cheeses 🧂, such as American, blue, Cheddar, Monterey, Swiss
Other	1 oz	Lunch meats 🧂, such as bologna, salami, pimento loaf
	1 oz	Sausage 🧂, such as Polish, Italian
	1 oz	Knockwurst, smoked
	1 oz	Bratwurst 🧂
	1 (10/lb)	Frankfurters 🧂 (turkey or chicken)
	1 tbs	Peanut butter (contains unsaturated fat)
Count as 1 high-fat meat plus 1 fat exchange:		
1 frank	(10/lb)	Frankfurters 🧂 (beef, pork, or combination)

🧂 400 milligrams or more sodium per exchange. Meats contribute no fiber to the diet.

[a]These items are high in saturated fat, cholesterol, and calories and should be used no more than three times per week.

Table C.3
U.S. Exchange System: Vegetable List

5 g carbohydrate, 2 g protein, 25 cal
All portion sizes, except as otherwise noted, are ½ c of any cooked vegetable or vegetable juice, 1 c of any raw vegetable.

Artichokes, ½ medium	Cabbage, cooked	Leeks	Spinach, cooked
Asparagus	Carrots	Mushrooms, cooked	Summer squash (crookneck)
Bean sprouts	Cauliflower	Okra	Tomatoes, 1 large
Beans (green, wax, Italian)	Eggplant	Onions	Tomato/vegetable juice (salt symbol)
Beets	Green peppers	Pea pods	Turnips
Broccoli	Greens (collard, mustard, turnip)	Rutabagas	Water chestnuts
Brussels sprouts	Kohlrabi	Sauerkraut (salt symbol)	Zucchini, cooked

Starchy vegetables such as corn, peas, and potatoes are found on the Starch/Bread List.
For free vegetables, see the Free Food List (Table D.7).

(salt symbol) 400 milligrams or more sodium per serving. Most vegetable servings contain 2 to 3 grams dietary fiber.

Table C.4
U.S. Exchange System: Fruit List

15 g carbohydrate, 60 cal
All portion sizes, unless otherwise noted, are ½ c fresh fruit or fruit juice, ¼ c dried fruit.

Amount	Food	Amount	Food
Fresh, Frozen, and Unsweetened Canned Fruit		**Dried Fruit**	
1	Apples, raw, 2" across	4 rings	Apples (fiber symbol)
½ c	Applesauce, unsweetened	7 halves	Apricots (fiber symbol)
4	Apricots, medium, raw	2½ medium	Dates
½ c (4 halves)	Apricots, canned	1½	Figs (fiber symbol)
½	Bananas, 9" long	3 medium	Prunes (fiber symbol)
¾ c	Blackberries, raw (fiber symbol)	2 tbs	Raisins
¾ c	Blueberries, raw (fiber symbol)	**Fruit Juice**	
⅓	Cantaloupe, 5" across	½ c	Apple juice/cider
1 c	Cantaloupe, cubes	⅓ c	Cranberry juice cocktail
12	Cherries, large, raw	⅓ c	Grape juice
½ c	Cherries, canned	½ c	Grapefruit juice
2	Figs, raw, 2" across	½ c	Orange juice
½ c	Fruit cocktail, canned	½ c	Pineapple juice
½	Grapefruit, medium	⅓ c	Prune juice
¾ c	Grapefruit, segments		
15	Grapes, small		
⅛	Honeydew melon, medium		
1 c	Honeydew melon, cubes		
1	Kiwis, large		
¾ c	Mandarin oranges		
½	Mangoes, small		
1	Nectarines, 1½" across (fiber symbol)		
1	Oranges, 2½" across		
1 c	Papayas		
1 (¾ c)	Peaches, 2¾" across		
½ c (2 halves)	Peaches, canned		
½ large or 1 small	Pears		
½ c (2 halves)	Pears, canned		
2	Persimmons, medium, native		
¾ c	Pineapple, raw		
⅓ c	Pineapple, canned		
2	Plums, raw, 2" across		
½	Pomegranates (fiber symbol)		
1 c	Raspberries, raw (fiber symbol)		
1¼ c	Strawberries, raw, whole		
2	Tangerines, 2½" across		
1¼ c	Watermelon, cubes		

(fiber symbol) 3 grams or more dietary fiber per serving. Average fiber contents of fresh, frozen, and dry fruits: 2 grams per serving.

C

Table C.5
U.S. Exchange System: Milk List

Nonfat and very low-fat milk = 12 g carbohydrate, 8 g protein, trace fat, 90 cal; low-fat milk = 12 g carbohydrate, 8 g protein, = g fat, 120 cal; whole milk = 12 g carbohydrate, 8 g protein, 8 g fat, 150 cal.

Amount	Food	Amount	Food
Nonfat and Very Low-Fat Milk		**Low-Fat Milk**	
1 c	Nonfat milk	1 c fluid	2% milk
1 c	½% milk	8 oz	Plain low-fat yogurt, with added nonfat milk solids
1 c	1% milk	**Whole Milk**	
⅓ c	Dry nonfat milk	1 c	Whole milk
½ c	Evaporated nonfat milk	½ c	Evaporated whole milk
1 c	Low-fat buttermilk	8 oz	Whole plain yogurt
8 oz	Plain nonfat yogurt		

Table C.6
U.S. Exchange System: Fat List

5 g fat, 45 cal

Amount	Food	Amount	Food
Unsaturated Fats		**Saturated Fats**	
⅛ medium	Avocados	1 slice	Bacon[a]
1 tsp	Margarine	1 tsp	Butter
1 tbs	Margarine, diet[a]	½ oz	Chitterlings
1 tsp	Mayonnaise	2 tbs	Coconut, shredded
1 tbs	Mayonnaise, reduced calorie[a]	2 tbs	Coffee whitener, liquid
	Nuts and seeds:	4 tsp	Coffee whitener, powder
6 whole	Almonds, dry roasted	1 tbs	Cream (heavy, whipping)
1 tbs	Cashews, dry roasted	2 tbs	Cream (light, coffee, table)
20 small or 10 large	Peanuts	2 tbs	Cream (sour)
2 whole	Pecans	1 tbs	Cream cheese
2 tsp	Pumpkin seeds	¼ oz	Salt pork[a]
1 tbs	Other nuts	Two tablespoons of low-calorie salad dressing is a free food.	
1 tbs	Seeds, pine nuts, sunflower seeds (without shells)		
2 whole	Walnuts		
1 tsp	Oil (corn, cottonseed, safflower, soybean, sunflower, olive, peanut)		
10 small or 5 large	Olives[a]		
1 tbs	Salad dressing, all varieties[a]		
2 tsp	Salad dressing, mayonnaise type		
1 tbs	Salad dressing, mayonnaise type, reduced calorie		
2 tbs	Salad dressing, reduced calorie		

[a]If more than one or two servings are eaten, these foods provide 400 milligrams or more sodium.

400 milligrams or more sodium per serving.

Table C.7
U.S. Exchange System: Free Foods

A free food is any food or drink that contains less than 20 cal/serving. People with diabetes are advised to eat as much as they want of those items that have no serving size specified. They may eat two or three servings per day of those items that have a specific serving size. It is suggested that they spread the servings out through the day.

Amount	Food
Drinks	Bouillon, low-sodium
	Bouillon or broth without fat
	Carbonated drinks, sugar-free
	Carbonated water
	Club soda
1 tbs	Cocoa powder, unsweetened
	Coffee/tea
	Drink mixes, sugar-free
	Tonic water, sugar-free
Nonstick Pan Spray	
Fruit	
½ c	Cranberries, unsweetened
½ c	Rhubarb, unsweetened
Vegetables (raw, 1 c)	Cabbage
	Celery
	Chinese cabbage
	Cucumbers
	Green onions
	Hot peppers
	Mushrooms
	Radishes
	Zucchini
Salad Greens	Endive
	Escarole
	Lettuce
	Romaine
	Spinach
Sweet Substitutes	Candy, hard, sugar-free
	Gelatin, sugar-free
	Gum, sugar-free
2 tsp	Jam/jelly, sugar-free
1 to 2 tbs	Pancake syrup, sugar-free
	Sugar substitutes (saccharin, aspartame, acesulfame-K)
2 tbs	Whipped topping

Amount	Food
Condiments	
1 tbs	Catsup
	Horseradish
	Mustard
	Pickles, dill, unsweetened
2 tbs	Salad dressing, low-calorie
1 tbs	Taco sauce
	Vinegar
Seasonings	Basil, fresh
	Celery seeds
	Chili powder
	Chives
	Cinnamon
	Curry
	Dill
	Flavoring extracts (almond, butter, lemon, peppermint, vanilla, walnut, etc.)
	Garlic
	Garlic powder
	Herbs
	Hot pepper sauce
	Lemon
	Lemon juice
	Lemon pepper
	Lime
	Lime juice
	Mint
	Onion powder
	Oregano
	Paprika
	Pepper
	Pimento
	Soy sauce
	Soy sauce, low-sodium ("lite")
	Spices
¼ c	Wine, used in cooking
	Worcestershire sauce

3 grams or more dietary fiber per serving.
400 milligrams or more sodium per serving.

C

Table C.8

U.S. Exchange Combination Foods

Much of the food we eat is mixed together in various combinations. These combination foods do not fit into only one exchange list. It can be quite hard to tell what is in a certain casserole dish or baked food item. This is a list of average values for some typical combination foods. This list will help you fit these foods into your meal plan. Ask your dietitian for information about any other foods you'd like to eat. The *American Diabetes Association/American Dietetic Association Family Cookbooks* and the *American Diabetes Associates Holiday Cookbook* have many recipes and further information about many foods, including combination foods. Check your library or local bookstore.

Food	Amount	Exchanges
Casseroles, homemade	1 c (8 oz)	2 starch, 2 medium-fat meat, 1 fat
Cheese pizza, thin crust	¼ of 15 oz or ¼ of 10"	2 starch, 1 medium-fat meat, 1 fat
Chili with beans, commercial	1 c (8 oz)	2 starch, 2 medium-fat meat, 2 fat
Chow mein, without noodles or rice	2 c (16 oz)	1 starch, 2 vegetable, 2 lean meat
Macaroni and cheese meat, 2 fat	1 c (8 oz)	2 starch, 1 medium-fat
Soups:		
Bean	1 c (8 oz)	1 starch, 1 vegetable, 1 lean meat
Chunky, all varieties	10¾-oz can	1 starch, 1 vegetable, 1 medium-fat meat
Cream, made with water	1 c (8 oz)	1 starch, 1 fat
Vegetable or broth	1 c (8 oz)	1 starch
Spaghetti and meatballs, canned	1 c (8 oz)	2 starch, 1 medium-fat meat, 1 fat
Sugar-free pudding, made with nonfat milk	½ c	1 starch
If beans are used as a meat substitute:		
Dried beans, peas, lentils	1 c (cooked)	2 starch, 1 lean meat

400 milligrams or more sodium per serving.

3 grams or more dietary fiber per serving.

Table C.9

U.S. Exchange Foods for Occasional Use

The following list includes average exchange values for some foods high in sugar and fat. People are advised to use them only occasionally and in moderate amounts.

Food	Amount	Exchanges
Angel food cake	1/12 cake	2 starch
Cake, no icing	1/12 cake or a 3" square	2 starch, 2 fat
Cookies	2 small, 1¾" across	1 starch, 1 fat
Frozen fruit yogurt	⅓ c	1 starch
Gingersnaps	3	1 starch
Granola	¼ c	1 starch, 1 fat
Granola bars	1 small	1 starch, 1 fat
Ice cream, any flavor	½ c	1 starch, 2 fat
Ice milk, any flavor	½ c	1 starch, 1 fat
Sherbet, any flavor	¼ c	1 starch
Snack chips, all varieties	1 oz	1 starch, 2 fat
Vanilla wafers	6 small	1 starch, 1 fat

If more than one serving is eaten, these foods have 400 milligrams or more sodium.

APPENDIX D

D

Table of Intentional Food Additives

Table D.1
A guide to Intentional Food Additives

Abbreviation	Type of Additive	Uses
AA	Anticaking agents	Keeps dry powders and crystals from clumping together (e.g., salt, powdered sugar).
C	Colors	Synthetic (laboratory-made) vegetable and fruit concentrates and other substances used to color foods (e.g., soft drinks, frosting).
E	Emulsifiers	Used to make oil and water mix (e.g., salad dressings, sauces).
EX	Extenders	"Fillers" such as fruit pulp and texturized protein (e.g., fruit drinks, hamburger).
FA	Flavoring agents	Used to add particular flavors to food (e.g., pudding, rye bread).
N	Nutrients	Used to add vitamins or minerals to foods (e.g., breakfast cereals, skim milk).
P	Preservatives	Used to keep food from spoiling (e.g., breads, breakfast cereals).
S	Sweeteners	Used to sweeten foods. Some are "artificial," such as aspartame and saccharin, and some are extracted from plants, such as sugar cane and sugar beets (e.g., soft drinks, catsup).
T	Texturizers	Used to improve the texture of food by stabilizing moisture content, dryness, volume, tenderness, or hardness (e.g., cakes, breads).
TH	Thickeners	Used to improve the consistency of foods (e.g., low-fat salad dressings, low-calorie jams).

Common Food Additives and Their Primary Function

Alginates (T)
Alpha tocopherol (vitamin E) (N)
Alpha tocopheryl acetate (vitamin E) (N)
Ascorbic acid (vitamin C) (P)
Aspartame (NutraSweet™) (S)
Baking powder (T)
Beet juice (C)
Beet sugar (S)
Beta-carotene (N)
BHA (P)
BHT (P)
Calcium carbonate (calcium) (N)
Calcium pantothenate (pantothenic acid, a B vitamin) (N)
Calcium propionate (P)
Calcium silicate (AA)
Cane sugar (S)
Carotene (C)
Carrageenan (E, TH)
Cellulose gum (TH)
Chromium chloride (N)
Citric acid (FA, P)
Corn syrup (S)
Cupric oxide (copper) (N)
Cyanocobalamin (vitamin B_{12}) (N)
Dextrin (TH)
Dextrose (S)
Dibasic calcium phosphate (calcium phosphorus) (N)
Diglycerides (E)
EDTA (P)
Extracts (FA)
FD&C blue no. 1 (C)
FD&C red no. 3 (use is being phased out) (C)

continued

D

Table D.1
A guide to Intentional Food Additives, continued

Common Food Additives and Their Primary Function

FD&C yellow no. 5 (C)
Ferrous fumarate (iron) (N)
Ferrous sulfate (iron) (N)
Fructose (S)
Fruit pulp (EX)
Gelatin (TH)
Glycerol (T)
Glyceryl abietate (E, T)
Guar gum (T, TH)
High-fructose corn syrup (S)
Honey (S)
Hydrolyzed protein (EX)
Lecithin (E)
Magnesium oxide (magnesium) N
Maltodextrin (S, TH)
Manganese sulfate (manganese) (N)
Modified food starch (TH)
Monoglycerides (E)
Monosodium glutamate (MSG) (FA)
Natural flavorings (FA)
Niacinamide (niacin, vitamin B_3) (N)
Nitrates (P)
Nitrites (P)
Paprika (C)
Pectin (TH)
Phosphoric acid (P)
Phytonadione (vitamin K) (N)
Polysorbates (P)
Potassium benzoate (P)
Potassium bicarbonate (T)
Potassium chloride (N)
Potassium iodite (iodine) (N)
Potassium metabisulfite (P)
Potassium sorbate (P)
Propyl gallate (P)
Propyleneglycol (T)
Pyridoxine hydrochloride (vitamin B_6) (N)
Reduced iron (iron) (N)
Retinol (vitamin A) (N)
Riboflavin (vitamin B_2) (N)
Saccharin (S)
Salt (FA, P)
Silicondioxide (AA)
Sodium aluminum phosphate (T)
Sodium ascorbite (vitamin C) (N)
Sodium benzoate (P)
Sodium bicarbonate (baking soda) (T)
Sodium bisulfite (P)
Sodium chloride (P, FA)
Sodium citrate (P)
Sodium erythorbate (P)
Sodium hexametaphosphate (P, T)
Sodium metabisulfite (P)
Sodium molydate (molybdenum) (N)
Sodium nitrate (P)
Sodium nitrite (P)
Sodium propionate (P)
Sodium saccharin (S)
Sodium selenate (selenium) (N)
Sodium stearolyn-2-lactylate (P, E)
Sodium sulfite (P)
Sorbitan monostearate (E)
Sorbitol (S, T)
Sorghum (S, P)
Starches (TH)
Sucrose (S)
Sugar (P)
Sulfur dioxide (P)
Sweeteners (FA)
Texturized protein (EX)
Thiamin hydrochloride (thiamin, B_1) (N)
Thiamin mononitrate (thiamin, B_1) (N)
Tocopherols (P, N)
Turmeric (C)
Vitamin A palmitate (vitamin A) (N)
Vitamin C (ascorbic acid) (N, P)
Vitamin E (N, P)
Xanthan gum (T)
Xylitol (S)
Yeast (T)
Zinc oxide (zinc) (N)

APPENDIX E

Cells

This appendix presents an overview of the basic structure and functions of ***cells*** in the human body. Cell structure and function are central to discussions of nutrition for it is within cells that nutrients are utilized to sustain life and health. The life-sustaining processes that take place within each of the more than one hundred trillion cells in the human body are maintained by the nutrients we consume in our diet. Chemical reactions within the cells produce energy from carbohydrates, proteins, and fats; cause proteins to be broken down or built up; and in thousands of other ways keep us in a state of health.

➡ **cell**
The basic unit of life, of which all living things are composed. Every cell is surrounded by a membrane and contains cytoplasm, within which are organelles and a nucleus; the cell nucleus contains chromosomes.

A cell is a basic unit of life. Any substance that does not consist of one or more cells cannot be alive. Cells are the "building blocks" of tissues (such as muscles and bones), organs (such as the kidneys and liver), and systems (the respiratory and digestive systems, for example). Normal cell health and functioning are maintained when a state of nutritional and environmental utopia exists within and around the cells. A disruption in the availability of nutrients or the presence of harmful substances in the cell's environment can initiate disorders that eventually affect our health or growth. Health problems in general begin with disruptions in the normal activity of cells.

The types and amounts of food and supplements people consume affect the cells' environment and their ability to function normally. Excessive or inadequate supplies of nutrients and other chemical substances disrupt cell functions and result in health problems. Humans remain in a state of health as long as their cells do.

Cell Structures and Functions

A generalized diagram of a human cell is shown in Illustration F.1. Although all cells have some structures and functions in common, the specific functions performed, and the structures that support those functions, can vary a good deal from cell to cell. Cells lining the esophagus, stomach, and intestines, for example, are specialized to produce and secrete mucus that helps food pass through the digestive tract. Red blood cells are specially formed to transport oxygen and carbon dioxide.

Every cell is surrounded by a cell membrane that helps move nutrients into and out of the cell. Inside the cell membrane lies the cytoplasm, a fluid material that contains many organelles, tubes, and particles. Among the organelles in the cytoplasm are ribosomes, mitochondria, and lysosomes. Each of these "little organs" is encased in a cell membrane and performs specific functions. Ribosomes assemble amino acids into proteins following the instructions of DNA and its messenger RNA. The mitochondria are made of intricately folded membranes that bear thousands of highly organized sets of enzymes on their surfaces. These enzymes are actively involved in the production of energy and are found in particularly dense quantities in muscle cells. The lysosomes are like packets of enzymes. The enzymes are used to break down old cell particles that are being recycled and to destroy substances that are harmful to the body.

Cytoplasm also contains a highly organized system of membranes called the endoplasmic reticulum. When these membranes are dotted with ribosomes, they are called "rough endoplasmic reticulum." When ribosomes are absent, the endoplasmic reticulum is referred to as "smooth." Some membranes within the

E

cytoplasm form tubes that collect certain types of cellular material and transport it out of the cell. These membranous tubes are called "Golgi apparatus." The rough and smooth endoplasmic reticulum are continuous with the Golgi apparatus, so secretions produced throughout the cell can be collected and transported to its exterior.

Within each cell is a nucleus covered by a two-layer membrane. The nucleus contains chromosomes and the genetic material DNA. DNA encodes all of the instructions a cell needs to conduct protein synthesis and to replicate life.

All cells within the body are part of a complex communication system that uses hormones, electrical impulses, and other chemical messengers to link each cell to the others. No cell is an island that operates independently from the others.

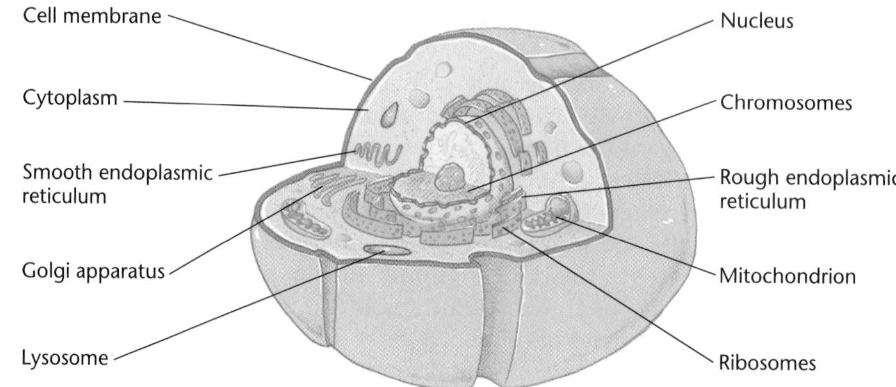

Illustration E.1
Generalized structure of a human cell.

cell membrane: the membrane that surrounds the cell and encloses its contents; made primarily of lipid and protein.

chromosomes: a set of structures within the nucleus of every cell that contain the cell's genetic material, DNA, associated with other materials (primarily proteins).

cytoplasm (SIGH-toe-plazm): the cell contents, except for the nucleus.
cyto = cell
plasm = a form

Golgi (GOAL-gee) **apparatus:** a set of membranes within the cell where secretory materials are packaged for export.

lysosomes: cellular organelles; membrane-enclosed sacs of degradative enzymes.
lysis = dissolution

mitochondria (my-toe-KON-dree-uh; *singular* **mitochondrion):** the cellular organelles responsible for producing ATP aerobically; made of membranes (lipid and protein) with enzymes mounted on them.
mitos = thread (referring to their slender shape)
chondros = cartilage (referring to their external appearance)

nucleus: a major membrane-enclosed body within every cell, which contains the cell's genetic material, DNA, embedded in chromosomes.
nucleus = a kernel

organelles: membrane-bound subcellular structures such as ribosomes, mitochondria, and lysosomes.
organelle = little organ

ribosomes: protein-making organelles in cells; composed of RNA and protein.
ribo = containing the sugar ribose
some = body

rough endoplasmic reticulum (en-doh-PLAZ-mic reh-TIC-you-lum): intracellular membranes dotted with ribosomes, where protein synthesis takes place.
endo = inside
plasm = the cytoplasm

smooth endoplasmic reticulum: smooth intracellular membranes bearing no ribosomes.

Contents

APPENDIX F

Canada: Recommendations, Choice System, and Labels

This appendix presents details for Canadians on Canada's exchange system (called the choice system in Canada), and food guides. Appendix C includes addresses of Canadian governmental agencies and professional organizations that may provide additional information.

Choice System for Meal Planning

The *Good Health Eating Guide* is the Canadian choice system of meal planning.[1] It contains several features similar to those of the U.S. exchange system including the following:

- Foods are divided into lists according to carbohydrate, protein, and fat content.
- Foods are interchangeable within a group.
- Most foods are eaten in measured amounts.
- An energy value is given for each food group.

Tables F.1 through F.9 present the Canadian choice system.

Table F.1

Average Energy Requirements for Canadians

Age	Sex	Average Height (cm)	Average Weight (kg)	Requirements[a]					
				(kcal/kg)[b]	(MJ/kg)[b]	(kcal/day)	(MJ/day)	(kcal/cm)	(MJ/cm)
Infants (months)									
0–2	Both	55	4.5	120–100	0.50–0.42	500	2.0	9	0.04
3–5	Both	63	7.0	100–95	0.42–0.40	700	2.8	11	0.05
6–8	Both	69	8.5	95–97	0.40–0.41	800	3.4	11.5	0.05
9–11	Both	73	9.5	97–99	0.41	950	3.8	12.5	0.05
Children and Adults (years)									
1	Both	82	11	101	0.42	1100	4.8	13.5	0.06
2–3	Both	95	14	94	0.39	1300	5.6	13.5	0.06
4–6	Both	107	18	100	0.42	1800	7.6	17	0.07
7–9	M	126	25	88	0.37	2200	9.2	17.5	0.07
	F	125	25	76	0.32	1900	8.0	15	0.06
10–12	M	141	34	73	0.30	2500	10.4	17.5	0.07
	F	143	36	61	0.25	2200	9.2	15.5	0.06
13–15	M	159	50	57	0.24	2800	12.0	17.5	0.07
	F	157	48	46	0.19	2200	9.2	14	0.06
16–18	M	172	62	51	0.21	3200	13.2	18.5	0.08
	F	160	53	40	0.17	2100	8.8	13	0.05
19–24	M	175	71	42	0.18	3000	12.6		
	F	160	58	36	0.15	2100	8.8		
25–49	M	172	74	36	0.15	2700	11.3		
	F	160	59	32	0.13	1900	8.0		
50–74	M	170	73	31	0.13	2300	9.7		
	F	158	63	29	0.12	1800	7.6		
751	M	168	69	29	0.12	2000	8.4		
	F	155	64	23	0.10	1500	6.3		

[a]Requirements can be expected to vary within a range of ±30 percent.

[b]First and last figures are averages at the beginning and end of the three-month period.

Source: Health and Welfare Canada, *Nutrition Recommendations: The Report of the Scientific Review Committee* (Ottawa: Canadian Government Publishing Centre, 1990), Tables 5 and 6, pp. 25, 27.

F

Table F.2
Canadian Choice System: Starch Foods

1 starch choice = 15 g carbohydrate (starch), 2 g protein, 290 kJ (68 kcal)

Food	Measure	Mass (Weight)
Breads		
Bagels	½	30 g
Bread crumbs	50 mL (¼ c)	30 g
Bread cubes	250 mL (1 c)	30 g
Bread sticks	2	20 g
Brewis, cooked	50 mL (¼ c)	45 g
Chapati	1	20 g
Cookies, plain	2	20 g
English muffins, crumpets	½	30 g
Flour	40 mL (2½ tbs)	20 g
Hamburger buns	½	30 g
Hot dog buns	½	30 g
Kaiser rolls	½	30 g
Matzo, 15 cm	1	20 g
Melba toast, rectangular	4	15 g
Melba toast, rounds	7	15 g
Pita, 20-cm (8") diameter	¼	30 g
Pita, 15-cm (6") diameter	½	30 g
Plain rolls	1 small	30 g
Pretzels	7	20 g
Raisin bread	1 slice	30 g
Rice cakes	2	30 g
Roti	1	20 g
Rusks	2	20 g
Rye, coarse or pumpernickel	½ slice	30 g
Soda crackers	6	20 g
Tortillas, corn (taco shell)	1	30 g
Tortilla, flour	1	30 g
White (French and Italian)	1 slice	25 g
Whole-wheat, cracked-wheat, rye, white enriched	1 slice	30 g
Cereals		
Bran flakes, 100% bran	125 mL (½ c)	30 g
Cooked cereals, cooked	125 mL (½ c)	125 g
Dry	30 mL (2 tbs)	20 g
Cornmeal, cooked	125 mL (½ c)	125 g
Dry	30 mL (2 tbs)	20 g
Ready-to-eat unsweetened cereals	125 mL (½ c)	20 g
Shredded wheat biscuits, rectangular or round	1	20 g
Shredded wheat, bite size	125 mL (½ c)	20 g
Wheat germ	75 mL (⅓ c)	30 g
Cornflakes	175 mL (⅔ c)	20 g
Rice Krispies	175 mL (⅔ c)	20 g
Cheerios	200 mL (¾ c)	20 g
Muffets	1	20 g
Puffed rice	300 mL (1¼ c)	15 g
Puffed wheat	425 mL (1⅔ c)	20 g

[1] The tables for the Canadian choice system are adapted from the *Good Health Eating Guide Resource,* copyright 1994, with permission of the Canadian Diabetes Association.

continued

Table F.2
Canadian Choice System: Starch Foods, continued

1 starch choice = 15 g carbohydrate (starch), 2 g protein, 290 kJ (68 kcal)

Food	Measure	Mass (Weight)
Grains		
Barley, cooked	125 mL (½ c)	120 g
Dry	30 mL (2 tbs)	20 g
Bulgur, kasha, cooked, moist	125 mL (½ c)	70 g
Cooked, crumbly	75 mL (⅓ c)	40 g
Dry	30 mL (2 tbs)	20 g
Rice, cooked, brown & white (short & long grain)	125 mL (½ c)	70 g
Rice, cooked, wild	75 mL (⅓ c)	70 g
Tapioca, pearl and granulated, quick cooking, dry	30 mL (2 tbs)	15 g
Couscous, cooked moist	125 mL (½ c)	70 g
Dry	30 mL (tbs)	20 g
Quinoa, cooked moist	125 mL (½ c)	70 g
Dry	30 mL (2 tbs)	20 g
Pastas		
Macaroni, cooked	125 mL (½ c)	70 g
Noodles, cooked	125 mL (½ c)	80 g
Spaghetti, cooked	125 mL (½ c)	70 g
Starchy Vegetables		
Beans and peas, dried, cooked	125 mL (½ c)	80 g
Breadfruit	1 slice	75 g
Corn, canned, whole kernel	125 mL (½ c)	85 g
Corn on the cob	½ medium cob	140 g
Cornstarch	30 mL (2 tbs)	15 g
Plantains	⅓ small	50 g
Popcorn, air-popped, unbuttered	750 mL (3 c)	20 g
Potatoes, whole (with or without skin)	½ medium	95 g
Yams, sweet potatoes, (with or without skin)	½	75 g

Food	Choices per serving	Measure	Mass (Weight)
NOTE: Food items found in this category provide more than 1 starch choice:			
Bran flakes	1 starch + ½ sugar	150 mL (⅔ c)	24 g
Croissant, small	1 starch + 1½ fats	1 small	35 g
Large	1 starch + 1½ fats	½ large	30 g
Corn, canned creamed	1 starch + ½ fruits and vegetables	12 mL (½ c)	113 g
Potato chips	1 starch + 2 fats	15 chips	30 g
Tortilla chips (nachos)	1 starch + 1½ fats	13 chips	20 g
Corn chips	1 starch + 2 fats	30 chips	30 g
Cheese twists	1 starch + 1½ fats	30 chips	30 g
Cheese puffs	1 starch + 2 fats	27 chips	30 g
Tea biscuit	1 starch + 2 fats	1	30 g
Pancakes, homemade using 50 mL (¼ c) batter (6" diameter)	1½ starches + 1 fat	1 medium	50 g
Potatoes, french fried (homemade or frozen)	1 starch + 1 fat	10 regular size	35 g
Soup, canned* (prepared with equal volume of water)	1 starch	250 mL (1 c)	260 g
Waffles, packaged	1 starch + 1 fat	1	35 g

*Soup can vary according to brand and type. Check the label for Food Choice Values and Symbols or the core nutrient listing.

F

Table F.3
Canadian Choice System: Fruits and Vegetables

1 fruits and vegetables choice = 10 g carbohydrate, 1 g protein, 190 kJ (44 kcal)

Food	Measure	Mass (Weight)
Fruits (fresh, frozen, without sugar, canned in water)		
Apples, raw (with or without skin)	½ medium	75 g
Sauce unsweetened	125 mL (½ c)	120 g
Sweetened	see *Combined Food Choices*	
Apple butter	20 mL (4 tsp)	20 g
Apricots, raw	2 medium	115 g
Canned, in water	4 halves, plus 30 mL (2 tbs) liquid	110 g
Bake-apples (cloudberries), raw	125 mL (½ c)	120 g
Bananas, with peel	½ small	75 g
Peeled	½ small	50 g
Berries (blackberries, blueberries, boysenberries, huckleberries, loganberries, raspberries)		
Raw	125 mL (½ c)	70 g
Canned, in water	125 mL (½ c), plus 30 mL (2 tbs) liquid	100 g
Cantaloupe, wedge with rind	¼	240 g
Cubed or diced	250 mL (1 c)	160 g
Cherries, raw, with pits	10	75 g
Raw, without pits	10	70 g
Canned, in water, with pits	75 mL (⅓ c), plus 30 mL (2 tbs) liquid	90 g
Canned, in water, without pits	75 mL (⅓ c), plus 30 mL (2 tbs) liquid	85 g
Crabapples, raw	1 small	55 g
Cranberries, raw	250 mL (1 c)	100 g
Figs, raw	1 medium	50 g
Canned, in water	3 medium, plus 30 mL (2 tbs) liquid	100 g
Foxberries, raw	250 mL (1 c)	100 g
Fruit cocktail, canned, in water	125 mL (½ c), plus 30 mL (2 tbs) liquid	120 g
Fruit, mixed, cut-up	125 mL (½ c)	120 g
Gooseberries, raw	250 mL (1 c)	150 g
Canned, in water	250 mL (1 c), plus 30 mL (2 tbs) liquid	230 g
Grapefruit, raw, with rind	½ small	185 g
Raw, sectioned	125 mL (½ c)	100 g
Canned, in water	125 mL (½ c), plus 30 mL (2 tbs) liquid	120 g
Grapes, raw, slip skin	125 mL (½ c)	75 g
Raw, seedless	125 mL (½ c)	75 g
Canned, in water	75 mL (⅓ c), plus 30 mL (2 tbs) liquid	115 g
Guavas, raw	½	50 g
Honeydew melon, raw, with rind	⅒	225 g
Cubed or diced	250 mL (1 c)	170 g
Kiwis, raw, with skin	2	155 g
Kumquats, raw	3	60 g
Loquats, raw	8	130 g
Lychee fruit, raw	8	120 g
Mandarin oranges, raw, with rind	1	135 g
Raw, sectioned	125 mL (½ c)	100 g
Canned, in water	125 mL (½ c), plus 30 mL (2 tbs) liquid	100 g
Mangoes, raw, without skin and seed	⅓	65 g
Diced	75 mL (⅓ c)	65 g
Nectarines	½ medium	75 g
Oranges, raw, with rind	1 small	130 g

continued

F

Table F.3
Canadian Choice System: Fruits and Vegetables, continued

1 fruits and vegetables choice = 10 g carbohydrate, 1 g protein, 190 kJ (44 kcal)

Food	Measure	Mass (Weight)
Raw, sectioned	125 mL (½ c)	95 g
Papayas, raw, with skin and seeds	¼ medium	150 g
Raw, without skin and seeds	¼ medium	100 g
Cubed or diced	125 mL (½ c)	100 g
Peaches, raw, with seed and skin	1 large	100 g
Raw, sliced or diced	125 mL (½ c)	100 g
Canned in water, halves or slices	125 mL (½ c), plus 30 mL (2 tbs) liquid	120 g
Pears, raw, with skin and core	½	90 g
Raw, without skin and core	½	85 g
Canned, in water, halves	1 half plus 30 mL (2 tbs) liquid	90 g
Persimmons, raw, native	1	30 g
Raw, Japanese	¼	50 g
Pineapple, raw	1 slice	75 g
Raw, diced	125 mL (½ c)	75 g
Canned, in juice, diced	75 mL (⅓ c), plus 15 mL (1 tbs) liquid	55 g
Canned, in juice, sliced	1 slice, plus 15 mL (1 tbs) liquid	55 g
Canned, in water, diced	125 mL (½ c), plus 30 mL (2 tbs) liquid	100 g
Canned, in water, sliced	2 slices, plus 15 mL (1 tbs) liquid	100 g
Plums, raw	2 small	60 g
Damson	6	65 g
Japanese	1	70 g
Canned, in apple juice	2, plus 30 mL (2 tbs) liquid	70 g
Canned, in water	3, plus 30 mL (2 tbs) liquid	100 g
Pomegranates, raw	½	140 g
Strawberries, raw	250 mL (1 c)	150 g
Frozen/canned, in water	250 mL (1 c), plus 30 mL (2 tbs) liquid	240 g
Rhubarb	250 mL (1 c)	150 g
Tangelos, raw	1	205 g
Tangerines, raw	1 medium	115 g
Raw, sectioned	125 mL (½ c)	100 g
Watermelon, raw, with rind	1 wedge	310 g
Cubed or diced	250 mL (1 c)	160 g
Dried Fruit		
Apples	5 pieces	15 g
Apricots	4 halves	15 g
Banana flakes	30 mL (2 tbs)	15 g
Currants	30 mL (2 tbs)	15 g
Dates, without pits	2	15 g
Peaches	½	15 g
Pears	½	15 g
Prunes, raw, with pits	2	15 g
Raw, without pits	2	10 g
Stewed, no liquid	2	20 g
Stewed, with liquid	2, plus 15 mL (1 tbs) liquid	35 g
Raisins	30 mL (2 tbs)	15 g
Juices (no sugar added or unsweetened)		
Apricot, grape, guava, mango, prune	50 mL (¼ c)	55 g
Apple, carrot, papaya, pear, pineapple, pomegranate	75 mL (⅓ c)	80 g
Cranberry (see Sugars Section)		
Clamato (see Sugars Section)		

continued

F

Table F.3
Canadian Choice System: Fruits and Vegetables, continued

1 fruits and vegetables choice = 10 g carbohydrate, 1 g protein, 190 kJ (44 kcal)

Food	Measure	Mass (Weight)
Grapefruit, loganberry, orange, raspberry, tangelo, tangerine	125 mL (½ c)	130 g
Tomato, tomato-based mixed vegetables	250 mL (1 c)	255 g
Vegetables (fresh, frozen, or canned)		
Artichokes, French, globe	2 small	50 g
Beets, diced or sliced	125 mL (½ c)	85 g
Carrots, diced, cooked or uncooked	125 mL (½ c)	75 g
Chestnuts, fresh	5	20 g
Parsnips, mashed	125 mL (½ c)	80 g
Peas, fresh or frozen	125 mL (½ c)	80 g
Canned	75 mL (⅓ c)	55 g
Pumpkin, mashed	125 mL (½ c)	45 g
Rutabagas, mashed	125 mL (½ c)	85 g
Sauerkraut	250 mL (1 c)	235 g
Snow peas	250 mL (1 c)	135 g
Squash, yellow or winter, mashed	125 mL (½ c)	115 g
Succotash	75 mL (⅓ c)	55 g
Tomatoes, canned	250 mL (1 c)	240 g
Tomato paste	50 mL (¼ c)	55 g
Tomato sauce*	75 mL (⅓ c)	100 g
Turnips, mashed	125 mL (½ c)	115 g
Vegetables, mixed	125 mL (½ c)	90 g
Water chestnuts	8 medium	50 g

*Tomato sauce varies according to brand name. Check the label or discuss with your dietitian.

Table F.4
Canadian Choice System: Milk

Type of Milk	Carbohydrate (g)	Protein (g)	Fat (g)	Energy
Nonfat (0%)	6	4	0	170 kJ (40 kcal)
1%	6	4	1	206 kJ (49 kcal)
2%	6	4	2	244 kJ (58 kcal)
Whole (4%)	6	4	4	319 kJ (76 kcal)

Food	Measure	Mass (Weight)
Buttermilk (higher in salt)	125 mL (½ c)	125 g
Evaporated milk	50 mL (¼ c)	50 g
Milk	125 mL (½ c)	125 g
Powdered milk, regular	30 mL (2 tbs)	15 g
Instant	50 mL (¼ c)	15 g
Plain yogurt	125 mL (½ c)	125 g

Food	Choices per Serving	Measure	Mass (Weight)
NOTE: Food items found in this category provide more than 1 milk choice:			
Milkshake	1 milk + 3 sugars + ½ protein	250 mL (1 c)	300 g
Chocolate milk, 2%	2 milks 2% + 1 sugar	250 mL (1 c)	300 g
Frozen yogurt	1 milk + 1 sugar	125 mL (½ c)	125 g

Table F.5
Canadian Choice System: Sugars

1 sugar choice = 10 g carbohydrate (sugar), 167 kJ (40 kcal)

Food		Measure	Mass (Weight)
Beverages			
Condensed milk		15 mL (1 tbs)	
Flavoured fruit crystals*		75 mL (⅓ c)	
Iced tea mixes*		75 mL (⅓ c)	
Regular soft drinks		125 mL (½ c)	
Sweet drink mixes*		75 mL (⅓ c)	
Tonic water		125 mL (½ c)	
*These beverages have been made with water.			
Miscellaneous			
Bubble gum (large square)		1 piece	5 g
Cranberry cocktail		75 mL (⅓ c)	80 g
Cranberry cocktail, light		350 mL (1⅓ c)	260 g
Cranberry sauce		30 mL (2 tbs)	
Hard candy mints		2	5 g
Honey, molasses, corn & cane syrup		10 mL (2 tsp)	15 g
Jelly bean		4	10 g
Licorice		1 short stick	10 g
Marshmallows		2 large	15 g
Popsicle		1 stick (½ popsicle)	
Powdered gelatin mix (Jello®) (reconstituted)		50 mL (¼ c)	
Regular jam, jelly, marmalade		15 mL (1 tbs)	
Sugar, white, brown, icing, maple		10 mL (2 tsp)	10 g
Sweet pickles		2 small	100 g
Sweet relish		30 mL (2 tbs)	
The following food items provide more than 1 sugar choice:			
Brownie	1 sugar + 1 fat	1	20 g
Clamato juice	1½ sugars	175 mL (⅔ c)	
Fruit salad, light syrup	1 sugar + 1 fruits & vegetables	125 mL (½ c)	130 g
Aero® bar	2½ sugars + 2½ fats	1 bar	43 g
Smarties®	4½ sugars + 2 fats	1 box	60 g
Sherbet	3 sugars + ½ fat	125 mL (½ c)	95 g

F

Table F.6
Canadian Choice System: Protein Foods

1 protein choice = 7 g protein, 3 g fat, 230 kJ (55 kcal)

Food	Measure	Mass (Weight)
Cheese		
Low-fat cheese, about 7% milk fat	1 slice	30 g
Cottage cheese, 2% milkfat or less	50 mL (¼ c)	55 g
Ricotta, about 7% milkfat	50 mL (¼ c)	60 g
Fish		
Anchovies (see *Extras,* Table I-9)		
Canned, drained (e.g., mackerel, salmon, tuna packed in water)	(⅓ of 6.5 oz can)	30 g
Cod tongues, cheeks	75 mL (⅓ c)	50 g
Fillet or steak (e.g., Boston blue, cod, flounder, haddock, halibut, mackerel, orange roughy, perch, pickerel, pike, salmon, shad, snapper, sole, swordfish, trout, tuna, whitefish)	1 piece	30 g
Herring	⅓ fish	30 g
Sardines, smelts	2 medium or 3 small	30 g
Squid, octopus	50 mL (¼ c)	40 g
Shellfish		
Clams, mussels, oysters, scallops, snails	3 medium	30 g
Crab, lobster, flaked	50 mL (¼ c)	30 g
Shrimp, fresh	5 large	30 g
Frozen	10 medium	30 g
Canned	18 small	30 g
Dry pack	50 mL (¼ c)	30 g
Meat and Poultry (e.g., beef, chicken, goat, ham, lamb, pork, turkey, veal, wild game)		
Back, peameal bacon	3 slices, thin	30 g
Chop	½ chop, with bone	40 g
Minced or ground, lean or extra-lean	30 mL (2 tbs)	30 g
Sliced, lean	1 slice	30 g
Steak, lean	1 piece	30 g
Organ Meats		
Hearts, liver	1 slice	30 g
Kidneys, sweetbreads, chopped	50 mL (¼ c)	30 g
Tongue	1 slice	30 g
Tripe	5 pieces	60 g
Soyabean		
Bean curd or tofu	½ block	70 g
Eggs		
In shell, raw or cooked	1 medium	50 g
Without shell, cooked or poached in water	1 medium	45 g
Scrambled	50 mL (¼ c)	55 g

continued

F

Table F.6
Canadian Choice System: Protein Foods, continued

Food	Choices per Serving	Measures	Mass (Weight)
Note: The following choices provide more than 1 protein exchange:			
Cheese			
Cheeses	1 protein + 1 fat	1 piece	25 g
Cheese, coarsely grated (e.g., cheddar)	1 protein + 1 fat	50 mL (¼ c)	25 g
Cheese, dry, finely grated (e.g., parmesan)	1 protein + 1 fat	45 mL	15 g
Cheese, ricotta, high fat	1 protein + 1 fat	50 mL (¼ c)	55 g
Fish			
Eel	1 protein + 1 fat	1 slice	50 g
Meat			
Bologna	1 protein + 1 fat	1 slice	20 g
Canned lunch meats	1 protein + 1 fat	1 slice	20 g
Corned beef, canned	1 protein + 1 fat	1 slice	25 g
Corned beef, fresh	1 protein + 1 fat	1 slice	25 g
Ground beef, medium-fat	1 protein + 1 fat	30 mL (2 tbs)	25 g
Meat spreads, canned	1 protein + 1 fat	45 mL	35 g
Mutton chop	1 protein + 1 fat	½ chop, with bone	35 g
Paté (see *Fats and Oils* group, Table I-8)			
Sausages, garlic, Polish or knockwurst	1 protein + 1 fat	1 slice	50 g
Sausages, pork, links	1 protein + 1 fat	1 link	25 g
Spareribs or shortribs, with bone	1 protein + 1 fat	1 large	65 g
Stewing beef	1 protein + 1 fat	1 cube	25 g
Summer sausage or salami	1 protein + 1 fat	1 slice	40 g
Weiners, hot dog	1 protein + 1 fat	½ medium	25 g
Miscellaneous			
Blood pudding	1 protein + 1 fat	1 slice	25 g
Peanut butter	1 protein + 1 fat	15 mL (1 tbs)	15 g

F

Table F.7
Canadian Choice System: Fats and Oils

1 fat choice = 5 g fat, 190 kJ (45 kcal)

Food	Measure	Mass (Weight)	Food	Measure	Mass (Weight)
Avocado*	⅛	30 g	Nuts (continued):		
Bacon, side, crisp*	1 slice	5 g	Sesame seeds	15 mL (1 tbs)	10 g
Butter*	5 mL (1 tsp)	5 g	Sunflower seeds		
Cheese spread	15 mL (1 tbs)	15 g	Shelled	15 mL (1 tbs)	10 g
Coconut, fresh*	45 mL (3 tbs)	15 g	In shell	45 mL (3 tbs)	15 g
Coconut, dried*	15 mL (1 tbs)	10 g	Walnuts	4 halves	10 g
Cream, Half and half			Oil, cooking and salad	5 mL (1 tsp)	5 g
(cereal), 10%*	30 mL (2 tbs)	30 g	Olives, green	10	45 g
Light (coffee), 20%*	15 mL (1 tbs)	15 g	Ripe black	7	57 g
Whipping, 32 to 37%*	15 mL (1 tbs)	15 g	Pâté, liverwurst,	15 mL (1 tbs)	15 g
Cream cheese*	15 mL (1 tbs)	15 g	meat spreads		
Gravy*	30 mL (2 tbs)	30 g	Salad dressing: blue,	10 mL (2 tsp)	10 g
Lard*	5 mL (1 tsp)	5 g	French, Italian,		
Margarine	5 mL (1 tsp)	5 g	mayonnaise,		
Nuts, shelled:			Thousand Island	5 mL (1 tsp)	5 g
Almonds	8	5 g	Salad dressing,	30 mL (2 tbs)	30 g
Brazil nuts	2	10 g	low-calorie		
Cashews	5	10 g	Salt pork, raw	5 mL (1 tsp)	5 g
Filberts, hazelnuts	5	10 g	or cooked*		
Macadamia	3	5 g	Sesame oil	5 mL (1 tsp)	5 g
Peanuts	10	10g	Sour cream		
Pecans	5 halves	5 g	12% milkfat	30 mL (2 tbs)	30 g
Pignolias, pine nuts	25 mL (5 tsp)	10 g	7% milkfat	60 mL (4 tbs)	60 g
Pistachios, shelled	20	10 g	Shortening*	5 mL (1 tsp)	
Pistachios, in shell	20	20 g			
Pumpkin and	20 mL (4 tsp)	10 g			
squash seeds					

*These items contain higher amounts of saturated fat.

Table F.8
Canadian Choice System: Extras

Extras have no more than 2.5 g carbohydrate, 60 kJ (14 kcal)

Vegetables 125 mL (½ c)

Artichokes
Asparagus
Bamboo shoots
Bean sprouts, mung or soya
Beans, string, green, or yellow
Bitter melon (balsam pear)
Bok choy
Broccoli
Brussels sprouts
Cabbage
Cauliflower
Celery
Chard
Cucumbers
Eggplant
Endive
Fiddleheads
Greens: beet, collard, dandelion, mustard, turnip, etc.
Kale
Kohlrabi
Leeks
Lettuce
Mushrooms
Okra
Onions, green or mature
Parsley
Peppers, green, yellow or red
Radishes
Rapini
Rhubarb
Sauerkraut
Shallots
Spinach
Sprouts: alfalfa, radish, etc.
Tomato wedges
Watercress
Zucchini

Free Foods (may be used without measuring)

Artificial sweetener, such as cyclamate or aspartame
Baking powder, baking soda
Bouillon from cube, powder, or liquid
Bouillon or clear broth
Chowchow, unsweetened
Coffee, clear
Consommé
Dulse
Flavorings and extracts
Garlic
Gelatin, unsweetened
Ginger root
Herbal teas, unsweetened
Horseradish, uncreamed
Lemon juice or lemon wedges
Lime juice or lime wedges
Marjoram, cinnamon, etc.
Mineral water
Mustard
Parsley
Pimentos
Salt, pepper, thyme
Soda water, club soda
Soya sauce
Sugar-free Crystal Drink
Sugar-free Jelly Powder
Sugar-free soft drinks
Tea, clear
Vinegar
Water
Worcestershire sauce

Condiments

Food	**Measure**
Anchovies	2 fillets
Barbecue sauce	15 mL (1 tbs)
Bran, natural	30 mL (2 tbs)
Brewer's yeast	5 mL (1 tsp)
Carob powder	5 mL (1 tsp)
Catsup	5 mL (1 tsp)
Chili sauce	5 mL (1 tsp)
Cocoa powder	5 mL (1 tsp)
Cranberry sauce, unsweetened	15 mL (1 tbs)
Dietetic fruit spreads	5 mL (1 tsp)
Maraschino cherries	1
Nondairy coffee whitener	5 mL (1 tsp)
Nuts, chopped pieces	5 mL (1 tsp)
Pickles	
unsweetened dill	2
sour mixed	11
Sugar substitutes, granular	5 mL (1 tsp)
Whipped toppings	15 mL (1 tbs)

F

Table F.9
Canadian Choice System: Combined Food Choices

Food	Choices per serving	Measure	Mass (Weight)
Angel food cake	½ starch + 2½ sugars	1/12 cake	50 g
Apple crisp	½ starch + 1½ fruits & vegetables + 1 sugar + 1–2 fats	125 mL (½ c)	
Applesauce, sweetened	1 fruits & vegetables + 1 sugar	125 mL (½ c)	
Beans and pork in tomato sauce	1 starch + ½ fruits & vegetables + ½ sugar + 1 protein	125 mL (½ c)	135 g
Beef burrito	2 starches + 3 proteins + 3 fats		110 g
Brownie	1 sugar + 1 fat	1	20 g
Cabbage rolls*	1 starch + 2 proteins	3	310 g
Caesar salad	2–4 fats	20 mL dressing (4 tsp)	
Cheesecake	½ starch + 2 sugars + ½ protein + 5 fats	1 piece	80 g
Chicken fingers	1 starch + 2 proteins + 2 fats	6 small	100 g
Chicken and snow pea Oriental	2 starches + ½ fruits & vegetables + 3 proteins + 1 fat	500 mL (2 c)	
Chili	1½ starches + ½ fruits & vegetables + 3½ protein	300 mL (1¼ c)	325 g
Chips			
Potato chips	1 starch + 2 fats	15 chips	30 g
Corn chips	1 starch + 2 fats	30 chips	30 g
Tortilla chips	1 starch + 1½ fats	13 chips	
Cheese twist	1 starch + 1½ fats	30 chips	30 g
Chocolate bar			
Aero®	2½ sugars + 2½ fats	bar	43 g
Smarties®	4½ sugars + 2 fats	package	60 g
Chocolate cake (without icing)	1 starch + 2 sugars + 3 fats	1/10 of a 8" pan	
Chocolate devil's food cake (without icing)	2 starches + 2 sugars + 3 fats	1/12 of a 9" pan	
Chocolate milk	2 milks 2% + 1 sugar	250 mL (1 c)	300 g
Clubhouse (triple-decker) sandwich	3 starches + 3 proteins + 4 fats		
Cookies			
chocolate chip	½ starch + ½ sugar + 1½ fats	2	22 g
oatmeal	1 starch + 1 sugar + 1 fat	2	40 g
Donut (chocolate glazed)	1 starch + 1½ sugars + 2 fats	1	65 g
Egg roll	1 starch + ½ protein + 1 fat		75 g
Four bean salad	1 starch + ½ protein + 1 fat	125 mL (½ c)	
French toast	1 starch + ½ protein + 2 fats	1 slice	65 g
Fruit in heavy syrup	1 fruits & vegetables + 1½ sugars	125 mL (½ c)	
Granola bar	½ starch + 1 sugar + 1–2 fats		30 g
Granola cereal	1 starch + 1 sugar + 2 fats	125 mL (½ c)	45 g
Hamburger	2 starches + 3 proteins + 2 fats	junior size	
Ice cream and cone, plain flavour			
ice cream	½ milk + 2–3 sugars + 1–2 fats		100 g
cone	½ sugar		4 g
Lasagna			
regular cheese	1 starch + 1 fruits & vegetables + 3 proteins + 2 fats	3" x 4" piece	
low-fat cheese	1 starch + 1 fruits & vegetables + 3 proteins	3" x 4" piece	

* If eaten with sauce, add ½ fruits & vegetables exchange.

continued

F

Table F.9
Canadian Choice System: Combined Food Choices, continued

Food	Choices per serving	Measure	Mass (Weight)
Legumes			
Dried beans (kidney, navy, pinto, fava, chick peas)	2 starches + 1 protein	250 mL (1 c)	180 g
Dried peas	2 starches + 1 protein	250 mL (1 c)	210 g
Lentils	2 starches + 1 protein	250 mL (1 c)	210 g
Macaroni and cheese	2 starches + 2 proteins + 2 fats	250 mL (1 c)	210 g
Minestrone soup	1½ starches + ½ fruits & vegetables + ½ fat	250 mL (1 c)	
Muffin	1 starch + ½ sugar + 1 fat	1 small	45 g
Nuts (dry or roasted without any oil added).			
Almonds, dried sliced	½ protein + 2 fats	50 mL (¼ c)	22 g
Brazil nuts, dried unblanched	½ protein + 2½ fats	5 large	23 g
Cashew nuts, dry roasted	½ starch + ½ protein + 2 fats	50 mL (¼ c)	28 g
Filbert hazelnut, dry	½ protein + 3½ fats	50 mL (¼ c)	30 g
Macadamia nuts, dried	½ protein + 4 fats	50 mL (¼ c)	28 g
Peanuts, raw	1 protein + 2 fats	50 mL (¼ c)	30 g
Pecans, dry roasted	½ fruits & vegetables + 3 fats	50 mL (¼ c)	22 g
Pine nuts, pignolia dried	1 protein + 3 fats	50 mL (¼ c)	34 g
Pistachio nuts, dried	½ fruits & vegetables + ½ protein + 2½ fats	50 mL (¼ c)	27 g
Pumpkin seeds, roasted	2 proteins + 2½ fats	50 mL (¼ c)	47 g
Sesame seeds, whole dried	½ fruits & vegetables + ½ protein + 2½ fats	50 mL (¼ c)	30 g
Sunflower kernel, dried	½ protein + 1½ fats	50 mL (¼ c)	17 g
Walnuts, dried chopped	½ protein + 3 fats	50 mL (¼ c)	26 g
Perogies	2 starches + 1 protein + 1 fat	3	
Pie, fruit	1 starch + 1 fruits & vegetables + 2 sugars + 3 fats	1 piece	120 g
Pizza, cheese	1 starch + 1 protein + 1 fat	1 slice (⅛ of a 12")	50 g
Pork stir fry	½ to 1 fruits & vegetables + 3 proteins	200 mL (¾ c)	
Potato salad	1 starch + 1 fat	125 mL (½ c)	130 g
Potatoes, scalloped	2 starches + 1 milk + 1–2 fats	200 mL (¾ c)	210 g
Pudding, bread or rice	1 starch + 1 sugar + 1 fat	125 mL (½ c)	
Pudding, vanilla	1 milk + 2 sugars	125 mL (½ c)	
Raisin bran cereal	1 starch + ½ fruits & vegetables + ½ sugar	175 mL (⅔ c)	40 g
Rice krispie squares	½ starch + 1½ sugars + ½ fat	1 square	30 g
Shepherd's pie	2 starches + 1 fruits & vegetables + 3 proteins	325 mL (1⅓ c)	
Sherbet, orange	3 sugars + ½ fat	125 mL (½ c)	
Spaghetti and meat sauce	2 starches + 1 fruits & vegetables + 2 proteins + 3 fats	250 mL (1 c)	
Stew	2 starches + 2 fruits & vegetables + 3 proteins + ½ fat	200 mL (¾ c)	
Sundae	4 sugars + 3 fats	125 mL (½ c)	
Tuna casserole	1 starch + 2 proteins + ½ fat	125 mL (½ c)	
Yogurt, fruit bottom	1 fruits & vegetables + 1 milk + 1 sugar	125 mL (½ c)	125 g
Yogurt, frozen	1 milk + 1 sugar	125 mL (½ c)	125 g

Figure F.1
Example of a Food Label

Kellogg's
Raisin Bran*

OUR COMMITMENT TO QUALITY

Kellogg's is committed to providing foods of outstanding quality and freshness. If this product in any way falls below the high standards you've come to expect from Kellogg's, please send your comments and both top flaps to:
Consumer Affairs
KELLOGG CANADA INC.
Etobicoke, Ontario M9W 5P2

IF IT DOESN'T SAY Kellogg's* ON THE BOX, IT'S NOT Kellogg's* IN THE BOX.
SI LE NOM Kellogg's* N'EST PAS SUR LA BOÎTE, CE N'EST PAS Kellogg's* DANS LA BOÎTE.

• HIGH IN FIBRE
• LOW IN FAT
• PRESERVATIVE FREE
• SOURCE ÉLEVÉE DE FIBRES
• FAIBLE EN MATIÈRES GRASSES
• SANS AGENT DE CONSERVATION

NUTRITION INFORMATION
APPORT NUTRITIONNEL

	Per 40 g serving cereal (175 mL, ¾ cup) Par ration de 40 g de céréale (175 mL, ¾ tasse)	Per 40 g serving cereal with 125 mL Partly Skimmed Milk (2%) Par ration de 40 g de céréale avec 125 mL de lait partiellement écrémé (2,0 %)	
ENERGY	130Cal	195Cal	ÉNERGIE
	540kJ	810kJ	
PROTEIN	3.0g	7.3g	PROTÉINES
FAT	0.4g	2.9g	MATIÈRES GRASSES
CARBOHYDRATE	32g	38g	GLUCIDES
SUGARS*	11g	18g	*SUCRES
STARCH	16g	16g	AMIDON
DIETARY FIBRE	4.6g	4.6g	FIBRES ALIMENTAIRES
SODIUM	235mg	300mg	SODIUM
POTASSIUM	240mg	440mg	POTASSIUM

% of Recommended Daily Intake
% de l'apport quotidien conseillé

VITAMIN A	0%	7%	VITAMINE A
VITAMIN D	0%	23%	VITAMINE D
VITAMIN B1	62%	66%	VITAMINE B1
VITAMIN B2	3%	16%	VITAMINE B2
NIACIN	13%	18%	NIACINE
VITAMIN B6	13%	16%	VITAMINE B6
FOLACIN	11%	14%	FOLACINE
VITAMIN B12	0%	25%	VITAMINE B12
PANTOTHENATE	9%	15%	PANTOTHÉNATE
CALCIUM	1%	15%	CALCIUM
PHOSPHORUS	12%	23%	PHOSPHORE
MAGNESIUM	20%	27%	MAGNÉSIUM
IRON	38%	39%	FER
ZINC	16%	22%	ZINC

*Approximately half of the sugars occur naturally in the raisins.
Environ la moitié des sucres se retrouvent à l'état naturel dans les fruits.

Canadian Diabetes Association Food Choice Values: 40 g (175 mL, ¾ cup) cereal. Système des choix d'aliments de l'Association canadienne du diabète : 40 g (175 mL, ¾ tasse) céréale = 1 [■] + ½ [◆] + ½ [✱] choices/choix

INGREDIENTS / INGRÉDIENTS

WHOLE WHEAT, RAISINS (COATED WITH SUGAR, HYDROGENATED VEGETABLE OIL), WHEAT BRAN, SUGAR/GLUCOSE-FRUCTOSE, SALT, MALT (CORN FLOUR, MALTED BARLEY), VITAMINS (THIAMIN HYDROCHLORIDE, PYRIDOXINE HYDROCHLORIDE, FOLIC ACID, d-CALCIUM PANTOTHENATE), MINERALS (IRON, ZINC OXIDE).

BLÉ ENTIER, RAISINS SECS (ENROBÉS DE SUCRE, D'HUILE VÉGÉTALE HYDROGÉNÉE), SON DE BLÉ, SUCRE/GLUCOSE-FRUCTOSE, SEL, MALT (FARINE DE MAÏS, ORGE MALTÉ), VITAMINES (CHLORHYDRATE DE THIAMINE, CHLORHYDRATE DE PYRIDOXINE, ACIDE FOLIQUE, PANTOTHÉNATE DE d-CALCIUM), MINÉRAUX (FER, OXYDE DE ZINC).

Made by / Produit par
KELLOGG CANADA INC.
ETOBICOKE, ONTARIO
CANADA M9W 5P2
*Registered trademark of /
*Marque déposée de
KELLOGG CANADA INC. © 1994
00094

Food Labels in Canada

Consumers can gather a lot of information from a nutrition label. Figure F.1 demonstrates the reading of a food label and Table F.10 defines terms.

WHAT YOU WILL FIND ON A LABEL:

Nutrition Claims

- in Canada, it is optional for a company to decide to use claims,
- when claims appear on a label, they must follow government laws

Nutrition Information

- gives detailed nutrition facts about the product, including serving size and core list
- does not have to appear by law on food products in Canada
- refers to the food as packaged, so if you add milk, eggs or other food, the nutritional content of the food you eat can be very different

Serving Size

- the amount of food for which the information is given
- check the serving size: the serving size on the label may not be the same as the serving size you would actually eat (for example, the serving size of cereal may be ¾ cup, much smaller than your regular serving)

Core List

- the energy (in Calories and kilojoules), grams of protein, fat and carbohydrate for each serving
- some products break down fat into monounsaturates, polyunsaturates, saturates, and cholesterol (to find out what these mean, look at the Fats & Oils section)
- carbohydrates may include the amount of sugars, starch and fibre, or may list these items separately

Sodium and Potassium (in milligrams)

Vitamins and Minerals (as percent of your recommended daily intake)

Canadian Diabetes Association Food Choice Values and Symbols

- the Values and Symbols are tools to help you fit the food into your meal plan, they are not an endorsement by CDA
- it is up to the food company to decide if they want their foods analyzed and assigned symbols
- when they are on a label, they have been assigned by a dietitian working for CDA, so you can be sure the information is correct

Ingredients

- must be found on all food labels by law
- ingredients are listed in decreasing order by weight, so what you see first is what you get the most of

Table F.10
Terms on Food Labels

F

Energy

kcalorie reduced: 50% or fewer kcalories than the regular version.

light: term may be used to describe anything (for example, light in colour, texture, flavour, taste, or kcalories); read the label to find out what is "light" about the product.

low kcalorie: kcalorie-reduced and no more than 15 kcalories per serving.

Fat and Cholesterol

low cholesterol: no more than 3 mg of cholesterol per 100 g of the food and low in saturated fat; *does not* always mean low in total fat.

low fat: no more than 3 g of fat per serving; *does not* always mean low in kcalories.

lower fat: at least 25% less fat than the comparison food; be aware that 80% fat-free still means the food is 20% fat.

Carbohydrates: Fibre and Sugar

carbohydrate reduced: not more than 50% of the carbohydrate found in the regular version; *does not* always mean the product is lower in kcalories because other ingredients such as fat may have increased.

source of dietary fibre: a product that provides 2–4 g of fibre.

high source of dietary fibre: a product that provides 4–6 g of fibre.

very high source of fibre: a product that provides 6 g (or more) of fibre.

sugar free: low in carbohydrates and kcalories; can be used as an extra food in the exchange system.

unsweetened or no sugar added: no sugar was added to the product; sugar may be found naturally in the food (for example, fruit canned in its own juice).

Glossary

Note to the reader: If you have a customized book, this glossary may include some entries that will not be found in your book.

A

Absorption The process by which nutrients and other substances are transferred from the digestive system into the blood for transport throughout the body.

Adequate diet A diet consisting of foods that together supply approximately the RDA (Recommended Dietary Allowance) for protein, vitamins, and minerals and enough calories to meet a person's need for energy.

Adequate Intakes (AIs) "Tentative" RDAs (Recommended Daily Allowances). AIs are based on less conclusive scientific information than are the RDAs.

Aerobic fitness A state of respiratory and circulatory health as measured by the ability to deliver oxygen to muscles, and the capacity of muscles to use the oxygen for physical activity.

Aflatoxins Naturally occurring toxins produced by molds that may form on nuts, grains, seeds, and other plant products during storage.

Alcohol A substance formed by the fermentation of carbohydrates.

Alcoholism An illness characterized by a dependence on alcohol and a level of alcohol intake that interferes with health, family and social relations, and job performance.

Alcohol sugars Simple sugars that contain an alcohol group in their molecular structure. The most common alcohol sugars are xylitol, mannitol, and sorbitol.

Allergen *See* Food allergen.

Allergy *See* Food allergy.

Amino acids Specific nitrogen-containing molecules that serve as the building blocks for protein.

Amylophagia Laundry starch or cornstarch eating.

Anaphylactic (an-ah-fa-lac-tic) **shock** Reduced oxygen supply to the heart and other tissues due to the body's reaction to an allergen in food. Symptoms of anaphylactic shock include paleness, a flush discoloration of the lips and fingertips, a "faceless" expression, weak and rapid pulse, and difficulty in breathing.

Anorexia nervosa An eating disorder characterized by extreme weight loss, poor body image, and irrational fears of weight gain and obesity.

Antibodies Proteins the body makes to combat allergens.

Antioxidants Chemical substances that prevent or repair damage to cells caused by exposure to oxidizing agents such as automobile and factory emissions, smoke, ozone, and oxygen. Beta-carotene, vitamin E, and vitamin C function as antioxidants.

Appetite The desire to eat; a pleasant sensation that is aroused by thoughts of the taste and enjoyment of food.

Atherosclerosis "Hardening of the arteries" due to a buildup in plaque.

B

Balanced diet A diet that provides neither too much nor too little nutrients and other components of food such as fat and fiber.

Basal metabolism Energy used to support the body's ongoing processes that support growth, health, tissue repair and maintenance, and other functions. Assessed while at rest, basal metabolism includes energy the body expends for breathing, the pumping of the heart, the maintenance of body temperature, and other life-sustaining, ongoing functions.

Binge eating The consumption of a large amount of food in a small amount of time.

Binge eating disorder An eating disorder characterized by periodic binge eating, which normally is not followed by vomiting or the use of laxatives. People must experience eating binges twice a week on average for over six months to qualify for the diagnosis.

Bioavailability The amount of a nutrient consumed that is absorbed and available for use by the body.

Biotechnology As applied to food products, biotechnology is the process of modifying the composition and properties of foods by biologically altering their genetic makeup. Foods genetically modified are sometimes referred to as "transgenic."

Bulimia nervosa An eating disorder characterized by recurrent episodes of rapid, uncontrolled eating of large amounts of food in a short period of time. Episodes of binge eating are often followed by purging.

C

Calorie A unit of measure of the amount of energy supplied by food. Also known as *kilocalories,* or the *large Calorie* with a capital "C."

Cancer A group of diseases in which abnormal cells grow out of control and can spread throughout the body. Cancer is not contagious and likely has many causes.

Carbohydrates Substances in food that consist of a single sugar molecule or of multiples of them in various forms.

Cardiovascular disease Disorders related to plaque buildup in arteries of the heart, brain, or other organs or tissues.

Celiac disease An intestinal malabsorption disorder caused by a genetically-based intolerance to gluten in wheat, rye, and, to a lesser extent, barley and oats.

Cholesterol Fat-soluble, colorless liquid found in animals but not plants. It can be manufactured by the liver.

Chromosones Structures in the nuclei of cells which contain genes. Humans have 23 pairs of chromosomes; half of each pair comes from the mother and half from the father.

Chronic diseases Slow-developing, long-lasting diseases that are not contagious (e.g., heart disease, cancer, diabetes).

Cirrhosis of the liver Degeneration of the liver, usually caused by excessive alcohol intake over a number of years.

Coenzyme A chemical substance that activates an enzyme.

Colon The large intestine, or the last part of the intestines.

Colostrum The milk produced during the first few days after delivery. It contains more antibodies, protein, and certain minerals than the mature milk that is produced later. It is thicker than mature milk and has a yellowish color.

Complementary protein sources Plant sources of protein that together provide sufficient quantities of the nine essential amino acids.

Complete proteins Proteins that contain all of the essential amino acids in amounts needed to support growth and tissue maintenance.

Complex carbohydrates The form of carbohydrate found in starchy vegetables, grains, and dried beans and in many types of dietary fiber. The most common form of starch is made of long chains of interconnected glucose units.

Critical period A specific interval of time during which cells of a tissue or organ are genetically programmed to multiply. If the supply of nutrients needed for cell multiplication is not available during the specific time interval, the growth and development of the tissue or organ are permanently impaired.

D

Daily Reference Values (DRVs) Scientifically agreed-upon daily levels of fat, saturated fat, cholesterol, carbohydrate, dietary fiber, and protein intake compatible with health. DRVs are intended for use on nutrition labels only and are listed in the Nutrition Facts panel. "% Daily Value" on Nutrition Facts panels is calculated as the percentage of each DRV supplied by a serving of the labeled food.

% Daily Value (%DV) Scientifically agreed-upon standards of daily intake of nutrients from the diet developed for use on nutrition labels. The "% Daily Value" listed in nutrition labels represents the percentages of the standards obtained from one serving of the food product.

Development Processes involved in enhancing functional capabilities. For example, the brain grows, but the ability to reason develops.

Diabetes A disease characterized by abnormally high blood glucose levels. There are two major types: insulin-dependent and noninsulin-dependent diabetes mellitus.

Dietary fiber Complex carbohydrates and carbohydrate-like substances in food that cannot be broken down by human digestive enzymes.

Dietary Reference Intakes (DRIs) The general term for dietary intake standards.

Dietary supplements Any products intended to supplement the diet, including vitamin and mineral supplements; proteins, enzymes, and amino acids; fish oils and fatty acids; hormones and hormone precursors; and herbs and other plant extracts. Such products must be labeled "Dietary Supplement."

Dietary thermogenesis (thermogenesis means the production of heat) Energy expended during digestion of food and the absorption and utilization of nutrients. Some of the energy escapes as heat. It accounts for approximately 10% of the body's total energy need. Also called *specific dynamic action* of foods and *diet-induced thermogenesis.*

Digestion The breakdown of food in the digestive system.

Disaccharides (*di* = two, *saccharide* = sugar) Simple sugars consisting of two molecules of monosaccharides linked together. Sucrose, maltose, and lactose are disaccharides.

DNA (deoxyribonucleic acid) Genetic material contained in cells that initiates and directs the production of proteins in the body.

DNA fingerprinting The process of identifying specific individuals by their DNA. This is possible because no two individuals have the same genetic makeup. Differences among individuals are due to variation in the sections of DNA molecules that do not transmit genetic information.

Double-blind, placebo-controlled food challenge A test used to determine the presence of a food allergy or other adverse reaction to a food. In this test, neither the patient nor the care provider knows whether a suspected offending food or a placebo (a "sugar pill") is being tested.

E

Edema Swelling due to an accumulation of fluid in body tissues.

Electrolytes Minerals such as sodium and potassium that carry a charge when in solution. Many electrolytes help the body maintain an appropriate amount of fluid.

Empty-calorie foods Foods that provide an excess of calories in relation to nutrients. Soft drinks, candy, sugar, alcohol, and fats, such as butter, margarine, and oil, are considered empty-calorie foods.

Energy nutrients Carbohydrates, proteins, and fats are the "energy nutrients." Alcohol also provides energy, but it is not considered a nutrient.

Enrichment The replacement of thiamin, riboflavin, niacin, and iron that are lost when grains are refined.

Environmental trigger An environmental factor, such as inactivity, a high-fat diet, or a high sodium intake, that causes a genetic tendency toward a disorder to be expressed.

Enzymes Protein substances that speed up chemical reactions. Enzymes are found throughout the body but are present in particularly large amounts in the digestive system.

Ergogenic aids Substances that increase the capacity for muscular work.

Essential amino acids Amino acids that cannot be synthesized in adequate amounts by humans and therefore must be obtained from the diet. They are sometimes referred to as "indispensable amino acids."

Essential fatty acids Components of fat that are a required part of our diet. They are a component of certain hormones and cell membranes. Linoleic and alpha-linolenic acid are identified as essential fatty acids.

Essential hypertension Hypertension of no known cause; also called *primary* or *idiopathic hypertension,* it accounts for over 95% of all cases of hypertension.

Essential nutrients Substances the body requires for normal growth and health but cannot manufacture in sufficient amounts; they must be obtained in the diet.

F

Fat *See* Lipids.

Fats Substances in food that are soluble in fat and not water.

Fatty acids Fat-soluble components of fats.

Fermentation The process by which carbohydrates are converted to ethanol by the action of the enzymes in yeast.

Fetus A baby in the womb from the eighth week of pregnancy until birth. (Before then, it is referred to as an *embryo.*)

Fluorosis Fluoride overdose characterized by discolored ("mottled") teeth in children.

Food additive Any substance added to food that becomes part of the food or that affects the characteristics of the food. The term applies to substances added both intentionally and unintentionally to food.

Food allergen A substance in food (usually a protein) that is identified as harmful by the body and elicits an allergic reaction by the immune system.

Food allergy Adverse reaction to a normally harmless substance in food that involves the body's immune system. (Also called *food sensitivity.*)

Foodborne illness An illness related to consumption of foods or beverages containing disease-causing bacteria, viruses, parasites, toxins, or other contaminants.

Food insecurity The limited or uncertain availability of safe, nutritious foods.

Food intolerance Adverse reaction to a normally harmless substance in food that does *not* involve the body's immune system.

Food security Access at all times to a sufficient supply of safe, nutritious foods.

Fortification The addition of one or more vitamins and/or minerals to a food product.

Free radicals Chemical substances (usually oxygen or hydrogen) that are missing an electron and consequently act as oxidizing agents by stealing electrons from nearby atoms or molecules.

Fruitarian A form of vegetarian diet where fruits are the major ingredient. Such diets provide inadequate amounts of a variety of nutrients.

Functional foods A marketing term for foods that contain added ingredients intended to benefit health. They do not have to be demonstrated to benefit health.

G

Genes The basic units of heredity that occupy specific places (loci) on chromosomes. Genes consist of large DNA molecules, each of which contains the code for the manufacture of a specific enzyme or other protein.

Genome Combined term for "genes" and "chromosomes." It represents all the genes and DNA contained in an organism and principally located in chromosomes. The human genome consists of about 100,000 genes.

Geophagia Clay or dirt eating.

Gestational diabetes Diabetes first discovered in pregnancy.

Gingivitis A form of peridontal disease characterized by a minor inflammation of the gums.

Glycerol A syrupy, colorless liquid component of fats that is soluble in water. It is similar to glucose in its chemical structure.

Glycogen The body's storage form of glucose. Glycogen is stored in the liver and muscles.

Growth A process characterized by increases in cell number and size.

H

Heartburn A condition that results when acidic stomach contents are released into the esophagus, usually causing a burning sensation.

Heart disease One of a number of disorders that result when circulation of blood to parts of the heart is inadequate. Also called *coronary heart disease.* ("Coronary" refers to the blood vessels at the top of the heart. They look somewhat like a crown.)

Hemochromatosis A disorder affecting 1 in 200 people that occurs due to a genetic deficiency of a protein that helps regulate iron absorption. Individuals with hemochromatosis absorb more iron than normal and have excessive levels of body iron.

Hemoglobin The iron-containing protein in red blood cells.

Hemorrhoids Swelling of veins in the anus or rectum.

Histamine A substance released in allergic reactions. It causes blood vessels to dilate and blood pressure to drop and stimulates the release of stomach acids and other fluids. Antihistamines neutralize the effects of histamine and are used in the treatment of allergies.

Homocysteine A compound produced when the amino acid methionine is converted to another amino acid, cysteine. High blood levels of homocysteine increase the risk of hardening of the arteries, heart attack, and stroke.

Hunger Unpleasant physical and psychological sensations (weakness, stomach pains, irritability) that lead people to acquire and ingest food.

Hydrogenation The addition of hydrogen to unsaturated fatty acids.

Hypertension High blood pressure. It is usually defined as blood pressure that typically exceeds 140/90 millimeters of mercury.

Hypoglycemia A disorder resulting from abnormally low blood glucose levels. Symptoms of hypoglycemia include irritability, nervousness, weakness, sweating, and hunger. These symptoms are relieved by eating foods that provide calories.

I

Immune system Body tissues that provide protection against bacteria, viruses, and other substances identified by cells as harmful.

Incomplete proteins Proteins that are deficient in one or more essential amino acids.

Initiation The start of the cancer process; it begins with the alteration of DNA within cells.

Insoluble fiber Dietary fiber that does not dissolve in water. It "holds" water and is found in plant cell walls.

Insulin-dependent diabetes mellitus (IDDM) This type of diabetes is due to the destruction of insulin-producing cells in the pancreas, which leads to a lack of insulin. Also called *type 1 diabetes.*

Insulin resistance A condition in which cells "resist" the action of insulin in facilitating the passage of glucose into cells.

Intentional food additives Substances added to food to improve certain properties such as shelf life, color, nutritional value, texture, flavor, and consistency. Intentionally added sugars and salt account for 90% of our intake of food additives. Also called *"direct" food additives.*

Iron deficiency A disorder characterized by weakness, fatigue, short attention span, poor appetite, increased susceptibility to infection, and irritability that results from a depletion of iron stores in the body.

Iron-deficiency anemia A condition that results when the content of hemoglobin in red blood cells is reduced due to a lack of iron. It is characterized by the signs of iron deficiency plus paleness, exhaustion, and a rapid heart rate.

K

Kwashiorkor (kwa-she-or-kor) A deficiency disease primarily caused by a lack of complete protein in the diet. It usually occurs after children are taken off breast milk and given solid foods that contain protein of low quality.

L

Lactose intolerance The term for gastrointestinal symptoms (flatulence, bloating, abdominal pain, diarrhea, and "rumbling in the bowels") resulting from the consumption of more lactose than can be digested with available lactase.

Lactose maldigestion A disorder characterized by reduced digestion of lactose due to the low availability of the enzyme lactase.

Life expectancy The average length of life of people of a given age.

Lipids Compounds that are insoluble in water and soluble in fat. Triglycerides, saturated and unsaturated fats, and essential fatty acids are examples of lipids, or "fats."

Lymphatic system A network of vessels that absorb some of the products of digestion and transport them to the heart, where they are mixed with the substances contained in blood.

M

Malnutrition "Poor" nutrition resulting from a lack or excess of calories or nutrients needed by the body.

Marasmus A condition of severe body wasting due to deficiencies of both protein and calories. Also called *protein-energy malnutrition* and *protein-calorie malnutrition."*

Maximal oxygen consumption The highest amount of oxygen that can be delivered to, and utilized by, muscles for physical activity. Also called VO_2 *max* and *maximal volume of oxygen.*

Metabolism The chemical changes that take place in the body. The conversion of glucose to energy or to body fat is an example of a metabolic process.

Minerals In the context of nutrition, minerals are specific, single atoms that perform particular functions in the body.

Monosaccharides (*mono* = one, *saccharide* = sugar) Simple sugars consisting of one sugar molecule. Glucose, fructose, and galactose are monosaccharides.

Myoglobin The iron-containing protein in muscle cells.

N

Neural tube defects Malformations of the spinal cord and brain. They are among the most common and severe fetal malformations, occurring in approximately one in every 1000 pregnancies. Neural tube defects include spina bifida (spinal cord fluid protrudes through a gap in the spinal cord), anencephaly (absence of the brain or spinal cord), and encephalocele (protrusion of the brain through the skull).

Nonessential amino acids Amino acids that can be readily produced by humans from components of the diet. Also referred to as *dispensable amino acids.*

Nonessential nutrients Nutrients required for normal growth and health that can be manufactured in sufficient quantities by the body from other components of the diet. We do not need a dietary source of nonessential nutrients.

Noninsulin-dependent diabetes mellitus (NIDDM) In this type of diabetes, insulin is produced by the body but is utilized poorly. Also called *type 2 diabetes.*

Nutrient-dense foods Foods that contain relatively high amounts of nutrients compared to their calorie content. Broccoli, collards, bread, cantaloupe, and lean meats are examples of nutrient-dense foods.

Nutrients Chemical substances found in food that are used by the body for growth and health. The six categories of nutrients are carbohydrates, proteins, fats, vitamins, minerals, and water.

Nutrition The study of foods, their nutrients and other chemical constituents, and the effects of food constituents on health.

O

Obesity A condition characterized by excess body fat. Excess fat in men is defined as body fat levels greater than 20% of total weight. The corresponding figure for women is 25%.

Osteomalacia Condition in which a loss of bone mineral leads to a softening of the bones; adult counterpart of rickets.

Osteoporosis (*osteo* = bones, *poro* = porous, *osis* = abnormal condition) A condition characterized by porous bones: it is due to the loss of minerals from the bones.

Overweight Body weights that are higher than normal but lower than those used to designate obesity. Overweight in men and women is defined as BMI (body mass index) of approximately 25 to 30.

P

Pagophagia Ice eating.

pH A measure of the acidity of a solution, based on a scale from 0 to 14: a pH of 7 is neutral; greater than 7 is alkaline; less than 7 is acidic.

PKU (Phenylketonuria) A very rare disorder caused by the lack of the enzyme phenylalanine hydroxylase. Lack of this enzuyme causes phenylalanine, an essential amino acid, to build up in the blood.

Physical fitness The condition of the body as measured by muscular strength, endurance, and flexibility in the conduct of physical activity.

Phytochemicals Chemical substances in plants, some of which perform important functions in the human body. Phytochemicals give plants color and flavor and provide protection against insects and diseases.

Pica The regular consumption of nonfood substances such as clay or laundry starch.

Plant stanols Substances in corn, wheat, oats, rye, olives, wood, and some other plants that are similar in structure to cholesterol, but are not absorbed by the body. They decrease cholesterol absorption.

Plaque Deposits of cholesterol, other fats, calcium, and cell materials in the lining of the inner wall of arteries. On teeth, a soft, white sticky material formed by bacteria.

Plumbism Lead (primarily from old paint flakes) eating.

Polysaccharides (*poly* = many, *saccharide* = sugar) Carbohydrates containing many molecules of sugar linked together. Starch, glycogen, and dietary fiber are the three major types of polysaccharides.

Prebiotics "Intestinal fertilizer." Certain types of nondigestible carbohydrates added to some dairy products that support the growth of beneficial bacteria in the gut.

Precursor In nutrition, a nutrient that can be converted into another nutrient. Beta-carotene is a precursor of vitamin A.

Probiotics Strains of *Lactobacillus acidophilus* and *Lactobacillus johnsnii* added to some dairy products that limit the growth of harmful bacteria.

Progression The uncontrolled growth of abnormal cells.

Promotion The period in cancer development when the number of cells with altered DNA increases.

Prostate A gland located above the testicles in males. The prostate secretes a fluid that surrounds sperm.

Proteins Substances in food made of amino acids.

Provitamins Precursors of vitamins that can be converted to vitamins in the body (e.g., beta-carotene).

Purging The use of self-induced vomiting, laxatives, or diuretics (water pills) to prevent weight gain.

R

Recommended Dietary Allowances (RDAs) Daily levels of nutrient intake that meet the needs of practically all healthy persons while decreasing the risk of certain chronic diseases.

Rectum The last five inches of the colon.

Reference Daily Intakes (RDIs) Standards by which the nutrient composition of food and supplement products are assessed and reported on labels. They are set by the Food and Drug Administration and are based on the 1968 RDA table. (Formerly called the "U.S. RDAs.")

Registered dietitian An individual who has acquired knowledge and skills necessary to pass a national registration examination and participates in continuing professional education. Qualification is conferred by the Commission of Dietetic Registration.

Remodeling The breakdown and buildup of bone tissue.

Restrained eating The purposeful restriction of food intake below desired amounts in order to control body weight.

S

Salt sensitivity A genetically determined condition in which a person's blood pressure rises when high amounts of salt or sodium are consumed.

Satiety A feeling of fullness or of having had enough to eat.

Saturated fats The type of fat that tends to raise blood cholesterol levels and the risk for heart disease. They are solid at room temperature and are found primarily in animal products such as meat, butter, and cheese.

Simple sugars Carbohydrates that consist of a glucose, fructose, or galactose molecule; or a combination of glucose and either fructose or galactose:
glucose + fructose = sucrose
glucose + glucose = maltose
glucose + galactose = lactose
High-fructose corn syrup and alcohol sugars are also considered simple sugars. Simple sugars are often referred to as *sugars*.

Single gene defect A disorder resulting from one abnormal gene. Also called "inborn error of metabolism." Over 800 single gene defects have been cataloged; most are very rare.

Soluble fiber Dietary fiber that dissolves in water and produces a gel.

Stomach ulcer An open sore in the lining of the stomach.

Stroke The event that occurs when a blood vessel in the brain suddenly ruptures or becomes blocked, cutting off the blood supply to a portion of the brain. Stroke is usually associated with "hardening of the arteries" in the brain. (Also called a *cerebral vascular accident.*)

Sulfites Sulfur-containing compounds used as preservatives in foods to eliminate bacteria, preserve freshness, prevent browning, and increase storage life; can cause acute asthma attacks, and even death, in people who are sensitive to them.

T

Thermogenesis *See* Dietary thermogenisis.

Tolerable Upper Intake Levels (ULs) Upper limits of nutrient intake compatible with health. The ULs do not reflect desired levels of intake. Rather, they represent total daily levels of nutrient intake from food, fortified food products, and supplements that should not be exceeded.

Tooth decay The disintegration of teeth due to acids that are produced by bacteria in the mouth and feed on sugar. Also called *dental caries* or *cavities.*

Trans fats Unsaturated fatty acids in fats that contain atoms of hydrogen attached to opposite sides of carbons joined by a double bond.

Triglyceride A lipid containing glycerol and three fatty acids. Also called *triacylglyceride.*

Trimester One-third of the normal duration of pregnancy. The first trimester is 0 to 13 weeks, the second 13 to 26 weeks, and the third trimester is 26 to 40 weeks.

U

Underweight A condition characterized by an inadequate supply of body fat. For adult females, less than 20% body fat is considered underweight; for males the figure is 12%.

Unintentional food additives Substances that inadvertently end up in foods due to production, storage, or processing methods. Food packaging particles, pesticides used on crops, and antibiotics given to livestock are examples of unintentional food additives. Unintentional food additives are found in very small amounts in foods and generally are not listed on food ingredient labels. Also called *"inadvertent," "incidental,"* and *"indirect" food additives.*

Unsaturated fats Unsaturated fats are found primarily in vegetable products and tend to lower blood cholesterol.

V

Vitamin B complex All known water-soluble vitamins except C; includes thiamin (B_1), riboflavin (B_2), pyridoxine (B_6), niacin, folic acid, cobalamin (B_{12}), pantothenic acid, and biotin.

Vitamins Chemical substances that perform specific functions in the body.

W

Water balance The ratio of the amount of water outside cells to the amount inside cells; this balance is needed for normal cell functioning.

Index

Note to the reader: If you have a customized book, this index may include some entries that will not be found in your book.

Daily Values for Food Labels

The Daily Values are standard values developed by the Food and Drug Administration (FDA) for use on food labels. Daily Values for protein, vitamins, and minerals reflect average allowances based on the RDA. Daily Values for nutrients and food components, such as fat and fiber, that do not have an established RDA but do have important relationships with health are based on recommended calculation factors as noted.

Proteins, Vitamins, and Minerals

Nutrient	Amount
Protein[a]	50 g
Thiamin	1.5 mg
Riboflavin	1.7 mg
Niacin	20 mg NE
Biotin	300 µg
Pantothenic acid	10 mg
Vitamin B_6	2 mg
Folate	400 µg
Vitamin B_{12}	6 µg
Vitamin C	60 mg
Vitamin A	5000 IU
Vitamin D	400 IU
Vitamin E	30 IU
Vitamin K	80 µg
Calcium	1000 mg
Iron	18 mg
Zinc	15 mg
Iodine	150 µg
Copper	2 mg
Chromium	120 µg
Selenium	70 µg
Molybdenum	75 µg
Manganese	2 mg
Chloride	3400 mg
Magnesium	400 mg
Phosphorus	1000 mg

[a]The Daily Values for protein vary for different groups of people: pregnant women, 60 g; nursing mothers, 65 g; infants under 1 year, 14 g; children 1 to 4 years, 16 g.

Nutrients and Food Components

Food Component	Amount	Calculation Factors
Fat	65 g	30% of kcalories
Saturated fat	20 g	10% of kcalories
Cholesterol	300 mg	Same regardless of kcalories
Carbohydrate (total)	300 g	60% of kcalories
Fiber	25 g	11.5 g per 1000 kcalories
Protein	50 g	10% of kcalories
Sodium	2400 mg	Same regardless of kcalories
Potassium	3500 mg	Same regardless of kcalories

Note: Daily Values were established for adults and children over 4 years old. The values for energy-yielding nutrients are based on 2000 kcalories a day.

GLOSSARY OF NUTRIENT MEASURES

kcal: kcalories; a unit by which energy is measured.

g: grams; a unit of weight equivalent to about 0.03 ounces.

mg: milligrams; one-thousandth of a gram.

µg: micrograms; one-millionth of a gram.

IU: international units; an old measure of vitamin activity determined by biological methods (as opposed to new measures that are determined by direct chemical analyses). Many fortified foods and supplements use IU on their labels.

- For vitamin A, 1 IU = 0.3 µg retinol, 3.6 µg β-carotene, or 7.2 µg other vitamin A carotenoids.
- For vitamin D, 1 IU = 0.025 µg cholecalciferol.
- For vitamin E, 1 IU = 0.67 natural α-tocopherol (other conversion factors are used for different forms of vitamin E).

mg NE: milligrams niacin equivalents; a measure of niacin activity.

- 1 NE = 1 mg niacin.
  = 60 mg tryptophan (an amino acid).

µg DFE: micrograms dietary folate equivalents; a measure of folate activity.

- 1 µg DFE = 1 µg food folate.
  = 0.6 µg fortified food or supplement folate.
  = 0.5 µg supplement folate taken on an empty stomach.

µg RAE: micrograms retinol activity equivalents; a measure of vitamin A activity.

- 1 µg RE = 1 µg retinol.
  = 12 µg β-carotene.
  = 24 µg other vitamin A carotenoids.